MANAGEMENT OF WHEAT AND BARLEY DISEASES

MANAGEMENT OF WHEAT AND BARLEY DISEASES

Edited by
Devendra Pal Singh, PhD, MBA

Apple Academic Press Inc.
3333 Mistwell Crescent
Oakville, ON L6L 0A2 Canada

Apple Academic Press Inc.
9 Spinnaker Way
Waretown, NJ 08758 USA

© 2018 by Apple Academic Press, Inc.

First issued in paperback 2021

No claim to original U.S. Government works

ISBN 13: 978-1-77-463661-9 (pbk)
ISBN 13: 978-1-77-188546-1 (hbk)

Library and Archives Canada Cataloguing in Publication

Management of wheat and barley diseases / edited by Devendra Pal Singh, PhD, MBA.

Includes bibliographical references and index.
Issued in print and electronic formats.
ISBN 978-1-77188-546-1 (hardcover).--ISBN 978-1-315-20753-7 (PDF)

1. Wheat--Diseases and pests. 2. Barley--Diseases and pests. I. Singh, Devendra Pal, editor

SB608.W5M36 2017   633.1'194   C2017-902381-0   C2017-902382-9

Library of Congress Cataloging-in-Publication Data

Names: Singh, Devendra Pal, editor.
Title: Management of wheat and barley diseases / editor: Devendra Pal Singh, PhD, MBA.
Description: Waretown, NJ : Apple Academic Press, 2017. | Includes bibliographical references and index.
Identifiers: LCCN 2017015091 (print) | LCCN 2017017929 (ebook) | ISBN 9781315207537 (ebook) | ISBN 9781771885461 (hardcover : alk. paper)
Subjects: LCSH: Wheat--Diseases and pests. | Barley--Diseases and pests.
Classification: LCC SB608.W5 (ebook) | LCC SB608.W5 M328 2017 (print) | DDC 638/.155--dc23
LC record available at https://lccn.loc.gov/2017015091

Apple Academic Press also publishes its books in a variety of electronic formats. Some content that appears in print may not be available in electronic format. For information about Apple Academic Press products, visit our website at **www.appleacademicpress.com** and the CRC Press website at **www.crcpress.com**

ABOUT THE EDITOR

Devendra Pal Singh, PhD, MBA

Dr. Devendra Pal Singh is the Principal Scientist (Plant Pathology) and Principal Investigator (Crop Protection Programme) at ICAR—Indian Institute of Wheat and Barley Research (ICAR—IIWBR) Karnal, India. He was the Coordinator at the Rice Research Station, Guyana Rice Development Board, Guyana, Specialist in Zambia on a UNDP/FAO Project (United Nations Development Programme / Food and Agriculture Organization) and a Visiting Scientist at the Université Catholique de Louvain, Louvain-la-Neuve, Belgium. Dr. Singh has published over 110 research papers and edited two books. He is fellow of the Indian Phytopathological Society, the Indian Society of Mycology and Plant Pathology, and the Indian Society of Plant Pathologists and sits on the editorial boards and panels of reviewers of several international journals in plant pathology, including the journal of the American Phytopathological Society. Dr. Singh earned his PhD in Plant Pathology from the G. B. Pant University of Agriculture and Technology, Pantnagar, India, and his MBA from Indira Gandhi National Open University (IGNOU), New Delhi, India.

CONTENTS

LIST OF CONTRIBUTORS

Gazala Ameen
North Dakota State University, 306 Walster Hall, 1402 Albrecht Boulevard, Fargo, North Dakota, 58108, USA. E-mail: gazala.ameen@ndsu.edu

Mohammed Shamshul Qumor Ansari
Department of Mycology and Plant Pathology, Institute of Agricultural Sciences, Banaras Hindu University, Varanasi 221005, Uttar Pradesh, India. E-mail: md.samshul37@gmail.com

Ritu Bala
Department of Plant Breeding and Genetics, Punjab Agricultural University, Ludhiana 141004, India. E-mail: rituraje2010@pau.edu

S. Bandyopadhyay
Uttar Banga Krishi Viswavidyalaya (North Bengal Agriculture University), Pundibari, Coochbehar 736165, West Bengal, India. E-mail: badyopadhyaysekhar@yahoo.co.in

Ashwani Kumar Basandrai
CSKHPAU, Rice and Wheat Research Centre, Malan District, Kangra 176047, Himachal Pradesh, India. E-mail: ashwanispp@gmail.com

Daisy Basandrai
CSKHPAU, Rice and Wheat Research Centre, Malan District, Kangra 176047, Himachal Pradesh, India. E-mail: bunchy@rediffmail.com

Subhash C. Bhardwaj
ICAR-IIWBR, Regional Station, Flowerdale, Chota Shimla, Shimla 171002, Himachal Pradesh, India. E-mail: scbfdl@hotmail.com

P. M. Bhattacharya
Uttar Banga Krishi Viswavidyalaya (North Bengal Agriculture University), Pundibari, Coochbehar 736165, West Bengal, India. E-mail:pmb_ubkv@yahoo.co.in

S. P. Bishnoi
Division of Nematology, Rajasthan Agricultural Research Institute, SKN University of Agriculture, Durgapura, Jaipur 302018, Rajasthan, India. E-mail: satyapal_bishnoi@yahoo.com

Robert Saxon Brueggeman
North Dakota State University, 306 Walster Hall, 1402 Albrecht Boulevard, Fargo, North Dakota 58108, USA. E-mail: robert.brueggeman@ndsu.edu

Ramesh Chand
Department of Mycology and Plant Pathology, Institute of Agricultural Sciences, Banaras Hindu University Varanasi 221005, Uttar Pradesh, India. E-mail: rc_vns@yahoo.co.in

A. K. Chowdhury
Department of Plant Pathology, Uttar Banga Krishi Viswavidyalaya, Pundibari 736165, Coochbehar, West Bengal, India. E-mail: akc_ubkv@rediffmail.com

T. Dhar
Regional Research Sub-Station, Uttar Banga Krishi Viswavidyalaya (North Bengal Agriculture University), Manikchak, Malda 732203, West Bengal, India. E-mail:tapamay_ubkv@yahoo.co.in

Om P. Gangwar
ICAR-IIWBR, Regional Station, Flowerdale, Chota Shimla, Shimla 171002, Himachal Pradesh, India.
E-mail: gangwarop@gmail.com

R. P. Ghasolia
Rajasthan Agricultural Research Institute, SKN University of Agriculture, Durgapura, Jaipur 302018,
Rajasthan, India. E-mail: rpghasolia.ppath@sknau.ac.in

P. Jayaprakash
ICAR-Indian Agricultural Research Institute, Regional Station, Wellington 643231, Tamil Nadu, India.
E-mail: jpsarit@gmail.com

A. K. Joshi
International Maize and Wheat Improvement Center (CIMMYT) G-2, B-Block, NASC Complex, DPS
Marg, New Delhi 110012, India. E-mail: a.k.joshi@cgiar.org

Jaspal Kaur
Department of Plant Breeding and Genetics, Punjab Agricultural University, Ludhiana 141004, India.
E-mail: jassu75@pau.edu

Hanif Khan
ICAR-IIWBR, Regional Station, Flowerdale, Chota Shimla, Shimla 171002, Himachal Pradesh, India.
E-mail: hanif.gene@gmail.com

Jagdish Kumar
ICAR- National Institute of Biotic Stress Management, Baronda, Raipur 493225, Chhattishgarh, India.
E-mail: moola01@yahoo.com

Subodh Kumar
ICAR-IIWBR, Regional Station, Flowerdale, Chota Shimla, Shimla 171002, Himachal Pradesh, India.
E-mail: subodhfdl@gmail.com

V. K. Mishra
Department of Genetics and Plant Breeding, Institute of Agricultural Sciences, Banaras Hindu University, Varanasi221005, Uttar Pradesh, India. E-mail: vkmbhu@gmail.com

Dharam Pal
ICAR-IARI, Regional Station, Tutikandi Centre, Shimla 171004, Himachal Pradesh, India. E-mail:
dpwalia@rediffmail.com

Anju Pandey
Department of Mycology and Plant Pathology, Institute of Agricultural Sciences, Banaras Hindu University, Varanasi 221005, Uttar Pradesh, India. E-mail: pandeyanju931@gmail.com

B. M. Patel
Wheat Research Station, Sardarkrushinagar Dantiwada Agricultural University, Vijapur 382870, District-Mehsana, Gujarat, India. E-mail: bmpatelxp@gmail.com

S. I. Patel
Wheat Research Station, S. D. Agricultural University, Vijapur 382870, Gujarat, India

Madhu Patial
ICAR-IARI, Regional Station, Tutikandi Centre, Shimla 171004, Himachal Pradesh, India. E-mail:
mcaquarian@gmail.com

Pramod Prasad
ICAR-IIWBR, Regional Station, Flowerdale, Chota Shimla, Shimla 171002, Himachal Pradesh, India.
E-mail: pramoddewli@gmail.com

Anju Rani
Department of Bioscience & Biotechnology, Banasthali University, Rajasthan, India

Jonathan Kertz Richards
North Dakota State University, 306 Walster Hall, 1402 Albrecht Boulevard, Fargo, North Dakota 58108, USA. E-mail: jonathan.richards@ndsu.edu

M. S. Saharan
ICAR-Indian Agricultural Research Institute, New Delhi 110012, India. E-mail: mssaharan7@yahoo.co.in

Siddanna Savadi
ICAR-IIWBR, Regional Station, Flowerdale, Chota Shimla, Shimla 171002, Himachal Pradesh, India. E-mail: siddannasavadi@gmail.com

Indu Sharma
ICAR-Indian Institute of Wheat and Barley Research, Karnal, Aggarsain Marg, Haryana 132001, India. E-mail: ispddwr@gmail.com

T. R. Sharma
ICAR- NRCPB, Lal Bahadur Shashtri Building, Pusa, New Delhi 110012, India. E-mail: pdnrcpb@gmail.com

P. S. Shekhawat
Division of Plant Pathology, Rajasthan Agricultural Research Institute, SKN University of Agriculture, Durgapura, Jaipur, Rajasthan 302018, India. E-mail: pdsingh87@yahoo.co.in

Devendra Pal Singh
Crop Protection Programme, ICAR-Indian Institute of Wheat and Barley Research, Karnal 132001, Aggarsain Marg, Haryana, India. E-mail: dpkarnal@gmail.com

M. Sivasamy
ICAR-Indian Agricultural Research Institute, Regional Station, Wellington 643231, Tamil Nadu, India. E-mail: iariwheatsiva@rediffmail.com

Shyam Solanki
North Dakota State University, 306 Walster Hall, 1402 Albrecht Boulevard, Fargo, North Dakota 58108, USA. E-mail: shyam.solanki@ndsu.edu

V. A. Solanki
Department of Plant Pathology, N. M. College of Agriculture, Navsari Agricultural University, Navsari 396450, Gujarat, India. E-mail: ppnmca@yahoo.co.in

S. S. Vaish
Department of Mycology and Plant Pathology, Institute of Agricultural Sciences, Banaras Hindu University, Varanasi 221005, India. E-mail: shyam_saran@rediffmail.com

S. P. Val-Moraes
UNESP Univ. Estadual Paulista/College of Agricultural and Veterinarian Sciences, Technology Department-Via de Acesso Prof. Paulo Donato Castellane, s/n 14884900, Jaboticabal, SP, Brazil. E-mail: valmoraes.silvania@gmail.com

V. K. Vikas
ICAR-Indian Agricultural Research Institute, Regional Station, Wellington 643231, Tamil Nadu, India. E-mail: vkvikaswtn@gmail.com

R. S. Yadav
ICAR-Directorate of Groundnut Research, Post Box-5, Ivnagar Road, Junagadh 362001,Gujarat, India. E-mail: yadavrs2002@gmail.com; yadavrs2002@nrcg.res.in

LIST OF ABBREVIATIONS

AFLP	amplified fragment length polymorphisms
AICRP	All India Coordinated Research Project
AICW & BIP	All India Coordinated Wheat and Barley Improvement Project
AICWP	All India Coordinated Wheat Improvement Programme
AMOVA	analysis of molecular variance
AMP	antimicrobial proteins
APR	adult plant resistance
AUDPC	area under disease progress curve
AVg	avirualence genes
BaMMV	barley mild mosaic virus
BGRI	Borlaug Global Rust Initiative
Bgt	*Blumeria graminis* f. sp. *tritici*
BMGF	Bill & Melinda Gates Foundation
BMV	brome mosaic virus
BPI	Black Point Index
BYDV	barley yellow dwarf virus
CA	conservation agriculture
CABI	Centre for Agriculture and Bioscience International
CAPS	cleaved amplified polymorphic sequence
CCN	cereal cyst nematode
CDD	cumulative degree-days
CFU	colony forming units
CGIAR	Consultative Group on International Agricultural Research
CIM	composite interval mapping
CIMMYT	International Maize and Wheat Improvement Center
CLAES	Central Laboratory of Agricultural Expert System
CRI	crop research institute
CRISPR	clustered regularly interspaced short palindromic repeat
CRN	crinkling and necrosis
CSEPs	candidate secreted effector proteins
CVT	common varietal trial
CZ	central zone
DAC & FW	department of agriculture, cooperation and farmers' welfare

DArT	diversity arrays technology
DAS	days after sowing
DBCP	dibromochloropropane
DNA	deoxyribonucleic acid
DON	deoxynivalenol
DRRW	durable rust resistance in wheat
DWR	directorate of wheat research
EC	experiments conducted
ELISA	enzyme-linked immunosorbent assay
ESTs	expressed sequence tags
ETI	effector triggered immunity
FAO	Food and Agriculture Organization of the United Nations
FHB	*Fusarium* head blight
FYM	farm-yard manure
GCRM	global cereal rust monitoring system
GRI	global rust initiative
GS	genomic selection
HIGS	host induced gene silencing
HLB	*Helminthosporium* leaf blight
HR	hypersensitive response
HYSPIT	HYbrid single-particle lagrangian integrated trajectory
I	index
IARI	Indian Agricultural Research Institute
ICAR	Indian Council of Agricultural Research
ICARDA	International Centre for Agricultural Research in Dry Areas
IIWBR	Indian Institute of Wheat and Barley Research
IDM	integrated disease management
IMTEH	Institute of Microbial Technology
IRs	infection responses
ISBP	insertion site-based polymorphism
IWF	intracellular wash fluids
KALRO	Kenya Agricultural and Livestock Research Organization
KB	Karnal bunt
KVK	KrishiVigyan Kendra
LOD	logarithm of the odds
LR	leaf rust
LS	loose smut
LRR	leucine rich repeat
MAMPs	microbe associated molecular patterns
MAPK	mitogen-activated protein kinase

MARS	marker-assisted recurrent selection
MAS	marker-assisted selection
MEs	mega environments
MLO	mildew resistance locus
MoT	*Magnaporthe oryzae* pathotype *Triticum*
MR	moderately resistant
MS	moderately susceptible
NEPZ	North-eastern plains zone
NFNB	net form net blotch
NGS	next generation sequencing
NGSN	national genetic stock nursery
NHR	non-host resistance
NHZ	Northern hills zone
NILs	near isogenic lines
NIV	nivalenol
NLR	NOD-like receptors
NOAA	National Oceanic and Atmospheric Administration
NWPZ	North-western plains zone
PAMPs	pathogen associated molecular patterns
PBC	pseudo black chaff
PBI	Plant Breeding Institute
PCR	polymerase chain reaction
PDA	potato dextrose agar
PGPR	plant growth promoting rhizobacteria
Pgt	*Puccinia graminis tritici*
PINs	puroindoline proteins
PM	powdery mildew
pt	pathotype
PTGS	post-transcriptional level
Ptr	*Pyrenophora tritici repentis*
pts.	pathotypes
PZ	peninsular zone
QTL	quantitative trait loci
R	resistance
R^2	mean phenotypic variation
RAMP	randomly amplified microsatellite polymorphism
RAPD	random amplified polymorphic DNA
RCTs	resource-conserving technologies
RFLP	restriction fragment length polymorphism
RH	relative humidity

RIL	recombinant inbred line
RIP	ribosome-inactivating protein
RLKs	receptor-like kinases
RNAi	RNA interference
ROS	reactive oxygen species
S	susceptible
SA	salicylic acid
SAARC	South Asian Association for Regional Cooperation
SAUs	State Agricultural Universities
SBWV	wheat soil-borne mosaic virus
SCAR	sequence characterized amplified regions
scFv	single chain variable fragment
SFNB	spot form net blotch
SHZ	Southern hills zone
SNP	single nucleotide polymorphism
SPAR	simple primer amplification reaction
SPS	sanitary and phytosanitary
Sr	stem rust
SRAP	sequence-related amplified polymorphism
SRT	seedling resistance test
STS	sequence tagged sites
TALENs	transcription activator-like effector nucleases
TILLING	targeting induced local lesions in genomes
TPN	trap plot nursery
Ug99	Uganda 99 stem rust race
USDA	United States Department of Agriculture
VIGS	virus induced gene silencing
WSMV	wheat streak mosaic virus
Yr	yellow rust
ZFNs	zinc-finger nucleases

FOREWORD 1

Numerous plant diseases and pests pose major threats to food security and safety and cause significant losses if proper diagnostics and timely decisions on control strategies are not implemented. Wheat and barley crops being grown on millions of hectares on diverse geographies and environmental conditions are no exception to this and are also conducive for disease epidemics. Evolution, selection, and migration of new and more virulent/ aggressive as well as fungicide-resistant/tolerant races of pathogens evolving and rate of evolution may accelerate further under changing climatic and injudicial fungicidal use situations. It is important to reduce losses due to diseases in these crops to harness full potential of genetic gain in yield and quality as well as maintaining high produce quality for different market niche through sound management strategies. This book, **Management of Wheat and Barley Diseases,** edited by Dr. Devendra Pal Singh, brings together a wealth of updated information, new knowledge, improved disease identification, and management technologies from many expert scientists who are actively engaged in managing a diverse range of wheat and barley diseases.

—**Dr. Ravi P. Singh**
Distinguished Scientist
Head—Wheat Improvement & Rust Research
Global Wheat Program
International Maize and Wheat Improvement Center (CIMMYT)
Carretera México-Veracruz Km. 45, El Batán,
Texcoco, C.P. 56237, México

FOREWORD 2

Three species of wheat, *Triticum aestivum, T. durum,* and *T. dicoccum,* are grown in India. *T. aestivum* is grown in all the six agro-ecological regions. *T. durum* is cultivated in central and peninsular zones. *T. dicoccum* is grown in some parts of the peninsular zone. Barley has been traditionally a feed crop grown under low fertility and drier regions. Increased demand of barley for the malt industries in recent years has resulted in its cultivation under high fertility and irrigated conditions in Northwestern and Eastern plains zones in India. Rice–wheat is a major cropping system in the Indo Gangetic plains in Southern Asia. The enhanced cultivation of wheat in nontraditional areas like warm and humid climate in Eastern and Far Eastern regions, caused changes in the cropping system, crop cultivation and tillage practices, and the climate also brought changes in the spectrum and severity of diseases of wheat and barley in India as well as in other countries of the world. The plant pathogens also keep evolving with the changes in the genetic constitution of disease-resistant varieties of wheat and barley and rends these susceptible to diseases, thus posing a threat to food security. Strategic survey and surveillance, diagnosis, use of effective management practices at critical stages of disease cycles, and integrated pest management practices are therefore need to be updated with time to effectively manage the diseases.

I take this opportunity to congratulate Dr. Devendra Pal Singh, editor of the book and all the contributors of chapters for bringing out new knowledge and technologies for disease management in wheat and barley. The book will be of much use to different stakeholders of wheat and barley industry. The publication is quite timely, much required, relevant, and is an important milestone in achieving a hunger-free world using an agro-ecological approach and growing more with less.

—**Dr. Jeet Singh Sandhu**
Deputy Director General (Crop Sciences)
Indian Council of Agricultural Research,
Krishi Bhavan, New Delhi-110001, India

INTRODUCTION

Wheat and barley are important cereal crops in the world and contribute significantly in the nutrition of human beings. Barley is also used in the malt industries and in animal feed. Three types of wheat are grown commercially. The most popular wheat is "Bread wheat" (*Triticum aestivum*) used for making bread, roti, cookies, and pastries. It accounts for nearly 90% of global wheat production. Its grain is semi-hard in nature and thus is suitable for flour production. The hexaploid wheat is generated through natural crossing between tetraploid wheat and grass (AABBDD). The next is "Durum wheat" (*T. durum*), which is mainly grown for pasta production. Its grain is comparative by harder and larger. It is nearly 10% of total wheat production in world. It is used for the production of semolina. It is tetraploid in nature, developed due to crossing of two grasses (AABB). The "Emmer wheat" or hulled wheat is a type of awned tetraploid wheat. The domesticated species are *Triticum turgidum* sub sp. *dicoccum* and *Triticum turgidum* conv. *durum*. "Triticale" is a hexaploid and is a result of a human-made cross between durum and rye. It is competitive as feed grain or green forage, especially in marginal areas. It has comparatively fewer disease problems than wheat and is more competitive in problem soils. The grains generally are not suitable for bread making.

Wheat is grown by itself or in cropping systems with rice, soybean, sugarcane, maize, cotton, etc. It is quite important to maintain the good health of these crops to feed the global human population, which is growing every day. The task is becoming difficult due to diminishing resources, decreasing farmland, adverse climate; poor soil health, and biotic stresses. Diseases account for a sizable share of these losses in these crops, which varies from one agro-climatic zone to another, with degree of susceptibility of crop varieties to diseases, with new pathotypes, and with lack of management of these at the farm level.

Wheat and barley are known to be affected due to rust diseases and, there had been epidemics of rusts in some countries. Being a compound interest disease, it is difficult to manage in field in situations of susceptible cultivars, favorable weather, and virulent pathotypes. The deployment of resistant varieties against rusts is the most practical, simple, effective, ecofriendly, and economic method to keep away the rusts. However, there had been a race between pathotypes of rusts, and wheat and barley breeders and farmer keep evolving thus rendering resistant varieties susceptible after a gap of a few years. There is continuous need to breed new varieties with effective rust-resistant genes.

The powdery mildew disease is also gaining importance due to use of defeated genes, lack of laboratories for maintaining pathotypes for use in evaluation of new varieties and breeding material, and overemphasis of using rust-resistant cultivars without proper tolerance to powdery mildew. The smut also causes losses in yield of wheat and barley but is easy to manage using fungicides. Bunts are also becoming important globally due to strict quarantines and thus causing impediments in global wheat trade. The seedborne nature of smuts and bunts is helping these to transmit through infected seeds to newer areas.

The hemibiotrophic pathogens such as *Magnaporthe oryzae* pathotype *Triticum, Bipolaris sorokiniana* (causing spot blotch of wheat and barley), and tan spot caused by *Pyrenophora tritici repentis* and necrotrophic diseases like Septoria blight, Fusarium head blight, foot rot, take-all disease, etc., are gaining importance in different countries. Wheat blast is considered another such disease that may affect rice wheat cropping system. The nematodes are mostly inhabitants of soil and affect the yield of wheat and barley by reducing the vigor of plants. The resource conservation technologies being adopted in most common rice–wheat cropping systems generated a lot of interest on their effects on the crop health of wheat. Likewise, cultivation of malt barley under high irrigated and fertility conditions demands different approaches for crop health management. The emergence of new pathotypes like Ug99 of stem rust, new diseases like wheat blast, the introduction of exotic pathogens in newer areas, zero tolerance to some of the plant pathogens in grain trade, and changes in the pattern of severity of different diseases under changing climate posed the a need of bringing out this book on management of diseases of both wheat and barley.

The chapters are divided into three sections of approaches. These are (i) disease identification and management practices, (ii) diseases in diverse agroecological conditions, and (iii) pathogenic variability and its management. The first chapter gives an introduction of different strategies of management of diseases in wheat and barley along with a list of diseases and their symptoms, whereas, the rest of the chapters deals with reviews of the latest information generated on the diseases like leaf, stripe and stem rusts; their pathotypes including Ug99 pathotype of stem rust; other foliar diseases like powdery mildew, spot blotch; wheat blast and net blotch, plant pathogenic nematodes; host resistance to different diseases in wheat including molecular markers and gene pyramiding; host pathogen relationship; smuts and bunts; black point disease of wheat; survey and surveillance of wheat diseases; disease management in wheat and barley in differentagroecological conditions; and impact of resource conservation technologies on soil and crop health. The book gives up-to-date accounts of information and technologies available for management of diseases of two important cereals, wheat and barley; and aimes to benefit the researchers, students, teachers, farmers, seed growers, traders, and other stakeholders dealing with these commodities. It also advances our knowledge in the field of plant pathology, plant breeding, and plant biotechnology, agronomy, grain quality, and pesticide industries. The book is expected to serve as a reference book of disease management technologies to contain the losses in wheat and barley yields, quality, increase in yield, and quality of produce with reduced cost of cultivation, and thus aid in ensuring food security at the global level.

The genuine and time-bound contributions made by actively engaged internationally reputed wheat and barley scientists are greatly acknowledged. The editor is also thankful to heads of organizations like ICAR-IIWBR, Karnal, Hon. Secretary, DARE & DG and DDG (CS) of ICAR, New Delhi, India; HE Mr. V. Mahalingam, High Commissioner, High Commission of India, Georgetown, Guyana; and Mr. N. Hassan, GM, GRDB, Georgetown, Guyana, for allowing me to use their resources for bringing out this book. Special thanks to Dr. Subhash C. Bhardwaj, Wheat and Barley Rust Specialist, ICAR-IIWBR RS Flowerdale, Shimla, India, and his team members, Dr. Robert Saxon Brueggeman, South Dakota, USA, and his team, Dr. S. P. Val-Moraes, Brazil, and Dr. A. K. Joshi, CIMMYT, India, and his team for readily agreeing to contribute and for their strong commitments for the project.

Appreciation to family members, Mrs. Sadhna Singh, Mr. Anubhav Singh, and Ms. Surbhi Singh, for their regular help, love and encouragement to the editor while he was working on the assignment. The editor is highly grateful to Dr. J. S. Sandhu, DDG (CS), ICAR, and Dr. Ravi P. Singh, Distinguished Scientist and Head, Wheat Improvement and Rust Research Global Program, CIMMYT, Mexico, for contributing forewords to the book. Thanks are due to Ms. Sandra Jones Sickels, Vice President, Editorial and Marketing of Apple Academic Press as well as Mr. Ashish Kumar and Mr. Rakesh Kumar for working together to publish the book.

PART I
Disease Identification and Management Practices

CHAPTER 1

STRATEGIC DISEASE MANAGEMENT IN WHEAT AND BARLEY

DEVENDRA PAL SINGH*

ICAR-Indian Institute of Wheat and Barley Research, Karnal 132001, Haryana, India

Corresponding author. E-mail: dpkarnal@gmail.com

CONTENTS

ABSTRACT

Wheat and barley suffer mostly from diseases of fungal origin. Amongst these leaf, stripe and stem rusts, spot blotch, powdery mildew, and loose smut are common but are caused by different species of fungi. Karnal bunt occurs in wheat only and is of importance in international grain trade. Covered smut is an important disease of barley only in cooler regions. Cereal cyst nematode is important in wheat and barley whereas ear cockle nematode occurs in wheat only. All diseases are not important in different agro-ecological regions. Stripe rust occurs in cooler and humid agro-ecological zones whereas leaf and stem rusts like warmer climate. Powdery mildew, smuts, and bunts are of importance in cooler and humid climate. Spot blotch is quite important disease on both crops in warm and humid climate. The diseases and nematodes affect the wheat and barley crop at different growth stages and plant parts. The diseases like rusts are multiple cycle in nature whereas loose smut and flag smut are simple interest disease and easy to manage. The rusts are quite evolved for developing newer pathotypes thus posing continuous threat to wheat cultivars. Powdery mildew and wheat blast pathogens are known to develop resistance to fungicides in a faster way than other pathogens. The chapter deals with diagnosis and strategies to manage the important diseases of wheat and barley under different agro-ecological zones.

1.1   INTRODUCTION

The diseases may affect the wheat and barley yields from 1 to 100% depending on the susceptibility of varieties, virulence of pathotypes of pathogens, growth stage of crop at infection, favorable weather conditions, and time of availability of inocula and nutrients. The typical symptoms of diseases may appear in field as leaf edging, lesions, and discoloration of plant parts. The damage due to diseases generally initiates in the parts of field and in a gradual manner. The disease problems in bread wheat (*Triticum aestivum*), durum wheat (*Triticum durum*), emmer wheat (*Triticum dicoccum*), and in triticale are almost identical. In barley also, the diseases like rusts, powdery mildew, smuts, and spot blotch are mostly similar to wheat in appearance but are caused by different pathogens or their pathotypes. The yield of affected crop is reduced due to low numbers of spikes per unit area, total grains per spike, and thousand grain weights. The reduced plant stand happens due to reduced seed viability, poor seedling vigor, poor tillering, and poor seed setting.

1.1.1 LOSSES

Spring wheat losses due to fusarium head scab in South Dakota were 5,800,000 bu in wheat alone, with an estimated loss in value of $17,800,000 during epidemic of 1999. The yield losses mainly due to leaf and stripe rusts, tan spots, root rot, crown rot and Septoria blight complex during 2003 were up to 22.3% in South Dakota, USA (Ruden, 2016). Appel et al. (2016) reported 8.4% losses due to different wheat diseases in Kansas, USA, on an average of 20 years. The incidence, severity, and yield loss caused by 41 pathogens were assessed from a survey in Australia. The losses due to diseases were of $913 × 10^6/year or 19.5% in the decade from 1998–1999 to 2007–2008. The three most important pathogens were *Pyrenophora tritici-repentis, Puccinia striiformis,* and *Phaeosphaeria nodorum,* with the average annual losses of $212 × 10^6, $127 × 10^6, and $108 × 10^6, respectively. Cultural methods (rotation, paddock preparation) used for 10 pathogens > 50% of the control. Breeding and the use of resistant cultivars contributed more than 50% of control for seven pathogens and pesticides for three pathogens. The relative importance of pathogens varied between regions and zones (Murray & Brennan, 2009). The losses due to leaf rust, powdery mildew, spot blotch, and scab were in the range of 18–20, 8, 14–19, and 8%, respectively, depending on wheat cultivar, in Brazil. The overall losses due to different foliar diseases were 19–42% (Luz, 1984). Rana et al. (2006) estimated losses in wheat yield due to powdery mildew in Himachal Pradesh, India during 1994–1997 and these were in the range of 9–47%. The cereal cyst nematode (CCN) caused yield losses in wheat in Northern India ranging from 8 to 49% during 1974–1976 (Bhatti et al., 1981). The losses due to spot blotch were up to 63.4% in wheat yield in Haryana, India (Malik et al., 2008b). Singh et al. (2004) recorded higher losses in popular wheat cultivars due to leaf blight in warmer regions of India as compared to northwestern plain zone. The timely sown crop suffered higher losses. The losses were in the range of 2–39%. In case of susceptible varieties to spot blotch, the losses in yield were up 51% in peninsular, 41% in Northeastern, and 27% in Northwestern zones of India (Singh et al., 2002). The losses due to spot blotch were recorded in panicle weight, number of grains per panicle, and grain weight per panicle in case of spot blotch of wheat and reductions were in the range of 17–48, 22–59, and 25–56%, respectively, in case of six genotypes. Infection of *Bipolaris sorokiniana* was high in case of discolored seeds (90–100%) than that in black pointed and healthy looking seeds (20–60%). The losses in seed germination were high in discolored seeds (90–100%) (Singh & Kumar, 2008). The yield losses due to leaf rust in susceptible wheat varieties in Northern India were in the range of 11–26% with average coefficient of infection (ACI) of 71 in unsprayed control (Singh, 1999).

In barley, the yield losses due to stripe disease were in the range of 20–70% during 1992–1993 (Kumar et al., 1998). Net blotch is second only to scald as a cause of yield loss in barley, which can range between 20 and 30%. Net blotch is more common than scald in the warmer, drier regions of Canada. The losses due to sport blotch and net blotch in barley in Haryana, India were 53% in case of susceptible cultivars (Singh, 2004a).

1.1.2   DISEASE TRIANGLE

The appearance of a plant disease is mainly as a result of interaction between susceptible host, a virulent pathogen, and an environment favorable for disease development. These components are therefore quite important in disease management and one of these components is eliminated to manage plant diseases effectively.

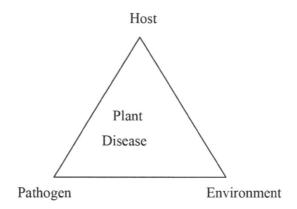

FIGURE 1.1   Plant disease triangle.

The terrestrial plants have little thermal storage capacity and are immobile. The immune system is also absent in plants. Therefore, genetic constitution of plants is quite important for managing diseases which are mainly caused by fungi, and are highly dependent on environment. A host may be immune or susceptible to a disease. The high degree of pathogen virulence and favorable environmental conditions are important factors for disease development (Fig. 1.1). The fourth factor representing human activities like altered crop cultivation practices, uniformity of genetic makeup of cultivar covering large acreages, genetic manipulation, conservation agriculture, irrigation methods, green house cultivation, hydroponics, etc. also contributes in change in crop environment, expression of host resistance, survival of pathogen, and vector population (Francl, 2016).

1.2 DISEASES OF WHEAT AND BARLEY

Wheat and Barley crops suffer from by fungi, bacteria, viruses, and nematodes. The fungi are major pathogens whereas nematodes, generally remain in soil and weaken the vigor of plant. The fungal diseases are known to cause epidemics in wheat and barley. The major diseases and their pathogens of wheat and barley are shown in Table 1.1 (Mathre, 2016; Murray, et al., 2016).

TABLE 1.1 Diseases of Wheat and barley along with Their Pathogens.

Diseases	Pathogens	
	Wheat	**Barley**
Fungal diseases		
Barley stripe	-	*Drechslera graminea* (Rabenh.) Shoemaker
Leaf (brown) rust	*Puccinia triticina* Eriks. Syn *P. recondita* Roberge ex Desmaz. f. sp. *tritici* (Eriks. & E. Henn.) D. M. Henderson, *P. tritici-duri* Viennot-Bourgin	*Puccinia hordei* Otth
Stripe (yellow) rust	*P. striiformis* Westend. anamorph *Uredo glumarum* J. C. Schmidt	*P. striiformis* Westend.
Stem (black) rust	*Puccinia graminis* Pers.: Pers. Syn. *P. graminis* Pers.:Pers. f. sp. *tritici* Eriks. & E. Henn.	*P. graminis* Pers.:Pers.
Crown rot	-	*Puccinia coronata* Corda
Powdery mildew	*Erysiphe graminis* DC. f. sp. *tritici* Em. Marchal, *Blumeria graminis* (DC.) E. O. Speer, Syn. *E. graminis* DC., *Oidium monilioides* (Nees) Link [anamorph]	*Erysiphe graminis* DC. f. sp. *hordei* Em. Marchal, Syn. *B. graminis* (DC.) E. O. Speer, *Oidium monilioides* (Nees) Link [anamorph]
Alternaria leaf blight	*Alternaria triticina* Prasada & Prabhu	-
Halo spot	-	*Pseudoseptoria donacis* (Pass.) Sutton, Syn. *Selenophoma donacis* (Pass.) R. Sprague & A. G., Johnson Syn. *Septoria donacis*
Spot blotch	*B. sorokiniana* (Sacc.) Shoemaker Syn. *Helminthosporium sativus*	*Cochliobolus sativus* (Ito & Kuribayashi) Drechs. ex Dastur, *B. sorokiniana* (Sacc.) Shoemaker [anamorph]
Net Blotch	-	*Drechslera teres* (Sacc.) Shoemaker, *Pyrenophora teres* Drechs. [teleomorph]

TABLE 1.1 *(Continued)*

Diseases	Pathogens	
	Wheat	**Barley**
Stripe disease	-	*Drechslera graminea* (Rabenh.) Shoemaker
Stagonospora blotch	-	*Stagonospora avenae* f. sp. *triticea* T. Johnson, *Phaeosphaeria avenaria* f. sp. *triticea* T. Johnson [teleomorph], *Stagonospora nodrum* (Berk.) Castellani & E. G. Germano, Syn. *Septoria nodorum* (Berk.) Berk. in Berk. & Broome, *P. nodorum* (E. Muller) Hedjaroude [teleomorph]
Covered smut	-	*Ustilago hordei* (Pers.) Lagerh.
False loose smut	-	*Ustilago avenae* (Pers.) Rostr., Syn. *U. nigra* Tapke
Speckled snow mold		*Typhula idahoensis* Remsberg
Leaf spot	-	*Ascochyta hordei* K. Hara (United States, Japan), *A. graminea* (Sacc.) R. Sprague & A. G. Johnson, *A. sorghi* Sacc., *A. tritici* S. Hori & Enjoji
Scald	-	*Rhynchosporium secalis* (Oudem.) J. J. Davis
Scab (head) blight	*Fusarium* spp., *Gibberella zeae* (Schwein.) Petc *Fusarium graminearum* Schwabe, Group II [anamorph], *G. avenacea* R. J. Cook *F. avenaceum* (Fr.:Fr.) Sacc. [anamorph], *F. culmorum* (Wm. G. Sm.) Sacc. *F. nivale* Ces ex Berl. & Voglino,	*Fusarium* spp., *F. graminearum* Schwabe
Stagonospora blotch	*Phaeosphaeria avenaria* (G. F. Weber) O. Eriksson f. sp. *triticea* T. Johnson, *Stagonospora avenae* (A. B. Frank) Bissett f. sp. *tritica*, T. Johnson [anamorph], *Septoria avenae* A. B. Frank f. sp. *triticea* T. Johnson, *P. nodorum* (E. Müller) Hedjaroude, *Stagonospora nodorum* (Berk.) Castellani & E. G. Germano [anamorph], Syn. *Septoria nodorum* (Berk.) Berk. in Berk. & Broome	*Stagonospora avenae* f. sp. *triticea* T. Johnson, *Phaeosphaeria avenaria* f. sp. *triticea* T. Johnson [teleomorph], *Stagonospora nodrum* (Berk.) Castellani & E. G. Germano = *Septoria nodorum* (Berk.) Berk. in Berk. & Broome, *Phaeosphaeria nodorum* (E. Muller) Hedjaroude [teleomorph]
Septoria speckled leaf blotch	-	*Septoria passerinni* Sacc., *Stagonospora avenae* f. sp. *triticea* T. Johnson
Sharp eyespot	-	*Rhizoctonia cerealis* Van der Hoeven *Ceratobasidium cereale* D. Murray & L. L. Burpee [teleomorph]
Flag smut	*Urocystis agropyri* (G. Preuss) Schrot.	
Karnal (partial) bunt	*Tilletia indica* Mitra syn. *Neovossia indica* (M. Mitra) Mundk.	

TABLE 1.1 *(Continued)*

Diseases	Pathogens	
	Wheat	**Barley**
Loose smut	*Ustilago tritici* (Pers.) Rostr. Syn. *U. segetum tritici, U. segetum nuda, U. segetum avena*	*Ustilago tritici* (Pers.) Rostr., Syn. *U. nuda* (C. N. Jensen) Rostr., nom. nud.
Covered smut	-	*Ustilago hordei* (Pers.) Lagerh.
False loose smut	-	*Ustilago avenae* (Pers.) Rostr. Syn. *U. nigra* Tapke
Tan spot, yellow leaf spot, red smudge	*P. tritici-repentis* (Died.) Drechs., *Drechslera tritici-repentis* (Died.) Shoemaker [anamorph]	*P. tritici-repentis* (Died.) Drechs., Syn. *P. trichostoma* (Fr.) Fuckel. *Drechslera tritici-repentis* (Died.) Shoemaker [anamorph] Syn. *Helminthosporium tritici-repentis* Died.
Septoria blotch	*Septoria tritici* Roberge in Desmaz., *Mycosphaerella graminicola* (Fuckel) J. Schröt. In Cohn [teleomorph]	-
Zonate eyespot	*Drechlera gigantean* (Heald & Wolf) Ito	-
Common bunt, stinking smut	*T. tritici* (Bjerk.) G. Wint. in Rabenh. Syn. *Tilletia caries* (DC.) Tul. & C. Tul. *T. laevis* Kühn in Rabenh., *T. foetida* (Wallr.) Liro	-
Crown rot, foot rot, seedling blight, dry land root rot	*Fusarium* spp., *F. pseudograminearum* O'Donnell et. T. Aoki sp. nov., *Gibberella zeae* (Schwein.) Petch, *F. graminearum* Schwabe, Group II [anamorph], *G. avenacea* R. J. Cook, *F. avenaceum* (Fr.:Fr.) Sacc. [anamorph], *F. culmorum* (W. G. Smith) Sacc.	-
Pythium root rot	*Pythium aphanidermatum* (Edison) Fitzp., *P. arrhenomanes* Drechs., *P. graminicola* Subramanian, *P. myriotylum* Drechs, *P. volutum* Vanterpool & Truscot	*Pythium* spp., *P. arrhenomanes* Drechs., *P. graminicola* Subramanian *P.tardicrescens* Vanderpool
Common root rot	*B. sorokiniana* (Sacc.) Shoemaker, *Cochliobolus sativus* (Ito & Kurib-ayashi) Drechs. Ex Dast., [teleomorph]	*Bipolaris sorokiniana* (Sacc.) Shoemaker, *Fusarium culmorum* (Wm. G. Smith) Sacc., *F. graminearum* Schwabe, *Gibberella zeae* (Schwein.) Petch [teleomorph]
Anthracnose	*Colletotrichum graminicola* (Ces.) G. W. Wils., *Glomerella graminicola* Politis [teleomorph]	*Colletotrichum graminicola* (Ces.) G. W. Wils., *Glomerella graminicola* Politis [teleomorph]
Ascochyta leaf spot	*Ascochyta tritici* S. Hori & Enjoji	-
Aureobasidiu-m decay	*Microdochium bolleyi* (R. Sprague) DeHoog & Hermanides-Nijhof Syn. *Aureobasidium bolleyi* (R. Sprague) Arx	-
Black head molds, sooty molds	*Alternaria* spp., *Cladosporium* spp., *Epicoccum* spp., *Sporobolomyces* spp., *Stemphylium* spp., and other genera	-

TABLE 1.1 *(Continued)*

Diseases	Pathogens	
	Wheat	**Barley**
Cephalosporiu-m stripe	*Hymenula cerealis* Ellis & Everh. Syn. *Cephalosporium gramineum* Nisikado & Ikata in Nisikado *et al*	*Hymenula cerealis* Ellis & Everh. (*Cephalosporium gramineum* Nisikado & Ikata in Nisikado *et al.*)
Cottony snow mold	*Coprinus psychromorbidus* Redhead & Traquair	-
Downy mildew, crazy top	*Sclerophthora macrospora* (Sacc.) Thirumalachar et al.	*Sclerophthora rayssiae* Kenneth *et al.*
Dilophospora leaf spot, twist	*Dilophospora alopecuri* (Fr.) Fr.	-
Dwarf bunt	*Tilletia controversa* Kühn in Rabenh	*Tilletia controversa* Kuhn in Rebenh.
Take-all	*Gaeumannomyces graminis* (Sacc.) Arx & D. Olivier var. *tritici* J. Walker, *G. graminis* (Sacc.) Arx & D. Olivier var. *avenae* (E. M.Turner) Dennis	*Gaeumannomyces graminis* var. *tritici* J. Walker
Ergot	*Claviceps purpurea* (Fr.:Fr.) Tul., *Sphacelia segetum* Lév. [anamorph]	*Claviceps purpurea* (Fr.:Fr.) Tul., *Sphacelia segetum* Lev. [anamorph]
Eyespot, foot rot, straw breaker	*Tapesia yallundae* Wallwork & Spooner, *Ramulispora herpotrichoides* (From) Arx [anamorph], Syn. *Pseudocercosporella herpotrichoides* (Fron) Deighton, *T. acuformis* (Boerema, Pieters & Hamers) Crous, *Ramulispora acuformis* (Nirenberg) Crous [anamorph], Syn. *Pseudocerco-sporella herpotrichoides* var. *acuformis* Nirenberg	*Pseudocercosoporella herpotrichoides* (Fron) Deighton, *Tapesia yallundae* Wallwork & Spooner [teleomorph]
False eyespot	*Gibellina cerealis* Pass.	-
Halo spot	*Pseudoseptoria donacis* (Pass.) Sutton, Syn. *Selenophoma donacis* (Pass.) R. Sprague & A. G. Johnson	*Pseudoseptoria donacis* (Pass.) Sutton (*Selenophoma donacis* (Pass.) R. Sprague & A.G. Johnson), *Septoria donacis*
Foot rot, dry land foot rot	*Fusarium* spp.	-
Leptosphaeria leaf spot	*Phaeosphaeria herpotrichoides* (De Not.) L. Holm, Syn. *Leptosphaeria herpotrichoides* De Not., *Stagonospora* sp. [anamorph]	-
Microscopica leaf spot	*Phaeosphaeria microscopica* (P. Karst.) O. Eriksson, Syn. *Lepto-sphaeria microscopica* P. Karst	-
Phoma spot	*Phoma* spp., *P. glomerata* (Corda) Wollenweb. & Hochapfel, *P. sorghina* (Sacc.) Boerema et al., Syn. *P. insidiosa* Tassi	-

TABLE 1.1   *(Continued)*

Diseases	Pathogens	
	Wheat	**Barley**
Pink snow mold, Fusarium patch	*Microdochium nivale* (Fr.) Samuels & I. C. Hallett, Syn. *Fusarium nivale* Ces. ex Berl. & Voglino, *Monographella nivalis* (Schaffnit) E. Müller [teleomorph]	*Microdochium nivale* (Fr.) Samuel & I. C. Hallett (*Fusarium nivale* (Fr.) Sorauer, *Monographella nivalis* (Schaffnit) E. Müller [teleomorph]
Platyspora leaf spot	*Clathrospora pentamera* (P. Karst.) Berl. Syn. *Platyspora pentamera* (P. Karst.) Wehmeyer	-
Rhizoctonia root rot	*Rhizoctonia solani* Kühn, *Thanatephorus cucumeris* (A. B. Frank) Donk [teleomorph]	*Rhizoctonia solani* Kuhn, *Thanatephorus cucumeris* (A. B. Frank) Donk [teleomorph]
Ring spot, wirrega blotch	*Pyrenophora semeniperda* (Brittlebank & Adam) Schoemaker, Syn. *Drechslera campanulata* (Lev.) Sutton, *D. wirreganesis* Wallbork, Lichon & Sivanesan	-
Sclerotinia snow mold, snow scald	*Myrioslerotinia borealis* (Bubák & Vleugel) L. M. Kohn Syn. *Sclerotinia borealis* Bubák & Vleugel	-
Sclerotium wilt	*Sclerotium rolfsii* Sacc., *Athelia rolfsii* (Curzi) Tu & Kimbrough [teleomorph]	-
Sharp eyespot	*Rhizoctonia cerealis* Van der Hoeven, *Ceratobasidium cereale* D. Murray & L. L. Burpee [teleomorph]	-
Snow scald (Sclerotinia snow mold)	-	*Myrioslerotinia borealis* (Bubak & Vleugel) L. M. Kohn (*Sclerotinia borealis* Bubak & Vleugel)
Snow rot	*Pythium* spp., *P. aristosporum* Vanterpool, *P. iwayamai* Ito, *P. okanoganense* Lipps	*Pythium iwayamai* Ito, *P. okanoganense* Lipps, *P. paddicum* Harane
Southern blight, sclerotium base rot	*Sclerotium rolfsii* Sacc., *Athelia rolfsii* (Curzi) Tu & Kimbrough [teleomorph]	*Sclerotium rolfsii* Sacc. (India, California, Puerto Rico), *Athelia rolfsii* (Curzi) Tu & Kimbrough [teleomorph]
Kernel blight (black point)	-	*Alternaria* spp. *Arthrinium arundinis* Dyko. & Sutton, *Apiospora montagnei* Sacc. [teleomorph], *Cochliobolus sativus* (Ito & Kuribayashi) Drechs. Ex Dastur, *Fusarium* spp.
Speckled snow mold, gray snow mold or typhula blight	*Typhula idahoensis* Remsberg, *T. incarnata* Fr., *T. ishikariensis* Imai, *T. ishikariensis* Imai var. *canadensis* Smith & Arsvoll	*Typhula idahoensis* Remsberg
Snow molds	-	*Typhula incarnata* Fr., *T. ishikariensis* Imai
Storage molds	*Aspergillus* spp., *Penicillium* spp., and others	-

TABLE 1.1   *(Continued)*

Diseases	Pathogens	
	Wheat	**Barley**
Zoosporic root rot	*Lagena radicicola* Vant & Ledingham, *Ligniera pilorum* Fron & Gaillat, *Olpidium brassicae* (Woronin) Dang., *Rhizophydium graminis* Ledingham	-
Tar spot	*Phyllachora graminis* (Pers.:Fr.) Nitschke, *Linochora graminis* (Grove) D. G. Parbery [anamorph]	-
Verticillium wilt	-	*Verticillium dahliae* Kleb. (Idaho)
Wheat blast	*Magnaporthe oryzae pathotype Triticum (*Hebert) Barr (anamorph *Pyricularia grisea* (Cooke) Sacc.)	-
Wirrega blotch	-	*Drechslera wirreganensis* Wallwork *et al.* (Australia)
Bacterial diseases		
Bacterial leaf blight	*Pseudomonas syringae* subsp. *syringae* van Hall	*Pseudomonas syringae* pv. *syringae* van Hall
Bacterial kernel blight	-	*Pseudomonas syringae* pv. *syringae* van Hall
Black chaff and bacterial streak	-	*Xanthomonas translucens* pv. *translucens* (ex Jones *et al.* Vauterin *et al.*)
Bacterial stripe	-	*Pseudomonas syringae* pv. *striafaciens* (Elliott) Young *et al.*
Basal glume rot	-	*Pseudomonas syringae* pv. *atrofaciens* (McCulloch) Young *et al.*
Bacterial mosaic	*Pseudomonas fuscovaginae* (ex Tanii *et al.*) Miyajima *et al.*	-
Bacterial sheath rot	*Pseudomonas fuscovaginae* ex Tanii *et al.* Miyajima *et al.*	-
Black chaff, bacterial streak	*Xanthomonas translucens* pv. *undulosa* (Smith *et al.*) Vauterin et al.	-
Pink seed	*Erwinia rhapontici* (Millard) Burkholder emend. Hauben *et al.*	-
Stem melanosis	*Pseudomonas cichorii* (Swingle) Stapp	-
Spike blight	*Rathayibacter tritici* (Carlson & Vidaver ex Hutchinson) Zgurskaya et al.	-
Gumming	*Rathayibacter iranicus* (Carlson & Vidaver ex Scharif) Zgurskaya et al.	-
Viral diseases and virus like agents		
Agropyron mosaic	*Rymovirus, Agropyron mosaic virus* (AgMV)	-

TABLE 1.1   *(Continued)*

Diseases	Pathogens	
	Wheat	**Barley**
African cereal streak	-	African cereal streak virus (Ethiopia, Kenya)
Chloris striate mosaic	*Monogeminivirus, Chloris striate mosaic virus* (CSMV)	-
Barley mild mosaic	-	genus *Bymovirus, Barley mild mosaic virus* (BaMMV) (Europe, Japan)
Barley mosaic	-	Barley mosaic virus (Japan)
Barley yellow mosaic	-	*Bymovirus, Barley yellow mosaic virus* (BaYMV) (Europe, Japan)
Barley yellow streak mosaic	-	Barley yellow streak mosaic virus
Barley yellow striate mosaic	*Cytorhabdovirus, Barley yellow striate mosaic virus* (BYSMV)	-
Barley yellow dwarf	*Luteovirus, Barley yellow dwarf virus* (BYDV)	*Luteovirus, Barley yellow dwarf virus (BYDV)*
Barley stripe mosaic	*Hordeivirus, Barley stripe mosaic virus* (BSMV)	*Hordeivirus, Barley stripe mosaic virus* (BSMV)
Brome mosaic	-	*Bromovirus, Brome mosaic virus* (BMV)
Cereal northern mosaic	*Cytorhabdovirus, Cereal northern mosaic virus* (NCMV)	-
Cereal tillering	-	*Reovirus, Cereal tillering disease virus* (CTDV) (Italy, Sweden, Finland)
Barley yellow stripe	-	Virus-like agent (Turkey)
Brome mosaic	*Bromovirus, Brome mosaic virus* (BMV)	-
Cereal northern mosaic (barley yellow striate mosaic)	-	genus *Cytorhabdovirus, Northern cereal mosaic virus* (NCMV) (Japan, Korea, Pakistan, Siberia)
Chloris striate mosaic	-	*Monogeminivirus, Chloris striate mosaic virus* (CSMV) (Australia)
Cocksfoot mottle Maize streak	*Sobemovirus, Cocksfoot mottle virus* (CoMV) *Monogeminivirus, Maize streak virus* (MSV)	-
Eastern wheat striate	-	Eastern wheat striate virus (India)
Enanismo	-	Virus like agent (Columbia, Equador)
Hordeum mosaic	-	*Rymovirus, Hordeum mosaic virus* (HoMV)
Northern cereal mosaic	*Cytorhabdovirus, Cereal northern mosaic virus* (NCMV)	-

TABLE 1.1 *(Continued)*

Diseases	Pathogens	
	Wheat	**Barley**
Oat sterile dwarf	*Fijivirus, Oat sterile dwarf virus* (OSDV)	-
Oat blue dwarf	-	*Marafivirus, Oat blue dwarf virus* (OBDV)
Oat pseudorosette	-	*Tenuivirus, Oat pseudorosette virus* (Siberia)
Oat sterile dwarf	-	*Fijivirus, Oat sterile dwarf virus* (OSDV)
Rice black-streaked dwarf	*Fijivirus, Rice black-streaked dwarf virus* (RBSDV)	*Fijivirus, Rice black-streaked dwarf virus* (RBSDV) (Japan, Korea, China, Russia)
Rice stripe	-	*Tenuivirus, Rice stripe virus* (RSV) (Japan, Korea, China, Russia)
Rice hoja blanca	*Tenuivirus, Rice hoja blanca virus* (RHBV)	-
Russian winter wheat mosaic	*Cytorhabdovirus, Russian winter wheat mosaic virus* (WWRMV)	-
Tobacco mosaic	*Tobamovirus, Tobacco mosaic virus* (TMV)	-
Wheat American striate mosaic	*Nucleorhabdovirus, Wheat American striate mosaic virus* (WASMV)	-
Wheat yellow leaf	-	*Closterovirus, Wheat yellow leaf virus* (WYLV) (Japan)
Russian winter wheat mosaic	-	Winter wheat Russian mosaic virus (WWRMV)
Wheat dwarf	*Monogeminivirus, Wheat dwarf virus* (WDV)	*Monogeminivirus, Wheat dwarf virus* (WDV)
Wheat European striate mosaic	*Tenuivirus, Wheat European striate mosaic virus* (EWSMV)	-
Wheat soilborne mosaic	*Furovirus, Wheat soilborne mosaic virus* (SBWMV)	*Furovirus, Wheat soil-borne mosaic virus* (SBWMV)
Wheat streak mosaic	*Rymovirus, Wheat streak mosaic virus* (WSMV)	*Ryemovirus, Wheat streak mosaic virus* (WSMV)
Wheat yellow leaf	*Closterovirus, Wheat yellow leaf virus* (WYLV)	-
Nematode disease		
CCN	*Heterodera avenae* Wollenweber	*Heterodera avenae* Wollenweber, *H. filipjevi* (Madzhidov) Stelter, *H. latipons* Franklin
Root-knot nematode	*Meloidogyne* spp., *M. naasi* Franklin, *M. chitwoodi* Santos	*Meloidogyne* spp. *M. naasi* Franklin, *M. artiellia* Franklin *M. chitwoodi* Golden *et al.*
Seed gall (wheat gall) nematode	*Anguina tritici* (Steinbuch) Chitwood	*Subanguina radicicola* (Greeff) Paramonov

TABLE 1.1 *(Continued)*

Diseases	Pathogens	
	Wheat	**Barley**
Bulb and stem nematode	*Ditylenchus dipsaci* (Kühn) Filepjev	-
Dagger, American nematode	*Xiphinema americanum* Cobb	-
Grass cyst nematode	*Punctodera punctata* (Thorne) Mulvey & Stone	-
Lesion nematode	*Pratylenchus* spp., *P. thornei* Sher & Allen, *P. neglectus* (Rensch et al.) Filepjev	*Pratylenchus* spp.
Ring nematode	*Criconemella* spp.	-
Root gall nematode	*Subanguina* spp., *S. radicicola* (Greeff) Paramonov	-
Spiral nematode	*Helicotylenchus* spp.	-
Stubby-root nematode	*Paratrichodorus* spp., *P. minor* (Colbran) Siddiqi	-
Stunt nematode	*Merlinius brevidens* (Allen) Siddiqi	*Merlinius brevidens* (Allen) Siddiqi, *Tylenchorhynchus dubius* (Butschli) Filipjef, *T. maximus* Allen
Parasitic higher plants		
Witchweeds	*Striga* spp., *S. lutea* Lour. Syn. *S. asiatica* (L.) Kuntze	-
Phytoplasmal diseases		
Aster yellows	Aster yellows phytoplasma	Aster yellows phytoplasma

1.3 DIAGNOSIS OF DISEASES IN FIELD

The first step in diagnosis of diseases is to look the growth of crop in field. If crop has patches, then plants having patches are compared with good-looking crop. Any change in number and size of leaves, color of spikes, and developing grains is observed. Besides diseases, the other factors affecting crop condition may be soils (alkaline, saline), sunlight (scorching, fog), temperature (too low, too high), frost, crop husbandry (crop canopy, establishment, nutrients, residue of previous crop in cropping system, lodging, tillage practices, water quality and irrigation methods, and weeds). The insects, rodents, and birds may also affect both underground and upper parts of plants and seed storage. The nutrients may be deficient (NPK, Cu, Zn, S, Mn, Mg, Cu, Fe, Mo, B, and Ca) and in excess thus causing toxicities (Al, B, Mn, Fe, P, and salts).

The important viral diseases of wheat are barley yellow dwarf virus (BYDV), brome mosaic virus (BMV), and soil borne wheat mosaic virus (SBWMV) and wheat streak mosaic virus (WSMV). Bacterial stripe, basal glume rot, and bacterial spike blight are diseases caused due to bacteria. The major fungal diseases of wheat are, root rot, crown rot, bare patch, pythium root rot, sclerotium wilt, take-all, stem rust, leaf rust, stripe rust, eye spot, sharp eye spot, flag smut, powdery mildew, spot blotch, septoria blight, tan spot, alternaria blight, fusarium head blight, downy mildew, common and dwarf bunt, loose smut, ergot, black molds, Karnal bunt, wheat blast, and black point. The important nematodes are CCN, root knot nematode, root lesion nematode, and seed gall or ear cockle nematode. The important insects are, aphids, sting bugs, army worms, cut worms, cereal leaf beetle, thrips, hessian fly, wheat stem maggot, saw fly, grasshoppers, white grubs, wire worms, mites, etc. Barley also gets similar diseases but which are caused by different pathogens and sub species. The diseases like covered smut, scald, and barley stripe only occur on barley (Mathre, 1997).

1.4 SYMPTOMS OF WHEAT AND BARLEY DISEASES

The diseases of wheat and barley may be identified looking at the affected plants carefully for timely management of crop health. The characteristic symptoms of different wheat and barely diseases are listed below.

1.4.1 LEAF RUST

Infection sites are primarily found on the upper surfaces of leaves and leaf sheaths, and occasionally on the neck and awns. The pustules are circular or slightly elliptical. These are smaller than those of stem rust, usually do not coalesce, and contain masses of orange to orange brown urediospores (Fig. 1.2).

1.4.2 STRIPE RUST

The pustules contain yellow to orange-yellow urediospores, and form narrow stripes on the leaves. It may also develop on leaf sheaths, necks, and glumes. The disease develops in patches initially in field and spreads later in whole field. The yellow powder of urediospores may be seen on soil under the severely infected plants (Fig. 1.3).

(a)   (b)

FIGURE 1.2   Symptoms of leaf rust of (a) wheat and (b) barley.

(a)   (b)

FIGURE 1.3   Stripe rust on (a) wheat and (b) barley.

1.4.3   STEM RUST

Pustules are dark reddish brown in color, and may occur on both sides of the leaves, on the stems, and on the spikes. With light infections, the pustules are usually separate and scattered, but with heavy infections they may coalesce. The infected leaves show "flecks" initially. Before the spore masses break through the epidermis, the infection sites feel rough to the touch; as the spore masses break through, the surface tissues take on a ragged and torn appearance (Fig. 1.4).

(a)   (b)

FIGURE 1.4   Urediopustules of stem rust on (a) wheat leaf and (b) on stem of barley.

1.4.4   SPOT BLOTCH

The pathogen develops water-soaked specks within 48 h of inoculation which later turn in to dark brown spots. Later, these develop in to elongate to oval shaped, dark brown color lesions. The lesions may be or may not be surrounded by light yellow colored halo. These when coalesce forms bigger spots and leaf blight. The symptoms generally develop first on lower leaves and moves upward and spikes may be blighted under highly favorable conditions. The seeds develop in these spikes are shriveled, thin, and discolored (Figs. 1.5 & 1.6).

FIGURE 1.5 Spot blotch infected (a) leaf, (b) spike, (c) eye spots on seeds, (d) discolored seeds of wheat.

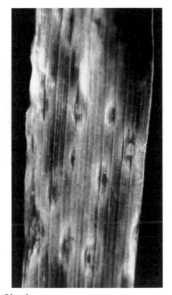

FIGURE 1.6 Spot blotch of barley.

1.4.5 WHEAT BLAST

The lesions develop on the spikelet as light brown elongated spots. Spike infection is the most notable symptom of the disease. Wheat blast is mainly a spike disease but can also produce lesions on all the above ground parts of the plant under certain conditions (Kohli et al., 2011) (Fig. 1.7). It was first reported from Brazil and now reached in Bangladesh.

FIGURE 1.7   Blast infected wheat spike, Source: Dr. Pawan Kumar Singh (CIMMYT)

1.4.6   NET BLOTCH OF BARLEY

The plant parts, leaves, stems, panicles, and seeds are affected due to net blotch. The pathogen infects healthier leaves more quickly than drier, older leaves. Initially, brown spots or blotches appear near the tips of the seedling leaves. Each spot has a darker brown stripe running at 90° to each other within the lighter brown area of the spot. The spots form net-like appearance and often surrounded by areas of chlorosis and areas of dead tissue. The blotches later grow in length to form narrow streaks. These may coalesce forming long brown stripes with irregular darker margins; dark green water-soaked spots; narrow brown blotches with netted appearance, surrounding tissue yellow; and stripes running the length of leaf (Fig. 1.8).

FIGURE 1.8   Net blotch of barley.

1.4.7   BARLEY STRIPE

The infected seedlings show small yellow spots on leaves; yellow to tan stripes along leaf blade before heading. The strips later result death of diseased tissue. The plant dies before emergence of panicle (Fig. 1.9).

FIGURE 1.9   Barley stripe disease.

1.4.8   TAN SPOT OF WHEAT

At first, lesions appear as tan to brown flecks, which expand into large, irregular, oval to lens shaped tan blotches with a yellow or chlorotic margin. As these spots coalesce, large blotches are formed. The development of a dark brown to black spot in the center of the lesion is the characteristic of the disease. As the disease progresses, entire leaves, spikes, and even whole plants may be killed.

1.4.9   ALTERNARIA LEAF BLIGHT OF WHEAT

Small, chlorotic, oval, or elliptical shaped lesions appear and, as they enlarge, these lesions become irregular in shape. The chlorotic borders of the lesions may become diffuse and turn light to dark brown in color (Fig. 1.10). Infection usually starts on the lower leaves, but symptoms can be found on all plant parts. The disease appears in the field along with spot blotch and difficult to identify by seeing the symptoms.

FIGURE 1.10   Symptoms of Alternaria blight of wheat produced in glass house conditions.

1.4.10   SEPTORIA BLOTCH

Initial infection sites tend to be irregular in shape, oval to elongated chlorotic spots or lesions. As these sites expand, the centers of the lesions become pale, straw colored, and slightly necrotic, often with numerous small black dots (pycnidia). The lesions of septoria blotch tend to be linear and restricted laterally. Heavy infection can kill leaves, spikes, or even the entire plant.

1.4.11   FLAG SMUT OF WHEAT

Masses of black teliospores are produced in narrow strips just beneath the epidermis of leaves, leaf sheaths, and occasionally the culms. Diseased plants often are stunted, tiller profusely, and the spikes may not emerge. A severe infection usually induces the leaves to roll, producing an onion-type leaf appearance. The epidermis of older diseased plants tends to shred, releasing the teliospores (Fig. 1.11).

FIGURE 1.11   Flag smut infected wheat plant.

1.4.12   LOOSE SMUT

The entire inflorescence, except the rachis, is replaced by masses of smut spores. These black teliospores often are blown away by the wind, leaving only the bare rachis and remnants of other floral structures. Smutted spikes often emerge several days before the healthy spikes and in some wheat varieties like "Sonalika" showed pale colored flag leaf even in case of emergence of smutted spike. The delicate seed membranes rupture shortly after head emergence, exposing masses of dark brown to black spores (Fig. 1.12). Seed infected with the loose smut fungi appears healthy.

(a)   (b)

FIGURE 1.12   Loose smut infected spikes of (a) wheat and (b) barley.

1.4.13   COVERED SMUT OF BARLEY

The growth of affected tillers is stunted. The spike emergence is late. The kernels in infected spikes are replaced with grey fungal masses. These masses are broken into small balls and powdery teliospores are dispersed and adhere to seed coat of healthy seeds (Fig. 1.13).

FIGURE 1.13   Covered smut of barley.

1.4.14 KARNAL BUNT OF WHEAT

Karnal bunt is difficult to detect prior to harvest. Only few kernels per spike may be affected by the disease. Following harvest, diseased kernels can be easily detected by visual inspection: a mass of black teliospores replaces a portion of the endosperm, and the pericarp may be intact or ruptured. The infection may be slight to severe in grain. Diseased kernels give off a fetid or fishy odor when crushed (Nagarajan et al., 1997) (Fig. 1.14).

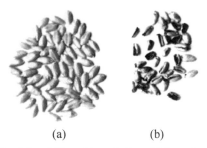

(a)   (b)

FIGURE 1.14   Healthy (a) and Karnal bunt infected (b) seeds.

1.4.15 POWDERY MILDEW

The initial symptoms appear as white to pale gray, fuzzy, or powdery colonies of mycelia, and conidia on the upper surfaces of leaves and leaf sheaths, and sometimes on the spikes. Older fungal tissue is yellowish grey. Host tissue beneath the fungal material becomes chlorotic or necrotic and, with severe infections, the leaves may dry and die. Eventually, black spherical fruiting structures (cleistothecia) may develop in the mycelia and can be seen without magnification (Fig. 1.15).

(a)   (b)

FIGURE 1.15   Powdery mildew of wheat on (a) spikes and (b) leaf.

1.4.16   TAKE-ALL

The affected field has poorly defined patches. The infected plants are stunted, have few tillers, and display drought symptoms. White heads are produced after flowering where plants produce no seed. Whole plants are affected. The rotting of the roots and lower stems may occur. The basal stem, leaf sheath tissues, and roots may turn a shiny black color.

1.4.17   COMMON ROOT ROT

The sub crown internodes, coleoptiles, and leaf sheaths are dark or brown in color. Brown streaks may also appear on the basal internodes of tillers. The tillering is reduced with slight root rotting. Individual plants or groups of plants may lodge. Infection early in the crop development can cause pre- or post-emergence "damping off" of seedlings.

1.4.18 SCLEROTIUM WILT

It results pre- or post-emergence "damping off" of seedlings. White, fluffy fungal mycelia on the surface may appear on affected plant parts. Later, the crop is affected due to rotted culms, crowns, and roots and such plants die. It leads to "white heads." Sclerotia are commonly found on the crown tissues, culms, or near the soil surface. Young sclerotia are whitish and turn brown to dark brown with age.

1.4.19 BYDV

The affected plants show a yellowing or reddening of leaves, stunting, and an upright posture of thickened stiff leaves, reduced root growth, delayed heading, and a reduction in yield. The spikes of affected plants tend to remain erect and become black and discolored during ripening due to colonization by saprophytic fungi.

1.4.20 BACTERIAL STRIPE OR BLACK CHAFF

Black chaff occurs primarily on the glumes. Bacterial stripe appears primarily on the leaves and leaf sheaths. Narrow chlorotic, water-soaked lesions or stripes with droplets of sticky yellowish exudates are formed during rain or dew. The exudates dry to form crusty droplets. If infection occurs early in the crop cycle, the spike may be infected. The infected spikes are sterile under severe disease conditions and may kill entire plant.

1.4.21 BACTERIAL LEAF BLIGHT

The leaves, culms, and spikes of wheat and triticale can be infected. Initially small, dark green, water-soaked lesions develop that turn dark brown to blackish in color later. The spikes are affected at bases of the glumes and may eventually extend over the entire glumes. Diseased glumes have a translucent appearance when held toward the light. Dark brown to black discoloration occurs with age. Under wet or humid conditions, whitish grey bacterial ooze may be present. Dark discoloration of the stem occurs.

1.4.22 DOWNY MILDEW

Diseased plants tiller profusely; have short, erect, irregular, or crooked yellowish green culms; and the leaves are thick, erect, and usually in whorls. Tillers die prematurely or spikes never develop. If formed, the spikes may be branched, and some of the floral tissues grow into leaf like structures.

1.4.23 COMMON AND DWARF BUNT

The "bunt balls" develops in spike instead of kernels and these are filled with black teliospores. The bunt balls of common bunt are caused by *Tilletia tritici* and *Tilletia laevis*. The dwarf bunt, caused by *Tilletia controversa*, is more nearly spherical. When bunt balls are crushed, they give off a fetid or fishy odor. Infected spikes tend to be bluish green in color, and the glumes tend to spread apart slightly; the bunt balls often become visible after the soft dough stage.

1.4.24 ERGOT

At flowering, infected florets produce yellowish, sticky, and sweet exudates that are visible on the glumes. As the spike matures, kernels of infected florets are replaced by brown to purplish black fungal structures.

1.4.25 CCN

The patches of unthrifty yellowed and stunted plants are seen in case of CCN. The infected roots become "knotted," "ropey," and swollen (Fig. 1.16). Development of root systems is retarded and shallow. In spring, characteristic "white cysts" which are about the size of a pin head can be seen with the naked eye if roots are carefully dug and washed free of soil.

FIGURE 1.16   CCN infected roots.

1.4.26   EAR COCKLE NEMATODE

The nematode causes stunted plants and distorted leaves. The seeds are transformed into galls which contain a dried mass of nematodes (Fig. 1.17).

FIGURE 1.17   Ear cockle infected wheat grains.

1.4.27   FUSARIUM HEAD BLIGHT

Symptoms of fusarium head blight include tan or light brown lesions encompassing one or more spikelets. An orange fungal mass may be seen on glume. The infected grains are often shriveled and have a white chalky appearance. Some kernels may have a pink discoloration.

1.4.28   SOOTY HEAD MOLDS

Dark green or black mold growth develops on the surface of mature wheat heads. Sooty molds are most common when mature wheat is subjected to repeated rains and delayed harvest. The sooty head mold growth is normally superficial. Its effect on grain is thought to be minor, but it can make for dusty harvest operations. Sooty molds can contribute to a discoloration of the grain called "black point."

1.4.29   BARLEY YELLOW DWARF

Stunted growth of plants; yellow green blotches at leaf tip, leaf margin, or leaf blade; leaves turn bright yellow, red, or purple.

1.4.30   COMMON ROOT ROT

Brown lesions on leaves nearest soil extend to stem; resembles drought; death of lower leaves; and rotting of roots.

1.4.31   BASAL GLUME ROT

Brown discoloration at base of the glumes; dark line where glumes attach to spike; water-soaked spots on leaves; yellow and necrotic spots on leaves.

1.4.32   BACTERIAL LEAF BLIGHT, BLACK CHAFF

Water-soaked spots on foliage; shriveling dead leaves; glossy yellow or brown streaks; plant appears stunted; due to slow plant growth.

1.5   GROWTH STAGES AND DISEASES

Not all diseases are important at throughout the growth period of plants. The management of diseases therefore, begins from selection of field, seed, varieties, and continues till grain filling stage. The diseases like foot rot, take-all, loose smut, covered smut, flag smut, common bunt, seedling blight, etc. may be easily managed through using clean field, clean seeds of resistant varieties, and pre-sowing seed treatment with appropriate fungicides (Maloy, 2003). The foliar diseases like powdery mildew, rusts, blights, blotches, blast etc. appear on plants right from seedling stage till physiological maturity. These may be therefore managed by the use of resistant varieties along with the use of foliar sprays of fungicides, biocontrol agents, and balanced use of fertilizers. The infection of bunts and smuts takes place during boot leaf to flowering stage and while threshing of grains. Timely harvest is important to avoid seed discoloration by weaker fungal pathogens.

1.6   STRATEGIES FOR MANAGING DISEASES

It is important to get the information on diseases of wheat and barley during previous growing seasons, to determine the strategies that can be applied. Field location, topography, and wind-direction are also important considerations especially in case of spot blotch of wheat and barley and net blotch of barley, since pathogens of these diseases are windborne. The use of good quality, pathogen-free, non-damaged seed with good germination and vigor accounts for an effective and holistic disease management. The disease free seeds from loose smut, covered smuts, common bunt, dwarf bunt, Karnal bunt, spot blotch, net blotch, blast etc. may avoid early season disease problems, and reduce or eliminate the risk of introducing a seed-borne pathogen to new areas. These measures, along with seed treatment with fungicides, are of much important in case of diseases susceptible varieties in case of wheat and barley (Turkington et al., 2011). Proper seed germination and plant establishment are important aspects for good crop yields. Wegulo (2016) developed strategies for strategic disease management in wheat in Nebraska, USA.

1.6.1   CULTURAL CONTROL

Deep seeding weakens the seedlings and these become vulnerable to attack by soil borne pathogens. Destruction of infested barley residues minimizes buildup of leaf spot epidemics. The adoption of conservation tillage has not led to a dramatic increase in barley leaf spot diseases in Canada. Take-all of wheat (caused by *Gaeumannomyces graminis*) is easily managed by short rotation of one year, out of susceptible crops, which may include susceptible weed hosts such as grasses.

1.6.2 RESISTANT CULTIVARS

The resistant cultivars may be used to contain losses due to different diseases. For this, one must be aware of level of resistance in wheat and barley varieties against different diseases and severity of occurrence of diseases in the agro climatic zone. The varieties grown in Northwestern plain zones are not yielding good in Northeastern plain zones due to heavy attack on these by spot blotch pathogen under warmer conditions in India. Likewise, varieties bred for other zones may suffer badly due to attack of stripe rust in Northern hills and plains zones in India. Singh (2008) reported that barley varieties RD 2508, RD 2035, DWR UB 52, RD 2552, and RD 2624 have multiple disease resistance in India. Out of these, RD 2035 is resistant to CCN also.

Under the All India Coordinated Wheat and Barley Improvement project (AICW&BIP), the breeding materials of wheat and barley at pre-coordinated as well as coordinated yield trials stage are evaluated at hot spot locations and under artificial epiphytotic conditions. Hot spots can be defined as specific sites or locations where the pest populations are endemic in nature and carry adequate variability spectrum. Therefore, evaluating genotypes at these sites gives reliable information on the level of host resistance to diseases, or pests in general (Sharma et al., 2001). The data generated at multilocations are used to promote the resistant genotypes in yield trials and identify resistant varieties.

Planting at least 5–6 varieties of wheat and barley in each agro ecological zone will reduce the chances of developing epidemics. Because of genetic differences, wheat varieties will react differently to diseases and some varieties will mature sooner or later than others. Wheat and barley improvement program in India and in few other countries is actively using new resistant sources, genes and their combinations, molecular techniques, and proper evaluation before release of varieties and therefore avoided epidemics of rusts and other diseases in spite of growing of wheat areas in different agro ecological zones. The resistant wheat and barley varieties are regularly tested under AICWBIP and listed in the annual reports of AICWBIP, Karnal (www.dwr.res.in)

1.6.3   CHEMICAL CONTROL

The use of chemicals in wheat and barley is more common as seed treatment. These may be used as foliar sprays under disease epidemic situation in case of susceptible cultivar especially in case of wheat blast, Karnal bunt, rusts, powdery mildew, head scab and spot blotch.

1.6.3.1   SEED TREATMENT

Seed treatment with fungicides and biocontrol agents is quite beneficial for control of seed-borne infection, getting higher germination and proper seedling vigor. It is the cheapest method in chemical control. However, farmers are sometimes reluctant to follow it due to hidden infection of diseases like loose smut, hill bunt in wheat, and covered smuts of barley in or on seed which results losses in field. Singh (1998) reported effective control of covered smut of barley using seed treatment of carboxin @ 0.25%, tebuconazole @ 0.15%, and *Trichoderma viride*. Wet treatment of fungicides was effective at half dose of fungicides as dry seed treatment. The seed treatment of carboxin and thiram @ 0.25 and 3.0%, respectively, was found most effective in reducing the seed infection of *B. sorokiniana* and increasing germination of seeds (Singh & Kumar, 2008). The infection of loose smut of wheat was controlled fully with the seed treatment of tebuconazole at 0.05% when infected seeds were soaked in the suspension of fungicide overnight at room temperature of 20–25°C. Even lower doses of 0.025 and 0.0125% of tebuconazole could reduce seed infection up to 98 and 82%, respectively (Singh, 2004b). It is lower than the doses of dry seed treatment of (0.1–0.15%) of tebuconazole recommended by Sinha and Singh (1996).

1.6.3.2   FOLIAR SPRAYS

The diseases like Karnal bunt are properly controlled with the application of foliar sprays of fungicides like propiconazole @ 0.1% at boot leaf stage (Goel et al., 2000) and tebuconazole (Sharma et al., 2005). Fungicidal sprays are also advocated to control early infection of rusts in case of susceptible varieties and favorable weather, powdery mildew, and spot blotch in wheat and barley. In Europe, foliar sprays are used a few times in crop season to manage rusts, powdery mildew, head scab, and leaf blotches. It is, however, quite costly and difficult operation and one has to plan the sowing of wheat accordingly to minimize expenses and facilitate effective in foliar sprays. Care may be taken to avoid sprays at grain maturity to reduce residue problem and development of resistance in fungal pathogens. Malik et al. (2008a) found two foliar sprays of propiconazole and defenoconazole at 0.1 and 0.2% respectively, effectively lowered the disease severity of spot blotch in wheat in India and gave increase in grain yield in the range of 40–46% over untreated check plots. Singh et al. (2010) worked out an integrated pest management in barley and use of seed treatment with insecticide and vitavax power @ 3 g/kg of seed, higher dose of K fertilizer (40 kg/ha), and foliar sprays of propiconazole and insecticide resulted good control of diseases and insect pests and yield increase was up to 30%. Sharma et al. (2005) reported effectiveness of new fungicides, tebuconazole (folicur) at 0.2% and contaf 5 EC at 0.1%. These fungicides gave 90% control of Karnal bunt in field.

1.7   MAJOR WHEAT AND BARLEY DISEASES IN DIFFERENT ZONES OF INDIA AND THEIR STRATEGIC MANAGEMENT

Successful management of wheat and barley diseases is achieved in India during past five decades. The diseases vary in occurrence and severity in different agro ecological zones in India. Leaf rust and foliar blights are present in all six zones later are more damaging in warmer and dry zones. The important diseases and their strategic management are given in Table 1.2.

TABLE 1.2   Major Diseases of Wheat and Barley in Different Agro Climatic Zones of India and their Strategic Management.

Agro ecological zone	Important diseases rank wise		Management strategies
	Wheat	**Barley**	
Northwestern plains zone	1. Stripe rust	1. Stripe rust	• Resistant varieties
	2. Leaf rust	2. Leaf rust	• Seed Treatment with systemic fungicides
	3. Powdery mildew	3. Foliar blight complex	
	4. Karnal bunt	4. Loose smut	• Foliar Sprays of fungicides
	5. Loose smut	5. Covered smut	• Crop rotation
	6. Flag smut	6. CCN	
	7. Foliar blight		
	8. CCN		

TABLE 1.2   *(Continued)*

Agro ecological zone	Important diseases rank wise		Management strategies
	Wheat	**Barley**	
Northeastern plains zone	1. Leaf rust	1. Leaf rust	• Resistant varieties
	2. Foliar blight	2. Foliar blight complex	• Seed Treatment
	3. Karnal bunt	3. Loose smut	• Foliar Sprays of fungicides
	4. Loose smut	4. Covered smut	• Crop rotation
	5. Ear cockle nematode	5. CCN	
Central zone	1. Leaf rust	1. Leaf rust	• Resistant varieties
	2. Stem rust	2. Stem rust	• Seed treatment
	3. Foot rot	3. Foliar blight	• Foliar sprays of fungicides
	4. Foliar blight		
Peninsular zone	1. Leaf rust	1. Leaf rust	• Resistant varieties
	2. Stem rust	2. Stem rust	• Foliar sprays of fungicides
	3. Foot rot	3. Foliar blight	
	4. Foliar blight		
Northern hills zone	1. Stripe rust	1. Stripe rust	• Resistant varieties
	2. Leaf rust	2. Leaf rust	• Seed Treatment with systemic fungicides
	3. Powdery mildew	3. Loose smut	
	4. Karnal bunt	4. Covered smut	
	5. Loose smut		• Foliar Sprays of fungicides
	6. Common bunt, Hill bunt		
Southern hill zone	1. Stripe rust	1. Stripe rust	• Resistant varieties
	2. Leaf rust	2. Leaf rust	• Seed Treatment
	3. Stem rust	3. Stem rust	• Foliar Sprays of fungicides
	4. Powdery mildew	4. Loose smut	
	5. Loose smut	5. Covered smut	

The management practices followed in integrated management of diseases is given in Table 1.3.

TABLE 1.3 Management Practices for Effective Diseases Management at Different Stages of Crop Growth.

S. No.	Crop growth stage	Practice to be followed
1.	Pre-sowing	Summer plowing in May–June to reduce inoculum of CCN in endemic areas.
		Use clean and healthy seeds which are harvested from a healthy crop. Also seed should not have nematode galls. The galls are removed by dipping seeds in brine solution (2–5% salt solution in water).
2.	Sowing	Use seeds of newly released disease resistant varieties are recommend.
		Following varieties also have in built resistance to major diseases:
		NHZ (hills and foothills of Northern India): HS507, HS542, HPW42, VL404, VL616, VL832.
		NWPZ (Punjab, Haryana, Rajasthan, and Western UP): HD3086, WB02, HD2687, Raj3765, PBW723, HD2967, DBW71, DBW91, WH1142, WH1105, WH1080, KRL210, KRL213, PBW 660
		NEPZ (Eastern UP, Bihar, West Bengal, Assam and other NE states): HD2967, HD2888, K0307, K1006, K9107
		CZ (Kota region of Rajasthan, Madhya Pradesh, Gujarat): GW322, GW173, GW273, GW1139, Raj 1555, DL788-2, HD8498(d), HI8737, MPO1215(d), UAS 428(d), UAS440(d)
		PZ (Maharashtra, Karnataka, Tamil Nadu): DWR 1006 (d) MACS 2467 (d), HD 2189, HD 2501, NI 5439, NIDW 15 (d), AKDW2997-16 (d), DBW 93, MACS 6222, MACS 6478, NIAW 34, NIAW 1415, NIDW 205(d), Raj 4083
		SHZ: (Nilgiris and Pallini hills of Tamil Nadu): HW 1085, HW 2004, HW 5216.
		In North Indian conditions, the seed should be treated with fungicides like carboxin or carbendazim WP @ 2.5 g/kg of seed or tebuconazole @ 1.25 g/kg of seed to control loose smut infection, if the seed was taken from infected field. The seed treatment should be done one day before sowing.
3.	Boot leaf stage	Use foliar spray of propiconazole @ 0.1% in case of favorable weather (cold temperature with humidity and rains coincide with ear emergence) to control the infection of Karnal bunt especially in fields meant for seed production in Northern hills zone and Northwestern plains zone.
4.	Storage	Do not store the seed with high moisture > 14% and dry it properly before storing to prevent losses due to storage insect pests and molds

The wheat crop grown in India is, up to larger extent, free from the use of pesticides and these are rarely used as foliar sprays except in seed crop.

1.8   CONCLUDING REMARKS

The well-defined and targeted strategic disease management in wheat and barley derives full advantage of the resistance breeding, gene deployment, use of chemical, cultural, biological control measures and their integration. Survey and surveillance program is the key for deciding the strategies and planning for timely actions to be taken in next crop season as well as in the crop of same season, right from field selection, preparations, seed selection, pre-sowing seed treatment, foliar sprays and proper harvesting, threshing and storage. Proper diagnosis as well as growth stage on which a disease appears is important aspects for taking decisions on best methods to be adopted for disease management. Mere presence of disease does not qualify for the use of chemicals. Likewise, late appearance of disease after grain filling has much implications to yield and grain quality. Some diseases which are also transmitted through seed, need careful eradication of inocula in seed before sowing. While use of resistant varieties is most appropriate among methods of disease management, incorporation of resistance in one cultivar to major diseases with other desirable agronomic characters is almost impossible. Further evolution of new pathotypes of pathogens possess extra burden on breeders over time and susceptible varieties need to be replaced with resistant varieties. An integrated approach is therefore, adopted. Efforts are required to keep the wheat and barley crops free of the pesticide residues, keeping in view of proper human and animal health. This approach need to be recast for different agro ecological regions and may be refined further depending on the levels of resistance to major diseases in that region.

KEYWORDS

- **wheat**
- **barley**
- **strategic disease management**
- **losses**
- **symptoms**
- **rusts**
- **smuts**

- other foliar diseases
- Karnal bunt
- host resistance
- chemical control
- biological control

REFERENCES

Appel, J. A., et al. Kansas Cooperative Plant Disease Survey Report Preliminary 2014 Kansas Wheat Disease Loss Estimates August 14, 2014. 2016, pp 1–3. (Source: http://agriculture. ks.gov/docs/default-source/PP-Disease-Reports-2014/2014-ks-wheat-disease-loss-estimates.pdf).

Bhatti, D. S., et al. Estimation of Loss in Wheat Yield Due to the Cereal Cyst Nematode *Heterodera avenae. Trop. Pest Manage.* **1981,** *27,* 375–378.

Francl, L. J. APS, Education, Instructor Communication, Teaching Articles. The Disease Triangle: A Plant Pathological Paradigm Revisited. 2016, (Source: http://www.apsnet.org/edcenter/instcomm/TeachingArticles/Pages/DiseaseTriangle.aspx).

Goel, L. B., et al. Evaluation of Tilt against Karnal Bunt. *Indian Phytopathol.* **2000,** *53,* 301–302.

Kohli, M. M., et al. Pyricularia Blast-a Threat to Wheat Cultivation. *Czech J. Genet. Plant Breed.* **2011,** *47,* S130–S134.

KSU. Kansas State University, Wheat Blast. 2016. KSU, Manhattan, KS 66506, USA (Source: https://www.k-state.edu/wheatblast/).

Kumar, V., et al. Estimation of Yield Losses in Barley Due to *Drechslera graminae* the Causal Agent of Stripe Disease. *Indian Phytopathol.* **1998,** *51* (4), 365–366.

Luz, W. C. D. Yield Losses Caused by Fungal Foliar Wheat Pathogens in Brazil. *Phytopathology.* **1984,** *74,* 1403–1407.

Malik, V. K.; Singh, D. P.; Panwar, M. S. Losses in Yield Due to Varying Severity of Leaf Blight Caused by *Bipolaris sorokiniana* in Wheat. *Indian Phytopath.* **2008b,** *61,* 526–527.

Malik, V. K.; Singh, D. P.; Panwar, M. S. Management of Spot Blotch of Wheat (*Triticum aestivum*) Caused by *Bipolaris sorokiniana* Using Foliar Sprays of Botanicals and Fungicides. *Indian J. Agric. Sci.* **2008a,** *78,* 646–648.

Maloy, O. C. *Plant Disease Control: Principles and Practice;* John Wiley and Sons, Inc: New York, 1993; p 346.

Mathre, D. E. *Diseases of Barley (Hordeum vulgare L.);* 2nd ed.; American Phytopathological Society (APS): St. Paul, MN, 1997; p 120.

Mathre, D. E. *Common Names of Plant Diseases, Diseases of Barley (Hordeum vulgare L.);* APS Publications: St. Paul, MN, 2016. (Source: http://www. apsnet.org/publications/common names/Pages/Wheat.aspx).

Murray, G. M.; Brennan, J. P. Estimating Disease Losses to the Australian Wheat Industry. *Australas. Plant Pathol.* **2009,** *38,* 558–570.

Murray, T. D. et al. APS Publications, Common Names of Plant Diseases, Diseases of Wheat (*Triticum* spp. L.). 2016. (Source: http://www. apsnet.org/publications/common names/ Pages/Wheat.aspx).

Nagarajan, S., et al. Karnal Bunt (*Tilletia indica*) of Wheat-a Review. *Rev. Plant Pathol.* **1997,** *76,* 1207–1214.

Rana, S. K., et al. Estimation of Losses Due to Powdery Mildew of Wheat in Himachal Pradesh. *Indian Phytopathol.* **2006,** *59,* 112–114.

Ruden, B. Crop Profile for Wheat (Spring) in South Dakota. Pesticide Impact Assessment Program SDSU Cooperative Extension Service. 2016. (Source: https://ipmdata.ipmcenters. org/documents/cropprofiles/SDwheat-spring.pdf, 2016).

Sharma, A. K., et al. Efficacy of Some New Molecules against Karnal Bunt of Wheat. *Indian J. Agric. Sci.* **2005,** *75,* 369–370.

Sharma, A. K.; Singh, D. P.; Kumar, J. Concept and Use of Multilocation Hot Spot Testing for Identifying Potential Donor Lines/Varieties in Wheat. In *Role of Resistance in Intensive Agriculture;* Kalyani Publishers: Ludhiana, 2001; pp 223–232.

Sinha, V. C.; Singh, D. P. Raxil (tebuconazole) in the Control of Loose Smut of Wheat. *J. Mycol. Pl. Pathol.* **1996,** *26,* 279–281.

Singh, D. P. Assessment of Losses Due to Brown Rust in Two Popular Cultivars of Wheat. *Pl. Dis. Res.* **1999,** *14,* 60–62.

Singh, D. P. Chemical and Biological Control of Covered Smut of Barley. *J. Mycol. Pl. Pathol.* **1998,** *29,* 256–257.

Singh, D. P. Evaluation of Barley Genotypes against Multiple Diseases. *SAARC J. Agric.* **2008,** *6,* 117–120.

Singh, D. P.; Pankaj Kumar. Role of Spot Blotch (*Bipolaris sorokiniana*) in Deteriorating Seed Quality in Different Wheat Genotypes and its Management Using Fungicidal Seed Treatment. *Indian Phytopathol.* **2008,** *61,* 49–54.

Singh, D. P. Assessment of Losses Due to Leaf Blights Caused by *Bipolaris sorokiniana* (Sacc.) Shoemaker and *Helminthosporium teres* (Sacc.) in Barley. *Pl. Dis. Res.* **2004a,** *19,* 173–175.

Singh, D. P. Effect of Reduced Doses of Fungicides and Use of *Trichoderma viride* during Seed Activation Stage in Controlling the Wheat Loose Smut. *J. Mycol. Pl. Pathol.* **2004b,** *34,* 396–398.

Singh, D. P., et al. Losses Caused Due to Leaf Blight in Wheat in Different Agroclimatic Zones of India. *Pl. Dis. Res.* **2002,** *17,* 313–317.

Singh, D. P., et al. Integrated Pest Management in Barley (*Hordeum vulgare*). *Indian J. Agric. Sci.* **2010,** *80,* 437–442.

Singh, D. P., et al. Assessment of Losses Due to Leaf Blight in Popular Varieties of Wheat under Different Sowing Conditions and Agroclimatic Zones in India. *Indian J. Agric. Sci.* **2004,** *74,* 110–113.

Turkington, T. K., et al. Foliar Diseases of Barley: Don't Rely on a Single Strategy from the Disease Management Toolbox. *Prairie Soils Crops J.* **2011,** *4,* 142–150.

Wegulo, Stephen. University of Nebraska–Lincoln, National Institute of Agriculture and Natural Resources Crop Watch, Fall Strategies for Managing Wheat Diseases. 2016. (Source: http://cropwatch. unl.edu/fall-practices-to-counter-wheat-disease).

CHAPTER 2

MANAGEMENT OF RUST DISEASES IN WHEAT AND BARLEY: NEXT GENERATION TOOLS

SIDDANNA SAVADI, PRAMOD PRASAD,
SUBHASH C. BHARDWAJ*, OM P. GANGWAR, HANIF KHAN,
and SUBODH KUMAR

ICAR-Indian Institute of Wheat and Barley Research, Regional Station, Flowerdale, Shimla 171002, Himachal Pradesh, India

Corresponding author. E-mail:scbfdl@hotmail.com

CONTENTS

ABSTRACT

The rust diseases are the destructive diseases of cereal crops. Development of host resistance is an effective means of managing these diseases. Adequate understanding of plant pathogen interactions will facilitate speeding up breeding for disease resistance in plants. Recent advancements in molecular biology and biotechnology have provided molecular tools which facilitate designing novel strategies for developing resistance in crop plants. Strategies based on transgenic and genomics approaches can be deployed for understanding the disease resistance mechanisms and development of resistant cultivars. Transgenic expression of antimicrobial peptides, enzymes that synthesize antifungal metabolites, growth inhibitors, proteins that inhibit fungal virulence proteins, and proteins that induce natural plant defenses are produced in plants upon attack by the fungal pathogens. Besides, studies have shown that animal immunity can be engineered into plants to provide resistance against pathogens. Further, multiple or combination of R genes could be used to develop wheat transgenics with durable and broad spectrum resistance. Furthermore, introduction of non-host resistance (NHR) is another effective means of rust resistance and it needs exploration of molecular mechanisms controlling the NHR. RNAi-based host plant mediated/induced pathogen gene silencing (HIGS) is another strategy recently deployed to explore for fungal resistance. Genomic selection is a novel molecular breeding approach provides opportunity for manipulation of complex traits like yield, drought tolerance, and disease resistance. Very recently, *in vivo* genome editing tools like the clustered regularly interspaced short palindromic repeat (CRISPR)/Cas facilitate precise manipulation of genomes and understanding of disease resistance mechanisms and development of resistant cultivars. Apart from enhancing host plant resistance, other disease management strategies such as integrated disease management (IDM), disease epidemics prediction system, use of new generation fungicides with novel mode of action and nanotechnology based diagnostic devices are also critical for strategic management of diseases.

2.1　INTRODUCTION

The world population is projected to double from six billion to twelve billion in next 50 years if the population continues to grow at the current growth rate (FAO, 2009; Tilman et al., 2011). Feeding such a huge population with increasing diversification of food habit requires food production at a much higher rate than today from the dwindling land and water resources. However, increasing incidences of biotic and abiotic stresses is a major limiting factor in achieving a sustained increase in food production. The rapidly evolving pathogens such as rusts, pose a major obstacle to sustainable food production due to break down of resistance in popular crop cultivars and results in epidemics. In recent years, this kind of breakdown of resistance to diseases and epidemics has been observed in wheat as well as in other field crops (Wellings et al., 2011; Chakraborty et al., 2011).

Among the biotic stresses, the rust diseases (yellow or stripe, brown or leaf and black or stem rusts) are the major threat to sustainable wheat production in the world since these are air borne in nature, able to travel long distances, and evolve new virulent races faster (McDonald & Linde, 2002). In recent times, wheat rust epidemics have caused severe losses in wheat yields in many of the wheat growing countries (Singh et al., 2011; Figlan et al., 2014). The new strain of stem rust, Ug99 is virulent on most of the popular wheat varieties in East Africa, West and South Asia (Rutkoski et al., 2011). Therefore, there is a need for continuous watchfulness and preparedness to prevent repercussions of rust epidemics for ensuring food security.

Among the four general disease control methods, that is, exclusion, eradication, protection, and resistance, the most effective and sustainable in means of rust management is the deployment of resistant cultivars. In the past, efforts of wheat breeders and plant pathologists have had notable successes in breeding rust-resistant wheat varieties. However, now there is an increased need to breed for rust resistance in wheat at faster pace due to changes in global climatic conditions, faster rate of evolution of new rust races, need for reduction in use of fungicides due to environmental and human health concerns, and an emphasis on using sustainable wheat production systems (ICARDA, 2011). Faster development of resistant varieties requires a comprehensive knowledge of genetics of rusts and wheat system interaction. Recent advancements in molecular biology of resistance and high throughput technologies have provided newer tools for understanding plant pathogen interactions in a comprehensive manner which would allow hastening of the varietal development for disease resistance in wheat.

2.2   GENETICS OF RUST RESISTANCE

Genetic resistance in wheat is based on two classes of genes, resistance (R genes), and adult plant resistance (APR) genes. Resistance genes generally provide seedling and race specific resistance, and thus short-lived because single genes resistance are easily defeated by frequent mutations in the pathogen population. In contrast, APR genes function mainly at the adult stage and provide long lasting resistance (Ellis et al., 2014). Both R and APR genes are named as *Lr*, *Sr*, and *Yr* for leaf, stem, and stripe rust resistance, respectively, without distinction between the two classes. Since durable resistance is more desirable, breeders and pathologists are putting more emphasis on discovery, characterization and use of APR genes in wheat breeding programs (Ellis et al., 2014). Genes conferring APR such as *Sr2*, *Sr55*, *Sr56*, *Sr57*, and *Sr58* for stem rust, *Lr34* for leaf rust and *Yr36* for stripe rust have been characterized. In addition, many quantitative trait loci (QTLs) conferring APR have been identified (Ellis et al., 2014; Singh et al., 2015).

2.3   TOOLS AND STRATEGIES FOR RUST RESISTANCE BREEDING

Knowledge derived from basic research on a model plant and pathogen systems on disease resistance mechanisms has been valuable in understanding the basic defense mechanism in plants. However, an adequate understanding of species-specific interactions in disease resistance mechanisms of crop plants and their pathogens will help further in speeding up breeding for disease resistance. In the recent years, advancement in molecular biology and biotechnology has provided tools for molecular level studies hitherto difficult to study crop and pathogen interactions. The molecular level studies will facilitate designing better strategies for developing disease resistance in crop plants. Tools and strategies based on transgenic and genomics approaches can be deployed for understanding the wheat-rust pathogen interaction, and subsequently applying the derived knowledge for development of rust resistant wheat cultivars in short time to manage the fast-evolving races of rust pathogens.

2.4   TRANSGENIC APPROACHES FOR RUST RESISTANCE BREEDING

Transgenic expression of useful traits is regarded as a speedy plant breeding approach to improve the productivity of crops. To date, only a few crops such as soybean, maize, and cotton, and a few countries have benefited from transgenic improvements, but due to high adoption rate, the area under transgenic crops has increased tremendously in the last decade (James, 2010) leading to a significant increase in crop productivity. However, due to extensive regulatory measures and negative public perception, application of transgenic technology to other crops and in other parts of the world remains restricted (Tester & Langridge, 2010). In the past decade, many recombinant DNA technology strategies have been tried in attempts to enhance plant disease resistance to pathogens (Bieri et al., 2003; Sohn et al., 2006; Jia et al., 2010a, 2010b; Huang et al., 2012; Kouzai et al., 2013). A range of products, which include antimicrobial peptides, enzymes that synthesize antifungal metabolites, growth inhibitors, proteins that inhibit fungal virulence proteins, and proteins that induce natural plant defenses are produced in plants upon attack by the fungal pathogens (Punja, 2001). Genes for some of these defense products have been cloned and expressed in plants to achieve disease resistance in susceptible plants (Bliffeld et al., 1999; Osusky et al., 2000; Wally et al., 2009; Huang et al., 2012). Apart from these strategies, studies have shown that animal immunity can be engineered into plants to provide resistance against pathogens (Schillberg et al., 2001; Li et al., 2008; Yajima et al., 2010; Cheng et al., 2015a; Tran et al., 2015). Further, multiple or combination of R genes and APR genes could be used to develop wheat transgenics with durable and broad spectrum resistance (Singh et al., 2015).

2.4.1 R AND APR GENES MEDIATED RESISTANCE

Some host genes responsible for disease resistance have been cloned and characterized in wheat (Huang et al., 2003; Feuillet et al., 2003; Cloutier et al., 2007; Krattinger et al., 2009). Attempts have been made to utilize such genes for engineering disease resistance in wheat crop (Horvath et al., 2002; Brunner et al., 2011). Transgenic approach was explored in wheat using the *Pm3* resistance alleles for powdery mildew resistance. The *Pm3* alleles from different wheat lines were transformed into a susceptible wheat line under the control of maize *Ubiquitin* promoter. Overexpression of most of these alleles conferred improved resistance to powdery mildew in wheat (Brunner et al., 2011). Moreover, resistance to powdery mildew disease was improved when a mixture of these transgenic lines with different functional *Pm3* alleles were grown in the field like the classical multiline approach (Brunner et al., 2012). Likewise, wild relatives of wheat possess a great degree gene or allelic diversity for resistance to different diseases. Introgression of useful genes from wild and related species into cultivated varieties by classical breeding methods is often tedious, time consuming and are also associated with transfer of unwanted traits. Thus, transgenic technology along with a rapid discovery of the relevant genes in the wild germplasm provides a promising way to rapidly develop pre-breeding material for further use in wheat breeding programs (Keller et al., 2015). A leaf rust resistance gene *Lr10*, cloned by map-based approach was transformed into wheat. Transgenic wheat overexpressing *Lr10* gene showed enhanced resistance against leaf rust (Feuillet et al., 2003). A stem rust susceptible barley cultivar was transformed with the *Rpg1* genomic clone containing promoter, gene and 3' non-transcribed sequences. Transgenic barley lines showed improved resistance to *Pgt-MCC*, pathotype of the stem rust fungus *Puccinia graminis* f. sp. *tritici* (Horvath et al., 2003). Allele mining by sequencing of the stem rust resistance gene *Rpg1* in barley germplasm accessions identified a significant polymorphism among the accessions as well as gene regions having relevance in disease resistance. This study also revealed the presence of other stem rust resistance gene/s apart from *Rpg1* (Mirlohi et al., 2008). Expression analysis of two barley iso-lines, a susceptible cultivar and its resistant *Rpg1* transgenic, challenged with two *Pgt* pathotypes (MCC and QCC) identified different signaling components involved in *Rpg1* gene mediated stem rust resistance in barley (Zhang et al., 2008). Transgenic wheat lines carrying *Lr34*, an APR gene showed leaf rust resistance at both seedling as well as adult stage, and in one genetic background it also increased the cold tolerance (Risk et al., 2012). *Sr33* gene introgressed from *Aegilops tauschii* into wheat was cloned by map-based approach and transformed into a susceptible wheat variety. Transgenic lines carrying *Sr33* were resistant to Ug99 race (Periyannan et al., 2013).

2.4.2 *PATHOGEN DERIVED RESISTANCE*

Plants can be protected from diseases with transgenes that are derived from the pathogens themselves, a concept referred to as pathogen-derived resistance. For example, plant viral transgenes can protect plants from infection by the virus from which the transgene was derived. Effectors are molecules secreted by pathogens that manipulate host cell function, thereby aiding infection or triggering defense responses (Kamoun, 2006). Effectors subvert the plant defense by deregulating functions of plant cell in different compartments. Some effectors interact with host targets that are involved in plant defense while others act in self-defense to guard the pathogen from the anti-microbial compounds produced by host plant (Rovenich et al., 2014). Some effectors can be used in crop disease resistance breeding, such as to accelerate R gene cloning and utilization (Vleeshouwers & Oliver, 2014). Many effectors are translocated inside host cells, and they can cause severe physiological changes inside host cells even at very low concentrations. Hence, these fungal effectors can be utilized for generating transgenic plants with resistance to different diseases (Zhang et al., 2015). Recently, expression of *Phytophthora sojae* CRN (crinkling and necrosis) effector encoding gene, *PsCRN115*, in *Nicotiana benthamiana*, significantly improved disease resistance without affecting the plant growth and development, suggesting plant defense could be improved through ectopic expression of effector genes in plants (Zhang et al., 2015).

Broad-spectrum resistance can also be engineered by synchronized expression of an R gene and a cognate *Avr* transgene, driven by a pathogen-inducible promoter (De Wit, 1992). This strategy enables induction of defense by multiple pathogens without pyramiding numerous *R* transgenes. Furthermore, this system might avoid the frequent break down of resistance in crop varieties, which is generally observed with *R* genes because selection pressure on pathogens for forming new strains through mutations is very low in *Avr* transgene compared to *R* genes. The crux of this strategy lies in choosing an appropriate pathogen inducible promoter that drives the *Avr* gene rapidly to a wide variety of pathogens without causing burden on plant metabolism due to leaky expression of the *Avr* transgene. Therefore, it is necessary to identify such pathogen inducible promoters or design synthetic promoters by combining *cis* regulatory elements associated with defense against multiple pathogens to engineer resistance based on *Avr* protein (Rushton et al., 2002). Attempts to design synthetic promoters have resulted in some configurations that are strongly induced by pathogens, but quiescent under disease free conditions. Designing of such synthetic promoters might provide important tools for engineering *R/Avr* resistance for durable resistance in plants (van der Biezen, 2001).

So far, only six rust effector proteins have been characterized: AvrP123, AvrP4, AvrL567, AvrM, RTP1, and PGTAUSPE-10-1. However recently, genomic and transcriptomic sequence analyses of wide range of rust fungi have led to discovery of hundreds of small secreted proteins which are considered as rust candidate secreted effector proteins (CSEPs) (Petre et al., 2014). Functional characterization of these candidates will allow their utilization in rust pathogen resistance in wheat and barley.

2.4.3 ANTIMICROBIAL PROTEINS (AMPS) MEDIATED DISEASE RESISTANCE

Antimicrobial proteins are produced in plants apart from proteins that are specifically induced in resistant plants during a defense response against pathogens. In general, AMPs counter microbial infections through the hypersensitive response (HR) which pertains with rapid death of the plant cells resulting in isolation of pathogen and a consequence of it restricting the disease spread to other parts (Wang &Wang, 2004; Mur et al., 2008). The antifungal activity of AMP is believed to be through their role in fungal cell lysis or interference with cell wall synthesis. Chitin, a component of fungal cell wall component has been implied as target for AMPs (Boman, 2000; Nawrot et al., 2014), and binding of AMPs to chitin induces permeabilization or pore formation in fungal membrane (van der Weerden, 2010; Nawrot et al., 2014). AMPs have been classified into different families such as defensins, thionins, lipid transfer proteins, puroindolines, snakins, cyclotides, and hevein-like proteins (Nawrot et al., 2014).

Many AMPs have been isolated from different plant species and viewed to be involved in either constitutive or induced resistance to various pathogens (Sun et al., 2008; Yang et al., 2008). Introduction of different AMPs in many plant species through plant transformation have enhanced resistance to a wide range of pathogens (Hancock &Lehrer, 1998; Cary et al., 2000). Transgenic overexpression of *Beauveria bassiana* chitinase gene, *Bbchit1*, in poplar enhanced resistance to *Alternaria alternata* and *Colletotrichum gloeosporioides* (Jia et al., 2010a). Overexpression of *Leonurus japonicus* AMP gene encoding a plant non-specific lipid transfer protein, *LJAMP2*, in tobacco significantly increased resistance to fungal and bacterial pathogens (Yang et al., 2008). The *LJAMP2* gene was also introduced into Chinese white poplar (*Populus tomentosa*), and transgenic plants showed increased resistance against infection caused by pathogenic fungi (Jia et al., 2010b). Recently, a barley *chit-2* and a wheat *ltp* were introduced singly and in combination into carrot by *Agrobacterium*-mediated transformation. The level of disease resistance in plants expressing both genes was double that of single gene transformants against *Botrytis* and *Alternaria* (Jayaraj & Punja, 2007). In another study, two genes *Bbchit1* and *LJAMP2* were introduced singly and in combination into *P. tomentosa*. The level of disease reduction in double-transgenic lines was between 82 and 95%, compared to 65 and 89% in single-gene transformants carrying either *LJAMP2* or *Bbchit1* (Huang et al., 2012). These results indicate that genes deployed in combination give better resistance against fungal pathogens than the single genes.

Egorov et al. (2005) carried out analysis of cationic anti-microbial peptides from seeds of *Triticum kiharae*, a synthetic allopolyploid of *Triticum timopheevii* and *A. tauschii*, which has species status. *T. kiharae* is a hardy crop and is highly resistant to most of the wheat pathogens (Dorofeev et al., 1979). A total of 24 novel and widely diverse AMPs were isolated and characterized (Egorov et al., 2005). Such diverse group of AMPs from wild and related species of wheat will provide suitable arsenals to thwart attacks of rust pathogens in wheat by introducing them through plant transformation.

The puroindoline proteins (PINs) are proteins found in *Triticeae* tribe which affect the wheat grain quality, and also display antibacterial and antifungal properties. Antimicrobial activity of the PINs (PINA and PINB) against fungal pathogens has been reported under *in vitro* conditions (Dubreil et al., 1998). The antimicrobial property of PINs is thought to be due to the hydrophobic tryptophan domain (TRD), which has lipid-binding properties (Dubreil et al., 1998; Bolar et al., 2000). Two tetraploid wheat cultivars were transformed with *pin*A sequences and it increased antifungal activity against the *Puccinia triticina* in the transgenic lines of tetraploid wheat (Luo et al., 2008). Alfred et al. (2013) have shown spraying of the PuroA, Pina-R39G, and PuroB peptides adversely affect the morphology and number of the stripe rust spores (*Puccinia striiformis* f. sp. *tritici*) on wheat seedlings, suggesting these peptides to be effective against the rust diseases of wheat. Above cited strategies and AMPs active against rust pathogens could be effectively used to develop rust resistant wheat transgenics.

2.4.4 PLANTIBODY-MEDIATED RESISTANCE

Plant defense mechanisms against pathogen attack are different from animals; there is no intrinsic immune system in plants as in animals. Antibodies present in animals as a part of the vertebrate adaptive immune system, bind to specific foreign antigens, and eliminate them from the body. Plants can be engineered to express an antibody against a pathogen protein essential for pathogenesis to provide resistance to the pathogen. In other words, an animal strategy of defense is imported to the plant kingdom to fight the challenges posed by the pathogens. This approach is described as plantibody-mediated disease resistance. With comprehensive knowledge of the pathogen life cycle and products, it is possible to counter any disease by designing expression constructs so that pathogen-specific antibodies accumulate at high levels in appropriate sub-cellular compartments. Plantibodies mediated disease resistance concept was first developed to tackle viral pathogens, now it has been extended to a diverse range of plant pathogens. In addition, it has been deployed for detection of proteins involved in plant–pathogen interactions (Safarnejad et al., 2011).

Whole antibodies and their fragments produced in a plant can bind to specific pathogen antigens that come in contact with the host and inactivate them. Thus, sequestration of pathogen proteins necessary for completion of the infection cycle by antibodies will inactivate the pathogen and prevent disease development (Safarnejad et al., 2011). Antibodies can also be deployed for the immune modulation of endogenous targets, for example, inhibiting the activity of a host enzyme or metabolite that may be involved in the pathogenesis and this strategy can also be used in basic research to find out the roles of host factors involved in the pathogenesis (Nölke et al., 2006; Pfalz et al., 2011).

Due to the complexity in cloning the entire antibody molecules, and improper folding and assembly of antibodies due to differences in post-translational modifications between plants and mammals, whole antibody production in plants has not been successful (Hiatt et al., 1989; De Neve et al., 1993). In contrast, expression of only the variable regions of the antibodies that are involved in antigen binding has been successful (Bird et al., 1988). These smaller antibody different diseases molecules are less prone to folding and other post translation modifications, and subcellular targeting related problems in plants. Single chain variable fragment (scFv) antibodies have been successfully expressed in plants and bacteria (Chaudhary et al., 1990; Galeffi et al., 2002).

Antibodies mediated resistance in plants was first time explored in model plant *Arabidopsis* as a strategy to prevent fungal diseases by Peschen et al. (2004). They expressed a fusion protein comprising a recombinant scFv recognizing a surface protein of *Fusarium oxysporum* f.sp. *matthiola,* linked to an antifungal protein from *Aspergillus giganteus*. *Arabidopsis* transgenic plants expressing the antibody fusion protein showed a high level of resistance when challenged with the pathogen (Peschen et al., 2004). In a complementary study the same fusion protein was expressed in transgenic wheat, and the transgenic plants showed strong resistance against *Fusarium asiaticum* (Li et al., 2008). Recently, gene encoding a scFv recognizing the pathogen *Sclerotinia scerotiorum* was expressed in rapeseed (*Brassica napus*). Transgenic plants expressing the scFv showed less severe symptoms than wild type plants (Yajima et al., 2010).

To achieve broad-spectrum and durable resistance, use of antibody stacking, that is, co-expression of two or more different scFvs with different target specificities in a single transgenic line was suggested by Schillberg et al. (2001). However, broad-spectrum and durable resistance can be easily achieved by constructing diabodies or dimerized antibodies because multivalent antibodies have greater avidity and can bind specifically to a unique range of pathogens (Safarnejad et al., 2011). Thus, identification of monoclonal antibodies against rust pathogens, followed by single scFv, multiple scFv stacking, and multivalent antibody transgenic strategies can be deployed to develop rust resistant wheat varieties.

2.4.5   SYNTHETIC R GENES MEDIATED RESISTANCE

Increase in genome sequencing projects and functional genomics studies will facilitate cloning of *R* genes and other resistance related genes in crop plants including wheat. The availability of large number of cloned *R* genes will permit investigation of domains contributing disease resistances. Domains contributing to resistance against different pathogens could be utilized to design synthetic *R* genes providing broad and durable resistance. Further, the gene-shuffling approach could be utilized for generating a library of recombinants, and used for the production of libraries of potential synthetic *R* genes (Jones, 2001).

Two groups investigating domains responsible for different disease resistances in *Cf-4* and *Cf-9* genes found that 10–18 amino acids in leucine rich repeat (LRR) domain are most important for different recognition capacities (Van der Hoorn et al., 2001; Wulff et al., 2001). The substitution of two amino acids in particular domain of *Pm3* alleles enhanced their capacity to induce a HR in *Nicotiana*. The same substitutions in *Pm3f* allele in wheat not only improved disease resistance but also broadened the resistance spectrum. The results of this study suggest that one can enhance the resistance spectrum of an existing gene via minimal targeted modifications in the specific domains. Further, this also allows the designing of synthetic gene configurations with newer specificities and more durable type of resistance from the available major genes (Keller et al., 2015) (Fig. 2.1).

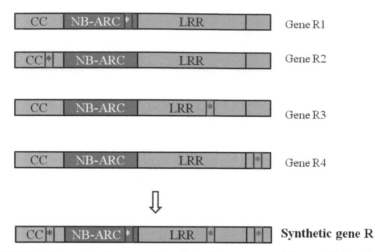

FIGURE 2.1   Schematic diagram of synthetic R gene created by combining the critical regions of different gene forms conferring resistances to different races of pathogen.

*Indicates gene region critical for conferring resistance against particular race of pathogen.

2.4.6   INTRODUCTION OF NON-HOST RESISTANCE FROM RICE

The ability of all the species of a plant to resist infection by all isolates of a pathogen species is termed non-host resistance (NHR). For instance, all species of rice are resistant to all known rust diseases, although other cereals are susceptible. Hence, rice has NHR against rust pathogens. NHR involve preformed physical or chemical barriers to stop establishment of infection structures and induced defense mechanisms (Heath, 2000; Mysore & Ryu, 2004). However, molecular mechanisms controlling the NHR remain relatively unexplored.

NHR against the powdery mildews in *Arabidopsis* is demonstrated to involve two multi component defense systems, firstly, *PEN* gene mediated resistance acting prior to pathogen invasion and is followed by *EDS1/PAD4/ SAG101* genes controlled resistance activated after the invasion by pathogen (Collins et al., 2003; Lipka et al., 2005; Stein et al., 2006; Lipka et al., 2010). Understanding molecular mechanisms controlling NHR against rust pathogens has been slower in contrast to that against powdery mildew pathogens (Boyajyan et al., 2014).

Histological and cytological studies in *Puccinia*-Gramineae and *Uromyces*-dicotyledons patho-systems have demonstrated that invasion of non-host species by most of the rust pathogens is halted rapidly once the first haustorium mother cell is formed (Heath, 1981; Niks et al., 1983; Hoog-kamp et al., 1998). Recent molecular studies conducted to study non-host-rust patho-systems such as growth of *P. graminis, P. triticina, P. striiformis, P. hordei, Melampsora lini* on rice (Ayliffe et al., 2011a, 2011b) emphasized that the NHR to rust fungi is under polygenic control.

Arabidopsis is a non-host to *Phakopsora pachyrhizi* causing Asian Soybean Rust in soybean. Recently, transcriptional profiling of *pen* mutants of *Arabidopsis* has allowed the identification of genes involved in post-invasion induced NHR, *PINGs* such as *bright trichomes 1* (*BRT1*) (Langenbach et al., 2013). Transformation of Arabidopsis *BRT1* resulted in reduction of Asian Soybean Rust disease symptoms in soybean transgenic lines (Conrath et al., 2013). Likewise, identifying the mechanism controlling the NHR against rust in rice and transferring it to wheat and other rust susceptible cereals can provide broad and durable resistance against the rust pathogens. Efforts are on for identifying the NHR mechanism in rice using different molecular biology and genomics tools (Ayliffe et al., 2011; Li et al., 2012a).

2.4.7 IN VIVO GENOME EDITING MEDIATED DISEASE RESISTANCE

Genome editing is precise modification of targeted endogenous nucleic acid molecules to bring about desirable changes in organisms. Recently, development of tools based on site-specific nucleases (SSNs), such as zinc-finger nucleases (ZFNs), transcription activator-like effector nucleases (TALENs), and the clustered regularly interspaced short palindromic repeat (CRISPR)/ Cas, have facilitated targeted *in vivo* gene editing in many organisms (Gaj et al., 2013). Some of the successful examples of application of these genome editing methods in plants include creation of fragrant rice by knocking out *OsBADH2* (Shan et al., 2015) and creation of Celiac-safe wheat by elimination of gluten by ectopic expression of gluteinases in wheat (Wen, 2014).

Among the genome editing methods, TALENs are fusion proteins created by linking the DNA recognition repeats of TAL effectors (proteins secreted by bacterial type III secretion system) with the DNA cleavage domains of Type II restriction endonucleases (Christian et al., 2010; Li et al., 2011). Recent studies have shown that custom TALEs can be created to recognize specific DNA sequences in different types of cells of plant and animals (Zhang et al., 2011). Such TALEs fused with endonucleases can be used to create site-specific double strand break in endogenous genes to facilitate homologous recombination for genome engineering applications (Miller et al., 2011; Mahfouz et al., 2011). TALENs have been recognized for the *in vivo* modifications of the host R genes to modify their level and range of resistance against pathogens. Recently, TALENs have been used to engineer disease resistance in crop plants (Li et al., 2012b; Wang et al., 2014).

TALENs were used to edit bacterial blight susceptibility gene, *Os11N3*, in rice to provide resistance against the pathogen, *Xanthomonas oryzae. X. oryzae* pv. *oryzae* using its TAL effectors *AvrXa7* activates *Os11N3* gene (encoding a member of the SWEET sucrose-efflux transporter family), and thus redirect nutrients from the plant cell so as to satisfy its nutritional needs (Antony et al., 2010). Two pairs of designer TALENs were used to induce mutations in *Os11N3* promoter to interfere with the virulence function of *AvrXa7*, but not the developmental function of *Os11N3* (Li et al., 2012b). Recently, TALEN technology was used to introduce site-specific mutations in the three homoeoalleles of mildew resistance locus (*MLO*) gene encoding MLO proteins in wheat. TALEN-induced mutation in three *TaMLO* homoeologs of same plant conferred heritable and broad-spectrum resistance to powdery mildew (Wang et al., 2014). Further, they used CRISPR technology to generate transgenic wheat plants that carry mutations in the *TaMLO-A1* allele which conferred resistance to powdery mildew in transgenic plants (Wang et al., 2014).

TALENs could also be utilized for precise transgene stacking by combining it with the uni-directional site-specific recombination systems. This strategy could be deployed for stacking a combination of R genes required for durable resistance at a specific locus in the plant genome. Site-specific stacking allows all inserted R genes to inherit together in progeny and also their efficient removal as a unit, if required. In addition, such a stacking method could be used to target transgenes to a "safe spot" in a plant genome that have decreased potential for transgene escape into wild relatives (Yau et al., 2013).

2.4.8   RNA INTERFERENCE (RNAI) TECHNOLOGY

RNA silencing is a gene expression regulatory mechanism found in eukaryotic organisms, and is referred to as gene quelling in fungi, co-suppression in plants and RNA interference (RNAi) in animals. RNA silencing is a nucleotide sequence-specific degradation of mRNA or inhibition of m-RNA translation at the post-transcriptional level (PTGS in plants) or epigenetic modification at the transcriptional level by RNA-directed DNA methylation (RdDM in plants) (Baulcombe, 2004).

Using RNAi strategy one can down regulate the pathogen genes involved in invasion, growth, and pathogenicity in plant system by introducing dsRNA sequences specific to those genes. This strategy is now widely regarded as host plant mediated/induced pathogen gene silencing (HIGS). In the past, HIGS has been mainly utilized to produce virus resistant plants and recently it has been explored for fungal pathogen resistance as well (Baulcombe, 2015). Interactions between fungal pathogens and their host cells is different from the viral particles because in case of fungus interaction occurs via a specialized cell called a haustorium, unlike viral pathogens, which multiplies inside the infected plant cells. The extrahaustorial matrix is formed between plant cell and fungal haustorium membrane, which acts as interface for signal exchange and nutrient uptake from host cell (Panstruga, 2003). Similarly, if dsRNA or siRNA silencing signals generated in the host cell can cross this barrier, gene silencing may be triggered in haustorial cells and possibly interfere with pathogenicity or other metabolic processes.

Recently, RNAi has been used as a tool to silence the expression of fungal or *Oomycete* genes (Nowara et al., 2010; Tinoco et al., 2010; Helber et al., 2011; Pliego et al., 2013; Panwar et al., 2013a, 2013b; Vega-Arreguı́n et al., 2014) as well as target genes essential for virulence of a pathogen to engineer resistance in plants (Koch et al., 2013; Govindarajulu et al., 2015; Hu et al., 2015; Cheng et al., 2015b). Apart from it, one can also deploy abundant and diverse plant siRNAs involved in silencing of endogenous genes that are targets of fungal pathogens for host invasion for engineering resistance in crop plants (Baulcombe, 2015). HIGS via transgenic expression of the dsRNA of *Blumeria graminis* genes conferred resistance to powdery mildew disease in barley (Nowara et al., 2010; Tinoco et al., 2010). It was observed that transgenics producing dsRNAs of *B. graminis* genes reduced powdery mildew symptoms significantly, but in transgenic control plants that had lost the hairpin RNAi cassette were as susceptible as wild type control plants suggesting trafficking of dsRNA or siRNA from host plants into *B. graminis* (Nowara et al., 2010). Although *B. graminis* pathogen is not closely related to rust fungi, but rusts are similar to *B. graminis* and the other powdery mildews in their biotrophic habit and formation of haustorial cells and extrahaustorial matrix (Szabo & Bushnell, 2001; Mendgen & Hahn, 2002). Hence, a proof of HIGS concept in conferring resistance against *B. graminis* is an encouraging sign for application HIGS in engineering rust resistance in wheat. Recently, a repertoire of small RNAs induced in wheat stripe rust fungus (*Puccinia striiformis* f.sp. *tritici*) during pathogenesis were deciphered (Mueth et al., 2015), which provide sRNA-targets for HIG based engineering of stem rust resistance in wheat. Yin et al. (2011) have demonstrated RNAi silencing of some *P. striiformis* f. sp. *tritici* genes using virus induced gene silencing (VIGS) system suggesting (Fig. 2.2) (Table 2.1).

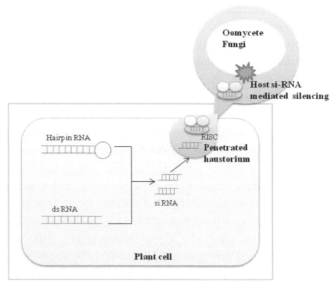

FIGURE 2.2   Schematic diagram showing host induced silencing (HIGS) of virulent genes in *Oomycete* fungi.

VIGS could be used as a powerful reverse genetics tool for analyses of gene function in rust fungi haustorial cells. In addition VIGS will be instrumental in designing gene constructs for engineering RNAi mediated defenses in wheat plants (Yin et al., 2011). VIGS has been used for functional analysis of *Lr21*-mediated leaf rust (Scofield et al., 2005) and *Sr33*-mediated stem rust resistance (Periyannan et al., 2013) in wheat.

Apart from genes contributing to resistance, plants also have genes that are required for susceptibility to certain pathogens. Vogel et al. (2002) identified a gene, *PMR6,* required for susceptibility to powdery mildew in Arabidopsis Col-0. *PMR6* encodes a pectate lyase-like protein with a novel C-terminal domain. Mutation of *PMR6* conferred resistance to powdery mildew in Arabidopsis. However, *pmr6*-mediated resistance needed neither salicylic acid (SA) nor jasmonic acid or ethylene, for signal transduction, and thus does not require the activation of well-described defense pathways. From the results of Vogel et al. (2002) study can be inferred that RNAi strategy can be employed to silence genes in host that are responsible for its susceptibility to certain pathogens.

TABLE 2.1 Examples of Using the RNAi Technology for Engineering Resistance against Fungal Pathogens.

Host plant	Target gene	Pathogen species	Effects	Reference
Wheat	*MLO*	*B. graminis* f. sp. *tritici*	*MLO9*	Riechen (2007)
Barley	*Avra10*	*B. graminis*	Reduced fungal development in the absence of the cognate R gene *MLO10*	Nowara et al. (2010)
Tobacco	*GUS*	*Fusarium verticillioides*	Proof of concept	Tinoco et al. (2010)
Barley and wheat	*PSTha12J12* (haustorial Pst transcript)	*Puccinia striiformis* f. sp. *tritici* or *P. graminis* f. sp. *tritici*	No obvious reductions in rust development or sporulation	Yin et al. (2011)
Wheat	*PtMAPK1* (MAP kinase), *PtCYC1* (cyclophilin) and *PtCNB* (calcineurin B)	*P. triticina*, *P. graminis* and *P. striiformis*	Disease suppression, reducedfungal growth and sporulation	Panwar et al. (2013a, 2013b)
Arabidopsis and barley	*CYP51A, CYP51B* and *CYP51C*	*Fusarium graminearum*	Resistance	Koch et al. (2013)

2.5   GENOMIC APPROACHES FOR RUST RESISTANCE BREEDING

In wheat several kinds of molecular markers have been developed and located on genetic maps; and many of these markers have been used to tag the genomic regions contributing to qualitative and quantitative traits and some of which are being routinely used in marker-assisted breeding programs. A large number of expressed sequence tags (ESTs) have been generated for wheat and have been used for development of functional markers, generation of transcript maps, and construction of microarrays. In recent years, the availability of wheat genome sequences has further increased the genomic resources in wheat. These genomic resources combined together with new approaches such as expression analyses, association mapping, allele mining, along with bioinformatics tools have potential to identify and clone genes responsible for a trait and to explore their use in classical as well as transgenic wheat breeding (Keller et al., 2015). A major problem in map-based isolation of genes is the insufficient marker coverage in a targeted region. This problem can now be approached more efficiently using chromosome sorting technique combined with next generation sequencing, with this one can successfully increase the number of single nucleotide polymorphism (SNP) markers in a specific chromosomal region (Shatalina et al., 2013).

2.5.1   MARKER-ASSISTED SELECTION

Marker-assisted selection (MAS) provides opportunities for enhancing the response from selections in breeding programs because molecular markers can be applied for selection even at the seedling stage with high precision and at reduced cost of its application. Numerous genes conferring race specific resistances and QTL for disease resistances have been reported in wheat (Miedaner & Korzun, 2012). MAS can be deployed in introgressing both single gene controlled resistance as well as quantitative disease resistance.

Resistance to biotrophs like rust pathogens controlled by a single gene is unlikely to be durable. In the past, many race-specific resistances of wheat to powdery mildew (*B. graminis*) and stripe rust (*P. striiformis*) failed when introgressed alone into commercial cultivars. This was even true for genes like *Yr11*, *Yr13*, and *Yr14* that express a partial resistance in adult plants (Johnson, 1992). Therefore, it might be worthwhile to select for quantitative resistances against these pathosystems, a strategy that was successful for powdery mildew in wheat (Miedaner & Flath, 2007), leaf rust in barley (*P. hordei*) (Parlevliet, 1995), and common rust in maize (*P. sorghi*) (Stuthman et al., 2007). Identification of novel genes/QTLs controlling resistance to wide range of pathogens, that is, broad-spectrum resistance will improve the prospect of using MAS in resistance breeding.

A few monogenically inherited resistance genes confer broad and durable resistance, for example, *Lr34* gives resistance to leaf rust, stripe rust, and powdery mildew (Lillemo et al., 2008), and *Yr36* provides high-temperature dependent durable resistance against a range of stripe rust races (Uauy et al., 2005). These genes have been cloned (Fu et al., 2009; Krattinger et al., 2009) and perfect markers directly drawn from the gene sequence are available (Fig. 2.3) (Table 2.2).

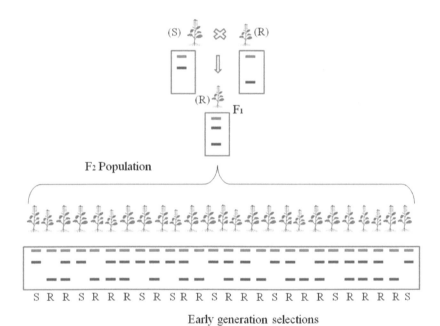

FIGURE 2.3 Schematic representation of MAS procedure for disease resistance in plants.

Success of MAS program depends on the availability of markers that are tightly linked to or derived from the gene governing the trait of interest. In the past, availability of such markers was scarce and therefore, there are only a few cases of MAS where durable resistance has been success-fully transferred into elite breeding material. Examples in which MAS has been successfully applied to practical breeding are the wheat rust resistance

TABLE 2.2 Examples of Marker-Assisted Backcrossing and Gene Pyramiding for Fungal Disease Resistance in Cereals.

Crop	Trait(s)	Gene/QTLs	References
Barley	Leaf rust	*Rphq6*	van Berloo et al. (2001)
	Stripe rust	QTLs on 4H and 5H	Toojinda et al. (1998)
	Stripe rust	*Rspx Rspx*, QTLs 4, 7 QTL 5	Castro et al. (2003)
Rice	Blast	*Pi1*	Liu & Anderson (2003)
	Disease resistance	Blast QTLs and quality loci	Toojinda et al. (2005)
	Sheath blight, bacterial blight, yellow stem borer	*RC7* chitinase gene, *Xa21, Bt*	Datta et al. (2002)
	Blast disease	*Pi1, Piz-5, Pi1, Pita*	Hittalmani et al. (2000)
Wheat	Powdery mildew	22 *Pm* genes	Zhou et al. (2005)
		Pm2, Pm4a	Liu et al. (2000)

genes *Lr34* (Lillemo et al., 2008) and *Yr36* (Uauy et al., 2005), and two
QTL for resistance to *Fusarium* head blight in wheat (*Fhb1* and *Qfhs.ifa-5A*)
(Miedaner et al., 2006). The main advantage of MAS over phenotypic selection during introgression of *Fhb1* and *Qfhs.ifa-5A* QTLs was halving the
cycle length of selection because crossing, selfing, and selection by MAS
could be done in two years, whereas phenotypic selection needed four years
due to the necessity of adult-plant testing in the field. Nonetheless, the best
way to breed for cultivars with durable disease resistance is to apply MAS
followed by phenotypic selection (Miedaner & Korzun, 2012). However,
due to constraints, like lack of tightly linked or perfect markers, and the existence of QTL background effects hinder the wider application of MAS. With
development of high throughput sequencing and genotyping platforms; and
recently the introduction of genomic selection will reduce the problems associated with deployment of MAS in practical breeding programs of durable
and broad spectrum resistance.

2.5.2 GENOMIC SELECTION (GS)

Genomic selection is a new molecular breeding approach with which it
is possible to develop new varieties in short time, and is particularly suitable for manipulation of complex traits like yield, drought tolerance, and
disease resistance. GS is a type of MAS, that is, plant lines are selected
on the basis of genotyping done using a large number of markers spread
across genome compared to few specific markers used in conventional
MAS (Heffner et al., 2009; Desta & Ortiz, 2014). Conventional MAS has
not been successful for improving complex traits because MAS allows
identification of few large-effect QTLs, whereas GS allows identification
of large number of QTLs having both small- and large-effects thereby
facilitate complete dissection of complex traits (Heffner et al., 2009;
Desta & Ortiz, 2014). Although association-mapping approach is effective
compared to MAS, but it is still inadequate to estimate the true QTL effects
on trait (Desta & Ortiz, 2014).

GS involves prediction of breeding values of genotypes to be tested
based on statistical models developed in reference genotypes, whose thorough genotyping and phenotyping data are available (Heffner et al., 2009).
Prediction of breeding values for each of test lines is done using only the
genotypic data without the phenotype. Hence, breeders can know the worth
of their breeding lines even before they are tested in the field condition.
Based on simulation and empirical studies in plants (Wong & Bernardo,

2008; Zhong et al., 2009), it is found that GS would lead to greater gains per unit time than phenotypic selection and MAS. GS showed a greater response to selection compared to phenotypic selection and marker assisted recurrent selection (MARS) for all three kinds of heritability values (Bernardo & Yu, 2007; Heslot et al., 2014).

The genome-wide marker coverage makes GS suitable to improve traits such as APR which are based on multiple genes. QTL mapping studies have indicated that APR to stripe rust, caused by *Puccinia striiformis* f. sp. *tritici* and leaf rust, *P. triticina*, is based on several genes with some having small effects (Mallard et al., 2005; Navabi et al., 2005; Rosewarne et al., 2008). It has been observed that different resistance sources have different APR genes for stripe rust (Navabi et al., 2004) and leaf rust resistance (Singh et al., 2005). Breeding efforts to pyramid 4–5 APR genes from different resistance sources have been successful in creating lines with high levels of resistance (Singh et al., 2000). Although, either MAS or GS can serve as important tools for combining several APR genes from different resistant sources, GS has the advantage that APR genes need not be mapped, and QTL with minor effects can also be selected (Rutkoski et al., 2011).

Rutkoski et al. (2011) of Cornell university have proposed GS model for APR to stem rust with genotypes having varying levels of stem rust APR. Screening for APR in locations such as Kenya where disease pressure is high and highly virulent races are present would provide informative phenotypic data. Such informative data would allow for the training of GS models to effectively select for APR. Once the GS model has been built, it can be implemented in selection of wheat lines for stem rust resistance (Rutkoski et al., 2011). Simulation results from a selected bulk scheme in wheat based on a combination of GS with MAS and phenotypic selection showed increased gains per unit time compared to MAS and phenotypic selection alone (Heffner et al., 2010). GS is a promising strategy for rapid improvements in quantitative traits, and is desperately needed as stem rust races such as Ug99 which continues to evolve rapidly and migrates toward major wheat producing regions. Thus, GS could be an important tool for achieving objective of developing durable stem rust resistance in wheat (Rutkoski et al., 2011). Recently, Daetwyler et al. (2014) have shown the possibility of using GS for resistance in wheat through the genomic prediction study in wheat landraces for rust resistance. Further, use of genotyping by sequencing, the ultimate genotyping platform, will enhance the efficiency of GS in breeding for multigenic traits like durable rust resistance in wheat crop (Poland et al., 2012).

2.5.3 TILLING AND ECO-TILLING

The large and polyploidy nature of wheat genome makes it hard to detect gene mutations based on phenotypic screening due to gene redundancy. Therefore, carrying out forward genetic studies in wheat is more difficult than in diploid plants (Dong et al., 2009). However, availability of genome sequences allows efficient use of high throughput genomic tools such as Targeting Induced Local Lesions in Genomes (TILLING) for functional characterization of genes. TILLING is a reverse genetics method which employs chemical mutagenesis with a high-throughput mutation detection. Since TILLING is reverse genetics approach, it identifies single base pair mutations in candidate genes responsible for the expected phenotype (McCallum et al., 2000; Henikoff & Comai, 2003) which is in contrast to the forward genetics approaches where mutants are first identified based on phenotypes.

While TILLING discovers mutations in EMS mutagenized populations, the modified form known as Eco-TILLING detects SNPs occurring in natural populations (Comai et al., 2004; Haughn & Gilchrist, 2006). Further, Eco-TILLING has additional applications in genetic mapping, breeding and genotyping, and also provides information concerning gene structure, linkage disequilibrium, population structure, or adaptation (Haughn & Gilchrist, 2006). Since TILLING directly introduces genetic variations into improved or elite genotypes, it avoids the need for introgression of alleles from non-adapted/wild type accessions into cultivated high-yielding varieties. Therefore, the introduction of agronomically undesirable traits due to linkage drag is avoided (Sestili et al., 2010) (Fig. 2.4).

In contrast to other reverse genetics approaches such as RNA interference and insertional mutagenesis, TILLING is non-transgenic and creates a series of allelic mutations, including knockouts, in the desired gene (Henikoff & Comai, 2003). The TILLING method is useful for both functional genomics as demonstrated in Arabidopsis (McCallum et al., 2000) and crop improvement as demonstrated in wheat (*Triticum aestivum* L.) (Slade et al., 2005), maize (*Zea mays* L.) (Till et al., 2004), rice (*Oryza sativa* L.) (Wu et al., 2005; Till et al., 2007), barley (*Hordeum vulgare* L.) (Caldwell et al., 2004) and soybean [*Glycine max* (L.) Merr.] (Cooper et al., 2008).

Wheat has the ability to tolerate mutations at much higher frequency than other plants, perhaps due to its allopolyploid nature (Slade et al., 2005) and hence, requires a relatively small mutant population for screening of desired mutants compared to other plants (Dong et al., 2009). Therefore, TILLING and Eco-TILLING are the appropriate tools that can be used to

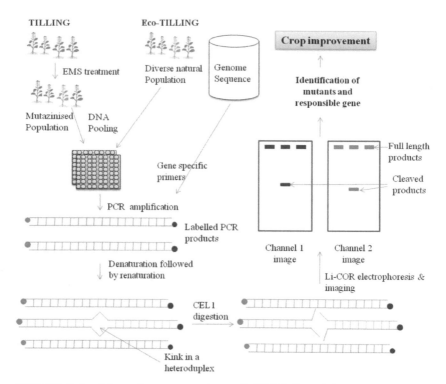

FIGURE 2.4 Schematic representations of TILLING and Eco-TILLING procedures (modified from Esfeld et al., 2009).

create mutations as well as search for naturally existing genetic variations for functional characterizations of *R* genes and development of resistant varieties for rust resistance in wheat (Table 2.3).

TILLING approach was used during map based cloning of *Yr36* gene in wheat. *Yr36* was mapped between *Xucw129* and *Xucw148* marker placed at 0.02 cM apart. This region has two pairs of duplicated IBR and WKS genes. Of these two pairs of gene, WKS genes contained domains associated with plant responses to pathogens in other species. So *WKS1* and *WKS2* were chosen for functional characterization. Screening of TILLED population for disease resistance showed that *WKS1* mutants produced susceptible reactions. In contrast, none of the *WKS2* mutants was susceptible suggesting that *WKS1* is *Yr36* (Fu et al., 2009). Further, the investigation of wheat natural diversity using allele mining tools allows for the identification of new functional disease resistance alleles. For instance, screening about 2000 landraces of wheat from different geographical regions resulted in discovery of more

TABLE 2.3 Examples of Using the TILLING Strategy in Allele Mining for Disease Resistance in Plants.

Crop	Target trait	Target genes	Numbers of identified mutations	References
Hordeum vulgare	Immunity to fungus	*MLO9 (mildew resistance locus 9)*	4	Gottwald et al. (2009)
	Virus resistance	*eIF4E (eukaryotic translation initiation factor 4E)*	5	Talamè et al. (2008)
	Immunity to fungus	*Rpg1 (barley stem rust resistance protein gene 1)*	4	
Solanum lycopersicum	Virus resistance	*eIF4E1 (eukaryotic translation initiation factor 4E)*	7	Piron et al. (2010)
Triticum aestivum	Grain hardness	*Pina, Pinb (puroindoline a, b)*	18	Feiz et al. (2009)

number of functional alleles of *Pm3* (powdery mildew resistance gene). The allele number increased from seven to 16 (Bhullar et al., 2009, 2010).

2.6   OTHER MANAGEMENT STRATEGIES FOR WHEAT AND BARLEY RUSTS

Apart from tools used for developing host plant resistance, other disease management strategies such as integrated disease management (IDM), disease epidemics prediction system, use of new generation fungicides with novel mode of action, use of nanotechnology for development of diagnostic devices and novel fungicides are also critical for strategic management of fast evolving races of rust pathogens to sustain the wheat production in the coming years.

2.6.1   *EFFICACIOUS MICROBIAL BIO-PESTICIDES*

Microbial pesticides are the integral part of the IDM. Microbial bio-pesticides used against fungal pathogens include *Trichoderma harzianum*, which is an antagonist of *Rhizoctonia*, *Pythium*, *Fusarium,* and other soil-borne pathogens (Harman, 2005) and specific strains of *Bacillus subtilis*, *Pseudomonas fluorescens,* and *Pseudomonas aureofaciens* are being used against a range of pathogens including damping-off and soft rots (Kloepper et al., 2004; Haas & Défago, 2005; Berg, 2009; Choudhary & Johri, 2009). Biological control of rusts on geranium (Rytter et al., 1989), bean (Mizubuti et al., 1995; Yuen et al., 2001), and coffee (Haddad et al., 2009; Ward et al., 2012) has been previously explored. Recently, an endophytic *B. subtilis* strain E1R-j has been shown effective as biological control agent against wheat stripe rust in both greenhouse and field trials (Li et al., 2012c). Certain strains of rhizosphere bacteria, called plant growth promoting rhizobacteria (PGPR), and soil inhabiting fungi, called Arbuscular mycorrhizal fungi promote plant growth directly by affecting the availability of nutrients (Saikia et al., 2012; El-Amri et al., 2013), while other strains of them suppress soil-borne pathogens, or by stimulating plant natural defenses, through induced systemic resistance (Tortora et al., 2012; El-Amri et al., 2013). Combined application of arbuscular mycorrhizal fungi and *Azospirillum amazonense* improved wheat plant growth, yield, and quality, in addition to the reduction in rust disease severity (Ghoneem et al., 2015).

However these micropesticides suffer from certain limitations such as low persistence, low efficacy, low virulence, specific host range, and slow mechanism of action, making their use difficult and ineffective to manage diseases under field conditions. Taking into consideration their potential as bio-control agents efforts are made to overcome the drawbacks while retaining the advantageous properties (Glare et al., 2012). Efforts to improve the efficacy of micropesticides through conventional approaches like strain selection from different geographical regions as well as through induced mutation variation for higher virulence, persistence, and greater efficiency have had limited success. Recently genetic engineering techniques have been employed for increasing virulence, persistence under field conditions, and broadening the host specificity of micropesticides. Although the scope of employing genetic engineering for improvement of biopesticides is huge, due to strict regulatory approaches, and negative public perception about GM organisms, use of GM based biopesticides are not promoted (Glare et al., 2012).

2.6.2 MODELING WHEAT RUST DISEASES EPIDEMICS AND EXPERT SYSTEMS FOR CROP PROTECTION

In recent years frequent wheat rust epidemics occurred in different parts of the world especially in developing countries due to lack of epidemics warning systems and thus appropriate preventive measures could not be taken at right time to prevent epidemics. Development of appropriate models for prediction of wheat rust disease epidemics will enable prevention of wheat yield losses due to rust diseases.

Most of the simulation models are developed to predict the dynamics of disease intensity in particular wheat diseases. Examples include models for leaf rust (Rossi et al., 1997; Räder et al., 2007) and for *Septoria* blotches (Djurle & Yuen, 1991; El Jarroudi et al., 2009). Most of these models are very elaborate and specific to a single disease, and may address small regional areas (Savary et al., 2015). However, while developing long-term robust strategies for disease management one should consider a set of diseases that may cause epidemics over a large region. Therefore, in the case of wheat, one should think of considering a group of 5–15 diseases (Duveiller et al., 2007) and considering large spatial scales (Savary et al., 2006), which is required for robust estimates of epidemic patterns. These requirements can be fulfilled by the use of a simulation model able to handle a wide range of diseases, that would be common across the diseases, easy to share, and use with frugal inputs and few parameters, and that would offer robust estimates of epidemic intensity at large spatial scales. Such a model, EPIRICE, was developed for rice (Savary et al., 2012), whose structure accommodates fungal, bacterial, and viral pathogens, which can develop lesions on small fractions of leaves, entire leaves, entire tillers, or whole plants. Likewise, EPIWHEAT model is developed to provide a generic, robust modeling structure for a wide array of wheat pathogens such as leaf rust (*P. triticina*) and *Septoria* blotch (*Zymoseptoria tritici*) that may cause epidemics. EPIWHEAT simulated epidemics were comparable to observations at the field, national (France), and European scales suggesting model has better predicting potential of epidemics of wheat foliar diseases at large scales (Savary et al., 2015).

Proper dissemination of outputs of predictive systems to farmers is also important for effective disease management. It can be achieved through Expert Systems (Es). Es are the frontiers, which combines and integrate both science of plant pathology and art of diagnosis and disease management. Es are computer programs that emulate the logic and problem solving proficiency of human expert (Travis & Latin, 1991). The Es are programmed to review a consultation and provide user with an explanation. PLANT/ds was the first expert system in plant pathology developed by Michalski et al. (1983) to diagnose 17 soybean diseases in Illinois. The PLANT/ds expert system offers advice about soybean diseases on the basis of symptoms communicated by the user. Later on many expert systems have been built for plant disease diagnosis (Blancard et al., 1985; Plant et al., 1989; Latin et al., 1990; Gent & Ocamb, 2009) and management (Roach et al., 1985; Ramon & Roland, 1994). Recent developments in imaging processing will greatly reduce error prone communication between system and user arising as they reduce errors in diagnosis of diseases (Patil & Kumar, 2011). Mohammad Ei-Helly et al. (2004) developed Central Laboratory of Agricultural Expert System (CLASE) by incorporating image analysis into diagnostic expert system and used it to manage cucumber crop. To diagnose a disorder from a leaf image, four image-processing phases were used: enhancement, segmentation, feature extraction, and classification. It was tested for diagnosis of three different disorders such as leaf miner, powdery and downy diseases. They found a great reduction in diagnosis errors diagnosis and miscommunication between system and user (Ei-Helly et al., 2004). The usefulness and impact of predictive systems in disease management also depends in part on a social network for information exchange and its implementation (Gent et al., 2013).

2.6.3 NEW GENERATION FUNGICIDES

Fungicides are in common use for the control of plant diseases world over since the last century and have become an integral component of our crop production system. Despite their important role in controlling plant disease, conventional fungicides have certain shortcomings, such as development of fungicide resistance in pathogen, hazardous effect on health of environment, and non-target organisms. The development of fungicides has passed through several stages. Generally, the type of materials developed corresponded with the knowledge of disease etiology and also the chemical and biological properties of the compounds. Recently new generation fungicides with novel modes of action have been developed (Thind, 2006). These new fungicide compounds are expected to provide improved disease control options, and are ecologically safe and have a good efficacy even at much lower doses. These require fewer treatments per season compared to earlier compounds. Since they possess novel modes of action, there are less chances of resistance development to these fungicides. These are easily degraded and cause less risk to the environment. There is considerable improvement in their formulations and are safer to the crops (Thind, 2006).

2.6.4 SYNTHETIC DEFENSE ELICITORS FOR INDUCING RESISTANCE

Plant defense responses can be induced by chemical elicitors. A chemical can be considered as a defense elicitor/activator if it triggers resistance to pathogens by inducing the host defense mechanism in the same or similar fashion as induced by pathogen stimuli and it should not be directly toxic to the pathogen (Kessmann et al., 1994). Treatment of plants the chemicals such as SA, 2,6-dichloroisonicotinic acid (INA), and acibenzolar-S-methyl benzo (1, 2, 3) thiadiazole-7-carbothioic acid S-methyl ester (BTH) has been shown to activate the natural immune responses in plants (Métraux et al., 1991; Uknes et al., 1992; Schob et al., 1997; Knoth et al., 2009). The compounds that elicit a plant's innate immune response have been proposed as a safe alternative to the toxic pesticides for controlling diseases (Uknes et al., 1992; Bektas & Eulgem, 2014).

Apart from using in crop protection, they can also be used in studying the molecular basis of plant defense (Bektas & Eulgem, 2014). In the recent times, as an alternative to forward and reverse genetic approaches, chemical approaches have gained importance for analysis of biological pathways and networks. Use of small bioactive molecules can allow researchers to study the functions and interactions that were previously difficult to study by genetic analyses due to gene redundancy or lethality of mutations. Synthetic elicitors are small drug-like molecules that induce plant defense responses, but are distinct from known natural elicitors of plant immunity. Recent advances in combinatorial chemical synthesis (Blackwell & Zhao, 2003; Stockwell, 2004; Raikhel & Pirrung, 2005) have resulted in a large number of compounds with defense inducing properties. These wide-ranging collections of new synthetic elicitors along with powerful reverse genetics tools and resources available for model plants and crop systems, will allow investigation of the intricacies of plant defense signaling pathways and networks in a comprehensive manner. A prominent example highlighting potency of synthetic elicitors in basic research on plant immunity is the role of INA in the discovery of NPR1 as a central regulator of SA-dependent immune responses in Arabidopsis (Knoth et al., 2009).

2.6.5   NANOTECHNOLOGY

Nanotechnology is a modern technology dealing with nanoscale materials and processes. Nanotechnology offers novel tools and products for plant protection in agriculture (Ghormade et al., 2011). Nanotechnology based tools such as sensors, labeling agents, *polymerase chain reaction* (PCR), and Enzyme-linked immunosorbent assay (ELISA) have the potential to improve the sensitivity, selectivity, speed, cost, and convenience of disease diagnosis. Nanoscale labeling agents, such as quantum dots, facilitate intracellular labeling and visualization allowing study of interacting molecules. Lab-on-a-chip developed using nanotechnology having functionality such as PCR and ELISA units into handheld devices will facilitate on site diagnosis of diseases (Khiyami et al., 2014). Researchers are developing nanosensors for detection of plant pathogens and forecasting of diseases in crop systems. Recently a highly sensitive DNA based nanosensor was developed for detection of white leaf disease in sugarcane in Thailand (Wongkaew & Poosittisak, 2014). Further, nanosensors can be linked to a GPS system for real-time monitoring of disease under field conditions (Sharon et al., 2010).

Micronutrients have role in enhancing host defense mechanism in plants against pathogens. However, unavailability some of the metal, metal oxides, and carbon in soil are likely to affect efficacy of disease suppression in plants. Greater availability and translocation within plants is the one of the notable characteristics of nanoscale size metals and metal oxides. Hence the nanoformulations of metals and metal oxides have offer novel platforms for plant protection (Servin et al., 2015). Silver nanoemulsions have been shown to have the property of broad-spectrum fungicides. These silver-based nanofungicides have the potential to tackle most of the fungal pathogens of crop plants and hold promise for future. Research on nanofungicides is under progress. Simple cost effective nano-dispersed formulations can be prepared and are suited for developing new forms of fungicidal materials. For instance, Syngenta's Banner MAXX ™ is a micro-dispersed systemic fungicide offering broad-spectrum disease control in turf and ornamental plants (Abd-Elsalam, 2013). It provides excellent tank mix compatibility and stability.

Nanoparticles can also be used for genetic transformation of plants. Introduction of resistance genes in plant cells through nanoparticles-mediated transformation can be employed for development of resistant varieties (Mahendra et al., 2012).

2.7 CONCLUDING REMARKS

Pathogens have been the cause of concern for food production since the beginning of agriculture. Pathogens and plants have coevolved during the evolution, and due to their antagonistic association, there is continuous weapon race between the two to overcome each other's strategic moves. However, modern intensive and monoculture agricultural systems, changing climatic conditions and narrow genetic base in the wheat breeding programs have caused rapid evolution of rust pathogens. Currently, wheat rust diseases have become the main foci of resistance breeding than ever before due to frequent rust epidemics in the East and South African, and Middle East countries. Recent breakthroughs in omics and genetic modification approaches have provided resources and high throughput tools needed for rapid development of resistant wheat cultivars to cope with pace of the rapidly evolving virulent races of rust pathogens. Apart from resistance breeding, management practices based on modern technologies like remote science technology, information technology, and nanotechnology will facilitate efficient disease control and lessen the yield losses due to diseases.

KEYWORDS

- wheat
- barley
- rust management
- transgenic
- resistance
- genomics
- *R* gene
- nanotechnology
- *Puccinia*
- biopesticide
- RNAi
- TILLING
- TALEN

REFERENCES

Abd-Elsalam, K. A. Nanoplatforms for Plant Pathogenic Fungi Management. *Fungal Genom. Biol.* **2013,** *2,* 2.

Alfred, R. L., et al. Stability of Puroindoline Peptides and Effects on Wheat Rust. *World J. Microbiol. Biotechnol.* **2013,** *29* (8), 1409–1419.

Antony, G., et al. Rice *xa13* Recessive Resistance to Bacterial Blight is Defeated by Induction of the Disease Susceptibility Gene *Os-11N3*. *Plant Cell.* **2010,** *22,* 3864–3876.

Ayliffe, M., et al. Determining the Basis of Nonhost Resistance in Rice to Cereal Rusts. *Euphytica.* **2011a,** *179,* 33–40.

Ayliffe, M., et al. Nonhost Resistance of Rice to Rust Pathogens. *Mol. Plant Microbe Interact.* **2011b,** *24,* 1143–1155.

Baulcombe, D. C. VIGS, HIGS and FIGS: Small RNA Silencing in the Interactions of Viruses or Filamentous Organisms with their Plant Hosts. *Curr. Opin. Plant Biol.* **2015,** *23,* 141–146.

Baulcombe, D. RNA Silencing in Plants. *Nature.* **2004,** *43* (1), 356–363.

Bektas, Y.; Eulgem, T. Synthetic Plant Defense Elicitors. *Front. Plant Sci.* **2014,** *5,* 804.

Berg, G. Plant Microbe Interactions Promoting Plant Growth and Health: Perspectives for Controlled Use of Microorganisms in Agriculture. *Appl. Microbiol. Biotechnol.* **2009,** *84,* 11–18.

Bernardo, R.; Yu, J. Prospects for Genome Wide Selection for Quantitative Traits in Maize. *Crop Sci.* **2007,** *47,* 1082–1090.

Bhullar, N. K., et al. Wheat Gene Bank Accessions as a Source of New Alleles of the Powdery Mildew Resistance Gene *Pm3*: A Large Scale Allele Mining Project. *BMC Plant Biol.* **2010,** *10,* 88.

Bhullar, N. K., et al. Unlocking Wheat Genetic Resources for the Molecular Identification of Previously Undescribed Functional Alleles at the *Pm3* Resistance Locus. *Proc. Natl. Acad. Sci.* **2009,** *106,* 9519–9524.

Bieri, S.; Potrykus, I.; Futterer, J. Effects of Combined Expression of Antifungal Barley Seed Proteins in Transgenic Wheat on Powdery Mildew Infection. *Mol. Breed.* **2003,** *11,* 37–48.

Bird, R. E., et al. Single-Chain Antigen-Binding Proteins. *Science.* **1988,** *242,* 423–426.

Blackwell, H. E.; Yunde, Z. Chemical Genetic Approaches to Plant Biology. *Plant Physiol.* **2003,** *133,* 448–455.

Blancard, D.; Bonnet, A.; Coltno, A. TOM, un Systeme Expert en Maladies des Tomates. *P.H.M. Rev. Horticole.* **1985,** *261,* 714.

Bliffeld, M., et al. Genetic Engineering of Wheat for Increased Resistance to Powdery Mildew Disease. *Theor. Appl. Genet.* **1999,** *98,* 1079–1086.

Bolar, J. P., et al. Expression of Endochitinase from *Trichoderma harzianum* in Transgenic Apple Increases Resistance to Scab and Reduces Vigor. *Phytopathology.* **2000,** *90,* 72–77.

Boman, H. G. Innate Immunity and the Normal Microflora. *Immunol. Rev.* **2000,** *173* (1), 5–16.

Boyajyan, A., et al. Molecular Mechanisms and Mediators of the Immune Response in Plants. *J. Plant Sci.* **2014,** *2* (1), 23–30.

Brunner, S., et al. Transgenic *Pm3* Multilines of Wheat Show Increased Powdery Mildew Resistance in the Field. *Plant Biotechnol. J.* **2012,** *10,* 398–409.

Brunner, S., et al. Transgenic *Pm3b* Wheat Lines Show Resistance to Powdery Mildew in the Field. *Plant Biotechnol. J.* **2011,** *9,* 897–910.

Caldwell, D. G., et al. Structured Mutant Population for Forward and Reverse Genetics in Barley (*Hordeumvulgare* L.). *Plant J.* **2004,** *40,* 143–150.

Cary, J. W., et al. Transgenic Expression of a Gene Encoding a Synthetic Antimicrobial Peptide Results in Inhibition of Fungal Growth *in vitro* and *in planta. Plant Sci.* **2000,** *154,* 171–181.

Castro, A. J., et al. Mapping and Pyramiding of Qualitative and Quantitative Resistance to Stripe Rust in Barley. *Theor. Appl. Genet.* **2003,** *107,* 922–930.

Chakraborty, S., et al. Rust-Proofing Wheat for a Changing Climate. *Euphytica.* **2011,** *179,* 19–32.

Chaudhary, V. K., et al. A Rapid Method of Cloning Functional Variable Region Antibody Genes in *Escherichia coli* as Single-Chain Immunotoxins. *Proc. Nat. Acad. Sci. USA.* **1990,** *87,* 1066–1070.

Cheng, W., et al. Tissue-Specific and Pathogen-Inducible Expression of a Fusion Protein Containing a *Fusarium*-Specific Antibody and a Fungal Chitinase Protects Wheat against *Fusarium* Pathogens and *Mycotoxins. Plant Biotechnol. J.* **2015a,** *13,* 664–674.

Cheng, W., et al. Host-Induced Gene Silencing of an Essential Chitin Synthase Gene Confers Durable Resistance to *Fusarium* Head Blight and Seedling Blight in Wheat. *Plant Biotechnol. J.* **2015b,** *13* (9), 1335–1345. http://dx.doi.org/10.1111/ pbi.12352 (accessed Sep 30, 2015)

Choudhary, D. K.; Johri, B. N. Interactions of *Bacillus* spp. and Plants with Special Reference to Induced Systemic Resistance (ISR). *Microbiol. Res.* **2009,** *164,* 493–513.

Christian, M., et al. Targeting DNA Double-Strand Breaks with TAL Effector Nucleases. *Genetics.* **2010,** *186,* 757–761.

Cloutier, S., et al. Leaf Rust Resistance Gene *Lr1*, Isolated from Bread Wheat (*Triticum aestivum* L.) is a Member of the Large *psr567* Gene Family. *Plant Mol. Biol.* **2007,** *65,* 93–106.

Collard, B. C. Y., et al. An Introduction to Markers, Quantitative Trait Loci (QTL) Mapping and Marker-Assisted Selection for Crop Improvement: The Basic Concepts. *Euphytica.* **2005,** *142,* 169–196.

Collins, N. C., et al. SNARE-Protein -Mediated Disease Resistance at the Plant Cell Wall. *Nature.* **2003,** *425,* 973–977.

Comai, L., et al. Efficient Discovery of DNA Polymorphisms in Natural Populations by Ecotilling. *Plant J.* **2004,** *37,* 778–786.

Conrath, U., et al. Genes to Enhance the Disease Resistance in Crops. Patent WO/2013/093,738, June 27, 2013.

Cooper, J. L., et al. TILLING to Detect Induced Mutations in Soybean. *BMC Plant Biol.* **2008,** *8,* 9.

Daetwyler, H. D., et al. Genomic Prediction for Rust Resistance in Diverse Wheat Landraces. *Theor. Appl. Genet.* **2014,** *127,* 1795–1803.

Datta, K., et al. Pyramiding Transgenes for Multiple Resistance in Rice against Bacterial Blight, Yellow Stem Borer and Sheath Blight. *Theor. Appl. Genet.* **2002,** *106,* 1–8.

De Neve, M., et al. Assembly of an Antibody and its Derived Antibody Fragment in *Nicotiana* and *Arabidopsis. Transgenic Res.* **1993,** *2,* 227–237.

De Wit, P. J. G. M. Molecular Characterization of Gene-for-Gene Systems in Plant-Fungus Interactions and the Application of *Avirulence* Genes in Control of Plant Pathogens. *Annu. Rev. Phytopathol.* **1992,** *30,* 391–418.

Desta, Z. A.; Ortiz, R. Genomic Selection: Genome-Wide Prediction in Plant Improvement. *Trends Plant Sci.* **2014,** *19* (9), 592–601.

Djurle, A.; Yuen, J. E. A Simulation Model for *Septoria nodorum* in Winter Wheat. *Agril. Syst.* **1991,** *37,* 193–218.

Dong, C.; Dalton-Morgan, J.; Vincent, K.; Sharp, P. A Modified TILLING Method for Wheat Breeding. *Plant Genome.* **2009,** *2,* 39–47.

Dorofeev, V. F., et al. *Flora of Cultivated Plants;* Dorofeev, V. F., Korovina, O. N., Eds.; Kolos: Leningrad, 1979; pp 320–321.

Dubreil, L., et al. Spatial and Temporal Distribution of the Major Isoforms of Puroindolines (Puroindoline-a and Puroindoline-b) and Non Specific Lipid Transfer Protein (ns-LTPe1) of *Triticum aestivum* Seeds. Relationships with their *in vitro* Antifungal Properties. *Plant Sci.* **1998,** *138,* 121–135.

Duveiller, E.; Singh, R. P.; Nicol, J. M. The Challenges of Maintaining Wheat Productivity: Pests, Diseases, and Potential Epidemics. *Euphytica.* **2007,** *157,* 417–430.

Egorov, T. A., et al. Diversity of Wheat Anti-Microbial Peptides. *Peptides.* **2005,** *26*(11), 2064–2073.

Ei Helly, M., et al. *Integrating Diagnostic Expert System with Image Processing Via Loosely Coupled Technique;* Central Laboratory for Agricultural Expert System (CLAES): Imbaba, Giza, 2004.

El Jarroudi, M., et al. Assessing the Accuracy of Simulation Model for *Septoria* Leaf Blotch Disease Progress on Winter Wheat. *Plant Dis.* **2009,** *93,* 983–992.

El-Amri, S. M., et al. Role of *Mycorrhizal* Fungi in Tolerance of Wheat Genotypes to Salt Stress. *Afr. J. Microbiol. Res.* **2013,** *7* (14), 1286–1295.

Ellis. J. G., et al. The Past, Present and Future of Breeding Rust Resistant Wheat. *Front. Plant Sci.* **2014,** *5,* 641.

Esfeld, K.; Zerihun, T. The Improvement of African Orphan Crops through TILLING. *Inside Issue.* **2009,** *6,* 44.

FAO. *Global Agriculture Towards 2050.* FAO: Rome, 2009.

Feiz, L.; Martin, J. M.; Giroux, M. J. Creation and Functional Analysis of New Puroindoline Alleles in *Triticumaestivum. Theor. Appl. Genet.* **2009,** *118,* 247–257.

Feuillet, C., et al. Map-Based Isolation of the Leaf Rust Disease Resistance Gene *Lr10* from the Hexaploid Wheat (*Triticum aestivum* L.) Genome. *Proc. Natl. Acad. Sci. USA.* **2003,** *100,* 15253–15258.

Figlan, S. S.; Le Roux, C.; Terefe, T.; Botes, W.; Visser, B.; Shimelis, H.; Tsilo, T. Wheat Stem Rust in South Africa: Current Status and Future Research Directions. *Afr. J. Biotechnol.* **2014,** *13,* 4188–4199.

Fu, D., et al. A Kinase-START Gene Confers Temperature-Dependent Resistance to Wheat Stripe Rust. *Science.* **2009,** *323,* 1357–1360.

Gaj, T.; Gersbach, C. A.; Barbas III, C. F. Zfn, Talen, and CRISPR/Cas-Based Methods for Genome Engineering. *Trends Biotechnol.* **2013,** *31* (7), 397–405.

Galeffi, P.; Giunta, G.; Guida, S.; Cantale, C. Engineering of a Single Chain Variable Fragment Antibody Specific for the *Citrus tristeza* Virus and its Expression in *Escherichiacoli* and *Nicotiana tabacum. Eur. J. Plant Pathol.* **2002,** *108,* 479–483.

Gent, D. H.; Ocamb, C. M. Predicting Infection Risk of Hop by *Pseudoperonspora humuli. Phytopathology.* **2009,** *99,* 1190–1198.

Gent, D. H., et al. The Use and Role of Predictive Systems in Disease Management. *Annu. Rev. Phytopathol.* **2013,** *51,* 267–289.

Ghoneem, K. M., et al. Postulation and Efficiency of Leaf Rust Resistance Genes of Wheat and Biological Control of Virulence Formulae of *Puccinia triticina* Races. *Egypt. J. Biol. Pest Control.* **2015,** *25*(1), 23–31.

Ghormade, V.; Deshpande, M. V.; Paknikar, K. M. Perspectives for Nano-Biotechnology Enabled Protection and Nutrition of Plants. *Biotechnol. Adv.* **2011,** *29,* 792–803.

Glare, T., et al. Have Biopesticides Come of Age? *Trends Biotechnol.* **2012,** *30,* 250–258.

Gottwald, S., et al. TILLING in the Two-Rowed Barley Cultivar 'Barke' Reveals Preferred Sites of Functional Diversity in the Gene *HvHox1. BMC Res. Notes.* **2009,** *2,* 258.

Govindarajulu, M., et al. Host Induced Gene Silencing Inhibits the Biotrophic Pathogen Causing Downy Mildew of Lettuce. *Plant Biotechnol. J.* **2015,** *13* (7), 875–883.

Haas, D.; Défago, G. Biological Control of Soil-Borne Pathogens by Fluorescent *Pseudomonads. Nat. Rev. Microbiol.* **2005,** *3,* 307–319.

Haddad, F., et al. Biological Control of Coffee Rust by Antagonistic Bacteria under Field Conditions in Brazil. *Biol. Control.* **2009,** *49,* 114–119.

Hancock, R. E.; Lehrer, R. Cationic Peptides: A New Source of Antibiotics. *Trends Biotechnol.* **1998,** *16,* 82–88.

Harman, G. E. Overview of Mechanisms and Uses of *Trichoderma* spp. *Phytopathology.* **2005,** *96,* 190–194.

Haughn, G. W.; Gilchrist, E. J. TILLING in the Botanical Garden: A Reverse Genetic Technique Feasible for All Plant Species. In *Floriculture, Ornamental and Plant Biotechnology: Advances and Topical Issues;* 1st ed.; Global Science Books: London, 2006; pp 476–482.

Heath, M. C. Non-host Resistance and Nonspecific Plant Defenses. *Curr. Opin. Plant Biol.* **2000,** *3,* 315–319.

Heath, M. C. Resistance of Plants to Rust Infection. *Phytopathology.* **1981,** *71,* 971–974.

Heffner, E. L., et al. Plant Breeding with Genomic Selection: Gain Per Unit Time and Cost. *Crop Sci.* **2010**, *50,* 1681–1690.

Heffner, E. L.; Sorrellsa, M. E.; Jannink, J. Genomic Selection for Crop Improvement. *Crop Sci.* **2009**, *49,* 1–12.

Helber, N., et al. A Versatile Monosaccharide Transporter that Operates in the Arbuscular Mycorrhizal Fungus *Glomus* sp. is Crucial for the Symbiotic Relationship with Plants. *Plant Cell.* **2011**, *23,* 3812–3823.

Henikoff, S.; Comai, L. Single-Nucleotide Mutations for Plant Functional Genomics. *Annu. Rev. Plant Biol.* **2003**, *53,* 375–401.

Heslot, N., et al. Integrating Environmental Covariates and Crop Modelling into the Genomic Selection Framework to Predict Genotype by Environment Interactions. *Theor. Appl. Genet* .**2014**, *127,* 463–480.

Hiatt, A.; Caffferkey, R.; Bowdish, K. Production of Antibodies in Transgenic Plants. *Nature.* **1989**, *342,* 76–78.

Hittalmani, S., et al. Fine Mapping and DNA Marker-Assisted Pyramiding of the Three Major Genes for Blast Resistance in Rice. *Theor. Appl. Genet.* **2000**, *100,* 1121–1128.

Hoogkamp, T.; Chen, W. Q.; Niks, R. Specificity of Prehaustorial Resistance to *Puccinia hordei* and to Two Inappropriate Rust Fungi in Barley. *Phytopathology.* **1998**, *88,* 856–861.

Horváth, E., et al. *In vitro* Salicylic Acid Inhibition of Catalase Activity in Maize: Differences between the Isozymes and a Possible Role in the Induction of Chilling Tolerance. *Plant Sci.* **2002**, *163,* 1129–1135.

Horvath, H., et al. Genetically Engineered Stem Rust Resistance in Barley Using the *Rpg1* Gene. *Proc. Natl. Acad. Sci. USA.* **2003**, *100* (1), 364–369.

Hu, Z., et al. Down-Regulation of *Fusarium oxysporum* Endogenous Genes by Host-Delivered RNA Interference Enhances Disease Resistance. *Front. Chem.* **2015**, *3,* 1–10.

Huang, L., et al. Map-Based Cloning of Leaf Rust Resistance Gene *Lr21* from the Large and Polyploid Genome of Bread Wheat. *Genetics.* **2003**, *164,* 655–664.

Huang, Y., et al. Combined Expression of Antimicrobial Genes (*Bbchit1* and *LJAMP2*) in Transgenic Poplar Enhances Resistance to Fungal Pathogens. *Tree Physiol.* **2012**, *32,* 1313–1320.

ICARDA. *Strategies to Reduce the Emerging Wheat Stripe Rust Disease,* ICARDA: Aleppo, Syria, 2011, pp 1–32. Source: http://www.icarda.org/striperust2014/wp-content/uploads/2014/01/ Strategies_ to_reduce.pdf (accessed Sep30, 2015).

James, C. Global Status of Commercialized Biotech/GM Crops: 2010. ISAAA Brief No. 42. ISAAA: Ithaca, NY, 2010; Source: http://isaaa.org/resources/publications /briefs/42/executive summary/default.asp (accessed Sep 30, 2015).

Jayaraj, J.; Punja, Z. K. Combined Expression of *Chitinase* and Lipid Transfer Protein Genes in Transgenic Carrot Plants Enhances Resistance to Foliar Fungal Pathogens. *Plant Cell Rep.* **2007**, *26,* 1539–1546.

Jia, Z. C., et al. The *Chitinase* gene (*Bbchit1*) from *Beauveria bassiana* Enhances Resistance to *Cytospora chrysosperma* in *Populus tomentosa* Carr. *Biotechnol. Lett.* **2010a**, *32,* 1325–1332.

Jia, Z.; Gou, J.; Sun, Y.; Yuan, L.; Tang, Q.; Yang, X.; Luo, K. Enhanced Resistance to Fungal Pathogens in Transgenic *Populus tomentosa* Carr. by Overexpression of an nsLTP-like Antimicrobial Protein Gene from Motherwort (*Leonurus japonicus*). *Tree Physiol.* **2010b**, *13,* 1599–1605.

Johnson, R. Past, Present and Future Opportunities in Breeding for Disease Resistance, with Examples from Wheat. *Euphytica.* **1992**, *63,* 3–22.

Jones, J. D. Putting Knowledge of Plant Disease Resistance Genes to Work. *Curr. Opin. Plant Biol.* **2001,** *4,* 281–287.

Kamoun, S. A. Catalogue of the Effector Secretome of Plant Pathogenic *Oomycetes. Annu. Rev. Phytopathol.* **2006,** *44,* 41–60.

Keller, B., et al. Genomic Approaches Towards Durable Fungal Disease Resistance in Wheat. In *Advances in Wheat Genetics: From Genome to Field;* Springer: Tokyo, 2015; pp 369–375.

Kessmann, H.; Staub, T.; Hofmann, C.; Maetzke, T.; Heraozog, J.; Ward, E.; Uknes, S.; Ryals, J. Induction of Systemic Acquired Resistance in Plants by Chemicals. *Annu. Rev. Phytopathol.* **1994,** *32,* 439–459.

Khiyami, M. A., et al. Plant Pathogen Nanodiagnostic Techniques: Forthcoming Changes? *Biotechnol. Biotechnol. Equip.* **2014,** *28*(5), 775–785.

Kloepper, J. W.; Ryu, C. M.; Zhang, S. A. Induced Systemic Resistance and Promotion of Plant Growth by *Bacillus* spp. *Phytopathology.* **2004,** *94,* 1259–1266.

Knoth, C., et al. The Synthetic Elicitor 3, 5-Dichloroanthranilic Acid Induces NPR1-Dependent and NPR1-Independent Mechanisms of Disease Resistance in *Arabidopsis. Plant Physiol.* **2009,** *150,* 333–347.

Koch, A., et al. Host-Induced Gene Silencing of Cytochrome P450 Lanosterol C14Ademethylase-Encoding Genes Confers Strong Resistance to *Fusarium* Species. *Proc. Natl. Acad. Sci. USA.* **2013,** *110,* 19324–19329.

Kouzai, Y., et al. Expression of the Chimeric Receptor between the Chitin Elicitor Receptor CEBiP and the Receptor-Like Protein Kinase Pi-d2 Leads to Enhanced Responses to the Chitin Elicitor and Disease Resistance against *Magnaporthe oryzae* in Rice. *Plant Mol. Biol.* **2013,** *81,* 287–295.

Krattinger, S. G., et al. A Putative ABC Transporter Confers Durable Resistance to Multiple Fungal Pathogens in Wheat. *Science.* **2009,** *323,* 1360–1363.

Langenbach, C.; Campe, R.; Schaffrath, U.; Goellner, K.; Conrath, U.UDP-Glucosyltransferase *UGT84A2/BRT1* is Required for *Arabidopsis* Non-host Resistance to the Asian Soybean Rust Pathogen *Phakopsora pachyrhizi. New Phytol.* **2013,** *198,* 536–545.

Latin, R. X., et al. An Expert System for Diagnosing Muskmelon Disorders. *Plant Dis.* **1990,** *74,* 83–87.

Li, H. P., et al. Engineering *Fusarium* Head Blight Resistance in Wheat by Expression of a Fusion Protein Containing a *Fusarium*-Specific Antibody and an Antifungal Peptide. *Mol. Plant Microbe Interact.* **2008,** *21,* 1242–1248.

Li, H., et al. Microscopy and Proteomic Analysis of the Non-Host Resistance of *Oryza sativa* to the Wheat Leaf Rust Fungus, *Puccinia triticina* f. sp. *tritici. Plant Cell Rep.* **2012a,** *31,* 637–650.

Li, T., et al. High-Efficiency TALEN-Based Gene Editing Produces Disease-Resistant Rice. *Nat. Biotechnol.* **2012b,** *30,* 390–392.

Li, T., et al. TAL Nucleases (Talns): Hybrid Proteins Composed of TAL Effectors and Foki DNA-Cleavage Domain. *Nucleic Acids Res.* **2011,** *39* (1), 359–372.

Li, W. P., et al. Expression and Characterization of a Recombinant Cry1Ac Crystal Protein Fused with an Insect-Specific Neurotoxin ω-ACTX-Hv1a in *Bacillus thuringiensis. Gene.* **2012c,** *498,* 323–327.

Lillemo, M., et al. The Adult Plant Rust Resistance Loci *Lr34/Yr18* and *Lr46/Yr29* are Important Determinants of Partial Resistance to Powdery Mildew in Bread Wheat Line Saar. *Theor. Appl. Genet.* **2008,** *116,* 1155–1166.

Lipka, U., et al. Live and Let Die-Arabidopsis Non-host Resistance to Powdery Mildews. *Eur. J. Cell Biol.* **2010**, *89*, 194–199.

Lipka, V., et al. Pre- and Post-Invasion Defenses Both Contribute to Non-host Resistance in Arabidopsis. *Science.* **2005**, *310*, 1180–1183.

Liu, J., et al. Molecular Marker-Facilitated Pyramiding of Different Genes for Powdery Mildew Resistance in Wheat. *Plant Breed.* **2000**, *119*, 21–24.

Liu, S. X.; Anderson, J. A. Marker Assisted Evaluation of *Fusarium* Head Blight Resistant Wheat Germplasm. *Crop Sci.* **2003**, *43*, 760–766.

Luo, L., et al. Expression of Puroindoline a Enhances Leaf Rust Resistance in Transgenic Tetraploid Wheat. *Mol. Biol. Rep.* **2008**, *35*, 195–200.

Mahendra, R., et al. Strategic Nanoparticle-Mediated Gene Transfer in Plants and Animals-a Novel Approach. *Curr. Nanosci.* **2012**, *8*, 170–179.

Mahfouz, M. M., et al. *De novo*-Engineered Transcription Activator-Like Effector (TALE) Hybrid Nuclease with Novel DNA Binding Specificity Creates Double-Strand Breaks. *Proc. Natl. Acad. Sci. USA.* **2011**, *108*, 2623–2628.

Mallard, S., et al. Genetic Analysis of Durable Resistance to Yellow Rust in Bread Wheat. *Theor. Appl. Genet.* **2005**, *110*, 1401–1409.

McCallum, C. M., et al. Targeted Screening for Induced Mutations. *Nat. Biotechnol.* **2000**, *18*, 455–457.

McDonald, B. A.; Linde, C. Pathogen Population Genetics, Evolutionary Potential and Durable Resistance. *Annu. Rev. Phytopathol.* **2002**, *40*, 349–379.

Mendgen, K.; Hahn, M. Plant Infection and the Establishment of Fungal Biotrophy. *Trends Plant Sci.* **2002**, *7*, 352–356.

Métraux, J. P., et al. Induced Resistance in Cucumber in Response to 2, 6-Dichloroisonico-tinic Acid and Pathogens. In *Advances in Molecular Genetics of Plant-Microbe Interactions;* Hennecke, H., Verma, D. P. S., Eds.; Kluwer: Dordrecht, 1991; Vol. 1, pp 432–439.

Michalski, R. S., et al. A Computer Based Advisory System for Diagnosing Soybean Diseases in Illinois. *Plant Dis.* **1983**, *67*, 459–463.

Miedaner, T.; Flath, K. Effectiveness and Environmental Stability of Quantitative Powdery Mildew (*Blumeria Graminis*) Resistance among Winter Wheat Cultivars. *Plant Breed.* **2007**, *126*, 553–558.

Miedaner, T.; Korzun, V. Marker-Assisted Selection for Disease Resistance in Wheat and Barley Breeding. *Phytopathology.* **2012**, *102*, 560–566.

Miedaner, T., et al. Stacking Quantitative Trait Loci (QTL) for *Fusarium* Head Blight Resistance from Non-Adapted Sources in an European Elite Spring Wheat Background and Assessing their Effects on Deoxynivalenol (DON) Content and Disease Severity. *Theor. Appl. Genet.* **2006**, *112*, 562–569.

Miller, J. C., et al. A TALE Nuclease Architecture for Efficient Genome Editing. *Nat. Biotechnol.* **2011**, *29*, 143–148.

Mirlohi, A., et al. Allele Sequencing of the Barley Stem Rust Resistance Gene *Rpg1*Identifies Regions Relevant to Disease Resistance. *Phytopathology.* **2008**, *98*, 910–918.

Mizubuti, E. S. G., et al. Selection of Isolates of *Bacillus subtilis* with Potential for the Control of Dry Bean Rust. *Fitopatol. Bras.* **1995**, *20*, 540–544.

Mueth, N. A., et al. Small RNAs from the Wheat Stripe Rust Fungus (*Puccinia striiformis* f.sp. *tritici*). *BMC Genomics.* **2015**, *16*, 718.

Mur, L. A. J., et al. The Hypersensitive Response; the Centenary is upon us but how much do we know? *J. Exp. Bot.* **2008**, *59*, 501–520.

Mysore, K. S.; Ryu, C. M. Non-host Resistance: How much do we know? *Trends Plant Sci.* **2004**, *9*, 97–104.

Navabi, A., et al. Inheritance and QTL Analysis of Durable Resistance to Stripe and Leaf Rusts in an Australian Cultivar, *Triticum aestivum* 'Cook'. *Genome.* **2005**, *48*, 97–107.

Navabi, A.; Singh, R. P.; Tewari, J. P.; Briggs, K. G. Inheritance of High Levels of Adult-Plant Resistance to Stripe Rust in Five Spring Wheat Genotypes. *Crop Sci.* **2004**, *44*, 1156–1162.

Nawrot, R., et al. Plant Antimicrobial Peptides. *Folia Microbiol.* **2014**, *59*, 181–196.

Niks, R. Comparative Histology of Partial Resistance and the Nonhost Reaction to Leaf Rust Pathogens in Barley and Wheat Seedlings. *Phytopathology.* **1983**, *73*, 60–64.

Nölke, G.; Fischer, R.; Schillberg, S. Antibody-Based Metabolic Engineering in Plants. *J. Biotechnol.* **2006**, *124*, 271–283.

Nowara, D., et al. HIGS: Host-Induced Gene Silencing in the Obligate Biotrophic Fungal Pathogen *Blumeria graminis. Plant Cell.* **2010**, *22*, 3130–3141.

Osusky, M.; Zhou, G.; Osuska, L.; Hancock, R.; Kay, W.; Misra, S. Transgenic Plants Expressing Cationic Peptide Chimeras Exhibit Broad-Spectrum Resistance to Phytopathogens. *Nat. Biotechnol.* **2000**, *18*, 1162–1166.

Panstruga, R. Establishing Compatibility between Plants and Obligate Biotrophic Pathogens. *Curr. Opin. Plant Biol.* **2003**, *6*, 320–326.

Panwar, V.; McCallum, B.; Bakkeren, G. Endogenous Silencing of *Puccinia triticina* Pathogenicity Genes through *in Planta*-Expressed Sequences Leads to Suppression of Rust Diseases on Wheat. *Plant J.* **2013a**, *73*, 521–532.

Panwar, V.; McCallum, B.; Bakkeren, G. Host-Induced Gene Silencing of Wheat Leaf Rust Fungus *Puccinia triticina* Pathogenicity Genes Mediated by the Barley Stripe Mosaic Virus. *Plant Mol. Biol.* **2013b**, *81*, 595–608.

Parlevliet, J. E. Genetic and Breeding Aspects of Durable Resistance of Crops to Pathogens. *Afri. Crop Sci. J.* **1995**, *3*, 1–13.

Patil, J. K.; Kumar, R. Advances in Image Processing for Detection of Plant Diseases. *J. Adv. Bioinfo. Appl. Res.* **2011**, *2*(2), 135–141.

Periyannan, S., et al. The Gene *Sr33*, an Ortholog of Barley *Mla* Genes, Encodes Resistance to Wheat Stem Rust Race *Ug99. Science.* **2013**, *341*, 786–788.

Peschen, D., et al. Fusion Proteins Comprising a *Fusarium*-Specific Antibody Linked To Antifungal Peptides Protect Plants against a Fungal Pathogen. *Nat. Biotechnol.* **2004**, *22*, 732–738.

Petre, B.; David, L. J.; Sébastien, D. Effector Proteins of Rust Fungi. *Front. Plant Sci.* **2014**, *5*, 416.

Pfalz, M., et al. Metabolic Engineering in *Nicotiana benthamiana* Reveals Key Enzyme Functions in Arabidopsis Indole Glucosinolate Modification. *Plant Cell.* **2011**, *23*, 716–729.

Piron F., et al. An Induced Mutation in Tomato *Eif4e* Leads to Immunity to Two Potyviruses. *PLoS ONE.* **2010**, *5*(6), e11313.

Plant, R. E., et al. CALEX/Peaches, an Expert System for the Diagnosis of Peach and Nectarine Disorders. *Hort. Sci.* **1989**, *24*, 700.

Pliego, C., et al. Host Induced Gene Silencing in Barley Powdery Mildew Reveals a Class of Ribonuclease-Like Effectors. *Mol. Plant Microbe Interact.* **2013**, *26*, 633–642.

Poland, J., et al. Genomic Selection in Wheat Breeding Using Genotyping-by-Sequencing. *Plant Genome.* **2012**, *5*, 103–113.

Punja, Z. K. Genetic Engineering of Plants to Enhance Resistance to Fungal Pathogens-a Review of Progress and Future Prospects. *Can. J. Plant Pathol.* **2001**, *23*, 216–235.

Räder, T.; Racca, P.; Jörg, E.; Hau, B. PUCREC/ PUCTRI—a Decision Support System for the Control of Leaf Rust of Winter Wheat and Winter Rye. *EPPO Bull.* **2007,** *37,* 378–382.

Raikhel, N.; Michael, P. Adding Precision Tools to the Plant Biologists' Toolbox with Chemical Genomics. *Plant Physiol.* **2005,** *138,* 563–564.

Ramon, M. C. U.; Roland, F. An Expert Advisory System for Wheat Disease Management. *Plant Dis.* **1994,** *78,* 209–215.

Riechen, J. Establishment of Broad-Spectrum Resistance against *Blumeria graminis* f. sp. *tritici* in *Triticum aestivum* by RNAi-Mediated Knock-Down of *MLO*. *J. Verbrauch. Lebensm.* **2007,** *2,* 120.

Risk, J. M., et al. Functional Variability of the *Lr34* Durable Resistance Gene in Transgenic Wheat. *Plant Biotechnol. J.* **2012,** *10,* 477–487.

Roach, J. W., et al. POMME, a Computer-Based Consultation System for Apple Orchard Management using Prolog. *Expert Systems.* **1985,** *2,* 56–69.

Rosewarne, G. M., et al. Quantitative Trait Loci for Slow-Rusting Resistance in Wheat to Leaf Rust and Stripe Rust Identified with Multi-Environment Analysis. *Theor. Appl. Genet.* **2008,** *116,* 1027–1034.

Rossi, V., et al. A Simulation Model for the Development of Brown Rust Epidemics in Winter Wheat. *Euro. J. Plant Pathol.* **1997,** *103,* 453–465.

Rovenich, H.; Boshoven, J. C.; Thomma, B. P. Filamentous Pathogen Effector Functions: Of Pathogens, Hosts and Microbiomes. *Curr. Opin. Plant Biol.* **2014,** *20,* 96–103.

Rushton, P. J., et al. Synthetic Plant Promoters Containing Defined Regulatory Elements Provide Novel Insights into Pathogen-and Wound-Induced Signaling. *Plant Cell.* **2002,** *14,* 749–762.

Rutkoski, J. E., et al. Genomic Selection for Durable Stem Rust Resistance in Wheat. *Euphytica.* **2011,** *179,* 161–173.

Rytter, J. L.; Lukezic, F. L.; Craig, R.; Moorman, G. W. Biological Control of Geranium Rust by *Bacillus subtilis*. *Phytopathology.* **1989,** *79,* 367–370.

Safarnejad, M. R., et al. Antibody-Mediated Resistance against Plant Pathogens. *Biotechnol. Adv.* **2011,** *29,* 961–971.

Saikia, S. P.; Bora, D.; Goswami, A.; Mudoi, K. D.; Gogoi, A. A Review on the Role of *Azospirillum* in the Yield Improvement of Non-Leguminous Crops. *Afr. J. Microbiol. Res.* **2012,** *6,* 1085–1102.

Savary, S.; Nelson, A.; Willocquet, L.; Pangga, I.; Aunario, J. Modeling and Mapping Potential Epidemics of Rice Diseases Globally. *Crop Prot.* **2012,** *34,* 6–17.

Savary, S.; Stetkiewicz, S.; Brun, F.; Willocquet, L. Modeling and Mapping Potential Epidemics of Wheat Diseases-Examples on Leaf Rust and *Septoria tritici* Blotch Using EPIWHEAT. *Eur. J. Plant Pathol.* **2015,** *142,* 771–790.

Savary, S.; Teng, P. S.; Willocquet, L.; Nutter, F. W. Jr. Quantification and Modeling of Crop Losses: A Review of Purposes. *Annu. Rev. Phytopathol.* **2006,** *44,* 89–112.

Schillberg, S.; Zimmermann, S.; Zhang, M. Y.; Fischer, R. Antibody-Based Resistance to Plant Pathogens. *Transgenic Res.* **2001,** *10,* 1–12.

Schob, H.; Kunz, C.; Meins, F. Jr. Silencing of Transgenes Introduced into Leaves by Agro Infiltration: A Simple, Rapid Method for Investigating Sequence Requirements for Gene Silencing. *Mol. Gen. Genet.* **1997,** *256,* 581–585.

Scofield, S. R., et al. Development of a Virus-Induced Gene-Silencing System for Hexaploid Wheat and its Use in Functional Analysis of the *Lr21*-Mediated Leaf Rust Resistance Pathway. *Plant Physiol.* **2005,** *138,* 2165–2173.

Servin, A., et al. A Review of the Use of Engineered Nanomaterials to Suppress Plant Disease and Enhance Crop Yield. *J. Nanopart. Res.* **2015**, *17*, 92.

Sestili, F., et al. Production of Novel Allelic Variation for Genes Involved in Starch Biosynthesis through Mutagenesis. *Mol. Breed.* **2010**, *25*, 145–154.

Shan, Q.; Zhang, Y.; Chen, K.; Zhang, K.; Gao, C. Creation of Fragrant Rice by Targeted Knockout of the *OsBADH2*Gene Using TALEN Technology. *Plant Biotechnol. J.* **2015**, *13*, 791–800.

Sharon, M.; Choudhary, A. K.; Kumar, R. Nanotechnology in Agricultural Diseases and Food Safety. *J. Phytology.* **2010**, *2*, 83–92.

Shatalina, M., et al. Genotype-Specific SNP Map Based on Whole Chromosome 3B Sequence Information from Wheat Cultivars Arina and Forno. *Plant Biotechnol. J.* **2013**, *11*, 23–32.

Singh, R. P., et al. Emergence and Spread of New Races of Wheat Stem Rust Fungus: Continued Threat to Food Security and Prospects of Genetic Control. *Phytopathology.* **2015**, *105*, 872–884.

Singh, R. P.; et al. The Emergence of *Ug99* Races of the Stem Rust Fungus is a Threat to World Wheat Production. *Ann. Rev. Phytopathol.* **2011**, *49*, 465–481.

Singh, R. P.; Huerta-Espino, J.; Rajaram, S. Achieving Near-Immunity to Leaf and Stripe Rusts in Wheat by Combining Slow Rusting Resistance Genes. *Acta Phytopathol. Entomol. Hung.* **2000**, *35*, 133–139.

Singh, R. P.; Huerto-Espino, J.; William, H. M. Genetics and Breeding for Durable Resistance to Leaf and Stripe Rusts in Wheat. *Turk. J. Agr. For.* **2005**, *29*, 121–127.

Slade, A. J., et al. A Reverse Genetic, Nontransgenic Approach to Wheat Crop Improvement by TILLING. *Nat. Biotechnol.* **2005**, *23*, 75–81.

Sohn, K. H., et al. Expression and Functional Roles of the Pepper Pathogen-Induced Transcription Factor RAV1 in Bacterial Disease Resistance, and Drought and Salt Stress Tolerance. *Plant Mol. Biol.* **2006**, *61*, 897–915.

Stein, M., et al. Arabidopsis PEN3/PDR8, an ATP Binding Cassette Transporter, Contributes to Nonhost Resistance to Inappropriate Pathogens that Enter by Direct Penetration. *Plant Cell.* **2006**, *18*, 731–746.

Stockwell, B. R. Exploring Biology with Small Organic Molecules. *Nature.* **2004**, *432*, 846–854.

Stuthman, D. D.; Leonard, K. J.; Miller-Garvin, J. Breeding Crops for Durable Resistance to Disease. *Adv. Agron.* **2007**, *95*, 319–367.

Sun, J. Y., et al. Characterization and Antifungal Properties of Wheat Non-Specific Lipid Transfer Proteins. *Mol. Plant Microbe Interact.* **2008**, *21*, 346–360.

Szabo, L. J.; Bushnell, W. R. Hidden Robbers: The Role of Fungal *Haustoria* in Parasitism of Plants. *Proc. Natl. Acad. Sci. USA.* **2001**, *98*, 7654–7655.

Talamè, V., et al. TILLMore, a Resource for the Discovery of Chemically Induced Mutants in Barley. *Plant Biotechnol. J.* **2008**, *6*, 477–485.

Tester, M.; Langridge, P. Breeding Technologies to Increase Crop Production in a Changing World. *Science.* **2010**, *327* (5967), 818–822.

Thind, T. S. Significant Achievements and Current Status: Fungicide Research. In *One Hundred Years of Plant Pathology in India -an Overview;* Chahal, S. S., Khetarpal, R. K., Thind, T. S., Eds.; Scientific Publishers India: Jodhpur, India, 2006; pp 267–306.

Till, B. J., et al. Discovery of Chemically Induced Mutations in Rice by TILLING. *BMC Plant Biol.* **2007**, *7*, 19.

Till, B. J., et al. Mismatch Cleavage by Single-Strand Specific Nucleases. *Nucleic Acids Res.* **2004**, *32*, 2632–2641.

Tilman, D.; Balzer, C.; Hill, J.; Befort, B. L. Global Food Demand and the Sustainable Intensification of Agriculture. *Proc. Natl. Acad. Sci. USA.* **2011,** *108,* 20260–20264.

Tinoco, M. L., et al. *In vivo* Trans-Specific Gene Silencing in Fungal Cells by *in planta* Expression of a Double-Stranded RNA. *BMC Biol.* **2010,** *31,* 27.

Toojinda, T., et al. Introgression of Quantitative Trait Loci (QTLs) Determining Stripe Rust Resistance in Barley: An Example of Marker-Assisted Line Development. *Theor. Appl. Genet.* **1998,** *96,* 123–131.

Toojinda, T., et al. Molecular Breeding for Rainfed Lowland Rice in the Mekong Region. *Plant Prod. Sci.* **2005,** *8,* 330–333.

Tortora, M. L.; Díaz-Ricci, J. C.; Pedraza, R. O. Protection of Strawberry Plants (*Fragaria ananassa* Duch.) against Anthracnose Disease Induced by *Azospirillum brasilense. Plant Soil.* **2012,** *356,* 279–290.

Tran, D. T., et al. A Codon-Optimized Nucleic Acid Hydrolyzing Single-Chain Antibody Confers Resistance to *Chrysanthemums* against *Chrysanthemum stunt viroid* Infection. *Plant Mol. Biol. Rep.* **2015,** *34(*1), 221–232.

Travis, J. W.; Latin, R. X. Development, Implementation, and Adoption of Expert Systems in Plant Pathology. *Annu. Rev. Phytopathol.* **1991,** *29,* 343.

Uauy, C., et al. High-Temperature Adult-Plant (HTAP) Stripe Rust Resistance Gene*Yr36* from *Triticum turgidum* ssp. *dicoccoides* is Closely Linked to the Grain Protein Content Locus *Gpc-B1. Theor. Appl. Genet.* **2005,** *112,* 97–105.

Uknes, S., et al. Acquired Resistance in Arabidopsis. *Plant Cell.* **1992,** *4,* 645–656.

van Berloo, R.; Aalbers, H.; Werkman, A.; Niks, R. E. Resistance QTL Confirmed through Development of QTL- NILs for Barley Leaf Rust Resistance. *Mol. Breed.* **2001,** *8,* 187–195.

van der Biezen, E. A. Quest for Antimicrobial Genes to Engineer Disease-Resistant Crops. *Trends Plant Sci.* **2001,** *6,* 89–91.

van der Hoorn, R. A. L.; Roth, R.; De Wit, P. J. G. M. Identification of Distinct Specificity Determinants in Resistance Protein*Cf-4*Allows Constructions of a *Cf-9*Mutant that Confers Recognition of Avirulence Protein *AVR4. Plant Cell.* **2001,** *13,* 240–250.

van der Weerden, N. L.; Hancock, R. E. W.; Anderson, M. A. Permeabilization of Fungal Hyphae by the Plant Defensin *NaD1* Occurs through a Cell Wall-Dependent Process. *J. Biol. Chem.* **2010,** *285,* 37513–37520.

Vega Arreguı́n, J. C.; Jalloh, A.; Bos, J. I.; Moffett, P. Recognition of an *Avr3a*Homologue Plays A Major Role in Mediating Non-host Resistance to *Phytophthora capsici* in *Nicotiana* Species. *Mol. Plant Microbe Interact.* **2014,** *27,* 770–780.

Vleeshouwers, V. G. A. A.; Oliver, R. P. Effectors as Tools in Disease Resistance Breeding against Biotrophic, Hemibiotrophic, and Necrotrophic Plant Pathogens. *Mol. Plant Microbe Interact.* **2014,** *27,* 196–206.

Vogel, J. P., et al. PMR6, a Pectate Lyase–Like Gene Required for Powdery Mildew Susceptibility in *Arabidopsis. Plant Cell.* **2002,** *14,* 2095–2106.

Wally, O.; Jayaraj, J.; Punja, Z. Comparative Resistance to Foliar Fungal Pathogens in Transgenic Carrot Plants Expressing Genes Encoding for Chitinase, β-1, 3-Glucanase and Peroxidise. *Eur. J. Plant Pathol.* **2009,** *123,* 331–342.

Wang, Z.; Wang, G.; APD: The Antimicrobial Peptide Database. *Nucleic Acids Res.* **2004,** *32,* 590–592.

Wang, Y., et al. Simultaneous Editing of Three Homoeoalleles in Hexaploid Bread Wheat Confers Heritable Resistance to Powdery Mildew. *Nat. Biotechnol.* **2014,** *32,* 947–951.

Ward, N. A.; Robertson, C. L.; Chanda, A. K.; Schneider R. W. Effects of *Simplicillium lanosoniveum* on *Phakopsora pachyrhizi*, the Soybean Rust Pathogen, and its use as a Biological Control Agent. *Phytopathology.* **2012**, *102*, 749–760.

Wellings, C. R. Global Status of Stripe Rust: A Review of Historical and Current Threats. *Euphytica.* **2011**, *179*, 129–141.

Wen, N. Celiac-Safe Wheat, Reaching One Objective by Two Approaches, Gluten Elimination by Random and Site-Directed Mutagenesis, and Detoxification by Ectopic Expression of 'Glutenases'. Dissertation, Washington State University, Pullman, Washington, 2014.

Wong, C. K.; Bernardo, R. Genome Wide Selection in Oil Palm: Increasing Selection Gain Per Unit Time and Cost with Small Populations. *Theor. Appl. Genet.* **2008**, *116*, 815–824.

Wongkaew, P.; Poosittisak, S. Diagnosis of Sugarcane White Leaf Disease Using the Highly Sensitive DNA Based Voltammetric Electrochemical Determination. *Am. J. Plant Sci.* **2014**, *5*, 2256–2268.

Wu, J. L., et al. Chemical and Irradiation-Induced Mutants of *Indica* rice *IR64* for Forward and Reverse Genetics. *Plant Mol. Biol.* **2005**, *59*, 85–97.

Wulff, B. B.; Thomas, C. M.; Smoker, M.; Grant, M.; Jones, J. D. Domain Swapping and Gene Shuffling Identify Sequences Required for Induction of an Avr-Dependent Hypersensitive Response by the Tomato*cf-4* and *cf-9*Proteins. *Plant Cell.* **2001**, *13*, 255–272.

Yajima, W.; Verma, S. S.; Shah, S.; Rahman, M. H.; Liang, Y.; Kav, N. N. Expression of *Antisclerotinia scFv* in Transgenic *Brassica napus* Enhances Tolerance against Stem Rot. *N. Biotechnol.* **2010**, *27*, 816–821.

Yang, X. Y., et al. Characterization and Expression of an nsLTPs-Like Antimicrobial Protein Gene from Motherwort (*Leonurus japonicus*). *Plant Cell Rep.* **2008**, *27*, 759–766.

Yau, Y. Y.; Easterling, M.; Stewart, Jr. N. Precise Transgene Stacking in Planta through the Combined Use of Talens and Unidirectional Site-Specific Recombination Systems. *OA Biotechnol.* **2013**, *2*(3), 24.

Yin, C.; Jurgenson, J. E.; Hulbert, S. H. Development of a Host-Induced RNAi System in the Wheat Stripe Rust Fungus *Puccinia striiformis* f. sp. *tritici*. *Mol. Plant Microbe Interact.* **2011**, *24*, 554–561.

Yuen, G. Y.; Steadman, J. R.; Lindgren, D. T.; Schaff, D.; Jochum, C. Bean Rust Biological Control Using Bacterial Agents. *Crop Prot.* **2001**, *20*, 395–402.

Zhang, F., et al. Efficient Construction of Sequence-Specific TAL Effectors for Modulating Mammalian Transcription. *Nat. Biotechnol.* **2011**, *29*, 149–153.

Zhang, L., et al., Parallel Expression Profiling of Barley–Stem Rust Interactions. *Funct. Integr. Genomics.* **2008**, *8*, 187–198.

Zhang, M., et al. Two Cytoplasmic Effectors of *Phytophthora sojae* Regulate Plant Cell Death Via Interactions with Plant Catalases. *Plant Physiol.* **2015**, *167*, 164–175.

Zhong, S., et al. Factors Affecting Accuracy from Genomic Selection in Populations Derived from Multiple Inbred Lines: A Barley Case Study. *Genetics.* **2009**, *182*, 355–364.

Zhou, R. H., et al. Development of Wheat Near-Isogenic Lines for Powdery Mildew Resistance. *Theor. Appl. Genet.* **2005**, *110*, 640–648.

CHAPTER 3

HOLISTIC MANAGEMENT OF FOLIAR BLIGHT DISEASE OF WHEAT AND BARLEY

A. K. CHOWDHURY[1,*], P. M. BHATTACHARYA[1], S. BANDYOPADHYAY[1], and T. DHAR[2]

[1]*Department of Plant Pathology, Uttar Banga Krishi Viswavidyalaya, Pundibari 736165, Cooch Behar, West Bengal, India*

[2]*Regional Research Sub-Station, Uttar Banga Krishi Viswavidyalaya, Manikchak, Malda 732203, West Bengal, India*

Corresponding author. E-mail: akc_ubkv@rediffmail.com

CONTENTS

ABSTRACT

Wheat and barley, the earliest known commercial grains are grown throughout the world and has importance since Biblical and Indus Valley Civilization. They contribute about one-third of the world's total grain production. Wheat provides nourishment for humans as major source of carbohydrate. Beside the staple food, barley also contains many medicinal properties. However, the production and productivity of both the crops particularly in South East Asia remains almost stagnant for the last five years. As very little scope for increasing the area under wheat and barley cultivation due to growing urbanization, crop diversification, dwindling water resources, micronutrient deficiencies and soil health deterioration, the genetic improvement, and crop management are only options to break the yield barrier. There are numerous foliar blights either of seed borne and/or soil borne diseases reported on wheat and barley. Among them, spot blotch (*Bipolaris sorokiniana*), Alternaria blight (*Alternaria triticina*) and tan spot (*Pyrenophora tritici- repentis*) are recorded in most wheat growing areas of South East Asia where warm and humid climate persist. It takes a heavy toll by reduction of annual yields by 20% in the farmers' fields. The foliar blight in barley includes spot blotch (*Bipolaris sorokiniana*), net blotch (*Helminthosporium teres*), and speckled blotch (*Septoria passerinii*) diseases. Considering the importance of the crops in the economy of developing countries, the disease management by integrated way has been highlighted and reviewed.

3.1 INTRODUCTION

Wheat and barley are the major cereal crops in the word. The global statistics indicate that wheat and barley production in the world was 707.2 and 135.65 million ton, respectively, in 2014–2015 and contribute one-third of the world's total grain production (FAO, 2014; USDA, 2014). Wheat (*Triticum* spp.) was one of the first domesticated food crops and it has been the basic staple food of the major civilizations of Europe, West Asia, and North Africa over 8000 years. Now, wheat is grown on more land area than any other commercial crops and considered to be the most important food grain source for humans. It is regarded as the best of the cereal foods, most important source of carbohydrate throughout world and also provides more nourishment for humans than any other food source. It is a major diet crop because of its agronomic adaptability, easy properties of grain storage and also conversion of grain into flour for making edible, palatable, interesting, and satisfying foods. Doughs produced from bread wheat flour differ from other cereals in their unique viscoelastic properties. Wheat starch is easily digested as well as its protein. Beside these, it contains minerals, vitamins and fats (lipids), and is highly nutritious when small amount of animal or legume protein is added. It is reported that wheat-based diet is higher in fiber than a meat-based diet (Johnson et al., 1978).

Barley (*Hordeum vulgare* L.), is one of the first cultivated grains and now grown in almost all countries. In Tibet, barley grain is a staple food and is reported to be eaten widely by peasants in Medieval Europe. Beside this, it is used as animal fodder, source of fermentable material for beer and certain distilled beverages, and also as an additive of various health foods. The soups and stews prepared from barley are very popular in some countries. Malt prepared from barley grains are commonly made by traditional and ancient method of preparation. It has also some medicinal properties like lowering of blood sugar, blood pressure, and cholesterol and promoted as weight loss food. It is very popular food during digestive complaints like diarrhoea, stomach pain, and inflammatory bowel conditions.

The attacks of pests and diseases have a serious impact on the economic output of a farm. Farmers adopt different technologies to combat the pest and disease infestation depending on the crop grown, nature of the attack, weather conditions, and so forth, and they need to ensure the balance of pests and disease prevention and treatment methods against damage to the environment. Like other crops wheat and barley is also suffered from many diseases caused by fungi, bacteria, viruses, and so forth. Among the different diseases, foliar blight of wheat and barley is one of the important diseases particularly in warmer growing areas of eastern plains of South Asia which is characterized by an average temperature in the coolest month above 17°C (Dubin & Bimb, 1991; Dubin & van Ginkel, 1991; Duveiller & Gilchrist, 1994; Saari, 1998; Dubin & Duveiller, 2000; Duveiller, et al 2005; Joshi et al., 2007a).

Yield losses due to three foliar blight pathogens are variable but result in considerable crop loss. The destructive capacity of pathogen is evident from the reports around the world. In last two decades, spot blotch have emerged as serious concerns for wheat cultivation in the South East Asia. The disease causes significant yield losses from 22% to complete failure of crop under severe epidemics. Indian subcontinent has 10 mha of spot blotch-affected area (Nagarajan & Kumar, 1998) and the losses due to this pathogen may be 10–50% which can be more devastating for the farmers in the Eastern Gangetic plains, and it depends on the level of resistance in a cultivar against leaf blight and weather conditions. Besides, losses in grain yield, leaf blight causes seed discoloration, shriveled seeds, and loss in seed viability. On an average, it is estimated that a South Asian country loses 20% of total crop yield through leaf blight disease (Singh et al., 2004; Malik et al., 2008a, Chowdhury et al., 2013). In Nepal, it was shown that grain yield losses due to spot blotch were 52% under soil nutrient stress and 26% under optimum fertilization and it continues to cause substantial grain yield reductions under resource-limited farming conditions (Sharma & Duveillar, 2007). In Bangladesh, the average losses due to these foliar blights were estimated to be 15% in the farmers' fields (Alam et al., 1998). The pathogen also causes grain yield losses up to 10, 15, and 20% in Scotland, Canada, and Brazil, respectively, through common root rot and seedling blight. Grain yield losses due to foliar blight vary greatly, depending on wheat crop husbandry. In affected areas, yield losses depend upon genotypes, sowing time, year, location, and stress conditions (Rosyara et al., 2005). Diseased wheat plots in Mexico without fungicides yielded 43% less grains (Vilareal et al., 1995). In the United States, it has been reported to be a serious foliar disease which causes economic loss of wheat yield (Wegulo et al., 2009) and fungicide application increased yield ranging from 27 to 42% at Nebraska, United States. In favorable years the rates of grain infection by this fungus were detected to be as high as 70% (Sharma-Poudyal et al., 2005). At higher latitudes of Canada and United States prairies (Gonzalez & Trevathan, 2000; Fernandez & Jefferson, 2004) and in parts of Australia (Lehmensiek et al., 2004), *Bipolaris sorokiniana* is a dominant pathogen among fungi, causing common root rot and resulting in crop losses up to 19%. In Turkey, *B. sorokiniana* has been observed to be widespread and affects the sub crown internodes and crowns of wheat (Eken & Demirci, 1998).

3.2 FOLIAR BLIGHT DISEASE AND PATHOGENS INVOLVED

There are numerous foliar blights either of seed borne and/or soil borne diseases reported on wheat (Prescott et al., 1986; Mathur & Cunfer, 1993). The three blight diseases (spot blotch, tan spot, and *Alternaria* blight) either individually or in combination have been recorded in wheat growing areas of India (Joshi et al., 1978; Nema, 1986) Bangladesh (Alam et al., 1994), Nepal (Sharma, 1996), and Pakistan (Hafiz, 1986).

Typical symptoms caused by *B. sorokiniana* (Sacc.) Shoemaker, *Alternaria triticina* Prasada and Prabhu, *Pyrenophora tritici-repentis* (Died.) Drechs are found in regions where the spot blotch, leaf blight, and tan spot diseases occur separately. However, it is often impossible to recognize the causal organism of a blight symptom without the help of a microscope in wheat-growing areas where all the pathogens occur together. The term *Helminthosporium* leaf blight (HLB) is commonly used to refer to this disease complex.

3.2.1   WHEAT

3.2.1.1   SPOT BLOTCH CAUSED BY BIPOLARIS SOROKINIANA

Among the foliar blight pathogens, spot blotch caused by *B. sorokiniana* (Sacc.) Shoem is one of the most concerning diseases in relatively warm and humid growing regions of India and other south asian countries due to its wide spread prevalence and increasing severities. (Joshi et al., 2007b). The occurrence of *B. sorokiniana* as wheat pathogen in the Northwestern part of the Russian Federation (Smurova, 2008) suggests its potential to become a serious wheat pathogen in Europe. At present, under European conditions, *B. sorokiniana* causes yield losses mostly due to root rot (Rossi et al., 1995) and seed black point, which affects seed germination rates and also causes root rots in seedlings (Hudec & Muchova, 2008). Recently the wide spread use of conservation tillage practices and poor nutrient management may also be favorable for spot blotch incidence in the South East Asia (Duveiller & Sharma, 2009). It is the major biotic constraint in wheat in the Gangetic plain of South Asia, especially in the rice-wheat cropping system (Duveiller et al., 1998).

   B. sorokiniana is an aggressive pathogen that causes spot blotch, root and crown rots, node cankers, ear head and seedling blight (Zillinisky, 1983). Lesions in leaves appear as small, chlorotic, and oval shaped with dark centers. Later it becomes reddish brown centers with yellow margins and tapered ends. They may reach several centimeters before coalesce and induce the death of the leaf also. If spikelets are affected, it can result in shriveled grain and black point of the embryo at the end of the seed (Duveiller & Dubin, 2002).

3.2.1.2   ALTERNARIA BLIGHT CAUSED BY ALTERNARIA TRITICINA

A. triticina is the causal agent of *Alternaria* leaf blight of wheat, described from India during 1970s (Joshi et al., 1978). It is more prevalent especially on susceptible cultivars of durum and bread wheat and cause major damage under wet or humid conditions. *A. triticina* is one of several species of *Alternria* that have been isolated from wheat leaves and demonstrated to be pathogenic, while others appear to be primarily saprophytes.

The lesions of *A. triticina* are irregular in shape and dark brown to grey. Initially, the disease appears as small, oval discolored lesions, which coalesce with disease progression, resulting in leaf death (Singh & Srivastava, 1997).

3.2.1.3   TAN SPOT CAUSED BY PYRENOPHORA TRITICI-REPENTIS

Typically, the first foliar symptoms of tan spot (*P. tritici-repentis*) appears as small, light brown blotches that develop into oval shaped, light brown, necrotic lesions bordered with a yellow halo (Schilder & Bergstrom, 1993). Lesions coalesce with age and favor the early senescence of the entire leaf. Necrosis of leaf often starts near the tip and progresses toward the base of the leaf.

3.2.2   BARLEY

Among the different diseases, leaf blight (spot blotch by *B. sorokiniana* (Sacc.) Shoemaker, net blotch by *Drechslera teres* (Sacc.) Shoemaker teleomorph *Pyrenophora teres* (Drechs), speckled leaf blotch by *Septoria passerinii,* are important. Of diseases, blotches are considered to be very important because they are air borne and can spread very easily and cause epidemic.

3.2.2.1   SPOT BLOTCH

B. sorokiniana (Sacc.) Shoemaker (telomorph: *Cochliobolus sativus* (Ito & Kurib Drechsl ex Dastur) causes spot blotch of wheat and barley. The pathogen has been described as the most important fungal pathogen of barley (Arabi & Jawhar, 2004; El Yousfi & Ezzahiri, 2002; Valjavec-Gratian & Steffenson, 1997). The spot blotch fungus has a wide range of hosts and pathogenicity is variable in nature. The fungus has been reported to attack at least 28 species of cultivated and wild grasses. The spot blotch fungus has many physiologic races that differ greatly in virulence and ability to attack specific cereals and grasses. Yield losses caused by the pathogen have been estimated in average 10% in barley and 5% in spring wheat on a long-term, region-wide basis, with losses in individual fields in some years have been reported to be above 30% (Stack, 1991). In the United States, it has been reported that yield losses ranged from 16 to 33% in 1960s on barley (Valjavec-Gratian & Steffenson, 1997). The infection depends on climatic conditions and ranges from 10 to 100% infection by the fungus. Another problem is the spots enlargement as the leaf grows and spread along the entire leaf blade and can produce brown lesions. The producers have 100% losses because infected plant cannot produce normal heads (Teng, 1987; Agrios, 1988; Minella, 2001; Castro & Bach, 2004). The fungus overwinters in diseased crop residue of wild and cultivated grasses or cereals grains, on or within the seed, and on the seedling leaves of winter barley.

Spots (lesions) appear as chocolate brown-to-black discoloration near the soil line or at the base of the sheaths that cover the leaves of the seedlings. Infections may continue until the seedlings turn yellow and die, either before or after emergence which results in poor crop stand. The symptoms are more frequent in later cases, after emergence. Affected seedlings may be dwarfed, produce excess number of tillers, and have dark green leaves. Diseased seedlings commonly have weaker, dark brown, rotted crowns as well as roots. Tillers may be affected or killed. In extreme seedling infections, plants may be stunted, the panicles may emerge partially and the kernels are either fully or partially chaffy. Barley plants that is not infected at seedling stage usually appear normal up to about heading stage, when the typical leaf lesions, of various sizes and shapes, appear on the lower leaves in warm, humid weather. The center of every lesion is dark brown; the color gradually fades at the edge into the normal green color of the leaf. Spots are oblong or eye-shaped and the centers are lighter brown than the margins. Where numerous lesions are formed, they may coalesce producing larger irregular blotches. Heavily infected leaves dry out and die prematurely. The centers of older lesions on both living and dead leaves have an olive-green color caused by the growth of the fungus with abundant conidial production. Spot blotch starts on the older leaves and leaf sheaths which progresses upwards to the younger leaves. The lesions do not have the net like appearance like net blotch. Dark brown-to-black spots on the glumes and kernels may also be observed. The black point at the germination end of a kernel is a common and characteristic symptom. Kernels which are infected early become shriveled and are light weight. The market price is also reduced on the grains that contain 5% or more of black-point kernels. The fungus survives in diseased crop residue of wild and cultivated grasses, on or inside the seed of cereal grains, and also on the leaves of the seedling of winter barley.

3.2.2.2   NET BLOTCH

In severe epidemic, net blotch can cause significant reduction in yield and lower the quality of the grains from malting quality to cattle feed. In general the flag and flag-1 leaves must be infected for yield loss to occur, with losses generally ranging between 10 and 20%, and sometimes losses of more than 30% also reported. Net blotch, although common early in the season, rarely develops sufficiently in spring to cause significant yield loss. In normal infection 0–10% yield loss generally occurs by the pathogen but in severe infection the loss can exceed 20%. Net blotch mostly causes reductions in grain quality by reducing the grain size (http://www.croppro.com.au/crop_disease_manual/ch02s04. php).

Net blotch is caused by the fungus *Helminthosporium teres (P. teres)*. The blotch is basically a leaf disease and restricted to the *Hordeum* species of cultivated and wild barley. It is mostly found during cool and humid periods. Self-sown and volunteer barley plants are commonly infected late. The estimated annual loss caused by net blotch disease ranges from a trace to 0.5% of the potential yield.

The disease appears first on the seedling leaves of both self-sown and spring-sown barley as ovoid to oblong brown spots or blotches, often restricted by the veins and marked internally with a characteristic netting. The net like blotches, made up of narrow lines that are darker brown and crosshatched, are best seen by holding the leaves against light. Infections occur on the leaves from seedling stage until near maturity stage. When a huge number of lesions are formed, they enlarge and merge, forming long brown streaks with irregular margins that may cover most area of the leaf and affect its usefulness. The symptoms do not extend into the leaf sheath and the leaf tissue does not split. Small brown streaks, without the netted appearance, develop on the chaff (glumes), causing reduced yields and shriveled seed. At the base, infected kernels have indistinct brown lesions. At harvest, the stems may be a dull brown and may lack strength.

Net blotch fungus can survive on infected barley stubble as long as the stubble is present on the soil surface. However, the primary inoculum levels are significantly reduced after two years. Ascospores are produced by pseudothecia on the stubble residues spread by rain-splash or wind to infect neighbouring plants. Majority of these ascospores travel only short distances within the crop. Infection requires moist conditions with temperatures below 25°C but is most rapid at 25°C. Secondary infection is provided by conidia produced from lesions on leaves (Turkington et al., 2011).

3.2.2.3 SPECKLED LEAF BLOTCH

Besides spot and net blotch, speckled leaf blotch caused by *S. passerinii.* is also important. Majority of barley cultivars are susceptible to speckled leaf blotch. The pathogen overwinters on seeds, stubble, straw, and leaves. Spores infect the new crop during wet weather. Secondary infection to nearby plants results from spores produced on infected leaf spots which is carried away from one field to another by splashing rain and wind. Wet and windy weather favors this disease, whereas dry conditions reduce or prevent new infections and spore production on diseased plants. The successful infection and establishment by the spores inside the plant depend on wetness for six hours or more on the plant surface. The disease increases with higher level of nitrogenous fertilizers as it produces more dense foliage creating humid microclimate inside the crop.

Light brown, elongated spots which are surrounded by yellow tissues in the margins of the leaf are dried frequently. Several spots may coalesce. Eventually, pycnidia are formed as lines of very small dot like black structures in the brown tissue of the lesion. Lesions on the upper leaves and glumes significantly reduce photosynthetic activity of the plants as well as yield. Seed set is not affected but seed filling is hampered. The grains may become shriveled and these grains become chaffy at harvest (Tekauz, 2003).

3.3 MANAGEMENT OF FOLIAR BLIGHT DISEASE

3.3.1 WHEAT

The management of foliar blight by different methods draws attention of plant pathologists and developed along with the development of plant pathology. It becomes critical particularly under favorable environmental regime especially for fungal pathogens, as they are highly sporulating with shorter life cycle as well as easy dissemination properties. Generally multi-pronged strategy becomes essential and application of a single tactics results in obvious failure. Conventionally foliar spray has been a first choice to researchers, extension agents and to practicing farmers. Integrated disease management (IDM) with various stages of development in different countries has been attempted, however, not yet accepted in larger scale among the growers' community. Researches relevant to foliar blight management can be considered and categorized as component for developing IDM.

3.3.1.1   BREEDING FOR RESISTANCE

The best approach to manage foliar blight disease is to develop resistant varieties. Inheritance of its resistance has been reported by many workers. In many countries highly susceptible cultivars are quickly replaced by less susceptible cultivars to protect the disease. Sources of spot blotch resistance had been identified over the years and is governed by one or more genes. The origin of resistance was differentiated into three categories: Latin America, China and wild relatives of wheat or alien species (van Ginkel & Rajaram, 1998). The Latin American sources are mainly obtained from Brazil, and generally trace back to Italian ancestry. Among them, older two resistant Brazilian varieties are BH 1146 and CNT 1 (Mehta, 1985). The Chinese sources are mostly materials from Yangtze River Basin. Early Chinese sources of resistance used at International Maize and Wheat Improvement Center (CIMMYT) include Shanghai# 4, Suzhoe# 8, and Yangmai# 6.

Alien resistance sources are many. Alien sources used at CIMMYT had centred around crosses involving *Thinopyrum curvifolium* (Mujeeb-Kazi et al., 1996b; Vilareal et al., 1992). Two outstanding lines Mayoor and the Chirya series were developed by using these alien sources in combination with Chinese resistance sources. It may sometimes be difficult to accurately quantify or determine the exact contribution of alien genes in such multi-parent crosses. van Ginkel and Rajaram (1998) indicated that there is a component of alien-derived resistance present in certain advanced materials developed by A. Mujeeb-Kazi at CIMMYT. van Ginkel and Rajaram (1998) found *Aegilops squarrosa* crosses had shown good resistance to spot blotch in Mexico. The progress in transferring resistance genes from *T. curvifolium, Triticum tauschii,* and *Elymus curvifolius* to wheat germplasm has also been achieved to some extent (Mujeeb-Kazi et al., 1996a, 1996b). Among 250 synthetic hexaploid (2N=6X=42) amphiploid of wheat, some resistant stocks against *B. sorokiniana* have been identified (Mujeeb-Kazi, 1996b).

Significant progress of research has been done at Punjab Agricultural University, India, by using more than 14,000 entries of wheat and related wild species, representing 13 genera and 136 species, for spot blotch resistance (Dhaliwal et al., 1993; Singh & Dhaliwal, 1993). They identified several promising entries including *Aegilops triaristata, Triticum dicoccoides, Aegilops cylindrical* and *Triticum boeoticum.* Studies on the inheritance on spot blotch resistance with Latin American and Indian resistance sources had shown one or two genes to be involved (Srivastava et al., 1971; Srivastava, 1982; Adlakha et al., 1984). Velazquez Cruz (1994) observed six genes to be segregating in four moderately resistant to resistant lines (Gisuz,

Cugap, Chirya 1, and Sabut) developed at CIMMYT, with two to three genes providing good resistant levels. Conner (1990) and Gilchrist et al. (1991a, 1991b) had shown that foliar and seed infection by *B. sorokiniana* may be determined by different genetic systems. Heritabilities of resistance tend to be low and much influenced by the changing environment, as determined by temperature, humidity, inoculum pressure, planting date, observation date, and so forth. These environmental components and the variability in the pathogen make breeding for resistance difficult (Duveiller & Gilchrist, 1994).

In Zambia, spot blotch is the devastating disease under rainfed wheat production (Muyanga, 1994; Mukwavi, 1995). Most resistance cultivars in Zambia contain Brazilian ancestry. Brazilian sources of resistance were first introduced in 1979 from Passo Fundo (Raemaekers, 1985). In 1984, PF 7748 was released as Whydah, and in 1986 and 1988, Hornbil and Coucal were released (Raemaekers,1988) a CIMMYT cross, was released as Coucal in 1988 (Raemaekers, 1988). The release of these three varieties enhanced yield potential to increase from 1.6–1.7 t ha^{-1} to 2.7 t ha^{-1} (Mukwavi, 1995). More previous studies indicated that the resistance to spot blotch in wheat and barley is inherited and governed by quantitative trait and no major genes are known to involve (Mehta, 1993). Though, sources of higher levels of resistance had been achieved using conventional breeding methods, mainly because of the polygenic nature of this resistance. It becomes necessary to identify the quantitative trait loci (QTLs) with the expectation that once the genes corresponding to QTLs are cloned, transformation method may offer new options for combating the disease in future. Induced variation for higher levels of resistance is also expected through somatic embryogenesis (Vasil & Vasil, 1986; Mehta, 1996; Bohorova et al., 1995). Valjavec-Gratian and Steffenson (1997) found that both virulence in the pathogen and resistance in the host are controlled by monogene in progenies of barley cross Bowman/ND 5883 and *B. sorokiniana* isolate ND 90Pr. Resistance to *B. sorokiniana* in barley is reported to be conferred by a single gene at the seedling stage. On the basis of both qualitative and quantitative data analyses, this gene was mapped to the distal region of chromosome 1P. Two QTLs were identified for spot blotch resistance at the adult plant stage: The largest QTL mapped to chromosome 5P and the other mapped to chromosome 1P near the seedling resistance locus (Steffenson et al., 1996). Jana and Bailey (1995) reported that cultivated barley conserved *in situ* in the Middle East preserve resistance to *B. sorokiniana* in either single or multiple combinations in cultivated landrace population, but it was less effective in preserving the diversity for resistance than in wild barley. Bailey and Wolfe

(1994) suggested that a new breeding strategy for developing lines resistant to common root rot by involving high yield character to screen to eliminate the highly susceptible lines from crosses and then selecting the best yielding lines with the lowest disease. While analyzing resistance to common root rot and spot blotch in barley, the heritability to common root rot resistance of cross Fr926-77 × Deuce was found to range from 56 to 85%. Heritability for spot blotch resistance in this cross was 43–615. For another cross of Virden × Elice, the heritability to common root rot resistance ranged from 53 to 78%, while 73 to 78% for spot blotch. However, no association could be detected between heritability to common root rot and spot blotch resistance in barley (Kutcher et al., 1994). A high level of resistance to spot blotch was achieved by crossing Chinese hexaploid wheat with a moderately resistant commercial cultivar in Nepal by Sharma et al. (1997). They made selections for low area under disease progress curve (AUDPC) of spot blotch in segregating generations.

Recently, Singh et al. (2013, 2014) worked on three sets of mapping populations made for spot blotch resistance in wheat *viz.* Sonalika/BH1146 (set-I), Kanchan/Chiryal (set-II) and HUW234/YM#6 (set-III) in three hot spots in India and identified 23 resistant lines across the locations. They further reported that in population I (BH 1146 × Sonalika), through composite interval mapping (CIM) two stable QTLs were identified for the DS on chromosome 7B and 7D at logarithm of the odds (LOD) score above the threshold value of 2.4 that explained 11.6 and 9.8% mean phenotypic variation (R^2), respectively. These two stable QTLs were detected in three environments and three years consecutively. Similarly in population II (Chirya 1 × Kanchan), using CIM, one stable and consistent QTL was detected on chromosome 7D over the two years and two locations with a LOD score value of 6.4 explaining mean phenotypic variation (R^2) of 10.5%.

Arun et al. (2003) reported that somaclones (R_2, R_3, and R_4 generations) regenerated from immature embryos of two popular wheat varieties in Eastern India, HUW-206 and HUW-234 displayed improved earliness, enhanced resistance to spot blotch disease and yield over the parent. The superiority of variants for both yield traits and spot blotch resistance was established in R_4 generation. This confirmed the possibility of wheat improvement through somaclonal variation. Similar types of studies were conducted by Liatukas and Ruzgas (2012) on resistance of European wheat varieties against *B. sorokiniana* in Turkey.

Very recently, Singh et al. (2014b, 2015a, 2015b) evaluated a total of 1281 genotypes and varieties of wheat (*Triticum aestivum, Triticum durum,* and *Triticum dicoccum*) and also Triticale at 14 hot spot locations in India against *B. sorokiniana* and *A. triticina.* They observed that 11 genotypes *viz.* HD 3081, HS 513, HP 1913, KARAWANI/4NIF/ 3/SOTY//NAD63 / CHRIS, NIAW 1846, PBW 629, PBW 665, PDW 314, TL 2978, VL 971, and VL 972 were highly resistant to foliar blight pathogens. CIMMYT developed 56 genotypes for specific objective of spot blotch resistance and when evaluated in different hot spots of South East Asia, it was observed that eight genotypes were highly resistant against the disease across the locations (Singh et al., 2015a, 2015b).

3.3.1.2   CULTURAL PRACTICES

The reports from different countries on managing foliar blight through manipulation of agronomic practices suggest that different mineral nutrients may reduce foliar blight (Huber et al., 1987; Morh et al., 1987; Krupinsky & Tanaka, 2001).Although soil moisture and soil nutrient stress occur together in the wheat fields of South Asia, however, quantitative information is lacking on the effect of low soil moisture and poor soil fertility on foliar blight severity (Regmi et al., 2002). Factors such as minimum tillage or surface seeding, irrigation, late planting, and low soil fertility may be responsible for higher foliar blight severity in the wheat-based cropping systems of the Indo-Gangetic plains (Sharma & Duveiller, 2003). Application of higher nitrogen with lower disease severity was reported by Chaurasia and Duveiller (2006) and Sharma and Duveiller (2004). On the contrary, Singh et al. (1998) reported more disease infection on higher nitrogen application.

Seeding dates could be adjusted slightly to reduce spot blotch incidence on wheat during the post-flowering period (Hetzler et al., 1991). By adjusting the sowing date, it was possible to keep high air temperatures from coinciding with critical physiological growth stages (e.g., anthesis). The early seeding is recommended for reducing losses due to foliar blight in areas to avoid pre-monsoon showers result in high moisture at the end of the wheat season in northeast and southern parts of Eastern India, Bangladesh, and parts of South America. In Uttar Pradesh (Eastern India), Singh et al. (1998) reported that late-sown (December 30) wheat fields suffered more from foliar blight than plots sown on the optimal date (November 30).

Tan spot and *Stagonospora* blotch [*Phaeosphaeria avenaria* (G. F. Weber) O. Eriksson f. sp. *triticea* T. Johnson, anamorph *Stagonospora avenae* (A. B. Frank) Bissett f. sp. *tritica* T. Johnson] increased in no-tillage systems in wheat monoculture or wheat followed by fallow and reduced disease incidence occurred when wheat followed other crops (Bailey et al.,

2000; Fernandez et al., 1998). In some studies, conventional tillage increases crop residue mineralisation, reducing fungal inoculum (Sutton & Vyn, 1990). However, others (Krupinsky & Tanaka, 2001; Jorgensen & Olsen, 2007; Krupinsky et al., 2004, 2007) reported opposite results regarding the effect of no-tillage on necrotrophic wheat diseases, depending on the crop growth stage evaluated (early or late in the season) and the existing environment.

In long term trials with different nutrient management in rice-wheat cropping system, Gami et al. (2001) and Regmi et al. (2002) reported that declining soil fertility affects foliar blight severity. Factors that may alter the severity of wheat foliar blight caused by nonspecific pathogens include change in seeding date (Hetzler et al., 1991; Toledo & Guzman, 1998) and mineral nutrients (Huber et al., 1987; Krupinsky & Tanaka, 2001; Regmi et al. 2002). Sharma and Duveiller (2004) opined soil fertility stress has pronounced effect on foliar blight severity and/or disease levels than moisture stress. They observed that under low soil fertility, reduction in grain yield and 1000 kernel weight due to HLB were 52.1 and 15.5%, respectively, compared to 26.1 and 8.05 under optimum soil fertility and also recorded 22.1% yield reduction under low soil moisture compared to 11.85 under optimum soil moisture.

In 1960s, Ledingham (1961) showed that a longer time gap between two crops in the same season with different susceptible hosts may cause lowering of the rate of common root rot. The effect of tillage and crop rotation may not be consistent all over the world. Duczek et al. (1999) observed a low sporulation rate of *B. sorokiniana* in the soil when rotation is made with oil seed rape and red clover. As the frequent tillage of soil produces more infectivity of *B. sorokiniana* (Reis, 1991), lesser tillage or reduction in frequency of tillage is advocated to reduce infectivity of the soil.

The effect of cultural practices on leaf blotches of wheat were assessed in Brazil. Four soil management systems (no-tillage, minimum tillage, conventional tillage using a disk plough plus disk harrow and tillage using a mold board plough plus disk harrow) and three crop rotation systems, that is, system I (wheat/soybean), system II (wheat/soybean and common vetch/ maize or sorghum), and system III (wheat/soybean, black oats or white oats/ soybean, and common vetch/ maize or sorghum) were compared. Leaf blotch incidence was higher under monoculture and no-tillage condition than under crop rotation. Crop rotation under no-tillage controlled leaf blotch diseases and yield stability of wheat was also maintained (Prestes et al., 2002).

Ploughing or clearing the stubble, grass weeds and volunteer cereals reduce inoculum similar to crop rotation (Diehl et al., 1982). They further noted that eradicant fungicide treatment of the seed and crop rotation with non-host crops can control spot blotch. However, Cook and Veseth (1991) contradicted the views of Diehl et al. (1982) in reducing the blight incidence. Hossain (1998) observed that nutrient elements, especially application of micro-elements Cu, B, and Mo in soil along with normal fertilization with the objective of possible disturbance of the overall natural soil fertility balance reduced leaf spot incidence. Boron application increased grain formation and wheat yields also. An *in vitro* study revealed that boron has a strong inhibitory effect on *B. sorokiniana.*

Extensive studies were conducted in Nepal from 2000 to 2002, on experimental station and in farmers' fields of terai region to compare plots sprayed with fungicides with untreated check under recommended and no fertilizer conditions on HLB incidence and yield loss. It was observed that yield reduction of 34% when no chemical fertilizer was added and it came down to 17% at the recommended level of NPK (120:60:30 kg/ha). Further it was noted that losses due to disease decreased from 19 to 14% when supplementary irrigation was given and physiological stress was less (Sharma & Duveiller, 2004). In different locations and of Northeastern plains zone of India, application of higher doses of potash in combination with other nutrients (120:60:60 kg N, P, K/ha) reduced foliar blight incidences as compared to 120:60:40 kg N, P, K/ha (Singh et al., 2008; Mukherjee, 2008). Application of farmyard manure (FYM) 12.0 t/ha, led to decline of AUDPC by 30%, suggesting not only that FYM contains a significant amount of potassium, but increasing levels of organic matter impart beneficial role and reduce disease severity.

Crop rotation facilitates the decomposition of stubble on which pathogens carry over and natural competitive organisms, thereby reducing the pathogens on the remaining residue when unrelated crops are being grown (Cook & Veseth, 1991). Leaf spot disease severity were higher in monoculture compared with following an alternative crop when tan spot *(Drechslera tritici-repentis)* and *Stagonospora nodorum* blotch (*S. nodorum)* were the dominant diseases, Fernandez et al. (1998) observed that monoculture wheat had a higher disease severity than wheat grown after flax or lentil. It is interesting to note sometime that rotations have limited impact on wheat disease severity instead of reduction of pathogen populations (Bailey et al., 2000).

Mahto et al. (2006) computed the foliar blight incidence on three different wheat varieties, Rohini, BL 1473 and Bhrikuti under surface seeding conditions and conventional tillage. Surface seeding was particularly superior to conventional tillage in low land situation. The results indicate that the average foliar blight induced reductions in grain yield, 1000 grain weight, and kernel per spike were 16, 12, and 155, respectively, under conventional tillage whereas the corresponding reductions were 23, 13, and 195 under surface seeding. The same observation was noticed at Cooch Behar, West Bengal, India by using three different wheat varieties PBW 343, NW 2045, and NW 1012 after three years of continuous zero-tillage under zero-tillage and conventional tillage conditions (Chowdhury et al., 2014).

3.3.1.3 INDUCED RESISTANCE AND BIOLOGICAL CONTROL

Despite success in chemical management, yield losses in wheat due to foliar blight indicate the need to search for alternative strategies of disease control. On a global basis one of the promising alternative strategies is induced resistance. In the broadest sense, induced resistance means the inhibition of pathogen or its activity by a prior activation of the plants innate defense potential. Defense is activated by necrotizing pathogen as well as by chemicals mimicking factors of the natural defense systems, such as salicylic acid (Bochow et al., 2001; Sticher et al., 1997). Hait and Sinha (1986) reported that seed treatment with systemic acquired resistance (SAR) chemicals (heavy metals) provide protection in wheat seedlings from *Helminthosporium* infection.

Pre-treatment of wheat leaves with *Bipolaris oryzae* (inducer organism) reduced the number of spot blotch lesion and lesion size of *B. sorokiniana* (Sarhan et al., 1991). An antifungal substance in apoplastic fluids of "induced" plants were identified which able to prevent spore germination of *B. sorokiniana in vitro* (Chakraborty & Sinha, 1984). Plant growth-promoting fungal rhizosphere isolates of *Phoma* spp. from Zoysia grass (*Zoysia* sp.) suppressed *B. sorokiniana* due to the competitive root colonization (Shivanna et al., 1996). Successful antagonists against seed borne *B. sorokiniana* were identified and potential are *Chaetomium* sp., *Idriella bolleyi*, and *Gliocladium roseum* (Knudsen et al., 1995). The antagonistic activity of *Chaetomium globosum* against *Drechslera sorokiniana* was first observed by Mandal (1995) in India, which was further confirmed by Biswas et al. (2000). The effectiveness of *Trichoderma* spp. have been recorded in reducing the foliar disease severity on wheat plants (Hasan & Alam, 2007; Singh et al., 2008). The root exudates of *Brassica* species suppressed many soil-borne fungi, including *B. sorokiniana* and chemical identified as isothiocyanates (Kirkegaard et al., 1996). Biological control of leaf blight of wheat with botanicals has been successfully done by many scientists (Malik et al., 2008b; Islam et al., 2006; Hossain et al., 2005; Hossain & Schlosser, 1993; Ashrafuzzaman & Hossain, 1992). Islam et al. (2006) reported reduction in disease severity in flag leaf stage, flowering stage, milking stage, and hard dough stage which was achieved by seed treatment with *Allium sativum* extract and *Polygonum hydropiper* leaf and stem extract and yield enhancement to 29.74% over untreated control. Hossain et al. (2015) reported the extracts of thirteen plant species to control leaf blight of wheat (*B. sorokiniana*). Disease severity was best controlled by using seed treatment with *Curcuma longa,* rhizome extract and *A. sativum* extract. Among all the treatments, *C. longa* rhizome extract treated seeds resulted the highest grain yield (3.60 t/ha) which was 39.44 higher over untreated control.

3.3.1.4 CHEMICALS

Seed treatment with different fungicides and various other substances including non-conventional chemicals may provide variable protection against foliar blight incidence of wheat. Hait and Sinha (1986) reported that phytoalexin inducers like Cupric chloride and Ferric chloride at 10^{-3} M concentration protected the wheat seedlings from foliar blight infection. Mukherjee (2008) found that nickel chloride, salicylic acid, and cupric chloride at 10^{-3} M concentration were very much effective in controlling the foliar blight disease of wheat caused by *B. sorokiniana*. Results also showed that all the concentrations of different chemicals tested have positive effect. The best result was recorded with salicylic acid at 10^{-3} M followed by nickel chloride and these were at par with propiconazole.

The other effective seed treating fungicides includes captan, mancozeb, thiram, pentachloronitrobenzene, proline and triadimefon (Stack & McMuller, 1988; Mehta, 1993). The foliar pathogen can be controlled with seed treating fungicides like guazatine and guazatine+ imazalil (Schilder & Bergstrom, 1993). Pavlova et al. (2002) indicated that seed treatment with flutriafol + thiabendazole, carboxin + thiram, difenoconazole + cyproconazole, carbendazim, fludioxonil, triticonazole, tebuconazole, and diniconazole provided protection against root rot caused by *Fusarium, Helminthosporium, Bipolaris,* and *Rhizoctonia* spp. Domanov (2002) conducted field trials to test carbendazim + carboxin which gave good control of root rots and increased yield of wheat.

The efficacy of Raxil (tebuconazole) was studied on spring wheat cultivar and seed treatment with tebuconazole controlled root rots caused by

B. sorokiniana (*C. sativus*) and *Alternaria* sp. (Korobov & Korobova, 2002). Seed treatment with Vitavax 200 B and Bavistin increased seed germination by 43% and reduced seedling infection by *B. sorokiniana* in Nepal (Sharma et al., 2004a, 2004b). The seed treatment with newly developed fungicidal formulation, Vitavax 200 WS (carboxin + thiram 1:1) @ 2.0, 2.5, and 3.0 g/ kg reduced seedling mortality as well as incidence of foliar diseases at different wheat growing locations of India (Singh et al., 2007).

Mehta (1998) observed that foliar fungicides are effective in controlling spot blotch. Non-systemic and systemic foliar fungicides *viz.* dithiocarbamates (*viz.* mancozeb) and triazoles (*viz.* propiconazole, tebuconazole, flutriazol, prochloraz, and triadimenol) and dicarboxymides (*viz.* iprodione) are effective against-spot blotch. Foliar applications of systemic fungicides like tebuconazole, epoxiconazole, flutriafol, cyproconazole, flusilazole, epoxiconazole and metconazole, applied between heading and grain filing stages, have been proved cost effective. Under severe disease levels, a second spray can result in a grain yield increase by 38–61% (de Viedma & Kohli, 1998). Several fungicides were proved to be useful and economical in the control of tan spot and spot blotch.

In fungicide tests, manzate reduced disease severity by 25%. Osorio et al. (1998) observed the average disease severity on combined flag leaf (F) and (F-1) leaves by more than 75% and application of tebuconazole (Folicur) twice as spray significantly reduced the diseased severity by 86%. (Loughman et al., 1998). In Argentina, Annone (1998) observed positive response with foliar fungicides to reduce disease development under conditions of moderate to low disease pressure, but inconsistent results under high disease pressure. Among fungicides tested under field conditions, tebuconazole and propiconazole proved to be the most effective fungicides to control foliar blight, though results varied widely from 30 to 80%.

Tewari and Wako (2003) reported that the mixture of tebuconazole (Folicur) and Metacid applied as foliar spray at the boot stage of wheat crop effectively suppressed all the foliar diseases (brown and yellow rusts, powdery mildew and leaf blight) showing curative property with no phytotoxic effect on the plant. They also observed that sencor and propiconazole (Tilt) was incompatible and highly phytotoxic. They further noticed that the mixture of zinc sulphate and urea completely inhibited mycelia growth of *B. sorokiniana* and 78% growth inhibition of *A. triticina.* Rashid et al. (2001) reported propiconazole (Tilt 250 EC) to be very effective against foliar blight of wheat (*B. sorokiniana*). The different fungicides like flusilazole, prochloraz, propiconazole, and tebuconazole were effective against tan spot disease of wheat (Colson et al., 2003). Application of propiconazole, hexaconazole, prochloraz, and quintal @ 0.1% gave good control of leaf blight of wheat in the field (Chandrasekhar, 2003; Ramachandra & Kalappanavar, 2004). Singh et al. (2008) proposed three foliar applications of propiconazole @ 0.1% after appearance of the disease significantly reduce the disease and increase yield tested over several locations of India. Sarkar et al. (2010) reported the effectiveness of some newly synthesized organotin compounds against *B. sorokiniana* both *in vitro* and *in vivo*. The application of propiconazole and tebuconazole @ 0.1% on susceptible wheat varieties like HUW234 and DBW14 under field conditions was proved to be very effective in controlling spot blotch disease of wheat (Singh et al., 2014a).

The disease incidence percentage and disease severity (AUDPC) of spot blotch of wheat was found to be minimum by seed treatment with combined carboxin 37.5% + thiram 37.5%WS @ 2.5 g kg^{-1} seed with two sprays of propiconazole 25% EC (0.1%) at boot leaf stage and after 20 days of first spray. The 1000 grain weight and the grain yield were also highest in the above treatment, that is, 44.69 g and 4.35 t ha^{-1}, respectively, in comparison to other treatment combinations (Mahapatra & Das, 2013).

3.3.2   BARLEY

The growers should have proper knowledge of the pathogen and manipulation of inter related factors to achieve effective control of the pathogen. Modification of environment is very difficult but sowing of resistant varieties as well as following of proper cultural practices as well as use of chemicals must be done by the growers to minimize the disease. Although several strategies like genetic modification, use of chemicals and cultural methods are available for the control of foliar and other diseases of cereals, rarely the use of a single method provides complete protection against a disease. So, growers need to adopt an IDM strategy by combining a number of available methods for disease management to get most effective as well as long lasting protection against the foliar diseases.

3.3.2.1   RESISTANT VARIETIES

Growing of resistant varieties is the most easiest and effective way to check disease severity but very few varieties possess good resistance against major diseases. So, alternative way to minimize the disease should be explored and acceptable to the growers to reduce the risk. Such methods include rotation of barley crops with non-host crops like legumes, solanaceous crops or cruciferous crops; avoidance of sowing of barley continuously for number of years in the same field and maintenance of clean fallow land. Sowing the crop in off season favors disease development and inoculum can be built up early in the next growing season. The foliar diseases of barley are the main problem to retain good yield and quality of barley, and the problem becomes more severe if susceptible varieties are grown. As the pathogens have several numbers of strains or pathotypes (Mathre, 1997; Tekauz, 1990, 2003; Xi et al., 2002, 2003) and new strains can evolve very fast in response to change of varieties and/or modifications of production technologies, even "resistant" varieties may become susceptible and suffer from diseases. As a result of foliar blight, yield losses up to 20% have been reported by some workers across the region (Buchannon & Wallace, 1962; Ghazvini & Tekauz, 2004; McDonald & Buchannon, 1964; Skoropad, 1960; Orr et al., 1999; Kutcher & Kirkham, 1997). These studies have also shown the utility of different foliar fungicides in increasing yields, 1000-grain weight, and grain plumpness in affected crops.

3.3.2.2   CULTURAL PRACTICES

Different information on the incidence and severity of diseases in a particular locality collected from previous growing seasons help to develop a strategy to minimize the disease (Turkington et al., 2011). Early sowing of barley has been suggested to minimize the disease severity. Net blotch disease severity was reduced significantly in early sown crop (October 25) as compared to late sowings (November 14 and December 4) in India (Prasad et al., 2001).

The use of quality seeds with good germination and vigor and free from pathogen is another important tool of an effective and holistic disease management program (Turkington et al., 2011). The seeds should be treated with thiram or difenoconazole to reduce the carry over seed-borne inoculum of net blotch disease (http://www.croppro.com.au/crop_disease_manual/ch02s04. php). Diseased barley residue should be destroyed to eliminate the primary source of inoculum that can initiate disease in the next cropping season and may cause leaf spot epidemics (Turkington et al., 2011). Diversification of crops in a rotation helps to break-up the life cycles of many disease-causing organisms. Crop rotation of small grains and grasses with non-grass crops, preferably legumes-soybeans and forage legumes should be practiced (Turkington et al., 2011). Besides above, growing of resistant varieties with good agronomic quality as well as quality parameters may provide barley plants to protect from foliar diseases (Turkington et al., 2011).

3.3.2.3 BOTANICALS AND BIOCONTROL AGENTS

Spraying of some botanicals can reduce spot blotch of barley. Fifteen percent concentration of garlic extract can check the growth of *B. sorokiniana* by 67.5% (Hasan et al., 2012). Besides this, aqueous extracts from *Bauhinia forficata* showed protection of 92–100% to barley plants against spot blotch disease in local and systemic action (Bach et al., 2014). It was also observed that barley seeds induced systemic resistance on the plants to subsequent infection with *B. sorokiniana* when inoculated with the fungi *Idriella bolleyi* (Liljeroth & Bryngelson, 2002). Reduced symptoms could also be achieved, although not to the full extent, after spraying with the bacterial biocontrol agent *Pseudomonas chlororaphis* strain MA 342. Treated seeds could be stored dry for at least two years without losing the disease suppressing effect of the bacterial treatment (Johnson et al., 1998).

3.3.2.4 CHEMICALS

Besides this, application of some chemicals can reduce the blotch diseases of barley. Hundred percent mycelial growth of *B. sorokiniana* was inhibited with the application of propiconazole or hexaconazole or difenoconazole + propiconazole at all concentration starting from 100 to 500 ppm (Hasan et al., 2012). Treatment with fungicide, pyraclostrobin provided protection up to 60% against foliar diseases (Bach et al., 2014). Any azole fungicide can be sprayed to control all the blotch diseases of barley very effectively (Hasan et al., 2012).

3.4 CONCLUDING REMARKS

Foliar blight is probably the most serious disease of wheat and barley in the ME 5A mega environment characterized by high temperature (coolest month >17°C) and high relative humidity (van Ginkel & Rajaram, 1998). Though several strategies including cultural modifications, growing of resistant cultivars and use of chemicals are available to control foliar diseases, but the use of a single one of these does not provide complete protection. Different control strategies to decrease the severity of foliar blight epidemics, such as crop rotation, manipulation of sowing date, seeding rate or agronomic treatments during the growing season, use of resistant varieties, chemical and biological control are aimed at different stages in the development of epidemics. Clearly, implementing an effective, economic, and long lasting wheat and barley foliar disease management strategy is not a simple issue. There are many factors for producers to consider, some risks, and a number of potential solutions. However, as outlined here there are a variety of resources available to assist in developing and implementing such a plan, which, when combined with good crop management, should result in a platform for sustainable wheat and barley production.

KEYWORDS

- **foliar blight**
- **wheat**
- **barley**
- **spot blotch**
- *Alternaria blight*
- **tan spot**
- **net blotch**
- **specked leaf blotch**
- **disease management**

REFERENCES

Adlakha, K. L.; Wilcoxson, R. D; Raychaudhury, S. P. Resistance of Wheat to Leaf Spot Caused by *Bipolaris sorokiniana*. *Plant Dis.* **1984,** *68,* 320–321.

Agrios, G. N. *Plant Pathology;* Academic Press: Cambridge, MA, 1988; p 803.

Alam, K. B.; Banu, S. P.; Shaheed, M. A. In *Occurrence and Significance of Spot Blotch Disease in Bangladesh,* Proceedings of the International Workshop on *Helminthosporium* Blights of Wheat: Spot Blotch and Tan Spot, CIMMYT, El Batan, Mexico, Feb 9–14, 1997; Duveiller, E., Dubin, H. J., Reeves, J., McNab, A., Eds.; CIMMYT: El Batan, Mexico, 1998; pp 63–66.

Alam, K. B., et al. *Bipolaris* Leaf Blight (Spot Blotch) of Wheat in Bangladesh. In *Wheat in Heat-Stressed Environments: Irrigated, Dry Areas and Rice-Wheat Farming Systems,* Saunders, D. A., Hettel, G. P., Eds.; D. F. CIMMYT: Mexico, 1994; pp 339–342.

Annone, J. G. In *Tan Spot of Wheat in Argentina: Importance and Disease Management Strategies,* Proceedings of the International Workshop on *Helminthosporium* Blights of Wheat: Spot Blotch and Tan Spot, CIMMYT, El Batan, Mexico, Feb 9–14, 1997; Duveiller, E., Dubin, H. J., Reeves, J., McNab, A., Eds.; CIMMYT: El Batan, Mexico, 1998; pp 339–345.

Arabi, M. I. E.; Jawhar, M. Identification of *Cochliobolus sativus* (Spot Blotch) Isolates Expressing Differential Virulence on Barley Genotypes in Syria. *J. Phytopathol.* **2004,** *152,* 461–464.

Arun, B., et al. Wheat Somaclonal Variants Showing, Earliness, Improved Spot Blotch Resistance and Higher Yield. *Euphytica.* **2003,** *132,* 235–241.

Ashrafuzzaman, H.; Hossain, I. Antifungal Activity of Crude Extracts of Plants against *Rhizoctonia solani* and *Bipolaris sorokiniana. Proc. BAU. Res. Progr.* **1992,** *6,* 188–192.

Bach, E. E., et al. Control of Spot Blotch in Barley Plants with Fungicide and *Bauhinia variegata* Linn. Leaf Extract. *Emir. J. Food Agric.* **2014,** *26*(7), 630–638.

Bailey, K. L.; Wolfe, R. I. Genetic Relationships between Reaction to Common Root Rot and Yield in the Progeny of a Barley Cross. *Can. J. Plant Pathol.* **1994,** *16,* 163–169.

Bailey, K. L., et al. Managing Crop Losses from Diseases with Fungicides, Rotation, and Tillage in the Saskatchewan Parkland. *Can. J. Plant Sci.* **2000,** *80,* 169–175.

Biswas S. K.; et al. Antagonism of *Chaetomium Globosum* to *Drechslera sorokiniana,* the Spot Blotch Pathogen of Wheat. *Indian Phytopathol.* **2000,** *53,* 436–440.

Bochow, H., et al. Mechanisms of Interactions in the System of Plant, Parasite and Beneficial Organisms. *J. Plant Dis. Prot.* **2001,** *108,* 626–652.

Bohorova, N. E., et al. Tissue Culture Response of CIMMYT Elite Bread Wheat Cultivars and Evaluation of Regenerated Plants. *Cer. Res. Comm.* **1995,** *23,* 243–249.

Buchannon, K. W.; Wallace, H. A. H. Note on the Effect of Leaf Disease on Yield, Bushel Weight and Thousand-Kernel Weight of Parkland Barley. *Can. J. Plant Sci.* **1962,** *42,* 534–536.

Castro, O. L.; Bach, E. E. Increased Production of β-1, 3 Glucanase and Proteins in *Bipolaris Sorokiniana* Pathosystem Treated Using Commercial Xanthan Gum. *Plant Physiol. Biochem.* **2004,** *42,* 165–169.

Chakraborty, D.; Sinha, A. K. Similarity between the Chemically and Biologically Induced Resistances in Wheat Seedlings to *Drechslera sorokiniana. Z. Pflanzenkr. Pflanzenschutz.* **1984,** *91,* 59–64.

Chandrasekhar, M. E. Studies on Leaf Blight Disease of Barley (Hordium vulgare L.) Caused by Helminthsporium sativum Pam., King and Bakke. M.Sc. (Agri.) Thesis, University of Agricultural Sciences, Dharwad, 2003; p 103.

Chaurasia, P. C. P.; Duveiller, E. Management of Leaf Blight (*Bipolaris sorokiniana*) Disease of Wheat with Cultural Practices. *Nepal Agric. Res. J.* **2006,** *7,* 63–69.

Chowdhury, A. K.; Bhattacharya, P. M.; Singh, G. Foliar Blight: The Major Biotic Constraint of Wheat in Rice Wheat Systems of Eastern Gangetic Plains. *Rev. Plant Pathol.* **2014,** *6,* 437–472.

Chowdhury, A. K., et al. Spot Blotch Disease of Wheat a New Thrust Area for Sustaining Productivity. *J. Wheat Res.* **2013,** *5*(2), 1–11.

Colson, E. S.; Platz, G. J.; Usher, T. R. Fungicidal Control of *(Pyrenophora tritici repentis)* in Wheat. *Aus. Plant Pathol.* **2003,** *32,* 241–246

Conner, R. L. Interrelationship of Cultivar Reaction to Common Root Rot, Black Point, and Spot Blotch in Spring Wheat. *Plant Dis.* **1990,** *74,* 224–227.

Cook, R. J.; Veseth, R. J. *Wheat Health Management.* APS Press: St Paul, MN, 1991; p 152.

de Viedma, L. Q.; Kohli, M. M. In *Spot Blotch and Tan Spot of Wheat in Paraguay,* Proceedings of the International Workshop on *Helminthosporium* Blights of Wheat: Spot Blotch and Tan Spot, CIMMYT, El Batan, Mexico, Feb 9–14, 1997; Duveiller, E., Dubin, H. J., Reeves, J., McNab, A., Eds., CIMMYT: El Batan, Mexico, 1998; pp 126–133.

Dhaliwal, H. S., et al. Evaluation and Cataloguing of Wheat Germplasm for Disease Resistance and Quality. *In Biodiversity and Wheat Improvement;* Dhamania, A. B., Ed.; John Wiley and Sons: New York, 1993; pp 123–140.

Diehl, J. A., et al. The Effect of Fallow Periods on Common Root Rot of Wheat in Rio Grande do Sul, Brazil. *Phytopathology.* **1982,** *72,* 1297–1301.

Domanov, N. M. Kolfugo Duplet for Seed Treatment of Cereals. Zashchita Karantin Rasteni: Moscow, Russia, 2002; Vol. 6, p 29.

Dubin, H. J.; Bimb, H. P. Effects of Soil and Foliar Treatments on Yield and Diseases of Wheat in Lowland Nepal. *In Wheat in Heat-Stressed Environments: Irrigated, Dry Areas and Rice – Wheat Farming Systems,* Saunders, D. A., Hettel, G. P., Eds.; CIMMYT: Mexico, 1991; pp 484–485.

Dubin, H. J.; Duveiller, E. In *Helminthosporium Leaf Blights of Wheat: Integrated Control and Prospects for the Future,* Proceedings of the International Conference on Integrated Plant Disease Management for Sustainable Agriculture, New Delhi, Nov 10–15, 1997; Indian Phytopathological Society: New Delhi, India, 2000; pp 575–579.

Dubin, H. J.; van Ginkel, M. In *The Status of Wheat Diseases and Disease Research in the Warmer Areas,* Proceedings of the Wheat for Non-Traditional Warm Areas Conference, *Foz do Iguacu,* Brazil, Jul 29–Aug 3, 1990; Saunders, D. A., Hettel, G., Eds.; UNDP/CIMMYT: Mexico, 1991; pp 125–145.

Duczek, L. J., et al. Survival of Leaf Spot Pathogens on Crop Residues of Wheat and Barley in Saskatchewan. *Can. J. Plant Pathol.* **1999,** *21,* 165–173.

Duveiller, E.; Dubin, H. J. *Helminthosporium* Leaf Blights: Spot Blotch and Tan Spot. In *Improvement and Production, Plant Production and Protection Series No 30,* Curtis, B. C., et al., Eds.; FAO of United Nations: Rome, 2002; pp 285–299.

Duveiller, E.; Gilchrist, L. In *Productions Constraints Due to Bipolaris sorokiniana in Wheat: Current Situation and Future Prospects,* Proceedings of the CIMMYT/UNDP Workshop: Wheat in the Warmer Areas, Rice/Wheat Systems, Nashipur, Dinajpur, Bangladesh, Feb 13–16, 1993; Saunders, D. A., Hettel, G., Eds.; CIMMYT/UNDP: Nashipur (Dinajpur), Bangladesh, 1994; pp 343–352.

Duveiller, E.; Sharma, R. C. Genetic Improvement and Crop Management Strategies to Minimize Yield Losses in Warm Nontraditional Wheat Growing Areas Due to Spot Blotch Pathogen *Cochliobolus sativus. J. Phytopathol.* **2009,** *157,* 521–534.

Duveiller, E., et al. Epidemiology of Foliar Blights (Spot Blotch and Tan Spot) of Wheat in the Plains Bordering the Himalayas. *Phytopathology.* **2005,** *95* (3), 248–256.

Duveiller, E.; van Ginkel, M.; Dubin, H. J. In *Helminthosporium Diseases of Wheat: Summary of Group Discussions and Recommendations,* Proceedings of the International Workshop on *Helminthosporium* Blights of Wheat: Spot Blotch and Tan Spot, CIMMYT, El Batan, Mexico, Feb 9–14, 1997; Duveiller, E., Dubin, H. J., Reeves, J., McNab, A., Eds.; CIMMYT: El Batan, Mexico, 1998; pp 1–5.

Eken, C.; Demirci, E. The Distribution, Cultural Characteristics, and Pathogenesis of *Drechslera sorokiniana* in Wheats and Barley in Erzurum Region. *Turk. J. Agric. For.* **1998,** *22,* 175–180.

El Yousfi, B.; Ezzahiri, B. Net Blotch in Semi-Arid Regions of Morocco II Yield and Yield-Loss Modelling. *Field Crops Res.* **2002,** *73,* 81–93.

Food and Agriculture Organization of the United Nations (FAO). n.d. *"World Food Situation: FAO Food Price Index";* Retrieved May 13, 2014. (http://www.fao.org/worldfoodsituation/foodpricesindex/en/).

Fernandez, M. R. Jefferson, P. G. Fungal Populations in Roots and Crowns of Common and Durum Wheat in Saskatchewan. *Can. J. Plant Pathol.* **2004,** *26,* 325–334.

Fernandez, M. R., et al. Effect of Crop Rotations and Fertilizer Management on Leaf Spotting Diseases of Spring Wheat in South-Western Saskatchewan. *Can. J. Plant Sci.* **1998,** *78,* 489–496.

Gami, S. K., et al. Long Term Changes in Yield and Soil Fertility Status in a 20-Year Rice-Wheat Experiment in Nepal. *Bio. Fert. Soils.* **2001,** *34,* 73–78.

Ghazvini, H.; Tekauz, A. Yield Loss in Barley Inoculated with High and Low Virulence Isolates of *Bipolaris sorokiniana. Czech J. Genet. Plant Breed.* **2004,** *40,* 149.

Gilchrist, L. I.; Pfeiffer, W. H.; Rajaram, S. Progress in Developing Bread Wheats Resistant to *Helminthosporium sativum.* In *Wheat for the Nontraditional, Warm Areas,* Saunders, D. A. Ed.; CIMMYT: Mexico, 1991a; pp 469–472.

Gilchrist, L. I.; Pfeiffer, W. H.; Velaquez, C. Resistance to *Helminthosporium sativum* in Bread Wheat: Relationship among Infected Plant Parts and Associated with Agronomic Traits. In *Developing Sustainable Wheat Production Systems. The 8th Regional Wheat Workshop for Eastern, Central and Southern Africa;* Tanner, D. G., Ed.; CIMMYT: Addis Ababa, 1991b; pp 176–181.

Gonzalez, M. S.; Trevathan, L. E. Identity and Pathogenicity of Fungi Associated with Root and Crown Rot of Soft Red Winter Wheat Grown on the Upper Coastal Plain Land Resource Area of Mississippi. *J. Phytopathol.* **2000,** *148,* 77–85.

Hafiz, A. *Plant Diseases;* Pakistan Agricultural Research Council: Islamabad, 1986; p 552.

Hait, G. N.; Sinha, A. K. Protection of Wheat Seedlings from *Helminthosporium* Infection by Seed Treatment with Chemicals. *J. Phytopathol.* **1986,** *115,* 97–107.

Hasan, M. M.; Alam, S. Efficacy of *Trichoderma harzianum* Treated Seeds on Field Emergence, Seedling Disease, Leaf Blight Severity and Yield of Wheat cv. Gourab and Shourav Under Field Condition. *Intl. J. Biol. Res.* **2007,** *3*(6), 23–30.

Hasan, M. M., et al. *In Vitro* Effect of Botanical Extracts and Fungicides against *Bipolaris sorokiniana,* Causal Agent of Leaf Blotch of Barley. *J. Agrofor. Environ.* **2012,** *6*(1), 83–87.

Hetzler, J., et al. Interactions between Spot Blotch (*Cochliobolus sativus*) and Wheat Cultivars. In *Wheat for the Non-traditional Warm Areas;* Saunders, D. A., Ed.; CIMMYT: Mexico, 1991; pp 146–164.

Hossain, I.; Schlosser, E. Control of *Bipolaris sorokiniana* in Wheat with Neem Extracts. *Bangladesh J. Microbiol.* **1993,** *10,* 39–42.

Hossain, I. In *Controlling Leaf Spot of Wheat through Nutrient Management,* Proceedings of the International Workshop on *Helminthosporium* Blights of Wheat: Spot Blotch and Tan Spot, CIMMYT, El Batan, Mexico, Feb 9–14, 1997; Duveiller, E., Dubin, H. J., Reeves, J., McNab, A., Eds.; CIMMYT: El Batan, Mexico, 1998; pp 354–358.

Hossain, M. M., et al. Effect of Plant Extract on the Incidence of Seed-Borne Fungi of Wheat. *J. Agric. Rural Dev.* **2005,** *3*(1–2), 39–43.

Hossain, M. M.; Hossain, I; Khalequzzaman, K. M. Effect of Seed Treatment with Biological Control Agent against Bipolaris Leaf Blight of Wheat. *Int. J. Sci. Res. Agric. Sci.* **2015,** *2*(7), 151–158.

Huber, D. M., et al. Amelioration of Tan Spot Infected Wheat with Nitrogen. *Plant Dis.* **1987,** *71,* 49–50.

Hudec, K.; Muchova, D. Correlation between Black Point Symptoms and Fungal Infestation and Seedling Viability of Wheat Kernels. *Plant Prot. Sci.* **2008,** *44,* 138–146.

Islam, M. A., et al. Seed Treatment with Plant Extract and Vitavax 200 in Controlling Leaf Spot (*Bipolaris sorokiniana*) with Increasing Grain Yield of Wheat. *Int. J. Sustain. Agric. Tech.* **2006,** *2*(8), 15–20.

Jana, S.; Bailey, K. L. Responses of Wild and Cultivated Barley from West Asia to Net Blotch and Spot Blotch. *Crop Sci.* **1995,** *35,* 242–246.

Johnson, L.; Hokeberg, M.; Gerhardson, B. Performance of the *Pseudomonas chlororaphis* Biocontrol Agent MA 342 against Cereal Seed-Borne Diseases in Field Experiments. *Eur. J. Plant Pathol.* **1998,** *104,* 701–711.

Johnson, V. A., et al. Grain Crops. In *Protein Resources and Technology*, Milner, M., Scrimshaw N. S., Wang, D. I. C, Eds.; AVI Publishing: Westport, CT, 1978; pp 239–255.

Jorgensen, L. N.; Olsen, L. V. Control of Tan Spot (*Drechslera tritici-repentis*) Using Cultivar Resistance, Tillage Methods and Fungicides. *Crop Prot.* **2007,** *26,* 1606–1616.

Joshi, A. K., et al. Associations of Environments in South Asia Based on Spot Blotch Disease of Wheat Caused by *Cochliobolus sativus. Crop Sci.* **2007b,** *47,* 1071–1081.

Joshi, A. K., et al. Stay Green Trait: Variation, Inheritance and its Association with Spot Blotch Resistance in Spring Wheat (*Triticum aestivum* L.). *Euphytica.* **2007a,** *153,* 59–71.

Joshi, L. M., et al. *Annotated Compendium of Wheat Diseases in India.* ICAR: New Delhi, 1978; p 332.

Kirkegaard, J. A.; Wong, P. T. W.; Desmarchelier, J. M. *In Vitro* Suppression of Fungal Root Pathogens of Cereals by *Brassica* Tissues. *Plant Pathol.* **1996,** *45,* 593–603.

Knudsen, I. M. B.; Hockenhull, J.; Jensen, D. F. Bio-Control of Seedling Disease of Barley and Wheat Caused by *Fusarium culmorum* and *Bipolaris sorokiniana*: Effects of Selected Fungal Antagonists on Growth and Yield Components. *Plant Pathol.* **1995,** *44,* 467–477.

Korobov, V. A.; Korobova, L. N. Raxil in the South of West Siberia. *Zashchita Karantin Rasteni.* Novosivirec Agrarian University: Siberia, Russia, 2002; 29–30.

Krupinsky J. M., et al. Nitrogen and Tillage Effects on Wheat Leaf Spot Diseases in the Northern Great Plains. *Agronomy J.* **2007,** *99,* 562–569.

Krupinsky, J. M.; Tanaka, D. L. Leaf Spot Diseases on Winter Wheat Influenced by Nitrogen, Tillage and Haying after a Grass-Alfalfa Mixture in the Conservation Reserve Program. *Plant Dis.* **2001,** *85,* 785–789.

Krupinsky, J. M., et al. Leaf Spot Diseases of Barley and Spring Wheat as Influenced by Preceding Crops. *Agronomy J.* **2004,** *96,* 259–266.

Kutcher, H. R.; Kirkham, C. Fungicide Control of Foliar Leaf Spot Pathogens of Barley and Wheat at Melfort, Saskatchewan, in 1997. In *1997 Pest Management Research Report (PMRR): 1997 Growing Season;* **1997**; pp 274–275. Available from http://www.cpsscp.ca/pmrr/common/pmrr_1997. pdf (Accessed Dec 3, 2010).

Kutcher, H. R., et al. Heritability of Common Root Rot and Spot Blotch Resistance in Barley. *Can. J. Plant Pathol.* **1994,** *16,* 287–294.

Ledingham, R. J. Crop Rotations and Common Root Rot in Wheat. *Can. J. Plant Sci.* **1961,** *41,* 479–486.

Lehmensiek, A., et al. QTLs for Black-Point Resistance in Wheat and the Identification of Potential Markers for Use in Breeding Programmes. *Plant Breed.* **2004,** *123,* 410–416.

Liatukas, Z.; Ruzgas, V. Spot Blotch Resistance in Derivatives of European Winter Wheat *Turk. J. Agric. For.* **2012,** *36,* 341–351.

Liljeroth, E.; Bryngelson, T. Seed Treatment of Barley with *Idriella bolleyi* Causes Systematically Enhanced Defence against Root and Leaf Infection by *Bipolaris sorokiniana*. *Biocon. Sci. Technol.* **2002,** *12,* 235–249.

Loughman, R., et al. In *Crop Management and Breeding for Control of Pyrenophora tritici-repentis Causing Yellow Spot of Wheat in Australia*, Proceedings of the International Workshop on *Helminthosporium* Blights of Wheat: Spot Blotch and Tan Spot, CIMMYT, El Batan, Mexico, Feb 9–14, 1997; Duveiller, E., Dubin, H. J., Reeves, J., McNab, A., Eds.; CIMMYT: El Batan, Mexico, 1998; p 354.

Mahapatra, S.; Das, S. Efficacy of Different Fungicides against Spot Blotch of Wheat in Terai Region of West Bengal. *J. Wheat Res.* **2013,** *5*(2), 18–21.

Mahto, B. N.; Duveiller, E.; Sharma, R. C. Effect of Surface Seeding on Foliar Blight Severity and Wheat Performance. *Field Crop Res.* **2006,** *97*(2–3), 344–352.

Malik, V. K.; Singh, D. P.; Panwar, M. S. Losses in Yield Due to Varying Severity of Leaf Blight Caused by *Bipolaris sorokiniana* in Wheat. *Indian Phytopath.* **2008a,** *61*(4), 526–527.

Malik, V. K.; Singh, D. P.; Panwar, M. S. Management of Spot Blotch of Wheat (*Triticum aestivum*) Caused by *Bipolaris sorokiniana* Using Foliar Sprays of Botanicals and Fungicides. *Indian J. Agric. Sci.* **2008b,** *78,* 646–648.

Mandal, S. Effect of Some Antagonists on Drechslera sorokiniana, Causal Agent of Spot Blotch of Wheat. M.Sc. Thesis, I.A.R.I., New Delhi, 1995, p 74.

Mathre, D. E. *Compendium of Barley Diseases.* 2nd ed.; American Phytopathological Society Press: St. Paul, MN, 1997; p 87.

Mathur, S. B.; Cunfer, B. M. *Seed-Borne Diseases and Seed Health Testing of Wheat;* Danish Government Institute Seed Pathology for Developing Countries: Copenhagen, Denmark, 1993; p 168.

McDonald, W. C.; Buchannon, K. W. Barley Yield Reductions Attributed to Net Blotch Infections. *Can. Plant Dis. Surv.* **1964,** *44,* 118–119.

Mehta, Y. R. Spot Blotch (*Bipolaris sorokiniana*). In *Seedborne Diseases and Seed Health Testing of Wheat;* Mathur, S. B., Cunfer, B. M., Eds.; Institute of Seed Pathology for Developing Countries: Copenhagen, Denmark, 1993; pp105–112.

Mehta, Y. R. In *Breeding Wheats for Resistance to Helminthosporium Spot Blotch,* Proceedings of the International Symposium: Wheats for More Tropical Environments, Mexico DF, Sep 24–28, 1984; Villareal, R. L., Klatt, A. R, Eds.; CIMMYT: Mexico, 1985; pp 135–144.

Mehta, Y. R. In *Constraints on the Integrated Management of Spot Blotch of Wheat*, Proceedings of the International Workshop on *Helminthosporium* Blights of Wheat: Spot Blotch and Tan Spot, CIMMYT, El Batan, Mexico, Feb 9–14, 1997; Duveiller, E., Dubin, H. J., Reeves, J., McNab, A., Eds.; CIMMYT: El Batan, Mexico, 1998; pp 18–27.

Mehta, Y. R. Variacao Somaclonal Nas Plantas de Trigo Regeneradas Atraves de Embriogenese. *Fitopatol. Bras.* **1996,** 21 (Abstract).

Minella, E.; Barley for Beer: Agronomic Characteristics from Cultivars in Sul Region. Cevada Cervejeira: Características e Desempenho Agronômico das Cultivares Indicadas Para a Região Sul. Embrapa Trigo: Passo Fundo, Brazil, Circular Técnica Online. 2001; Vol. 4, pp 1–8. www.cnpt.embrapa.br/biblio/p_ci04. htm

Morh, R. M., et al. Effect of Chloride Fertilization on Bedford Barley and Katepwa Wheat. *Can. J. Soil Sci.* **1987,** *75,* 15–25.

Mujeeb-Kazi, A., et al. Registration of Five Wheat Germplasm Lines Resistant to *Helminthosporium* Leaf Blight. *Crop Sci.* **1996a,** *36,* 216–217.

Mujeeb-Kazi, A.; Rosas, V.; Roldan, S. Conservation of the Genetic Variation of *Triticum tauschii* (Coss) Schmalh *(Aegilops squarrosa osa auct non* L.) in Synthetic Hexaploid

Wheats (*T. turgidum* L s lat. *T. tauschii*= 42, AABBDD) and; 2n = 6 its Potential Utilization for Wheat Improvement. *Genet. Resour. Crop Evol.* **1996b,** *43,* 129–134.

Mukherjee, S. Studies on Foliar Blight Complex of Wheat with Particular Reference to Bipolaris sorokiniana. Ph. D. Thesis, Institute of Agricultural Sciences, Visva Bharati, 2008.

Mukwavi, M. V. In *Breeding for Disease Resistance with Emphasis on Durability,* Proceedings of a Regional Workshop for Eastern, Central and Southern Africa, Njoro, Kenya, Oct 2–6, 1994; Wageningen Agricultural University: The Netherlands, 1995; pp 143–145.

Muyanga, S. Wheat Production and Research in Zambia: Constraint and Sustainability. In *Developing Sustainable Wheat Production System.* The 8th Regional Workshop for Eastern, Central and Southern Africa, Addis Ababa, CIMMYT: Ethiopia, 1994; pp 57–64.

Nagarajan, S.; Kumar, J. In *Foliar Blights of Wheat in India: Germplasm Improvement and Future Challenges for Sustainable, High Yielding Wheat Production,* Proceedings of the International Workshop on *Helminthosporium* Blights of Wheat: Spot Blotch and Tan Spot, CIMMYT, El Batan, Mexico, Feb 9–14, 1997; Duveiller, E., Dubin, H. J., Reeves, J., McNab, A., Eds.; CIMMYT: El Batan, Mexico, 1998; pp 52–58.

Nema, K. G. Foliar Diseases of Wheat-Leaf Spots and Blights. In *Problems and Progress of Wheat Pathology in South Asia,* Joshi, L. M., Singh, D. V., Eds.; Malhotra Publishing House: New Delhi, 1986; pp 162–175.

Orr, D. D.; Turkington, T. K.; Kutcher, R. *Impact of Tilt Timing on Disease Management in Harrington Barley, Lacombe, Alberta and Melfort, Saskatchewan, 1999;* 1999 Pest Management Research Report (PMRR): 1999 Growing Season, Compiled for The Expert Committee on Integrated Pest Management by Southern Crop Protection and Food Research Centre (SCPFRC), Agriculture and Agri-Food Canada (AAFC): London, Ontario, Canada N5V 4T3, 1999; pp 322–324, Available from http://www.cpsscp.ca/pmrr/common/pmrr_1999. pdf (Accessed Dec 3, 2010).

Osorio, L., et al. In *Improving Control of Tan Spot Caused Pyrenophora tritici-repentis in the Mixteca Alta of Oaxaca, Mexico,* Proceedings of the International Workshop on *Helminthosporium* Blights of Wheat: Spot Blotch and Tan Spot, CIMMYT, El Batan, Mexico, Feb 9–14, 1997; Duveiller, E., Dubin, H. J., Reeves, J., McNab, A., Eds.; CIMMYT: El Batan, Mexico, 1998; pp 142–145.

Pavlova, V. V.; Dorofeeva, L. L.; Kozhukhovskaya, V. A. Effectiveness of Seed Treatments against Root Rots of Cereals. Moscow Russia; Izdatel stvokolos [Ru], All Russian Research Institute of Phytopathology: Russia, 2002; Vol. 8, pp 21–23. (CAB International).

Prasad, R.; Singh, H. C.; Singh, S. K. Effect of Sowing Date, Temperature and Relative Humidity on the Incidence of Net Blotch of Barley. *Indian Phytopathol.* **2001,** *54,* 304–306.

Prescott, J. M., et al. *Wheat Disease and Pests, A Guide for Field Identification;* CIMMYT: Mexico, 1986; p 135.

Prestes, A. M.; Santos, H. P.; Reis, E. M. Práticas Culturais e Incidência de Manchas Foliares em Trigo. *Pesq. Agropec. Bras.* **2002,** *37*(6), 791–797. http://www.scielo.br/pdf/pab/v37n6/10556. pdf

Raemaekers, R. H. In *Breeding Wheat with Resistance to Helminthosporium sativum in Zambia,* Proceedings of the International Symposium: Wheats for the More Tropical Environments, Mexico DF, Sep 24–28, 1984; Villareal, R. L., Klatt, A. R. Eds.; CIMMYT: Mexico, 1985; pp 145–148.

Raemaekers, R. H. *Helminthosporium sativum*: Disease Complex on Wheat and Sources of Resistance in Zambia. In *Wheat Production Constraints in Tropical Environments,* Klatt, A. R., Ed.; CIMMYT: Mexico, 1988; pp 175–185.

Ramachandra, C. G.; Kalappanavar, I. K. Evaluation of Fungicides against *Exserohillum hawaiiensis* Both Under Laboratory and Field Conditions. *Indian Phytopathol.* **2004,** *57,* 343.

Rashid, A. Q. M. B.; Sarker, K.; Khalequzzaman, K. M. Control of *Bipolaris* Leaf Blight of Wheat with Foliar Spray of Tilt 250 Ec. *Bangladesh J. Plant Pathol.* **2001,** *17,* 45–47.

Regmi A. P., et al. The Role of Potassium in Sustaining Yields in a Long-Term Rice-Wheat Experiment in the Indo-Gangetic Plains of Nepal. *Biol. Fertil. Soils.* **2002,** *36,* 240–247.

Reis, E. M. Integrated Disease Management the Changing Concept of Controlling Head Blight and Spot Blotch. In *Wheats for the Non Traditional Warm Areas,* Saunders, D. A., Ed.; CIMMYT: Bangkok, Thailand, 1991; pp 165–177.

Rossi, V., et al. Fungi Associated with Foot Rots on Winter Wheat in Northwest Italy. *J. Phytopathol.* **1995,** *143,* 115–119.

Rosyara, U. R., et al. Yield and Yield Components Response to Defoliation of Spring Wheat Genotypes with Different Level of Resistance to *Helminthosporium* Leaf Blight. *J. Inst. Agric. Anim. Sci.* **2005,** *26,* 43–50.

Saari, E. E. In *Leaf Blight Diseases and Associated Soil Borne Fungal Pathogens of Wheat in South and South East Asia,* Proceedings of the International Workshop on *Helminthosporium* Blights of Wheat: Spot Blotch and Tan Spot, CIMMYT, El Batan, Mexico, Feb 9–14, 1997; Duveiller, E., Dubin, H. J., Reeves, J., McNab, A., Eds.; CIMMYT: El Batan, Mexico, 1998; pp 37–51.

Sarhan, A. R. T., et al. Increased Levels of Cytotoxins in Barley Leaves Having the Systemic Acquired Resistance to *Bipolaris sorokiniana* (Sacc.) Shoemaker. *J. Phytopathol.* **1991,** *131,* 101–108.

Sarkar, B., et al. Crystal Structure, Antifungal Activity and Phytotoxicity of Diorganotin Compounds of Dihalo Substituted [(2-hydroxyphenyl) methylideneamino)] Thiourea. *Appl. Organomet. Chem.* **2010,** *24*(12), 842–852.

Schilder, A. M. C.; Bergstrom, G. Tan Spot. In *Seed Borne Diseases and Seed Health Testing of Wheat, Copenhagen;* Mathur, S. B., Cunfer, B. M., Eds.; Jordburgsforlaget: Denmark, 1993; pp 113–122.

Sharma, R. C., et al. Resistance to *Helminthosporium* Leaf Blight and Agronomic Performance of Spring Wheat Genotypes of Diverse Origins. *Euphytica.* **2004a,** *139,* 33–44.

Sharma, R. C.; Duveiller, E. Advancement toward New Spot Blotch Resistant Wheats in South Asia. *Crop Sci.* **2007,** *47,* 961–968.

Sharma, R. C.; Duveiller, E. Effect of *Helminthosporium* Leaf Blight on Performance of Timely Seeded Wheat Under Optimal and Stress Levels of Soil Fertility and Moisture. *Field Crop Res.* **2004,** *89,* 205–218.

Sharma, R. C.; Duveiller, E. In *Effect of Stress on Helminthosporium Leaf Blight in Wheat,* Proceedings 4th International Wheat Tan Spot and Spot Blotch Workshop, Bemidji, MN, USA, July 21–24, 2002; Rasmussen, J. B., Friesen, T. L., Ali, S., Eds.; Agricultural Experiment Station, North Dakota State University: Fargo, 2003; pp 140–144.

Sharma, R. C., et al. *Helminthosporium* Leaf Blight Resistance and Agronomic Performance of Wheat Genotypes across Warm Regions of South Asia. *Plant Breed.* **2004b,** *123,* 520–524.

Sharma, R. C., et al. Heritability Estimates of Field Resistance to Spot Blotch in Four Spring Wheat Crosses. *Plant Breed.* **1997,** *116,* 64–68.

Sharma, S. In *Wheat Diseases in Western Hills of Nepal,* Proceedings of the National Winter-Crops Technology Workshop, Kathmandu, Nepal, Sep 7–10, 1995; NARC & CIMMYT: Kathmandu, Nepal, 1996; pp 339–344.

Sharma-Poudyal, D.; Duveiller, E.; Sharma, R. C. Effects of Seed Treatment and Foliar Fungicides on *Helminthosporium* Leaf Blight and Performance of Wheat in Warmer Growing Conditions. *J. Phytopathol.* **2005**, *153*, 401–408.

Shivanna, M. B.; Meera, M. S.; Hyakumachi, M. Role of Root Colonization Ability of Plant Growth Promoting Fungi in the Suppression of Take-All and Common Root Rot of Wheat. *Crop Prot.* **1996**, *15*, 497–504.

Singh D. P., et al. Management of Leaf Blight Complex of Wheat (*Triticum aestivum*) Caused by *Bipolaris sorokiniana* and *Alternaria triticina* in Different Agroclimatic Zones Using an Integrated Approach. *Indian J. Agric. Sci.* **2008**, *78*, 513–517.

Singh, D. P., et al. Evaluation of Sources of Resistance to Leaf Blight (*Bipolaris sorokiniana* and *Alternaria triticina*) in Wheat (*Triticum aestivum*) and Triticale. *Indian Phytopathol.* **2015a**, *68*, 221–222.

Singh, D. P., et al. Optimum Growth Stage of Wheat and Triticale for Evaluation of Resistance against Spot Blotch. *Indian Phytopathol.* **2014b**, *67*, 423–425.

Singh, D. P., et al. Management of Spot Blotch of Wheat Caused *Bipolaris sorokiniana* in Wheat Using Fungicides. *Indian Phytopathol.* **2014a**, *67*, 308–310.

Singh, D. P. Assessment of Losses Due to Leaf Blights Caused by *Bipolaris sorokiniana* (Sacc.) Shoemaker and *Helminthosporium teres* (Sacc.) in Barley. *Pl. Dis. Res.* **2004**, *19*, 73–75.

Singh, D. P.; Chowdhury, A. K.; Kumar, P. Management of Losses Due to Seed-Borne Infection of *Bipolaris sorokiniana* and *Alternaria triticina* in Wheat (*Triticum aestivum*) Using Seed Treatment with Vitavax 200WS. *Indian J. Agric. Sci.* **2007**, *77*(2), 101–103.

Singh, D. V.; Srivastava, K. D. Foliar Blights and *Fusarium* Scab of Wheat: Present Status and Strategies for Management. *In Management of Threatening Plant Disease of National Importance*, Agnihotri, A. P., Sarvoy, A. K., Singh, D. V., Eds.; Malhotra publishing House: New Delhi, 1997; pp 1–16.

Singh, G., et al. Phenotypic and Marker Aided Identification of Donors for Spot Blotch Resistance in Wheat. *J. Wheat Res.* **2014**, *6*(1), 98–100.

Singh, G., et al. Phenotyping of Recombinant Inbred Lines for HLB Resistance in Wheat. *Wheat Barley Newslett.* **2013**, *7*(1), 9–10.

Singh, P. J.; Dhaliwal, H. S. Resistance to Foliar Blight of Wheat in *Aegilops* and Wild *Triticum* Species. *Indian Phytopathol.* **1993**, *46*, 941–943.

Singh, P. K., et al. Development and Characterization of the 4th CSISA-Spot Blotch Nursery of Bread Wheat. *Eur. J. Plant Pathol.* **2015b**, *143* (3), 595–605. https://www.research-gate.net/publication/280324049_Development_and_characterization_of_the_4th_CSISA-spot_blotch_nursery_of_bread_wheat

Singh, R. V., et al. In *Influence of Agronomic Practice on Foliar Blight and Identification of Alternate Hosts in the Rice-Wheat Cropping System*, Proceedings of the International Workshop on *Helminthosporium* Blights of Wheat: Spot Blotch and Tan Spot, CIMMYT, El Batan, Mexico, Feb 9–14, 1997; Duveiller, E., Dubin, H. J., Reeves, J., McNab, A., Eds.; CIMMYT: El Batan, Mexico, 1998; pp 346–348.

Skoropad, W. P. Barley Scald in the Prairie Provinces of Canada. *Commonw. Phytopathol. News.* **1960**, *6*, 25–27.

Smurova, S. G. *A New Sources and Donors of Wheat Resistance to Cochliobolus sativus Drechs. ex Dastur.* Ph.D. Thesis, All-Russia Institute of Plant Protection, Sankt-Petersburg, Russian Federation (in Russian), 2008.

Srivastava, O. P. Genetics of Resistance to Leaf Blight in Wheat. *Indian J. Genet.* **1982**, *42*, 140–141.

Srivastava, O. P.; Luthra, J. K.; Narula, P. N. Inheritance of Seedling Resistance to Leaf Blight of Wheat. *Indian J. Genet. Plant Breed.* **1971,** *31,* 209–211.

Stack, R. W.; McMullen, M. Root and Crown Rots of Small Grains. In *NSDU Extension Service Bulletin,* Fargo, North Dakota, USA, 1988; p 785.

Stack, R. W. In *Yield Losses in Barley Due to Common Root Rot in North Dakota,* Proceedings of International Conference on Common Root Rot, Saskatoon, Saskatchewan, Canada, Aug 11–14, 1991; Tinline, R. D., et al., Eds.; Agriculture, Saskatoon, Sask, Canada, 1991; pp 1–5.

Steffenson, B. J.; Hayes, P. M.; Kleinhofs, A. Genetics of Seedling and Adult Plant Resistance to Net Blotch (*Pyrenophora teres f. teres*) and Spot Blotch (*Cochliobolus sativus*) in Barley. *Theor. Appl. Genet.* **1996,** *92,* 552–558.

Sticher, L.; Mauch-Mani, B.; Métraux, J. P. Systemic Acquired Resistance. *Annu. Rev. Phytopathol.* **1997,** *35,* 235–270.

Sutton, J. C.; Vyn, T. J. Crop Sequences and Tillage Practices in Relation to Diseases of Winter Wheat in Ontario. *Can. J. Plant Pathol.* **1990,** *12,* 358–368.

Tekauz, A. Characterization and Distribution of Pathogenic Variation in *Pyrenophora teres* f. *teres* and *P. teres* f. *maculata* from Western Canada. *Can. J. Plant Pathol.* **1990,** *12,* 141–148.

Tekauz, A. Diseases of Barley. In *Diseases of Field Crops in Canada,* Bailey, K. L., et al., Eds.; Canadian Phytopathological Society: Moncton, 3rd ed.; 2003; pp 30–53.

Teng, P. S. *Crop Loss Assessment and Pest Management.* APS Press: St. Paul, MN, 1987; p 58.

Tewari, A. N.; Wako, K. Effect of Some Agrochemicals on Foliar Disease of Wheat. *Pl. Dis. Res.* **2003,** *18,* 39–43.

Toledo, J. B.; Guzman, E. A. In *Importance of Spot Blotch Caused by Bipolaris sorokiniana in Bolivia,* Proceedings of the International workshop on *Helminthosporium* Blights of Wheat: Spot Blotch and Tan Spot, CIMMYT, El Batan, Mexico, Feb 9–14, 1997; Duveiller, E., Dubin, H. J., Reeves, J., McNab, A., Eds.; CIMMYT: El Batan, Mexico, 1998; pp 146–149.

Turkington, T. K., et al. Foliar Diseases of Barley: Don't Rely on a Single Strategy from the Disease Management Toolbox. *Prairie Soils Crops J.* **2011,** *4,* 142–150. (www.prairiesoilsandcrops.ca).

United States Department of Agriculture (USDA) Global Barley Production. United States Department of Agriculture, USA, 2014.

Valjavec-Gratian, M.; Steffenson, B. J. Pathotypes of *Cochliobolus sativus* on Barley in North Dakota. *Plant Dis.* **1997,** *81,* 1275–1278.

van Ginkel, M.; Rajaram, S. In *Breeding for Resistance to Spot Blotch in Wheat: Global Perspective,* Proceedings of the International Workshop on *Helminthosporium* Blights of Wheat: Spot Blotch and Tan Spot, CIMMYT, El Batan, Mexico, Feb 9–14, 1997; Duveiller, E., Dubin, H. J., Reeves, J., McNab, A., Eds.; CIMMYT: El Batan, Mexico, 1998; pp 162–170.

Vasil, I.; Vasil, V. Regeneration in Cereal and Other Grass Species. In *Cell Culture and Somatic Cell Genetics of Plants;* Vasil, I., Vasil, V., Eds.; Academic Press: New York, 1986; Vol. 3, pp 121–150.

Velazquez Cruz, C. *Genetica de la Resistencia a Bipolaris sorokiniana in Trigos Harineros.* Ph.D Thesis, Montecillo, Mexico, 1994, p 84.

Vilareal, R. L., et al. In *Advanced Lines Derived from Wheat (Triticum aestivum L.) and Thinopyrum curvifolium Resistant to Helminthosporium sativum.* Proceedings of International

Crop Science Congress, 1992; Buxton, D. R., Shibles, S., Forsberg, R. A., Eds.; Crop Science Society of America/ Iowa State University: Ames, IA, 1992.

Vilareal, R. L., et al. Yield Losses to Spot Blotch in Spring Bread Wheat in Warm Non Traditional Wheat Production Areas. *Plant Dis.* **1995,** *79,* 893–897.

Wegulo, S. N.; Breathnach, J. A.; Baenziger, P. S. Effect of Growth Stage on the Relationship between Tan Spot and Spot Blotch Severity and Yield in Winter Wheat. *Crop Prot.* **2009,** *28,* 696–702.

Xi, K., et al. Distribution of *Rhynchosporium secalis* Pathotypes and Cultivar Reaction on Barley in Alberta. *Plant Dis.* **2003,** *87,* 391–396.

Xi, K., et al. Pathogenic Variation of *Rhynchosporium secalis* in Alberta. *Can. J. Plant Pathol.* **2002,** *24,* 176–183.

Zillinisky, F. J. Common Diseases of Small Grain Cereals. A Guide to Identification. CIMMYT: Mexico, 1983; p 141.

CHAPTER 4

OVERCOMING STRIPE RUST OF WHEAT: A THREAT TO FOOD SECURITY

OM P. GANGWAR, SUBHASH C. BHARDWAJ*, SUBODH KUMAR, PRAMOD PRASAD, HANIF KHAN, and SIDDANNA SAVADI

ICAR-Indian Institute of Wheat and Barley Research, Regional Station, Flowerdale, Shimla 171002, Himachal Pradesh, India

Corresponding author. E-mail: scbfdl@hotmail.com

CONTENTS

ABSTRACT

Wheat yield is affected by various biotic and abiotic stresses. Of the biotic stresses, yellow rust is most important disease especially in cooler regions of the world. During the past two decades, a number of virulent races/pathotypes of wheat yellow rust pathogen have emerged, causing concerns to wheat production. Adaptations and spread of new virulent races to the regions of warmer climate, is further posing the threat to world food security. Recently the new race defeating *Yr27* has caused severe losses in many countries of Africa and Asia. Growing resistant cultivars of wheat is most economical and effective in managing the yellow rust. Fungicides may be resorted in case of sudden outbreak of disease but it is not a pragmatic approach. Development and distribution of rust resistant varieties through accelerated breeding assisted by modern biotechnological tools, strengthening the research capacity and awareness of the farmers, continuous surveillance and strategic planning at country level, and effective global collaboration are essential for containing yellow rust and ensuring world food security.

4.1   INTRODUCTION

With more than 215 mha planted annually, wheat (*Triticum aestivum* L.) is the most widely cultivated cereal in the world. It is the most important protein source and provides around 20% of global calories for human consumption. The annual global wheat trade is around 130 million tons, and higher than the combined volume of trade of maize and rice. The developing countries are major producers of wheat and amounts more than 60% of wheat produced globally. China and India are top producing countries of wheat and together produce nearly twice as much wheat as the USA and Russia combined. In North Africa, West and Central Asia, wheat is the dominant staple food crop and contributes in 40–50% of the total calories needed (www.wheatinitiative.org). Over the next four decades, the human population in world may increase by 2 billion people to exceed 9 billion people by 2050. Recent Food and Agriculture Organization (FAO) estimates indicate that to meet the projected demand, global agricultural production will have to increase by 60% than what was produced during 2005–2007 (FAO, 2013). Again, the increased amount of food has to be produced from lesser land area and resources. It may be possible through genetic, physiological, and agronomic interventions, precision breeding for improved elasticity in varieties to mitigate the climate change, and monitoring for climate change and crop modeling for advance yield forecasts (Sharma et al., 2015). Climate change is considered as driving force of new stripe rust problems in wheat in cooler environments. The changes are facilitating the new races to spread across the borders of many countries. Global food security is one of the most pressing issues for humanity, and agricultural production is critical for achieving this. Stripe rust disease has long been among the major biotic threats to wheat productivity particularly in cooler climate in Central, and West Asia as well as in North Africa. Due to the evolution of new virulent races/pathotypes and their dissemination by winds, stripe rust disease emerged as recurrent threat to wheat production. Recently, the emergence of *Yr27* virulent race of *Puccinia striiformis* f. sp. *tritici* (*Pst*) affected wheat production in a number of countries of the world (FAO, 2015) (http://www.fao.org/agriculture/crops/wheatrust/en/). The emphasis has been given to have robust surveillance for stripe rust pathogen, introgression of stripe rust resistance from wild relatives, conserved germplasm/collections, educating the wheat farmers about the recurrence of rusts, symptoms, and disease management as well as skill up gradation of young wheat breeders and pathologist which will go a long way in proper stripe rust management and ensure food security.

4.2   AN OVERVIEW OF THE PATHOGEN, *PUCCINIA STRIIFORMIS* F. SP. *TRITICI*

Wheat stripe or yellow rust pathogen (*Puccinia striiformis* Westend. f. sp. *tritici* Eriks. & Henn.) belong to genus *Puccinia*, family Pucciniaceae, order Pucciniales (erstwhile Uredinales), class Pucciniomycetes, subphylum Pucciniomycotina, and phylum Basidiomycota (Bauer et al., 2006). Stripe rust disease of wheat was first described by Gadd and Bjerkander in 1777 but Schmidt (1827) designated the stripe rust fungus infecting barley glumes as *Uredo glumarum* (Humphrey et al., 1924). Westendorp (1854) used *Puccinia striaeformis* for stripe rust collected from rye (Stubbs, 1985). Later on, Eriksson and Henning (1896) showed that stripe rust resulted from a separate pathogen, which they named *Puccinia glumarum*. This name of stripe rust pathogen was in the vogue until Hylander et al. (1953) revived the name *P. striiformis* Westend. Nowadays, the *formae speciale* is added after the scientific name and currently written as *Pst*.

It is an obligate parasite and highly specialized fungus. Although, Schroeter (1879) first observed the phenomenon of physiological specialization (specialization of parasitism) in *Puccinia graminis* but the identification of races/pathotypes were done on 12 differential cultivars by Gassner and

Straib (1932). The heteroecious nature of stripe rust pathogen was unknown until 2010 when Jin et al. (2010) elucidated the mystery of the life cycle of *Pst*. They unequivocally demonstrated that barberry (*Berberis* spp.) (especially *Brachyponera chinensis, Berberis vulgaris*) serves as alternate hosts for the wheat stripe rust pathogen. Zhao et al. (2013) also showed that *P. striiformis* f. sp. *tritici* can infect some *Berberis* spp. under natural conditions, and the sexual cycle of the fungus may contribute to the diversity of *P. striiformis* f. sp. *tritici* in China. An analysis of linkage disequilibrium and genotypic diversity indicated a strong regional heterogeneity in levels of recombination, with clear signatures of recombination in the Himalayan (Nepal and Pakistan) and near-Himalayan regions (China) and a predominant clonal population structure in other regions. The existence of a high genotypic diversity, a high ability for sexual reproduction as well as the independent maintenance of strongly differentiated populations in the Himalayan region suggests this region as the putative centre of origin of *Pst* (Ali et al., 2014).

On the contrary, Wang and Chen (2015) reported that barberry does not function as an alternate host for *Pst* in regions of U.S. Pacific Northwest. The teliospores of *Pst* need a minimum of 32 h continual dew-forming conditions to infect barberry, and infection reaches at the peak after incubation of inoculated plants for 88 h. The lack of required period of leaf wetness conditions during the season of telial maturity effectively negates *Pst* infection of barberry plants in the U.S. Pacific Northwest. The survival and perpetuation of *Pst* in off season have been surreptitious in many countries of the world. Kumar et al. (2013) observed that latent infection (dormant mycelia) in winter wheat plays a pivotal role in overwintering and transmission of inoculum to subsequent spring crops in the regions of Central Alberta. They observed dormant mycelia intermittently in infected winter wheat leaves using polymerase chain reaction (PCR) assay from winter to early spring. Mutation, somatic recombination, parasexuality, selection, and sexual recombination are considered to be mechanisms determining the genetic variability of *Pst* that result in evolution of new aggressive races or pathotypes (Hovmoller et al., 2011; Jin, 2011; Knott, 1989; Stubbs, 1985; Park & Wellings, 2012).

4.3   DISEASE SYMPTOMS AND SIGN OF THE PATHOGEN

Stripe rust is considered as "polio of agriculture" as it spreads by releasing trillions of spores in the wind, each of which can cause new infection, and may end up in massive crop failures within a few weeks after attack especially under conducive weather conditions. Stripe rust pathogen infects almost all green parts of the wheat plant. However, leaves are mainly affected and infection can occur anytime from the one-leaf stage to till crop remains green, during crop cycle. The visible symptoms appear about a week of infection, and sporulation begins in about two weeks, under optimum temperature conditions (Chen, 2005). The initial symptoms are long, yellow to orange-colored fine stripes on leaves, usually between veins in horizontal manner. The symptoms also develop on leaf sheaths, glumes, and awns under favorable weather conditions on susceptible wheat plants. The yellow powdery mass of uredospores sticks on tips of fingers if these are rub on infected leaves as well as it falls on the soil surface. Resistant wheat cultivars are characterized by various infection types from no visual symptoms to small hypersensitive flecks to uredia surrounded by chlorosis or necrosis with restricted uredospore production (Chen et al., 2014). On seedlings, uredia produced by the infection of a single uredospore are not confined by leaf veins, but progressively emerge from the infection site in all directions, potentially covering the entire leaf surface (Fig. 4.1).

FIGURE 4.1   Stripe/yellow rust disease on leaf, glume, and awns.

4.4　STRIPE RUST OF WHEAT—A THREAT TO FOOD SECURITY

"Rust never sleeps" warned the Nobel Laureate Norman Borlaug, referring to the crop-destroying fungus that rank among humankind's most formidable agricultural foes (Global Food Security, 2015) (http://www.foodsecurity. ac.uk/research/current/defeating-wheat-disease.html). Stripe rust disease is continuing a threat to wheat production. It is capable of reducing the grain quality and yield in susceptible wheat cultivars, significantly. Stripe rust is principally a disease of cooler climate (2–15 °C) at higher elevation, northern latitude and cooler years (Roelfs et al., 1992). However, the recent outbreaks have defied this assumption and present races of *Pst* are also adapted to high temperatures, and hence becoming a problem in countries closer to the equator also. The *Pst* has remained a significant threat in the majority of wheat-growing regions of the world with potential to inflict regular regional crop losses ranging from 0.1 to 5%, with rare events giving losses of 5–25%. Regions with current vulnerability to *Pst* include the United States (particularly Pacific Northwest), East Asia (China North-West and South-West), South Asia (India, Pakistan, and Nepal), Oceania (Australia, New Zealand), East Africa (Ethiopia, Kenya), the Arabian Peninsula (Yemen), and Western Europe (East England) (Wellings, 2011). In 2009–2010, stripe rust resistant gene *Yr*27, succumbed to an aggressive new race of *Pst* that caused significant yield losses in Azerbaijan, Ethiopia, Iran, Iraq, Kenya, Morocco, Syria, Turkey, and Uzbekistan, and threatening the food security and livelihood of resource-poor farmers and their communities (ICARDA, 2014) (http://www.icarda.org/striperust2014/wp-content/uploads/2014/01/Strategies_to_reduce.pdf). Today's food security situation is being worsened by new aggressive pathotypes of wheat rust pathogen that are emerging more frequently and spreading much faster in new areas. It is further fuelled by climate change and conducive environments in increasingly fragile ecosystems. Under conducive circumstances, the stripe may cause yield losses ranging from 50 to 100% due to damaged plants and shriveled grain. Chen (2005) estimated yield losses ranging from 10 to 70% in the situations of where infection occurs earlier, the disease develops fast, duration is longer and the cultivars are susceptible.

During last 30 years, stripe rust caused severe economic losses to crops in North America, Europe, Australia, Central Asia, West Asia, South Asia, and North Africa. The losses ranged from 30 to 50% of the expected grain yield of a country's wheat production. Epidemics of stripe rust in the 1980s caused by a race that overcomes resistance in many wheat varieties spread from East Africa to the Middle East, Turkey, Iran, Afghanistan, Pakistan, and

India. They affected major wheat-growing regions and hit the livelihoods of millions of farmers. In the following years, researchers released varieties of wheat that were resistant to this group of pathotypes in most of these areas. However, many of these varieties fell susceptible to the new pathotypes of stripe rust. Some of the aggressive new pathotypes of $Yr27$ and other genes are tolerant to higher temperatures and adapt rapidly to new environmental conditions that result in severe crop loss in new geographical areas (Sci Dev Net, 2015) (http://www.scidev.net/global/policy/news/stripe-rust-threat-to-wheat-worse-than-predicted.html). Word's major epidemics of stripe rust are listed in Table 4.1. The innovative strategies for stripe rust management and planned actions can avert the future incurring endemic losses caused by Pst to wheat crop.

TABLE 4.1   World's Major Stripe Rust Epidemics with Comments on Severity and Losses.

Region	Nature of epidemic and losses
United Kingdom	Rothwell Perdix epidemic in 1966, Joss Cambier epidemic in 1996, and Sleijpner, Hornet epidemic during 1988–1989
Australia	Losses up to 80% during 1983–1986
	Annual fungicide expenditure AUD $ 40–90 million (2003–2006) incurred due to rust epidemics during 2002–2010
New Zealand	60% crop losses (60%) during 1980–1981
Iran	National losses of 1.5 million tons of wheat in 1993
Chile	Regular epidemics during 1976–1988, and 2001
USA	Widespread rust epidemic in 10 states during 1957–1958
	A yield reduction worth US$ 15–30 million tons during 1960–1964
	Widespread in 20 states in 2000
	National losses 3.4% or 11.7 million tons occurred in wheat yield in 2003
China	Loss of 6.00, 3.20, 2.65, and 1.40 tons of wheat due to stripe rust epidemics in 1954, 1964, 1990, and 2002, respectively.
Spain, North Africa	Siete Cerros epidemic in 1978
Republic of South Africa	Annual fungicide expenditure of ZAR 5–28 m during 1996–1999
India	High incidence of stripe rust was witnessed during 1994–2004, the breakdown of resistance due to $Yr27$ gene in most widely cultivated wheat variety, PBW 343 resulted high incidence of stripe rust

TABLE 4.1   *(Continued)*

Region	Nature of epidemic and losses
Pakistan	US$ 100 million losses were reported in NWFP in 2005
Italy	Undetermined; widespread epidemic were witnessed on susceptible cultivars during 1977–1978
Czechoslovakia	A loss of 30% in wheat yield was reported in susceptible cultivars in 1977

Source: Wellings (2011).

4.5   STRATEGIES FOR OVERCOMING THE STRIPE RUST

4.5.1   SURVEY AND SURVEILLANCE

Annual pathogen monitoring surveys are conducted to detect new pathotypes with potential to overcome the resistance genes that are deployed *vis-a-vis* overall distribution of pathotypes in a given area. In many cases, there is a lag period from the time of detection of a pathotype with that when its population reached to the levels of causing significant crop losses. It gives an opportunity for promotion of alternative cultivars. It may be achieved by meaningful selection, increase in seed, and replacement of susceptible cultivars (McIntosh & Brown, 1997). Increased surveillance, both at international and regional levels involve testing and tracking rust types using geospatial tools, monitoring the wheat varieties they attack, and determining stripe rust resistant resources. The use of monitoring and prediction systems for the management of wheat rust diseases has not been very widely practiced, however, under an extreme threat of breakdown of resistance, accurate prediction is useful to devise counter measures. All countries need robust food security strategies that include sharing information on crop breeding across regions.

Surveillance of wheat rust pathogens, including assessments of rust incidence and virulence characterization via either trap plots or pathotype surveys, has provided information which is fundamental in formulating and adopting appropriate national and international policies, investments and strategies in plant protection, plant breeding, seed systems, and in rust pathogen research (Park et al., 2011). For instance, in Iran, a monitoring network of agricultural extension specialists and researchers gathered reports of new pathotypes that prompted the country's plant protection authorities to stock up on fungicide, and establish a nursery, testing rust samples sent by wheat breeders. This helped in identification of 10 varieties resistant to stripe rust over the past three years, and distribution of seeds of these to the farmers. Wheat producing countries are also encouraged to design agricultural systems that enable new varieties to be released and faster multiplication of their seeds. In Egypt, for example, such a system has led to the production of new resistant varieties that can cover 30% of the country's wheat growing area in just three years. These varieties often offer better quality and higher yields, a powerful incentive for farmers to adopt them.

Wang et al. (2008) reported that the application of the PCR assay may be useful for rapid and reliable detection of dormant mycelia of *Pst* in infected leaves of overwintering wheat plants to determine the initial inoculum potential and thus to predict early outbreak for facilitating effective management of the disease. Field pathogenomics is another approach that adds highly informative data to surveillance module by enabling fast evaluation of pathogen variability, population structure, and host genotype. Pathogenomics provides the data on host and pathogen needed for deployment of effective resistance genes (Fig. 4.2) (Derevnina & Michelmore, 2015).

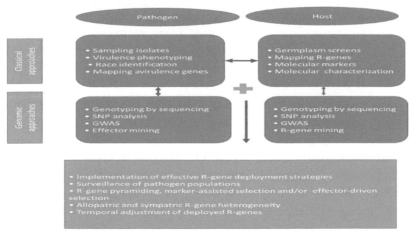

FIGURE 4.2   Field pathogenomics, genomic DNA-based approaches, and deployment of resistance genes are driven by knowledge of pathogen population genomics. Abbreviations: GWAS-Genome-wide association studies (From Derevnina, L.; Michelmore, R. W. Wheat Rusts Never Sleep but Neither Do Sequencers: Will Pathogenomics Transform the Way Plant Diseases are Managed? Genome Biol. 2015, 16, 44. DOI 10.1186/s13059-015-0615-3). © Derevnina and Michelmore; licensee BioMed Central. 2015. http://creativecommons.org/licenses/by/4.0).

Production: You can get the original from here: https://genomebiology.biomedcentral.com/articles/10.1186/s13059-015-0615-3.

4.5.2   HOST GENETIC RESISTANCE

Genetic resistance is the primary tool to protect wheat crops from stripe rust disease and it is most effective, economic, and most environmentally safe means of curbing the devastating effect of *Pst* on wheat crop. However, achieving and maintaining adequate resistance is difficult. Wheat researchers and breeders are scrambling to protect wheat from stripe rust and developing wheat varieties with several rust resistance genes, so that if the fungus mutates to outwit one defense, there are others to take on the pathogen. A systematic nomenclature for the genes conferring stripe rust resistance (*Yr*-gene catalog) was first proposed by Lupton and Macer (1962) and they suggested that new, and genetically distinct, differential host varieties should be produced for the identification of physiologic races of *P. glumarum*. Line (2002) published a broad description on the identification and designation of stripe rust resistance genes. Because of the efforts made by researchers, so far seventy-six stripe rust resistance genes with official name *Yr* followed by a number or provisional, have been identified and designated.

Breeding strategies are built on stacking or pyramiding pathotype specific genes and the utilization of minor genes from tested genetic resources. Resistance pyramids coupled with diversity for resistance like non-race-specific resistance and adult plant resistance are targeted at durable resistance. While for immediate use, stacking of major genes could be considered, for longer-term breeding solutions, durable host plant resistance is very useful.

A few historic cultivars imparted resistance to stripe rust pathogen for many years *viz.* Wilhelmina, Capelle-Desprez, Manella, Juliana, and Carstens VI (Stubbs, 1985). Most cultivars have remained resistant for five years or more, which is about the agronomic lifespan of a cultivar where an active breeding program exists. However, some cultivars have rusted even before they were grown on more than a fraction of the cultivated acreage. In most, if not all the cases, the failures have been due to inadequate knowledge of the virulences present in the pathogen population. In other cases, mutations or perhaps a recombination of existing virulence combinations occurred and rendered the host susceptible. Scientists have shown that wild relatives of cultivated wheat such as goat grass (*Aegilops* spp.) exhibit resistance to a number of fungal diseases including the stripe rust of wheat. Rust threatening wheat crops worldwide could be thwarted with genetics, new pathotypes of wheat stripe rust have been mutating and evolving constantly to overcome unfavorable weather conditions and the efforts of plant scientists are focused to breed resistance into wheat crops. Anticipatory resistance breeding is the process of predicting future pathotypes and producing resistant germplasm to avert future losses. This is supported by genetic analyses to catalog the identity and distribution of resistance genes in current cultivars. A national germplasm enhancement program ensuring both currently effective and potentially new sources of resistance are available in a wide range of adapted genotypes enables rapid cultivar replacement before or soon after the detection of new pathotypes. The policy of recommending only rust-resistant cultivars in the more rust-prone areas has resulted in significant reductions in pathogen population size and variability (McIntosh & Brown, 1997). There is a need to "stack" different types of resistance genes in order to protect wheat plants against the diseases effectively. There are two main types of resistance, major gene resistance defends the wheat plant against infection by specific race of stripe rust and adult plant resistance or minor gene resistance, on the other hand, retards the fungal infection and limits its nutrient intake from the host wheat plants. While adult stage resistance is only partial, it protects against multiple pathotypes of stripe rust pathogen and lasts for decades before a mutated form of pathogen overcomes it. The problem is that it can take years to achieve this goal using traditional breeding methods. In the meantime, the fungus continues to evolve resistance to wheat crops' natural defenses. Dr. Ravi Singh at Centro Internacional de Mejoramiento de Maíz y Trigo (CIMMYT), Mexico speaks of a gene cassette, which is a type of mobile genetic element that contains a gene or may be genes and a recombination site/s. Gene cassettes often carry resistance genes. One other way to hasten the development of a long lasting stripe rust resistant wheat variety is to engineer plants' DNA to carry resistance genes, creating what are known as genetically modified (GM) wheat. In the coming years genetic engineering would also make it easier to pull in genes from wild relatives, a rich source of rust resistance, without dragging along potentially unappealing agronomic qualities. Even if the researchers surmount the technical hurdles of isolating the resistance genes and inserting them into the wheat genome, they will undoubtedly face considerable hurdles in bringing the product to market (The Scientist, 2015) (http://www.the-scientist.com/?articles.view/articleNo/40085/title/Putting-Up-Resistance/).

4.5.3 CULTURAL, CHEMICAL, AND BIOLOGICAL ENDEAVORS

These strategies facilitate in reducing the intensity of an epidemic, restrict the spread of the pathogen, or provide long-term partial management. A farmer can harvest benefits of diverse resistance gene deployment if more than one cultivar with diverse resistance is used. In some areas, regulating the timing, frequency, amount of irrigation, and fertigation applications can aid in stripe rust disease management. Zadoks and Bouwman (1985) emphasized the importance of "green bridges" in carrying inoculum from one crop to the next. The green bridge may consist of volunteer plants, crops grown successively in one area, or wild accessory hosts. Removal or avoidance of these bridges is helpful where the inoculum is endogenous. Fungicides application for stripe rust disease management has been successfully used in Europe, permitting high yields (6–7 tons/ha) and where prices for wheat are supported (Buchenauer, 1982). In Australia, foliar fungicide spraying has increased due to the breakdown of resistance to stripe rust in wheat varieties (Murray & Brennan, 2009). Northwestern India has also 10 mha stripe rust-prone area. For controlling initial load of inoculum or under high stripe rust incidence in India, need based fungicides belonging to triazole group such as Propiconazole 25% EC (Tilt), Tebuconazole 25% EC (Folicur), and Triadimefon 25% EC (Bayleton) have been used effectively at the rate of 0.1% for the management of stripe rust of wheat in some pockets in Northern India. Fungicides can be applied for the management of wheat rust in cases of emergency but this option is not generally affordable for resource poor farmers and fungicides are not environment friendly.

Darluca filum is one of the more aggressive hyperparasites which are capable of infecting a range of rust fungi including *P. graminis* f. sp. *tritici*. However, this and other hyperparasites appear less promising because of the wide and rapid dispersion of the stripe rust pathogen and difficulties in accumulating the hyperparasite in requisite proportion under field conditions. However, Li et al. (2013) reported that *Bacillus subtilis* strain E1R-j effectively manage stripe rust in the field.

4.5.4 SUSTAINED INVESTMENT FOR RESEARCH

To protect the wheat crop from the dreadful fungus that causes stripe rust, countries need to take pre-emptive action with sustained investment in research, surveillance, a strategy to boost crop diversity, and policies to encourage farmers to adopt disease resistant crop varieties. Developing countries need help with crop surveillance and the development of rust resistant varieties of wheat. Donor governments, development agencies, and the international research community must increase their attention and support to low income countries striving to develop strategies to prevent wheat rust. Climate change is likely to increase the spread and severity of stripe rust disease, further threatening food security. To combat stripe rust, greater investments in research and regional coordination are essential. Simply maintaining the biological capital, and the beneficial production and economic outcomes it bestows, requires continual reinvestment in new crop defense. In his research findings, Beddow et al. (2015) estimated that 5.47 million tons of wheat are lost due to the stripe rust pathogen each year, equivalent to a loss of US$ 979 million per year. Comparing the cost of developing stripe rust resistant varieties of wheat with the cost of stripe rust induced yield losses by using a probabilistic Monte Carlo simulation model, a sustained annual research investment of at least US$ 32 million into stripe rust resistance is economically justified. Most of the countries do not spend even 1% of the money, the vibrant programs of wheat rust management are saving in the form of avoidable losses.

4.5.5 COLLABORATION, CAPACITY BUILDING, AND COMMUNICATION

For over 40 years, a collaborative network of publicly funded international wheat scientists has made a significant contribution to food security in the developing world. Thousands of modern wheat varieties were made available for the farmers to use in both favorable and marginal environments on well over 50 ha. Millions of small-scale farmers in the developing world have benefited (Reynolds & Borlaug, 2006). Networks of scientists and agriculture specialists should exchange information for an early warning of rust incidence in their area, scientists use slow-rusting genes to extend the time that varieties can resist the disease in a bid to slow down the progression of epidemics and farmers avoid covering large areas with wheat varieties with similar genetic backgrounds and degrees of resistance. The resources deployed to contain the worst effects of *Pst* will need to train a new generation of breeders and pathologists in host-pathogen genetics. *Plant Breeding Institute* (PBI), Sydney and CIMMYT, Mexico are doing an effective job at international level, while ICAR-IIWBR, Shimla extends similar help to the South Asian Association for Regional Cooperation (SAARC) countries. Extension services and farmer education are given little attention in most of the countries threatened by new rust races. Participatory farmer education methods, such as farmer field schools, have proven to be extremely effective, empowering farmers with strong observation and decision-making abilities. Properly trained farmers will be a major support to the implementation of national contingency plans. They could play a role in early recognition and reporting of changes in disease severity and virulence in their fields. Thus, they need to be trained to understand the risks associated with virulent pathotypes and the importance of the various field management practices (planting dates, planting periods, choice of varieties, responsible fungicide use, etc.) for disease management and yield improvement. This system has proved very effective in Northwestern India where ICAR-IIWBR is working in tandem to tackle the stripe rust effectively. Government's awareness to growers at the list of potential, recommended wheat varieties from the provincial system, departments of agriculture and choose a variety with good to very good resistance is a steering factor. An approach of fast seed multiplication and diversification of resistant varieties to reduce the impact of stripe rust disease is depicted in Figure 4.3.

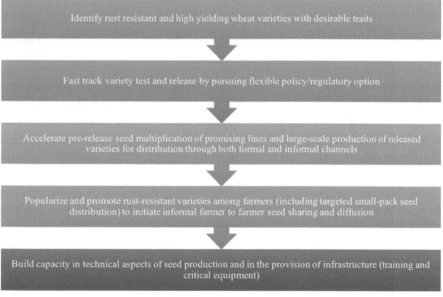

FIGURE 4.3   An approach to seed multiplication and diversification of new stripe rust resistant varieties (modified from source: ICARDA, 2014).

4.6   CONCLUDING REMARKS

Yellow rust pathogen will continue to mutate and overcome crop resistance. Sustained investment in agricultural research and preparedness to manage rust problems in the long-term basis in different countries is needed. Farmers may be educated and encouraged to adopt stripe rust resistant varieties. The use of fungicides as foliar sprays may be done as and when it is necessary especially in susceptible varieties. A combined strategy between low income countries and developmental partners is required which may include, regular and strategic disease surveillance, development and dissemination of new resistant varieties, strengthening research capacity, and ensuring adoption and multiplication of new seeds. The advantage of new molecular tools like gene cassettes, genetic engineering of resistance, transgenics, and so forth may be exploited to manage stripe rust especially in situation where, conventional practices are not yielding desirable results.

KEYWORDS

- **stripe rust**
- ***Puccinia striiformis* f. sp. *tritici***
- **wheat**
- **virulence**
- **surveillance**
- **resistance**
- **food security**
- **aggressiveness**

REFERENCES

Ali, S., et al. Origin, Migration Routes and Worldwide Population Genetic Structure of the Wheat Yellow Rust Pathogen *Puccinia striiformis* f. sp. *tritici*. *PLoS Pathog.* **2014,** *10*(1), 1–12.

Bauer, R., et al. The Simple-Septate Basidiomycetes: A Synopsis. *Mycol. Prog.* **2006,** *5,* 41–66.

Beddow, J. M., et al. Research Investment Implications of Shifts in the Global Geography of Wheat Stripe Rust. *Nat. Plants.* **2015,** *1*(10), 15132.

Buchenauer, H. Chemical and Biological Control of Cereal Rust. In *The Rust Fungi;* Scott, K. J., Chakravorty, A. K., Eds.; Academic Press: London, 1982; pp 247–279.

Chen, W.; Wellings, C.; Chen, X.; Kang, Z.; Liu, T. Wheat Stripe (Yellow) Rust Caused by *Puccinia striiformis* f. sp. *tritici. Mol. Plant Pathol.* **2014,** *15*(5), 433–446.

Chen, X. M. Epidemiology and Control of Stripe Rust (*Puccinia striiformis* f. sp. *tritici*) on Wheat. *Can. J. Plant Pathol.* **2005,** *27,* 314–337.

Derevnina, L.; Michelmore, R. W. Wheat Rusts Never Sleep but Neither Do Sequencers: Will Pathogenomics Transform the Way Plant Diseases are Managed? *Genome Biol.* **2015**, *16*, 44 (DOI 10.1186/s13059-015-0615-3).

Eriksson, J.; Henning, E. *Die Getreiderosle. Ihre Geschichte und Nalur sowie Massregein Gegen Dieselben;* PA Norstedt & Söner AB: Stockholm, Sweden, 1896; p 463.

FAO. *Feeding the World. FAO Statistical Yearbook 2013;* World Food and Agriculture Organization: Rome, 2013; pp 123–158.

FAO. Wheat Rusts: A Constant Threat to Wheat Crops Around Globe. 2015. http://www.fao.org/agriculture/crops/wheatrust/en/ (accessed on 30 Nov. 2015).

Gassner, G.; Straib, W. Die Bestimmung der Biologischen Rassen des Weizengelbrostes *(Puccinia glumasum* f. sp. *tritici* (Schmidt.) Erikss. und Henn.). *Arbeiten Biol. Reichsans. Land- Forstw.* **1932**, *20*, 141–163.

Global Food Security. Joining Forces to Defeat Wheat Disease. 2015. http://www.foodsecurity.ac.uk/research / current/defeating-wheat-disease.html (accessed on 30 Nov. 2015).

Hovmoller, M. S.; Sorensen, C. K.; Walter, S.; Justesen, A. F. Diversity of *Puccinia striiformis* on Creals and Grasses. *Annu. Rev. Phytopathol.* **2011**, *49*, 197–217.

Humphrey, H. B.; Hungerford, C. W.; Johnson, A. G. Stripe Rust *(Puccinia glumarum)* of Cereals and Grasses in the United States. *J. Agric. Res.* **1924**, *29*, 209–227.

Hylander, N.; Jorstad. I.; Nannfeldt, J. A. Enumeratio uredionearum Scandinavicarum. *Opera Bot.* **1953**, *1*, 1–102.

ICARDA. Strategies to Reduce the Emerging Wheat Stripe Rust Disease. 2014. http://www.icarda.org/striperust2014/wp-content/uploads/2014/01/Strategies_to_reduce.pdf.

Jin, Y. Role of *Berberis* spp. as Alternate Hosts in Generating New Races of *Puccinia graminis* and *P. striiformis. Euphytica.* **2011**, *179*, 105–108.

Jin, Y.; Szabo, L. J.; Carson, M. Century-Old Mystery of *Puccinia striiformis* Life History Solved with the Identification of *Berberis* as an Alternate Host. *Phytopathology.* **2010**, *100*, 432–435.

Knott, D. R. *The Wheat Rusts-Breeding for Resistance;* Springer-Verlag: Berlin, Germany, 1989; p 201.

Kumar, K.; Holtz, M. D.; Xi, K.; Turkington, T. K. Overwintering Potential of the Stripe Rust Pathogen *(Puccinia striiformis)* in Central Alberta. *Can. J. Plant Pathol.* **2013**, *35*(3), 304–314.

Li, H.; Zhao, J.; Feng, H.; Huang, L.; Kang. Z. Biological Control of Wheat Stripe Rust by an Endophytic *Bacillus subtilis* Strain E1R-j in Greenhouse and Field Trials. *Crop Prot.* **2013**, *43*, 201–206.

Line, R. F. Stripe Rust of Wheat and Barley in North America: A Retrospective Historical Review. *Annu. Rev. Phytopathol.* **2002**, *40*, 75–118.

Lupton, F. G. H.; Macer, R. C. F. Inheritance of Resistance to Yellow Rust *(Puccinia glumarum* Erikss. & Henn.) in Seven Varieties of Wheat. *Trans. Brit. Mycol. Soc.* **1962**, *45*(1), 21–45.

McIntosh, R. A.; Brown, G. N. Anticipatory Breeding for Resistance to Rust Diseases in Wheat. *Annu. Rev. Phytopathol.* **1997**, *35*, 311–326.

Murray, G. M.; Brennan, J. P. *The Current and Potential Costs from Diseases of Wheat in Australia*; Grains Research & Development Corporation: Barton, Australia, 2009.

Park, R. F.; Wellings, C. R. Somatic Hybridization in the Uredinales. *Annu. Rev. Phytopathol.* **2012**, *50*, 219–239.

Park, R., et al. International Surveillance of Wheat Rust Pathogens: Progress and Challenges. *Euphytica.* **2011**, *179*, 109–117.

Reynolds, M. P.; Borlaug, N. E. Impacts of Breeding on International Collaborative Wheat Improvement. *J. Agric. Sci.* **2006,** *144,* 3–17.

Roelfs, A. P.; Singh, R. P.; Saari, E. E. *Rust Diseases of Wheat: Concepts and Methods of Disease Management;* CIMMYT: Mexico, DF. 1992; p 81.

Sci Dev Net. Stripe Rust Threat to Wheat Worse Than Predicted. 2015. http://www.scidev. net/global/policy/news/stripe-rust-threat-to-wheat-worse-than-predicted.html.

Schroeter, J. Entwicklungsgeschichte Einiger Rost Pilze. *Beitrage zur Biologie der Pflanzen.* **1879,** *3,* 51–93.

Sharma, I., et al. Enhancing Wheat Production- A Global Perspective. *Indian J. Agric. Sci.* **2015,** *85*(1), 3–13.

Stubbs, R. W. Stripe Rust. In: *The Cereal Rusts: Diseases, Distribution, Epidemiology, and Control*; Roelfs, A. P., Bushnell, W. R., Eds.; Academic Press: Orlando, FL, 1985; Vol. 2, pp 61–101.

The Scientist. Putting up Resistance. 2015. http://www.the-scientist.com/?articles.view/articleNo/ 40085/title/Putting-Up-Resistance (accessed on 30 November, 2015).

Wang, M. N.; Chen, X. M. Barberry does not Function as an Alternate Host for *Puccinia striiformis* f. sp. *tritici* in the U. S. Pacific Northwest Due to Teliospore Degradation and Barberry Phenology. *Plant Dis.* **2015,** *99*(11), 1500–1506. http://dx.doi.org/10.1094/PDIS-12-14-1280-RE (accessed on 30 Sep. 2015).

Wang, X., et al. The Development of a PCR-Based Method for Detecting *Puccinia striiformis* Latent Infections in Wheat Leaves. *Eur. J. Plant. Pathol.* **2008,** *120,* 241–247.

Wellings, C. R. Global Status of Stripe Rust: A Review of Historical and Current Threats. *Euphytica.* **2011,** *179,* 129–141.

Zadoks, J. C.; Bouwman, J. J. Epidemiology in Europe. In: *The Cereal Rusts: Diseases, Distribution, Epidemiology, and Control;* Roelfs, A. P., Bushnell, W. R., Eds.; Academic Press: Orlando, FL, 1985; Vol. 2, pp 329–369.

Zhao, J., et al. Identification of Eighteen *Berberis* Species as Alternate Hosts of *Puccinia striiformis* f. sp. *tritici* and Virulence Variation in the Pathogen Isolates from Natural Infection of Barberry Plants in China. *Phytopathology.* **2013,** *103*(9), 927–934.

CHAPTER 5

POWDERY MILDEW OF WHEAT AND ITS MANAGEMENT

ASHWANI KUMAR BASANDRAI* and DAISY BASANDRAI

CSKHPAU, Rice and Wheat Research Centre, Malan, District Kangra, Himachal Pradesh, 176047, India

Corresponding author. E-mail: bunchy@rediffmail.com, ashwanispp@ gmail.com

CONTENTS

ABSTRACT

Powdery mildew caused by *Blumeria graminis* f. sp. *tritici* is the fourth most important disease of wheat after three rusts. It inflicts yield losses to the tune of 13–34% and 50–100% under low or moderate infestation and severe infestation, respectively. It is essential to understand the damaging trend, etiology, virulence pattern, host resistance, genetics of host resistance, and fungicidal management for its successful control. Molecular studies on host–parasite interaction of wheat-powdery mildew system are continuously being conducted resulting in identification and mapping of resistant genes. Currently, ~80 designated resistance alleles have been identified at 51 loci (*Pm1-Pm55, Pm18 = Pm1c, Pm22 = Pm1e, Pm23 = Pm4c, Pm31 = Pm21, Pm48 = Pm46*). In addition, <50 resistance genes have been located but carry temporarily designated nomenclature. Use of molecular markers such as simple sequence repeat (SSR), amplified fragment length polymorphism (AFLP), restriction fragment length polymorphism (RFLP), random amplified polymorphic DNAs (RAPD), and sequence tagged sites (STS) has contributed to identification and mapping of many resistant genes. Fungicidial sprays have been successfully applied as a shorterm alternative for its management. In this chapter damages caused by wheat powdery mildew, virulence structure of the pathogen in epimeological important areas, identification of resistant sources, postulation of racespecific resistant genes, major resistance genes and molecular markers flanking the resistant genes, and fungicdal management of the disease have been reviewed.

5.1   INTRODUCTION

Wheat (*Triticum aestivum* L.) is an extensively grown cereal in the most parts of the world. It is a staple food for more than 40% global population (Akhtar et al., 2011). Common bread wheat (*T. aestivum*) and durum wheat (*T. durum* Desf.) contributes 90% of the world's wheat production covering an area of more than 200 mha (Xin et al., 2012). In the last few decades, the area under the crop has increased globally. However, its productivity is less as compared to the potential yield of commonly cultivated varieties. Among various factors for low productivity, diseases take a heavy toll of the crop. Wheat powdery mildew (PM) caused by the fungus *Blumeria graminis* (DC) Speer f. sp. *tritici* Em Marchal, (*Bgt*) (syn. *Erysiphe graminis* (DC) f. sp. *tritici*) is the fourth most devastating disease after the three rusts. The disease can be partially managed by the use of fungitoxicants. However, development and cultivation of resistant cultivars is the effective, economically viable practical and environmentally safe method to manage this disease. The lifecycle of pathogen is such that evolution and spread of new virulences render the resistant varieties susceptible in a short period of time. Hence, effective virulence monitoring programs were carried out in epidemiologically important countries along with strong breeding programs leading to continuous development of resistant varieties. Efforts have also been made to transfer effective genes from wild species and related genera (Mwale et al., 2014). Till date, 54 PM resistant genes have been identified, catalogued, and designated as *Pm1* to *Pm54* with multiple alleles at some loci. In addition, the same number of genes has been temporarily designated. Recent DNA markers have been widely used for the characterization, location, and mapping of resistant genes. In the event of loss of resistance due to sudden and fast appearance of new virulences, effective fungicides have been identified and widely used. This chapter reviews the recent developments on various aspects of PM of wheat.

5.2 ECONOMIC IMPORTANCE AND DISTRIBUTION

Wheat PM is an important disease primarily in the northern hemisphere. Before the "Green Revolution" era, it was economically damaging in cooler, maritime or semi-continental climates. However, over the past several decades, with the cultivation of post green revolution varieties, the disease has gained importance even in some hotter and drier areas due to the adoption of intensive production techniques leading to thicker, more compact, and more humid plant canopies resulting from the use of semi dwarf cultivars, excessive use of nitrogen fertilizers and assured irrigation facilities. The disease causes crop yield losses of 13–34% and 50–100%, under low and severe infestation, respectively (Alam et al., 2011; Mwale et al., 2014). Rana et al. (2005) reported that PM of wheat caused yield losses from 8.7 to 41.3% in Himachal Pradesh. The disease causes economic losses in all the continents where, the wheat is grown and the distribution of the disease has been reviewed by Christina et al. (2012).

5.2.1 EUROPE

In Europe, PM is has wide incidence in the major cereal growing regions of Ireland, the United Kingdom and Northern Europe including France,

Sweden, Denmark, Germany, and Poland. Long distance mildew spore transport in large areas of Northern Europe has made it an epidemiologically important unit for this disease. The disease is sporadic in the United Kingdom whereas, it remained important in Scandinavian bread wheat growing regions. In Eastern Europe, PM is common in North-West Poland, Croatia, and Hungary. In Italy it takes a heavy toll of the crop on both durum and bread wheat. In Turkey, wheat PM caused 5–30% losses in the central region and was a major constraint to durum wheat production in the Aegean and Mediterranean coastal areas.

5.2.2   ASIA

PM is an important disease in the higher elevations of Asia and the coastal area around the Caspian Sea. It occurs in Russia, Northern Caucasus, Volga Basin, Central Chernozem, Ural, and Volga-Vyatka regions, Baltic States, and Trans Caucasia. It also affects areas north of Afghanistan and west of China, Kazakhstan, Kirgizstan, Turkmenistan, and Tajikistan. In China, during early eighties, epidemics of PM and yellow rust (YR) affected 6 mha area out of 28 mha mean annual national wheat area. During 2001, PM and YR resulted in 1.6 t/ha yields in Yunnan province which was below the mean national yield. Both the diseases lead to shifting of wheat area from PM prone north-central region to south. On an average, an area of 6.9 mha was affected annually by wheat PM between 2004 and 2009. During 2009, out of about 24 mha planted area, severe PM epidemic affected about 20% area in the North-West Chinese province of Ningxia. In Japan, losses are estimated at 20% and it is also considered an emerging problem in Pakistan. It is a sporadic disease in India, occurring mainly in the northwestern plains. The disease is considered one of the most important fungal diseases of wheat in Iran around the Caspian Sea.

5.2.3   AFRICA

In North Africa, PM is important in the coastal regions of the Mediterranean Sea on bread and durum wheat. The disease has increased and has become economically important in the cooler regions of East Africa.

5.2.4 NORTH AMERICA

In Canada, PM occurred in British Columbia and the eastern provinces. It was quite serious in the eastern provinces in the United States and it is economically damaging in the mild, humid mid-Atlantic states of Maryland, Virginia, North Carolina, and South Carolina.

5.2.5 LATIN AMERICA

In South America, the disease increased after the breakdown of the resistance gene *Pm8* and it is a significant factor contributing to yield instability in Brazil. Wheat PM epidemics are expected to be affected by climate change. Since, *B. graminis* is a biotroph therefore disease is influenced strongly by host plant stage and status of water and nitrogen in plants. PM severity on wheat plants increased with increasing levels of wheat shoot nitrogen and water contents. When water was moderately available, wheat plants grown in elevated CO_2 had lower levels of shoot nitrogen but higher water content than plants grown at ambient CO_2 and the disease severity remained unchanged. It is expected that climate change may influence PM distribution with regard to crop water availability.

5.3 SYMPTOMS

Disease symptoms may appear at any time after the plant emergence and the signs are most common on leaves. But these may also develop on other aerial parts of the plant. The disease manifests as white, cottony patches, or colonies of mycelium and asexually produced conidia (also called oidia) in long chains, which are easily dislodged by wind or rain. The patches merge and cover large areas of the stem, leaf surface or heads and awns of susceptible cultivars. Leaf tissues on the adaxial side of the colony become yellow turning tan or brown. The plants remain stunted under severe infections. Dark round cleistothecia (135–280 μm in diameter) develop in the fungal mass, as the pustules get older. Ascospores develop in the cleistothecia and serve as the long-term survival structures (http://www.fao.org/docrep/006/y 4011e/y4011e0l.htm# Mwale et al., 2014).

5.4   DISEASE DEVELOPMENT IN RELATION TO ENVIRONMENTAL CONDITIONS

The development of PM is favored by relatively cold weather, that is, air temperatures range of 14.5–18.2°C and relative humidity (RH) range of 50–80%. The conidia of PM of wheat germinates at temperature between 5 and 30°C with an optimum of 20°C in dark at 100% RH. After conidial landing, precipitation, high wind velocities, and both very low and high vapor pressure deficits of the air exhibited a negative effect upon the computed infection probability. Rainfall after inoculation inhibited PM development (Christina et al., 2012). RH, air temperature, and precipitation were the most important meteorological factors influencing disease development.

5.5   PHYSIOLOGICAL SPECIALIZATION

The primary inoculum of Bgt is thought to originate mainly from cleistothecia (chasmothecia, or sexual fruiting bodies). Subsequently, the pathogen undergoes multiple cycles of asexual reproduction via conidia. Sexual reproduction leads to shuffling of virulence gene combinations and allows selection of effective virulence combinations in fit backgrounds. Thus, depicting virulence phenotypes are a more suitable method of characterizing individuals than "races," because each specific virulence phenotype may be ephemeral and there may not be a direct line of descent between individuals possessing the same virulence phenotype. There are currently 70 identified alleles at 54 loci conferring resistance to wheat PM (Mwale et al., 2014; McIntosh et al., 2014). Extensive production of wheat cultivars with one or a few *Pm* genes places strong selective pressure on the pathogen population and abundant spore production provides ample potential for mutations to virulence, resulting in rapid appearance and increase of adapted isolates. Wind-aided conidial dispersal quickly spreads virulent individuals leading to low durability of resistance and long-distance spread of new virulence.

The pioneer work on pathogenic variation dates back to 1930 and 1933, when Waterhouse and Mains, reported races in Australian and American wheat PM populations, respectively (Wolfe & Schwarzbach, 1978). Numerous surveys of the *B. graminis* f. sp. *tritici* population virulence have been performed in United States and Europe (Moseman, 1966; Christina et al., 2012, 2014; Wolfe & Schwarzbach, 1978; Szunics et al., 2001; Lesovoi & Kol'obritskii, 1980; Solc & Paulech, 1977; Gromashevs'ka, 1983; Woznick,

1982; Stojanovic et al., 1990), Japan (Oku et al., 1987), China (Wu et al., 1983), and Argentina (Wolfe & Schwarzbach, 1978).

In India, scanty and localized information is available about race spectrum of Bgt. Arya (1962) reported three races (3, 4, and 10) from ten isolates collected from Jodhpur. Prabhu and Prasada (1963) indicated the prevalence of different races of Bgt in Shimla and Wellington (Goel et al., 1986; Bahadur & Aggarwal, 1997). It is evident that wheat PM populations are pathogenically variable. However, due to use of varied differentials, the exact nature of comparable race flora couldnot be ascertained.

Briggle (1969) established near isogenic lines, having single PM resistance genes in the background of cv. Chancellor and till date more than 54 cataloged PM resistance genes with multiple alleles at some locus and additional resistant lines with undesignated genes have been identified which have been extensively used as differential sets to study pathogenic variation in relation to host resistance.

Royer et al. (1984) distinguished single colony isolates of Bgt from Central Pennsylvania (United States) using nine *Pm* genes. Virulence was rare on lines with genes *Pm1*, *Pm3a*, and *Pm3b*. In North Carolinavirulences were detected against all the known PM genes, that is, *Pm1*, *Pm3*, *Pm2*(+), *Pm3a*, *Pm3b*, *Pm3c*, *Pm4*, *Pm6*, *Pm7*, *Pm8*, and Michigan Amber using mobile nursery technique (Leath & Murphy, 1985). Costamilan (2005) reported that out of 31 combinations of effective and ineffective resistance genes, gene *Pm4a*+ and *Pm6* remained totally and highly effective to all isolates, whereas genes *Pm3a* and *Pm8* were totally ineffective. The virulence formula *Pm1, 3c, 4a, 6, 1+?, 2+Mld, 4a+..., D2* (effective genes) /2, *3a, 8, D1* (ineffective genes) was most frequent with 15% occurrence. A mobile nursery comprising near-isogenic lines with genes *Pm1, Pm2, Pm3a, Pm3b, Pm3c, Pm4, and Pm5* was exposed to the field inoculum of *Bgt* in South Ontario, Canada and virulence pattern was different in various regions (Bailey & McNeill, 1983). Gene *Pm3a* was effective at most of the sites and single pustule isolates provided better estimate of the occurrence of virulence genes than collections of mass cultures (Menzies & McNeill, 1986). Parks et al. (2008) analyzed 207 single ascosporic Bgt isolates on a set of 16 mildew resistance (*Pm*) genes. Virulence to genes *Pm3a, Pm3c, Pm5a*, and *Pm7* was present in >90% isolates and 10% isolates were virulent on *Pm1a, Pm16, Pm17*, and *Pm25* whereas, 71–88% isolates possessed few multi locus virulence phenotypes. Virulence frequencies on individual *Pm* genes varied from place to place and genetic (phenotypic) distance between isolate sub populations increased significantly with increasing geographic separation. It may be due to deployment of different *Pm* genes and restricted gene flow in the pathogen population. According to Christina et al. (2015) since 1993, virulence profiles of U.S. *Bgt* populations revealed that genes *Pm1a*, *1b*, *4b*, *16*, and *36* remained effective and most of other *Pm* genes, that is, *Pm1-Pm25 were* widely defeated. Several resistance genes, that is, *lAG12*, *Pm25*, *Pm34*, *Pm35*, and *Pm37* introgressed from wild relatives in soft red winter wheat backgrounds adapted to the Mid-Atlantic and South-eastern United States were highly effective. Genes *Pm3a* and *Pm3b* remained very effective in areas west of the Appalachian Mountains (Kentucky, Nebraska, Oklahoma, and Texas). This discrepancy might have persisted due to newly reported migration patterns in the U.S. *Bgt* population. Two hundred thirty eight *Bgt* isolates, collected from 12 U.S. states from 2003 to 2010 were evaluated for local and regional population differences, linkage disequilibrium, and migration. Isolates from the South-east, Mid-Atlantic, and Great Lakes regions comprised a single large cluster, and were genetically separated from the populations in Kentucky, Oklahoma, and Texas. The evidence suggested annual re-establishment of PM was primarily from the local sources, resulting in a large-scale mosaic of overlapping local populations with some long-distance dispersal in a west-to-east direction.

In China, extensive studies have been undertaken on virulence structure of *Bgt*. Wu and Liu (1983) observed that genes *Pm2* and *Pm4* were very effective against the local *Bgt* populations in Jiangsu State. In Sichuan, among 67 isolates of Bgt, frequency of virulence genes v_1 and v_5 was 90% and it was 65.7, 7.5, 4.5, and 16.4 for v_8, v_{2x}, v_{Ba}, and v_{kg}, respectively. Cultivars, Baimain 3 (*Pm2x*), 81-7241(Ba), and KenGui-A (KG) remained highly resistant (Li & Huang, 1991). Chen et al. (2013) reported that during 2011 and 2012 Bgt races 411, 377, and 317 were dominant in the North-eastern spring wheat, winter wheat region of Shandong and Huberi, and Sichuan, respectively. The frequencies of virulent genes *v3b*, *v5*, *v7*, *v8*, *v1+2+9*, and *v17* were higher (more than 90%), and it was lower (<31.4%) for genes V2, V4a, V4b, V12, V13, V16, V18-V23, and V5+6. Zeng et al. (2013) characterized virulence and diversity of 1082 isolates of Bgt from eight major wheat-growing regions of China using 22 differential lines with known *Pm* genes. None of the isolates were virulent on gene *Pm 21* and <20.0% were virulent on *Pm13*. Virulence frequencies was >50.0% on genes *Pm1a*, *Pm3b*, *Pm3c*, *Pm3f*, *Pm5a*, *Pm6*, and *Pm8*. In all, 1028 pathotypes were detected, of which 984 were unique. Phenotypic diversity indices revealed a high level of diversity within populations and genetic distance between different populations correlated significantly with geographical distance (R-2 = 0.494, $P \leq 0.001$). Isolates from Xinjiang appeared to form a separate group. Virulence and diversity of the eight populations suggested that varieties with effective resistance gene combinations should be developed at regional level. Xu et al. (2014) studied molecular genetic variation in wild populations of *Bgt* in China. The frequency of V1a, V3a, V3c, V3e, V5a, V6, V7, V8, and V19 was high (>75%), indicating their wide distribution; whereas frequency of genes V1c, V5b, V12, V13, V16, V21, VXBD, V2+6, V2+Mld, and V4+8, with less distribution was lower (0–20%). The Nei's gene diversity (H), Shannon's information index (I) and the percentage of polymorphic loci (P) were 0.23, 0.35, and 67.65%, respectively, which revealed virulence diversity. The results from single nucleotide polymorphisms (SNPs) of 38 isolates showed that three housekeeping genes contained nine SNP sites. Ten haplotypes (H1–H10) were inferred from the concatenated sequences, with 1 haplotype (H1) comprising over 55% of Qinghai population. Phylogenic analysis did not show obvious geographical subdivision between the isolates. A multi locus haplotype network presented a radial structure, with H1 in the centre as an inferred ancestor. Using analysis of molecular variance (AMOVA), 1.63 and 98.63% variation was among and within populations, respectively, with a low fixations index ($F_{ST} = 0.01634$, $P < 0.05$). This revealed a relatively high genetic diversity but a low genetic divergence in Qinghai population. The molecular data on gene flow (Nm = 6.32) confirmed the migration of pathogen populations among areas in Qinghai Province. Liu et al. (2015) investigated 17 pathogenic populations in Sichuan and classified 109 isolates into high virulence (HV, 92 isolates) and low virulence (LV, 17 isolates) groups. Populations from Yibin (Southern region), Xichang (Western region), and Meishan (Middle region) showed lower virulence frequencies than populations from other regions. Many of the known resistance genes lost their resistance. Gene *Pm21* remained immune to all the PM populations in Sichuan whereas, genes *Pm13, Pm5b, Pm2+6*, and *PmXBD* maintained their resistance. AMOVA revealed significantly higher and lower levels of variation within and among regional populations. High levels of gene flow were detected among populations in the four regions. Closely related populations within each region were distinguished by cluster analyses using inter simple sequence repeat (ISSR) and sequence-related amplified polymorphism (SRAP) alleles and SRAP alleles were more informative. No significant association between these alleles and the virulence or pathogenicity was detected.

According to Lesovoi and Kol'Nobritskii (1980), genes *Pm1, Pm2, Pm3a, Pm3b, Pm3c, Pm4, Pm5*, and *M1r* were not effective against the predominant *Bgt* races in Ukraine and gene *Pm4* was highly resistant followed by genes *Pm2+Pm7* and *Pm2+Pm6*. Bogdanovich (2003) reported that in five regions of Ukraine, Bgt populations were virulent on gene *Pm5* and

Pm8 and Bgt populations in the vicinity of Kalush were more diverse with higher virulence. It was inferred that chemical pollution had influence on the virulence and race structure of PM populations (http://www.prague2003. fsu.edu/content/ pdf/544.pdf). Babayants et al. (2015) studied virulence frequency of 750 Bgt isolates in Ukraine using 23 *Pm*-genes and gene combinations during 2004–2013. They identified 78 previously known and 39 new pathotypes and 60–90% isolates were virulent on genes *Pm6*, *Pm8*, *Pm8+11*, *Pm2+4b+8*, *Pm3g*, *Pm10+15*, and *Pm10+14+15*, whereas, virulence frequency was variable for genes *Pm1a*, *Pm2*, *Pm3a*, *Pm3b*, *Pm3c*, *Pm5*, and *Pm7* and it varied from 1 to 8% on genes *Pm20*, *Pm37*, *Pm4a+*, and gene combinations *Pm3c+5a+35*. Breeding lines CN240/06, CN98/06, and CN158/06 derived from interspecific crosses were highly resistant.

Negulescu et al. (1978) reported that gene *Pm6* and *Pm4* was effective and partially effective, respectively, in Romania. In Switzerland, Streckeisen and Fried (1985) detected 43 pathotypes out of 162 isolates of Bgt using a set of 24 genotypes. No isolate could overcome the resistance of CWW 1645/5 and line 623/65. Sappo (*Pm2+Pm4b*) and Walter (*Pm2+Pm6*) lost their resistance. Mascher et al. (2012) monitored the pathogenic variation and virulence structure of Bgt populations in Switzerland following global analysis technique where the tester lines were directly planted in the field at 8–17 sites throughout the country during 2003–2010. It could enable to identify the upcoming of virulences during whole season. Little changes were observed among the dominating resistances and multiple virulences increased. The virulence structures of the populations changed over the years and sites which could be due to wheat varieties cultivated and environmental factors. Important micro evolutional processes have taken place in the wheat PM population over the last 30 years. There has been a considerable change in the race composition of the pathogen population and in the prevalent races. Of the 78 races identified, only 11 have "lived" for more than 15 years. Many races were only isolated in one or two years. The number of virulence genes rose from 2.03 in 1973 to 5.63 in 1993. Szunics et al. (2001) reported in Hungary that on the basis of race composition and virulence the wheat PM population between 1971 and 1999 could be divided into four distinct groups. A large proportion of the PM isolates are virulent to most of the resistance genes. The complete resistance was provided by gene *Pm4a* (Khapli) whereas, genes *Pm2 + Mld* (Halle st. 13471), *Pm4b+* (TP 315/2) and *Pm1+2+9* (Normandie) remained partially effective. The majority of cultivated varieties carry the resistance gene *Pm8* due to the presence of the 1B/1R translocation. Limpert et al. (1987) observed differences in virulence frequencies of single colony isolates of Bgt using differentials with genes

Pm2, Pm4b, Pm8, M1i, Pm2+Pm6, Pm4b+Pm5, and *Pm2+Pm4b+Pm8.*
Virulence frequency on genes *Pm2, Pm2+Pm6,* and *Pm4b* was higher in
England and decreased to east and it was reverse for gene *Pm8.* Virulence
was higher on gene *M1i.* Hovmoller (1987a) studied virulence structure
of Bgt in north west Europe using mobile nurseries and cvs Karka (*M1i*),
Disponent (*Pm8*) and Longbow (*Pm2*) were susceptible to 50% isolates and
Sleipner (*Pm2+Pm8+*) was highly resistant.

Stojanovic and Ponos (1990) analyzed 375 isolates of Bgt from 76 localities in South Eastern Yugoslavia, on *Pm* genes *Pm1, Pm2, Pm3a, Pm3b,
Pm3c,* and *Pm4a.* Thirty eight virulence gene combinations were detected
and majority of the isolates had 3–4 virulence genes. Isolates with virulence
formulae *Pm 2, 3b, 4a/1, 3a, 3c; 1, 2, 3b, 4a/pm3a, 3c,* and *1, 3b, 4b/2,
3a, 8* were the most common. Stojanovic et al. (1991) worked out virulence
pattern of 735 isolates during the year 1988–1989 in Serbia (Yugoslavia).
Virulence genes, *1, 2, 3a, 5, 8/4a,* and *1, 2, 3a, 3c, 5, 6, 8/3b; 5, 6, 8/3b,* and
4a were dominant in 1988 and 1989, respectively. Alleles, v-3b and v-4a,
were the least frequent. Most pathotypes had 5–8 virulent genes.

Vechet (2012) based on the observations from 1999 to 2010; at two locations in Czechoslovakia reported that virulence at crop research institute
(CRI) was higher to the gene *Pm1* whereas, at Humpolec it was higher on
gene *Pm17.* During the year 2010 in CRI, the virulence was the highest
on gene *Pm1* whereas, at Humpolec it was the highest on genes *Pm3f* and
Pm4a. Salari et al. (2003) reported that Bgt races 5, 11, 14, 19, 21, 24, 27,
28, 31, 32, 44, 50, 53, 58, 66, 73, and 84 were prevalent in Sistan province
of Iran during 1998–2001. These races were new to Iran and races 11, 53 and
73 were dominant. The isolates were compatible with genes *Pm8, Pm3a* and
Pm3b but were incompatible with gene *Pm4b.* Among 70 wheat cultivars,
cultivar Hirmand with probable gene *Pm4b* showed no symptoms under the
greenhouse and field conditions and cv. Chamran was highly resistant.

In India, the disease had been important in northern and southern hill
zone and foothills of the country. However, with the cultivation of post green
revolution high yielding, semi dwarf varieties responsive to high fertilization and extension of area under irrigation the disease has started causing
immense losses in Northern hills zone (NHZ) and North-western plains zone
(NWPZ). Moreover, maturation of cleistothecia from dry temperate zone of
Himachal Pradesh (Sharma et al., 1992) in ascospores and occurrence of the
disease throughout the year in NWPZ and NHZ (Basandrai et al., 1989) justified the virulence analysis in these epidemiologically important areas. Upadhyay and Kumar (1974) evaluated near-isogenic PM resistant lines having
genes *Pm1– Pm4* at Shimla and Wellington and *Pm2* (CI 12632 × Cc^8 and

Ulka × Cc[8]) was resistant at all the locations. In North India, strains were virulent on genes *Pm3a, Pm3b,* and *Pm3c* whereas, in Nilgiri hills on *Pm1, Pm3b,* and *Pm4* were prevalent. Singh and Sood (1977) grouped eight Bgt isolates collected from Kangra valley (Himachal Pradesh) as culture 1 (virulent on *Pm3b* and *Pm3c*) & 2 (virulent on *Pm3b, Pm3c,* and *Pm4*). Saharan et al. (1981) reported that Bgt collections from Himachal Pradesh were virulent on genes *Pm3a* and *Pm3c*. Virulence was abundant on gene *Pm2* and was absent on *Pm1* and *Pm4*. Sharma and Singh (1990a) studied the virulence structure of Bgt populations with mobile nurseries. Virulence was most common on genes *Pm (Ma)* and *Pm7* whereas, it was rare on *Pm1* and no virulence was trapped on genes *Pm2, Pm4,* and *Pm6*. Sharma et al. (1990) classified 20 single spore isolates of Bgt from Punjab and Himachal Pradesh into nine virulence combinations based on their reaction on lines *Pm1-Pm4* and *Pm (Ma),* and genes *Pm5, Pm6, Pm7,* and *Pm8*. All the pathotypes were virulent on genes *Pm3c, Pm5,* and *Pm (Ma)* and avirulent on genes *Pm1* and *Pm4a*. Sharma and Singh (1990b) identified 14 and 5 races from 37 conidial, 12 ascosporic isolates, respectively, using the differential set of Sharma et al. (1990), excluding line having gene *Pm5*. In general, races from conidia were more virulent than races from ascospores. Using the same differential set, Pathania et al. (1996) identified 26 and 14 pathotypes among 63 conidial and 21 ascosporic isolates collected from different agro climatic areas of Himachal Pradesh. Pathotype, 23 and 11, with virulence on eight *Pm* genes and *Pm (Ma)* were the most and least virulent, respectively. Gene *Pm1* was effective against all pathotypes, whereas virulence on genes *Pm2* and *Pm6* was rare. Paul et al. (2000) studied the virulence spectrum of 45 conidial and 11 ascosporic single colony isolates collected from different agroclimatic zones of Himachal Pradesh. The isolates were grouped into 29 virulence combinations (pathotypes). Pathotypes with intermediate virulence (on 3–6 genes) were more frequent. Virulence combination *Pm1,2 4a,2+6, 2+4b, 4b+8/Pm 3a, 3b, 3c, 5, 6, 8 (Ma)* was the most virulent and virulence combination *Pm1,2,3a,3b,3c,4a,5,6,7,8,2+6,2+4b,4b+8/Pm (Ma)* was the least virulent. The latter virulence along with virulence combination *Pm 1, 2, 3b, 4a, 2+6, 2+4b, 4b+8IPm30, 3c, 5, 6, 7, 8, Pm(Ma)* were the most prevalent ones. Genes *Pm2, Pm4a, Pm2+4b,* and *Pm4b+8* were effective against all the pathotypes of *E. graminis tritici* in the state. Virulence was rare on genes *Pm1* and *Pm 2+6*. Basandrai et al. (2007) studied the virulence structure of Bgt in HP during 1991 and 1992 using mobile nurseries. Virulence was frequent on *Pm3a, Pm3c, Pm5, Pm7, Pm8,* and *Pm(Ma)* and no virulence was detected on gene *Pm4a* and it was rare on genes *Pm1, Pm2,* and *Pm6*. Combination of gene *Pm1* with *Pm2, Pm3a, Pm3b, Pm3c, Pm4a,*

Pm5, Pm6, Pm7, Pm8, and *Pm(Ma)* and genes *Pm2* with *Pm4a, Pm2,* with *Pm6, Pm2* with *Pm8, Pm4a* with *Pm8* and *Pm6* with *Pm8* were expected to be effective against prevailing virulences in the State.

In India, the pathogenic variation was studied during early 1970s, 1980s, and 1990s. Thereafter, no systematic work has been undertaken. The virulence on genes, *Pm3a, Pm3b, Pm6, Pm7,* and *Pm(Ma)* was more prevalent during early eighties. The virulence on gene *Pm8* increased from 2% (Sharma & Singh, 1990; Sharma et al., 1990) during 1980s to more than 50% during the 1990s (Anonymous, 1998; Paul et al., 2000; Basandrai et al., 2003, 2007). Gene *Pm8* is linked with gene *Lr26 /Yr9* for leaf rust and YR resistance. Indian wheat program depended heavily on leaf rust and YR resistant genes *Lr26* and *Yr,* respectively. Varieties viz., PBW 343, WH 542, HPW 184, HPW 42, HS 240, HS 277, HUW 234 etc., were extensively grown over a large acreage throughout the country during the 1990s which resulted in selection pressure for corresponding virulences of leaf rust and YR and Bgt. Hence, there is an urgent need to undertake virulence analysis of Bgt in epidemiologically important areas.

5.6 HOST–PATHOGEN INTERACTION

Components of race specific incompatible reaction to Bgt have been widely investigated in wheat. In general, resistance was attributed to the morphological characters of the host, that is, number of surface hairs, cutical thickness, papilla deposition epidermal cell wall thickness and inability of fungus to penetrate host cells, haustorial abortion or restricted haustorial development in host epidermal cell, necrosis of tissues surrounding or underlying infected cell or cells and suppression of sporulation.

Lupton (1956) reported that conidia of *Bgt* germinated and produced appressoria but did not penetrate the epidermal wall on leaves of *T. carthlicum, T. dicoccum* var. *farrum* and *T. dicoccum* s. sp. *georgicum, T. timopheevii,* and *Aegilops caudata* but penetration was normal in *A. ovata.* In *A. speltoides,* the penetrating hyphae burst on entering the host cell and killed it. The penetration of susceptible, resistant, and highly resistant cultivars was 65, 45, and 27% respectively during first 36 h (Tao et al., 1982). Irregular appressoria were unable to invade host cells in wheat varieties having gene *Pm4* (Yao et al., 1998). Ghemawat (1969, 1979) reported that papillae formation hindered the penetration of the fungus and it became more effective with increasing age. Tosa et al. (1985) reported that unsuccessful conidial penetration was associated with papillae formation and increased resistance resulted in decreased normal primary haustorial formation and spores were unable to penetrate the cell wall and induced the hypersensitive response in resistant lines. According to Wen and Tao (1989), papillae induction in resistant lines ceased the conidia to produce appressoria. *T. timopheevii* exhibited papillae mediated resistance and epidermal hypersensitivity (Nashaat & Moore, 1991) and epidermal hypersensitivity was also exhibited by *T. carthlicum*. Hu et al. (1998) suggested that papilla-associated resistance depended on the time of initiation and speed of papilla formation.

Degeneration and death of haustoria and mycelium, three or four days after normal conidial penetration was reported in *A. ovata* (Lupton, 1956). Haustorial abortion leading to hypersensitive reaction (Ghemawat, 1969) and abnormal development of primary germ tube penetrations (Ghemawat, 1979) were reported in young leaves. Haywood and Ellingboe (1979) reported that in incompatible genotypes, viz., *P1a/Pm1a*, *P2a/Pm2a*, *P3a/Pm3a*, and *P4a/Pm4a* haustorial formation were 15, 66, 18, and 3%, respectively. More than 75% host cells showing successful penetrations were heavily stained indicating mesophyll collapse and necrogenic protoplasts. In compatible *Px/pmx* genotypes, 87% parasitic units formed 35–55 μm long haustoria within 30 h after inoculation. Delayed and smaller haustorial formation was observed in resistant cultivars (Tao et al., 1982; Basandrai et al., 2011). Wu et al. (1985) indicated that resistance was conferred by delay in early haustorial development, relatively slow increase in rate of haustorial formation, reduction in numbers and final size of haustoria and relatively more rapid distortion and disintegration of haustoria. Components of race- specific incompatible reaction to *E. graminis* f. sp. *tritici* (Egt) have been widely investigated in wheat (Hyde & Colhoun 1975; Nashaat & Moore, 1991). Basandrai et al. (2011) studied the infection process of Bgt pathotype 4 on seedlings of Pm lines and two commercially grown cvs. in comparison with universal susceptible cv. Agra Local. In incompatible host parasite interactions, like in Nora × Cc8 (*Pm1+4b+9*), conidial germination and appressorium formation were less. Germtubes appeared weak, shrivelled and distorted in Sappo (*Pm 2+4b*), and Kronjuwel (*Pm4b+8*). Haustorial abortion, less secondary haustorium production, and restricted growth of mycelial wefts, with no subsequent conidiophore formation were observed in TP 114 (*Pm 2+6*). In Norka × Cc8, Sappo and Kronjuwel, development of haustorium, and elongated secondary hypha was completely absent. In compatible host parasite combinations, the sequence of infection events was similar to the susceptible check, however incubation period and latent period were longer in Amigo (*Pm17*) and cvs., HS 240 and HS 295. In CS/Hope (*Pm5*), Kavkaz (*Pm8*), Amigo, and HS 295 colony size and colony number per unit area and sporulation index were reduced, whereas in Transec (*Pm7*) colony size and in HS 240 sporulation index were at par with the susceptible check. Sporulation capacity was also reduced in all the genotypes.

Hyde (1976) observed that the formation of elongated secondary hyphae (ESH), rate of colony growth and conidial production was slower on flag leaves than on the seedling leaves. Slesinski and Ellingboe (1969) hypothesized that a lower percentage of ESH was formed. The *P3/Pm3* genotypes (except *P3c/Pm3c*) inhibited only 60–70% of the applied parasitic units but *P2/Pm2* genotypes did not alter the efficiency of ESH production. Iliev (1988) observed less number of ESH and stronger inhibition of functional secondary hyphae was stronger in the resistant cultivars than the susceptible variety. Tao et al. (1982) reported that fewer and less densely sporulating colonies developed on resistant plants. It was also reported by Nashaat and Moore (1991), in *T. timopheevii* and CI 12644.

5.7   DISEASE MANAGEMENT STRATEGIES

5.7.1   HOST RESISTANCE

Reed (1912) reported immunity to Bgt in emmer wheat (*T. dicoccum* Schubl) varieties. Briggle and Scharen (1961) reported that wheat lines/varieties showing seedling susceptibility were resistant after jointing stage. Van Silfhout and Gerechter-Amitai (1988) discussed four different types of resistances, namely, true-seedling resistance, adult plant resistance (APR), overall resistance, and partial resistance to PM.

5.7.1.1   SEEDLING RESISTANCE

Varieties Klipkous and Wolkoring were immune whereas, Hope, Kleintrou, Marina, Regent, Rooi Wol, *T. durum,* and Wolbaard were highly resistant to Bgt (Moseman 1966). Moseman et al. (1984) reported that 49% of wild emmer was resistant to Bgt at seedling stage. Gerechter-Amitai and Van Silfhout (1984) evaluated 47 wild emmer wheat entries in six nurseries and discerned 11 markedly different reaction patterns indicating the presence of several different resistance genes. Liang et al. (1985) reported that out of 352 accessions cultivars Maris Huntsman (*Pm2, Pm6*), VPM × Mission 9.9.12.4.28.11, Yanxiaohei 2-2, and Qianhuan 2 were highly resistant while Ulka × Cc^8 (*Pm2*) was resistant. Suryanarayana et al. (1971) reported powdery mildew resistance in vars. Sharbati Sonora, Sonora-64, Chhoti Lerma, and Nadadores. Singh and Sood (1977) reported that four winter and 26 spring wheat cultivars were resistant to PM. Sharma and Singh (1990) reported that CPAN 1922 was resistant to race nine at seedling stage. Thirty-nine entries were resistant or moderately resistant to two *Bgt* isolates. Garcia-Sampedro et al. (1996) found seedling resistance in var Buck Charrna might have derived from the winter wheat Lovrin.

5.7.1.2 ADULT PLANT RESISTANCE

Newton and Cherwick (1947) reported that most of the varieties tested were more resistant at the adult than at the seedling stage. Smith and Blair (1950) inferred that resistance which retarded infection and growth of *Bgt* at adult plant stage was termed as APR. Resistance to *Bgt* in cultivars *Pm3a*, Hope, Michigan Amber, Indiana, Sonora, and gene *Pm6* was best expressed after three leaf stage (Briggle 1969; Moseman, 1966). Wheat varieties, viz., Axminister, Norka, Chul, Dixon, Huron, Erivan, Red Fern, Sonora, and Michigan Amber, several selections of Illinois no.1, Normandie, Russian selection, Hope, Regent, Renown selection RL 716.6, Thew, Kenya 744, Transec, Kavkaz, and Aurora (Goel et al., 1986; Moseman, 1966; Driscoll & Jensen, 1965) were resistant at adult plant stage. Bennett (1981) found that genotypes Flinor, Avalon, and Bounty were more resistant on leaves whereas, Iona, Wizard and Bounty were more resistant for ear infection. From time to time different workers have studied APR in different varieties of spring wheats (Goel et al., 1986; Moseman et al., 1984; Bennett, 1984; Bahadur & Aggarwal, 1997; Tomar & Menon, 2001; Alam et al., 2011), summer wheats (Bennett, 1984; Alam et al., 2011), winter wheats (Nass et al., 1995; Misic et al., 1997), and *T. durum* (Moseman, 1966; Moseman et al., 1984; Tomerlin et al., 1984; Casulli & Siniscalco, 1986).

From time to time, a large number of genotypes, that is, Weihenstephan M1, Zlatna Dolina, Sava, Sanja, Biserka, TP 114/65A, Maris Huntsman, Transec, Atlas 66, Klien Puntal, Gerardo, Valgerardo, Baccum, No. 28331, Kuban' 8, Sappo and Weibulls 121128, Erythrospermum 116180, Erythro-spermum 583/83, Lutescens 1276182, McNair 1789, CI 12633, ZG 2444/72, Idaed 59b, Arthur 71, Halle Stamm 1347, PI 170911, Fakir, Kenya Civet, Roazon, Sivka, Avalon, Purdue, Ave, Khapli/8 Chancellor, Peacock, ST 104-87, ST 204, SO 2392, Armur, NC97BGTD 7 and NC97BGTD 8 were documented as resistant sources for their utilization in breeding programs (Moseman, 1966; Negulescu et al., 1978; Babayants & Smilyanetes, 1991;

Murphy et al., 1999). Schneider and Heun (1988) reported that the diploid *Triticum* species were valuable source of qualitative and quantitative resistance genes and could be used to improve resistance in commercially cultivated wheats. New sources of resistance were developed and identified using resistant genotypes of *T. monococcum, T. boeticum, T. carthlicum, T. zhukovski* and *T. fungicidum, T. miguschovae, T. timopheevii, T. persicum, T. araraticum, Aegilop* species, *Hordeum bulbosum, Lolium multiflorum, Secale cereale,* and *H. chilense* (Tomerlin et al., 1984; Bennett, 1984; Tomar & Menon, 2001; Alam et al., 2011; Mwale et al., 2014).

Sources with APR to PM have been reported in India (Goel et al., 1986; Singh et al., 2005, 2016; Basandrai & Sharma, 1990). Varieties HS 1138-6-4, HB 117-107, Timgalin, and CPAN 718 were promising resistant sources (Upadhyay & Kumar, 1974). Resistant genotypes were identified by Singh et al. (2005). Basandrai et al. (1989) reported that 35 genotypes, *T. aestivum* (14), *T. durum* (4), *T. dicoccum* (1), *T. dicoccoides* (2), and triticale (11) were resistant to *Bgt* at three hot spot locations, that is, Dhaulakuan, Gurdaspur, and Palampur. Genotypes HD 2329, PBW 154, and SKML1 (Basandrai et al., 1991), CPAN 1922, CPAN 1676, PBW 34, and HB 208 (Sharma & Singh, 1990c), HD 2189 (Sharma et al., 1991a) showed adult plant resistance. Bahadur and Aggarwal (1997) identified sources of resistance to PM based their field reaction at hot spot locations.

Since the chances of losing resistance in varieties having single gene are more, commercial varieties with different gene combinations were evolved to prolong the life span of the resistant cultivars against mildew. Combination *Pm2+ Pm6* is the most widely used (Xiong et al., 1983; Li & Huang, 1991; Sharma & Singh, 1990b; Huszar, 1992; Kowalczyk et al., 1998). Gene combinations *Pm1+Pm2, Pm1+Pm6, Pm2+Pm4, Pm2+Pm8, Pm4+Pm6, Pm4+Pm8,* and *Pm6+ Pm7+ Pm8* (Sharma & Singh, 1990b) and *Pm1+Pm2, Pm1+Pm6, Pm1+Pm5, Pm1+Pm3c, Pm1+Pm7,* and *Pm1+Pm8* (Pathania et al., 1996) could confer complete resistance against the prevalent virulent strains of Bgt in Himachal Pradesh (India) and may be exploited for the effective management of PM. Sharma et al. (1991b) showed that gene combinations *Pm3c+Pm8, Pm3c+Pm (Ma),* and *Pm8+Pm (Ma)* reduced spores per colony and genes, viz., *Pm3a, Pm3b, Pm3c, Pm8,* and *Pm (Ma)* in various combinations increased the latent period.

Resistance at both the seedling and adult sage is often referred to as "overall" resistance (Zadoks, 1961). Sharma et al. (1979) reported that cvs. Vernal, Khapli, Thew, NP 200, TDW 1656, W 1656, V 156, and CC 422 showed overall resistance. Similarly, genotypes TL 365a/37, 667-7 and Maris Huntsman (Vershinina et al., 1984), Morroqui 588, Thew, CPAN 1922,

FRP 873, Arora, ISWYN 29, ISWYT 11, SWM 4123, and near-isogenic lines Axminster x Cc8, Yuma x Cc8, Khapli x Cc8 (Basandrai et al., 1991) *Pm1, Pm2, Pm4* genes and cultivars Thew, PBW 91, Yc1, Yc19 (Sharma et al., 1991a) possessed resistance throughout the growing season. Basandrai and Sharma (1990) indicated that 57 genotypes comprising accessions of *T. aestivum* (37), *T. durum* (4), *T. dicoccoides* (2), *T. dicoccum* (1) and triticale (13) were resistant at both the stages.

Combined effect of reduced colony formation and reduced spore producing capacity indicated slow spread of disease in cvs. Knox (Shaner, 1973; Shaner & Finney, 1977), Genesee (Ellingboe, 1975, 1976) and Red Coat (Rouse et al., 1980). Parleviet (1975) observed that plants with high infection type at all growth stages but with lower disease severity were having "partial" resistance, or "slow mildewing." Near-isogenic lines carrying *Pmx* genes showed partial resistance relative to Chancellor (Royer et al., 1984). Low infection rate of *Bgt* was observed on var HB 208 (Kapoor, 1990), CPAN 1676, HD 2204, and VL 421 (Sharma & Singh, 1990d) cv. HB 208 had the lowest disease severity, sporulation index and colony size (Sharma & Singh, 1990a) followed by CPAN 1922, PBW 34, CPAN 1676, HD 2285, and HD 2204 in comparison to Kalyansona and HS 86. Lines with gene *Pm3b* showed longer latent period and lines *Pm3a, Pm3b, Pm3c, Pm8,* and *Pm (Ma)* produced fewer colonies and fewer spores per colony (Sharma et al., 1991a).

Under All India Coordinated Research Project (AICRP) on wheat in India, stable sources of resistance have been identified based on mutilocational and multi-year testing and given below:

> *T. aestivum:* HPW 347, VL 930, HPW 317, VL 931, VL 943, VL 944, HUW 635, PBW 615, DBW 62, HS 522, HUW 629, KRL 250, NW 4091, PBW 635, HS 533, UAS 327, DBW 58, HI 1653, HI 1569, RSP 561, HS 534, HD 3013, HD 3048, HD 3052, DBW 52.

> *Triticum durum:* HI 8692, HI 8702, HI 8709, MACS 3742, MACS 3744, NIDW 577, HI 8708, UAS 432, PDW 315, PDW 317.

> Triticale: TL 2963.

> *T. dicoccum:* DDK 1037, DDK 1039.
> (http://hau.ernet.in/ research/pdf/lsharma.pdf)

Singh et al. (2016) evaluated 963 genotypes during cropping season from 2004–2005 to 2011–2012 in India and genotypes HD 2189, HI 977, HPW 236, HW 1095, HW 5013, MACS 1967, NW-RF-11, VL 858 (*T. aestivum*), NIDW 295 (*T. durum*), DDK 1009, DDK 1025, DDK 1029, SPL-DIC-02,

SPL-DIC-03 (*T. dicoccum*), and TL 2934, TL 2942, TL 2947 and TL 2949 (Triticale) with disease reaction "0-3" were resistant whereas, 57 genotypes showed differential response at various hot spot locations.

In India, most of the commercially grown varieties and advanced breeding material is susceptible. The disease appears late in the season, hence, APR with slow mildewing attribute may be deployed in disease prone areas.

5.7.1.3 POSTULATION OF RESISTANCE GENES

Gene-for-gene concept (Flor, 1942) has led to the discovery of infection-type matching technique which has been used by a number of workers in wheat *B. graminis tritici* system to postulate resistance genes. Pugsley and Carter (1953), without knowing the concept of infection-type-matching technique, grouped 12 wheat varieties into three groups based on their reaction with races P and P-1 of Bgt. They used this technique to identity six different PM resistant genes, designated as M1a, M1b, M1e, M1s, M1t, and M1u.

Wolfe (1965) grouped 26 wheat lines into 10 different groups, which formed the basis for the identification of races of PM in United Kingdom and other European countries. Based on the infection-type matching technique, Sebastian et al. (1983) attributed the resistance in lines IL-72-2919-1 and Va-66-54-10 to gene *Pm3a*. Moseman et al. (1984) evaluated PM resistance in 233 accessions of *T. dicoccoides* based on their reaction to four Bgt cultures having virulences corresponding to most of the known resistance genes. Race specific PM resistance genes were identified in 23 spring and 59 winter wheats having diverse virulence pattern, using 19 and 11 isolates of Bgt, respectively (Heun & Fischbeck, 1987a, 1987b). Most of the varieties carried single resistance gene whereas, resistance in some varieties was governed by combination of two and three genes. Similarly, race specific resistance genes were identified in soft red winter wheats in United States (Leath & Heun, 1990).

Postulation of race specific resistance genes in advanced Indian breeding material revealed the presence of genes *Pm3c, Pm5,* and *Pm8* individually or in combination with other known or unknown gene (s) (Basandrai et al., 2012, 2013b, c), in exotic and winter wheats (Anonymous, 1998) and in International Maize and Wheat Improvement Center (CIMMYT) material (Pathania et al., 1998; Paul et al., 1999). Presence of genes was confirmed by pedigree relationship and presence of gene *Pm8* was further confirmed by postulation of leaf rust resistance gene *Lr26/Yr9* (Nayar et al., 2001; Bhardwaj, 2011) which is closely linked with gene *Pm8*.

5.7.1.4   GENETICS OF RESISTANCE

5.7.1.5   DESIGNATED AND CATALOGUED GENES

Bgt undergoes continuous mutation leading to evolution of new and more virulent strains. Thus breeding for Bgt resistance using identified resistant sources and wild species has become a continuous process resulting in development of PM resistant or tolerant cultivars. Subsequently, resistant genes have been located and mapped on wheat chromosomes. Over 70 PM resistance genes or alleles have been designated on 54 loci (*Pm1–Pm54*). Six loci *Pm1 (Pm1a* to *Pm1e), Pm3 (Pm3a* to *Pm3r), Pm4 (Pm4a* to *Pm4d), Pm5 (Pm5a* to *Pm5e), Pm8/Pm17,* and *Pm24 (Pm24a* and *Pm24b)* have multiple resistant alleles (Table 5.1). Thirty-four designated PM resistance genes have been derived from *T. aestivum* and rest of the genes originated from species either closely related to or distant relatives of common wheat. The genes *Pm1b, Pm4d* and *Pm18 (T. monococcum), Pm4a (T. dicoccum), Pm4b* and *Pm33 (T. carthlicum), Pm5a, Pm5b (Mli), Pm5c, Pm5d* and *Pm5e* in cvs. Hope, Ibis, *T. sphaerococcum* var. *rotundatum,* CI 10904, and Fuzhuang 30, respectively, were identified and mapped on wheat chromosome 7BL. Genes *Pm16, Pm30, Pm31, Pm36, Pm41, Pm42, Pm49,* and *Pm50 were* located on chromosomes, 4A, 5BS, 6AL, 5BL, 3BS, 2BL, 2BS and 2AL, respectively, whereas genes *Pm26, Pm42, and Pm49 are* located on chromosome 2BS. Genes *Pm6, Pm27, Pm37,* and *Pm33* originated from chromosome 2G of *T. timopheevii* and were introgressed into chromosome 2BL of common wheat (Tao et al., 2000). *T. carthlicum* included genes *Pm4b* and *Pm33.* Genes *Pm12, Pm13, Pm19, Pm29, Pm32, Pm34,* and *Pm35* were transferred from *Aegilops* whereas, genes *Pm7, Pm8, Pm17,* and *Pm20* originated from *Secale.* Two genes *Pm40* and *Pm43* were introgressed into common wheat from *Elytrigia intermedium* while gene *Pm21* originated from *Haynaldia villosum* (Mwale et al., 2014). Different genes were derived from different T. spp. and allied genera viz., Genes *Pm 12, Pm 53 (A. speltoides), Pm13 (A. longissum), Pm19, Pm35 (A. tauschii),* genes *Pm21 (Pm31), Pm40,* and *Pm53* from *H. villosum, E. intermedium,* and *Thinopyrum paniticum,* respectively (Table 5.1).

5.7.1.6   TEMPORARILY DESIGNATED RESISTANCE GENES

Some of the recessive and dominant genes for PM resistance are located on the same chromosomes but are different from known resistant sources

TABLE 5.1 Designated Names, Chromosomal Positions, Cultivar/Line and Sources of Identified Resistance Genes to Powdery Mildew in Wheat.

Gene	Position	Cultivar/line	Source
Pm1a	7AL	Axminster	*T. aestivum*
Pm1b	7AL	MoczZlatka	*T. monococcum*
Pm1c/Pm18	*7AL*	Weihestephan M1N	*T. aestivum*
Pm1d	7AL	*Triticum spelta* var. *duhamelianum*	*T. spelta*
Pm1e (Pm22)	7AL	Virest	*T. aestivum*
Pm2	5DS	Ulka/XX 194	*T. aestivum/ Aegilops tauschii*
Pm3a	1AS	Asosan	*T. aestivum*
Pm3b	1AS	Chul	*T. aestivum*
Pm3c	1AS	Sonora	*T. aestivum*
Pm3d	1AS	Kolibri	*T. aestivum*
Pm3e	1AS	W150	*T. aestivum*
Pm3f	1AS	Michigan Amber	*T. aestivum*
Pm3g	1AS	Aristide	*T. aestivum*
Pm3h	1AS	Abessi	*T. durum*
Pm3i	1AS	N324	*T. aestivum*
Pm3j	1AS	GUS 122	*T. aestivum*
Pm3k	1AS	IG46439	*T. dicoccoides*
Pm4a	2AL	Khapli	*T. dicoccum*
Pm4b	2AL	Armada	*T. carthlicum*
Pm4c (Pm23)	2AL	81-7241	*T. aestivum*
Pm4d	2AL	T27d2	*T. monococcum*
Pm5	7BL	Xiaobaidong	*T. aestivum*
Pm5a	7BL	Hope	*T. dicoccum*
Pm5b	7BL	Ibis	*T. aestivum*
Pm5c	7BL	Kolandi	*T. aestivum* ssp. *T. aestivum* ssp. *T. sphaerococcum*
Pm5d	7BL	IGV 1-455	*T. aestivum*
Pm5e	7BL	Fuzhuang 30	*T. aestivum*
Pm6	2BL	TP 114	*T. timopheevii*
Pm7	4BS.4BL-2RL	Transec	*S. cereale*
Pm8	1RS.1BL	Disponent	*S. cereale*
Pm9	7AL	N14	*T. aestivum*
Pm10	1D	Norin 26	*T. aestivum*

TABLE 5.1   *(Continued)*

Gene	Position	Cultivar/line	Source
Pm11	6BS	Chinese spring	*T. aestivum*
Pm12	6BS-6SS.6SL	Trans. Line 31	*A. speltoides*
Pm13	3BL.3SS-3S 3DL.3SS-3S	C strans. Line	*Aegilops longissima*
Pm14	6BS	Norin 10	*T. aestivum*
Pm15	7DS	Norin 26	*T. aestivum*
Pm16	4A.	Norman rec. line	*T. dicoccoides*
Pm17	1RS.1AL	Amigo	*S. cereale*
Pm19	7D	XX 186	*A. tauschii*
Pm20	6BS.6RL	KS93WGRC28	*S. cereale*
Pm21(Pm31)	6VS.6AL	Yangmai 5 line	*Haynaldia villosa*
Pm23(Pm4c)	2AL	82-7241	*T. aestivum*
Pm24a	1DS	Chiyacao	*T. aestivum*
Pm24b (mlbhl)	1DS	Baihulu	*T. aestivum*
Pm25	1A	NC96BGTA5	*T. boeoticum*
Pm26	2BS	TTD140	*T. dicoccoides*
Pm27	6B-6G	146-155-T	*T. timopheevii*
Pm28	1B	Meri	*T. aestivum*
Pm29	7DL	Pova	*A. ovate*
Pm30	5BS	C20	*T. dicoccoides*
Pm31 (Pm21)	6AL	G-305-M/781	*T. dicoccoides*
Pm32	1BL.1SS	L501	*A. speltoides*
Pm33	2BL	PS5	*T. carthlicum*
Pm34	5DL	NC97BGTD7	*A. tauschii*
Pm35	5DL	NC96BGTD3	*A. tauschii*
Pm36	5BL	MG29896	*T. dicoccoides*
Pm37	7AL	NC99BGTAG11	*T. timopheevii*
Pm38	7DS	RL6058	*T. aestivum*
Pm39	1BL	Saar	*T. aestivum*
Pm40	7BS	GRY19	*E. intermedium*
Pm41	3BL	IW2	*T. dicoccoides*
Pm42	2BS	G-303-1M	*T. dicoccoides*
Pm43	2DL	CH5025	*Thinopyrum intermedium*
Pm44	3AS	Hombar	*T. aestivum*
Pm45	6DS	D57	*T. aestivum*

TABLE 5.1   *(Continued)*

Gene	Position	Cultivar/line	Source
Pm46	5DS	Tabasco	*T. aestivum*
Pm47	7BS	Hongyanglazi	*T. aestivum*
Pm49 (Ml5323)	2BS	MG5323	*T. dicoccum*
Pm50	2AL	K2	*T. dicoccum*
Pm 51	2BL	CH 7086	Putative *Thinopyrum ponticum*
			derivative
Pm 52	2BL	Liangxing 99	–
Pm 53	5BL	NC09BGTS16, PI669386 = Saluda*3/ TAU829	*A. speltoides*
Pm 54	6BL	Pioneer (®) variety 26R61 (shortened as 26R61	*T. aestivum*
		NAU421	
Pm 55	5AS		*Dasypyrum villosum*

Information on genes is per the paper of Mwale et al. (2014), Zhang et al. (2016), McIntosh et al. (2016), Hao et al. (2015), Zhan et al. (2014), Yin et al. (2014), Song et al. (2015), Peterson et al. (2014), Sharma (2012), Briggle (1969), Hsam et al. (1998), Singrun et al. (2003), Lutz et al. (1995), Briggle & Scharen (1961).

and wild relatives of wheat. Nine PM resistance genes have been mapped on chromosome 2B and four of these genes *MlZec1, PmY39, MlAB10,* and *PmPs5B*, carry temporarily designated names. Chhuneja et al. (2012) reported the presence of two temporarily designated genes *PmTb7A.1* and *PmTb7A.2* in accession PAU 5088 which were located and mapped on chromosome 7AL where 10 other temporary designated genes are also mapped. The source of resistance was from a diploid AbAb genome of *T. boeoticum* combines freely with the A genome of wheat. "Mzalenod Beer" (hexaploid triticale)/"Baofeng 7228"//"90 Xuanxi," carry recessive resistance gene, *PmLK906* and *TaAetPR5* which were located on chromosome 2AL. These appeared different from the known dominant alleles found on *Pm4* locus located on the same chromosome. Additionally, PM resistant genes/alleles located on the chromosome 2A include *PmDR147* in *T. durum* line DR147, *PmPS5A* from *T. carthlicum*. Recently, genes Mllw170, MLNCD1, *PmAS846, PmTm4, Pm, pmCH 89, Pm Tb7A1, PmTb7A.2, PmL962, MlHLT, Pm2b (Km 2939), MlUM 15MlHUBE, PmPB3558*, and *PmYB* etc. have also been identified (Table 5.2).

TABLE 5.2 Chromosomal Position, Cultivars/Lines, and Sources, of the Temporarily Designated Powdery Mildew Resistance Genes in Wheat.

Gene	Position	Cultivar/line	Source
Mld	4B	Maris Dove	*T. durum*
Ml-Ad	-	Adlungs Alemannen	*T. aestivum*
Ml-Br	-	Bretonischer	*T. aestivum*
Ml-Ga	-	Garnet	*T. aestivum*
MlRE	6AL	RE714	*T. aestivum*
Mljy	7B	Jieyan 94-1-1	*T. aestivum*
Mlsy	7B	Siyan 94-2-1	*T. aestivum*
mlRD30	7AL	RD 30	*T. aestivum*
PmDR147	2AL	DR 147	*T. durum*
MlZec1	2BL	Zecoi 1	*T. dicoccoides*
PmPs5A	2AL	Am 4	*T. carthlicum*
PmPs5B	2BL	Am9/3*Laizhou953	*T. carthlicum*
PmE	2AL	Xiaohan/4*Bainong3217	*E. orientale*
PmP		Xiaobing/3*Bainong 3217 Fuco/ Agropyron	*Agropyron* spp.
PmY39	2U(2B)	953*4/Am9	*Aegilops umbellulata*
PmH	7BL	Hongquanmang	*T. aestivum*
PmY150	6B/6S	*Ae. longissima*/ 3*Laizhou 953	*A. longissima*
PmM53	5DL	M53	*A. tauschii*
PmU	7AL	UR206/Laizhou	*Triticum urartu*
PmY201	5DL	Y201	*A. tauschii*
PmY212	5DL	Y212	*A. tauschii*
Mlm2033	7AL	TA2033	*T. monococcum*
Mlm80	7AL	M80	*T. monococcum*
PmE	7BS	TAI7047	*E. intermedium*
PmYU25	2DL	TAI7047	*E. intermedium*
PmAS846	5BL	N9134	*T. dicoccoides*
PmAeY2	C5DL	Y189	*A. tauschii*
PmY39-2	6AS	N9628-2	*A. umbellulata*
Pm2026	5AL	TA2026	*T. monococcum*
mlIW72	7AL	IW72	*T. dicoccoides*
PmTm4	7BL	Tangmai 4	*T. aestivum*
PmLK906	2AL	Lankao 90(6)	*T. aestivum*
PmYm66	2AL	Yumai 66	*T. aestivum*

TABLE 5.2   *(Continued)*

Gene	Position	Cultivar/line	Source
MlWE18	7AL	3D249	*T. dicoccoides*
MlAG12	7AL	NC06BGTAG12/ Jagger	*T. aestivum*
MlWE29	5BL	WE29	*T. dicoccoides*
TaAetPR5	2AL	EU082094	*A. tauschii*
Ml3D232	5BL	3D232	*T. dicoccoides*
MlAB10	2BL	NC97BGTAB10	*T. aestivum*
MlIw170			
PmG16	7AL	G18-16	*T. dicoccoides*
PmHNK	3BL	Zhoumai 22	*T. aestivum*
Pm07J126– 07		Jian126	*T. aestivum*
PmAs846	5BL	N9134 & N9738	*T. dicoccoides*
PmTb7A.1	7AL	pau5088	*T. boeoticum*
PmTb7A.2	7AL	pau5088	*T. boeoticum*
PmLX66	5DS	Liangxing 66	*T. aestivum*
PmG25	5BL	N0308	*T. dicoccoides*
PmZB90	2AL	ZB90	*T. aestivum*
MlIw170	2BS	IW170	*T. dicoccoides*
MLNCD1	7DL	NC96BGD1 PI 597348 = Sa-luda*3 /TA2570	*T. aestivum*
PmAS846	5BL	N9134 and N9738	*T. diccocoides*
PmTm4	7BL	Tangmai 4	*T. aestivum*
Pmx	2AL	Xiaohongpi	*T. aestivum*
MlLX99	2BL	Liangxing 99	*T. aestivum*
pmCH 89	4BL		*T. intermedium*
Pm Tb7A1	7AL		*T. boeticum*
PmTb7A.2	7AL		*T. boeticum*
PmL962	2BS	L 962	*T. intermedium*
MlHLT	1DS	*HULUTOU*	Chinese wheat land race
Pm 2b (Km 2939)	5DS	Km 2939	*Agropyrum christatum*
PmWFJ	5DS	Wanfengjian 34	*T. aestivum*
MlUM 15		*NC09BGTUM 15 (NC-UM15)*	
MlHUBEL	2D	*Hubel*	*T. speltoid* wheat
PmPB3558	5DS	PB 3558	*Agropyron cristatum*

TABLE 5.2 *(Continued)*

Gene	Position	Cultivar/line	Source
PmYB	5DS	-	-
Pm5055	2B	N0324	*T. dicoccoides*
			Aegilops tauschii
PmT	7D	Synthetic 43	
PML10103	6B	Shannong	*Aegillops squarrosa*
PmSe5785	2DL	SE 785	
Pm5VS	5DS	NAU415	*Dasypyrum villosum*

Information up to genes *PmZB 90* (Yi et al., *2013)* is per the paper of Mwale et al. (2014).

5.7.1.7 MOLECULAR MARKERS LINKED TO POWDERY MILDEW RESISTANCE GENES

Molecular markers are the tools which help to locate and identify parts of DNA positioned near a gene or genes of interest (Alam et al., 2011). Molecular identification of specific DNA sequences can be used to identify the presence or absence of *Pm* genes in wheat cultivars, their chromosomal location, the number of genes, and their inheritance. One or more markers may be used in location and mapping, to increase the precision and exact positioning of the identified genes. Markers are associated with a particular wheat chromosome and the region where polymorphism may occur hence, it becomes easier to map the identified gene following polymorphism of a marker using preferred gene mapping software. These molecular markers have helped to locate and map, over 54 designated and >50 temporarily designated PM genes on wheat chromosomes. Restriction fragment length polymorphisms (RFLP) was used to map genes *Pm1a, Pm1c, Pm2, Pm6, Pm13*, and *Pm26* whereas, temporary designated genes such as *MIRE, Mlm2033, Mlm80, Pm2026,* and *mlIW72* were located and mapped using RFLP in combination with other known markers (Mwale et al., 2014). Random amplified polymorphic DNA (RAPD) was used to map PM resistance genes viz., *Pm1a, Pm6, Pm13,* and *Pm25* and amplified fragment length polymorphisms (AFLP) was used to locate and map genes *Pm1c, Pm4a, Pm17, Pm24a,* and *Pm29* and temporary designated genes *mlRD30, MlZec1, PmP,* and *PmM53* were mapped using AFLP in combination with other markers. Simple sequence repeats (SSR) individually or in combination with other markers were used to identify and map genes *Pm46, Pm47, Pm49* and *Pm50, Pm1e, Pm5e, Pm24a, Pm24b, Pm27, Pm30, Pm31, Pm36, Pm40, Pm42, Pm43,* and *Pm45* and t *PmU*, *PmY201*, *PmY212*, *MIAB10*, *Pm07J126* (Mwale et al., 2014) *Pm51*, *Pm53*, and *Pm54* (Table 5.3). PM resistance genes namely *Pm1a*, *Pm2*, *Pm4a*, *Pm6*, *Pm13*, *Pm41*, *Pm42*, *Pm45*, *PmTb7A1*, and *PmTb7.2* were tagged and mapped using sequence tagged sites (STS) markers and other molecular markers. The molecular markers linked with different resistant genes have been summarized in Table 5.3.

5.7.1.8   ADULT PLANT RESISTANCE

Vertical resistance which is governed by single or few genes is race specific and it is overcome easily by pathogen on widespread cultivation of such cultivars which exert selection on the pathogen population to favor isolates with corresponding virulence genes.

Horizontal resistance to PM referred to as "slow mildewing," "adult plant resistance" and partial resistance is more durable type of resistance. By definition, horizontal resistance is not race-specific; however, race-specific partial and APR is known to exist (Christina et al., 2012). Many APR cultivars viz., Knox, Knox 62, Massey, Genesee, Redcoat, Diplomat, Est Mottin, and Maris Huntsman remained effective for many years and over large production areas. Many winter and spring wheat genotypes with partial or APR have been identified in China, Hungary, India, Denmark, Finland, Norway, Sweden, and Lithuania (Christina et al., 2012; Sharma et al., 1991b; Basandrai et al., 2011).

Shaner (1973a) defined the rate-reducing components (lower sporulation capacity and infection efficiency) of APR and outlined methods for evaluating it. Gustafson and Shaner (1982) defined adult plant or slow mildewing resistance as resistance that retarded infection, growth, and reproduction of the pathogen in adult plants but not in seedlings.

Several quantitative trait loci (QTLs) governing APR to PM have been mapped near loci of defeated major genes, for example, *Pm4*, *Pm5*, and *Pm6*. Further, a single locus could be responsible for both qualitative and quantitative expression of a trait and resistance conferred by such QTLs may be the result of the residual effects of defeated major genes. Partial resistance may be the result of alternate alleles at the R-gene loci or unique loci occurring as a part of a gene cluster. The effectiveness of APR depends on its expression during the plant growth stages at which mildew epidemics are initiated and develop. Expression of QTL governing APR to PM can vary with plant growth stage (Bougot et al., 2006), and Muranty et al. (2009) recommended pyramiding QTLs such as the ones on chromosomes 2B and

TABLE 5.3 Molecular Markers Linked to Powdery Mildew Resistance Genes in Wheat.

Gene	Position	Type of markers	Closest/flanking marker	Linkage distance/contribution	Mapping population
Pm1a	7AL	RAPD, STS, RFLP	UBC320420, UBC638550	Both co-segregate	F5, F2 lines, BSA
		RFLP	WHS178-9.4kb-EcoRI	2.8 ± 2.7 cM	F2 lines, NILs
		RFLP, STS	CDO347, mwg2062, cdo347, psr121, psr148, psr680, psr687, wir148, C607, STS638542, ksuh9	Co-segregate All Co-segregate	F2 lines, NILs F2 lines
Pm1c	7AL	RFLP, RAPD	WHS178-15kb-EcoRI, OPH-111900	4.4 ± 3.6 cM, 13 cM	F2 lines, BSA
		AFLP	S19M22-325/200	Co-segregate	F3 + F4 lines, BSA
			S14M20-137/138	Co-segregate	
Pm1e	7AL	SSR, AFLP	GWM344-null-S13M26-372	0.9cM, 0.2 cM	F2:3 lines, BSA
Pm2	5DS	RFLP	WHS350-6.5kb-	3.8 cM, 2.7 ± 2.6 cM	F2:3 lines, BSA
			EcoRV, WHS295		F2 lines, NILs, F2 lines, NILsF2 lines, NILs
		RFLP	BCD1871	3.5 cM	
		STS	STSwhs350		
Pm3a	1AS	RFLP	WHS179	3.3 ± 1.9 cM	DH, NILs
Pm3b	1AS	RFLP	BCD1434	1.3 cM	F2 lines, NILs
Pm3g	1AS	RFLP	Gli-A5	5.2 cM	DH
Pm4a	2AL	RFLP	BCD1231, CDO678	Co-segregate	F2 lines, NILs
		AFLP	4aM1	3.5 cM	F3+F4 lines, BSA
		STS	STSbcd1231-1.7kb	Co-segregate	NILs
Pm4d	2AL	STS	Xbarc122, Xgwm526	1.0 cM; 3.4 cM	F2:3 lines

TABLE 5.3 (Continued)

Gene	Position	Type of markers	Closest/flanking marker	Linkage distance/contribution	Mapping population
Pm5e	7BL	SSR	GWM1267-136	6.6 cM	F2:3 lines, BSA
Pm6	2BL	RFLP	BCD135-9kb-EcoRV	1.6 ± 1.5 cM	F2 lines, NILs
Pm8	1RS.1BL	RFLP	IAG95	Tightly linked	F2 lines, BSA
		RAPD	OPJ07-1200, OPR19-1350		Translocation lines
		STS	SEC-1b-412bp		Translocation lines
		STS	STSiag95-1050	Co-segregate	DH, F2:3 lines
Pm12	6BS- SS.6SL	RFLP	Psr10, psr106, Nor-2, Psr141, psr113, psr142, psr149, psr2	Co-segregate	F2 lines
Pm13	3BL.3SS- 3S	RFLP	psr305, psr1196		Recombinant
	3DL.3SS- 3S	RFLP, RAPD, STS	cdo460, utv135, OPV13800, UTV13, OPX12570, UTV14		Recombinant lines
Pm17	1RS.1AL	RFLP, AFLP	IAG95–CA/CT-355	1.5 cM	F2:3 lines
Pm21	6VS.6AL	RAPD	OPH171900	Co-segregate	F2 lines
		RAPD, SCAR	OPH171400, SCAR1265, SCAR1400	All co-segregate	F2 lines
Pm24a	1DS	AFLP, SSR	E34/M51-407, Xgwm337-204	Co-segregate, 2.4 ± 1.2 cM	F2:3 lines, BSA
		SSR	Xgwm1291	Co-segregate	F2:3 lines
Pm24b	1DS	SSR	Xgwm603/Xgwm789, Xbarc229	1.5–1.0 cM	F2:3
Pm25	1A	RAPD	OPA04950	12.8 cM	BC1F1 lines, BSA

TABLE 5.3 *(Continued)*

Gene	Position	Type of markers	Closest/flanking marker	Linkage distance/ contribution	Mapping population
Pm26	2BS	RFLP	wg516	Co-segregate	RSLs
Pm27	6B-6G	RFLP, SSR	psp3131	Co-segregate	F2 lines
Pm29	7DL	RFLP, AFLP	S24M13-233, S19M23-240, S22M26-192, S25M15-145, S13M23-442, S22M21-217, S17M25-226	All co-segregate	F2 lines, BSA
Pm30	5BS	SSR	Xgwm159-460, Xgwm159-500	5-6 cM	BC2F2 lines, BSA
Pm31	6AL	SSR	Xpsp3029	0.6 cM	BC2F2 lines, BSA
Pm36	5BL	SSR	BJ261635	Co-segregate	BC5F5
Pm40	7BS	SSR	Xgwm297	0.4 cM	F2 lines
Pm41	3BL	SSR, ISBP, STS	BE489472	Co-segregate	F2 lines
Pm42	2BS	SSR, AFLP-SCAR, EST-STS, RFLP-STS	BF146221	Co-segregate	F2 lines
Pm43	2DL	SSR	Xwmc41	2.3 cM	F3 and BC1 lines
Pm45	6DS	SSR, STS	Xmag6176	2.8 cM	F2 lines
Pm46	5DS	SSR	Xgwm205, Xcf81	18.9 cM	F2 and F2, 3 lines
Pm47	7BS	SSR, EST	Xgwm46, BE606897	1.7 cM, 3.6 cM	F2, 3 lines
Pm49	2BS	EST-SSR	CA695634	0.84-1.00 cM	F1+F2
Pm50	2AL	SSR	Xgwm294	2.9 cM	BC1
MlRE	6AL	SSR, RFLP	XksuD27		BSA, F3
		SSR, AFLP	Xgwm344, XE33M62-392,		BSA, F3 lines
mlRD30	7AL		XE39N58-414		

TABLE 5.3 *(Continued)*

Gene	Position	Type of markers	Closest/flanking marker	Linkage distance/ contribution	Mapping population
PmDR147	2AL	SSR, AFLP, SSR	Xgwm311, Xgwm382 E35M56-330, E35M57-56, E37M54-286, E38M54-207	5.9 cM, 4.9 cM, 2.0 cM	BC3F2 lines BSA, F3 lines
MlZec1	2BL		Xwmc356-2B		
PmPs5B	2BL	SSR	Xgwm356,	10.2 cM	BC3, BC4 lines
PmPs5B	2BL	SSR	Xgwm111, Xgwm382	1.1 cM	F2 lines
PmPs5B	2BL	SSR	Xgwm526, Xwmc317, Xgwm265, Xgwm311, Xgwm382	2.9 cM, 3.6 cM, 4.4 cM	F2 lines
PmE	2AL				
PmH	7BL	SSR	Xgwm611, Xpsp3033	5.9 cM, 13.2 cM	BC5F2 lines
PmP		AFLP SSR	XM55P66, XM55P37, Xgwm257, Xgwm296	0.8cM, 2.4 cM Co-segregate	F2 lines F2 lines
PmY39	2U(2B)	SSR	Xgwm319, Xgwm325, Xwmc382		F2 lines
PmY150	6B/6S	AFLP, SSR	Xwmc397 P16M16-109, P5M16-161, Xwmc289b, Xgwm583	1.0 cM, 3.0 cM, 20.0 cM, 33.0 cM, 24.0 cM	F2, F3 lines
PmM53	5DL		Xgwm292		
PmU	7AL	SSR	Xgwm273, Xpsp3003	2.2 cM, 3.8 cM	F1, F2 lines
PmY201	5DL	SSR	Xgwm174	5.2 cM	F2 lines
PmY212	5DL	SSR	Cfd57	5.6 cM	F2 lines

TABLE 5.3 *(Continued)*

Gene	Position	Type of markers	Closest/flanking marker	Linkage distance/ contribution	Mapping population
Mlm2033	7AL	RFLP, STS, EST	Xgwm344, Xmag2185	< 2 cM	F2, F3 lines
Mlm80	7AL	RFLP, STS, EST	Xgwm344, Xmag2185	< 2 cM	F2, F3 lines
PmE	7BS	SSR	Xgwm297	13.0 cM	F2 lines
PmYU25	2DL	SSR	Xgwm210	16.6 cM	F2 lines
PmAS846	5BL	SSR	Xgwm67	20.6 cM	F1, F2 lines
		SSR	Xgwm583, Xgwm174, Xgwm182, Xgwm271	25.7cM, 16.7 cM, 9.1 cM, 7 cM	F2 lines
PmAeY2	5DL				
PmY39-2	6AS	SSR, SSR, STS, RFLP	Xwmc553, Xwmc684, Xcfd39, Xgwm126, MAG1491, MAG1493	10.99 cM, 7.43 cM 0.9 cM, Co-segregate	F2, F3 lines F2, F3 lines
Pm2026	5AL	SSR, EST, RFLP	MAG1494, MAG2170 Xgwm344, Xcfa2040, Xcfa2240, Xcfa2257, Xwmc525, MAG2185		F2, F3 lines
mlIW72	7AL	SSR, EST-SSR, EST-STS	MAG1759 Xcau12, Xgwm611, PmTm4, XEST92, Xbarc1073, Xbarc82		F2, F3 lines, BSA
PmTm4	7BL	STS, SSR	Xwmc276 EST48, EST83 (EST84,		F2, F3 lines, BSA

TABLE 5.3 *(Continued)*

Gene	Position	Type of markers	Closest/flanking marker	Linkage distance/ contribution	Mapping population
	2AL		Xksum193, PmYm66	3.72cM, 6.15cM	
PmLK906	2AL	SSR	Xgwm265, Xgdm93		F2, F3 lines, BSA
MlAG12	7AL	SSR	Xwmc273, Xwmc346	8.3 cM, 6.6 cM	F2, F3 lines
MlIWE29	5BL	SSR, SSR, STS, EST-	Xgwm415, Xwmc75, Xwmc525, Xcfa2040, Xwmc273, XE13-2, Xmag1759, MlWE18	2.5cM, 17.6cM	F2, F3 lines
MlIWE18	7AL	STS	Xcfa2240		
Ml3D232	5BL	SSR, EST, STS		0.8 cM	F2, F3 lines
TaAetPR5	2AL	STS	p9-7p1, p9-7p2	7.62 cM	BSA
MlIAB10	2BL	SSR	Xwmc445	7 cM	F2:3 lines
PmG16	7AL	SSR, STS, DArT, CAPS	Xgwm344, wPt-9217, wPt-1424, wPt-6019	3.6 cM	Recombinant, F2:3 lines
PmHNK	3BL	SSR	Xwmc291, Xgwm108	3.8 cM, 10.3 cM	BC1, F2, F3 lines
PmAs846	5BL	STS	BJ261635, CJ840011	Co-segregate	F2, F2:3 lines
PmTb7A.1	7AL	SSR, RFLP, STS, DArT	wPt4553, Xcfa2019	4.3cM	Recombinant lines
PmTb7A.2	7AL	SSR, RFLP, STS, DArT	MAG2185, MAG1759		Recombinant lines
PmLX66	5DS	SCAR, SSR	SCAR203, Xcfd81	0.4, 2.8 cM	F2, F2:3 lines

TABLE 5.3 *(Continued)*

Gene	Position	Type of markers	Closest/flanking marker	Linkage distance/contribution	Mapping population
Pm07J126		SSR	Xbarc183	7.4cM, 9.4 cM, 1.2 cM,	F2 lines
		SSR	Xgpw7425, Xwmc75, Xgwm408, Xwmc810	22.3 M, 25.4 cM, 29.3, cM	F2 lines
PmG25	5BL	SRAP, RGA	Me5/Em5-650, Me8/Em16-600	12.9 cM, 9.7 cM	F2 lines
PmZB90	2AL				
Pm 51	2BL	SSR and EST	Xwmc 332	3.2cM	Thinopyrum ponticum
			Xwmc 317	28.6cM	
Pm 53	5BL	SSR and SNP	Xgwm 499 and Xwmc 759	iwa6024 (0.7cM proximal) and iwa 2454 (1.8cM distal)	A. speltoides
Pm 54 (26R61)	6BL	SSR	Xbarc 134		26R61
pmCH 89	4BL	SSR	xwmc 310 and xwmc 125	3.1cM and 2.7cM	T. intermedium
Pm Tb7A1	7AL	STS marker	Ta7AL 4556232	0.6cM	Triticum bioticum
			Ta7AL 4426363	6.0cM	
PmTb7A.2	7AL	STS marker	Ta7AL 4556232	0.6cM	T. bioticum
			Ta7AL 4426363	6.0cM	
PmL962	2BS	SSR	Xwmc 314, re443737	2.09cM, 3.74cM	T. intermedium
MHLT	1DS	BSA SSR markers	Xgwm337, xcfd 83/xcfd72	3.6cM	Chinese wheat land race HULUTOU
Pm 2b (Km 2939)	5DS	SCAR	SCAR 112	0.5cM	A. christatum
			SCAR 203	1.5cM	

TABLE 5.3 *(Continued)*

Gene	Position	Type of markers	Closest/flanking marker	Linkage distance/ contribution	Mapping population
MIUM 15		BSA SSR, STS, SNP	Xwmc525/IWA8057, MIUM 15, Xcfa2240	0.7cM 0.8cM 2.8cM	NC09BGTUM 15 (NC-UM15)
MIHUBEL	2D	BSA SSR EST	Xgwm 265	0.4cM	Hubel speltoid wheat
PmPB3558	5DS	SSR based on the SNP	*Xcfd81* and *Xbwm25*	–	*A. cristatum*
PmYB	5DS	SSR, SCAR	*SCAR112 -Cfd81*-SCAR marker SCAR203	0.9 and 1.9 cM	–
PmFG	5DS	SCAR	SCAR203	*Xbwm21/Xcfd81/ Xscar112* (distal) and *Xbwm25* (proximal) at 0.3 and 0.5 cM	-

Information up to genes *PmG 20* is as per the paper of Mwale et al. (2014).

5D to provide resistance throughout the plant development. APR in culti-vars, Houser, Massey, and Red coat reduced grain yield losses under PM epidemics throughout the plant-growing season. APR to PM is a heritable with heritability estimates higher than 0.75 and a predominance of additive genetic effects (Das & Griffey, 1995) have been reported. APR is quantitative in nature and as many as 18 QTLs governed resistance. Most studies have identified 1–4 QTLs have been identified having consistent major effects over environments. Additional sources of APR to PM have been identified, characterized, and mapped, which facilitated their use in marker-assisted breeding programs. The majority APR QTLs have been mapped in winter wheats; however, QTL genes have been identified in spring wheat also, that is, gene *Pm38* located at *Lr34/Yr18* locus and *Pm39* located at the *Lr46/Yr29* locus (Lillemo et al., 2008, 2010). Race-specific major genes *Pm3a, Pm3g, Pm4b, Pm5, Pm6,* and *MlRe* were identified in APR mapping and these contributed to seedling and/or APR. QTLs that were mapped at or near loci of the defeated genes, *Pm4, Pm5, Pm6, and MlRe,* contributed to APR even when exposed to virulent isolates or populations of *B. graminis.* Addition-ally, QTL on chromosome 4A derived from *T. militinae,* chromosome 5D derived from RE 714 and on chromosome 7B from RE 714 contributed to APR (Christina et al., 2012). Asad et al. (2014) mapped three QTL on chro-mosomes 2BS (*QPm.caas-2BS.2*), 3BS (*QPm.caas-3BS*), and 5AL (*QPm. caas-5AL*) with the resistance alleles contributed by Pingyuan 50 explaining 5.3, 10.2, and 9.1% of the phenotypic variances, respectively, and one QTL on chromosome 3BL (*QPm.caas-3BL*) derived from Mingxian 169 accounting for 18.1% of the phenotypic variance. *QPm.caas-3BS, QPm.caas-3BL,* and *QPm.caas-5AL* appeared to be the new PMAPR loci and *QPm.caas-2BS.2* and *QPm.caas-5AL* are possibly pleiotropic or closely linked resistance loci to stripe rust resistance QTL. Pingyuan 50 could be a potential genetic resource to facilitate breeding for improved APR to both PM and stripe rust. Herrera (2014) demonstrated that gene *Lr67/Yr46* has pleiotropic effect on stem rust and PM resistance and is associated with leaf tip necrosis. Genes are designated as *Sr55, Pm46,* and *Ltn3,* respectively. It is interesting that loci conferring APR to PM have been reported in each of the three home-ologous chromosome groups. Recently, Li et al. (2014) reviewed the genetic study and application of APR to PM. They have reported 119 PMAPR QTLs on 21 chromosomes and 11 loci which are located on chromosomes 1BS, 1BL, 2AL, 2BS (2), 2DL, 4DL, 5BL, 6AL, 7BL, and 7DS and showed pleio-tropic effects for resistance to leaf rust, stripe rust, and PM.

5.7.1.9   MORPHOLOGICAL AND CHEMICAL BASIS OF RESISTANCE

PM resistance genes at *Pm3* locus have been cloned and sequenced. Gene *Pm3*, located on wheat chromosome 1A, is a member of a large cluster of nucleotide binding site (NBS)/leucine-rich repeat (LRR) receptor-like genes (Christina et al. (2012). The receptor-like NBS-LRR family is the largest class of R genes. To date, 17 functional alleles have been identified at the *Pm3* locus; *Pm3a* to *Pm3g* and *Pm3k* to *Pm3t.* The allelic variability at this locus is believed to have arisen mainly following the domestication of bread wheat, with the high sequence conservation among *Pm3* alleles suggesting recent evolution from a mildew-susceptible ancestral sequence (Christina et al., 2012).

5.7.1.10   GENETIC ENGINEERING

Plant transformation can break interspecies crossing barriers and is a viable alternative to conventional breeding methods for disease resistance leading to expansion of available gene pool. The procedure is limited only to already cloned genes and requires extensive testing to ensure stability and heritability of the transgene and the transformation may have a negative effect on agronomic performance. Wheat is not amenable to *Agrobacterium* mediated transformation. However, particle bombardment was used successfully to generate transgenic wheat expression. Barley-seed class II chitinase and a tobacco β-2, 3-glucanase gene was transferred to wheat seedlings via *Agrobacterium* transformation where increased resistance to PM was reported. Wheat was transformed using antifungal proteins viz., Ag-AFP from *Aspergillus giganteus*, a barley class II chitinase, and type I ribosome-inactivating protein (RIP). It was observed that simultaneous expression of the Ag-AFP and the barley chitinase enhanced PM resistance quantitatively, whereas the RIP gene had no effect on this disease. An alternative strategy, that is, genetic engineering to defense signaling pathways was manipulated to activate multiple defense genes to induce systemic acquired resistance (SAR). The *NPR1* gene from *Arabidopsis*, a key regulator of SAR, was used to engineer wheat plants with improved resistance to Fusarium head blight. However, more research is needed to demonstrate the feasibility of manipulating SAR to produce PM resistant wheat with overall good agronomic performance (Christina et al., 2012).

5.7.2 FUNGICIDAL MANAGEMENT

Fungicides are widely used for the management of PM in cereals. Several systemic fungicides such as benlate (0.1%), karathane (0.1%), etc., can control PM through foliar sprays. Seed dressing and soil drenching with 0.01% calaxin was also found effective. One spray of propiconazole (Tilt 25EC @ 0.1%) on disease appearance is highly effective (http://nsdl.niscair. res.in/ jspui/bitstream/123456789/647/1/revised%20crop%20Diseases-%20 Formatted.pdf). Fungicides like morpholines (e.g., fenpropidin), triazoles (e.g., tebuconazole, propiconazole, hexaconazole, and cyproconazole), and strobilurin are used for its control.

5.7.2.1 SEED TREATMENT

The disease can be controlled by using systemic fungicides as seed treatment, demethylation inhibitor seed treating fungicides (DMIs), viz., difenoconazole, flutriafol and triadimenol were extensively used. Out of these triadimenol can protect wheat seedlings infection for up to 60 days after emergence (Marchioro & Franco, 2010). Moreover, seed treating efficacy of fungicides like triadimenol and difenoconazole is further supplemented as these also reduce smuts and other foliar pathogens.

5.7.2.2 FOLIAR FUNGICIDES

Foliar fungicides are effective but these should only be applied if the cultivar is susceptible and an economic return is assured, that is, in seed production plots (Leath & Bowen, 1989). Avoid applying fungicides too early so that these may be effective during the grain-filling period. In case of moderate pre flowering severity of PM, early season application (at Feekes scale 6–8) of the systemic fungicide triadimefon, maintained 8–17% higher yield compared to the control (Lipps & Madden, 1989). Singh and Ramesh (2000) reported that three sprays of Bayleton 25% WP (0.05%) at stem elongation, flag leaf and flowering stages resulted in 69.54% control over the unsprayed check followed by Tilt 25 EC (67.82%), Punch (62.47%), and Topas (55.75%) and were effective in increasing yield from 15.8 to 31.8% over control. Basandrai et al. (2013c) reported that 1–2 foliar sprays of fungicides propiconazole, hexaconazole, and triadimefon @ 0.1% and mancozeb 75WP @ 0.25% resulted in significantly less severity of leaf rust, YR and PM and resulted in significantly more yield as compared with no spray (check).

Combinations containing azol and morpholine derivatives, Fenpiclonil +difenoconazole, carbendazim+triadimefon, benomyl+polymarcin were advantageous against infection and development of *Bgt* and increasing the yield. Propiconazole+tridemorph, flutriafol+chlorothalonil, propiconazole+ chlorothalonil consistently increased yield. Fungicide mixtures Tilt turbo (propiconazole+tridemorph) and Dorin (triadimenol+tridemorph) decreased the incidence of Bgt strains less sensitive to triadimenol (Fletcher et al., 1987). Boiko and Pokova (1993) suggested alternate applications of Tilt (propiconazole) and Bayleton (tridemorph) to reduce chances of the development of mildew resistant populations of Bgt.

An integrated disease management system should be used with genetic resistance as the cornerstone of the program. Cultural management, including proper management of nitrogen fertilization, is essential to minimize risk of crop damage from PM (http://www.fao.org/ docrep/006/y4011e/ y4011 e01.htm). Fungicides should be used in conjunction with disease monitoring system starting from planting through flowering stage. However, the efficacy of fungicides varies largely due to the development of isolates tolerant to different fungicide groups viz., triazoles (flutriafol, propiconazole and triadimefon), Resistance to strobilurin fungicides has been detected in isolates from Bgt in England and Germany (Fraaije et al., 2002). Reduced effectiveness of the triazole fungicides triadimefon and propiconazole was found in the Netherlands following its intensive use (De Waard et al., 1986). A quantitative real-time PCR diagnostic procedure has been developed for the early detection of resistance genes at low frequency which would be invaluable for further resistance risk assessment and validation of anti-resistance strategies. Moreover, triazoles are combined with another active ingredient, viz., morpholine to increase efficacy and reduce pressure on the pathogen to produce triazole-tolerant isolates.

5.8 CONCLUDING REMARKS

Wheat PM caused by *B. graminis* f. sp. *tritici* has become a challenging disease of wheat throughout the world, as crop damage levels are high. Its incidence and severity is increasing in NWPZ, the grain bowl of the country. However, till date, much emphasis is given on breeding for rust resistance and no consistent efforts are being made to breed for PM resistance in the National Breeding Program. During the past four decades, wheat varieties with leaf rust and YR resistant genes *Lr26/Yr9* and leaf rust resistant gene *Lr23* were developed and widely cultivated. Gene *Lr26* is closely linked with gene *Pm8*, hence the pathotypes of Bgt virulent on *Pm8* appeared and spread making the resistant varieties susceptible. At present, all the commonly grown varieties of *T. aestivum* and *T. durum* are susceptible. Hence, there is need to undertake the work on virulence analysis of Bgt population in epidemiologically important areas. The discovery of more than 50 designated genes (*Pm1* to *Pm54*) with 70 alleles as well as over 50 additional non-designated genes, land races, allied species and allied genera, etc. may be evaluated under Indian conditions to identify PM resistant sources to identify suitable donors to develop resistant wheat varieties. The use of molecular markers such as SSR, AFLP, RFLP, RAPD, and STS has led to successful gene identification and mapping in over 33 resistant genes to wheat PM including the recently designated *Pm46* to *Pm54* genes and over 42 temporarily designated PM resistance genes. These markers may be utilized in marker-assisted selection (MAS) for precise and accelerated PM resistant varieties. The management of disease using fungicides has proved useful in situations where host resistance level is not satisfactory and in the event of sudden break down of resistance. However, fungicides like propiconazole, tebuconazole, etc. with multiple disease control and different mode of action should be preferred.

ACKNOWLEDGMENTS

The authors would like to thank the authors of various research and review papers and books on wheat powdery mildew from which literature have been used for in this chapter.

KEYWORDS

- wheat
- powdery mildew
- *Blumeria graminis* f sp. *tritici*
- virulence pattern
- pathogenic variability

- **fungal diseases**
- **epidemics**
- **host resistance**
- **management**
- **conidial landing**
- **phenotypes**

REFERENCES

Akhtar, S.; Anjum, F. M.; Anjum, M. A. Micronutrient Fortification of Wheat Flour: Recent Development and Strategies. *Food. Res. Int.* **2011,** *44,* 652–659.

Alam, M. A.; Xue, F.; Wang, C.; Ji, W. Powdery Mildew Resistance Genes in Wheat: Identification and Genetic Analysis. *J. Mol. Biol. Res.* **2011,** *1*(1), 20–39.

Alam. M. A.; Mandal, M. S. N.; Wang. C.; Ji, W. Chromosomal Location and SSR Markers of a Powdery Mildew Resistance Gene in Common Wheat Line N0308. *Afric. J. Micro. Res.* **2013,** 7(6), 477–482.

Anonymous. *"Monitoring of Virulence of Erysiphe graminis tritici in Himachal Pradesh and Their Utilisation in Characterisation of Resistance Sources in Triticum spp. ";* Final Technical Report of the Icar Adhoc Project: New Delhi, 1998, p 52.

Arya, H. C. Studies on the Physiologic Specialization and Varietal Reaction of Wheat to Powdery Mildew in India. *Indian Phytopathol.* 1962, *15,* 127–132.

Asad, M. A. Identification of QTL for Adult-Plant Resistance to Powdery Mildew in Chinese Wheat Landrace Pingyuan 50. *The Crop J.* **2014,** *2,* 308–314.

Babayants, L. T.; Smilyanetes, S. P. Race Composition of Powdery Mildew *Erysiphe graminis* f.sp. *tritici* March and Mildew Resistance of Wheat Varieties in the South Ukraine. *Milkol. Fitopatologiya.* 1991, *25,* 324–329.

Babayants, O. V., et al. Race Composition of *Blumeria graminis* (DC) Speer f. sp. *tritici* in the South of Ukraine and Effectiveness of *Pm*-Genes in 2004–2013. *Cereal Res. Commun.* **2015,** *43,* 449–458.

Bahadur, P.; Aggarwal, R. Powdery Mildew of Wheat: A Potential Disease in North Western India. In *Management of Threatening Plant Diseases of National Importance;* Agnihotri, V. P., Sarabhoy, A. K., Singh, D. V., Eds.; Malhotra Publishing House: New Delhi, 1997; pp 17–30.

Bailey, K.; McNeill, B. H. Virulence of *Erysiphe graminis* f sp. *tritici* in Southern Ontario in Relation to Vertical Resistance in Winter Wheat. *Can J. Plant Pathol.* **1983,** *5,* 148–153.

Basandrai Ashwani, K.; Sharma, B. K.; Basandrai, D. Efficacy of Triazole Fungicides for the Integrated Management of Yellow Rust, Leaf Rust and Powdery Mildew of Wheat. *Pl. Dis. Res.* **2013a,** *28,* 135–139.

Basandrai Ashwani, K.; Daisy, B.; Tyagi, P. D. Postulation of Resistant Genes for Powdery Mildew (*Blumeria graminis tritici*) in Indian Wheats. *Int. J. Plant Prot.* **2013b,** *6,* 171–176.

Basandrai Ashwani, K., et al. Race-Specific Resistance against Powdery Mildew in Advanced Indian Wheats and Some Commercially Grown Cultivars. *Crop Improv.* **2013c,** *40,* 58–64.

Basandrai Ashwani, K., et al. PathogenicVariation in Conidial Populations of *Erysiphe graminis* f. sp. *tritici* in Himachal Pradesh. *J. Mycol. Pl. Pathol.* **2003,** *33,* 80–83.

Basandrai Ashwani, K.; Pathania, N.; Paul R.; Tyagi, P. D. Virulence Dynamics of *Erysiphe graminis* f sp. *tritici. Res. Crops.* **2007,** *7,* 334–337.

Basandrai Ashwani, K.; Sharma, S. C. Jhooty, J. S. Diversity for Resistance to Powdery Mildew in Wheat. *Crop Improv.* **1989,** *16,* 98–100.

Basandrai, A. K.; Sharma, S. C. Sources of Powdery Mildew Resistance in Wheat and Triti-cale and Inheritance of Resistance in Selected*Triticum aestivum* L. and *T. durum* Desf. genotypes. *J. Genet. Breed.* **1990,** *44,* 253–258.

Basandrai, A. K., et al. Race Specific Resistance to *Blumeria graminis* f sp. *tritici* Causing Powdery Mildew in Some Indian Wheats. *Crop Improv.* **2012,** *39,* 183–188.

Basandrai, A. K.; Sharma, S. C.; Munshi, G. D. Resistance Behaviour of Some Wheats to Powdery Mildew. *Pl. Dis. Res.* **1991,** *6,* 103–106.

Basandrai, A. K., et al. Histological Studies of Powdery Mildew (*Erysiphe graminis* f. sp. *tritici*) Resistance in Wheat (*Triticum aestivum*). *Indian Phytopath.* **2011,** *64,* 229–234.

Bennett, F. G. A. Resistance to Powdery Mildew in Wheat: A Review of its Use in Agriculture and Breeding Programmes. *Plant Pathol.* **1984,** *33*(3), 279–300.

Bennett, F. G. A. The Expression of Resistance to Powdery Mildew Infection in Winter Wheat Cultivars. II. Adult Plant Resistance. *Ann. Appl. Biol.* **1981,** *98,* 305–317.

Bhardwaj, S. C., et al. *Resistance Genes and Adult Plant Resistance of Released Wheat Vari-eties of India*; Research Bulletin No. 5; Regional Research Station, Flowerdale, Directorate of Wheat Research: Shimla, India, 2011, p 31.

Boiko, N. I.; Pokova, O. V. Corbel on Winter Wheat. *Z. Rast.* **1993,** *8,* 14.

Bougot, Y., et al. A Major QTL Effect Controlling Resistance to Powdery Mildew in Winter Wheat at the Adult Plant Stage. *Plant Breed.* **2006,** *125,* 550–556. d-i:10.1111/j.1439-0523.2006.01308.x Boukhatem, N.

Briggle, L. W. Near-Isogenic Lines of Wheat with Genes for Resistance to *Erysiphe graminis* f. sp. *tritici. Crop Sci.* **1969,** *9,* 70–72.

Briggle, L. W.; Scharen, A. L. Resistance in *Triticum vulgare* to Infection by *Erysiphe graminis* f. sp. *tritici* as Influenced by Stage of Development of the Host Plant. *Pl. Dis. Reptr.* **1961,** *45,* 846–850.

Casulli, F.; Siniscalco, A. Observations on the Epidemiology of the Main Foliar Diseases of Wheat in three Localities of Southern Italy during 1983 and 1984. *Phytopath. Mediter-ranea.* **1986,** *25,* 61–67.

Chen Xiu-mei., et al. Analysis on the Race Population and Virulence Dynamics of *Blumeria graminis* f. sp *tritici* in Some Major Wheat Growing Regions of China during 2011–2012. *J. Triticeae Crops.* **2013**(http://en.cnki.com.cn/Article_en/CJFDTOTAL-MLZW201303034. htm).

Chhuneja, P., et al. Identification and Mapping of Two Powdery Mildew Resistance Genes in *Triticum boeoticum* L. *Theor. Appl. Genet.* **2012,** *124,* 1051–1058.

Chhuneja, P., et al. Fine Mapping of Powdery Mildew Resistance Genes PmTb7A.1 and PmTb7A.2 in *Triticum boeoticum* (Boiss.) using the Shotgun Sequence Assembly of Chro-mosome 7AL. *Theor. Appl. Genet.* **2015,** *128,* 2099–2111.

Christina, C., et al. Powdery Mildew. In *Disease Resistance in Wheat;* Indu S., Ed.; CAB International: Oxfordshire, 2012; pp 84–107.

Christina, C., et al. In *Structure and Regional Differences in U.S. Blumeria graminis f. sp. tritici Populations: Divergence, Migration, Fungicide Sensitivity, and Virulence Patterns,* Presented in 14th International Cereal Rusts and Powdery Mildews Conference, Helsingor, Denmark, Jul 5–8, 2015 2015 (http://emcrf.au.dk/fileadmin/EMCRF/Abstract_Pdf/5_Christina_Cowger.pdf).

Costamilan, L. M. Variability of the Wheat Powdery Mildew Pathogen*Blumeria graminis* f. sp. *tritici* in the 2003 Crop Season. *Fitopatol. Bras.* **2005,** *4,* 102–103.

Das, M. K.;Griffey, C. A. Gene Action for Adult-Plant Resistance to Powdery Mildew in Wheat. *Genome,* **1995,** *38*(2), 277–282.

De Waard, M. A., et al. Variation in Sensitivity to Fungicides which Inhibit Ergosterol Biosynthesis in Wheat Powdery Mildew. *Neth. J. Plant Pathol.* **1986,** *92,* 21–32.

Driscoll, C. J.; Jensen, N. F. Release of Wheat Rye Translocation Stock Involving Leaf Rust and Powdery Mildew Resistances. *Crop Sci.* **1965,** *5,* 279–280.

Elkot Ahmed, F. A., et al. Marker Assisted Transfer of Two Powdery Mildew Resistance Genes *PmTb7A.1* and *PmTb7A.2* from *Triticum boeoticum* (Boiss.) to *Triticum aestivum* (L.). *PLoS One.* **2015,** *10* (6), e0128297.

Ellingboe, A. H. Horizontal Resistance: An Artifact of Experimental Procedure? *Aust. Pl. Path. Soc. Newsl.* **1975,** *4,* 44–46.

Ellingboe, A. H. Genetics of Host Parasite Interactions. In*Physiological Plant Pathology;* Heitefuss, R., Williams, P. H., Eds.; Springer-Verlag: Berlin, 1976; pp 761–778.

Fletcher, J. T., et al. The Sensitivity of *Erysiphe graminis* f. sp. *tritici* to Various Fungicides Inhibiting Sterol Biosynthesis. *ISPP Chem. Control Newsl.* **1987,** *9,* 30–32

Flor, H. H. Inheritance of Pathogenicity in Melampsora Lini. Phytopathology. **1942,** *32,* 653–669.

Fraaije, B. A., et al. Following the Dynamics of Strobilurin Resistance in *Blumeria graminis* f. sp. *tritici* Using Quantitative Allele-Specific Real-Time PCR Measurements with the Fluorescent Dye SYBR Green I. *Plant Pathol.* **2002,** *51,* 45–54.

Garcia-Sampedro, C. Seedling Reaction to Powdery Mildew *Blumeria graminis* f. sp. *tritici* in Hexaploid Wheat Cultivars. *Rev. Fac. Agron. Vet. Univ. Buenos Aires.* **1996,** *16,* 133–139.

Gerechter-Amitai, Z. K.; Van Silfhout, C. H. Resistance to Powdery Mildew in Wild Emmer (*Triticum dicoccoides* Korn). *Euphytica.* **1984,** *33,* 273–280.

Ghemawat, M. S. Seeding and Mature Plant Resistance of Wheat to Powdery Mildew. *Diss. Abstr.* **1969,** *29,* 3990.

Ghemawat, M. S. Seedling Resistance of Wheat to Powdery Mildew Fungus, *Erysiphe graminis* f. sp. *tritici.* I. Resistance of First Leaves. *Mycopathologia.* **1979,** *68,* 131–137.

Goel, L. B.; Singh, S.; Sinha, V. C. The Mildews of Wheat. In *Problems and Progress of Wheat Pathology in South Asia;* Joshi, L. M., et al., Eds.; Malhotra Publishing House: New Delhi, India, 1986; pp 176–190.

Gromashevs' ka, Yu L. The Composition, Dynamics and Specialization of Races of the Pathogen of Powdery Mildew on Cultivated and Prospective Varieties of Wheat in Different Soil-Climatic Zones of the Ukraine. *Zakhist Roslin.* **1983,** *30,* 52–56.

Gustafson, G. D.; Shaner, G. Influence of Plant Age on the Expression of Slow-Mildewing Resistance in Wheat. *Phytopathology.* **1982,** *72,* 746–749.

Hao, Y., et al. Molecular Characterization of a New Powdery Mildew Resistance Gene *Pm54* in Soft Red Winter Wheat. *Theor. Appl. Genet.* **2015,** *128*(3), 465–476.

Haywood, M. J.; Ellingboe, A. H. Genetic Control of Primary Haustorial Development of *Erysiphe graminis* on Wheat. *Phytopathology.* **1979,** *69,* 48–53.

Herrera, F., et al. *Lr67/Yr46* Confers Adult Plant Resistance to Stem Rust and Powdery Mildew in Wheat. *Theor. Appl. Genet.* **2014,** *127,* 781–789.

Heun, M.; Fischbeck, G. Identification of Wheat Powdery Mildew Resistance Genes by Analyzing Host-Pathogen Interactions. *Plant Breed.* **1987b,** *98,* 124–129.

Heun, M.; Fischbeck, G. Genes for Powdery Mildew Resistance in Cultivars of Spring Wheat. *Plant Breed.* **1987a,** *99,* 282–288.

Hovmoller, M. S. Investigations of Virulence of Wheat and Barley Mildew in Denmark in 1987. *Tidsskr. Planteave.* **1987a,** *91,* 375–386.

Hu, D. W., et al. Ulra Structure and Cytochemistry of Papilla Response in Wheat against Attack by *Blumeria graminis* f. sp. *tritici. J. Zhejiang. Agri. Univ.* **1998,** *24,* 502–508.

Huang, J., et al. Molecular Detection of a Gene Effective against Powdery Mildew in the Wheat Cultivar Liangxing 66. *Mol. Breed.* **2012,** *30*(4), 1737–1745.

Huszar, J. The Variability of *Erysiphe graminis* f. sp. *tritici* in West Slovakia in 1989–1990. *Sbornik UVTIZ. Ochr. Rostl.* **1992,** *28,* 171–176.

Hyde, P. M. Comparative Studies of the Infection of Flag Leaves and Seedling Leaves of Wheat by *Erysiphe graminis. Phytopath. Z.* **1976,** *85,* 289–297.

Hyde, P. M.;Colhoun, J. Mechanisms of Resistance of Wheat to *Erysiphe graminis* f. sp. *tritici. Phytopath. Z.* **1975,** *82,* 185–206.

Iliev, I. V. Interaction between the Pathogen of Powdery Mildew (*Erysiphe graminis tritici*) and Wheat in Different Stages of the Infection Process. Formation of Initial Secondary Hyphae. *Rasteniev "dni Nauki.* **1988,** *25,* 20–23.

Kapoor, A. S. Evaluation of Slow Rusting and Slow Mildewing Resistance in Wheat under Field Conditions. *Him. J. Agric. Res.* **1990,** *16,* 56–58.

Kowalczyk, K. Identification of Powdery Mildew Resistance Genes in Common Wheat (*Triticum aestivum* L em Thell). XI. Cultivars Grown in Poland. *J. Appl. Genet.* **1998,** *39,* 225–236.

Leath, S.; Bowen, K. L. Effects of Powdery Mildew, Triadimenol Seed Treatment and Triadimefon Foliar Sprays on Yield of Winter Wheat in North Carolina. *Phytopathology.* **1989,** *79,* 152–155.

Leath, S.; Heun, M. Identification of Powdery Mildew Resistance Genes in Cultivars of Soft Red Winter Wheat. *Plant Dis.* **1990,** *74,* 747–752.

Leath, S.; Murphy, J. P. Virulence Genes of the Wheat Powdery Mildew Fungus, *Erysiphe graminis* f. sp. *tritici,* in North Carolina. *Plant Dis.* **1985,** *69,* 905.

Lesovoi, M. P.; Kol'nobritskii, N. I. Characteristics of Inheritance of Resistance of Wheat Cultivars to Powdery Mildew. *Selektsiya i Semenovodstvo.* **1980,** *11,* 17–18.

Li, C., et al. Characterization and Functional Analysis of Differentially Expressed Genes in Bgt-inoculated Wheat Near-Isogenic Lines by cDNA-AFLP and VIGS. *Crop Sci.* **2014,** *54*(5), 2214–2224.

Li, L. Y.; Huang, Y. J. Analysis of Six Virulence Genes of Wheat Powdery Mildew. *J. Southwest Agric. Univ.* **1991,** *13,* 473–476.

Liang, X. Y., et al. Identification of Resistance of Wheat Varieties to *Erysiphe graminis* f. sp. *tritici* and Screening Resistance Sources. *Zheijiang Nongye Kexue (Zhejiang Agri. Sci.).***1985,** *1,* 21–24.

Lillemo, M., et al. The Adult Plant Rust Resistance Loci *Lr34/Yr18* and *Lr46/Yr29* are Important Determinants of Partial Resistance to Powdery Mildew in Bread Wheat Line Saar. *Theor. Appl. Genet.* **2008,** *116* (8), 1155–1166. doi: 10.1007/s00122-008-0743-1.

Lillemo, M.; Singh, R. P.; Van Ginkel, M. Identification of Stable Resistance to Powdery Mildew in Wheat Based on Parametric and Nonparametric Methods. *Crop Sci.* **2010,** *50,* 478.

Limpert, E., et al. Analysis of Virulence in Populations of Wheat Powdery Mildew in Europe. *J. Phytopathol.* **1987,** *120,* 18.

Lipps, P. E.; Madden, L. V. Effect of Fungicide Application Timing on Control of Powdery Mildew and Grain Yield of Winter Wheat. *Plant Dis.* **1989,** *73,* 991–994.

Liu Na, Z., et al. Virulence Structure of *Blumeria graminis* f. sp. *tritici* and Its Genetic Diversity by ISSR and SRAP Profiling Analyses. *PLoS One.* **2015,** *10*(6), e0130881.

Lu, Y., et al. Genetic Mapping of a Putative *Agropyron cristatum*-Derived Powdery Mildew Resistance Gene by a Combination of Bulked Segregant Analysis and Single Nucleotide Polymorphism Array. *Mol. Breed.* **2015,** *35,* 96.

Lupton, F. G. H. Resistance Mechanisms of Species of *Triticum* and *Aegilops* and of Amphidiploids between them to *Erysiphe graminis* DC. *Trans. Brit. Mycol. Soc.* **1956,** *39,* 51–56.

Ma, P., et al. The Gene *PmYB* Confers Broad-Spectrum Powdery Mildew Resistance in the Multi-Allelic*Pm2* Chromosome Region of the Chinese Wheat Cultivar YingBo 700. *Mol. Breed.* **2015a,** *35,* 124.

Ma, P.; Xu, H.; Zhang, H., et al. The Gene *PmWFJ* is a New Member of the Complex *Pm2* Locus Conferring Unique Powdery Mildew Resistance in Wheat Breeding Line Wanfengjian 34. *Mol. Breed.* **2015b,** *35,* 210. doi:10.1007/s11032-015-0403-51.

Mascher, F., et al. Virulence Monitoring and the Structure of Powdery Mildew Populations between 2003 and 2010. *Rech. Agron. Suisse.* **2012,** *1,* 244–251.

Marchioro, V. S.; Franco, F. A. *Informações Técnicas Para Trigo e Triticale-Safra 2011. Cascavel PR.* COODETEC: Comissão Brasileira de Pesquisa de Trigo e Triticale, 2010.

McIntosh, R. A., et al. *Catalogue of Gene Symbols for Wheat: 2013–14 (Supplement),* 2014, http://maswheat.ucdavis.edu/ CGSW /2013–2014_Supplement.pdf

McIntosh, R. A., et al. *Catalogue of Gene Symbols for Wheat: 2015–16 (Supplement),* 2015–16, https://shigen.nig.ac.jp/wheat/komugi/genes/macgene/supplement2015.pdf

Menzies, J. G.; McNeill, B. L. Virulence of *Erysiphe graminis* f sp. *tritici* in Southern Ontarioin 1983, 1984, 1985. *Can J. Plant Path.* **1986,** *8,* 331–334.

Misic, T., et al. Prima, Very Early Dwarf Winter Wheat Variety. *Selekcija-i-Semenarstvo.* **1997,** *4,* 29–34.

Moseman, J. G., et al. Resistance of *Triticum dicoccoides* to Infection with *Erysiphe graminis tritici.* *Euphytica.* **1984,** *24,* 21–27.

Moseman, J. G. Genetics of Powdery Mildew. *Ann. Rev. Phytopathol.* **1966,** *4,* 269–290.

Muranty, H., et al. Two Stable QTL Involved in Adult Plant Resistance to Powdery Mildew in the Winter Wheat Line RE714 are Expressed at Different Times along the Growing Season. *Mol. Breed.* **2009,** *23,* 445–461.

Murphy, J. P. Registration of NC97BGTD 7 and NC97BGTD 8 Wheat Germplasm to Powdery Mildew. *Crop Sci.* **1999,** *39,* 884–885.

Mwale, V. M.; Chilembwe, E. H. C.; Uluko, H. C. Wheat Powdery Mildew (*Blumeria graminis* f. sp. *tritici*): Damage Effects and Genetic Resistance Developed in Wheat (*Triticum aestivum*). *Int. Res. J. Plant Sci.* **2014,** *5*(1), 1–16.

Nashaat, N. I.; Moore, K. The Expression of Components of Seedling Resistance to *Erysiphe graminis* f. sp. *tritici* in *Triticum timopheevi* and a Hexaploid Derivative. *Plant Pathol.* **1991,** *40,* 495–502.

Nass, H. G., et al. AC Winsloe Winter Wheat. *Can. J. Plant Sci.* **1995,** *75,* 905–907.

Nayar, S. K., et al. Revised Catalogue of Genes that Accord Resistance to *Puccinia* Species in Wheat. *Res. Bull.* **2001,** *3,* 48.

Negulescu, F.; Saulescu, N. N.; Ittu, G. Genes for Resistance to Powdery Mildew Used in the Wheat Breeding Programme at the ICCPT, Fundulea. *Probl. Genet. Teor. Apl.* **1978,** *10,* 13–25.

Newton, M.; Cherewick, W. J. *Erysiphe graminis* DC in Canada. *Can. J. Res.* **1947,** *25,* 73–93.

Oku, T.; Namba, S.; Yamashita, S.; Doi, Y. Physiologic Races of *Erysiphe graminis* f. sp. *tritici* in Japan. *Ann. Phytopath. Soc. Jpn.* **1987,** *53,* 470–477.

Parks, R. Virulence structure of the Eastern U.S. Wheat Powdery Mildew Population. *Plant Dis.* **2008,** *92,* 1074–1082.

Parleviet, J. E. Partial Resistance of Barley to Leaf Rust, *Puccinia hordei.* I. Effect of Cultivar and Development Stage on Latent Period. *Euphytica.* **1975,** *24,* 21–27.

Pathania, N.; Basandrai Ashwani, K.; Tyagi P. D. Postulation of Powdery Mildew Resistance Genes in some Wheat Stocks. *J. Mycol. Pl. Pathol.* **1998,** *28,* 11–14.

Pathania, N.; Tyagi, P. D.; Basandrai Ashwani, K. Virulence Spectrum of *Erysiphe graminis tritici* in Himachal Pradesh. *Pl. Dis. Res.* **1996,** *11,* 19–24.

Paul, R.;Basandrai Ashwani, K.; Tyagi, P. D. Identification of Resistance Genes against *Erysiphe graministritici* in Indian and Exotic Wheats. *Indian J Genet. Plant Breed.* **1999,** *59,* 125–134.

Paul, R.; Basandrai, A. K.; Tyagi, P. D. Virulence Spectrum of *Erysiphe graminis* f. sp. *tritici* in Himachal Pradesh. *Indian Phytopath.* **2000,** *53*(4), 415–418.

Peng, F. S., et al. Molecular Mapping of a Recessive Powdery Mildew Resistance Gene in Spelt Wheat Cultivar Hubel. *Mol. Breed.* **2014,** *34*(2), 491–500.

Petersen, S., et al. Mapping of Powdery Mildew Resistance Gene *Pm53* Introgressed from *Aegilops speltoides* into Soft Red Winter Wheat. *Theor. Appl. Genet.* **2014,** *128*(2), 303–312.

Prabhu, A. S.; Prasada, R. P. Physiologic Races of Wheat Powdery Mildew in Shimla and Nilgiri Hills. *Indian Phytopath.* **1963,** *16,* 201–204.

Pugsley, A. T.; Carter, M. V. The Resistance of Twelve Varieties of *Triticum vulgare* to *Erysiphe graminis tritici. Aust. J. Biol. Sci.* **1953,** *6,* 335–346.

Rana, S. K.; Sharma, B. K.; Basandrai, A. K. Estimation of Losses due to Powdery Mildew of Wheat in Himachal Pradesh. *Indian Phytopath.* **2005,** *59,* 112–114.

Reed, G. M. Infection Experiments with the Powdery Mildew of Wheat. *Phytopathology.* **1912,** *2,* 81–87.

Rouse, D. I., et al. Components of Rate-Reducing Resistance in Seedlings of Four Wheat Cultivars and Parasitic Fitness in Six Isolates of *Erysiphe graminis* f. sp. *tritici. Phytopathology.* **1980,** *70,* 1097–1100.

Royer, M. H., et al. Partial Resistance of Near-Isogenic Wheat Lines Compatible with *Erysiphe graminis* f. sp. *tritici. Phytopathology.* **1984,** *74,* 1001–1006.

Saharan G. S., et al. Virulence Pattern of Mass Isolates of Powdery Mildew Fungi of Wheat and Barley in Himachal Pradesh. *Indian J. Mycol. Pl. Pathol.* **1981,** *11,* 118–120.

Saidou, M., et al. Genetic Analysis of Powdery Mildew Resistance Gene Using SSR Markers in Common Wheat Originated from Wild Emmer (*Triticum dicoccoides* Thell). *Turk. J. Field Crops.* **2016,** *12,* 10–15. DOI: 10.17557/tjfc.83589*.

Salari, M., et al. Identification of Physiological Races of *Blumeria graminis* f. sp. *tritici* and Evaluation of Powdery Mildew Resistance in Wheat Cultivars in Sistan Province, Iran. *Commun. Agric. Appl. Biol. Sci.* **2003,** *68,* 549–553.

Schneider, D. M.; Heun, M. New Resistance Genes against Powdery Mildew in Diploid and Tetraploid *Triticum* Species. *Mitt. Biol. Bundesanst. Land Forstwirtsch. Berlin-Dahlem.* **1988,** *245,* 311.

Sebastian, S. A., et al. Inheritance of Powdery Mildew Resistance in Wheat Line IL-72–2219-I. *Plant Dis.* **1983**, *67*, 943–945.

Shaner, G. Evaluation of Slow Mildewing Resistance of Knox Wheat in the Field. *Phytopathology.* **1973b**, *63*, 867–72.

Shaner, G. Reduced Infectivity and Inoculum Production as Factors of Slow Mildewing in Knox Wheat. *Phytopathology.* **1973a**, *63*, 1307–1311.

Shaner, G.; Finney, R. E. The Effect of Nitrogen Fertilization on the Expression of Slow-Mildewing Resistance in Knox Wheat. *Phytopathology.* **1977**, *67*, 1051–1056.

Sharma, S. C.; Basandrai A. K.; Aulakh K. S. Virulence in *Erysiphe graminis* f sp. *Tritici* Causing Powdery Mildew in Wheat. *Pl. Dis. Res.* **1990**, *5*, 115–117.

Sharma, T. R.; Singh, B. M.; Basandrai A, K. Role of Cleistothecia in Annual Recurrence of Wheat Powdery Mildew in North India. *Indian Phytopath.* **1992**, *45*, 203–206.

Sharma, S. C., et al. Residual Resistance in Wheat to *Erysiphe graminis* f. sp. *tritici*. *Pl. Dis. Res.* **1991b**, *6*, 70–74.

Sharma, S. C., et al. Expression of Resistance of Wheat at Different Growth Stages of Powdery Mildew. *Pl. Dis. Res.* **1991a**, *6*, 14–18.

Sharma, S. K., et al. Performance of Wheat Cultivars against Powdery Mildew. *Indian Phytopath.* **1979**, *32*, 137–138.

Sharma, T. R.; Singh, B. M. Virulence Structure of *Erysiphe graminis tritici* in Himachal Pradesh. *Indian Phytopath.* **1990b**, *43*, 165–169.

Sharma, T. R.; Singh B. M. Development of Powdery Mildew on Different Wheat Cultivars. *Indian Phytopath.* **1990c**, *43*, 170–174.

Sharma, T. R.; Singh, B. M. Evaluation of Powdery Mildew Resistance in Some Indian Wheats. *Indian Phytopath.* **1990a**, *43*, 26–32.

Sharma, T. R.; Singh, B. M. Physiologic Races of *Erysiphe graminis tritici* in Himachal Pradesh. *Indian Phytopathol.* **1990d**, *43*, 165–169.

Shen, X. K. Identification and Genetic Mapping of the Putative *Thinopyrum intermedium*-Derived Dominant Powdery Mildew Resistance Gene *PmL962* on Wheat Chromosome Arm 2BS. *Theor. Appl. Genet.* **2015**, *128*, 517–528.

Singh, B. M.; Sood, A. K. Seedling Reaction of Spring and Winter Wheat Cultivars to Powdery Mildew. *Indian Phytopathol.* **1977**, *30*, 277–279.

Singh, D. P., et al. Powdery Mildew Resistant Genotypes in Wheat and Triticale. *Indian Phytopathol.* **2005**, *58*(1), 124.

Singh, D. P., et al. Identification of Resistant Sources . Against Powdery Mildew (*Blumeria graminis*) of Wheat. *Indian Phytopath.* **2016**, *69*, 413–415.

Singh, K. P.; Ramesh, P. Evaluation of Fungicides against Powdery Mildew of Wheat. *Indian Phytopath.* **2000**, *53*(2), 230–231.

Slesinski, R. S.; Ellingboe, A. H. The Genetic Control of Primary Infection of Wheat by *Erysiphe graminis* f. sp. *tritici*. *Phytopathology.* **1969**, *59*, 1833–1837.

Smith, H. C.; Blair, I. D. Wheat Powdery Mildew Investigations. *Ann. Appl. Biol.* **1950**, *37*, 570–583.

Solc, C.; Paulech, C. Physiological Races of the Fungus *Erysiphe graminis* f. sp. *tritici* Marchal Found in Slovakia in 1973–1974. *Pol'nohospodarstvo.* **1977**, *3*(10), 865–872.

Song, J., et al. Isolation and Identification of Differentially Expressed Genes Irom Wheat in Response to *Blumeria graminis* f. sp. *tritici* (Bgt). *Plant Mol. Biol. Report.* **2015**, *33*, 1371–1380

Stojanovic, S., Ponos, B. Virulence spectrum of *Erysiphe graminis* DC ex-Merat f.sp. *tritici* Em. Marchal Populations in S. East Part of Yugoslavia in 1986 and 1987. *Zast. Bilja.* **1990**, *41*, 41–47.

Stojanovic, S., et al. Virulence of *E. graminis* DC ex-Merat f. sp. *tritici* Em. Marchal Genotypes Proliferated by Sexual Reproduction. *Zast. Bilja.* **1991**, *42*, 7–19.

Stojanovic, S.; Stojanovic, J.; Ognjanovic, R. Review of Investigations of *E. graminis* DC ex. Merat f. sp. *tritici* Em. Marchal Population in Yugoslavia. *Zast. Bilja.* **1990**, *41*, 463–473.

Streckeisen, P.; Fried, P. M. Analysis of the Virulence of Wheat Powdery Mildew in Switzerland in 1981 to 1983. *Schweiz. Landwirtsch. Forsch.* **1985**, *24*, 261–269.

Suryanarayana, D.; Goel, L. B.; Sinha, V. C. Performance of Some of the Dwarf Wheats against Powdery Mildew Under Shimla Conditions. *Indian Phytopath.* **1971**, *24*, 605–607.

Szunics, L., et al. Dynamics of Changes in the Races and Virulence of Wheat Powdery Mildew in Hungary between 1971 and 1999. *Euphytica.* **2001**, *119*, 145–149.

Tao, J., et al. Varietal Resistance of Wheat to Powdery Mildew. *Acta Phytopath. Sin.* **1982**, *12*, 7–14.

Tao, W., et al. Genetic Mapping of the Powdery Mildew Resistance Gene *Pm6* in wheat by RFLP Analysis. *Theor. Appl. Genet.* **2000**, *100*, 564–568.

Tomar, S. M. S.; Menon, M. K. *Genes for Resistance to Rusts and Powdery Mildew in Wheat;* Indian Agricultural Research Institute: New Delhi, India, 2001; p 152.

Tomerlin, J. R., et al. Resistance to *Erysiphe graminis* f.sp. *tritici, Puccinia recondita* f.sp. *tritici* and *Septoria nodorum* in wild *Triticum*Species. *Plant Dis.* **1984**, *68*, 10–13.

Tosa, Y.; Mise, K.; Shishiyama, J. Recognition of Two Formae Speciales of *Erysiphe graminis*of Wheat Cells. *Annals Phytopath. Soc. Japan.* **1985**, *51*, 223–226.

Upadhyay, M. K.; Kumar, R. Field Reaction to Powdery Mildew Strains of Wheat Lines Possessing Known Genes for Resistance. *Indian J. Genet. Plant Breed.* **1974**, *34*, 150–155.

Van Silfhout, C. H.; Gerechter-Amitai, Z. K. A Comparative Study of Resistance to Powdery Mildew in Wild Emmer Wheat in the Seedling and Adult Plant Stage. *Neth. J. Pl. Path.* **1988**, *94*, 177–184.

Vechet, L. Incidence and Development of Powdery Mildew (*Blumeria graminis* f. sp. *tritici*) in the Czech Republic in the Years 1999–2010 and Race Spectrum of this Population. *J. Life Sci.* **2012**, *6*, 786–793.

Vershinina, V. A., et al. Sources of Powdery Mildew Resistance in Winter Bread Wheat in the North West of the RSFSR. *Nauchno tekhnicheskii Byulleten' Vsessoyuznogo Ordina Lenina i Ordena Druzhby Narodov Naucho-issledovatel'skogo Institute Rastenievodsta Imeni NI Vavilova.* **1984**, *142*, 47–49.

Wang. Z., et al. Genetic and Physical Mapping of Powdery Mildew Resistance Gene *MlHLT* in Chinese Wheat Landrace Hulutou. *Theor. Appl. Genet.* **2015**, *128*(2), 365–373.

Wang, Y. J., et al. Identification and Mapping of PmSE 5785, a New Recessive Powdery Mildew Resistance Locus in Synthetic Hexaploid Wheat. *Euphytica.* **2016**, *207*, 619–626.

Wen, C. J.; Tao, J. F. Relation between Papilla Formation and Infection of *Erysiphe graminis tritici* in Wheat. *Acta Phytopath. Sin.* **1989**, *19*, 17–20.

West, S. J. E. The Effect of Fungicides on Yield and Quality of Winter Wheat. *Aspects appl. Biol.* **1990**, *25*, 349–353.

Wolfe, M. S.; Schwarzbach, E. Patterns of Race Changes in Powdery Mildew. *Ann. Rev. Phytopathol.* **1978**, *16*, 159–180.

Wolfe, M. S. Physiological Specialization of *Erysiphe graminis* f. sp. *tritici* in United Kingdom. *Trans. Br. Mycol. Soc.* **1965**, *48*, 315–326.

Worthington, M., et al. *MlUM15*: An *Aegilops neglecta*-Derived Powdery Mildew Resistance Gene in Common Wheat. *Crop Sci.* **2014**, *54*, 1397–1406.

Wozniak-Strzembicka, A. Physiologic Differentiation of Wheat Mildew*Erysiphe graminis* f.sp. *tritici* in Poland in 1975–76 and 1978–79. *Hodowla Rosl. Aklim. Nasienn,* **1982,** *26,* 115–123.

Wu, O. S.; Liu, D. J. Physiological Specialization of *Erysiphe graminis* f. sp. *tritici* in Jiangsu Province. *J. Nainjing Agric. College.* **1983,** *4,* 9–15.

Wu, W., et al. Relation of Haustoria to Wheat Powdery Mildew Resistance. *Acta Phytopath. Sin.* **1985,** *15,* 31–35.

Xin, M., et al. Transcriptome Comparison of Susceptible and Resistant Wheat in Response to Powdery Mildew Infection. *Genomics Proteomics Bioinformatics.* **2012,** *10*(2), 94–106.

Xiong, E. H., et al. Assessment of Varietal Resistance to Powdery Mildew in Wheat. *Zhiwii Baohu (Plant Protection).***1983,** *3,* 5–7.

Xu, Z.; Duan, X.; Zhou, Y., et al. Population Genetic Analysis of *Blumeria graminis* f. sp. *tritici* in Qinghai Province, China. *J. Integ. Agric.* **2014,** *13*(9), 1952–1961.

Xue, Fei Wang.; Changyou, Li, Cong., et al. Molecular Mapping of a Powdery Mildew Resistance Gene in Common Wheat Landrace Baihulu and its Allelism with *Pm24. Theor. Appl. Genet.* **2012,** *125*(7), 1425–1432.

Yao, J. Q.; Shang H. S.; Li, Z. Q. On Irregular Appressoria of *Blumeria graminis. Acta Phytopath. Sin.* **1998,** *28,* 215–219.

Yu, S., et al. Localization of the Powdery Mildew Resistance Gene *Pm07J126* in wheat (*Triticum aestivum* L.). *Euphytica.* **2015,** *205*(3), 691–698.

Zadoks, J. C. Yelow Rust on Wheat, Studies in Epidemiology and Physiologic Specialization. *Neth. J. Plant Path.* **1961,** *67,* 69–256.

Zeng, F. S.; Yang, L. J.; Gong, S. J.; Shi, W. Q.; Zhang, X. J.; Wang, H.; Xiang, L. B.; Xue, M. F.; Yu, D. Z. Virulence and Diversity of *Blumeria graminis* f. sp. *tritici* Population in China. *J. Integr. Agric.* **2013.** Doi: 10.1016/S2095-3119(13)60669-3.

Zhan, H., et al. Chromosomal Location and Comparative Genomics Analysis of Powdery Mildew Resistance Gene *Pm51* in a Putative Wheat-*Thinopyrum ponticum* Introgression Line. *PLoS One,* **2014,** *9*(11), e113455.10.1371/journal.pone.0113455.

Zhang, R., et al. Development and Characterization of a *Triticum aestivum* –D. *Villosum* T5VS.5DL Translocation Line with Soft Grain Texture. *J. Cereal Sci.* **2010,** *51,* 220–225.

Zhang, R. et al. *Pm55,* A Developmental-Stage and Tissue-Specific Powdery Mildew Resistance Gene Introgressed from Dasypyrum Villosum into Common Wheat. *Theor. Appl. Genet.* **2016,** *129,* 1975–1984. doi: 10.1007/s00122-016-2753-8. *Epub* **2016** Jul. 15.

CHAPTER 6

MANAGEMENT OF KARNAL BUNT AND LOOSE SMUT DISEASES IN WHEAT

RITU BALA[1*], JASPAL KAUR[1], and INDU SHARMA[2]

[1]*Department of Plant Breeding and Genetics, Punjab Agricultural University, Ludhiana 141004, India*

[2]*ICAR-Indian Institute of Wheat and Barley Research, Karnal 132001, Haryana, India*

*Corresponding author. E-mail: rituraje2010@pau.edu

CONTENTS

ABSTRACT

The Karnal bunt of wheat caused by *Tilletia indica* Mitra is an important disease of wheat due to quarantine regulations imposed by several countries for import of Karnal bunt (KB) free wheat. The disease is difficult to manage due to its air borne, soil borne and seed borne nature. Several methods have been employed for its management but none of them gave 100% effectiveness. The loose smut (LS) of wheat caused by *Ustilago segatum tritici* is a menace to wheat as it has a potential to cause 100% loss of the yield. The disease can be effectively controlled by chemicals. But the management by chemicals is neither economic nor eco-friendly. Thus a review on the management of KB and LS is given here to devise effective, eco-friendly and economic strategies for the management of these diseases.

6.1 INTRODUCTION

Bunts and smuts of wheat are next to rusts in importance. The pathogens causing these diseases belong to order *Ustilaginales* of the class *Teliomycetes* and subdivision-*Basidiomycotina*. The important smuts and bunts associated with the wheat are Karnal bunt (KB), loose smut (LS), and flag smut. These pathogens produce characteristic thick walled brown or black spores—"brand spores"—which impart a sooty appearance. In most of the bunts and smuts except a few, the damaging part is grain; hence, apart from reducing yield, they also affect the quality of harvested grains. Hence, there is a dire need to avoid or manage these diseases in wheat crop. Out of all the smuts and bunts, a brief introduction about the KB and is their etiology, epidemiology with detailed management strategies are discussed in this chapter.

6.2 KARNAL BUNT

KB, kernel smut, or partial bunt is caused by a *Basidiomycetous* fungus, *Tilletia indica* (Syn. *Neovossia indica*). First observed by Mitra (1931) in experimental wheats grown at the Botanical station, Karnal in 1930 is now reported from several parts of India, Pakistan, Syria, Afghanistan, Mexico, Nepal, USA, Iran, and South Africa (Locke & Watson, 1955; Duran, 1972; Williams, 1983; Singh et al., 1989; Ykema et al., 1996; Torabi et al., 1996; Singh et al., 1998; Crous et al., 2000). In India, the disease is endemic to northwestern regions, foothills of north with sporadic occurrence in Bihar, West Bengal, Gujarat, and Madhya Pradesh (Swaminathan et al., 1971; Joshi et al., 1980; Singh et al., 1980; Gill et al., 1993; Pandey et al., 1994; Singh & Gogoi, 2011).

In Punjab, severity of the disease was highest during 2014–15 (2.24%) followed by 2005–06 (0.95%) and 1995–96 (0.62%) (Fig. 6.1). Earlier, the disease with high intensities was confined to the humid sub-mountainous districts like Gurdaspur, Ropar, and Hoshiarpur (Kang & Bedi, 1980; Aujla et al., 1986; Sharma et al., 1998, 2004). But during the last four years, disease is now becoming prevalent in all the districts of the Punjab. The factors contributing to the increase in KB incidence and prevalence are varietal susceptibility as all the cultivated varieties are susceptible and further the favorable environmental conditions for the pathogen multiplication and infection at heading stage of wheat (Kaur et al., 2015).

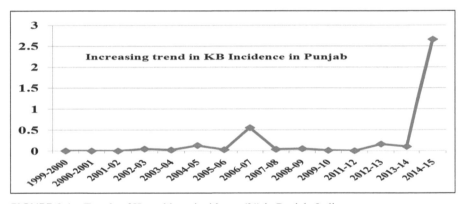

FIGURE 6.1   Trends of Karnal bunt incidence (%) in Punjab, India over years.

With minimum impact on crop yield, the KB greatly reduces quality of the flour and palatability of chapattis. Further, there was discoloration of the finished product. Besides this, the disease also had impact on seed weight, germination, and vigor depending upon severity of infection (Mehdi et al., 1973; Munjal, 1975; Aujla et al., 1983; Joshi et al., 1983; Bansal et al., 1984a; Warham, 1990). But, the most important losses due to KB are quarantine restrictions (both international and domestic) and the cost of grain fumigation. The total loss due to disease was estimated to be 7.02 million US dollars/ year (Brennan et al., 1990). The disease is rated as serious non-tariff barrier in world wheat trade. About 70 foreign countries enforce a quarantine restriction on wheat trade from the countries having

KB in import shipments (Lari et al., 2006). In the USA, appearance of the disease in the early 1996 led to the establishment of animal and plant health inspection service (APHIS) implementing an emergency quarantine, inspection, and certification program for wheat moving out of the infested areas, along with regulations on sanitizing machinery, and storage facilities.

A pest risk analysis (PRA) developed under the auspices of European commission rated KB as a serious disease and *T. indica* as a 1/A1 harmful pathogen organism under EU legislation (Sansford et al., 1998).

The fungal pathogen responsible for this disease was originally named as *T. indica* (Mitra, 1931) and later named as *N. indica* on the basis of formation of non-septate promycelium with a whorl of non-fusing primary sporidia at the apex and formation of apiculate teliospores (Mundkur, 1940). The controversies over the name of the fungus as *Neovossia* or *Tilletia* is still continued (Fischer & Holton 1957; Duran & Fischer, 1961; Khanna & Payak, 1968; Duran, 1972; Krishna & Singh, 1982). However, commonwealth mycological institute has retained the fungus as *T. indica* (Anonymous, 1983).

The disease becomes evident only when grains have fully developed. The infected spikes can be detected by the shiny silvery black spikelets with glumes spread apart and swollen ovaries. The pathogen converts the infected ovary into sorus where in a mass of dark brown colored teliospores are produced which remained covered by pericarp. Only a few kernels of some spikes are infected and usually only a portion of an infected kernel is replaced with fungal sorus. However, infection in kernels varies from small point of infection to completely bunted kernels and the embryo is not infected but gets shriveled under very severe infection (Fig. 6.2) (Mitra, 1935; Chona et al., 1961; Gill et al., 1981; Goates et al., 1988).

The initiation and development of the KB is highly dependent on suitable weather conditions during flowering, which is the most susceptible stage to infection. High-humidity (70%), low-temperature (19–23°C), continuous rainy/foggy, and cloudy weather for > 13 days from ear emergence to anthesis have been reported to favor the disease in different years at many places resulting into disease epiphytotics (Munjal, 1971; Aujla et al., 1977; Singh & Prasad, 1978). If at the time of anthesis, the temperature ranges from 8–10°C to 19–23°C coupled with slight shower, it results in well establishment of disease. Several other disease prediction models have been developed (Jhorar et al., 1992; Singh et al., 1996; Smiely, 1997).

FIGURE 6.2   Karnal bunt infected seeds.

The KB pathogen is soil borne and infection may take place if healthy seeds are sown in infested soil. Air borne nature of the fungus has also been proved (Mitra, 1937; Mundkur, 1943; Bedi et al., 1949). The telio-spores may be deposited into the soil at the time of harvesting, threshing, or winnowing, or they may become attached to the surface of the seed as an external contaminant. In India, the germination of teliospores took place from the middle of February to March when soil temperature and moisture are suitable. The survival of teliospores varied from two years in wheat straw, soil, and farmyard manure to four years in soil (Munjal, 1970; Aggarwal et al., 1977; Dhiman, 1982; Sharma & Nanda, 2002). Teliospores are diploid (2n) in nature and germinate giving rise to a long and sometimes unbranched promycelium bearing a whorl of monokaryotic haploid primary sporidia at the tip (Mitra, 1937; Mundkur, 1940; Holton, 1949; Krishna & Singh, 1981) under optimu m conditions of temperature (18–25°C) and rela-tive humidity (> 70%). These secondary sporidia are produced by primary sporidia. (Figs. 6.3-6.7). These secondary sporidia are air borne and lodge onto plant surfaces by air currents and rain splash or monkey jumping of sporidia from the soil surface to the ear heads. They may germinate and produce additional generations of secondary sporidia on the wheat heads and cause infection (Gill et al., 1981; Bedi, 1989; Bains & Dhaliwal, 1989; Dhaliwal & Singh, 1989; Gill et al., 1993; Nagarajan et al., 1997). Wheat spikes are susceptible from early head emergence to late anthesis (Duran & Cromarty, 1977; Warham, 1986; Bonde et al., 1997). *T. indica* is hetero-thallic and its successful infection and reproduction depend upon dikaryo-tization between airborne secondary sporidia of different mating types. The site of dikaryotization is not fully known, although apparent hyphal anasto-mosis has been observed on the glume surfaces. Sporidial germ tubes pene-trate through stomata of glumes, lemmas, and palea. In some cases, there may be direct penetration of immature seed (Aujla et al., 1988; Dhaliwal et al., 1988; Goates, 1988; Salzar- Huerta et al., 1990). A series of studies has been conducted on compatibility system of *T. indica* and both bipolar and tetrapolar compatibility systems have been proposed and heterothallism and pathogenicity in *T. indica* is controlled by multiple alleles at one locus (Krishna & Singh, 1983; Royer & Rytter, 1985; Aujla & Sharma, 1990; Chahal et al., 2003).

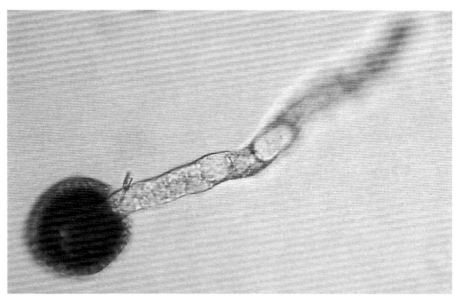

FIGURE 6.3   Germinating teliospore (GT) producing promycelium (PM).

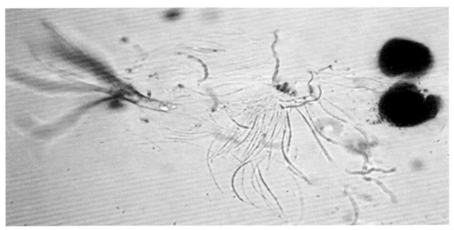

FIGURE 6.4   Promycelium (PM) bearing primary sporidia (PS).

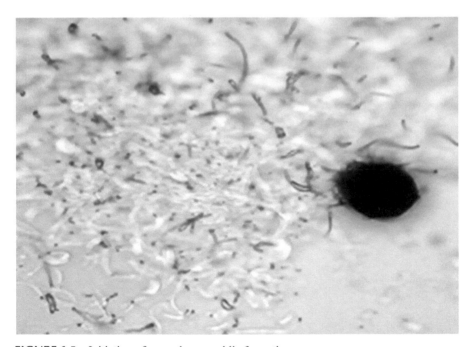

FIGURE 6.5   Initiation of secondary sporidia formation.

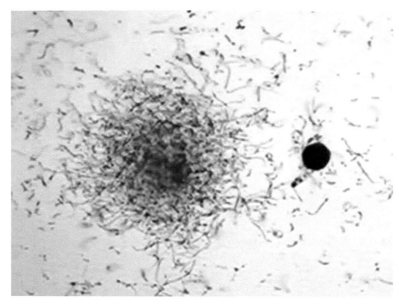

FIGURE 6.6   Development of colonies from germinating teliospores in cultures.

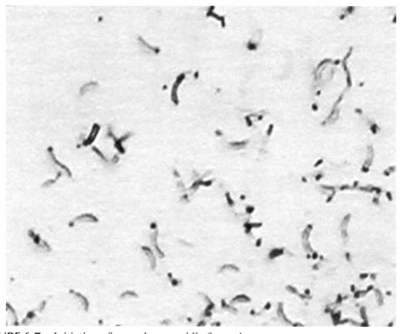

FIGURE 6.7   Initiation of secondary sporidia formation

T. indica is a highly variable fungus and this greater variation in *T. indica* is believed to be due to high level of outcrossing (Bonde et al., 1997). On the basis of teliospore morphology and teliospore size, races were observed (Mitra, 1935). The range of teliospore size from samples collected from different states and type specimen was very close so the conclusion was drawn that teliospore size was a slightly variable character and is influenced by the environmental factors, hence it should not be taken as a criterion to differentiate the different strains of *T. indica* (Bansal et al., 1984b). Further, no significant difference in the teliospore diameter, characteristics of primary and secondary sporidia were observed in the Indian and Mexican collections (Peterson et al., 1984).

The concept of races in the *T. indica* was declared to be controversial (Bonde et al., 1997) although; initially races/pathotypes were demonstrated in *T. indica* (Mitra, 1935; Munjal, 1970; Aujla et al., 1987). Thus, pathogenic variability in *T. indica* has not been assigned the race/pathotype approach as in each cycle nuclear fusion takes place heterothallic sporidia so variation in isolates can be classified as aggressiveness. Keeping this in view, different number of aggressive isolates have been demonstrated using a set of over-lapping differentials five aggressive (KB Ag) isolates have been identified by Singh et al. (1995). But, existence of gene for gene relationship was indicated in a *Triticum aestivum* line HD 29 using two isolates. Similarly, eight isolates of *T. indica* collected from different locations in Northern India were distinguished on the basis of differential reactions on 13 host lines distinguishing even the most resistant lines of durum (PDW 215), triticale (TL 1210), and wheat (HD 29) by the differential disease reactions with one or more isolates (Datta et al., 1999). Further, nine isolates of *T. indica* isolates were divided into six groups based on their pathogenic behavior on one triticale and 10 wheat varieties. They differed from each other with respect of percent teliospores germination and production of primary and secondary sporidia. A positive correlation was also observed between secondary sporidial production and resultant disease incidence and virulence (Pannu & Chahal, 2000). Pathogenic variability in *T. indica* isolates collected from North and Northwestern India (NW India) on 18 differential hosts identified three distinct aggressive types KB-AgI, KB-AgII, and KB-AgIII from different locations by Thirumalaisamy et al. (2006) and was further confirmed by Aggarwal et al., 2010. Furthermore, isozyme analysis has established fairly good evidence in favor of the existence of genetic variability (Bonde et al., 1985; Bonde et al., 1988, 1989; Kumar et al., 1995; Sharma et al., 1998a). But with the advent of DNA-based markers, a new option was available to investigate pathogen variability. Random amplified polymorphic DNA (RAPD) has been employed successfully to elucidate genetic relationships among different species of *Tilletia*, except *T. indica*. Different types polymerase chain reaction-based assay in conjunction with six short arbitrary RAPD primers to determine genetic variability among 15 isolates of *T. indica*, collected from different regions of Punjab found four RAPD groups with 17–35% similarity (Mishra et al., 2000). Datta et al. (2000) reported a genetic diversity analysis in Indian isolates of *T. indica*. A library of isolate Ni7 was constructed, and three repetitive elements (pnir9, pnir12, and pnir16) were identified for molecular analysis. Immuno-pathotyping of *T. indica* isolates was carried out using anti-mycelial antibodies (Varshney et al., 2003). These assays were employed for immune analysis of diversity among KB pathogen. Kumar et al. (2004) studied the pathogenic and molecular variations in a large number of isolates of *T. indica* by using PAGE and revealed the presence of three protein types. Monosporidial lines of *T. indica* were differentiated using universal rice primers thus proving the heterogeneity in the pathogen population (Aggarwal et al., 2010). Recently, pathogenic variability test revealed the existence of three different aggressive groups among the 34 selected isolates on a set of host differentials. A total of 64 MLs were developed from the isolates, which originated from six states, representing 19 locations in the northwestern plains zone of India. Self-paired MLs revealed two mating alleles. A total of 15 mating alleles were postulated using 46 MLs. This study confirmed the heterothallic nature of *T. indica* and the existence of multiple alleles controlling pathogenicity (Parveen et al., 2015).

6.2.1   DISEASE MANAGEMENT

The multiple modes of transmission of the pathogen and several attributes of the etiology, pathogen biology of KB along with the highly resistant and long-lived teliospore make its control very difficult. To manage KB, we need to first combat the disease in endemic areas, where the pathogen is already established and followed by preventing the entry of pathogen in disease free areas. Hence, management strategies for KB can be classified into four broad categories:

1. Management through cultural practices
2. Management through chemicals
3. Management through bio-agents
4. Management through deployment of host resistance

6.2.1.1   CULTURAL PRACTICES

The management through cultural methods usually includes alteration in the standard agronomic practices of the plant right from the seed sowing till storage of seed after harvesting. The objective is to modify the crop environment in a way so that it becomes unfavorable for the pathogen infection as well as disease development. Thus, the cultural control starts right from the selection of disease free seed followed by reduction in seed rate or planting density. The modification in crop environment may also be due to adjustments in sowing dates. In case of KB, both early and delayed sowing can manage the disease. The early sowing may result in disease escape due to non-availability of susceptible stage of host for the pathogen. While delayed planting avoid the coincidence between favorable weather conditions conducive for disease infection and the most susceptible stage of the host that is, heading. The effectiveness of delayed planting can be further enhanced if rainfall and optimum environmental conditions are avoided between awn emergence and the end of flowering. Reduction in nitrogen fertilizer amounts and alteration in irrigation schedule also help slightly in reducing the disease incidence. KB is most likely to be found in areas where lodging or water ponding have occurred. If we can avoid the irrigation at the time of heading, KB can be controlled to some extent. Burning of straw is another practice to reduce the KB. Intercropping of non-host crops and non-cultivation of wheat for two consecutive years can also reduce the disease by affecting the soil-borne primary inoculums (Singh & Singh, 1985). By changing rice-wheat-rice cropping system to Maize–wheat cropping system reduces the teliosporic counts in soil hence less disease. Certain cultural practices like crop rotation and disinfection of the soil help in reduction of teliospore load in the soil. The wheat can be rotated to durum wheat, triticale, or crops other than wheat. This practice can be practiced for at least five years if the soil has become infested. Soil solarization for 21–28 days in the month of May and June with polythene mulching was found effective in reduction of KB incidence by raising the soil temperature and thus rendering the teliospore non-viable due to high-soil temperature (Singh et al., 1998). Mulching of white plastic sheets to reduce the soil borne inoculum was also suggested. Amendments of soil with FYM, decomposable organic matters, different cakes, dust, wheat straw and biological mulches like chickpea, and sugarcane refuge also reduced the KB incidence (Mitra, 1937; Padwick, 1939; Singh & Singh, 1985; Singh et al., 1991a; Siddhartha, 1992; Gill et al., 1993; Singh et al., 1998; Sharma & Bassandrai, 1999). Zero tillage in rice-wheat cropping system along with tilling the highly infested soils can lower the disease incidence and effects dispersion of teliospores, respectively (Sharma et al., 2007; Allen et al., 2008). Cultural practices may be partially effective in controlling KB but cannot eliminate the disease completely. Most of the cultural practice used can suppress disease development but not eliminate the disease and further some of them will reduce the yield.

6.2.1.2 CHEMICALS

Management through chemicals gave a quick response for rusts and smuts but for bunts, this is a little tricky due to complexities involved in disease cycle. A number of chemicals have been used for the control of KB treating soil, seed, as well as foliar spray. A number of chemicals have been screened to be used as soil fumigants, seed dresser, and for spraying. However, the methyl bromide is the most effective soil fungicide which resulted in 99% reduction in teliospore germination (Singh & Krishna, 1980; Singh et al., 1985; Smilanick, 1986; Hoffman, 1986; Warham, 1986; Smilanick et al., 1987).

Several chemicals and plant extracts that inhibit teliospore germination are available. Ethyl mercury chloride, ethyl mercury, phosphate, fentin acetate, triphenyltin chloride, fentin hydroxide, oxycarboxin, benomyl, carbendazim, triadimenol, methfuroxan, tetramethylthiuram disulfide, indar, aureofungin, carboxin/thiram, difenoconazole/mefenoxam, PCNB 24% EC, and tebuconazole/thiram can be used as dry or slurry seed dressing fungicides (Gill et al., 1993). Among these, mercurial compounds (ethyl mercury chloride) can kill the teliospores but these are banned in most of the countries. Sodium phosphate salts have an inhibitory effect on teliospore germination. Timely applications of a phosphate compound to infested field soils may inhibit or delay teliospore germination during the wheat infection window and thereby reduce disease incidence. Some other available options are seed treatments with formaldehyde, ethanol, sodium hypochlorite, hot water treatment 60°C, dry heat treatment at 120–140°C for 10 min, common bleach, chlorine dioxide, quaternary ammonium solution, fumigation with methyl bromide, sulfur dioxide, and chloropicrin but again these also interfere with seed germination and are not very effective. The chemicals used for seed treatment have fungistatic rather than fungitoxic properties further seed treatments are partially ineffective. Use of ultrasonic vibration and radiation has also been employed, but it is also not so successful on the account of the feasibility to expose large quantities of seeds to kill seed borne teliospores and further there are adverse effect on seed germination (Singh et al., 1985; Prescott, 1986; Smilanick et al., 1987; Sharma & Nanda, 2002; Glenn & Peterson, 2005). Seed treatment does little to eliminate soil-borne inoculum. Further, seed treatment fungicides do not protect wheat plants from infection when seeds are planted in teliospore-infested soil, and do not persist long enough within the plant to inhibit the infection of florets. The foliar application of fungicides at anthesis stage is the most effective option (Smilanick et al., 1987). Mancozeb (0.25%), carbendazim (0.1%), triadimefon (0.2%), propiconazole (0.1%), cyproconazole (SAN 619), diniconazole (S308), fentin hydroxide, bitertanol, etc. have been reported effective against the KB. Folicur (0.2%), Contaf (0.1%), tilt (0.1%), and 100 g a.i. thifluzamide/ha have also been recommended (Goel et al., 2000b; Sharma et al., 2005). But the success for KB management has been seen only with the foliar applications of triazole group of chemicals, for example, tebuconazole, cyproconazole, propiconazole (tilt), and hexaconazole and the best example is in the form of a recommendation of a single spray of propiconazole (Tilt @ 0.01%) in Punjab was found highly effective in the management of the disease in seed production plots and the most successful chemical for KB management. Recently, 100% disease control and maximum yield was achieved due to the effect of seed treatment and two additional sprays of Tilt 25EC. Seed treatment and one foliar spray of Tilt25EC resulted in 89.96% disease control (Kumar & Singh, 2014). part from this, some weedicides like isoproturon, stomp, and 2,4-D inhibits the germination of teliospores and sporidia. Some other foliar fungicides are highly effective against KB, but none is currently registered for use (Smilanick et al., 1987b; ill et al., 1993; Goel, 2000, 2000a, 200b; Sharma et al., 1997; Singh et al., 1998; Sharma et al., 2005).

6.2.1.3 BIO-AGENTS

Several biocontrol agents including fungal, bacterial, and actinomycetes have been used to control the KB. Among these, the most exploited ones are the fungal bio-agents. Four BCA (fungi), viz., *Gliocladium deliquescens*, *G. virens*, *Trichoderma harzianum*, and *T. viride* were used to evaluate their antagonistic potential against the teliospores and sporidia of *T. indica* and on KB disease. There was a reduction in sporidia formation in cultural filtrate treated bits (Singh et al., 1991b). A large number of fungal, bacterial and actinomycetes isolates were tested for their antagonism against *T. indica* by dual culture method. Out of these, *T. koningii*, *T. lignorum*, *G.*

deliquescens, G. virens, bacteria isolates Bact-II, Bact-V, and actinomy-cetesisolates Act-V were observed to be either overgrowing or inhibiting the growth of *T. indica.* SEM observations revealed that the hyphal tips of *T. koningii* entered the hypha of *T. indica* by direct penetration without the formation of penetration peg or appressoria, while in *T. lignorum* the penetration was mediated through penetration pegs, which resulted in disintegration of mycelium of host fungus. The interaction between *G. deliquescens* and *T. indica* showed that the antagonist penetrated the host cell either through penetration pegs or formation of appressoria like structures while *G. virens* after penetration caused deformation and lysis of host mycelium. Bacteria isolates Bact-II and Bact-V showed clear inhibition zone causing lysis and deformation of *T. indica* mycelium. Actinomycetes isolates Act-V penetrated the host mycelium through penetration pegs and partially degraded the cell wall causing holes (Sharma et al., 1996). The antagonistic potential of *T. viride, T. harzianum,* and *G. deliquescens* on teliospore germination of *T. indica* using homogenized cultures resulted in more than 80% disease control in artificially inoculated ears by inhibiting the germination of teliospores of *T. indica* (Singh et al., 1998). *T. pseudo-koningii, T. lignorum, T. koningii, G. deliquescens,* and *G. virens* antago-nize teliospores and secondary sporidia and inhibit their germination while *T. lignorum* stimulate the growth of wheat plants (Amer et al., 1998). Some of the phytoextracts like dichloromethane (DCM) extract of *Chenopodium ambrosioides, Encelia farinosa,* and *Larrea tridentate* (500 mg/ml) reduced radial mycelial growth and found total inhibition of *T. indica.* They mark-edly differed in their fungitoxicity to *T. indica* (Rivera- Castaneda et al., 2001). Leaf extracts of *Vitex negundo, Cassia fistula, Azadirachta indica, Eucalyptus tereticornis,* and *Lantana camara* have been used and a65% of disease control has been achieved by spraying crude leaf extracts of neem, and lantana on wheat at boot leaf stage, 48 h after inoculation (Sharma & Basandrai, 1999, 2004).

6.2.1.4   HOST RESISTANCE

The management approaches mentioned above are only partially effec-tive. Cultural practices can reduce KB infection but also reduce the yield of the crop. Soil fumigations are not cost effective and can also reduce seed germination. Seed treatments are only effective when these are going to be sown in soils not infested with teliospores. Foliar fungicide sprays have achieved significant control of KB but more studies are needed to be carried out on timings of application particularly with natural infection. Further, residues of fungicides applied as foliar spray in the grain may pose a regulatory issue. Bio-control in field conditions offers several limitations. Hence, the most effective, eco-friendly, economic, and long-term strategy to minimize disease development is the identification and development of resistant varieties.

The process of identification of KB resistant sources was initiated in year 1949 in Gurdaspur, Punjab with the identification of KB free durum wheat and a wheat cultivar C 253 showing 9% KB infection (Bedi et al., 1949). Later on, resistance was identified in some of the aestivum-durum wheats and triticales (Chona et al., 1961; Munjal, 1971; Meeta et al., 1980). Till 1979, data on screening was collected from hot spot locations only under natural epiphytotics. In 1980s, with the development of "syringe inoculation method" at PAU, Ludhiana, the KB screening program included a large number of wheat germplasm and allied spp. under artificial epiphytotic conditions at PAU, AICWIP, CIMMYT, and Mexico (Aujla et al., 1980, 1982).

During the process of identification of KB resistant lines in the year 1980, 286 lines from AICWIP were evaluated at the Ludhiana center under artificial conditions using the isolates of the pathogen collected from WL 711, HD 2009, and WG 357 varieties grown in different agro climatic zones of Punjab. In this year, most of the lines scored high disease and only 10 lines showed up to 10% infection. In the first few years, none of the wheat genotypes showed highly resistant or immune reactions using "syringe inoculation method" (Aujla et al., 1980; Gill et al., 1981).

With more lines being brought under testing program, some of the lines remained disease free for 1–3 years before succumbing to a low level of susceptibility. Subsequently as a practical norm, lines showing up to 5% infection were rated as being resistant. As a result, a number of KB resistant varieties have been identified viz., DWL 1023, DWL 5010, HD 29, HD 2300, HD 2499, HD 4564, HD 4571, HS 346, HP 1531, HW 1045, HUW 453, MG 173-1-2-4, PBW 154, PBW 213, PBW 233, Raj 1707, Raj 2071, Raj 2296, WH 805 and WL7104 (Aujla et al., 1985; Gill et al., 1993, Singh et al., 1998). A number of resistant genotypes to KB were also identified through multilocation testing by Goel et al. (2000a).

Further at PAU, 835 lines found to show consistently low disease for 3–20 years identified by screening more than 45,000 genotypes obtained from national and international nurseries. These 835 KB resistant lines have been deposited with the NBPGR for long-term storage and dissemination (Sharma et al., 2002b). These lines have also been stored under natural conditions at

Keylong, Himachal Pradesh. KB resistant lines identified under the AICW & BIP are included in the National Genetic Stock Nursery (NGSN) for utilization by breeders. The most frequently used KB resistant stock, which was first reported from PAU is HD 29, HD 29 along with another resistant stock, HD 30 was registered under the AICWIP in 1999. Later, KB resistant lines of durum wheat (D 482, D 873, D 879, and D 895) and a triticale (TL 2807) from PAU Ludhiana were also registered with the NBPGR. Every year, the KB resistant lines are being identified under the AICWIP in India and added to the KB resistant material. Further, testing program of a large number of wheat lines in Mexico also found good resistance in several lines in all market classes (Singh & Rajaram, 2006).

Besides wheat (*T. aestivum*), resistance sources has also been identified in *T. durum*, triticale, wild spp. of *Triticum* and rye. Several accessions of wild wheats *Aegilops* and *Agropyron* were evaluated for resistance to KB and it was observed that 20 accessions of *T. urartu* and 22 accessions of *Ae. squarrosa* were resistant to KB. All the accessions of *T. urartu* remained free from KB infection. Among *Aegilops* spp. all accessions of *Ae. biuncialis*, *Ae. columaris*, *Ae. crassa*, *Ae. juvenalis*, *Ae. ovata*, and *Ae. speltoides* were found KB resistant (Warham, 1986; Gill & Aujla, 1986). Till 1980–1988, 407, 708, and 435 varieties of *T. aestivum*, *T. durum*, and triticale, varieties were found KB free, respectively; indicating that overall resistance was highest in triticale, followed by *T. durum* and *T. aestivum*. The additional D genome in *T. aestivum* was thought to be responsible for KB susceptibility (Aujla et al., 1990a, Aujla et al., 1992). Fuentes-Davilla & Rajaram (1994) evaluated wheat lines and cultivars derived from South American parents, Chinese parents, and Indian origin for KB resistance and found 13, 11, and 17 lines derived from South American parents, Chinese parents, and Indian origin, respectively, were KB resistant.

Apart from this, a number of KB resistant lines have been identified by screening of winter wheats. A method has been standardized for evaluating winter wheat lines against KB. It was cumbersome to evaluate winter wheats against the disease under controlled conditions maintaining plant growth vigor and seed set. Normally, high-spike sterility is encountered under such conditions and is likely to bias resistance evaluation. Thus, inoculations under normal field conditions are the best suited for KB screening provided the appropriate stage of inoculation is followed (up to the end of February/ early March). To ensure inoculations under this optimal period in winter and late flowering exotic lines, vernalization, and artificial extension of photoperiod may be necessary.

Chhuneja et al. (2004) screened US winter wheat lines and cultivars in green house specially constructed to simulate environmental conditions favorable for KB development and reported moderate resistance in few lines. Twelve more lines viz., NI98414/WESLEY, OGALLALA/ TX92D7960// BIG, DAWG/TX92U2317, 700-2, 701-14, 703-9, U5237 (2)-16, U5240 (1)-19), 95-934/Goldfield, U5240 (1)-19, KSU-08-27, and KSU-08-87 showed less than 1%KB infection and these winter wheat lines can be used as promising donor stocks for developing high-yielding KB resistant wheats (Sharma et al., 2009; Bala et al., 2015b).

6.2.1.4.1 Pyramiding of KB Genes for High Degree of KB Resistance

The zero tolerance limits of KB grains during the wheat trade necessitate the development of high degree of KB resistance. Keeping this in view, a high level of KB resistance was generated in synthetic hexaploid (SHW) wheat derived from *T. turgidum* and *T. tauschii*. SHs with 0% infection seem to aggregate resistance from the parental species. Four such synthetic wheat lines, SH 10, SH 12, SH 31, SH 46 were registered. However, the use of SHWs as donors is tedious due to hard threshing and unadapted derivatives (Villareal et al., 1994, 1996). Durum and triticales also had a low level of KB but transfer of these are not tedious due to involvement of minor genes thus, KB resistance in bread wheat is need of the hour which could be achieved by accumulation of diverse genes for raising the KB resistance. The presence of distinct resistance genes in donor stocks and the prevalence of additive gene action make this a viable option. Keeping this in the view, highly resistant stocks were crossed with each other from the identified resistant sources. Homozygous lines were derived from these resistant × resistant crosses. Stringent KB selection was done to obtain KB free plants. These KB free plants then planted in ear to row fashion for generation advancement and evaluated against mixture and individual isolates of *T. indica* collected from different agroclimatic zones of NWPZ. This led to the development of six KB free lines (KBRL 10, KBRL 13, KBRL 15, KBRL 18, KBRL 22, and KBRL 24) and three KB resistant wheats (W 7952, W 8086, and W 8618) by pyramiding of KB resistant genes and pedigree method, respectively. KBRL 10, KBRL 13, and KBRL 22 have been registered with NBPGR in 2001. Another KB resistant stock developed by PAU from ALDAN × H567.71 was KBRL 57. KBRL 22 along with KBRL 57 (ALDAN × H567.71) has also been used in incorporation of KB resistance in high-yielding variety, PBW 343 (Sharma et al., 2003; Bala et al., 2015a).

6.2.1.4.2 Marker Assisted Selection (MAS)/Chromosomal Location and Tagging of KB Resistance Genes

Screening of wheat genotypes for KB resistance is an essential exercise for identification of resistance donors. However, it is tedious because symptoms do not develop until the dough stage of plant development. Further it is cumbersome and labor intensive. Hence, MAS may be helpful in early selection in breeding program with the markers closely linked to the resistance genes.

Several gene-tagging studies have been carried out in India. But none of them reached a stage where MAS-worthy markers could be generated. Monosomic analysis in susceptible cultivar, WL 711 revealed the involvement of chromosomes 1D, 2A, 3B, 3A, 5B, and 7A for KB resistance/susceptibility. Chromosomes of WL 711 seemed to carry either a dominant gene(s) for susceptibility or a suppressor system, which inhibited the expression of resistance genes in WL711 (Gill, 1988). Analysis of monosomic series, aneuploides, ditelosomics, and nullitetrasomic compensating groups involving the D genome of the variety Chinese spring as well as addition and substitution lines of the D genome in *T. turgidum* cv. "Longdon" revealed that all homologous chromosomes were critical to KB resistance and susceptibility (Singh, 1989).

Later, Dhaliwal et al. (1990) identified 3D and 6D chromosomes of *Ae. squarrosa* carrying KB resistance genes. Gill et al. (1992) reported that the translocation of arm 4R and 6R of rye chromosome or small segment bearing genes for KB resistance to wheat homologous could be exploited. Since rye has been reported to be highly resistant to KB and due to cross pollination behavior, it can be readily used in KB resistance breeding program. Further, this is also feasible to transfer KB resistance from rye to wheat via triticale. Similarly, Datta et al. (1995) also reported that additional lines of chromosome 4R and 6R of *Secale cereale* cv. "*Imperial*" in *T. aestivum* cv. Chinese spring was completely free from KB resistance. Chromosomes of group 1, 3, 4, 6, and 7 were observed to carry genes for resistance and susceptibility (Dhaliwal & Singh, 1997).

With the advent of molecular markers, mapping of resistance genes became easier. Sukhwinder-Singh et al.(1994) used the restriction fragment length polymorphism (RFLP) for mapping a segregating population from HD 29 × WL 711, KB resistance gene(s) were observed to be located on chromosomes groups 4BL and 7BS of wheat. In a similar study based on RFLP mapping of recombinant inbred lines (RILs) from a cross between resistant synthetic wheat (*T. turgidum* ALTAR 84 ×*T. tauschii*) and the susceptible common cultivar Opata 85, regions on chromosomes arms 3BS and 5AL were shown to carry marker alleles (from durum parents) which were consistently associated with reduced kernel disease. The tagging strategy utilized four uniformly resistant and 13 uniformly susceptible F_3 families from a cross of KB resistant and (HD 29) and susceptible (WL 711) lines of *T. aestivum*. Bulked DNA of 8–10 plants in each family was tested by gel blot DNA hybridization. A total of 33% of the probes detected RFLP between R and S lines. Probe cxp 1 from chromosome 6 detected the polymorphic band in all the resistant bulk but was missing in all six susceptible families. This practice should reduce laborious disease screening requirements (Nelson et al., 1998). But in case of RFLPs, level of polymorphism was very low in wheat thus markers with high polymorphism was needed.

This led to the use of PCR-based markers AFLPs and microsatellite or simple sequence repeats (SSR) markers. SSRs have emerged as molecular markers of choice for genetic studies in major crop species. SSR belongs to a class of PCR-based markers with tandem repeats of a basic motif of fewer than six base pairs, are easier to use and more polymorphic. Identification of SSR markers linked to KB resistance offers the prospect of using MAS schemes in developing resistant wheat cultivars.

Singh et al. (1999) used amplified fragment length polymorphism (AFLP) to study polymorphism and their association with KB genes in wheat on RILs developed from a cross HD 29 × WL 711. Study of parents and a set of recombinant lines for AFLP with six primer combinations were analyzed and 336 loci scored, of these 77 were polymorphic between parents. Three of the polymorphic markers were associated with KB resistance genes.

In a study where resistant, *T. monococcum-T. durum* amphiploids were used in cross with susceptible cultivar WL 711, several genome specific microsatellite markers mapped on chromosomes 2AS, 3AS, 4AL, 5AL, and 6AL showed association with KB resistance. *T. monococcum*, diploid wheat has been found to be an excellent source of resistance against several wheat diseases. A set of microsatellite markers already mapped to the A genome of wheat was used to tag the KB resistance genes of *T. monococcum* transferred to *T. aestivum*, using F_3 population of the cross of synthetic amphiploid (*T. durum–T. monococcum*) ×*T. aestivum* cv. WL 711 (non-necrotic).

The *T. monococcum* specific alleles of the wheat microsatellite (WMS) loci *Xgwm0382*, *Xgwm0369*, *Xgwm0637*, *Xgwm0156*, and *Xgwm6617* mapped on 2AS, 3AS, 4AL, 5AL, and 6AL chromosomes, respectively, were found to be associated with KB resistance in different resistant derivatives (Vasu et al., 2000). When 90 SSRs and 81 AFLP loci were mapped on 130 RILs developed from WL 711 and HD 29, markers on chromosome 2A, 4B, and 7B accounted collectively for about one-third of the variation in the disease reaction. One SSR marker *Xgwm0538* was found closely related to KB resistance in wheat (Sukhwinder-Singh et al., 2003).

In a major mapping program at KSU (in collaboration with USDA, PAU, and CIMMYT), quantitative trait loci (QTL) associated with KB resistance were identified in three spring wheat RIL populations derived from WL 711 × HD 29, WH 542 × HD 29, and WH 542 × W 485. QTLs associated with KB resistance were located on chromosome 4B, explaining 19% of the phenotypic variation in WL 711/HD 29 population. Genomic regions on 4B, 5B, 6A, and 6B have consistent influence on the disease. Two new QTLs (*Qkb. ksu-5BL.1* and *Qkb.ksu-6BS.1*) with resistance alleles from HD 29 were identified and mapped in the intervals *Xgdm116–Xwmc235* on chromosome 5B (deletion bin 5BL9-0.76-0.79) and *Xwmc105–Xgwm88* on chromosome 6B (C-6BS5-0.76), respectively. Another QTL (*Qkb.ksu-4BL.1*) with a resistance allele from W485 mapped in the interval *Xgwm6–Xwmc349* on chromosome 4B (4BL5-0.86-1.00). *Qkb.ksu-6BS.1* showed pair wise interactions with loci on chromosomes 3B and 6A. Markers suitable for MAS are available for all three QTLs (Sukhwinder Singh et al., 2004, 2007). A number of molecular markers viz., *Xgwm337–1D*, *Xgwm637–4A*, *Xgwm538–4B*, and *Xgwm 6–4B5A*have shown apparent linkage with resistance to KB in two RILs developed from HD 29 × WH 542 and H 567.71 × WH 542 (Kumar et al., 2007, Kumar et al., 2015). These markers may be useful in MAS for KB resistance in wheat. In a more advanced approach, single nucleotide polymorphism (SNP) was developed from a co-dominant microsatellite marker, *Xgwm538*, associated with a QTL for KB resistance to improve gel-based resolution and amplification consistency. Amplification consistency is improved with *Xgwm538snp* since the amplification of competing non-target fragments is eliminated, and ambiguityis reduced since heterozygous plants are easily identified among backcross progeny (Brooks et al., 2006). So far, the gene tagging studies in KB resistance were done on RILs. The complex genetic control of KB resistance and difficulties associated with precise phenotyping render RILs less effective. Hence, use of KB resistant near isogenic lines (NILs) was done for genetic and molecular marker analysis. KB resistant NILs were developed in the background of variety PBW 343. Molecular marker analysis of these NILs showed the presence of donor alleles of four markers in; *Xgwm0099* (1AL), *Xgwm0149* (4BL), *Xgwm0174* (5DL), and *Xgwm0340* (3BL) in the resistant pool. Resistant and susceptible phenotypes showed association with three of the four markers viz., *Xgwm0099*, *Xgwm0174*, and *Xgwm0340* as indicated by the chi-square contingency test (Sehgal et al., 2008). Molecular marker analysis of KB resistant NILs showed association of two markers *Xbarc0045–3A 2B* and *Xgwm0261–2D* with KB resistance in two crosses WH 542 *5/ALDAN'S'/IAS58, and WH 542 *5/CMH77.308, respectively (Bala et al., 2010).

NILs could be used in molecular tagging of resistance with minimum genotyping efforts. It can also provide opportunity for precise genetic analysis of resistance genes within minimum background noise. Till date, a number of lines of wheat and allied species have been identified by screening under artificially epiphytotic conditions, thus, it was imperative to know the resistance mechanisms operating in these genotypes. Efforts have been made to identify morphological as well as genetic resistance against KB of wheat.

Morphologically, compact arrangement of the spikelets in *T. durum* and triticale and hairiness in durum wheat may impart KB resistance. But this type of resistance appeared to have very little role when variety was tested with "syringe inoculation method" (Warham, 1988). Further, the resistant lines of bread wheats, durum wheats, and triticales had comparatively more number of hairs/ mm^2 on glumes than the susceptible lines of respective groups (Aujla et al., 1990b; Sharma et al., 1998b; Grewal et al., 1999). A quantitative approach was followed initially for studying genetics of KB resistance indicating both additive and dominance gene effects (Chand et al., 1989; Gill et al., 1990; Sharma et al., 1991; Nanda et al., 1995). Later on, the approach was shifted to qualitative genetic analysis using segregating populations of several resistant × susceptible crosses. So, monogenic control of KB resistance was observed in a diallel cross involving four resistant (Weaver, W499, CRUZ ALTA, and K342) and two susceptible parents (LAJ 3302 and WL 3399). Weaver, W499, CRUZ ALTA were observed to possess different genes while CRUZ ALTA and K342 had same genes for resistance (Morgunov et al., 1994). Dominant to partially dominant nature of KB resistance was revealed in genetic analysis of segregating populations in the background of WL711. Progenies with accumulation of dominant/partially dominant genes were relatively more resistant than those with one gene (Singh et al., 1995a, 1995b; Fuentes- Davila et al., 1995). Further, single dominant gene, two dominant, and two complementary dominant genes were identified in SH cultivars Chen/*T. tauschii* and Chen/*T. tauschii*, Altar 84/*T. Tauschii* and Duergand/*T. tauschii*, respectively (Villareal et al., 1995). Sharma et al. (2004a) made an attempt to study the inheritance pattern of KB free trait in KB free stock KBRL 22. The KBRL 22 was derived from HD 29 and W 485 and used to introgress KB free trait in PBW 343.KB incidence was recorded on BC_1, BC_2, BC_3, and BC4 as well as on F_1 and F_2 plants after artificial inoculations. The segregation pattern in F_2 generation (3:1) indicated two independently segregating dominant genes in HD 29 and W 485 which jointly conferred resistance trait. Keeping in view the macro and micro environmental influences on KB incidence, genetic analysis has progressively shifted from segregating host populations to stable host populations like RILs. RILs are lines developed by single seed descent method of F_2 plants. These RILs are stable host populations which allow repetitive and replicated phenotyping. Nine loci were identified by screening 10 RIL populations using mixture of isolates representing genetically heterogeneous inoculum, derived from four KB resistant genotypes (ALDAN/IAS 58, H567.71/3*PAR, HD29, and W485) and one susceptible line, WH 542. Genetic analysis revealed that HD 29, W 485, and ALDAN/IAS 58 carried two genes whereas three genes were indicated in H 567.71/3*PAR. The six R × R RILs showed that genes in the four resistant parents were different and there may be nine loci carrying resistance in these four resistant donors (Sharma et al., 2005a).

As we know that the presence of oligogenic rather than monogenic resistance hampers precise genetic analysis. Thus besides stable host populations like RILs, an efficient screening system is also critical for reliable phenotyping of parents as well as host populations for genetic analysis. A screening system has two main components: (a) artificial inoculation techniques and (b) inoculum system. Artificial inoculation techniques as well as creation of artificial epiphytotic conditions have been standardized in case of KB of wheat (Aujla et al., 1982). However, type of pathogen population also contributes to the efficacy and reproducibility of a screening system. This aspect is important, particularly in case of KB of wheat where host–pathogen interaction is not simple and easy to understand like other *Basidiomycetous* pathogens. Further, *T. indica* is a heterothallic fungus and heterothallism is controlled by multiple alleles at one locus (Duran & Cromarty, 1977).

Screening of wheat genotypes for locating resistance to KB fungus, till date-involved use of inoculum system based on mixture of isolates. This approach suits in identifying genes for plant breeding lacking applied value.

Thus, single isolates were used for screening various segregating and RIL populations.

Singh et al. (1996) used the single isolate to study the inheritance pattern of KB resistance in S × R, S × S, and R × R crosses and found resistance to be a dominant character. Resistance was controlled by few major genes, along with minor genes. Some minor genes were present in otherwise susceptible parents. Differential reaction pattern was observed among resistant (HP 1531 more resistant than W 485 and susceptible (WL 711 more susceptible than HD 2329) groups to the isolate (P_{11}) used.

Harjit-Singh et al. (1999) used two isolates of *T. indica* for evaluating F_8RILs derived from a resistant (HD 29) and susceptible (WL 711) cross. Three genes were found to confer resistance to Ni 7 and two genes to Ni 8 whereas one gene was shown to be common against both the isolates. Similarly, a single isolate (Delhi isolate) was used to determine the nature and number of genes for resistance in three resistant genotypes of wheat viz., HD 29, HP 1531, and W 485. The F_1 hybrids between resistant and susceptible parents, namely WL 711, HD 2009, and HD 2285 were susceptible. The F_2 populations segregated in a ratio of 3S:1R. On the basis of segregation in F_3, F_2, and F_1, it was concluded that single recessive genes conferred resistance to Delhi isolate in three resistant parents (Bag et al., 1999). Single isolate inoculum system was also used on RIL derived from WH 542 × ALDAN 'S'/IAS 58. Three genes were observed to confer resistance against P_4 isolate in ALDAN 'S'/IAS 58 (Bodh, 2005). For more precise genetic analysis, attempts were made to develop genetically homogeneous inoculum system based on single compatible monosporidial pair on two sets of RIL populations derived from WH 542 × HD 29 and WH 542 × W485 and three additive genes were indicated in HD 29, W485, and ALDAN 'S'/IAS 58 (Sharma et al., 2006; Sirari et al., 2006; Sirari et al., 2008).

Generally, the use of homogenous inoculum is expected to simplify the genetic analysis but only for race specific major gene resistance as in rusts. As KB resistance in most of the studies discussed earlier is based on two or more genes with additive effects. Some element of specificity in KB—wheat system is however evident as exclusion of some pathogen population constituents led to change in the genes revealed by genetic analysis (Sirari et al., 2008). The element of specificity implies that a large number of compatible MLs will be required to represent the pathogen population of a particular geographical region, making the analysis of resistance, that is, operative in the field very laborious. Under these circumstances, the mixture of most virulent populations should be the preferred inoculum for revealing the comparatively smaller but effective genes, which are of relevance for transfer and commercial deployment.

A further advance in this regard would be the use of KB resistant NILs—a set of lines with diverse KB resistance genes introgressed into a susceptible genotype. It can also provide opportunity for precise genetic analysis of resistance genes within minimum background noise. Attempts were made in this regard for the development of KB resistant NILs developed using the stock KBRL 22 with high-KB resistance in the background of PBW 343 and six donor stocks ALDAN 'S'/IAS 58, CMH 77.308, H 567.71/3, HD29, HP1531, and W 485 in the background of WH 542. Genetic analysis of BC_5F_2 and BC_5F_3 populations of these NILs revealed the presence of two to three additive genes for KB resistance (Sehgal et al., 2008; Bala et al., 2011).

6.3   LOOSE SMUT

Loose smut of wheat is of common occurrence and first time illustrated in 1556 in Hieronymus Bock's Herbal and accurate symptomatology is described in Fabricius text of 1774 (Nielsen & Thomas, 1996). In India, the loose smut was first recorded from Punjab in 1897 (Anonymous, 2011). *Ustilago segetum* (Pers.) Roussel var. *tritici* Jons. (*U. nuda* var *tritici*/*U. nuda* f. sp. *tritici*) is known to be associated with this disease. The fungus produces smut spores/chlamydospores, which are spherical, finely echinulated, olivaceous brown in color, lighter on one side, and darker on another side (Fig. 6.8). Pathogen is internally seed borne fungus, survives as dormant mycelium inside the embryo of the seed. The teliospores are transported by wind from smutted heads to the florets of neighboring plants and germinate on the ovary. Dikaryotic infectious hyphae penetrate the ovary wall and grow toward the developing scutellum and embryo where they remain dormant until the next season when the seeds start to germinate. When the infected seed is sown, the mycelium inside the embryo, keep pace with the growing plant, and express symptoms at heading stage. The fungus colonizes the shoot apical meristem and ear primordia, resulting in the typical symptom of the smutted head (Malik & Batts, 1960a, b). Maddox (1896) firstly reported the embryo infection through wheat blossom. Mode of entry of the pathogen to the ovary via the stigma and further to the embryo was demonstrated by Lang (1910). During 1967, the release of a loose smut resistant variety Kalyansona led to the reduction in the incidence of loose

FIGURE 6.8   Loose smut symptoms in field.

smut disease, but again disease gained importance due to cultivation of susceptible variety like Sonalika (Swaminathan, 1971). Yield losses of about 2–4% due to loose smut of wheat have been estimated, but it has a potential to cause 100% losses (Joshi et al., 1985; Srivastava et al., 1992; Mathur & Barry, 1993). Persons (1954) reported 100% yield losses due to this disease in Georgia. The percent of yield loss caused by this pathogen is equal to the percent of smutted heads in the crop (Green et al., 1968). The degree of infection by loose smut in wheat varies greatly with the wheat varieties grown and with the environmental conditions prevalent during the infection stage (Saari & Wilcoxon, 1974). The black powdery mass of spores replaces the developing grains. Infected spike is initially covered with a delicate membrane which later on ruptures and releases spores. At the end, bare rachis is left behind. It is of more common occurrence in areas with cool and moist climate during the flowering. The seed infection occurs at 22–25°C and favored by frequent rains (7–21 mm) for 4–5 days with temperature range of 6–23°C. Since the disease is internally seed borne, seed testing and certification can ensure eradication of loose smut pathogen. Staining of embryo can help to get healthy seed. Wunderle et al. (2012) used seed lots of barley and wheat highly infected with loose smut, and studied the early establishment of the loose smut pathogens in the plant by fluorescence microscopy and also predicted loose smut infection by real-time PCR at the second leaf stage.

Physiological specialization in loose smut pathogen has been studied by different workers in India and abroad (Oort, 1947; Cherewick, 1952; Nielsen, 1987; Nielsen & Thomas, 1996; Menzies et al., 2009; Kaur et al., 2014). First study related to differential reaction of cultivars of wheat to *U. segetum* var. *tritici* was conducted by Tiemann (1925). Later on Grevel (1930) proposed a few principles from his observations, which are still valid today: (a) resistance to each race is monofactorial, and the genes can be readily transferred or combined, enabling for plant breeding for resistance, (b) the virulence pattern of a race does not depend on its geographic origin, but is determined by the cultivar on which it occurred, (c) cultivars select the races that are virulent on them, and (d) the same race may occur in different geographic regions. Gothwal and Pathak (1977) identified eight races of *U. segetum* var. *tritici* (LSR1-LSR8) on spring wheat genotypes of Canadian differential set. Saini et al. (1985) identified seven races. Rewal and Jhooty (1986) identified three races (T1, T10, andT11). Sharma et al. (1998) tested reaction of the TD lines and test cultivars to 11 pathotypes of loose smut pathogen (Fig. 6.9). The pathotypes gave resistant reaction on 13 TD lines (TD 1, TD 3, TD 5, TD 6, TD 8, TD 18, TD 12, TD 16, and TD 19), differential reaction on six lines (viz., TD 4, TD11, TD 13, TD 14, TD 15, and TD 17), and susceptible reaction on WL 711. Padmaja et al. (2006) grouped twenty-two isolates of *U. segetum* var. *tritici* collected from North India were into four distinct races based on the host pathogen reaction on a set of 19 Canadian differentials. Race T1 was detected in Himachal Pradesh; two races T7 and T11 were observed in Haryana isolates, races T7 and T4 were found in Rajasthan, and T1 and T11 were observed in Punjab. Knox and Menzies (2012) gave detailed account of races of *U. segetum* f. sp. *tritici* prevalent in different countries. Kaur et al. (2014) characterized 35 isolates of *U. segetum* f. sp. *tritici* collected from NW India for virulence studies on the standard Canadian differential set developed by Nielsen. Six races were identified and race T34 being the most common. Two new races T59 and T60 were also identified.

6.3.1 MANAGEMENT

Loose smut of wheat can be controlled by the use of clean seed, seed treatment with systemic fungicides or hot water, and host resistance (Bailey et al., 2003).

6.3.1.1 PHYSICAL METHOD

Control of loose smut by physical methods especially solar heat treatment is known since 1934 (Luthra & Sattar, 1934) who developed the solar heat treatment for eradicating seed borne inoculum. They suggested that in the plains of NW India, maximum temperature in the month of June soak in water for 4 h (8:00 a.m. to 12:00 noon) on bright summer days, then expose to sun for 4 h (12:00 noon to 4:00 p.m.) and finally dry for storage as in June temperature goes to 120°F in shade and 131°F in sun. Solar heat treatment was further modified by Patel et al. (1950) that the seed should be soaked in water from 8:00 a.m. to 12:00 noon, and then spread in thin layer on galvanized iron sheet in sun for one and a half to two hours to maintain contact with hot surface; seeds are stirred once or twice. Bedi (1957) observed that pre-soaking period of 4 h followed by one-hour exposure to sun light; under Punjab condition is enough to devitalize the intra-seminal mycelium. Rest of the exposure to sun rays constituted only the drying process. Even 5 min exposure of soaked seeds at 12:00 noon in Punjab was quite effective in reducing smut incidence. Till date, this method is being practiced for the management of loose smut at many places.

6.3.1.2 CHEMICAL

Many reports on the management of loose smut of wheat by using systemic fungicides are available in India and abroad. Soon after the discovery of oxathins, Hardison (1966) reported the efficacy of Vitavax against smuts and Hansing (1967, 1968) achieved complete control of loose smut in wheat cultivars Bishon, Kiowa, and Red Chief by using Vitavax 75 @ 202/c.wt. The systemic fungicides like carboxin (Vitavax 75 WP), carbendazim (Bavistin 50 WP), and tebuconazole (Raxil 2 DS) have been recommended in past decades (Goel et al., 2001; Kaur et al., 2002a; Maude & Shuring, 2008; Sinha & Singh, 1996) for the management of loose smut of wheat. Effective control of loose smut by seed treatments with Bavistin was achieved by Kulkarni and Chattannavar (1984); Verma (1984); Guldhe et al. (1985); Goel et al. (1995) evaluated a systemic fungicide "G 696" (Provax) as dry seed treatment for controlling loose smut of wheat and proved significantly superior in controlling the disease (giving 95 and 98% control, respectively) to the other treatments, except carboxin @ 0.3% which was on a par with it in effectiveness. Menzies (2008) reported carboxin tolerant strains of *U. tritici* in Canada.

Rana et al. (2000) worked out that wet seed treatment with propiconazole by dipping the seed in 0.01% solution for 6 h and dry seed treatment with Raxil (0.1%) are the most effective for the management of loose smut of wheat. Jones (1999) made attempts for using foliar spray of systemic fungicides to control loose smut of wheat and barley and achieved remarkable results. Sharma et al. (2001) tested the efficacy of thifluzamide (Pulsor 2F) against loose smut of wheat pathogen and achieved results at par with that of carboxin. Singh et al. (2015) evaluated the bio efficacy of new fungicide triticonazole 8% + pyraclostrobin 4% FS (Insure Perform 12% FS) @ 0.50, 0.75, and 1.00 ml per kg of seed against loose smut of wheat by taking infected seeds during 2011–2012 and 2012–2013 crop seasons at Karnal, India in comparison with recommended fungicide tebuconazole 2% DS. The test fungicide, triticonazole 8% + pyraclostrobin 4% FS (Insure Perform 12% FS) was highly effective in controlling the disease and gave more than 95% control that was at par with tebuconazole @ 1.0 g per kg of seed.

6.3.1.3   BIOLOGICAL

Gautam et al. (1995) studied the antagonistic effect of *Trichoderma* spp. on *U. segetum* var. *tritici* and their compatibility with fungicides and biofungicides. Maximum inhibitory effect on germination of chlamydospores of the smut pathogen was obtained *in vitro* with culture filtrate of *T. viride* (TV-5) followed by *T. koningii*, *T. hamatum*, *T. harzianum*, and *T. lignorum*. Antifungal compounds extracted from *T. viride* (TV-5) followed by *T. koningii* also suppressed mycelial growth of the host pathogen in the liquid medium. *T. koningii*, *T. harzianum*, and *T. lignorum* were compatible with carboxin at 200 and 500 ppm. Although neem oil and antifeedant neem formulations showed compatibility with all six isolates of *Trichoderma* but none had antifungal activity against the loose smut pathogen at 100 ppm except neem oil at 25 ppm. Attempts for the biological control of loose smut by using *T. viride*, *T. harzianum*, *Pseudomonas fluorescence* and *G. virens* in combination with Vitavax @ 0.125% for three years were made by Singh and Maheshwari (2001). They achieved better smut control even than full dose of Vitavax (0.25%). These treatments also enhanced the percent seed germination, seedling emergence in the field, and seed yield per plot. Singh et al. (2000) found lower incidence of loose smut in seeds treated with *T. viride*. The loose smut may be controlled effectively in case with much lower doses than recommended dose of systemic fungicide like tebuconazole is applied while seed is imbibed and mycelium of loose smut is activated before sowing. (Singh, 2004).

6.3.1.4 HOST RESISTANCE

Various studies have been performed for extensive evaluations of different wheat genotypes to collections or isolates of *U. tritici* (Atkins et al., 1947; Anderson, 1961; Calvo, 1978) (Fig. 6.9). Singh et al. (2001, 2002, 2008) identified confirmed sources of resistance in wheat and triticale against loose smut after evaluation of number of genotypes under artificially inoculated and disease epiphytotic conditions in India. Other sources of resistance have been reported through studies of inheritance or as reports of registered cultivars or germplasm. These included Downy, Auburn, E 6879, E 6878, E 6840, E 6824, E6160, E 6031, E 6006, E5070, CPAN 746, CPAN 744, Leeds, Wheaton, Manitou, Romany, Hawli, Davo, Canthatch, Morris, Manitou, Bezenchukskaya 98, DT 369, Giza 155, Giza 160, Giza 162, Giza 163, Giza1 64, Sakha 8, CGM 513, CGM 539, CM 33027, CM 39808, CM 70307, CM 48418, CM 43367, CM 64400, CGM 112, CM 32973, CM 64604, CM 59908, Saratovskaya (SAR) 29, Bezenchukskaya (BEK) 98, Selivanovsky Rusak (land race), Beloturka (durum land race), Thatcher, Myronivs'ka (MYR) 808, Odess'ka 16, Preson, Selkirk, Hope, Kawvale, Graecum 114, Kharkivs'ka (KHR) 8, KHR22, Florence/Aurore, K 32541, MYR 4, MIR 3, LUT 237H12, MIR 808, MYR yubilejna, Zhigulevskaya, Kazakhstanskaya 19, MN 81330, ND 597, ND596, ND607, SD8036,

FIGURE 6.9   Loose smut inoculation.

Chris, Ciano 67, Penewana, GW 1021, VL639, UP 2189, HW888, HW657, HW517, PBW65, Helios, Stettler, Goodeve, AC Cadillac, P8810-B5B3A2A2 (PI600683), L8800-CC7B1-B1D16 (PI 596348), P8802-C1*3A2C16 (PI 596351), DT676 (PI 650845), Unity, ML 521, W 59, W 1616, W 2484, W 2531, W 5915, W 6202, WL 1786, WL 2956, WL 3450, W 4461, W 5100, W 2615, WL 3951, WL 5907, W 2139, W 3899, W 4985, W 5450, and W 579 (Roberts et al., 1981; Patterson et al., 1982; Ghorpade, 1983; Busch et al., 1984; Troitskaya & Plakhotnik, 1986; McLeod et al., 1991; Sherif et al., 1991; Afonskaya et al., 1998; Beniwal et al., 1998; DePauw et al., 1998, 2007, 2009a, 2009b; Knox et al., 1998, 2000, 2008; Fox et al., 2010; Sharma et al., 2011). At Punjab Agricultural University, Ludhiana evaluation for loose smut resistance was initiated in the year 1976 under AICWIP and since then every year wheat lines are being evaluated against the disease and resistant stocks being maintained. From amongst these stocks, 635 bread wheat lines showing stable resistance from 10 to 30 years were stored at National Bureau of Plant Genetic Resources, New Delhi in the year 2004. The number has further increased to 762 lines. NILs for resistance have been developed that have potential for use in studies of the physiological and biochemical nature of resistance (Knox et al., 1998)

Resistance to loose smut is associated with several different chromosomes in wheat. A study by Dhitaphichit et al. (1989) assessed cytogenetic stocks of Hope chromosomes substituted for chromosomes in Chinese Spring. Nine different chromosomes were possess loose smut resistance to race T6. Similarly, when chromosomes of Thatcher were substituted into Chinese Spring, eight chromosomes were found to affect the level of loose smut. Chromosomes 4A, 7A and 5D were the only ones from both Hope and Thatcher that significantly reduced loose smut levels when substituted into Chinese Spring. Significant reductions in loose smut were associated with chromosomes 3A, 2D, 3D, and 4D in Hope. Significant reductions in loose smut were associated with chromosomes 7B, and 6D in Thatcher. Complete resistance to race T6 was associated with chromosome 7A from Hope, whereas complete resistance was associated with chromosome 7B of Thatcher. Chinese Spring had an intermediate level of resistance to race T6. Interestingly some chromosome substitutions significantly increased loose smut reaction of Chinese Spring. Using monosomic lines of the cultivar Cadet, Mathur et al. (1997) identified resistance to a mixture of races from India to be located on chromosomes 1B, 3D, and 7D. The 3D gene was characterized as a major gene while chromosomes 1B and 7D carried minor genes. Also using monosomic line crosses with resistant line PL, Heinrich (1970a) located resistance on chromosomes 5B and 4D. Knox and Howes (1994) determined cultivars Cadet, Kota, Thatcher, and TD 18 possess a resistance factor to race T19 on chromosome 6A using segregation within cytogenetic stocks and chromosome tracking with a monoclonal antibody marker. Procunier et al. (1997) located a gene from Biggar BSR to the long arm of chromosome 2B. Randhawa et al. (2009) determined the location of a loose smut resistance gene in durum wheat to be located on the short arm of chromosome 5B. The only chromosomes to show resistance in multiple investigators work were 3B and 4D (Heinrich, 1970b, Dhitaphichit et al., 1989) indicating that a diverse array of loose smut resistance genes are available.

DNA markers have been developed to a few genes for resistance to loose smut. Such markers not only aid selection for the resistance, but assist in the identifying genes by indicating their gene location in the case of mapped markers. Procunier et al. (1997) identified a RAPD and two RFLP markers to a gene for resistance to loose smut race T10 from the cultivar Biggar BSR. Knox et al. (2002) identified a sequence characterized amplified region (SCAR) marker to race T33 loose smut resistance in the durum wheat cultivar DT676 after AFLP analysis. Unfortunately, the resistance was not broad enough to cover all the virulence in the targeted region, leaving the authors to conclude another marker would need to be developed to resistance to the additional virulence. Randhawa et al. (2009) found AFLP markers to loose smut resistance in the durum cultivar D93213 and SSR markers in the cultivar P9162-BJ08*B. The resistance is likely the same between the two cultivars and DT676 because the SCAR marker developed by Knox et al. (2002) also co-segregated with resistance from D93213 and P9162-BJ08*B.

Many studies have been conducted to determine the inheritance of resistance to loose smut in wheat (Agrawal & Jain, 1965; Krivchenko & Bakhareva, 1985; Knox et al., 1999; Kassa et al., 2014). These studies revealed that one or two genes in common wheat controlled resistance to several races of *U. tritici*. Three genetic factors located on chromosomes 1B, 3D, and 7D in bread wheat Cadet. The durum wheats in general possess greater resistance to loose smut than bread wheats. Mau et al. (2004) conducted a study to determine the mode of inheritance of resistance to loose smut in the resistant durum wheat (*T. turgidum*) lines Orgaz, Tripolitico, and VIR53877 to identify a simply inherited resistance factor effective against multiple races of the pathogen. The resistant durum wheat lines were crossed with a susceptible durum wheat cultivar, "Sceptre", to produce F1, F2, and F3 generations. Plants of each generation were inoculated with races T26, T32, and T33 of *U. tritici*. The F1 plants from crosses Orgaz/Sceptre, Tripolitico/Sceptre, and VIR53877/Sceptre exhibited no smut, indicating dominant-gene expression. Segregation of resistance among the F2 and F3 progenies of the cross Orgaz/Sceptre fitted a one-gene ratio when tested with race T26 and a two-gene ratio to each of races T32 and T33. The genes controlling resistance to the races appeared to be independently inherited. In the cross Tripolitico/Sceptre, one gene controlled resistance to race T26, and two genes controlled resistance to each of races T32 and T33. These genes were not independently inherited. One gene conferring resistance to race T26, and two genes conferring resistance to each of T32 and T33, were also observed in the cross VIR53877/Sceptre. There appeared to be a common gene for resistance to both T32 and T33. Tripolitico appears to hold the greatest promise of providing broad resistance to *U. tritici* with simple inheritance. Sinha et al. (2000) identified PBW 65, HDR 70, VL 421, and WL 410 as resistant sources to loose smut of wheat after 8 years of screening. Randhawa et al. (2009) studied the genetic control of resistance to loose smut caused by *U. tritici* race T33 in two-durum RIL populations (DT662 × D93213 and Sceptre×P9162-BJ08*B) and identified molecular markers linked to the resistance. He suggested that SCAR and SSR markers could be used effectively for MAS to incorporate loose smut resistance into durum cultivars. Knox et al. (2008) conducted the study on "Glenlea" wheat and determined the effect of single gene in providing broad-spectrum resistance to multiple loose smut races representing virulences found on the Canadian prairies. He found that no one major gene provided complete or moderate resistance to all the races of *U. tritici* on the prairies (T2, T9, T10, and T39). Sharma et al. (2011) studied the genetics of loose smut pathogen in 20 crosses and concluded that resistance is controlled by two dominant complimentary genes, single dominant genes, two dominant genes with duplicate gene action, and single recessive gene in these 20 lines.

Kassa et al. (2014) studied the genetics of loose smut resistance in an F5-derived RIL population of 94 lines from the cross BW278/AC Foremost. The line AC Foremost is resistant and line BW278 is susceptible to *U. tritici* race T10. Phenotypic assessment revealed that a single gene, designated Ut6, segregated for resistance to race T10 in the RIL population. A modified bulked segregant analysis identified a microsatellite marker linked to Ut6. A linkage map was developed consisting of linked microsatellite loci and the resistance gene. The loose smut resistance gene Ut6 mapped to the long arm of chromosome 5B. Five microsatellite markers mapped within 6.7 cM of Ut6. The microsatellite markers gpw5029 and barc232 flanked Ut6 at distances of 1.3 and 2.8 cMon the distal and proximal sides, respectively. A diverse set of wheat lines was haplotyped for Ut6 using the linked micro-satellite markers *gpw5029* and *Xbarc232*. The haplotype analysis suggested that the microsatellite markers associated with Ut6 will be useful for MAS of loose smut resistant wheat lines.

6.4 CONCLUDING REMARKS

Many management strategies have been discussed for the management of KB and LS of the wheat. These strategies included cultural practices, chemical control, bio-control, seed treatment with hot water, and solar energy, seed treatment with fungicides and host resistance. Cultural methods are although cheaper but they are partially effective and further tend to reduce the yield of the crop. Chemical control gave a quick control but we cannot go with them due to environmental safety issues and residual effects. Biological control is not effective under field conditions. Thus, the cheapest and the most feasible method of management for KB and loose smut of wheat is the use of host resistance and breeding for varieties resistant to these diseases. However, an integrated approach is practicable and conductive for the better management of these diseases.

KEYWORDS

- **wheat**
- **Karnal bunt**
- *Tilletia indica*
- **loose smut**
- *Ustilago segetum* var. *tritici*
- **management**

REFERENCES

Afonskaya, E. Yu.; Rabinovich, S. V.; Dolhova, E. M.; Chernyaeva, I. N. In *The Genetic Nature of Group Resistance to Tilletia caries Tul. and Ustilago tritici Jens. in Some Culti-vars and Lines of Bread Wheat;* Raupp, W. J., Ed. Annual Wheat Newsletter, Contribution No. 98-430-D from the Kansas Agricultural Experiment Station, Kansas State University: Manhattan, 1998.

Aggarwal, R.; Tripathi, A.; Yadav, A. Pathogenic and Genetic Variability in *Tilletia indica* Monosporidial Culture Lines Using Universal Rice Primer-PCR. *Eur. J. Plant Pathol.* **2010,** *128,* 333–342.

Aggarwal, V. K.; Verma, H. S.; Khetrapal R. K. Occurrence of Partial Bunt on Triticale. *FAO Plant Prot. Bull.* **1977,** *25,* 210–211.

Agrawal, R. K.; Jain, K. B. L. Inheritance of Resistance of N.P. 790 Wheat to Loose Smut (*Ustilago Tritici*). *Indian J. Genet.* **1965,** *25,* 376–80.

Allen, T. W.; Workneh, F.; Steddom, K. C.; Peterson, G. L.; Rush, C. M. The Influence of Tillage on Dispersal of *Tilletia indica* Teliospores from a Concentrated Point Source. *Plant Dis.* **2008,** *92,* 351–356.

Amer, G. A. M.; Singh, D. V.; Aggarwal, R.; Dureja, P. Microbial Antagonism of *Neovossia indica*, Causing Karnal Bunt of Wheat. *Curr. Sci.* **1998,** *75,* 1393–1396.

Anderson, R. G. The Occurrence of Loose Smut Resistance in 42, 28 and 14 Chromosome Wheats. *Can. J. Plant Sci.* **1961,** *41,* 828–835.

Anonymous. *100 Years of Wheat Research in India- A Saga of Distinguished Achievements;* DWR: Karnal, Haryana, 2011; pp141–170.

Anonymous. *Description of Pathogenic Fungi and Bacteria*, *No. 748;* Commonwealth Agricultural Bureaux: Kew, 1983.

Atkins, I. M.; Hansing, E. D.; Bever, W. M. Reaction of Varieties and Strains of Winter Wheat to Loose Smut. *J. Am. Soc. Agron.* **1947,** *39,* 363–377.

Aujla, S. S., et al. Propiconazole a Promising Fungicide against Karnal Bunt of Wheat. *Pesticides.* **1988,** *14,* 35–38.

Aujla, S. S.; Gill, K. S.; Sharma, I.; Grewal, A. S. Prevalence of Karnal bunt in Punjab as Influenced by Varietal Susceptibility and Meteorological Factors. *Pl. Dis. Res.* **1986,** *1,* 51–54.

Aujla, S. S.; Grewal, A. S.; Gill, K. S.; Sharma, I. A Screening Technique for Karnal Bunt Disease of Wheat. *Crop Improv.* **1980,** *7,* 145–146.

Aujla, S. S.; Grewal, A. S.; Gill, K. S.; Sharma, I; Artificial Creation of Karnal Bunt Disease of Wheat. *Cereal Res. Commun.* **1982,** *10,* 171–176.

Aujla, S. S.; Grewal, A. S.; Sharma, I. Sources of Resistance in Wheat to Karnal Bunt. *Indian Phytopathol.* **1990a,** *43,* 91–93.

Aujla, S. S.; Grewal, A. S.; Sharma, I. Yield and Vigour Components of Wheat as Affected by Karnal Bunt Pathogen (*Neovossia indica*). *J. Res. Punjab Agric. Univ.* **1983,** *20,* 231–235.

Aujla, S. S.; Sharma, I. Compatibility System in *Neovossia indica.* *Indian Phytopath.* **1990,** *43,* 222–223.

Aujla, S. S.; Sharma, I.; Gill, K. S. Morphologic and Physiologic Resistance in Wheat to Karnal Bunt. *Pl. Dis. Res.* **1990b,** *5,* 119–121.

Aujla, S. S.; Sharma, I.; Gill, K. S. Stable Resistance in Wheat to Karnal Bunt (*Tilletia indica*). *Indian J. Agric. Sci.* **1992,** *62,* 171–172.

Aujla, S. S.; Sharma, I.; Gill, K. S.; Grewal, A. S. Prevalence of Karnal Bunt in Punjab as Influenced by Varietal Susceptibility and Meteorological Factors. *Pl. Dis. Res.* **1986,** *2,* 108–109.

Aujla, S. S.; Sharma, I.; Gill, K. S.; Grewal, A. S. In *Variable Resistance in Wheat Germplasm to Neovossia indica*, Proceedings of the 3rd National Seminar on Genetics and Wheat Improvement, IARI Regional Research Station, Flowerdale, Shimla, Himachal Pradesh, May 8–10, 1985; IARI: Himachal Pradesh, India, 1985.

Aujla, S. S.; Sharma, I.; Gill, K. S.; Grewal, H. S. Establishment of *Neovossia indica* in Wheat Kernel. *Pl. Dis. Res.* **1988,** *3,* 62–63.

Aujla, S. S.; Sharma, I.; Singh, B. B. Physiologic Specialization of Karnal Bunt of Wheat. *Indian Phytopathol.* **1987,** *40* (3), 333–336.

Aujla, S. S.; Sharma, K.; Chand, K.; Sawhney, S. S. Influence of Weather Factors on the Incidence and Epidemiology of Karnal Bunt of Wheat. *Indian J. Ecol.* **1977,** *8,* 175–179.

Bag, T. K.; Singh, D. V.; Tomar, S. M. S. Inheritance of Karnal Bunt Resistance in Some Indian Bread Wheat Lines and Cultivars. *J. Genet. Breed.* **1999,** *53,* 67–72.

Bailey, K. L., et al. *Diseases of Field Crops in Canada;* 3rd ed.; The Canadian Phytopathological Society: Winnipeg, 2003; p 290.

Bains, S. S.; Dhaliwal, H. S. Release of *Neovossia indica* Sporidia from Inoculated Wheat Spikes. *Pl. Soil.* **1989,** *115,* 83–87.

Bala, R. Molecular Tagging of Genes Conferring Resistance to Karnal Bunt *(Tilletia indica mitra)* in Near Isogenic Lines of Bread Wheat (*Triticum aestivum L.*). Ph.D. Dissertation, Punjab Agricultural University, Ludhiana, 2010.

Bala, R.; Kumar.; Bains, N. S.; Sharma, I. Development of Disease Resistant Bread Wheat (*Triticum aestivum*) Line in Background of PBW 343 and Genetics of Karnal Bunt Free Trait. *Indian Phytopathol.* **2015a,** *68* (1), 42–44.

Bala, R.; Sharma, I.; Bains, N. S. Genetic Analysis of Karnal bunt (*Tilletia indica* Mitra) in Near Isogenic Lines of Bread Wheat (*Triticum aestivum* L.). *J. Wheat Res.* **2011,** *3* (1), 59–62.

Bala, R.; Sharma, I.; Bains, N. S. Identification of New Donors of Karnal Bunt (*Tilletia indica*) Resistance in Winter Wheats. *Pl. Dis. Res.* **2015b,** *30* (1), 73–75.

Bansal, R.; Singh, D. V.; Joshi, L. M. Comparative Morphological Studies in Teliospores of *Neovossia indica*. *Indian Phytopathol.* **1984a,** *37,* 355–357.

Bansal, R.; Singh, D. V.; Joshi, L. M. Effect of Karnal-Bunt Pathogen (*Neovossia indica* (Mitra) Mundkur) on Weight and Viability of Wheat Seed. *Indian J. Agric. Sci.* **1984b,** *54* (8), 663–666.

Bedi, K. S. Further Studies on the Control of Loose Smut of Wheat in The Punjab. *Indian Phytopathol.* **1957,** *10,* 133–137.

Bedi, K. S.; Sikka, M. R.; Mundkur, B. B. Transmission of Wheat Bunt Due to *Neovossia indica* (Mitra) Mundkur. *Indian Phytopathol.* **1949,** *2,* 20–26.

Bedi, P. S. Impact of New Agricultural Technologies on Karnal Bunt of Wheat in Punjab. *Pl. Dis. Res.* **1989,** *4,* 1–9.

Beniwal, M. S; Karwasra, S. S; Gupta, A; Chhabra, M. L; Singh, R. Stable Sources of Resistance to Loose Smut of Wheat. *Ann. Biol.* **1998,** *14,* 231–232.

Bodh, R. Mating Behavior of Sporidia in *Tilletia indica* and Genetics of Karnal Bunt Resistance in Wheat. M.Sc. Thesis, Punjab Agricultural University, Ludhiana, India, 2005.

Bonde, M. R.; Peterson, G. L.; Dowler, W. M.; May, M. B. Comparison of *Tilletia indica* Isolates from India and Mexico by Isozyme Analysis. *Phytopathology.* **1985,** *75,* 1309.

Bonde, M. R.; Peterson, G. L.; Matsumoto, T. T. The Use of Isozymes to Identify Teliospores of *Tilletia indica*. *Phytopathology.* **1989,** *79,* 596–599.

Bonde, M. R.; Peterson, G. L.; Royer, M. Inheritance of Isozymes in the Smut Pathogen *Tilletia indica*. *Phytopathology.* **1988,** *78,* 1276–1279.

Bonde, M. R.; Peterson, G. L.; Schaad, M. W.; Smilanick, J. W. Karnal Bunt of Wheat. *Plant Dis.* **1997,** *81,* 1370–1377.

Brennan, J. P., et al. *Economic Losses from Karnal bunt of Wheat in Mexico*; CIMMYT Economic Working Paper, CIMMYT: Mexico, 1990, p 2.

Brooks, S. A.; See, D. R.; Guedira – Brown, G. SNP-Based Improvement of a Microsatellite Marker Associated with Karnal Bunt Resistance in Wheat. *Crop Sci.* **2006,** *46,* 1467–1470.

Busch, R.; McVey, D.; Rauch, T.; Baumer, J.; Elsayed, F. Registration of Wheat on Wheat. *Crop Sci.* **1984,** *24,* 622.

Calvo, J. A. Inheritance of Resistance of the Variety El Gaucho F. A. to Loose Smut (*Ustilago nuda* (Jens.) Rostr. *Revista de la Facultad de Agronomia* **1978,** *54,* 441–450.

Chahal, S. S.; Kaur, S.; Anita; Pannu, P. P. S. Compatibility in Sporidia of *Tilletia indica* and *Tilletia barclayana*. *Indian Phytopathol.* **2003,** *56,* 78–81.

Chand, K.; Gill, K. S.; Nanda, G. S.; Singh, G. Breeding for Karnal Bunt Resistance Through Intermating of Heat Cultivars with Low Coefficient of Infection. *Crop Improv.* **1989,** *16,* 178–179.

Cherewick, W. J. *Smut Diseases of Cultivated Plants in Canada;* Canada Department of Agriculture: Ottawa, Ontario, 1953; Vol. 887, pp 1–58.

Chhuneja, P.; Kaur, S.; Dhaliwal, H. S. In *Screening of US Winter Wheats Under a Screen House with Simulated Epidemiological Parameters for High Incidence of Karnal Bunt*, Proceedings of the XIVth Biennial Workshop on Smut Fungi, Idaho, USA, Jun 23–25, 2004.

Chona, B. L.; Munjal, R. L.; Adlakha, K. L. A Method for Screening Wheat Plants for Resistance to *Nevossia indica. Indian Phytopathol.* **1961,** *14,* 99–101.

Crous, P. W., et al. *Karnal Bunt Found in South Africa;* http: // www.sassp.co.2a/x6/newdiseasereports/January/feature/karnalbunt.htm. 2000.

Datta, R.; Rajebhosale, M. D.; Dhaliwal, H. S.; Singh, H.; Rajnekar, P. K.; Gupta, V. S. Interspecific Genetic Variability Analysis of *Neovossia indica* Causing Karnal Bunt of Wheat using Repetitive Elements. *Theor. Appl. Genet.* **2000,** *100,* 569–575.

Datta, R., et al. Gene-for- Gene Relationship for Resistance in Wheat to Isolates of Karnal Bunt, *Neovossia indica. Plant Breed.* **1999,** *118,* 362–364.

Datta, R.; Dhaliwal, H. S.; Gupta, S.; Multani, D. S. Transfer of Rye Chromosome Carrying Karnal Bunt Resistance to *Triticum aestivum* WL 711. *Wheat Inf. Serv.* **1995,** *80,* 20–25.

DePauw, R. M.; Knox, R. E.; Clarke, E. R.; Clarke, J. M; Fernandez, M. R.; McCaig, T. N. Helios Hard Red Spring Wheat. *Can. J. Plant Sci.* **2007,** *87,* 515–520.

DePauw, R. M.; Knox, R. E.; Clarke, E. R.; Clarke, J. M; McCaig, T. N. Stettler Hard Red Spring Wheat. *Can. J. Plant Sci.* **2009a,** *89,* 945–951.

DePauw, R. M.; Knox, R. E.; Thomas, J. B.; Smith, M.; Clarke, J. M.; Clarke, E. R., et al. Goodeve Hard Red Spring Wheat. *Can. J. Plant Sci.* **2009b,** *89,* 937–944.

DePauw, R. M.; Thomas, J. B.; Knox, R. E.; Clarke, J. M.; Fernandez, M. R.; McCaig, T. N., et al. AC Cadillac Hard Red Spring Wheat. *Can. J. Plant Sci.* **1998,** *78,* 459–462.

Dhaliwal, H. S., et al. Association of Different Traits with Monosomic Additional Lines of *Aegilopes squarrosa* in *Triticum aestivum. Wheat Inf. Serv.* **1990,** *71,* 14–15.

Dhaliwal, H. S.; Navarete, M. R.; Valdez, J. C. Scanning Electron Microscope Studies of Penetration Mechanism of *Tilletia indica* in Wheat Spikes. *Rev. Mex. Fitopathol.* **1988,** *7,* 150–155.

Dhaliwal, H. S.; Singh, D. V. Production and Inter- Relationship of Two Types of Secondary Sporidia of *Neovossia indica. Curr. Sci.* **1989,** *58,* 614–618.

Dhaliwal, H. S.; Singh, H. In *Breeding for Resistance to Bunts and Smuts, Indian Science,* Bunts and Smuts of Wheat, an International Symposium, Aug 17–20, 1997; North American Plant Protection Organization: North Carolina, USA, 1997.

Dhiman, J. S. *Epidemiology of Karnal Bunt of Wheat in Punjab.* Ph.D. Dissertation, Punjab Agricultural University, Ludhiana, 1982.

Dhitaphichit, P.; Jones, P.; Keane, E. M. Nuclear and Cytoplasmic Gene Control of Resistance to Loose Smut (*Ustilago tritici* (Pers.) Rostr.) in Wheat (*Triticum aestivum* L.). *Theor. App. Genet.* **1989,** *78,* 897–903.

Duran, R.; Cromarty, R. *Tilletia indica,* a Heterothallic Wheat Fungus with Multiple Alleles Controlling Incompatibility. *Phytopathology.* **1977,** *67,* 812–815.

Duran, R.; Fischer, G. W. *The Genus Tilletia;* Washington State University Press Co.: Pullman, WA, 1961.

Duran, R. Further Aspects of Teliospore Germination in North American Smut Fungi. *Canadian J. Bot.* **1972,** *50,* 2569–2573.

Fischer, G. W.; Holton, C. S. *Biology and Control of Smut Fungi;* Ronald Press Co.: New York, 1957, p. 622.

Fox, S. L.; McKenzie, R. I. H.; Lamb, R. J.; Wise, I. L.; Smith, M. A. H.; Humphreys, D. G., et al. Unity Hard Red Spring Wheat. *Can. J. Plant Sci.* **2010,** *90,* 71–78.

Fuentes- Davila, G.; Rajaram, S.; Singh, G. Inheritance of Resistance to *Tilletia indica* in Wheat. *Crop Prot.* **1995,** *13,* 20–24.

Fuentes-Davila, G.; Rajaram, S.; Sources of Resistance to *Tilletia Indica* in Wheat. *Genet.* **1994,** *50,* 205–209.

Fuentes-Davila, G.; Duran, R. *Tilletia indica,* Cytology and Teliospore Formation *in vitro* and in Immature Kernels. *Can. J. Bot.* **1986,** *64,* 1712–1719.

Gautam, M.; Srivastava, K. D.; Agarwal, R. Antagonistic Effect of *Trichoderma* spp. on *Ustilago segetum* var. *tritici* and their Compatibility with Fungicides and Biocides. *Indian Phytopathol.* **1995,** *48,* 466–470.

Ghorpade, D. S. Study of Genetic Stock Nurseries of Wheat. *J. Maharashtra Agric. Univ.* **1983,** *8,* 118–119.

Gill, K. S. In *Research on Karnal Bunt of Wheat Germplasm Screening, Breeding Strategies and Future Research Priorities,* Proceedings of the 5th Biennial Smut Workers' Workshop, CIMMYT: Cuidad Obregon, Sonora, Mexico, 1988; p 28.

Gill, K. S., et al. Breeding Wheat Varieties Resistant to Karnal Bunt. *Crop Improv.* **1981,** *8,* 73–80.

Gill, K. S., et al. Study of Gene Effects for Karnal Bunt (*N. indica* Mitra) Resistance in Bread Wheat (*Triticum aestivum*). *Indian J. Genet.* **1990,** *50,* 205–09.

Gill, K. S., et al. *Genetic Evaluation and Utilization of Wheat Germplasm for Breeding Varieties, Resistant to Karnal Bunt;* Final Report, PL 480 Project, 1992.

Gill, K. S.; Aujla, S. S.; Sharma, I. *Karnal Bunt and Wheat Production;* Punjab Agricultural University: Ludhiana, Punjab, 1993; pp 1–153.

Gill, K. S.; Aujla, S. S. In *Research on Breeding for Karnal Bunt Resistance in Wheat,* Proceedings of the 5th Biennial Smut Workers Workshop, Sonora, Mexico, Apr 28–30, 1986; CIMMYT: Mexico, 1986; p 13.

Glenn, D. L.; Peterson, G. L. Effect of Sodium Phosphate Salts and pH on Teliospore Germination and Basidiospores Production of *T. indica. Phytopathology.* **2005,** *95,* 535.

Goates, B. In *Histopathology of Karnal Bunt Infection of Wheat,* Proceeding of the5th Biennial Smut Workers Workshop, Ciudad Obregon, Sonora, Mexico, CIMMYT: Mexico, 1988.

Goel, L. B., et al. Efficacy of Raxil (Tebuconazole) for Controlling the Loose Smut of Wheat Caused by *Ustilago segetum* var. *tritici. Indian Phytopath.* **2001,** *54,* 270–271.

Goel, L. B., et al. Evaluation of Tilt against Karnal Bunt. *Indian Phytopathol.* **2000b,** *53,* 301–302.

Goel, L. B., et al. Incidence of Karnal Bunt on the Wheat Varieties of Northwestern India. *Pl. Dis. Res.* **2000a,** *15,* 116–118.

Goel, L. B.; et al. Evaluation of G 696 Fungicide for Controlling Loose Smut of Wheat *(Triticum aestivum)* Caused by *Ustilago segetum* var. *tritici. Indian J. Agric. Sci.* **1995,** *65* (7), 536–538.

Goel, L. B. Evaluation of Tilt against Karnal Bunt of Wheat. *Indian Phytopathol.* **2000,** *53,* 301–302.

Gothwal, B. D.; Pathak, V. N; Pathogenic Races of *Ustilago tritici* in India. *Indian Phytopath.* **1977,** *30,* 311–314.

Green, G. J., et al. The Experimental Approach in Assessing Disease Losses in Cereals, Rusts and Smuts. *Can. Plant. Dis. Surv.* **1968,** *48,* 61–64.

Grevel, F. K. Untersuchungen Ueber das Vorhandensein Biologischer Rassen des Flugbrandes des Weizens (*Ustilago tritici*). *Phytopathol. Z.* **1930,** *2,* 209–234.

Grewal, T. S.; Sharma, I.; Aujla, S. S. Role of Stomata and Hairs in Resistance, Susceptibility of Wheat to Karnal Bunt. *J. Mycol. Pl. Pathol.* **1999,** *29,* 217–221.

Guldhe, S. M.; Rant, J. G.; Wangikar, P. D. Control of Loose Smut Infection in Wheat by Physical and Chemical Methods of Seed Treatment. *PKV Res. J.* **1985,** *9,* 56–58.

Hansing, E. D. A Systemic Oxathiin Fungicide to Control Loose Smut, Seed Rots and Seedling Blights of Wheat, Barley, Oats and Sorghum. *USDA Comp.* **1968,** *CR.* 42–68.

Hansing, E. D. Systemic Oxathiin Fungicides for Control of Loose Smut (*Ustilago tritici*) of Winter Wheat. *Phytopathology.* **1967,** *57,* 814.

Hardison, J. Systemic Activity of Two Derivatives 1, 4-Oxathiin against Smut and Rust Diseases of Grasses. *Pl. Dis. Reptr.* **1966,** *50,* 624.

Harjit-Singh, Grewal; T. S; Pannu; P. P. S; Dhaliwal, H. S. Genetics of Resistance to Karnal Bunt Disease of Wheat. *Euphytica.* **1999,** *105,* 125–131.

Heinrich, J. Studies of Genetic Resistance of Wheat to Loose Smut, *Ustilago tritici* (Pers.) Rostr. Hodowla Roslin, Aklimatyzacja i Nasiennictwo, **1970a,** *14,* 393–404.

Heinrich, J. The Use of Monosomic Lines of the Wheat *Triticum aestivum* L. in Genetic Studies on Resistance to Loose Smut of Wheat, *Ustilago tritici* (Pers.) Rostr. Zeszyty Naukowe Wyzszej Szkoly Rolniczej w Krakowie **1970b,** *58,* 195–204.

Hoffman, J. A. In *Chemical Seed Treatment for Karnal Bunt*, Proceeding of the 5th Biennial Smut Worker's Workshop, Sonora, Mexico, Apr 28–30, 1986; CIMMYT; Mexico, 1986.

Holton, C. S. Observations on *Neovossia indica. Indian Phytopathol.* **1949,** *2,* 1–5.

Jhorar, O. P., et al. A Biometerological Model for Forecasting Karnal Bunt Disease of Wheat. *Pl. Dis. Res.* **1992,** *7,* 204–209.

Jones, P. Control of Loose Smut (*Ustilago nuda* and *U. tritici*) Infections in Barley and Wheat by Foliar Application of Systemic Fungicides. *Eur. J. Plant Pathol.* **1999,** *105,* 729–732.

Joshi, L. M., et al. Karnal Bunt, a Minor Disease that is Now a Threat to Wheat. *Bot. Rev.* **1983,** *49,* 309–330.

Joshi, L. M.; Singh, D. V.; Srivastava, K. D. Status of Rusts and Smuts during Wheat Era in India. *RACHIS.* **1985,** *4,* 10–16.

Joshi, L. M.; Singh, D. V.; Srivastava, K. D. Present Status of Karnal Bunt in India. *Indian Phytopathol.* **1980,** *33,* 147–148.

Kang, M. S.; Bedi, P. S. Prevalence and Incidence of Karnal Bunt of Wheat in the Punjab. *Indian J. Mycol. Pl. Pathol.* **1980,** *10,* 81 (abstr.).

Kassa, M. T.; Menzies, J. G.; McCartney, C. A. Mapping of the Loose Smut Resistance Gene Ut6 in Wheat (*Triticum aestivum L.*). *Mol. Breed.* **2014,** *33,* 569–576.

Kaur, J.; Bala, R.; Pannu, P. P. S.; Jindal, M.; Sharma, I. In *Status of Wheat Diseases in Punjab- A Decade Under Study*, Proceedings of the National Symposium on "Climate Challenges: Status and Management of Plant Diseases", Sri Konda Lakshman Telangana, Horticultural University, Rajendranagar, Hyderabad, Dec1–3, 2015; Indian Society of plant Pathologists: India, 2015.

Kaur, A., et al. Relative Efficacy of Systemic Fungicides as Seed Dresser against Loose Smut of Wheat. *J. Mycol. Pl. Pathol.* **2002a,** *32* (1), 110.

Kaur, G.; Sharma, I.; Sharma, R. C. Characterization of *Ustilago segetum tritici* Causing Loose Smut of Wheat in Northwestern India. *Can. J. Pl. Pathol.* **2014,** *36* (3), 360–366.

Khanna, A.; Payak, M. M. Teliospore Morphology of Some Smut Fungi. I. Electron Microscopy. *Mycologia.* **1968,** *58,* 562–569.

Knox, R. E., et al. Genetics of Resistance to *Ustilago tritici* in 'Glenlea' Wheat (*Triticum aestivum*). *Can. J. Pl. Pathol.* **2008,** *30,* 267–276.

Knox, R. E., et al. Inheritance of Loose Smut (*Ustilago tritici*) Resistance in Two Hexaploid Wheat (*Triticum aestivum*) Lines. *Can. J. Pl. Pathol.* **1999,** *21,* 174–180.

Knox, R. E.; Menzies, J. G. Resistance in Wheat to Loose Smut. In *Disease Resistance in Wheat (CABI Plant Protection Series);* Sharma, I., Ed.; CABI: Wallingford, Oxon, **2012**; pp 160–189.

Knox, R. E.; Howes, N. K. A Monoclonal Antibody Chromosome Marker Analysis Used to Locate a Loose Smut Resistance Gene in Wheat Chromosome 6A. *Theor. App. Genet.* **1994,** *89,* 787–793.

Knox, R. E.; Campbell, H. L.; DePauw, R. M.; Clarke, J. M.; Gold, J. J. Registration of P8810- B5B3A2A2 White-Seeded Spring Wheat Germplasm with *Lr35* Leaf and *Sr39* Stem Rust Resistance. *Crop Sci.* **2000,** *40,* 1512–1513.

Knox, R. E.; Fernandez, M. R.; Thomas, J. B.; Townley-Smith, T. F.; Campbell, H.; DePauw, R. M. Registration of Two Pairs of Wheat Genetic Stocks Near-Isogenic For Loose Smut Resistance: L8800-CC7B1B1D16 and L8800-CC7B1B1C1S, and P8802-C1*3A2A2U and P8802-C1*3A2C16. *Crop Sci.* **1998,** *38,* 557.

Knox, R. E.; Menzies, J. G.; Howes, N. K.; Clarke, J. M.; Aung, T.; Penner, G. A. Genetic Analysis of Resistance to Loose Smut and an Associated DNA Marker in Durum Wheat Doubled Haploids. *Can. J. Pl. Pathol.* **2002,** *24,* 316–322.

Krishna, A.; Singh, R. A. Multiple Alleles Controlling the Incompatibility in *Neovossia indica. Indian Phytopathol.* **1983,** *36,* 746–748.

Krishna, A.; Singh, R. A. Aberrations in the Teliospore Germination of *Neovossia indica. Indian Phytopathol.* **1981,** *34,* 260–262.

Krishna, A.; Singh, R. A. Taxonomy of Karnal Bunt Fungus, Evidence in Support of Genus *Neovossia. Indian Phytopath.* **1982,** *35,* 544–545.

Krivchenko, V. I; Bakhareva, Z. A Genetic Analysis of the Resistance of Spring Wheat to Loose Smut. *Soviet Genetics.* **1985,** *20,* 1079–1084.

Kulkarni, S.; Chattannavar, S. N. Efficacy of Fungicides against Loose Smut of Wheat. *Pl. Pathol. Newslett.* **1984,** *2,* 7–8.

Kumar, J., et al. Pathogenic and Molecular Variation among Indian Isolates of *Tilletia indica* Causing Karnal Bunt of Wheat. *Indian Phytopathol.* **2004,** *57,* 144–149.

Kumar, J.; Nagarajan, S.; Goel, L. B. Characterization of *Neovossia indica* by Electrophoresis of Mycelial Proteins. *Indian J. Mycol. Pl. Path.* **1995,** *25,* 105.

Kumar, M.; Luthra, O. P; Yadav, N. R; Chaudhary, L; Saini, N; Kumar, R; Sharma, I; Chawla, V. Identification of Micro Satellite Markers on Chromosomes of Bread Wheat Showing an Association with Karnal Bunt Resistance. *Afr. J. Biotechnol.* 2007, *6* (14), 1617–1622.

Kumar, S., et al. Identification and Validation of SSR Markers for Karnal Bunt (*Neovossia indica*) Resistance in Wheat (*Triticum aestivum*). *Indian J. Agric. Sci.* **2015,** *85* (5), 712–717.

Kumar, S.; Singh, D. Integrated Management of Karnal Bunt of Wheat. *J. Industrial Pollution Control.* **2014,** *30 (*2), 247–250.

Lang, W. Die Bluteninfektion Beim Weigen Flugbrand. *Zol. Bakt.* **1910,** *25,* 86.

Lari, M.; Castlebury, L. A.; Goates, B. J. Nonsystemic Bunt Fungi-*Tilletia indica* and *T. horrida,* A Review of History, Systematics and Biology. *Ann. Rev. Phytopathol.* **2006,** *44,* 113–133.

Locke, C. M.; Watson, A. J. Foreign Plant Diseases Intercepted in Quarantine Inspections. *Pl. Dis. Reptr.* **1955,** *39,* 518.

Luthra, J. C.; Sattar, A. Some Experiments on the Control of Loose Smut (*Ustilago tritici* Press.) Jens, of Wheat. *Indian J. Agric. Sci.* **1934,** *4,* 177–199.

Maddox, F. Smut and Bunt. *Agric. Gaz. Tasmania.* **1896,** *4,* 92–95.

Malik, M. M. S.; Batts, C. C. V. The Development of Loose Smut of Barley (*Ustilago nuda*) in the Brley Plant, with Observations on Spore Formation in Nature and in Culture. *Trans. British Mycol. Soc.* **1960b**, *43*, 126–131.

Malik, M. M. S.; Batts, C. C. V. The Infection of Barley by Loose Smut (*Ustilago nuda* [Jens.] Rostr.). *Trans. British Mycol. Soc.* **1960a**, *43*, 117–125.

Mathur, S. B.; Barry, M. C. *Seed Borne Diseases and Seed Health Testing of Wheat;* Jordburgsforlaget frededriksberg: Copenhagen, Denmark, 1993.

Mathur, H. C.; Chaudhary, H. B.; Singh, S. R. Identification of Chromosomes Carrying Genes for Resistance to Loose Smut of Bread Wheat (*Triticum aestivum* L.) in India. *Indian J. Genet.* **1997**, *57*, 115–119.

Mau, Y. S.; Fox, S. L.; Knox, R. E. Inheritance of Resistance to Loose Smut (*Ustilago tritici*) in Three Durum Wheat Lines. *Can. J. Pl. Pathol.* **2004**, *26*, 555–562.

Maude, R. B.; Shuring, C. G. Seed Treatments with Vitavax for the Control of Loose Smut of Wheat and Barley. *Annl. Appl. Biol.* **2008**, *64*, 259–263.

McLeod, J. G.; Townley-Smith, T. F.; DePauw, R. M.; Clarke, J. M.; Lendrum, C. W. B.; McCrystal, G. E. Registration of DT369 High-Yielding, Semi-Dwarf Durum Wheat Germplasm. *Crop Sci.* **1991**, *31*, 1717.

Meeta, M.; Dhiman, J. S.; Bedi, P. S.; Kang, M. S. Incidence and Pattern of Karnal Bunt Symptoms on Some Triticale Varieties Under Adaptive Trials in Punjab. *Indian J. Mycol. Pl. Pathol.* **1980**, *10*, LXXXI L (abstr.).

Mehdi, V.; Joshi, L. M.; Abrol, Y. P. Studies on 'Chapatis' Quality, VI Effect of Wheat Grains with Bunts on the Quality of 'Chapatis'. *Bull. Grain Tech.* **1973**, *11*, 195–197.

Menzies, J. G. Carboxin Tolerant Strains of *Ustilago nuda* and *Ustilago tritici* in Canada. *Can. J. Plant Pathol.* **2008**, *30*, 498–502.

Menzies, J. G.; Turkington, T. K.; Knox, R. E. Testing for Resistance to Smut Diseases of Barley, Oats and Wheat in Western Canada. *Can. J. Plant Pathol.* **2009**, *31*, 265–279.

Mishra, A.; Kumar, A.; Garg, G. K.; Sharma, I. Determination of Genetic Variability among Isolates of *Tilletia indica* using Random Amplified Polymorphic DNA Analysis. *Plant Cell Biotech. Mol. Biol.* **2000**, *1*(1–2), 29–36.

Mitra, M. A New Bunt on Wheat in India. *Ann. App. Biol.* **1931**, *18*, 178–179.

Mitra, M. Studies on the Stinking Smut (bunt) of Wheat in India. *Indian J. Agric. Sci.* **1937**, *4*, 467–468.

Mitra, M. Stinking Smut (Bunt) of Wheat with Special Reference to *Tilletia indica* Mitra. *Indian J. Agric. Sci.* **1935**, *5*, 51–74.

Morgunov, A.; Montoya, J.; Rajaram, S. Genetic Analysis of Resistance to Karnal Bunt in Bread Wheat. *Euphytica.* **1994**, *74*, 41–46.

Mundkur, B. B. A Second Contribution towards Knowledge of Indian Ustilaginales. *Trans. Br. Mycol. Soc.* **1940**, *24*, 312–336.

Mundkur, B. B. Studies in Indian Cereal Smut. V. Mode of Transmission of Karnal Bunt of Wheat. *Indian J. Agric. Sci.* **1943**, *13*, 54–58.

Munjal, R. L. Status of Karnal Bunt (*Neovossia indica*) of Wheat in North India during 1968–69 and 1969–70. *Indian J. Mycol. Pl. Pathol.* **1975**, *5*, 185–187.

Munjal, R. L. Studies on Karnal Bunt of Wheat. Ph.D. Thesis, Punjab University, Chandigarh, India. **1970**, p 125.

Munjal, R. L. In *Epidemiology and Control of Karnal Bunt of Wheat,* Proceedings of Symposium on Epidemiology, Forecasting and Control of Plant Disease, University of Lucknow, Lucknow, India, Jan 18–20, 1971; INSA: New Delhi, India, 1971.

Nagarajan, S; Aujla, S. S; Nanda, G. S; Sharma, I; Goel, L. B; Kumar, J. Singh, D. V; Karnal Bunt (*Neovossia indica*) of Wheat-a Review. *Rev. Pl. Pathol.* **1997**, *12*, 2–9.

Nanda, G. S.; Chand, K.; Sohu, V. S.; Sharma, I. Genetic Analysis of Karnal Bunt Resistance in Wheat. *Crop Improv.* **1995**, *22*, 189–193.

Nelson, J. C; Autrique, J. E; Fuentes- Davila G; Sorrells, M. E.; Chromosomal Location of Genes for Resistance to Karnal Bunt in Wheat. *Crop Sci.* **1998**, *38*, 231–236.

Nielsen, J. Races of *Ustilago tritici* and Techniques for their Study. *Can. J. Pl. Pathol.* **1987**, *9*, 91–105.

Nielsen, J.; Thomas, P. Loose Smut. In *Bunts and Smut Diseases of Wheat, Concepts and Methods of Disease Management;* Wilcoxson, R. D., Saari, E. E., Eds.; CIMMYT: Mexico, DF, 1996.

Oort, A. J. P. Specialization of Loose Smut of Wheat-a Problem for the Breeder. *Tijdschr. Plantenz.* **1947**, *53*, 25–43.

Padmaja, N.; Srivastava, K. D.; Aggarwal, R.; Singh, D. V. Pathogenic Variability in India Isolates of *Ustilago segetum* f. sp. *tritici* of Causing Loose Smut of Wheat. *Ann. Pl. Protec. Sci.* **2006**, *14*, 146–150.

Padwick, G. W. *Report of the Imperial Mycologist;* Scientific Report of the Agricultural Research Institute: New Delhi, India, 1939; pp 105–112.

Pandey, R. N., et al. Occurrence of Karnal Bunt in the North Gujarat. *Indian J. Mycol. Pl. Pathol.* **1994**, *24*, 234.

Pannu, P. P. S.; Chahal, S. S. Variability in *Tilletia indica*, the Incitant of Karnal Bunt of Wheat. *Indian Phytopathol.* **2000**, *53*, 279–282.

Parveen, S.; Saharan, M. S.; Verma, A.; Sharma, I. Pathogenic Variability among *Tilletia indica* Isolates and Distribution of Heterothallic Alleles in the Northwestern Plains Zone of India. *Turk. J. Agric. For.* **2015**, *39*, 63–73.

Patel, M. K.; Dhande, G. W.; Kulkarni Y. S. A Modified Treatment against Loose Smut of Wheat. *Curr. Sci.* **1950**, *19*, 324–325.

Patterson, F. L.; Shaner, G. E.; Ohm, H. W.; Finney, R. E.; Gallun, R. L.; Roberts, J. J., et al. Registration of Auburn Wheat (Reg. No. 652). *Crop Sci.* **1982**, *22*, 161–162.

Persons, T. D. Destructive Outbreak of Loose Smut in a Georgia Wheat Field. *Pl. Dis. Reptr.* **1954**, *38*, 422.

Peterson, G. L.; Bonde, M. R.; Dowler, W. M.; Royer, M. H. Morphological Comparisons of *Tilletia indica* (Mitra) from India and Mexico. *Phytopathology.* **1984**, *74*, 757 (abstr.).

Prescott, J. M. In *Karnal Bunt Research in Mexico*, Proceeding of the 5th Biennial Smut Workers Workshop, Sonora, Mexico, Apr 28–30, 1986; CIMMYT: Mexico, 1986; pp 13.

Procunier, J. D.; Knox, R. E.; Bernier, A. M.; Gray, M. A.; Howes, N. K. DNA Markers Linked to a T10 Loose Smut Resistance Gene in Wheat (*Triticum aestivum* L.). *Genome.* **1997**, *40*, 176–179.

Rana, S. K.; Basandrai, A. K.; Sood, A. K.; Sharma, B. K. Economical Management of Loose Smut of Wheat with Propiconazole. *Indian J. Agric. Sci.* **2000**, *70* (3), 163–164.

Randhawa, H. S., et al. Genetics and Identification of Molecular Markers Linked to Resistance to Loose Smut (*Ustilago tritici*) Race T33 in Durum Wheat. *Euphytica.* **2009**, *169*, 151–157.

Rewal, H. S.; Jhooty, J. S. Physiologic Specialization of Loose Smut of Wheat in the Punjab State of India. *Plant Dis.* **1986**, *70*, 228–230. doi, 10.1094/ PD-70–228.

Rivera-Castaneda, G., et al. *In vitro* Inhibition of Mycelial Growth of *Tilletia indica* by Extracts of Native Plants from Sonora, Mexico. *Revizta-Mexicano-de-Fitopatologia.* **2001**, *19*, 214–217.

Roberts, J. J.; Gallun, R. L.; Patterson, F. L.; Finney, R. E.; Ohm, H. W.; Shaner, G. E. Registration of Downy Wheat (Reg. No. 641). *Crop Sci.* **1981,** *21*, 350.

Royer, M. H.; Rytter, J. Artificial Inoculation of Wheat with *Tilletia indica* from Mexico and India. *Plant Dis.* **1985,** *69*, 317–319.

Saari, E. E.; Wilcoxon, R. D. Plant Disease Situation of High Yielding Dwarf Wheat in Asia and Africa. *Ann. Rev. Phytopath.* **1974,** *12*, 49–68.

Saini, R. G.; Gupta, A. K.; Sethi, D.; Sharma, S. C. Pathogenic Variation in Some Isolates of *Ustilago tritici. Indian J. Plant Pathol.* **1985,** *3*, 171–173.

Salzar-Huerta, F. I.; Osada, K. S.; Gilchrist, S. L.; Fuentes- Davilla, G. Evaluation de la Resistenicia de seis Genotipos de Trigo (*Triticum vulgare* L.) al Carbon Parcial Causado por el Hongo *Tilletia indica* Mitra en Invernadero. *Rev. Mex. Fitopathol.* **1990,** *8*, 145–152.

Sansford, C E. In *An Assessment of the Significance of the Initial Detection of Tilletia Indica Mitra in the USA in Early 1996 and the Potential Risk to the United Kingdom (And The European Union),* In Bunts and Smuts of Wheat: An International Symposium, Malik, V. S., Mathre, D. E., Eds.; North American Plant Protection Organization: Ottawa, Canada, 1998; pp 273–302.

Sehgal, S. K.; Kaur, G.; Sharma, I.; Bains, N. S. Development and Molecular Marker Analysis of Karnal Bunt Resistant Near Isogenic Lines in Bread Wheat Variety PBW343. *Indian J. Genet. Plant Breed.* **2008,** *68* (1), 21–25.

Sharma, A. K., et al. Efficacy of Some New Molecules Against Karnal Bunt of Wheat. *Indian J. Agric. Sci.* **2005,** *75*, 369–370.

Sharma, A. K., et al. Efficacy of Thifluzamide in the Control of Loose Smut of Wheat caused by *Ustilago segetum* var. *tritici. Indian J. Agric. Sci.* **2001,** *71* (10), 648–649.

Sharma, A. K.; Babu, K. S.; Sharma, R. K.; Kumar, K. Effect of Tillage Practices on *Tilletia indica* (Karnal Bunt Disease of Wheat) in a Rice-Wheat Rotation of the Indo-Gangetic Plains. *Crop Prot.* **2007,** *26* (6), 818–821.

Sharma, B. K.; Basandrai, A. K. Efficacy of Fungicides and Plant Extract for the Control of Karnal Bunt of Wheat (*Neovossia indica* Mitra) Mundkar. *J. Mycol. Pl. Pathol.* **2004,** *34*, 102–104.

Sharma, I., et al. Identifying Chromosome Regions Conferring Karnal Bunt Resistance in Wheat Using Deletion Stocks and Source of Resistance in Winter. *Pl. Dis. Res.* **2009,** *24* (1), 9–11.

Sharma, I.; Bains, N. S.; Sohu, V. S.; Sharma, R. C. Eight loci for Resistance to *Ustilago tritici* Race T11 Indicated in 20 Wheat Lines. *Cereal Res. Commun.* **2011,** *39*, 376–385.

Sharma, I.; Bains, N. S.; Bodh, R.; Sirari, A.; Sharma, R. C. Genetics of Karnal Bunt Resistance: Use of *Tilletia indica* Population at Different Levels of Heterogeneity. *Czech J. Plant Breeding.* **2006,** *42*, 6–11.

Sharma, I.; Bains, N. S.; Nanda, G. S. Inheritance of Karnal Bunt-Free Trait in Bread Wheat. *Plant Breed.* **2004a,** *123*, 96–97.

Sharma, I.; Bains, N. S.; Nanda, G. S.; Satija, D. R. Incorporation of Karnal Bunt Resistance in Wheat Varieties PBW 343. *Crop Improv.* **2003,** *30*, 21–24.

Sharma, I.; Bains, N. S.; Singh, K.; Nanda, G. S. Additive Genes at Nine Loci Govern Karnal Bunt Resistance in a Set of Common Wheat Cultivars. *Euphytica.* **2005a,** *142*, 301–307.

Sharma, I.; Nanda G S.; Singh, H.; Sharma, R. C. Status of Karnal Bunt Disease of Wheat in Punjab (1994–2004). *Indian Phytopath.* **2004b,** *57*, 435–439.

Sharma, I.; Nanda G. S.; Sharma, S.; Kaloty, P. K. Preliminary Studies on the use of Bioagents in the Management of Karnal Bunt of Wheat. *Pl. Dis. Res.* **1996,** *11*, 12–18.

Sharma, I.; Nanda, G. S. Factors Affecting Teliospore Viability of *Neovossia indica*. *J. Res. Punjab Agric. Univers.* **2002a**, *39*, 15–27.

Sharma, I.; Nanda, G. S.; Chand, K.; Sohu, V. S. Status of Karnal Bunt (*Tilletia indica*) Resistance in Bread Wheat (*Triticum aestivum*). *Indian J. Agric. Sci.* **2002b**, *71* (12), 794–796.

Sharma, I.; Nanda, G. S.; Kaloty, P. K . Ciperconazole-an Effective Fungicide for the Control of Karnal Bunt of Wheat. *Pl. Dis. Res.* **1997**, *12*, 35–36.

Sharma, I.; Nanda, G. S.; Kaloty, P. K. Karnal Bunt Resistance in Wheat in Relation to Hairyness, Waxiness and Period of Flowering. *Crop Improv.* **1998b**, *25*, 201–208.

Sharma, I.; Nanda, G. S.; Kaloty, P. K. Variability in *Neovossia indica*, Based on Pathogenicity and Isozyme Analysis. *Trop. Agric. Res. Extn.* **1998a**, *1*, 159–161.

Sharma, I.; Nanda, G. S.; Kaloty, P. K.; Grewal, A. S. Prevalence of Karnal Bunt of Wheat in Punjab. *Seed Res.* **1998**, *26*, 155–160.

Sharma, R. C.; Sharma, I.; Samra, J. S. In *Achievements in Disease Free Seed Production of Wheat in Punjab (India)*, Proceeding of the 2nd Asian Conference on Plant Pathology, National University of Singapore, Singapore, June 25–28, 2005; National University of Singapore: Singapore, 2005b; p 34.

Sharma, S. K., et al. Gene Action for Karnal Bunt Resistance in Wheat. *Indian J. Genet.* **1991**, *57*, 374–375.

Sharma, B. K.; Basandrai, A. K. Efficacy of Some Plant Extracts for the Management of Karnal Bunt (*Neovossia indica*) of Wheat. *Indian J. Agric. Sci.* **1999**, *69*, 837–839.

Sherif, S.; Ghanem, E. H.; Shafik, I.; Mostafa, E. E.; Abdel-Aleem, M. M. Integrated Control of Wheat Loose Smut in Egypt. *Assiut J. Agric. Sci.* **1991**, *22*, 153–163.

Siddhartha, V. S. *Epidemiological studies on Karnal bunt of wheat*. Ph.D. Dissertation, Indian Agricultural Research Institute, New Delhi, India, 1992.

Singh, D. V.; Aggarwal, R.; Nagarajan, S. Varibility in *Neovossia indica* in the Context of Host Resistance and Disease Management. In *Genetic Research and Education Current Trends and the Next Fifty Years;* Sharma. B., et al. Eds.; Indian Society of Genetics Plant Breeding: New Delhi, India, 1995; Vol. I, pp 314–324.

Singh, D. V.; Srivastava K. D.; Aggarwal, R. Karnal Bunt; Constraints to Wheat Production and Management. In *Bunts and Smuts of Wheat,* An International Symposium, 1998; Malik, V. S., Mathre, D. E., Sharma, Eds.; North American Plant Protection Organization: Ottawa, 1998; pp 201–222.

Singh, D. V. Bunts of Wheat in India. In *Problems and Progress of Wheat Pathology in South Asia*, Joshi, L. M., Singh, D. V., Srivastava, K. D., Eds.; Malhotra Publishing House: New Delhi, India, 1986; p 410.

Singh, A.; Prasad, R. Date of Sowing and Meteorological Factors in Relation to Occurrence of Karnal Bunt of Wheat in U.P. Tarai. *Indian J. Mycol. Pl. Pathol.* **1978**, *8*, 2.

Singh, D. P. Effect of Reduced doses of Fungicides and Use of *Trichoderma viride* During Seed Activation Stage in Controlling the Wheat Loose Smut. *J. Mycol. Pl. Pathol.* **2004**, *34*, 396–398.

Singh, D. P., et al. Efficacy of *Trichoderma viride* in Controlling of Loose Smut of Wheat Caused by *Ustilago segetum* var. *tritici* at Multilocation. *J. Biol. Contr.* **2000**, *14*, 35–38.

Singh, D. P., et al. Nature of Resistance in Wheat and Triticale to Loose Smut. *Indian Phytopathol.* **2008**, *61*, 528–529.

Singh, D. P., et al. Resistant Lines to Loose Smut (*Ustilago segetum* var. *tritici*) in Wheats (*Triticum aestivum, T. durum, T. dicoccum*) and Triticale. *Indian J. Agric. Sci.* **2002**, *72*, 308–310.

Singh, D. P., et al. Post Harvest Survey of Wheat Grains for the Presence of Karnal Bunt and Black Point Diseases in Different Agroclimatic Zones of India. *Indian J. Agric. Res.* **2003a,** *37,* 264–268.

Singh, D. P., et al. Bio-Efficacy of Triticonazole 8%+Pyraclostrobin 4% FS, Pyraclostrobin 20% FS, Triticonazole 2.5% against Loose Smut of Wheat. *African J. Crop Prot. Rural Sociol.* **2015,** *2,* 54–56.

Singh, D. P.; Sharma, A. K.; Grewal, A. S. Loose Smut Resistant Lines in Wheat with Combined Resistance to Karnal Bunt, Rusts, Powdery Mildew and Leaf Blight. *Wheat Inf. Serv.* **2001,** *92,* 27–29.

Singh, D. V., et al. First Report of *Neovossia indica* on Wheat in Nepal. *Plant Dis.* **1989,** *73,* 277.

Singh, D. V., et al. Mulching as a Means of Karnal Bunt Management. *Pl. Dis. Res.* **1991a,** *6,* 115–116.

Singh, D. V.; Gogoi, R. Karnal Bunt of Wheat (*Triticum* spp.)-A Global Scenario. *Indian J. Agric. Sci.* **2011,** *81*(1), 3–14.

Singh, D. V.; Srivastava, K. D.; Aggarwal, V. K. *Results of Co-ordinated Wheat Experiments;* AICWIP-ICAR: IARI, New Delhi, India, 1991b; pp 76–80.

Singh, D. V.; Srivastava, K. D.; Joshi, L. M.; Verma, B. R. Evaluation of Some Fungicides for the Control of Karnal Bunt of Wheat. *Indian Phytopathol.* **1985,** *38,* 571–73.

Singh, D., et al. Relation between Weather Parameters and Karnal Bunt in Wheat. *Indian J. Agric. Sci.* **1996,** *56,* 522–525.

Singh, D.; Maheshwari, V. K. Biological Seed Treatment for the Control of Loose Smut of Wheat. *Indian Phytopathol.* **2001,** *54* (4), 457–460.

Singh, D. V.; Srivastava, K. D.; Joshi, L. M. Occurrence and Spread of Karnal Bunt in India. *Indian Phytopathol.* **1980,** *33,* 249–254.

Singh, G.; Rajaram, S.; Monotoya, J.; Fuentes Davila, G. Genetic Analysis of Karnal Bunt Resistance in 14 Mexican Bread Wheat Genotype. *Plant Breed.* **1995a,** *81,* 117–120.

Singh, G.; Rajaram, S.; Monotoya, J.; Fuentes Davila, G. Genetic Analysis of Resistance To Karnal Bunt (*Tilletia indica* Mitra) in Bread Wheat. *Euphytica.* **1995b,** *81,* 117–120.

Singh, K. Cytogenetic Analysis of Karnal Bunt Resistance Involving Wild Donors and Cultivated Species Of Wheat. Ph. D. Dissertation, Punjab Agricultural University, Ludhiana, India, 1989.

Singh, R. A.; Krishna, A. In *Some Studies on Karnal Bunt of Wheat,* Proceedings of 19th All India Wheat Research Worker's Workshop, All India Co-ordinated Wheat Improvement Programme, Gujarat Agricultural University, Anand, Gujarat, Aug 11–14, 1980; IARI: New Delhi, India, 1980.

Singh, R. P.; Rajaram, S. Breeding for Resistance in Wheat. 2006. http//www.fao.org. DOCREP/006/Y4011E//y4011eOb.htm.

Singh, S. L.; Singh, P. P. Cultural Practices and Karnal Bunt Incidence. *Indian Phytopathol.* **1985,** *38,* 594.

Singh, S., et al. Identification of Amplified Fragment Length Polymorphism Markers Associated with Karnal Bunt (*Neovossia indica*) Resistance in Bread Wheat. *Indian J. Agric. Sci.* **1999,** *69,* 497–501.

Sinha, V. C., et al. Sources of Resistance in Wheat and Triticale Against Loose Smut Caused by *Ustilago segatum* var. *tritici. Indian Phytopathol.* **2000,** *53* (1), 76–79.

Sinha, V. C.; Singh, D. P. Raxil (Tebuconazole) in the Control of Loose Smut of Wheat. *J. Mycol. Pl. Pathol.* **1996,** *26,* 279–281.

Sirari, A; Sharma, I; Bains, N. S; Raj, B; Singh, S; Bowden, R. L. Genetics of Karnal Bunt Resistance in Wheat: Role of Genetically Homogenous *Tilletia indica* Inoculum. *Indian J. Genet.* **2008,** *68* (1), 10–14.

Sirari, A. Compatibility System in Tilletia Indica and Genetics of Karnal Bunt Resistance in Wheat. Ph.D Dissertation, Punjab Agricultural University, Ludhiana, India, 2006.

Smilanick, J. L. et al. Evaluation of Seed and Foliar Fungicides for Control of Karnal Bunt of Wheat. *Plant Dis.* **1987,** *71,* 94–96.

Smilanick, J. L. In *Fungicide Studies in the Field*, Proceedings of the 5th Biennial Smut Workers' Workshop. CIMMYT, Sonora, Mexico, Apr 28–30, 1986; CIMMYT: Mexico, 1986; pp 18–19.

Smiley, R. W. Risk Assessment for Karnal Bunt Occurrence in the Pacific Northwest. *Plant Dis.* **1997,** *81,* 689–892.

Srivastava, K. D., et al. Occurrence of Loose Smut and its Sources of Resistance in Wheat. *Indian Phytopathol.* **1992,** *45,* 111–112.

Sukhwinder Singh, et al. In *Molecular Characterization of Karnal Bunt Resistance in Wheat*, Proceedings of theXIVth Biennial Workshop on Smut Fungi, Idaho, USA, 2004; p12.

Sukhwinder-Singh, Brown-Guerdia, G. L; Grewal, T. S; Dhaliwal, H. S; Nelson, J. C; Singh, H.; Gill, B. S. Mapping of a Resistance Gene Effective Against Karnal Bunt Pathogen of Wheat. *Theor. Appl. Genet.* **2003,** *106,* 287–292.

Sukhwinder-Singh, Gill, K. S; Dhaliwal, H. S; Singh, H; Gill, B. S. Towards Molecular Tagging of Karnal Bunt Resistance Gene (s) in Wheat. *J. Pl. Biochem. Biotech.* **1994,** *43,* 79–83.

Sukhwinder-Singh, Sharma, I; Sehgal, S. K.; Bains, N. S.; Guo .Z.; Nelson, J. C.; Bowden, R. L. Molecular Mapping of QTLs for Karnal Bunt Resistance in Two Recombinant Inbred Populations of Bread Wheat. *Theor. Appl. Genet.* **2007,** *116,* 147–154.

Swaminathan, M. S. et al. *Incidence of Black Point and Karnal Bunt in North Western Plains Zones during 1969–70;* Wheat Newsletter 4, Indian Agriculture Research Institute: New Delhi, India, 1971; pp 1–4.

Thirumalaisamy, P. P.; Singh, D. V.; Aggarwal, R.; Srivastava, K. D. Pathogenic Variability in *Tilletia indica,* the Causal Agent of Karnal Bunt of Wheat. *Indian Phytopathol.* **2006,** *59,* 22–26.

Tiemann, A. Untersuchung euber die Empfanglichkeit des Sommerweizens fur Ustilago Tritici und der Einfluss des Aussern Bedingungen Dieser Krankheit. *Kuhn-Archiv.* **1925,** *9,* 405–467.

Torabi, M.; Mardoukhi, V.; Jaliani, N. First Report on the Occurrenceof Partial Bunt on Wheat in the Southern Parts of Iran. *Seed Plant.* **1996,** *12,* 8–9.

Troitskaya, L. A.; Plakhotnik, V. V. Intraspecific Differentiation of the Loose Smut Pathogen in Spring Wheat. *Selektsiya i Semenovodstvo. USSR.* **1986,** *3,* 26–28.

Varshney, G. K.; Singh, U S.; Mishra D. P.; Kumar, A. Immuno Pathotyping of Karnal Bunt (*Tilletia indica*) Isolates of Wheat using Anti-Mycelial Antibodies. *Indian J. Exper. Biol.* **2003,** *41* (3), 255–261.

Vasu, K.; Singh, H.; Chhuneja, P.; Singh, S. Dhaliwal, H. S. Molecular Tagging of Karnal Bunt Resistance Genes of *Triticum monococcum* L. *Crop Improv.* **2000,** *27,* 33–42.

Verma, H. S.; Singh, A.; Agarwal, V. K. Application of Seedling Crown Test for Screening of Systemic Fungicides against Loose Smut of Wheat. *Seed Res.* **1984,** *12,* 56–60.

Villareal, R. L.; Mujeeb-Kazi, A.; Fuentes, D. G.; Rajaram, S. Inheritance of Resistance to *Tilletia indica* (Mitra) in Synthetic Hexaploid Wheat x *Triticum aestivum* cross. *Plant Breed.* **1995,** 114, 547–548.

Villareal, R. L.; Mujeeb-Kazi, A.; Rajaram, S.; Inheritance of Threshability in Synthetic Hexaploid (*Triticum turgidum x T. tauschiii*) by *T. aestivum* crosses *Plant Breed.* **1996,** *115,* 407-409.

Villareal, R. L.; Mujeeb Kazi, A.; Fuentes Davila, G.; Rajaram, S.; Del Toro, E. Resistance to Karnal Bunt (*Tilletia indica* Mitra) in Synthetic Hexaploid Wheats Derived from *Triticum turgidum* x *T. tauschii. Plant Breed.* **1994,** *112,* 63–69.

Warham, E. J. Karnal Bunt Disease of Wheat – A Literature Review. *Trop. Pest Mgmt.* **1986,** *32,* 229–242.

Warham, E. J. Teliospore Germination Patterns in *Tilletia indica. Trans. Br. Mycol. Soc.* **1988,** *90,* 318–321.

Warham, E. J. Effect of *Tilletia indica* Infection on Viability, Germination and Vigour of Wheat Seed. *Plant Dis.* **1990,** *74,* 130–132.

Williams, P. C. Incidence of Stinking Smut (*Tilletia* spp.) on Commercial Wheat Samples in Northern Syria. *Rachis: Barley, Wheat, Triticale, Newslet.* **1983,** *2,* 21.

Wunderle, J., et al. Assessment of the Loose Smut Fungi (*Ustilago nuda* and *U. tritici*) in Tissues of Barley and Wheat by Fluorescence Microscopy and Real-Time PCR. *Eur. J. Plant Pathol.* **2012,** *133,* 865–875.

Ykema, R. E., et al. First Report of Karnal Bunt of Wheat in the United States. *Plant Dis.* **1996,** *80,* 120.

CHAPTER 7

FLAG SMUT OF WHEAT AND ITS MANAGEMENT PRACTICES

DEVENDRA PAL SINGH*

ICAR-Indian Institute of Wheat and Barley Research, Karnal 132001, Haryana, India

Corresponding author. E-mail: dpkarnal@gmail.com

CONTENTS

ABSTRACT

Flag smut caused by *Urocystis agropyri* is a seed and soilborne disease of wheat and capable of losses in situations where farmers' uses their own produced seeds without proper seed treatment. The disease is favored under sandy light soil conditions and cooler temperature. The infected plants may not develop spikes and thus disease may cause losses up to 100%. The teliospores remain viable in soil up to seven years and therefore crop rotation may play important role in management. Use of fungicidal seed treatment effectively manages the disease in case of susceptible wheat cultivars in cooler climate.

7.1 INTRODUCTION

Flag smut caused by *Urocystis agropyri* (Preuss) A. A. Fisch. Waldh. is an important disease of wheat in cooler climate and under sandy soil conditions. The first reports of flag smut's presence came from Australia in 1868. It spreads to other countries mainly through infected seeds. It produces basidiospores and teliospores. This pathogen is found globally, but is most problematic in Australia and India (CABI, 2016). Relatives of *U. agropyri*, infect other grasses. It causes 100% losses in infected tillers. The losses in yield due to flag smut are estimated on average greater than AUS$50 million. India reported yield losses from flag smut in the 1940s through the 1970s, up the tune of 15,000 tons each year. In India, the disease was initially reported from Lyallpur in Punjab, (now in Pakistan) by Butler (1918) and later from Punjab, Haryana, Himachal Pradesh, Uttar Pradesh, Delhi, and Rajasthan states in moderate to severe form (Goel et al., 1977). Beniwal (1992) reported 23–65% yield losses from flag smut infection in nine commercial wheat cultivars in the Indian state of Haryana. Tillering was reduced by 15–45%, plant height by 37–62%, earhead length by 28–46%, and 1000-grain weight by 19–37%.

Presently, the disease is occurring in Northwestern Plains Zone and Northern Hills Zone (Bansandrai et al., 1993). Flag smut is of no human or animal health concern and has no impact on grain quality.

7.2 SYMPTOMS

Flag smut is a systemic disease. It starts in young tissues. Initial symptoms are "leprous" spots and bending or twisting of coleoptiles. As plant grows further, the leaves develop white striations that eventually turn silvery gray, which is evidence of the pathogen's impending sporulation. In addition to it, the infected plants may have stunted growth, excess foliages, without any ear heads. Young sori are covered by the silvery host epidermis, which later ruptures to expose the spore mass, splitting the leaves into ribbons (Mordue & Waller, 1981). The inflorescence is stunted, distorted, and frequently sterile with the rachis bearing black streaks at maturity. Spike development usually stops before the head emerges from the leaf whorl, so that infected plants do not produce seeds. In cereals, especially in resistant varieties, only some tillers may show symptoms (Purdy, 1965; Bhatnagar, et al., 1978.). The dark gray colored spore mass falls on ground from infected leaves once these are dry. The leaf discoloration is due to fungal structures called sori. During sporulation, these sori burst through leaves releasing teliospores. The spores are reddish brown with smooth rounded texture. These tend to be in clumps of 5–6 with sterile cells around them. These are thus called "spore balls" and they measure about 20–50 microns. When in large numbers, spores look like brown or black dust. All three cultivated wheat species, *Triticum aestivum, T. durum,* and *T. dicoccum*, are affected due to flag smut (Fig. 7.1)

FIGURE 7.1   Flag smut symptoms of different stages.

7.3   DISEASE CYCLE

U. agropyri is soil borne and externally seed borne. The teliospores are wind dispersed or distributed through soils via machinery or animals. In soil, a dikaryotic teliospore germinates, meiosis is followed by mitosis, which in turns produces up to four basidiospores, each containing a single nucleus. Basidiospores germinate on seedlings, and each hypha undergoes plasmogamy with a compatible hypha. During this process, one nucleus transfers to the other hypha, thus reestablishing the dikaryotic state of the pathogen.

These hyphae form appressoria. The appressoria penetrate the coleoptile of an emerging seedling through the epidermal tissue, and hyphae grow between vascular tissue of the leaves. The hyphal cells give rise to smut sori, having teliospores. These emerge through the leaf tissue for wind dispersal. Teliospores remain in soils, and give rise to basidiospores under favorable conditions. Alternatively, teliospores may adhere to seeds during threshing and germinate to cause new infection (Duran, 1972). Teliospores overwinter in the soil, senescent plant tissues, and with the seeds. These spores remain viable for 3–7 years.

7.4 ENVIRONMENTAL CONDITIONS

This pathogen prefers arid summers, moderate temperatures, and mild winters. The teliospores germinate in dry soils when the temperature ranges from 4.4 to 26.6°C. The crop cultivation practices that leave plant debris on soil surfaces favor disease. Mild winters improve the pathogen's ability to establish infections for seeds sown in autumn or winter; spring plantings give the fungus less opportunity to establish.

7.5 PATHOGENIC VARIABILITY AND GENETICS OF HOST RESISTANCE

U. agropyri has shown the least tendency toward pathogenic specialization as compared to other cereal smuts (Fischer & Holton, 1957). According to Hafiz (1986), only six races exist in this pathogen. Some peculiar features in the life cycle of this fungus (e.g., production of fewer sporidia compared to related smut fungi, the absence of secondary sporidia, and sporidial fusions *in situ*) are probably responsible for limiting its variability (Nelson & Duran, 1984). As a consequence, resistance to flag smut has remained effective in Australia and elsewhere (Platz & Rees, 1980; Goel, 1992) and new races have not been detected (Line, 1998). Purdy (1965) and Line (1998) attribute disappearance of flag smut as a problem on wheat in parts of the United States and low incidence of disease in India at present is primarily due to the use of resistant varieties.

According to Johnson (1984), there is no evidence of genotype-specific pathogenicity of the type indicating a gene-for-gene relationship. The incorporation of resistance into commercial cultivars is considered as an effective disease management practice (Goel, 1992).

7.6   DISEASE MANAGEMENT

These include, use of disease resistant cultivars, chemical seed treatments, and crop rotation to reduce amount of inocula present. *U. agropyri* spores prefer to germinate in dry soils; therefore, maintaining soils wet reduces the infection. Presowing fungicidal seed treatment is an effective way to break the disease cycle. Carboxin is a commonly used fungicide on seeds. Dry seed dressing with non-systemics such as copper carbonate, and systemics such as carboxin, oxycarboxin, and pyracarbolid have also been used (Neergaard, 1977). In addition, fenfuram, triadimefon, triadimenol, and tebuconazole provide control of *U. agropyri* in the Indian subcontinent (Goel & Jhooty, 1985; Tariq et al., 1992). Seed treatment with carboxin (Vitavax 75 WP) or carbendazim (Bavistin 50 WP) @ 2.5 g/kg seed or tebuconazole (Raxil-2 DS) @ 1 g/kg seed was carried out to control seed borne diseases. Seed treatment with tebuconazole at 1.0, 1.5, and 2.0 g/kg, or carboxin and carbendazim at 2.0 and 2.5 g/kg seeds, reduced disease incidence in wheat (Kumar & Singh, 2004). Tetramethylthiuram disulphide (Thiram 75% DS) at 3 g/kg seed, thiophanate methyl (Topsin-M 70WP) at 1 and 2 g/kg seed, or a combination of tetramethylthiuram disulphide and carboxin were also found effective against flag smut. (Shekhawat et al., 2011).

Shallow sowing of seeds in soil also helps to reduce disease occurrence. Crop rotation with non-host crops such as soybeans, sorghum, and corn, reduces the risk of the disease emergence. Continuous wheat cultivation and use of disease susceptible varieties often favors and creates ideal conditions for flag smut to flourish. The effects of fertilizers on flag smut of wheat are complicated (Purdy, 1965). The increased levels of nitrogen are known to favor the disease in turf grasses (Smiley et al., 2005). In India, Kumar and Singh (2004) found that higher levels of nitrogen and phosphorus fertilizers, as well as the addition of poultry manure to the soil, reduced the level of the disease in wheat, but still one has to use fungicidal seed treatments to control it fully. The multilocation testing of genotype under All India coordinated Wheat and Barley Improvement Programme identified bread wheat varieties: WH 283, WH 291, HD 2329, PBW 65, HS 207, HS 277, and R 2184 resistant to flag smut disease. likewise, *durum* wheat: WH 896, WH 912, Raj 1555, PBW 34, and DWL 5023 were resistant/immune whereas, triticales: DT-46, DT-18, and TL 2597 were resistant to flag smut disease. Singh et al. (2016). Goel and Gupta (1990) also evaluated 163 commercial Indian wheat cultivars, the important donors in the Indian wheat breeding program and some international elite wheat lines evaluated for resistance to flag smut under artificial seed inoculation, 32 lines were found to be immune, 16 highly resistant, and 28 moderately resistant to flag smut. All the 13 tetraploid wheats tested were immune. Reaction of NI 5439 and WW 15 to flag smut differed from that observed at other places while that of Federation, Gabo, Gamenya, and Oxley was similar, suggesting the presence of different virulences in the pathogen. Varieties, CI 299417 (Brewster), CPAN 1235, CPAN 1830, CPAN 1869, CPAN 1884, CPAN 1885, CPAN 1922, CPAN 1929, CPAN 1973, CPAN 1994, CPAN 2019, CPAN 1922, CPAN 1929, CPAN 1973, CPAN 1994, CPAN 2019, CPAN 6051, HD 2278, Raj 1972, Raj 1973, and WG 2109, which were resistant to leaf as well as stripe rusts, can serve as promising donors of multiple disease resistance (Goel & Gupta 1990). Allan (1976) recorded the reaction to flag smut (*Urocystis agropyri* (Preuss) Schroet.) of F_3 and F_4 lines of 10 wheat (*Triticum aestivum* L. em Thell.) crosses involving the parents "Spinkcota," "lowin" (highly resistant); "Games," "Hurt," "Golden," "Dickson 114" (resistant); "Ridit" (moderately resistant); and CI 13749 (susceptible) was studied in a field near Bickleton, Washington. Segregation ratios in five crosses between highly resistant and resistant parents suggested that different large-effect genes governed resistance to flag smut; no completely susceptible lines were recovered, which indicated that the parental combinations tested had one or more small-effect genes for resistance in common with one another. Evidence suggested that "lowin" has two large-effect genes for resistance that differ from those possessed by "Spinkcota," "Dickson 114," and "Gaines." "Ridit" probably has one gene for moderate resistance that is not common to "Spinkcota," "Dickson 114," and "Gaines." The segregation patterns for reaction of two crosses involving the susceptible parent, CI 13749, suggested complementary resistance. Results indicated that a minimum of four different large-effect and three small-effect genes for flag smut resistance were represented among the eight parents studied. No associations were found between percentage of flag smut and plant height, stripe rust reaction, maturity, and spike type. Percentage of flag smut was associated with chaff color in two of seven crosses and with an expression in one of five crosses.

7.7 CONCLUDING REMARKS

Flag smut of wheat may become important disease especially in situations where soils are sandy, climate is cool, and seeds of susceptible wheat varieties are sown without fungicidal seed treatment year after year. The infected plant spreads the inoculum to healthy seeds during threshing operations The teliospore fall on soil and remain viable for years. Therefore, an integrated approach is suggested to manage disease in case of disease susceptible varieties.

KEYWORDS

- flag smut
- *Urocystis agropyri*
- wheat
- management

REFERENCES

Allan, R. E. Flag Smut Reaction in Wheat: Its Genetic Control and Associations with Other Traits. *Crop Sci.* **1976,** *16*(5), 685–687.

Bansandrai, A. K.; Sood, A. K.; Sud, A. K. Prevalence of Flag Smut of Wheat (*Urocystis agropyri*) in Himachal Pradesh. *Pl. Dis. Res.* **1993,** *8*(1), 85.

Beniwal, M. S. Effect of Flag Smut on Yield and Yield Components of Wheat Varieties. *Crop Res. (Hisar).* **1992,** *5*(2), 348–351.

Bhatnagar, G. C.; Gupta, R. B. L.; Mishra, V. L. Effect of Flag Smut, Caused by *Urocystis agropyri* on Yield Components of Wheat Cultivars in Rajasthan, India. *Pl. Dis. Report.* **1978,** *62*(4), 348–350.

Butler, E. J. *Fungi and Disease in Plants;* Thacker Spink and Co.: Calcutta, India, 1918; p 547.

CABI. *Urocystis agropyri* (Flag Smut of Wheat). http://www.cabi.org/isc/datasheet/55784. 2016.

Duran, R. Further Aspects of Teliospore Germination in North American Smut Fungi. *Can. J. Bot.* **1972,** *50,* 2569–2573.

Fischer, G. W.; Holton, C. S. *Biology and Control of the Smut Fungi;* The Ronald Press Company: New York, 1957.

Goel, R. K.; Gupta, A. K. Reaction of Some Wheat Cultivars and Lines to Flag Smut. *Indian J. Genet. Pl. Breed.* **1990,** *50,* 185–188.

Goel, L. B.; Singh, D. V.; Srivastava, K. D.; Joshi, L. M.; Nagarajan, S. *Smuts and Bunts of Wheat in India;* IARI: New Delhi, India, 1977; p 38.

Goel, R. K.; Jhooty, J. S. Chemical Control of Flag Smut of Wheat Caused by *Urocystis agropyri. Indian Phytopath.* **1985,** *38*(4), 749–751.

Goel, R. K. Flag Smut of Wheat. *Indian J. Mycol. Pl. Pathol.* **1992,** *22*(2), 113–124.

Hafiz, A. *Plant Diseases;* Pakistan Agricultural Research Council: Islamabad, Pakistan, 1986.

Johnson, R. A. Critical Analysis of Durable Resistance. *Ann. Rev. Phytopathol.* **1984,** *22,* 309–330.

Kumar, V. R.; Bhim S. Management of Flag Smut in Wheat. *Ann. Pl. Prot. Sci.* **2004,** *12*(1), 92–95.

Line, R. F. Quarantines for the Control of Flag Smut of Wheat: Are They Effective or Necessary? In *Bunts and Smuts of Wheat,* An International Symposium, Raleigh, North Carolina, Aug 17–20, 1997; Malik, V. S., Mathre, D. E., Eds.; North American Plant Protection Organization: Ottawa, Canada, 1998; pp 49–60.

Mordue, J. E. M.; Waller, J. M. *CMI Descriptions of Pathogenic Fungi and Bacteria, No. 716;* CAB International: Wallingford, 1981.

Neergaard, P. *Seed Pathology;* Macmillan Press Ltd.: London, 1977; Vol. 1, p 839.

Nelson, B. D. Jr.; Duran, R. Cytology and Morphological Development of Basidia, Dikaryons and Infective Structures of *Urocystis agropyri* from Wheat. *Phytopathology.* **1984,** *74*(3), 299–304.

Platz, G. J.; Rees, R. G. Flag Smut Resistance in Selected Wheat Cultivars and Lines. *Queensland J. Agric. Animal Sci.* **1980,** *37*(2), 141–143.

Purdy, L. H. Flag Smut of Wheat. *Bot. Rev.* **1965,** *31,* 565–606.

Shekhawat, P. S., et al. Relative Efficacy of Different Seed Dressing Fungicides against Flag Smut of Wheat. *Indian Phytopathol.* **2011,** *64*(3), 303–304.

Singh, D.P. et al. AICW&BIP Crop Protection Report, IIWBR, 2016, p. 221.

Smiley, R. W., et al. *Compendium of Turfgrass Diseases;* 3rd ed.; American Phytopathological Society: St. Paul, MN, 2005; p 167.

Tariq, A. H.; Khan, S. H.; Saleem, A. Control of Flag Smut (*Urocystis agropyri*) through Screening of Varieties and Systemic Seed-Dressing Fungicides. *Pakistan J. Phytopathol.* **1992,** *4*(1–2), 32–36.

CHAPTER 8

BLACK POINT OF WHEAT CAUSED BY *BIPOLARIS SOROKINIANA* AND ITS MANAGEMENT

MOHAMMED SHAMSHUL Q. ANSARI[1], ANJU PANDEY[1],
V. K. MISHRA[2], A. K. JOSHI[3], and R. CHAND[1*]

[1]*Department of Mycology and Plant Pathology, Institute of Agricultural Sciences, Banaras Hindu University, Varanasi 221005, India*

[2]*Department of Genetics and Plant Breeding, Banaras Hindu University, Varanasi 221005, India*

[3]*International Maize and Wheat Improvement Center (CIMMYT), G-2, B-Block, NASC Complex, DPS, New Delhi-110092, India*

Corresponding author. E-mail: rc_vns@yahoo.co.in

CONTENTS

ABSTRACT

Black point of wheat is prevalent in most of the wheat growing regions of the world. It reduces grain quality and germination. Symptoms of black point are easily recognizable by deposition of black fungal mass at the embryo end of the seed. Out of a number of fungal pathogens involved in the black point, *B. sorokiniana* dominates in the warm humid climate of South Asia. Black point appears due to invasion of seed surface by direct penetration of hyphae to outer layers of the pericarp or through stigma to the pericarp. This disease can be artificially produced by inoculation of spore suspension in the floral cavity between the lemma and palea. A scale has been developed for monitoring the quality of grain and black point severity. Warmer temperature and high relative humidity during grain filling period favors black point. Severity of black point increases many fold when rains occur during grain filling period. Genotypic variation for black point severity has been recorded in wheat germplasm. Foliar spray of sterol inhibiting fungicides was found effective in controlling this disease.

8.1 INTRODUCTION

Black point or seed discoloration of wheat is gaining importance to climate change mostly in the warm and humid climate of many countries (Kumar et al., 2002; Duczek and Flory, 1993; Chaurasia et al., 2000). This disease is caused by several fungi in combinations or alone. However, among these, the pathogenic fungi usually *Bipolaris sorokiniana* dominate. The spread of wheat cultivation in the warm humid climate after green revolution found the association of black point caused by the *Bipolaris sorokiniana* (Chaurasia et al., 2000). Black point is also influenced by the various environmental factors like low temperature, higher rainfall and humidity that make grain more vulnerable for the black point (Wang et al., 2003). This affects the quality of grain thereby market price and also poor seed quality. High incidence of black point reduced the milling quality and flour made out of such grains is poor quality (Lehmensiek et al., 2004; Solanki et al., 2006). Management of black point is lacking and black point resistant genotypes not yet identified. This review discuss the various aspects of black point aiming to emphasize need of regular wheat breeding program for black point management using resistant varieties.

8.2 DISTRIBUTION

Diseases, insects, draught, and frost are the main constrains to wheat production (Rajaram and van Ginkel, 1996; Miller and Pike, 2002). Higher requirements for the quality of the wheat seed is important to avoid spread of seed-borne diseases, weed seeds, and insect pests (Diekmann, 1993; Van Gastel et al., 2002). Among the pathogens, *B. sorokiniana* causal agent of spot blotch and black point of wheat emerged as a major problem in the warm humid tropic encompassing countries like India, Bangladesh, Nepal, Brazil, Argentina, and Peru (Kumar et al., 2002; Mathre, 1997; Duczek and Flory, 1993).

8.3 TAXONOMY OF THE PATHOGEN

The generic name *Bipolaris* for *Helminthosporium* species with fusoid, straight, or curved conidia germinating by one germ tube from each end (Bipolar germination) was proposed by Shoemaker (1959). *B. sorokiniana* is characterized by thick-walled, elliptical conidia (60–120 μm × 12–20 μm) with 5–9 cells. The fungus belongs to the sub-division—Ascomycotina, class—Loculoascomycetes, order—Pleosporales and family—Pleosporaceae. The *Cochliobolus sativus* (sexual stage) (Dastur, 1942) rarely occurs in nature (Rees, 1984) therefore *B. sorokiniana* (asexual stage) is the causal organism of spot blotch of wheat.

The members of Graminaceae family are generally infected by the *B. sorokiniana*, with rare chances of migration of spore from one crop to other (Bakonyi et al., 1997). This pathogen is seed and soil borne and can affect the seed germination and seedling emergence significantly in Bangladesh (Hossain and Hossain, 2001) and in China (Song et al., 2001). The seedling blight, root rot (Bhatti and Ilyas, 1986; Hafiz, 1986) and spot blotch (Iftikhar et al., 2006) of wheat due to *B. sorokiniana* is reported from Pakistan also.

The susceptibility of wheat to the pathogen increases at growth stage 56 (Zadoks et al., 1974) when approximately 75% of the head emerged. Under favorable weather conditions, for example, high temperature and relatively high humidity, disease become severe, especially in high yielding varieties. Under these conditions, spikelets of wheat may be infected causing grain shriveling and black point at embryo end (Nagarajan and Kumar, 1998; Singh et al., 1998).

8.4   SYMPTOMS

Black point is common in all wheat-growing regions of the world (Lorenz, 1986) and characterized by a dark discoloration of the embryo sides of the wheat and barley grains (Mak et al., 2006). Diseased kernels are discolored, and are black at the ends of the seed. Embryos are often shriveled and brown to black in color (Jain et al., 2012). When, the seed moisture content exceeds 20%, coupled with the relative humidity above 90%, the amount of black point increases dramatically. Seed with black point is more likely to have seedling blight and root rot problems. In most countries where cereals are grown commonly, black point can result in reduced grain quality and value (Wang et al., 2003). The disease can affect grain quality since food products made from infected kernels have displeasing odor and color. The black growth under microscope revealed the involvement of different fungi in different region. However, in warn humid climate, *B. sorokiniana* dominate with black point.

8.5   IMPORTANCE

Black point in wheat is an important seed-borne disease in all wheat-growing regions of the world including India (Hasabnis et al., 2006; Solanki et al., 2006). Khanum et al. (1987) reported that the black point in wheat is characterized by brown to black discoloration usually restricted to the germ end of the seeds. Black point symptom may also develop on the endosperm, ventral crease or even on the brush end of the grain. In case of severe infection, particularly when *B. sorokiniana* is involved, the grain may be completely discolored and shriveled. Impaired seed germination, reduced germination rate, reduced number of embryonic roots and coleoptile length, delayed seedling emergence, significant reduction in seedling vigor and grain/seed yield due to black point infection have been reported (Malaker and Mian, 2002; Ozer, 2005). The disease is also reported to affect luster and plumpness of grain and reduce its market value (Lehmensiek et al., 2004; Solanki et al., 2006). Mitchell (1985) reported that the severity of black point infection is probably more important to the milling and baking industry, than prevalence of the disease. Discoloration which is localized in the germ and does not extend beyond the bran layer is removed during milling. However, when discoloration extends into the crease and into the endosperm, it is not easily removed. The color of the flour milled from such wheat and the bread baked from the flour are darker, which is objectionable.

8.6 INFECTION BY *B. SOROKINIANA*, CAUSE OF BLACK POINT

Cytological study of wheat spike infection by *B. sorokiniana* was conducted by Qingmei et al., (2010). The wheat spikes were inoculated by injecting 10 μl of a conidial suspension of *B. sorokiniana* (5×10^5 conidia/ml) into the floral cavity between the lemma and palea of the first floret on spikelets using a pipette. Control plants were inoculated with distilled water. Three days after inoculation, typical symptoms of dark brown spots appeared on the lemma and glume of the inoculated spikelets. In spike infection, the hyphae extending between the epidermal cell wall layers or underneath the cuticular layer did not penetrate through the host cell wall. Hyphal growth was restricted to the intercellular space and no hyphae were found within the host cells. The fungus appeared to invade the surface of seeds in two ways:

1. Direct invasion of hyphae through the outer cell wall layers of the pericarp. Host cell death proceeded hyphal spreading.
2. Hyphae entered through the stigma in to the pericarp cells and extensively developed between the outer and inner cell layers of the pericarp of the seed embryo seven days after inoculation. Hyphae spread intracellularly only in the pericarp cells. *B. sorokiniana* was not able to colonize the aleuronic cells as well as the endosperm and embryo cells. This study demonstrates the differences in penetration and spread of the pathogen in glumes and seed tissues.

8.7 DISEASE ASSESSMENT

Based on infection, black point-affected seeds were categorized into six different grades on the basis of severity of black point infection. The grading was done according to 0–5 rating scale (Figure 8.1) as suggested by Gilchrist (1985) where

Grade 0 = Grains free from any discoloration (apparently healthy);
Grade 1 = Tip of the embryo brown to blackish;
Grade 2 = Discoloration covering the whole embryo;
Grade 3 = Embryo with 1/4 of the grain discolored;
Grade 4 = Embryo with 1/2 of the grain discolored;
Grade 5 = Embryo with more than 1/2 of the grain discolored and shriveled.

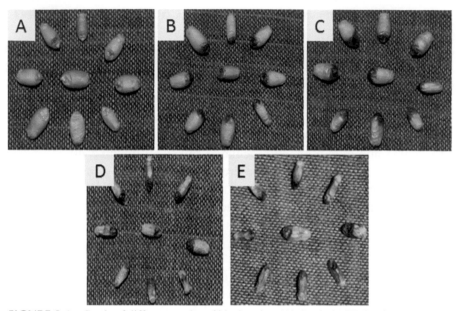

FIGURE 8.1   Seeds of different grades of black point. (A) Grade 0, (B) Grade 1, (C) Grade 2, (D) Grade 3, (E) Grade 4.

8.8   BLACK POINT INCIDENCE AND SEVERITY

Black point incidence and severity can be assessed by using black point severity index given by Harvey (1985). Harvey divides the black point-infected grains with different scores. Grains were given the following severity scores:

Score 0 = No black point symptoms;
Score 1 = Slight smudging, especially around the embryo and along the ventral crease;
Score 2 = Distinct darkening at embryo end and along the ventral crease;
Score 3 = Extensive blackening over at least one third of the grain.

Based on the above scores, the black point severity was calculated by using the formula

$$\text{Black point severity index} = \sum \frac{(\text{severity score} \times \text{number of grains in score}) \times 100}{\text{Total number of grains} \times \text{Maximum Score}}$$

8.9 LOSSES

8.9.1 SEED GERMINATION

Black point pathogen *B. sorokiniana* is seed and soil borne, and in nature, is seed transmitted (Neergaard 1977; Mehta 1993; Pandey et al., 2005). The pathogen was reported in seed samples of wheat by Shrestha et al. (1997). Seed infection by *B. sorokiniana* can adversely affect germination and root system development and can kill seedlings within a few days (Mehta, 1993).

Impaired seed germination and significant reduction in seedling vigor and grain yield to black point infection have been reported by Khanum et al. 1987; Rahman and Islam, 1998; Singh, 2007). Malaker and Mian (2002) found reduction in grain weight and seed germination related with the severity of black point infection. Increased seedling mortality and reduced seedling vigor and grain yield are also experienced when black pointed seeds are used for sowing. Hossain (2000) correlated the black point infection with leaf severity of wheat at flowering and milk-ripening stages under field conditions, when different black point-infected seeds were sown. *In vitro* results of Sultana and Rashid (2012) reported impairment in seedling caused by pathogen after sprouting of the seed. From this study, they found *B. sorokiniana* become more aggressive on coleoptiles, root and seed rot during transmission from seed to plant in wheat. In our own study also, the *in vitro* germination of seeds of wheat affected by black point varied in germination (Fig. 8.2). The more the infection, the less was germination. The infected seeds reduce the market value as well as the planting value also.

The experimental results of Al-Sadi and Leonard (2010) demonstrated the crown and root diseases on seven barley and three wheat varieties. Crown and root rot symptoms developed on barley and wheat cultivars following germination of infected seeds in sterilized growing medium. *B. sorokiniana* was only pathogen consistently isolated from crown and roots of emerging seedlings. This proved crown and root rot disease induced mostly due to the seed-borne inoculums of *B. sorokiniana*. Rashid et al. (2004) found minimum foliar infection at seedling, milking and grain filling stages of plant growth by sowing of healthy-looking seeds, while the black-pointed and shriveled seeds gave the minimum apparently healthy-looking grains and maximum black-pointed shriveled grains after harvest. Spike infection was significantly and negatively correlated with apparently healthy-looking grains but significantly and positively correlated with black-pointed and shriveled grains.

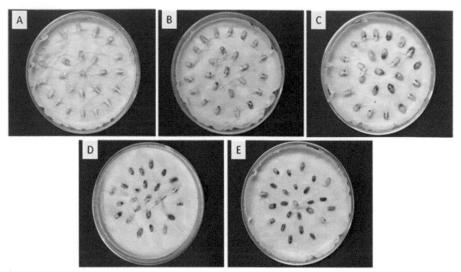

FIGURE 8.2 *In vitro* germination of different grades of seeds by blotter paper method: (A) Grade 0, (B) Grade 1, (C) Grade 2, (D) Grade 3, (E) Grade 4.

8.9.2 GRAIN QUALITY

Influence of black point on grain protein, fat, dry matter, and ash content have been observed. The disease grading is done according to 0–5 rating scale as suggested by Gilchrist (1985). The experimental data of Malaker and Mian (2002) revealed that the protein content was lowest in apparently healthy grains (Grade 0), which was statistically similar to that recorded under the black point-affected grains of Grade 1. The maximum protein content was found in grains under Grade 5, which was followed by Grade 4, Grade 3 and Grade 2. The parameter under Grade 1–5 was significantly different. The fat content was minimal in grains under Grade 0, which was statistically similar to those of Grade 1 and 2. The fat contents of black point-affected grains under Grade 4 and 5 were statistically similar, but significantly higher as compared to grains under the other four categories. The dry matter and ash contents of apparently healthy grains (Grade 0) and black point-affected grains under Grade 1 were statistically similar, but significantly higher when compared to those of the other grades of black point-affected grains. The lowest dry matter and ash contents were observed under Grade 5 which was followed by Grade 4, Grade 3, and Grade 2.

8.10   EFFECT OF TEMPERATURE, RAINFALL, AND HUMIDITY

Severity of black point disease is largely dependent on environmental conditions *viz.*, low temperature and frequent rainfall during kernel development/ grain filling (Wang et al., 2002). Black point-induced infection extended to embryo and ventral surface of the kernels also (Conner and Kuzyk, 1988).

Temperature and humidity play an important role for leaf blight severity of wheat. In Dinajpur district, temperature rapidly increases after third week of February to first week of March and it reaches about 25–30°C which is very favorable for *Bipolaris* infection. Therefore, maximum leaf blight infections of wheat occur in the period of late February to March. At that time, wheat passes its grain formation stage and *Bipolaris* has an opportunity to infect the grain resulting black point infected seeds.

From experimental findings, it has been observed that with the increase in the rainfall intensity during grain filling, the mean severity of black point increased. During 2014, the mean rainfall for the months January, February, and March was 59.5, 74.4, and 0.5 mm, respectively, and BPI was 1.6%. During 2015, it was 72.3, 1.0 and 24.7 mm, respectively, and BPI 2% (Fig. 8.3a,b). During March, the highest rainfall was observed in 2015 as

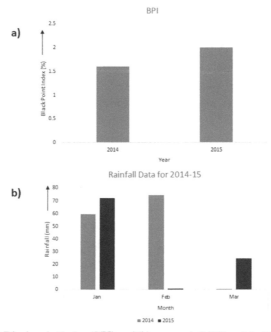

FIGURE 8.3   (a) Black point index (BPI) and (b) mean rainfall (Jan–Mar) for year 2014–2015.

compared to 2014. Present experimental findings revealed that increased rainfall intensity during grain filling increased the black point severity. The mean rainfall data of March, a peak period for grain filling, during 2014 and 2015 clearly demonstrated this fact.

8.11 MANAGEMENT

The disease may be managed by using seed dressing fungicides, foliar sprays, and use of resistant cultivars. Cecilia et al. (2004) observed that black point is a brownish or black discoloration of wheat kernels, and biological control is a complementary strategy to manage the disease.

8.11.1 REACTIONS OF WHEAT GENOTYPES AGAINST BLACK POINT

The variation in level of resistance to black point in wheat genotypes was found in field experiments conducted at Varanasi, IAS farm during 2014–15 and 2015–16. However, there was no association between spot blotch resistance and black point severity and both were independent. The spot blotch susceptible genotype Sonalika showed higher spot blotch severity score of 948 but the BPI was (3.4%) with higher germination of 93%. Similarly resistant genotypes TILHI and BCN/RIALTO showed lower spot blotch severity of 207 and 181, respectively, but the BPI was 8.35 and 1.97%, respectively. The seed germination was 82 and 90% in TILHI and BCN/RIALTO, respectively, indicating the independent nature of AUDPC and BPI (Table 8.1).

As the grade and number of black point count increases, the severity also increased linearly. There was a wide variation among the genotypes for grade and severity. The genotypes that fall under Grade 1 have lowest severity index of 0.43, 0.98 and 1.42 whereas, the genotypes with Grade 3 showed maximum severities of 6.03, 7.63 and 7.07, respectively (Table 8.1).

There is a positive and highly significant correlation among grade and black point index. Whereas, with germination percentage, both grade and black point index exhibits significant negative correlation. AUDPC showed negative correlation with grade and black point index and positive with germination percentage but not significantly. Germination rate, seedling emergence, and seedling establishment under the field conditions were reduced by black-pointed seeds. This result is in accordance with Hudec

TABLE 8.1 Pedigree of Genotypes and its Variation for Black Point Grade, Black Point Index, Germination Percentage and Area Under Disease Progress Curve.

S. No.	Pedigree of genotype	Grade	Black point index	Germination %	AUDPC
1	YAV_3/SCO//JO69/CRA/3/YAV79/4/AE.SQUARROSA (498)/5/2*OPATA	2.00	4.83	90.0	182.10
2	TILHI/PALMERIN F2004	2.00	2.10	90.0	233.33
3	ATTILA/3*BCN//BAV92/3/TILHI	2.00	2.10	90.0	240.43
4	NL 750	1.50	2.30	100.0	202.16
5	ASTREB/OAX93.10.1//SOKOLL	2.00	4.37	90.0	293.21
6	ASTREB/OAX93.10.1//SOKOLL	2.00	4.04	85.0	247.84
7	ALTAR 84/AEGILOPS SQUARROSA (TAUS)//OPATA	2.50	6.50	80.0	199.69
8	PBW343*2/KUKUNA//PBW343*2/TUKURU/3/PBW343	2.75	7.07	80.0	213.28
9	TILHI/SOKOLL	2.00	3.17	90.0	186.11
10	CMH79A.955/4/AGA/3/4*SN64/CNO67//INIA66/5/NAC/6/RIALTO	2.50	6.03	85.0	202.16
11	CHIRYA.3	1.50	1.37	95.0	193.21
12	SW89-5124*2/FASAN	2.00	5.49	90.0	207.72
13	NL748/NL837	1.75	0.85	100.0	257.41
14	BCN/RIALTO	2.00	1.97	90.0	181.79
15	ALTAR 84/AE. SQUARROSA (219)//OPATA/3/WBLL1/ FRET2//PASTOR	1.00	0.43	100.0	233.34
16	SURUTU-CIAT	2.25	4.95	87.5	229.32
17	W462//VEE/KOEL/3/PEG//MRL/BUC	1.50	1.12	95.0	228.40
18	CROC_1/AE. SQUARROSA (205)//KAUZ/3/SASIA/4/TROST	2.00	2.65	87.5	272.53
19	UP2338*2/4/SNI/TRAP#1/3/KAUZ*2/TRAP/KAUZ/5/MILAN/KAUZ//CHIL/ CHUM18/6/UP2338*2/4/SNI/TRAP#1/3/KAUZ*2/TRAP//KAUZ	1.50	1.65	97.5	181.18
20	GAN/AE.SQUARROSA (897)//OPATA/3/BERKUT	2.00	2.30	90.0	244.44
21	CNDO/R143//ENTE/MEXI_2/3/AEGILOPS SQUARROSA (TAUS)/4/ WEAVER/5/PASTOR	1.75	2.02	95.0	161.42

TABLE 8.1 *(Continued)*

S. No.	Pedigree of genotype	Grade	Black point index	Germination %	AUDPC
22	CNDO/R143//ENTE/MEXI_2/3/AEGILOPS SQUARROSA (TAUS)/4/ WEAVER/5/2*KAUZ	3.00	7.63	80.0	203.09
23	BECARD	1.50	1.85	97.5	197.22
24	PBW 343/PASTOR	1.50	3.90	95.0	201.24
25	ATTILA/3*BCN//BAV92/3/TILHI/4/SHA7/VEE#5//ARIV92	1.50	2.55	95.0	254.01
26	CIANO T 79	1.50	3.65	90.0	888.58
27	MILAN/KAUZ/3/URES/JUN//KAUZ/4/CROC_1/AE.SQUARROSA (224)// OPATA	1.75	6.08	90.0	225.31
28	JUPARE C 2001	1.50	2.05	95.0	182.10
29	PFAU/MILAN//TROST/3/PBW65/2*SERI.1B	1.25	1.42	97.5	226.24
30	CROC_1/AE. SQUARROSA (205)//KAUZ/3/ENEIDA	2.00	3.20	87.5	202.16
31	TILHI	2.25	8.35	82.5	207.72
32	VORB/4/D67.2/PARANA 66.270//AE. SQUARROSA (320)/3/ CUNNINGHAM	2.50	3.60	85.0	146.92
33	Sonalika	1.75	3.43	92.5	948.15
34	Yangmai 6- Sonalika RIL	1.50	3.63	92.5	720.37
35	Yangmai 6- Sonalika RIL	1.50	3.30	95.0	262.97
36	Yangmai 6- Sonalika RIL	2.50	4.99	85.0	309.57
37	Yangmai 6- Sonalika RIL	1.75	0.98	92.5	591.36
38	Yangmai 6	1.75	1.67	92.5	723.15

LSD (0.05%) Grade = 0.45; BPI = 1.85; Germination% = 4.3; AUDPC = 82.21

(2007) who reported that fungal species were able to kill or reduce embryo vigor. According to the results of our study, sowing rate should be increased for black-pointed seed lots to obtain optimum seedling rate in field conditions.

8.11.2 SEED TREATMENT

In an experiment conducted by Malaker & Milan (2009), seed treatment with either Vitavax-200 or Homai-80WP had no effect on black point incidence. However, both the fungicides were equally effective in increasing plant population, number of spikes and grain yield. The seed treatment with Vitavax-200 @ 2, 2.5 and 3 g/kg was effective against seed-borne pathogen, *B. sorokiniana*. Fungicide treatment also increased seed germination, seedling vigor and grain yield and reduced seedling mortality (Singh et al., 2007; Singh and Pankaj, 2008).

8.11.3 FOLIAR APPLICATION

Fungicides registered for foliar disease control can be applied only up to 30–45 days before harvest (Saskatchewan Ministry of Agriculture, 2010), and thus applications during susceptible kernel development stages would infringe on the pre-harvest application interval. When applied at the end of flowering, Conner and Kuzyk (1988) showed that fungicides were not consistently effective in reducing black point incidence. Foliar application of fungicide Tilt 250 EC reduced the severity of black point significantly as reported by Rashid et al. (2001).

Ansari (2015) conducted an experiment using treated seeds with Azoxystrobin and control. The black point-infected and black point-free seeds (100) from each genotype was soaked in 3.5 µl Azoxystrobin/100 seeds overnight (25 g a.i./100 kg seeds) and observed that the seed which was highly infected (Grades 3 and 4) had no effect of treatment compared to seeds of Grade 1 and 2. This was due to the fact that in case of severe infection, the whole grain may have become discolored and shriveled and embryo may have been killed. The severely infected seeds were not able to germinate even after treatment with fungicide, but those seeds which are partially infected (Grades 1 and 2) had good chance of seedling establishment because their embryo was intact.

8.12   FUTURE PROSPECTS

Black point is likely to increase in future due to changes in temperature and humidity because of climate change in the warm humid environment of South Asia. Therefore, the screening of wheat genotypes for black point with other important diseases is gaining importance. More systemic investigations about the mechanism of resistance in wheat genotypes are necessary for breeding black point resistance in new wheat cultivars. Despite the limited impact of agronomic practices, integrated methods for minimizing the development of kernel discoloration should be a future research objective for this disease.

KEYWORDS

- *Bipolaris sorokiniana*
- black point
- wheat
- management
- losses in seed quality

REFERENCES

Adlakha, K. L.; Joshi, L. M. Black Point of Wheat. *Indian Phytopath.* **1974,** *27,* 41–44.

Al-Sadi, A. M.; Leonard, M. Influence of Seed borne *Cochliobolus sativus* (Anamorph *Bipolaris sorokiniana*) on Crown Rot and Root Rot of Barley and Wheat. *J. Phytopathol.* **2010,** *158,* 683–690.

Ansari, M. S. Q. Effect of Azoxystrobin on the Biology of *Bipolaris sorokiniana* and Black Point of Wheat. Master Thesis. Banaras Hindu University: Varanasi, India, 2015.

Bakonyi, J.; Aponyi, I.; Fischl, G. Diseases Caused by *Bipolaris sorokiniana* and *Drechslera tritici* Repentis in Hungary. In *Helminthosporium Blights of Wheat: Spot Blotch and Tan Spot;* Duveiller, E., Dubin, H. J., Reeves, J., McNab, A., Eds.; CIMMYT: Mexico, D. F., 1997; pp 80–88.

Bhatti, M. A. R.; Ilyas, M. B. Wheat Diseases in Pakistan. In *Problems and Progress of Wheat Pathology in South Asia;* Joshi, L. M., Singh, D. V., Srivastava, K. D., Eds.; Malhotra Publishing House: New Delhi, India, 1986; pp 20–30.

Cecilia, M.; Marina, S.; Analía, P. Preliminary Studies on Biological Control of the Blackpoint Complex of Wheat in Argentina. *World J. Microbiol. Biotechnol.* **2004,** *20,* 285–290.

Chaurasia, S.; Chand, R.; Joshi, A. K. Relative Dominance of *Alternaria triticina* Pras. Et. Prab. and *Bipolaris sorokiniana* (Sacc.) Shoemaker in Different Growth Stage of Wheat (*T. aestivum* L.), *J. Plant Dis. Protect.* **2000**, *107,* 176–181.

Conner, R. L.; Kuzyk, A. D. Effectiveness of Fungicides in Controlling Stripe Rust, Leaf Rust, and Black Point in Soft White Spring Wheat. *Can. J. Plant Pathol.* **1988**, *10,* 321–326.

Dastur, J. F. Notes on Some Fungi Isolated from 'Black Point' Affected Kernels in the Central Provinces. *Indian J. Agric. Sci.* **1942**, *12,* 731–742.

Diekmann, M. *Seed-borne Diseases in Seed Production.* ICARDA: Aleppo, Syria, 1993.

Duczek, L. J.; Jones-Flory, L. L. Relationships between Common Root Rot, Tillering and Yield Loss in Spring Wheat and Barley. *Can. J. Plant Pathol.* **1993**, *15,* 153–158.

Gilchrist, L. I. In *CIMMYT Methods for Screening Wheat for Helminthosporium sativum Resistance,* Proceedings of the International Symposium: Wheat for More Tropical Environments, Mexico, Sept 24–28, 1984; CIMMYT: Mexico, D. F., 1985; pp 149–151.

Hafiz, A. *Plant Diseases.* Pakistan Agriculture Research Council: Islamabad, 1986; pp 552.

Hanson, E. W.; Christensen, J. J. *The Black Point Disease of Wheat in the United States;* The University of Minnesota Agricultural Experiment Station Technical Bulletin 206; University of Minnesota, Agricultural Experiment Station: St. Paul, MN, 1953; p 30.

Harvey, I. C. *Blackpoint of Wheat.* New Zealand Ministry of Agriculture and Fisheries Pamphlet: Lincoln, New Zealand, 1985.

Hasabnis, S. N.; Iihe, B. M.; Shinde, V. K. Incidence of Black Point Disease in Wheat Varieties. *J. Maharashtra Agric. Univ.* **2006**, *31,* 114–115.

Hossain, I.; Hossain, M. M. Effect of Black Pointed Grains in Wheat Seed Samples on Germination, Seedling Vigor and Plant Stand. *Pakistan J. Phytopathol.* **2001**, *13,* 1–7.

Hossain, M. M. Effect of Different Level of Black Pointed Seed on Germination, Seedling Vigor, Plant Stand and Seed Quality of Wheat. M. S. Thesis, Department of Plant Pathology, Bangladesh Agricultural University, Mymensingh, 2000; p 82.

Hudec, K. Influence of Harvest Date and Geographical Location on Kernel Symptoms, Fungal Infestation and Embryo Viability of Malting Barley. *Int. J. Food Microbiol.* **2007**, *113,* 125–132.

Iftikhar, S., et al. Prevalence and Distribution of Foliar Blight Pathogens of Wheat in Different Agro Ecological Zones of Pakistan with Especial Reference to *Bipolaris sorokiniana.* *Pakistan J. Bot.* **2006**, *38,* 205–210.

Jain, S.; Jindal, M.; Mohan, C. Status of Black Point Disease of Wheat in Punjab. *Plant Dis. Res.* **2012**, *27,* 28–33.

Khanum, M.; Nigar, Y.; Khanzada, A. K. Effect of Black Point Disease on the Germination of Wheat Varieties. *Pakistan J. Agric. Res.* **1987**, *84,* 467–473.

Kumar, J., et al. *Bipolaris sorokiniana,* a Cereal Pathogen of Global Concern: Cytological and Molecular Approaches towards Better Control. *Mol. Plant Pathol.* **2002**, *3,* 185–195.

Lehmensiek, A., et al. QTL's for Black Point Resistance in Wheat and the Identification of Potential Markers for Use in Breeding Programs. *Plant Breed.* **2004**, *123,* 410–416.

Lorenz, K. Effects of Black Point on Grain Composition and Baking Quality of New Zealand Wheat. *New Zeal. J. Agr. Res.* **1986**, *29,* 711–718.

Mak, Y., et al. Black Point is Associated with Reduced Levels of Stress, Disease- and Defence-Related Proteins in Wheat Grain. *Mol. Plant Pathol.* **2006**, *7,* 177–189.

Malaker, P. K.; Mian, I. H. Effect of Seed Treatment and Foliar spray with Fungicides in Controlling Black Point Disease of Wheat. *Bangladesh J. Agric. Res.* **2009**, *34:* 425-434.

Mathre, D. E. Compendium of Barley Diseases. *Am. Phytopathol. Soc.* **1997**, *35,* 17–21.

Mehta, Y. R. Spot Blotch (*Bipolaris sorokiniana*). In *Seed Borne Disease and Seed Health Testing of Wheat;* Mathur, S. B., Cunfer, B. M., Eds.; Institute of Seed Pathology for Developing Countries: Copenhagen, 1993; pp 105–112.

Miller, R. H.; Pike, K. S. Insects in Wheat-Based Systems. In *Bread Wheat: Improvement and Production;* Curtis, B. C., Rajaram, S., Gomez Macpherson, H., Eds.; Food and Agricultural Organization of the United Nations: Rome, Italy, 2002. ISBN9251048096.

Mitchell, T. A. Black Point. *Cereal News.* **1985,** *10,* 7–8.

Nagarajan, S.; Kumar, J. In *Foliar Blights of Wheat in India: Germplasm Improvement and Future Challenges for Sustainable, High Yielding Wheat Production,* Proceedings of the International Workshop on *Helminthosporium* Blights of Wheat: Spot Blotch and Tan Spot, Mexico, Feb 9–14, 1997; CIMMYT: El Batan, Texcoco, 1998; pp 52–58.

Neergaard, P. *Seed Pathology.* McMillan Press: London, 1977; Vol. 1.

Ozer, N. Determination of the Fungi Responsible for Black Point in Bread Wheat and Effects of the Disease on Emergence and Seedling Vigor. *Trakya Univ. J. Sci.* **2005,** *6,* 35–40.

Pandey, S. P., et al. Sources of Inoculum and Re-Appearance of Spot Blotch of Wheat in Rice-Wheat Cropping Systems in Eastern India. *Eur. J. Plant Pathol.* **2005,** *111,* 47–55.

Qingmei, H., et al. Cytological Study of Wheat Spike Infection by *Bipolaris sorokiniana. J. Phytopathol.* **2010,** *158,* 22–29.

Rahman, G. M. M.; Islam, M. R. Effect of Black Point of Wheat on Some Qualitative Characters of Its Grain and Seed Vigour. *Bangladesh J. Agric. Res.* **1998,** *23,* 283–287.

Rajaram, S.; van Ginkel, M. *A Guide to the CIMMYT Bread Wheat Section;* Wheat Program Special Report No. 5; CIMMYT: Mexico, D. F., 1996.

Rashid, A. Q.; Dhar, R. C.; Khalequzzaman, K. M. Association of *Bipolaris sorokiniana* in Wheat Seed and Its Effect on Subsequent Plant Infection at Different Growth Stages. *J. Agric. Rural Dev.* **2004,** *2,* 67–72.

Rashid, A.Q.; Sarker, K.; Khalequzzaman, K. M. Control of Bipolaris Leaf Blight of Wheat with Foliar Spray of Tilt 250 EC. *Bangladesh J. Plant Pathol.* **2001,** *17,* 45–47.

Rees, R. D.; Martin, D. I.; Law, D. P. Blackpoint in Bread Wheat: Effects on Quality and Germination, and Fungal Associations. *Aust. J. Exp. Agric. Anim. Husb.* **1984,** *24,* 601–605.

Saskatchewan Ministry of Agriculture. 2010 Guide to Crop Protection. Saskatchewan Ministry of Agriculture: SK, 2010.

Shoemaker, R. A. Nomenclature of *Drechlera* and *Bipolaris,* Grass Parasites Segregated from *Helminthosporium. Can. J. Bot.* **1959,** *37,* 879–887.

Shrestha, K. K.; Timila, R. D.; Mahto, B. N.; Bimb, H. P. In *Disease Incidence and Yield Loss due Foliar Blight of Wheat in Nepal,* Proceedings of the International Workshop on *Helminthosporium* Blights of Wheat: Spot Blotch and Tan Spot, CIMMYT, El Batan, Mexico, Feb 9–14, 1997; Duveiller, E., Dubin, H. J., Reeves, J., McNab, A., Eds.; CIMMYT: Mexico, D. F.,1997; pp 67–72.

Singh, D. P.; Choudhury, A. K.; Kumar, P. Management of Losses Due to Seedborne Infection of *Bipolaris sorokiniana* and *Alternaria triticina* in Wheat Using Seed Treatment with Vitavax 200 WS. *Indian J. Agric. Sci.* **2007,** *77,* 101–103.

Singh, D. P. Measurement of Ear Head Blight Due to *Bipolaris sorokiniana* in Wheat and Variations in Yield Components between Apparently Healthy Looking and Diseased Grains. *Indian Phytopath.* **2007,** *60,* 527–529.

Singh, D. P.; Pankaj, K. Role of Spot Blotch (*Bipolaris sorokiniana*) in Deteriorating Seed Quality in Different Wheat Genotypes and Its Management Using Fungicidal Seed Treatment. *Indian Phytopath.* **2008,** *61,* 49–54.

Singh, R. P.; Singh, A. K.; Singh, S. P. In *Distribution of Pathogens Causing Foliar Blight of Wheat in India and Neighboring Countries,* Proceedings of the International Workshop on *Helminthosporium* Blights of Wheat: Spot Blotch and Tan Spot, CIMMYT, El Batan, Mexico, Feb 9–14, 1997; Duveiller, E., Dubin, H. J., Reeves, J., McNab, A., Eds.; CIMMYT: Mexico, D. F., 1998; pp 59–62.

Solanki, V. A.; Augustime, N.; Patel, A. A. Impact to Black Point on Wheat Trade and Its Management. *Indian Phytopath.* **2006,** *59,* 44–47.

Song, Y. L.; He, W. L.; Yang, G. Q.; Liu, H. Y. Occurrence of Black Point of Wheat Seed and Its Control. *Acta Agric. Bor. Sin.* **2001,** *16,* 76–79.

Sultana, A.; Rashid, A. Q. M. B. Impact of Seed Transmission of *Bipolaris sorokiniana* On the Planting Value of Wheat Seeds. *J. Environ. Sci. Nat. Res.* **2012,** *5,* 75–78.

Van Gastel, A.; Bishaw, Z.; Gregg, B. *Wheat seed production. Bread Wheat: Improvement and Production.* Curtis, B. S., Rajaram, S., Gomez Macpherson, H., Eds.; Food and Agricultural Organization of the United Nations: Rome, Italy, 2002; pp 463–481.

Wang, H., et al. Effects of Foliar Fungicides on Kernel Black Point of Wheat in Southern Saskatchewan. *Can. J. Plant Pathol.* **2002,** *24,* 287–293.

Wang, H., et al. Kernel Discoloration and Downgrading in Spring Wheat Varieties in Western Canada. *Can. J. Plant Pathol.* **2003,** *25,* 350–361.

Zadoks, J. C.; Chang, T. T.; Konzak, C. F. A Decimal Code for the Growth Stages of Cereals. *Weed Res.* **1974,** *14,* 415–421.

CHAPTER 9

IMPORTANT NEMATODE PESTS OF WHEAT AND BARLEY AND THEIR MANAGEMENT

SHYAM SARAN VAISH*

Department of Mycology and Plant Pathology, Institute of Agricultural Sciences, Banaras Hindu University, Varanasi 221005, Uttar Pradesh, India

Corresponding author. E-mail: shyam_saran@rediffmail.com

CONTENTS

ABSTRACT

Wheat and barley are grown during Rabi season in the most parts of India. Wheat plays a significant role in Global Food Security. However, barley also getting status of a healthy food crop for human consumption due to rising health consciousness. The plant diseases are major biotic constraints that affect its production to a great extent. In addition to many fungal, bacterial and viral diseases, several plant parasitic nematodes, viz., cereal cyst nematodes (*Heterodera* spp.), root-knot nematodes (*Meloidogyne* spp.), wheat seed gall nematode (*Anguina tritici*), root lesion nematodes (*Pratylenchus* spp.), stem and bulb nematode (*Ditylenchus* spp.) and stunt nematode (*Tylenchorhynchus* spp.) are also considered to be one of the important constraints in their production. Hence, the efforts are made to describe symptoms caused by plant parasitic nematodes, their identification characteristics and biology. In addition to this, attempts are also made to give important effective methods for their management.

9.1   INTRODUCTION

Nematodes are tiny, complex, and un-segmented/undivided roundworms/ eelworms that are anatomically differentiated for feeding, digestion, loco-motion, and reproduction. These small animals occur worldwide in all environments. They are the most numerous multi-cellular animal life forms on earth. Many nematodes feed on soil microorganisms and contribute beneficially to the cycling of nutrients. Thus, nematodes make important contributions to organic matter decomposition and the food chain. By this means, most species are beneficial to agriculture. However, some species feed on the roots of plants and referred as plant-parasitic nematodes that have a retractable, hollow, spear-like structure ("stylet") which they use to break the walls of root cells and extract the contents for their nutrition. Endoparasitic nematode species enter into the root tissues by using their stylet and head to physically break root cell walls and by excreting enzymes that dissolve cell walls. Further, they are migratory or sedentary. Ectoparasitic nematodes remain outside the roots and use their stylet to feed on root hairs and surface cells of the root.

Keeping in view global food security, plant-parasitic nematodes that cause reduction in agricultural production considered to be of great concern. Although plant-parasitic nematodes are among the most widespread pests, and are frequently one of the most insidious and costly (Webster, 1987), data on their economic impact remain less than comprehensive, especially for crops produced in resource poor areas. In the tropical and sub-tropical climates, crop production losses attributable to nematodes were estimated at 14.6% compared with 8.8% in developed countries. Perhaps more importantly, only ~0.2% of the crop value lost to nematodes is used to fund nematological research to address these losses (Sasser & Freckman, 1987). One difficulty with assessing nematode impact is that damage resulting from nematode infection is often less obvious than that caused by many other pests or diseases. More recently, Handoo (1998) estimated global crop losses due to nematode attack in the region of $80 billion.

Population of plant-parasitic nematodes and losses caused by them governed by nematode species, feeding habit and habitat of nematode, susceptibility of host, availability of host from season to season, cropping pattern, cultural practices, soil properties, environmental conditions, use of inorganic fertilizers and pesticides and status soil organic matter, and so forth. A common feature of plant-parasitic nematodes on annual crops is that they have an uneven distribution within a field and the symptoms of damage, normally associated with high population densities, occur in patches (McSorley, 1998). Where a susceptible annual crop is replanted year after year, the nematodes spread and the patches increase in size and eventually coalesce (Brown, 1987; Swarup & Sosa-Moss, 1990). Mono-cropping/cropping system of rice–wheat enhances the horizontal spread of the common nematodes and may lead to their more uniform distribution. One factor likely to interfere with the uniform spread of the nematodes is soil texture as this can influence their abundance and distribution (Cadet et al., 1994; Norton, 1989; Upadhyay et al., 1972). For example, species of *Meloidogyne* are frequently more numerous and more pathogenic in light textured soils, giving rise to greater symptom expression in the sandier areas within the field (Prot & van Gundy, 1981).

In addition to the immediate concerns surrounding global food security issues, there is growing concern for pest and disease management under the predicted climate changes and the threat of the emergence of new pests, including nematodes. The Intergovernmental Panel on Climate Change (IPCC) assessments (IPCC, 2007) have concluded that, even if concentrations of all greenhouse gases had been kept constant at the levels present in 2000, a further overall warming of ~0.1°C per decade would be expected, due to the slow response of the oceans. About twice as much warming would be expected if emissions are within the range of scenarios used in IPCC assessments. Resulting changes would include an increase in frequency of heat extremes, heat waves, and heavy precipitation; changes in wind, precipitation, and temperature patterns; precipitation increases at high latitudes and decreases in most sub-tropical land regions. This would impact on species range shifts; water scarcity and drought risk in some regions of the dry tropics and sub-topics; and coastal damage from floods combined with sea-level rise. In an alternative example the rice root-knot nematode, *Meloidogyne graminicola*, can be maintained under damaging levels through good water management. However, with reduced availability of water following climatic changes and/or competition for urban use, reduced quality of water management, or the introduction of water saving mechanisms such as direct wet seeding is favoring the development of high populations of *M. graminicola*, drastically raising its economic significance as a damaging pest (De Waele & Elsen, 2007). Consequently, we can be sure that nematodes will continue to emerge as new or more aggressive pests of crops as farming practices adapt to fashion, as climate change occurs and as cropping systems intensify in response to an increasing global demand for food. In a world of limited means for nematode management, focus on plant-parasitic nematodes as a significant affliction of crop production is highly pertinent.

Wheat is grown as Rabi crop after rice in most of wheat growing area in India. However, wheat is also grown after cotton, maize, sorghum, pearl millet, groundnut, sesame, and other crops grown during Kharif season (crops are sown during June–July and harvested in September and October). At some places barley is grown after Kharif crops especially in Rajasthan, Eastern Utter Pradesh, and Madhya Pradesh. Among the plant-parasitic nematodes, cyst nematode (*Heterodera* spp.), lesion nematode (*Pratylenchus* spp.), root-knot nematode (*M. graminicola*), wheat seed gall nematode (*Anguina tritici*), stunt nematode (*Tylenchorhynchus* spp.), spiral nematode (*Helicotylenchus* spp.), bulb nematode (*Ditylenchus dipsaci*), and dagger nematode (*Xiphinema* spp.) are found associated with nematodes. However, cyst nematode (*Heterodera* spp.), lesion nematode (*Pratylenchus* spp.) root-knot nematode (*M. graminicola*), wheat seed gall nematode (*A. tritici*) and stunt nematode (*Tylenchorhynchus* spp.) are known to cause considerable damage to wheat crop. As far as barley is concerned with respect to plant-parasitic nematodes, cyst nematode (*Heterodera* spp.), lesion nematode (*Pratylenchus* spp.), and root-knot nematode (*M. graminicola*) are known to cause huge losses under favorable environmental conditions, soil properties, cropping system, and presence of susceptible varieties. This chapter focuses on important nematode species associated with wheat and barley, their key symptoms, biology, survival, and different management measures. Most of the developing countries including India lie in tropical or sub-tropical regions where climate is suitable for activity and multiplication of nematodes almost throughout the year. Sandy and warm soils in India and other developing countries in the arid zone are very favorable for nematode infection, especially in irrigated areas which are used for crop production continuously (Taylor, 1967).

In general, plants infected with plant-parasitic nematode do not produce dramatic symptoms as produced by fungal, bacterial, and viral pathogens. The most pronounced symptoms produced by plant-parasitic nematodes are stunting, yellowing, and wilting which might be confused with symptoms of nutritional deficiencies, presence of hardpan, problem of salinity and alkalinity, termite infection, and other factors. Therefore, to look into symptoms caused by plant-parasitic nematode, one needs special expertise and adequate long-term experience associated with the symptoms development in agricultural fields. Sometimes, when these symptoms of stunting, yellowing, and wilting in cereal crops caused by plant-parasitic nematodes are baffled with deficiency of nitrogen and one applies urea to recover these symptoms. Definitely it works in recovery from these symptoms, which might be due to management of pathogen by application of the nutrients to the relative disadvantage of the pathogen. But, the fact is that when one goes for the application of the nutrient repeatedly, it provides source of energy from plants to plant-parasitic nematodes, which ultimately leads to build up of nematode population with time and then time comes when even seed is not able germinate in these soils. Further, excessive doses of urea may lead to disruption of natural biological equilibrium. Therefore, one should be very clear about the symptoms produced by plant-parasitic nematodes.

9.2   CYST NEMATODES (*HETERODERA* SPP.)

Nematodes belonging to the genus *Heterodera* are called cereal cyst nematodes (CCN) because at the time of death the mature females get converted into a brown cyst containing eggs. Cysts may be light brown to dark reddish-brown and brown cysts are typically ovate to spheroid. The cyst nematode, *Heterodera avanae* was reported to be an important nematode problem of wheat and barley in Rajasthan, Madhya Pradesh, Eastern Uttar Pradesh, Bihar, Ladakh region of Jammu and Kashmir (Vaish et al., 2011), and Himachal Pradesh locally known to cause "molya" disease.

The Mediterranean cereal cyst nematode (MCCN), *Heterodera latipons*, is considered the most common species of cyst nematodes that limit the production of wheat and barley in several Mediterranean countries such as Cyprus, Jordan, Palestine, Italy, Libya, Spain, and Turkey (Franklin, 1969; Cohn & Ausher, 1973; Mor et al., 1992; Yousef & Jacob, 1994; Philis, 1995,

Rumpenhorst et al., 1996; Al-Abed et al., 2004). The use of germplasm that possesses resistant genes against CCN is considered the most effective control measures.

The chief symptoms of cyst nematode infected wheat and barley include stunting, yellowing, withering, and wilting. Further, such roots of such plants show presence of small, round white structure, which are female. Later, these females become brownish in color and dislodged from the roots and mixed in soil. These structures are referred as cysts containing several hundreds of eggs. The intensity of body color depends upon the maturity stage of the young female or cyst. In case of cyst nematode matrix with eggs may also be found attached to the posterior region of the female body.

The mature female bodies are found attached to roots by their head end embedded almost in the stele. The site of feeding is modified into a syncytium similar to that found in case of *Meloidogyne*. On the surface of infected roots, white to brown bodies of females can be discerned with naked eyes. The intensity of body color depends upon the maturity stage of the young female or cyst. In case of cyst nematode matrix with eggs may also be found attached to the posterior region of the female body.

The male is worm like, about 1.3 mm long by 30–40 μm in diameter. Fully developed females are lemon shaped, 0.6–0.8 mm in length and 0.3–0.5 mm in diameter. Approximately 21–30 days is required for the completion of life cycle of the cyst nematode. White females can be seen in the root system by 4–6 week old plants. In the absence of host crop the nematode multiplies on a number of graminaceous weeds.

9.3   ROOT-KNOT NEMATODES (*MELOIDOGYNE* SPP.)

M. graminicola (Golden & Birchfield, 1968) that causes root-knot disease is one of the most damaging pests of rice, wheat, and barley (Sharma & Rahaman, 1998; Singh et al., 2006; Singh & Singh, 2009; Pandey, 2011; Vaish & Pandey, 2012). Recently, this nematode has emerged as a notorious pest of wheat and barley grown after rice owing to late onset of winter and early onset of summer in rice growing regions. The climate change and environmental threats pose serious risk to the agriculture due to change in temperature, precipitation, natural biological equilibrium and distribution, and prevalence of agricultural pest and plant pathogens. The impact of climate change on disease for a given plant species depends on the nature of the effects climate change has on both the host and its pathogens (Boland et al., 2004). The root-knot nematode is polyphagous known to infect plants of almost all botanical families. It is also polycyclic pathogen produces its several cycles within the cropping season. The number of cycles depends on environmental conditions, genotype, and species of the nematode. Further infection by *M. graminicola* predisposes rice, wheat, and barley to foliar diseases particularly blast and leaf spot of rice and foliar blight or spot blotch of wheat and barley.

After rice, wheat *(Triticum aestivum* L.) is an important staple food for nearly 40% of the world population covering at least 43 countries and provides 20% of food calories to mankind. Globally, India is the second largest producer of wheat and rice next to China. Barley (*Hordeum vulgare* L.) is one of the fourth most important cereal crops around the world after rice, wheat, and maize. Barley is the oldest crop of the world has been cultivated since Neolithic period. It has been known as a major sustaining food played vital role in the evaluation of humankind and development of many civilizations.

Barley is sown as a Rabi crop in India. It is mainly grown in Rajasthan, Uttar Pradesh, Madhya Pradesh, Bihar, Himachal Pradesh, Haryana, and Punjab. Some cultivation is also undertaken in Bihar, Himachal Pradesh, and Uttaranchal. In 2011, the production was estimated at 1.41 million tons. With an increased area in cultivation, it is believed that barley cultivation would accelerate in the near future. Although the feed portion would remain stable, the food, seed, and industrial use would go up at a substantial rate.

9.3.1 SYMPTOMS

The main symptoms of root-knot disease caused by *Meloidogyne* spp. is presence of galls/knots on the root system of the infected plants. However, such plants also show the symptoms of yellowing, stunting, and wilting. The degree of symptom manifestation differs with time of infection, age of the plants, inoculum levels, and so forth. Root-knot disease is characterized by presence of galls/knots on root system of the infected plants. These plants exhibited yellowing, dwarfing, and wilting. However, the galls varied in their shapes and noticed from small to big in size. Barley roots show galls that are horseshoe or a complete spiral/curling, pyriform, round, oblong, and capitate/hooked in shape and varied in size and color. The most pronounced type of galls in early infection were curling type, which were seen in infected plants after 15 days of sowing when plants attained the growth stage 12. Further, these plants showed presence of 40–65 galls per plant along with acute stunting, yellowing, and wilting. However, wilting occurred in such severely infected plants after 2–3 weeks after infection. Early infection at the time of germination resulted in drying and death of these plants within 21–28 day after sowing. In these plants galls were round and some of the galls were brown to dark brown in color. Change in color of the galls was found associated with development of nematode and increase in its number within these galls. In some plants, roots were heavily laden with white to light brown round to pyriform galls 27 days after sowing. The reduction in height of plants in infected fields was recorded up to 66%. In fields, root-knot infected plants were observed in patches showing stunting, yellowing, and wilting with reduced tillering. Such plants recorded root galls in range of 30–97 galls/plant. In these patches, plants were weak with no ear or with a very short ear. Ear length was reduced greatly. Wheat roots are also show galls that vary from horseshoe or a complete spiral/curling, pyriform, round, oblong, and capitate/hooked in shape.

9.3.2 DIFFERENT GROWTH STAGES OF MELOIDOGYNE GRAMINICOLA

The root-knot nematode of wheat and barley was collected from infected roots and slides for different stages *viz.* eggs, second stage juveniles, males, and females were prepared for detailed observations of the morphological character and their measurements. For perennial pattern, posterior region of the females were cut with sharp blade in few drops of cotton blue and trimmed for final observations. Morphological details were similar to *M. graminicola* originally described by Golden and Birchfield (1968). Details of different stages are given below.

9.3.2.1 FEMALE (N = 25)

Female body is white, globose to pear shaped with varying neck length (50–215 μm). Female size is variable, 490–910 μm long and 300–590 μm wide. Body cuticle annulated, head not distinctly set off from neck and without annules. Cephalic framework is not prominent. Cephalids not observed. Stylet delicate, 11.2 μm long with rounded knobs. Esophagus well developed with elongate, cylindrical procorpus, and large, rounded meta-carpus provided with heavily sclerotized valve. Orifice of the dorsal esopha-geal gland 3.0 μm is posterior to base of stylet. Excretory pore very distinct, generally located about one and one-half stylet length or more from base of stylet. Ovaries two, convoluted. Vulva and anus terminally located. Perennial pattern found prominent with distinct and characteristic striations. Egg deposited in gelatinous matrix but egg sacs absent. Absence of egg sac is characteristic feature of this nematode.

Cuticular annulations are distinct, each annule measuring 2.0 μm in width. Lateral field wide consisting of 4–8 lines depending on young or older specimens at mid-body. Stylet robust, 15.4 μm long with rounded knobs. Opening of dorsal esophageal gland is nearly 3.5 μm posterior to base of stylet.

Medium bulb elongate with well-developed sclerotized valve. Length of esophagus (from anterior end to base of esophagus) measures 224.0 μm distinct nerve ring encircling isthmus just posterior to medium bulb. Cephalids located in anterior portion. Spicules are arcuate, 27.0 μm long. Testis 1 occasionally reflexed a short distance from its anterior end. Gubenaculum is usually 6.1 μm in length. Tail end is round, 11.2 μm long.

9.3.2.2   SECOND STAGE JUVENILES

Second stage juveniles are vermiform, body cylindrical, tapering more prominently toward posterior end. Head not offset from body with slight cephalic framework, bearing three faint post labial annules. Body annulated, cuticular annulations distinct and fine, each annule measuring about 1 μm in width. Lateral line is variable but four or more up to eight. Stylet is small 11.1 μm with rounded knobs, sloping posteriorly. Dorsal esophageal gland opening is nearly 3.0 μm posterior to base of the knob. Metacarpus spherical with prominent sclerotized valve. Tail is 71.0 μm long, hyaline tail terminal 18.0 μm in length, without distinct annulations.

9.3.2.3   EGGS (N = 25)

Eggs are cylindrical 92.5 μm long and 39.6 μm wide with hyaline shell culture marking with different embryonic stages. First molt larvae commonly seen in a population which give rise to second stage larvae after hatching.

9.3.2.4   PERINEAL PATTERN OF M. GRAMINICOLA

Perineal pattern with characteristics striation was similar to *M. graminicola* given by Golden and Birchfield (1968).

9.3.2.5   HOST RANGE OF MELOIDOGYNE

Root-knot nematode is well known for its polyphagous nature and wide host range besides, rice, wheat, and barley (Birchfield, 1965; Manser, 1971; Buangsuwon et al., 1971; Roy, 1977, 1979; Yik & Birchfield, 1979; MacGowan & Langdon, 1989).

9.3.3   SURVIVAL AND SPREAD

M. graminicola can sustain in flooded ill drained fields as egg masses or juveniles can withstand for long times under such circumstances. Roy (1982) reported that the population of *M. graminicola* drop quickly after four months but some egg masses can remain sustainable for at least 14 months in wet soil (Roy, 1982). Further, Bridge and Page (1982) found that *M. graminicola* can survive in flooded soils up to a depth of 1 m for at least five months. However, it cannot attack rice in waterlogged circumstances but swiftly attacks when infested soils are drained (Manser, 1968). Since root nematode of wheat and barley occurs in flooded rice also, there is further possibility of spread by irrigation and run-off water.

9.3.4   IMPACT OF RESOURCE CONSERVATION TECHNIQUES (RCTS) ON POPULATION DYNAMICS OF MELOIDOGYNE GRAMINICOLA

Study on effect of different soil parameters *viz.*, organic carbon content, available nitrogen, and pH of the soils from the fields of various resource conservation techniques (RCTs) and conventional techniques under rice-wheat cropping system revealed that organic carbon content and available nitrogen were higher in soils collected from the field which had intervention of zero tillage and double zero tillage as compared to conventional techniques. Population of *M. graminicola* was found to be higher in soils of those fields which had intervention of zero and double zero tillage showing positive correlation with organic carbon and available nitrogen of soil. Upadhyay and Vaish, (2014) observed that there was no relation found between the pH and initial population of nematode.

The higher population of second stage juvenile was obtained from fields with intervention of zero tillage. Whereas, conventional and zero tillage with *Sesbania* showed low population. Maximum stunting occurred with respect to zero tillage, rice-zero tillage wheat with residue, whereas, soil of conventional tilled transplanted puddle rice-conventional tilled broadcasted wheat with and without residues exhibited minimum reduction in plant height. Decline of shoot weight noticed to be related with advancement of stunting in most of the RCTs.

Progression in stunting was also to be found related with root-knot index (5.0) with respect to different RCTs over zero tillage rice-zero tillage wheat with *Sesbania*. Whereas, root-knot index was less than three with respect to conventional practice. It was established that the initial population of nematode, stunting, and shoot weight reduction was much more under RCT in compassion to conventional method, whereas, *Sesbania* incorporation in RCT was observed as superior option for management of the root-knot nematode (Upadhyay, 2012; Upadhyay et al., 2014a, 2014b).

9.4 WHEAT GALL NEMATODE OR EAR COCKLE OF WHEAT (*ANGUINA TRITICI*)

Needham (1743) reported *A. tritici* (Steinbuch) Filipjev from England as the first plant-parasitic nematode causing ear-cockle disease of wheat. Infective juveniles of *A. tritici* feed on young leaves as ectoparasite and become endoparasitic invading inflorescence and developing seeds. In infected wheat plants, seeds are transformed into the galls each containing around 10,000 or more J_2s as the dried mass. Wheat seed gall nematode is widely distributed and reported from Ethiopia, Hungary, India, Iran, Iraq, Afghanistan, Australia, Brazil, Bulgaria, China, Egypt, Israel, Lithuania, New Zealand, Pakistan, Poland, Romania, Russian Federation, Russian Far East, Syria, Switzerland, Turkey, and the Yugoslavia (Swarup & Sosa-Moss, 1990; Maqbool, 1988; Sikora, 1988). Early records of the nematode detection made in the United States include North and South Carolina, California, Georgia, Maryland, New York, Virginia, and West Virginia. *A. tritici* is a damaging pest in many developing countries. However, it is a pest of paramount regulatory importance in the developed countries and causes up to 70% yield losses (Luc et al., 1990). The disease has been eradicated in some developed countries by modern seed sanitation technique like gravity table seed processing or by sieving and freshwater flotation, however, it could easily be introduced. Wheat gall nematode has the historical significance of being the plant-parasitic nematode that was seen and demonstrated to cause a plant disease. The nematode is reported from all the important wheat-growing regions of the world.

Singh et al. (1953) had reported that the nematode causes a loss of 30% in the annual production of wheat in India. Nematode infected young plants exhibit slight enlargement at the basal portion of stem. Such infected plants are mostly stunted in appearance. The leaves are twisted and curled preventing normal emergence of ears.

Seed galls, which are green in the beginning, later turn hard, dark brown to black (Fig. 9.1 a-d). These seed galls measure about 3–5 mm in length and 2–3 mm in width. Each seed gall may contain 1000–30,000 or more nematode juveniles in a quiescent state. Even 28 years old seed galls have been found to produce viable juveniles.

FIGURE 9.1   Symptoms of ear cockle nematode of wheat, (a-b) infected spikes, (c) seed galls, (d) healthy seads.

The adult female is 2.64–4.36 mm long and adult males are 2.04–2.4 mm long, straighter than females, and more active. Anterior and posterior portions are slender while the middle portion is swollen. Life history, effect of sowing time, and disease record of *A. tritici* on wheat have been reported by various workers (Koshy & Swarup, 1971; Swarup & Gupta, 1971; Bhatti & Dalal, 1976; Paruthi & Bhatti, 1985; Gokte & Swarup, 1987; Singh et al., 1991). The wheat gall nematode or ear cockle of wheat (*A. tritici*) on wheat was reported first time from farmer's field of Jammu (Singh et al., 2011).

9.5 TUNDU DISEASE

Apart from ear cockle disease, the nematode is also involved in the yellow ear rot or yellow slime disease of wheat which is commonly known as 'tundu' disease in India. The association of *A. tritici* with bacterium *Corynebacterium (Clavibacter) tritici* results in the development of yellow slime disease of wheat.

9.6 ROOT LESION NEMATODES (*PRATYLENCHUS* SPP.)

The root lesion nematodes (RLN) are most economically important phytonematodes, and cosmopolitan in maize fields. However, more than 350 hosts have been recorded. Assessment of exact loss by lesion nematodes has not been made possible under field conditions due to presence of mixed population of the nematode in the field. Lesion nematodes have wide host range which can affect the selections of crop used to control the nematode in crop rotation sequences. Soil type tillage operations have also been reported to affect lesion nematodes' population dynamics. The plants show chlorosis, stunting, and general lack of vigor resulting into wilt. The plants form patches or zones in the field. Lesion nematodes reduce or inhibit root development by forming local lesions on young roots, which may then rot because of secondary fungi and bacteria. Crop yields are reduced.

There are two important species of root-lesion nematode in the northern grain region, *Pratylenchus thornei* (the most commonly found species) and *Pratylenchus neglectus*. Root-lesion nematodes feed, move, and reproduce inside plant roots (migratory endoparasitic nematodes). The damage to the plant roots leads to yield loss in crops such as wheat and chickpea. These nematodes can also build-up numbers under many other crops. Root-lesion nematodes in the northern grain region can be found deep in the soil profile and in some cases peak populations occur 30–60 cm deep in the soil.

Root-lesion nematodes are microscopic and cannot be seen with the naked eye in the soil or in plants. The most reliable way to confirm the presence of root-lesion nematodes is to test your farm soil. Nematodes are extracted from the soil for identification and to determine their population size. Look out for tell-tale signs of nematode infection in the roots and symptoms in the plant tops and if seen submit soil and root samples for nematode assessment.

Root-lesion nematodes invade the root tissue resulting in light browning of the roots or localized deep brown lesions. However, these lesions can be difficult to see on roots. The damage to the roots and the appearance of the lesions can be made worse by fungi and bacteria also entering the wounded roots. Roots infected by root-lesion nematodes are poorly branched, lack root hairs and do not grow deeply into the soil profile. Such root systems are inefficient in taking up soil nutrients (particularly nitrogen, phosphorus, and zinc under northern region conditions) and soil water.

When root-lesion nematodes are present in very high numbers the lower leaves of the wheat plants are yellow and the plants are stunted with reduced tillering. There is poor canopy closure so that the wheat rows appear more open. The tops of the plants may exhibit symptoms of nutrient deficiency (nitrogen, phosphorus, and zinc) when the roots are damaged by root-lesion nematodes.

Infected crops can wilt prematurely, particularly when conditions become dry later in the season because the damaged root systems are inefficient at taking-up stored soil moisture. With good seasonal rainfall, wilting is less evident and plants may appear nitrogen deficient. Deteriorating wheat yields over several years, called "wheat sickness," may also indicate a nematode problem.

The adult root-lesion nematodes are nearly all self-fertile females. Females are identified by the presence of vulva more than 70% length from head toward posterior region. They lay eggs inside the roots and pass through a complete life cycle in about six weeks under favorable conditions (warm, moist soil) and so pass through several generations in the life of one host crop. The newly hatched nematodes can remain in the plant roots, or leave the plant to seek another root system to attack. The nematodes survive through fallow periods, particularly in the subsoil where they escape the hot, drying conditions of the surface soil. In drought, the nematodes can dehydrate (anhydrobiosis) to further aid their survival until favorable conditions return. Both male and female of these nematodes are wormlike, 0.4–0.7 mm long and 20–25 μm in diameter. The life cycle of the various species of *Pratylenchus* is completed within 45–65 days.

9.7   STEM AND BULB NEMATODE (*DITYLENCHUS* SPP.)

Stem and bulb nematode occurs worldwide but is particularly prevalent and destructive in areas with temperate climate. It is one of the most destructive plant-parasitic nematodes and attacks a large number of host plants. On most crops, stem nematode causes heavy losses by killing seedlings, dwarfing plants, destroying bulbs, by causing the development of distorted, swollen, and twisted stems and foliage, and, by reducing yields greatly. The nematode feeds on stems, leaves, and bulbs and is rarely found in soil. The nematode is 1.0–1.3 mm long and about 30 μm in diameter. The females lay 200–500 eggs, mostly after fertilization. The total duration of life cycle ranges from 19 to 25 days. Reproduction continues throughout the year.

9.8   STUNT NEMATODE (*TYLENCHORHYNCHUS* SPP.)

The stunt or stylet nematode as member of the genus, *Tylenchorhynchus* Cobb, 1913 has been known to occur worldwide (O'Bannon et al., 1991). Many species, *viz.*, *Tylenchorhynchus vulgaris, Tylenchorhynchus eremicolus, Tylenchorhynchus indicus, Tylenchorhynchus brassicae, Tylenchorhynchus mashhoodi, Tylenchorhynchus capitatus, Tylenchorhynchus claytoni, Tylenchorhynchus brevidens, Tylenchorhynchus abuncus, Tylenchorhynchus nudus, Tylenchorhynchus maximus,* and *Tylenchorhynchus robustus* are known to parasitize numerous plant species of agronomic and horticultural importance. *T. vulgaris, T. eremicolus, Tylenchorhynchus persicus, T. indicus, T. brassicae,* and *T. mashhoodi* are found associated with wheat and barley in India. *Tylenchorhynchus* species are primarily considered to be migratory ectoparasitic feeders, feeding along the root surface penetrating epidermal cells of roots and root hairs. Occasionally, they feed endoparasitically, confined to the outer cortical layers of roots. They have been reported to feed in large aggregations at the root tip, causing mechanical breakdown of epidermal, cortical, and undifferentiated vascular tissue at this site. While this causes a reduction in main root growth, it is compensated for by increased lateral root growth. Stunt nematodes are common in sandy soils and well feed on roots of most plants. They do not usually cause noticeable damage to crops unless numbers are excessive >5000/l soil. Stunt nematodes are particularly prone to desiccation, as a result numbers can drop significantly after a dry summer. *T. persicus is* potentially an important nematode species in Punjab because of its polyphagous nature (Sakhuja et al., 2000). After several years of studying specimens and pertinent literature of all described stunt nematode species, Agricultural Research Service microbiologist Zafar A. Handoo—an expert on identifying nematodes at the ARS Nematology Laboratory, Beltsville, MD—recently completed an identification key. This is the first accurate, all-inclusive guide to diagnose and identify all known stunt nematode species. Handoo examined and evaluated all the information on these species contained in the United States Department of Agriculture (USDA) Nematode Collection, one of the world's largest and most valuable archives of these worms. Handoo's key is based on the overall morphology—the external features—of females, since males are not known in several species. In some cases, he used the differences in male reproductive organs. In his key, he identified the main characteristics useful in distinguishing species, such as the shape of the lip region and shape of the tail. Further studies are needed of the worm's morphology, including scanning electron microscopy to magnify male and female nematodes from a broader range of habitats.

9.9 MANAGEMENT OF NEMATODES

Nematode management is not simply to identify a specific nematode and use of effective chemical for its management but, to find out ecologically safe and sustainable means especially cultural practices of their management. Effectiveness of nematicides depends on environmental conditions. Unfavorable weather conditions reduce the effectiveness of nematicides negatively.

Management of plant-parasitic nematodes is problematical by the intricacy of the soil environment (Norton, 1978). Chemical, biological, and cultural methods along with the use of host plant resistance include management strategies that have reduced the risk of damage by many nematode species (Hague & Gowen, 1987; Heald, 1987; Kerry, 1987; Halbrendt & LaMondia, 2004; Starr & Roberts, 2004). None of these techniques is without challenges *viz.*, faith on chemical control has come into question because chemical pesticides can have negative environmental and human health effects and their practice is under review (Schierow, 2000; Martin, 2003). Usefulness of chemical suppression also can be restricted by regional production conditions. For example, there are only narrow windows of opportunity for pest management inputs if fields are occupied most of the time for suitable plant growth. In particular, narrow crop sequences with host crops of a specific plant-parasitic nematode or infestations with multiple nematode species can complicate the development of management strategies. Integrated methods for the management of plant-parasitic nematodes have been proposed for the production of high- and low-value crops (Brown, 1987; McKenry, 1987). These strategies partially benefit from physical, chemical, and biological factors in soils that constrain activities of plant-parasitic nematodes and may reduce their damage potential (Stirling, 1991; Robinson, 2004). An integrated management strategy requires in-depth evaluation of the cropping system. Key questions need to be answered before a successful holistic approach to nematode management is designed. How are plants established? How do vegetation and cropping period overlap? What is the type of cropping system? Are there cover crop periods? What is the crop sequence? What is the tillage system? Economical, biological, practical, and logistical feasibility of the potential management strategy need to be addressed.

9.9.1 CULTURAL MANAGEMENT

The cultural practices include crop rotation, fallowing, flooding, sanitation, ploughing during summer season, mulching, organic manure, spacing of plants in the field, time of sowing, resistant varieties, and so forth. Nematode management through inter cropping and crop rotation is based on the fact that some species of nematodes are host specific and can be managed by crop rotation with non-host crop. The period of flooding appears to vary with several factors such as kind of soil, season, and so forth. Warm conditions are said to reduce the period of time required for control. Juveniles are more easily killed by flooding than eggs. The period of flooding needs to be worked out for each condition. Fallow periods in cropping sequences can also reduce nematode populations. The summer ploughing 2–3 times during hottest period of the year help to expose nematodes to the drying action of sun and wind and reduce the population. The intensity of symptoms was directly related to the inoculum levels. Soil fertility and texture have a significant influence on nematode abundance and diversity. Plant-parasitic nematodes (e.g., *Meloidogyne, Rotylenchulus, Heterodera, Hoplolaimus* etc.) generally prefer sandy loam, loam, or sandy soils (Wallace, 1974; Singh & Singh, 2003). Some (e.g., *Hirschmaniella* and *Pratylenchus*) are however, more prevalent in heavy soil.

Since CCN is host specific, rotation with non-cereals offers good potential to reduce nematode density. However, as *Pratylenchus* is largely polyphagous, rotational options for these nematodes are far fewer (Nicol & Rivoal, 2007). Yield losses can also become very high in two-year rotations of cereals with summer fallow and in three-year rotations such as winter wheat, spring cereal, and a non-host broadleaf crop or fallow. Crop rotations that include broadleaf crops, corn, fallow, and resistant wheat, barley, or oat varieties can greatly reduce the nematode density. In addition, growing susceptible hosts <50% of the time in heavy-textured soils and <25% of the time in light textured soils can dramatically reduce numbers of *Heterodera avenae*.

Sanitation covers a wide range of cultural practices, including weed control, crop residue destruction, and disinfestations of farm equipment before moving it from heavily infested fields to uninfected fields. In monocultures, eliminating the weed hosts can be important in reducing the populations of plant-parasitic nematodes. Soil temperature plays crucial role in the activities of plant-parasitic nematodes, the time during which crop is planted is important (Ayaub, 1980). The addition of inorganic fertilizers alone without organic manure usually increases the nematode population and disease intensity. NH_4-N reduces the disease incidence while NO_3-N may increase the same. Particular forms as well as dose and proportion of NPK may also reduce or increase the incidence of disease. Use of tolerant/resistant varieties is most practical approach for the management of nematode diseases. Crop cultivars resistant to phytonematodes can be the most useful and cheapest means of nematode control for the small-scale farmers. Nematicidal plants with roots containing nematicidal substances have been investigated. These toxic substances reduce the population level of some nematode species. African marigolds (*Tagetes* spp.), asparagus, crotalaria, mustard, and several cruciferous plants have been reported to produce toxic substances. Various organic materials including agricultural and industrial by products, most of which are wastes, have been experimented for the control of phytonematodes infecting crops. Soil amendments with green manure, compost, oil cakes of neem, mahua, mustard, groundnut, cotton, linseed, karanj, sawdust, and so forth have been found to reduce nematode populations. Neem, karanj, and groundnut cakes incorporated into soil at the rate of 500 kg/ha give good control of plant-parasitic nematodes and could be practiced wherever possible. Apart from encouraging the multiplication of natural enemies like nematode trapping fungi, the decomposition products of these organic amendments are toxic to nematodes.

Use of organic manures is of great value since the decomposition products and promotion of natural enemies decrease nematode populations (Mankau, 1963; Goswami & Swarup, 1971; Singh & Sitaramaiah, 1973; Mishra & Prasad, 1974; Haseeb et al., 1984a; Singh & Singh, 1991, 1992, 2001; Alam, 1990; Singh, 2006, 2008). The amount of oil cakes or any organic matter to be incorporated depends on various factors like, soil type and texture, crop to be planted, the predominant nematode fauna of the soil and the amount of soil moisture present in the soil. In trap crops when grown in infested soils, the nematodes penetrate into the root system and start multiplying. Before the nematodes complete its life cycle, the plants are uprooted and destroyed. Applications of botanicals are easy, environmentally safe, and having no phytotoxic effect on crops. The inhibition of root-knot development may be due to the accumulation of toxic byproducts of decomposition to increased phenolic contents resulting in host resistance (Alam et al., 1979, 1980), or to changed physical and chemical properties of soil inimical to the nematodes (Ahmad et al., 1972). Use of resistance is one of the most economical and effective ways to manage plant-parasitic nematodes. The success has been achieved by identifying a dominant *Mi* gene governing resistance against *Meloidogyne* and its linkage to an *acid phosphatase* gene. Flooding also reduces numbers of nematode pests. However, availability of water is an important issue to practice flooding.

9.9.2 BIOLOGICAL MANAGEMENT

Application of natural enemies of plant-parasitic nematodes for controlling nematode population is an essential component of eco-friendly management. Manipulation of biotic agents present in soil or introduction of such agents in soil, for reduction of nematode populations, has been one of the most important research areas engaging the attentions of the plant protection scientists all over the world. Plant nematodes have suffered by many natural enemies, for example, fungi, bacteria, and predacious nematodes. Fungi have been the fore most amongst such agents. Certain fungi capture and kill nematodes in the soil. *Arthrobotrys* spp., *Dactylaria* spp., *Dactylella* spp., *Catenaria* spp., and *Trichothecium* spp. are the genera most commonly represented. The biological control of root-knot nematode on tomato under greenhouse condition by using predaceous fungi has been reported (Singh et al., 2001; Bandyopadhyay et al., 2001). Some fungi capture nematode by adhesion, but many employ specialized devices that include networks of adhesive branches, stalked adhesive knobs, non-constricting rings, and constricting rings. The surface of the nematode is penetrated and the fungus hyphae grow throughout the nematode body, digesting and absorbing its contents. Under favorable conditions, large numbers of nematodes may be captured, and killed especially by those fungi that form adhesive networks or hyphal loops. *Trichoderma harzianum*, *Trichoderma virens*, *Aspergillus niger*, *Paecilomyces lilacinus*, and *Pochonia chlamydosporia* are found promising biocontrol agents. Now mycorrhiza is not restricted to its use only as biofertilizers, its potential role in the biological control of plant-parasitic nematodes is reported by many workers. Sikora (1979) found that prior presence of vesicular-arbuscular mycorrhizas (VAM) fungi, *Glomus mosseae* has resulted into an increase in plant resistance against *Meloidogyne* spp. A bacterial parasite of nematodes, *Pasteuria penetrans*, has received much attention and research effort in recent years; *P. penetrans* is probably the most specific obligate parasite of nematodes, with a life cycle remarkably well adapted to parasitism of certain phytonematodes. It directly parasitizes juvenile nematode, thus affects penetration and reproduction. *P. penetrans* can survive several years in air dried soil apparently without loss of viability. Seed bacterization, soil drenching, and bare root dip applications with *Pseudomonas fluorescens*, *P. penetrans*, *Bacillus subtilis*, and *Bacillus polymyxa* effectively control plant-parasitic nematodes (Walia, 1994; Walia & Dalal, 1994; Singh et al., 1998a, 1998b; Rojas Miranda & Marban-Mendoza, 2000; Ravichandra & Reddy, 2008). Among the predatory nematodes, monarchs may be proved efficient predators because of stronger predatory potential, high rate of predacity, and high strike rate (Bilgrami, 1998).

9.9.3   CHEMICAL MANAGEMENT

There are two major groups of nematicides distinguished on the basis of the manner in which they spread or diffuse through the soil. Soil fumigants are available in the form of cylinders or liquids and diffuse in form of gases from the point of their injection into the soil. Non-fumigant nematicides *viz.*, carbamate or organophosphates which are water soluble compounds applied to the soil as liquid or granules. Recently, large number of non-fumigant and systemic nematicides are available which are safe on plant. Dibromochloropropane (DBCP) (nemagon) has been found very effective for standing crops against root-knot nematodes but its use has been suspended due to adverse effect on human beings. Carbofuran 3G (furadon), phorate

10G (thimet), fenamiphos (nemacur), fensulphothion (dasanit), and oxamyl have been recommended for the management of nematodes (Prasad,1990; Singh & Singh, 1991). The development of the *H. cabana* nematode was also delayed in pot condition experiment which was treated with carbofuran, phorate, and aldicarb nematicides (Singh & Singh, 1998). The dose and method of application would vary with crop. Bone (1987) reported that synthetic pheromones to disrupt nematode reproduction and chemicals that directly interfere with nematode chemoreceptors (Janson, 1987) should be exploited for nematode population management. Seed treatment with nematicides has been used successfully to control root-knot nematodes in various crops (Prasad & Chawla, 1991; Rahman & Das, 1994; Singh et al., 1998a). *A. tritici* nematode infested seeds are placed in 20% salt solution in suitable container and stirred vigorously for some time. The galls being lighter float on the surface and can be easily skimmed off and destroyed. The seeds are then thoroughly washed in fresh water, dried, and then can be sown or stored. Systemic nematicides like carbofuran and phorate are also very effective.

9.9.4   REGULATORY MANAGEMENT

Numerous attempts have been made to prevent the introduction of nematodes into countries or provinces by means of quarantine. Quarantines are established by legislative action in parliament and usually give quarantine authorities power to make and enforce regulations to accomplish the purpose. Such regulations usually prohibit bringing infected seeds into protected areas where similar crops might become infected.

9.9.5   PHYSICAL MANAGEMENT

Soil solarization with transparent polyethylene sheet has been attended as a means of raising the soil temperature to lethal levels to control soil pathogens (Sharma & Nene, 1990). Soil solarization with double transparent sheets caused the maximum reduction of nematode population (84.11%) whereas with double black, single transparent, and single black sheets reduced nematode population 80.96, 75.51, and 70.46%, respectively. The scientific principle involved in thermotherapy is that the pathogens present in soil and plant materials are inactivated or eliminated at temperatures nonlethal for the host tissues. Physical means of nematode control includes heat treatment of soil, solar drying, steam sterilization, hot water treatment, and soil solarization. Soil solarization, a method of pasteurization can effectively suppress most species of nematode along with other microbes and weeds under field conditions (Katan, 1981; Kumar et al., 1993; Sharma, 1985a; Sharma & Nene, 1985; Singh, 2008a, 2008b). Generally, sheets of 50–100 μm thickness are most suitable for raising the soil temperature. The use of transparent polyethylene had yielded better results than black sheets, since transparent sheet transmit most of the incident radiation to soil (Mehrer, 1979). Additions of salt, sugar, charcoal, and so forth create osmotic stress on nematodes and can be used for controlling nematodes. Ganguly et al. (1996) observed that mulching significantly increased rice seedling growth; the plant-parasitic nematodes were reduced by 50–87% in the solarized soil.

9.9.6 INTEGRATED MANAGEMENT OF NEMATODES

The development of pest management program for economic control of any pest implies the judicious selection of those available control techniques which will reduce the effects of the pest with minimum negative impact on the environment and with overall economic soundness. Integrated nematode management is based upon the system approach, follows location specific principle and is environment specific (Pankaj & Sharma, 1998). Utilization of the best combination of available management strategies for the pest complex at hand (nematodes, insect pests, disease organisms, weeds etc.) constitutes an integrated crop protection system. Resistant cultivars, crop rotation, pesticides, and sanitary and cultural practices can all be employed to the best possible advantage. An integrated management strategy prevents the excessive build up of any single nematode, insect, or disease population and minimizes the development of pest resistance to any single tactic. Integrated pest management systems require flexibility and depend upon the specific pest problem and locally available management options. A fixed set of recommendations may keep a pest complex in check for a limited period of time, but as the pest population shifts, recommendations will have to change also. Therefore, system development takes into account many factors including the species and race of pests present, the availability of resistant host plants, the longevity of the pest and the crops, cropping systems, and climate of the geographical region.

9.9.7 APPLICATION OF MOLECULAR BIOLOGY AND BIOTECHNOLOGY

Nematode diagnostics is essential for success of any nematode management program even more when the control method is highly specific to the control species/pathotypes/race. A number of new techniques for analyzing nucleic acids, proteins, carbohydrates, and lipids can be helpful in the identification of pests of those allozymes; monoclonal antibody and DNA based systems are most well developed for nematodes. Poly acrylamide gel electrophoresis (PAGE) has mainly focused on cyst and root-knot nematodes. *Polymerase chain reaction* (PCR) techniques or restriction fragment length polymorphism (RFLP) analysis used as supplementary tools wherever necessary. Nematode resistant transgenic plants can be designed by various approaches. The simple method is to introduce resistant genes effective against plant-parasitic nematodes from wild species to commercial cultivars.

CONCLUDING REMARKS

Nematodes parasitize the roots of our carefully nurtured crops, reduce their ability to produce yields, make them weak and perhaps vulnerable to many pests. Crop resistance offers great importance in this directivity. In addition, an intensive search of plant germplasm collection for natural resistance to nematodes and their races is also required. An in depth understanding is needed of the molecular basis of how and why plants are susceptible to nematodes. New progress is being made in studying changes in gene expression during the infection of plants in nematode host interactions where feeding sites are formed. There is a need to reduce these avoidable yield losses by developing new environment friendly, economically acceptable, and ecologically based management strategies as the current options become ineffective or unacceptable. The research on identifying the promising biological control agents, their mass production, application techniques, and their behavior in the soil under varying agro climatic conditions need to be intensified. Current cropping systems are often vulnerable to serious problems with plant-parasitic nematodes. Ecological and human health concerns and economic considerations require a holistic approach to a truly integrated pest management system. The benefits of novel management tools and agronomic practices in improving sustainable agricultural production require further exploitation to realize the goal of sustainable production with minimal environmental impacts. Detailed evaluation of the entire cropping system, though more cumbersome than single crop-approaches, is essential to integrate novel tools and strategies successfully into these systems. Chemical, biological, and cultural methods along with the use of host plant resistance comprise management strategies that have decreased the risk of damage by many nematode species. Future nematode management must employ sustainable agricultural practices that take into account beneficial, detrimental, and other nematode species in the rhizosphere and in soil.

KEYWORDS

- nematode pests
- wheat and barley
- nematode management
- cyst nematodes (*Heterodera* spp.)
- root-knot nematodes (*Meloidogyne* spp.)
- wheat gall nematode or ear cockle of wheat (*Anguina tritici*)
- root lesion nematodes (*Pratylenchus* spp.)
- stem and bulb nematode (*Ditylenchus* spp.)
- stunt nematode (*Tylenchorhynchus* spp.)

REFERENCES

Ahmad, R.; Khan, A. M; Saxena, S. K. In *Changes Resulting from Amending the Soil with Oil Cakes and Analysis of Oil Cakes (Abstract),* Proceedings of the 59th Session of the Indian Science Congress, Calcutta, India, Part III, 1972; Indian Science Congress Association: Calcutta, India, 1972; pp 1–164.

Al-Abed, A.; Al-Momany A.; Al Banna, L. *Heterodera latipons* on Barley in Jordan. *Phytopathol. Mediterranea.* **2004,** *43,* 311–317.

Alam, M. M. Control of Plant Parasitic Nematodes with Organic Amendments and Nematicides in Nurseries of Annual Plants. *J. Bangladesh Acad. Sci.* **1990,** *14,* 107–113.

Alam, M. M.; Khan, A. M.; Saxena, S. K. Mechanism of Control of Plant Parasitic Nematodes as a Result of the Application of Organic Amendments to the Soil Role of Phenolic Compounds. *Indian J. Nematol.* **1979,** *9,* 136–142.

Ayaub, S. M. *Plant Nematology: An Agricultural Training Aid;* Nema Aid Publication: Sacramento, CA, 1980; p 195.

Bandyopadhyay, P., et al. Eco-Friendly Management of Root-Knot Nematode of Tomato by *Arthrobotrys oligospora* and *Dactylaria brochopaga. Indian J. Nematol.* **2001,** *31,* 153–156.

Bhatti, D. S.; Dalal, M. R. Susceptibility of Some Common Wheat Varieties to the Seed Gall Nematode *Anguina tritici. Haryana Agric. Univ. J. Res.* **1976**, *6,* 162–163.

Bilgrami, A. L. Predatory Nematodes and Protozoans as Biopesticides of Plant Parasitic Nematodes. In: *Plant Nematode Management- a Biological Approach;* Trivedi, P. C. Ed.; CBS Publisher: New Delhi, India, 1998; pp 4–23.

Birchfield, W. Host Parasite Relations and Host Range Studies of a New *Meloidogyne* Species in Southern USA. *Phytopathology.* **1965,** *55,* 1359–1361.

Boland, G. J., et al. Climate Change and Plant Diseases in Ontario. *Can. J. Plant Pathol.* **2004,** *26,* 335–350.

Bone, I. W. Pheromone Communications in Nematodes. In *Vistas on Nematology: A Commemoration of the Twenty Fifth Anniversary of the Society of Nematologists*; Veech, J. A., Dickson, D. W., Eds.; Society of Nematologists: Hyattsville, MD, 1987; pp 147–152.

Bridge, J.; Page, S. L. J. The Rice Root-Knot Nematode, *Meloidogyne graminicola,* on Deep Water Rice (*Oryza sativa* sub. sp. *indica*). *Rev. de Nematol.* **1982,** *5,* 225–232.

Brown, R. H. Control Strategies in Low-Value Crops. In *Principles and Practice of Nematode Control in Crop;* Brown, R. H., Kerry, B. R., Eds.; Academic Press: Sydney, 1987; pp 351–388.

Buangsuwon, D., et al. Nematodes. In *Rice Diseases and Pests of Thailand;* English ed.; Rice Protection Research Centre, Rice Department, Ministry of Agriculture: Bangkok, Thailand, 1971; pp 61–67.

Cadet, P., et al. Relationships between Ferrisol Properties and the Structure of Plant Parasitic Nematode Communities on Sugarcane in Martinique (French West Indies). *Acta Ecologica.* **1994,** *15,* 767–780.

Cohn, E.; Ausher, R. *Longidorus cohni* and *Heterodera latipons,* Economic Nematode Pests of Oats in Israel. *Plant Dis. Rep.* **1973,** *57,* 53–54.

De Waele, D.; Elsen, A. Challenges in Tropical Plant Nematology. *Annu. Rev. Phytopathol.* **2007,** *45,* 457–485.

Franklin, M. T. *Heterodera latipons* n. sp., A Cereal Cyst Nematode from the Mediterranean Region. *Nematologica.* **1969,** *15,* 535–542.

Ganguly, A. K.; Pankaj, A.; Sirohi, A. Effect of Soil Solarization of Rice Nursery Beds to Suppress Plant Parasitic Nematodes. *Intl. Rice Res. Notes.* **1996,** *21,* 80–81.

Gokte, N.; Swarup, G. Studies on Morphology and Biology of *Anguina tritici. Indian J. Nematol.* **1987,** *17,* 306–307.

Golden, A. M.; Birchfield, W. Rice Root-Knot Nematode (*Meloidogyne graminicola*) as a New Pest of Rice. *Plant Dis. Rep.* **1968,** *52,* 423.

Goswami, B. K.; Swarup, G. Effect of Oil Cake Amended Soil on the Growth of Tomato and Root Knot Nematode Population. *Indian Phytopath.* **1971,** *24,* 491.

Hague, N. G. M.; Gowen, S. R. Chemical Control of Nematodes. In *Principles and Practice of Nematode Control in Crops;* Brown, R. H., Kerry, B. R., Eds.; Academic Press: Sydney, 1987; pp 131–178.

Halbrendt, J. M.; LaMondia, J. A. Crop Rotation and other Cultural Practices. In *Nematology Advances and Perspectives: Nematode Management and Utilization;* Chen, Z. X., et al., Eds.; CAB International: Oxfordshire, 2004; Vol. 2, pp 909–930.

Handoo, Z. A. Plant-Parasitic Nematodes. 1998. http://www.ars.usda.gov/Services/docs.htm.

Haseeb, A.; Alam, M. M.; Khan, A. M. Control of Plant Parasitic Nematodes with Chopped Plant Leaves. *Indian J. Plant Pathol.* **1984a,** *2,* 180–181.

Heald, C. M. Classical Nematode Management Practices. In *Vistas on Nematology;* Veech, J. A., Dickson, D. W., Eds.; Society of Nematologists: Hyattsville, MD, 1987; pp 100–105.

IPCC. Climate Projections. http://www.wmo.int/pages/themes/climate/climage projections. Php. 2007.

Janson, H. B. Receptors and Recognition in Nematodes. *Vistas on Nematology: A Commemoration of the Twenty Fifth Anniversary of the Society of Nematologists;* Veech, J. A., Dickson, D.W., Eds.; Society of Nematologists: Hyattsville, MD, 1987; pp 153–158.

Katan, J. Solar Heating (Solarization) of Soil for Control of Soil Borne Pests. *Ann. Rev. Phytopathol.* **1981,** *19,* 211–236.

Kerry, B. R. Biological Control. In *Principles and Practice of Nematode Control in Crops;* Brown, R. H., Kerry, B. R., Eds.; Academic Press: Sydney, 1987; pp 223–263.

Koshy, P. K.; Swarup, G. Distribution of *H. avenae, H. zeae, H. cajani,* and *Anguina tritici* in India. *Indian J. Nematol.* **1971,** *1*(2), 106–111.

Kumar, B. N., et al. Effect of Soil Solarization on Weeds and Nematodes Under Tropical Indian Conditions. *Weed Res.***1993,** *33,* 423–429.

Luc, M.; Sikora R. A.; Bridge, J. *Plant Parasitic Nematodes in Subtropical and Tropical Agriculture.* CABI Publisihing: Wallingford, 1990.

MacGowan, G. B.; Langdon, K. R. Hosts of the Rice Root-Knot Nematode, *Meloidogyne graminicola.* Nematology Circular, Gainesville No. 172, 1989.

Mankau, R. Effect of Organic Soil Amendments on Nematode Population. *Phytopathology.* **1963,** *53,* 881–882.

Manser, P. D. *Meloidogyne graminicola* a Cause of Root-Knot of Rice. *Pl. Protec. Bull. FAO.* **1968,** *16,* 11.

Manser, P. D. Notes on the Rice Root-Knot Nematode in Laos. *Pl. Protec. Bull. FAO.* **1971,** *19,* 138–139.

Maqbool, M. A. Present Status of Research on Plant Parasitic Nematodes in Cereals and Food and Forage Legumes in Pakistan. In *Nematodes Parasitic to Cereals and Legumes in Temperate Semi-Arid Regions;* Saxena, M. C., et al., Eds.; ICARDA: Aleppo, Syria, 1988; pp 173–180.

Martin, F. N. Development of Alternative Strategies for Man- Agement of Soilborne Pathogens Currently Controlled with Methyl Bromide. *Annu. Rev. Phytopathol.* **2003,** *41,* 325–350.

McKenry, M. V. Control Strategies in High-Value Crops. In *Principles and Practice of Nematode Control in Crops;* Brown, R. H., Kerry, B. R., Eds.; Academic Press: Sydney, 1987; pp 330–349.

McSorley, R. Population Dynamics. In *Plant and Nematode Interactions;* Barker, K. R., et al., Eds.; Madison Publishers: Madison, WI, 1998; pp 109–133.

Mehrer, Y. Prediction of Soil Temperature of a Soil Mulched with Transparent Polyethylene. *J. Appl. Meteorol.* **1979,** *18,* 1263–1267.

Mishra, S. D.; Prasad, S. K. Effect of Soil Amendments on Nematodes and Crop Yields. *Indian J. Nematol.* **1974,** *4,* 1–19.

Mor, M.; et al. Phenology, Pathogenicity and Pathotypes of Cereal Cyst Nematodes, *Heterodera avenae* and *Heterodera latipons* (Nematoda: Heteroderidae) in Israel. *Nematologica.* **1992,** *38,* 494–501.

Needham, T. A. Letter Concerning Certain Chalky Tubulous Concretions Called Malm; with Some Microscopical Observations on the Farina of the Red Lily, and of Worms Discovered in Smutty Corn. *Philos. Trans. R. Soc. London.* **1743,** *42,* 634–641. http://rstl.royalsocietypublishing.org/content/42/462-471/634.

Nicol, J. M.; Rivoal, R. Integrated Management and Biocontrol of Vegetable and Grain Crops Nematodes. In *Global Knowledge and its Application for the Integrated Control and Management of Nematodes on Wheat;* Ciancio, A., Mukerji, K. G., Eds.; Springer: The Netherlands, 2007; pp 243–287.

Norton, D. C. *Ecology of Plant-Parasitic Nematodes;* John Wiley & Sons: New York, 1978; p 268.

Norton, D. C. Abiotic Soil Factors and Plant-Parasitic Nematode Communities. *J. Nematol.* **1989,** *21,* 299–307.

O'Bannon, J. H., et al. Tylenchorhynchus Species as Crop Damaging Parasitic Nematodes. Nematology Circular No. 190, Florida Department of Agriculture and Consumer Service Division of Plant Industry, Florida, 1991.

Pandey, S. Studies on Root-Knot Disease of Barley *(Hordeum vulgare L.).* M.Sc (Ag) Thesis, Banaras Hindu University, Varanasi, India, 2011.

Pankaj, B.; Sharma, H. K. IPM Strategies for Nematode Management. In *Potential IPM Tactics;* Prasad, D., Gautam, R. D., Eds.; Westville Publishing House: Paschim Vihar, New Delhi, India, 1998; pp 131–145.

Paruthi, T. J.; Bhatti, D. S. Estimation of Loss in Yield and Incidence of *A. tritici* on Wheat in Haryana. *Int. Nematol. Network Newsl.* **1985,** *2*(3), 13–16.

Philis, J. An Up-Dated List of Plant Parasitic Nematodes from Cyprus and their Economic Importance. *Nematol. Mediterranea.* **1995,** *23,* 307–314.

Prasad, D. Control of Root Knot Nematode *Meloidogyne arenaria* on Groundnut by Chemical Seed Pelleting. *Curr. Nematol.* **1990,** 1, 31–34.

Prasad, D.; Chawla, M. L. Seed Dressing with Carbosulfan for the Management of Parasitic Nematodes on Groundnut. *Indian J. Nematol.* **1991,** *21,* 19–23.

Prot, J. C.; van Gundy, S. D. Effect of Soil Texture and the Clay Component on Migration of *Meloidogyne incognita* Second-Stage Juveniles. *J. Nematol.* **1981,** *13,* 213–217.

Rahman, M. F.; Das, P. Seed Soaking with Chemicals for Reducing Infestation of *Meloidogyne graminicola* in Rice. *J. Agri. Sci. Soci. North East India.* **1994,** 7, 107–108.

Ravichandra, N. G.; Reddy, B. M. R. Efficacy of *Pasteuria penetrans* in the Management of *Meloidogyne incognita* Infecting Tomato. *Indian J. Nematol.* **2008,** *38,* 172–175.

Robinson, A. F. Nematode Behavior and Migrations through Soil and Host Tissue. In *Nematology Advances and Perspectives: Nematode Morphology, Physiology, and Ecology;* Chen, Z. X., et al., Eds.; CAB International: Oxfordshire, 2004; Vol. 1, pp 330–405.

Rojas Miranda, T. P.; Marban-Mendoza, N. *Pasteuria penetrans*: Adherence and Parasitism in *Meloidogyne incognita* and *Meloidogyne arachibicida. Nematropica.* **2000,** *29,* 233–240.

Roy, A. K. Weed Hosts of *Meloidogyne graminicola. Indian J. Nematol.* **1977,** 7, 160–163.

Roy, A. K. Survival of *Meloidogyne graminicola* Eggs Under Different Moisture Conditions *In Vitro. Nematol. Mediterranea.* **1982,** *10*(2), 221–222.

Roy, T. K. Histochemical Studies of Hydrolytic Enzymes in *Meloidogyne incognita* (Nematoda: Tylenchidae) and Infected Host *Lycopersicum esculentum* and their Role in Host-Parasitic Relationship. *Indian J. Exp. Biol.* **1979,** *17,* 1357–1362.

Rumpenhorst, H. G., et al. The Cereal Cyst Nematode *Heterodera filipjevi* (Mazhidov) in Turkey. *Nematol. Mediterranea.* **1996,** *24,* 135–138.

Sakhuja, P. K., et al. Distribution and Management of Nematodes in Rice-Wheat-Legume Cropping Systems in Punjab. In *Nematode Pests in Rice-Wheat-Legume Cropping Systems,* Proceedings of the Review and Planning Meeting and Training Workshop, Indian Agricultural Research Institute, New Delhi, India, Apr 5–10, 1999; Sharma, S. B., Pankaj, Pande,

S., Johansen, C., Eds.; Rice-Wheat Consortium Paper Series 7; ICRISAT: Patancheru, Andhra Pradesh, India, 2000; pp 7–8.

Sasser, J. N.; Freckman, D. W. A World Perspective on Nematology: The Role of the Society. In *Vistas in Nematology;* Veech, J. A., Dickson, D. W., Eds.; Society of Nematologists: Hyattsville, ML, 1987; pp 7–14.

Schierow, L. J. FQPA: Origin and Outcome. *Choices Magaz. Food Farm Resour. Issues.* **2000,** *15,* 18–21.

Sharma, S. B.; Rahaman, P. F. Nematode Pests in Rice and Wheat Cropping System in the Indo-Gangetic Plain. In *Nematode Pests in Rice- Wheat- Legume Cropping System,* Proceeding of a Regional Training Course, Haryana Agricultural University, Haryana, Sep 1–5, 1997; Sharma, S. B., Johanson, C., Midha, S. K., Eds.; Rice-Wheat Consortium for Indo- Gangetic Plain; IARI: New Delhi, India, 1998.

Sharma, S. B. *Nematode Diseases of Chickpea and Pigeon Pea*; Pulse Pathology Progress Report No. 43. International Crops Research Institute for the Semi Arid Tropics: Patancheru, India, 1985a, p 103.

Sharma, S. B.; Nene, Y. L. Effect of Presowing Solarization on Plant Parasitic Nematodes in Chickpea and Pigeon Pea Fields. *Indian J. Nematol.* **1985,** *15,* 277–278.

Sharma, S. B.; Nene, Y. L. Effects of Soil Solarization on Nematodes Parasitic to Chickpea and Pigeon Pea. *J. Nematol.* **1990,** *22,* 658–664.

Sikora, R. A. Predisposition to *Meloidogyne* Infection by the Endo Tropic *Mycorrhizal* Fungus *Glomus mosseae.* In *Root-Knot Nematode Meloidogyne spp. Systematic Biology and Control;* Lamberti, F., Taylor, C., Eds.; Academic Press: London, 1979; pp 399–404.

Sikora, R. A. Plant Parasitic Nematodes of Wheat and Barley in Temperature and Temperate Semi-Arid Regions-A Comparative Analysis. In *Nematodes Parasitic to Cereals and Legumes in Temperate Semi-Arid Regions;* Saxena, M.C., et al., Eds.; ICARDA: Aleppo, Syria, 1988; pp 46–48.

Singh S. K.; Singh, K. P. A First Report of Root Knot Nematode (*Meloidogyne graminicola*) on Wheat in Utter Pradesh, India. *J. Mycol. Pl. Path.* **2009,** *39*(2), 340–341.

Singh, J., et al. In *Control of Root-Knot Nematode in Rice Nursery by Seed and Soil Treatment with Nematicides and a Neem Production,* International Symposium of Afro Asian Society of Nematologists, Coimbatore, April, 16–19, **1998**a.

Singh, K. P.; Bandyopadhyay, P.; Singh, V. K. Performance of two Predacious Fungi for Control of Root-Knot Nematode of Brinjal. *Mycopathol. Res.* **2001,** *39,* 95–100.

Singh, B.; Singh, J.; Mathur, S. C. Earcockle or Sehun disease of wheat. *Agric. Anim. Hush. Uttar Pradesh.* **1953,** *3,* 7-9.

Singh, K. P.; Singh, V. K. Nematicidal Natures of Arjun Bark *Terminelia arjuna* on Cyst Nematode, *Heterodera cajani* of Pigeon Pea. *New Agric.* **1991,** *2,* 77–78.

Singh, K. P.; Singh, V. K. *Terminelia arjuna* Leaf Powder Reduces Population Density of *Heterodera cajani. Intl. Pigeon Pea Newsl.* **1992,** *16,* 17–18.

Singh, K. P., et al. Biomass of Nematodes and Associated Roots: A Determinant of Symptom Production in Root-Knot Disease of Rice (*Oryza sativa* L.). *J. Phytopathol.* **2006,** *154,* 676–682.

Singh, K. P.; Singh, V. K.; Singh, L. P. Concomitant Effect of Sowing Time and Inoculation of *Anguina tritici* on wheat. *Curr.Nematol.* **1991,** *2,*155–158.

Singh, M.; Samar, R.; Sharma, G. L. Effect of Various Soil Types on the Development of *Pasteuria penetrans* on *Meloidogyne incognita* in Brinjal Crop. *Indian J. Nematol.* **1998b,** *28,* 35–40.

Singh, R. S.; Sitaramaiah, K. Control of Plant Parasitic Nematode with Organic Amendments of Soil Research Bulletin No. 6; G. B. Pant University of Agriculture Science and Technology: Pantnager, India, 1973, pp 1–289.

Singh, V. K. Effect of Soil Solarization for Management of Plant Parasitic Nematodes. *Ann. Plant Prot. Sci.* **2008,** *16,* 541–542.

Singh, V. K. Eco-Friendly Management of Plant Parasitic Nematodes in Vegetable Crops. *In Insect Pest and Disease Management;* Prasad, D., Ed.; Daya Publishing House: New Delhi, 2008; pp 440–447.

Singh, V. K.; Singh, K. P. Effect of Granular Nematicides on the Biology of *Heterodera cajani* of Pigeon Pea. *Indian J. Nematol.* **1998,** *28,* 168–173.

Singh, V. K.; Singh, K. P. Effect of Some Medicinal Plant Leaves on the Biology of *Heterodera cajani. Indian J.Nematol.* **2001,** *31,* 143–147.

Singh, V. K.; Singh, K. P. Penetration and Development of *Heterodera cajani* on Pigeon Pea in Relation to Different Types of Soil. *Curr. Nematol.* **2003,** *14*(1–2), 17–24.

Singh, V. K.; Dwivedi, M.C.; Wali, P. Occurrence of Wheat Ear Cockle Nematode (Seed Gall Nematode) *Anguina tritici* in Jammu. *Indian J. Nematol.* **2011,** *41*(1), 109–110.

Starr, J. L.; Roberts, P. A. Resistance to Plant-Parasitic Nematodes. In *Nematology Advances and Perspectives: Nematode Management and Utilization;* Chen, Z. X., et al., Eds.; CAB International: Oxfordshire, 2004; Vol. 2, pp. 879–907.

Stirling, G. R. *Biological Control of Plant-Parasitic Nematodes: Progress, Problems, and Prospects;* CAB International: Wallingford, 1991; pp 282.

Swarup, G.; Gupta, P. On the Ear Cockle and Tundu Diseases of Wheat II. Studies on *A. tritici* and *Corynebacterium tritici. Indian Phytopath.* **1971,** *24,* 359–365.

Swarup, G.; Sosa-Moss, C. Nematode Parasites of Cereals. In *Plant Parasitic Nematodes in Subtropical and Tropical Agriculture;* Luc, M., Sikora, R. A., Bridge, J., Eds.; CAB International: Wallingford, 1990; pp 109–136.

Taylor, A. *Introduction to Research on Plant Nematology: An FAO Guide to the Study and Control of Plant Parasitic Nematodes;* FAO: Rome, 1967; pp 133.

Upadhyay, R. S.; Oostenbrink, M.; Khan, A. M. The Effect of Different Soil Types on the Density of Nematode Populations. *Indian J. Nematol.* **1972,** *2,* 42–53.

Upadhyay, V. Association of Resource Conservation Techniques with Root Knot Diseases Under Rice-Wheat Cropping System. M.Sc.(Ag) Thesis, Banaras Hindu University, Varanasi, India, 2012.

Upadhyay, V.; Vaish, S. S. Effect of Resource Conservation Practices and Conventional Practice on *Meloidogyne graminicola* in Relation to Different Soil Parameters. *Environ. Ecol.* **2014,** *32*(2), 575–578.

Vaish, S. S.; Pandey, S. K. First Report of the Root Knot Disease of Barley Caused by *Meloidogyne graminicola* from India. *Curr. Nematol.* **2012,** *23*(1–2), 77–80.

Vaish, S. S.; Ahmed, S. B.; Prakash, K. First Documentation on Status of Barley Diseases from the High Altitude Cold Arid Trans-Himalyan Ladakh Region of India. *Crop Protect.* **2011,** *30,* 1129–1137.

Walia, R. K.; Dalal, M. R. Efficacy of Bacterial Parasite, *Pasteuria penetrans* Application as Nursery Soil Treatment in Controlling Root Knot Nematode *Meloidogyne javanica* on Tomato in Green House. *Pest Manag. Econ. Zool.* **1994,** *2,* 19–21.

Walia, R. K. Assessment of Nursery Treatment with *Pasteuria penetrans* for the Control of *Meloidogyne javanica* on Tomato in Green House. *J. Biol. Control.* **1994,** *8,* 68–70.

Wallace, H. R. The Influence of Root-Knot Nematode, *Meloidogyne javanica,* on Photosynthesis and Nutrient Demand by Roots of Tomato Plants. *Nematologica.* **1974,** *20,* 27–33.

Webster, J. M. Introduction. In *Principles and Practice of Nematode Control in Crops;* Brown, R. H., Kerry, B. R., Eds.; Academic Press: Melbourne, 1987, pp 1–12.

Yik, C. P.; Birchfield, W. Host Studies and Reactions of Cultivars to *Meloidogyne graminicola. Phytopathology.* **1979,** *69,* 497–499.

Yousef, D. M.; Jacob, J. J. A Nematode Survey of Vegetable Crops and Some Orchards in the Ghor of Jordan. *Nematol. Mediterranea.* **1994,** *22,* 11–15.

CHAPTER 10

DISEASE RESISTANCE BREEDING IN WHEAT: THEORY AND PRACTICES

HANIF KHAN, SUBHASH C. BHARDWAJ*, PRAMOD PRASAD,
OM P. GANGWAR, SIDDANNA SAVADI, and SUBODH KUMAR

*ICAR-Indian Institute of Wheat and Barley Research Regional Station,
Flowerdale, Shimla 171002, Himachal Pradesh, India*

Corresponding author. E-mail:scbfdl@hotmail.com

CONTENTS

ABSTRACT

Globally a number of diseases affect wheat include three rusts, foliar blights, Fusarium head scab, the loose smut, flag smut, and powdery mildew. Genetic resistance is the most effective, least expensive, and environmentally safe means of rust diseases management. A large number of resistance genes originating from wheat and related species has been cataloged. The key points for the management of wheat diseases particularly rusts has always been to avoid large scale planting of single genotype/similar resistance and deploy varieties with diverse resistance, if possible then resistance based on more than one effective gene. The long-term success of breeding for disease resistance is influenced by three important factors: (a) the nature of pathogen and diversity of virulence, (b) availability and type of resistance, and (c) screening methodology and selection environment for tracking resistance. Diversity and durability of the resistance are two most important feature of the breeding for disease resistance in wheat. Resistance based on additive genes and gene complex providing resistance to multiple diseases have been successfully exploited by wheat breeders. CIMMYT's spring wheat breeding program has carefully defined its target production environments into six mega-environments. DNA markers closely linked with resistance have been used successfully used by the concerted efforts from breeders, pathologists and molecular biologists for improving selection efficiency for disease resistance and other traits in wheat improvement programs.

10.1   INTRODUCTION

Wheat, one of mankind's important staple foods after rice, is grown on about 225 million hectare worldwide. Hexaploid bread wheat (*Triticum aestivum* L.) is a grass from the "Fertile Crescent" region of the Near East (the Karacadag Mountains in South Eastern Turkey) and presently cultivated worldwide. Nearly 24 species of *Triticum* are found in nature. Einkorn wheat (*Triticum monococcum*) and emmer/durum wheat (*Triticum turgidum* ssp. *durum* L.) date back to 8000–9000 sc. Einkorn wheat is diploid (AA) and durum is derived from wild emmer *Triticum dicoccoides* (AABB) which resulted by natural selection from the hybridization of *Triticum urartu* and *Aegilops speltoides*. Bread wheat (AABBDD) evolved from either wild or domesticated emmer hybridized naturally with another diploid grass *Aegilops cylindrica*. Synthetic hexaploid (SH) wheats have evolved through crossing wild goat grass *Aegilops tauschii* (D genome sp.) with durum wheat to create more genetic diversity.

The major biotic stresses of wheat existing under intensive agriculture are stem or black rust, leaf or brown rust, and stripe or yellow rust caused by *Puccinia graminis* f. sp. *tritici* (Pgt), *Puccinia triticina* (Ptr), and *Puccinia striiformis* f. sp *tritici* (Pst), respectively, which cause significant losses of grain production (McIntosh et al. 1995) and are quite important in breeding program. Other important diseases affecting wheat yield are leaf blight complex (spot blotch: *Bipolaris sorokiniana*; tan spot: *Pyrenophora tritici-repentis;*Septoria glume blotch: *Septoria nodorum*), powdery mildew (*Blumeria graminis* f. sp. *tritici*), Karnal bunt (*Tilletia indica*), loose smut (*Ustilago tritici* Pers.), flag smut (*Urocystis agropyri*), hill bunt (*Tilletia caries* and *Tilletia foetida*), cereal cyst nematode (*Heterodera* spp.), and ear cockle disease (*Anguina tritici*). Rusts are global pathogens and historically known to cause yield loss in all the wheat growing regions. Wheat rust management using host resistance is preferred keeping in view of economic and environmental concerns faced by farmers mostly belong to developing world. Further, repeated use of fungicides often results in development of resistance in rust pathogens to fungicides (Oliver, 2014). Therefore breeding for rust resistance has been a major thrust for wheat breeders worldwide. In India also breeders in almost all the major wheat breeding centers are engaged seriously in rust resistance breeding. All three rusts are not important in all the agro-climatic zones in India with the exception of Southern hill zone. However, among three ruts leaf rust is of prime importance throughout Indian subcontinent whereas stem rust is known to cause yield losses in central and peninsular India. The stripe rust is only important in cool weather prevails during early crop season in North Western plain zone and most of the growing season in Northern hill zone (Joshi et al., 1985).

The manipulation of inherent potential of crop plants in the form of resistant varieties is considered an economic viable and environmentally safe alternative to minimize losses in both produce quality and quantity due to diseases. Development of the resistant variety requires a thorough understanding of the coevolution and coexistence of crop plants and their pathogens. The development of disease resistant varieties involves manipulations of genetics of host plant and the pathogens in tandem with each other especially in the light of the interactions that nature has established for each combination of host and pathogen. The resistance breeding has become very precise and focused with the greater understanding of definite host–rust pathogen interaction (Bhardwaj, 2013).

10.2   GENETICS OF HOST–PATHOGEN INTERACTION

Physiological races do occur in case of obligate parasites like rusts and powdery mildew in wheat. These differ in their infection on differential hosts. While working on flax and *Melampsora lini* system Flor (1955) showed that the inheritance of both resistance in the host and the ability of the parasite to cause disease are controlled by pairs of matching genes. One of such gene is resistance (*R*) gene in host where its avirulence (*Avr*) gene is in parasite. Plants producing a specific *R* gene product are resistant toward a pathogen that produces the corresponding *Avr* gene product. According to this hypothesis for each gene in the host that conditions its response to the pathogen there is corresponding gene for pathogenicity in the pathogen. Resistance or low infection type (L) occurs when *R* gene in the host is matched by an *Avr* gene (P) in the pathogen. All other combinations produce a high infection type (H) or susceptible reaction.

Simplified and most common interaction model of host pathogen interaction in flax rust is presented in Table 10.1.

TABLE 10.1   Interaction between Host and Pathogen at the Pair of Corresponding Genes.

Pathogen	Host		
	RR	**Rr**	**rr**
PP	L	L	H
Pp	L	L	H
pp	H	H	H

Gene-for-gene relationships are widespread and very important aspect of plant disease resistance and evolved through a series of steps in evolution of each. Therefore if a host is resistant to pathogen a virulent mutant would have an advantage over avirulent. Likewise, if a host is susceptible to a pathogen a resistant mutant in host would be at advantage. Person (1959) studied plant pathosystem ratios rather than genetics ratios in host–parasite systems. In doing so he discovered the differential interaction that is common to all gene-for-gene relationships and is now known as the Person's differential interaction. Physiological races of the rusts were originally identified on the basis of their infection types on the wheat varieties with a combination of genes. However, with the understanding of the genetic basis they are presently identified in relation to the known genes for the resistance and are called "pathotypes." A vital outcome of this work is the assemblage of a collection of well-characterized pathogen isolates which over a period of time can provide a basis for predicting the presence of genes and gene combinations in wheat stocks of unknown resistance genotype—the classic "gene postulation studies."

Biffen (1905), working with yellow rust, was first to prove that resistance to a pathogen could be governed by a single gene which inherits in a Mendelian fashion. He demonstrated that resistance to yellow rust in rivet wheat was controlled by one recessive gene. Subsequently several other studies showed that resistance to various diseases is monogenically determined but case of duplicate, complementary, additive and other interactions have been reported. Much of the early work on the genetics of rust resistance in wheat was done in concurrence with wheat breeding without taking into consideration the physiological specialization of the pathogen which can substantially influence the conclusion drawn. Since the early studies were done in isolation different workers using different wheat cultivars and different races it was largely impossible to relate one study to another. Moreover many of them involved field studies with unknown mixture of races and results could vary from year to year. The results were complicated by the fact that a gene might provide resistance to one race and not to others (Knott, 1989). It is now recognized that disease resistance may show following three modes of inheritance: (a) oligogenic, (b) polygenic, and (c) cytoplasmic.

Genetic analyses have indicated that resistance to diseases in wheat is controlled generally by dominant (and few recessive genes) and virulence in pathogen is generally due to recessive genes (or *Avris* controlled by dominant genes). Resistance and Avr result from an active interaction between active gene products from resistant host and avirulent pathogen. Wheat rust resistance genes can be divided into two categories; seedling resistance genes or all stage resistance (*R* gene) which confers resistance during the life of plant and adult plant resistance (APR) genes that normally are not expressed in seedlings but become effective as the plant reaches the adult stage. Seedling resistance genes are characterized by a hypersensitive response (HR) that includes chlorosis or necrosis surrounding the site of infection and a reduction in uredinium size. There are two classes of APR genes: (a) those that produce a HR like that found in seedling genes and which may or may not be race specific; and (b) those that confer quantitative resistance that is presumed to be non race specific. Seedling genes may confer a differential response based on the particular virulence of the race or may confer resistance that is broad spectrum and does not discriminate between races of the fungus. This is also true of the hypersensitive-type APR genes. Partial or quantitative APR is characterized by reduced receptivity, smaller uredinia, and increased latent period.

Rust resistance gene of R class mostly conforms to Flor's gene-for-gene hypothesis (Flor, 1971). Two key genes are necessary for expression of resistance; the R gene in the host and the corresponding Avr effector gene in the rust pathogen. Each R gene confers resistance to pathogen strains carrying the matching *Avr effector* gene. In other words the efficacy of R genes is pathogen strain dependent. The ability of the pathogen to overcome resistance derives from mutation of the *Avr* gene resulting in loss of recognition by the corresponding R gene. R genes have been identified from many plant species including wheat and encode receptor proteins that either directly or indirectly distinguish pathogen Avr proteins. Many *Avr* genes have been isolated from various plant pathogens and they typically encode proteins that are secreted into the host to promote infection. Some R genes in crops have been referred to as "broad spectrum" because they confer resistance to all tested pathotypes of a single pathogen species (Ellis et al., 2014).

Wheat rust resistance genes of both R and APR classes are designated *Lr, Sr* and *Yr* (for leaf, stem, and stripe or yellow rust resistance, respectively) without distinction between R or APR classes and with increasing numbers to accommodate newly discovered genes. Rust resistance genes have been identified progressively in wheat and there are currently 75, 76, and 58 genes for stripe, leaf, and stem rust resistance that have been designated (McIntosh et al., 2013, personal comm.). For powdery mildew resistance in wheat so far 53 *Pm* genes have been designated (McIntosh et al., 2014). In addition many uncharacterized resistances await further genetic investigation before formal designations can be assigned. The usual steps taken in characterizing new sources of rust resistance are: (a) determine the number of genes and mode of inheritance; (b) generate single gene lines; (c) determine the chromosomal location and genetic map position; and (d) perform allelism tests with known rust resistance genes found in the same chromosomal region.

Early studies of genetics of disease resistance were complemented with the application of aneuploid techniques resulting in a progressive elucidation of resistance loci including their characteristic low infection type chromosomal location and linkage to traits of interest to wheat breeders. Progress became more rapid with the application of molecular marker technology and the development of mapping populations that permitted replicated studies across a range of environments. A large proportion of designated rust resistance genes have been shown to be pathotype specific including seedling effective genes and APR genes. In general genes for APR confer a partial often slow rusting phenotype (Singh et al., 2011). There is frequent evolution and selection of virulence in pathogen overcoming the deployed race-specific resistance genes. Although the life of effective race-specific resistance genes can be prolonged by using gene combinations and an alternative approach being implemented at CIMMYT is to deploy varieties that possess APR based on combinations of minor slow rusting genes (Singh et al., 2014). *APR* genes individually provide low levels of resistance and combinations of three or more genes are essential to express commercially adequate levels of resistance (Bariana & McIntosh, 1993). Currently there is a view among some wheat breeders and pathologists that more importance should be placed on discovery, characterization, and use of *APR* genes for durable resistance with an implied proposition that less emphasis be given to using resistance (R) genes because their lack of durability.

10.3   SCREENING METHODOLOGY AND SELECTION ENVIRONMENTS

Epidemics of several wheat diseases are naturally occurring in many wheat growing regions of the world. Moist conditions and dew formation will promote infection of susceptible lines. In some cases irrigation is used to promote disease development. Epidemics in some regions are sporadic in occurrence and severity. This is often due to either the absence or the scarcity of inoculum. Exogenous inoculum can be applied to wheat plants in the field using a variety of methods. Disease spores are generally very hydrophobic so do not mix readily with water; however, water-based spore suspensions can be injected into elongating wheat stems to infect plants in the field without the need for exogenous moisture. In case of rust, uredospores can be mixed with carriers such as talcum powder or mineral oil. Suspensions of spores in mineral oil can be applied efficiently to plants using sprayers of various kinds; typically hand-held sprayers are used. To distinguish new resistance sources from previously identified genes whether named or unnamed controlled-environment screenings with individual isolates or pathotypes are often used.

The evaluation of resistance in adult plants is generally conducted in the field; however, greenhouse screening is also possible. Infector rows of the susceptible varieties for particular disease are planted as spreaders in different arrangements depending on the nature of the field trial and it is advisable to inoculate infectors artificially to initiate epidemics. The infector rows are planted around the block and at every 15–20 m distance in breeding nurseries where large plots of segregating populations are grown for selection. Infectors should be planted as hills on one side of each plot in the pathway and around the block to establish a uniform disease pressure. This planting arrangement is often followed for phenotyping varieties, breeding materials and mapping populations. Humidity should be ensured for initial few days after inoculation.

Selection from early generations in screening nurseries would be expected to favor major gene resistances as these are largely convenient to phenotype. Observation of transgressive segregation helps in selecting genotypes with diverse and durables disease resistance.

10.4   RESISTANCE BREEDING

10.4.1   DURABLE RUST RESISTANCE

The durability of genetic resistance against rust diseases in wheat still remains a major challenge and is of special concern to wheat breeders and farmers. The semi-dwarf and dwarf varieties developed at CIMMYT, Mexico in the early days of green revolution (Penjamo 62, Pitic 62, Lerma Rojo 64, Sanora 64, and Siete Cerros, etc.) had been responsible for yield break-through in India, Pakistan, Turkey, Afghanistan, and many other parts of the world. The life time of most of these Mexican varieties was short as appearance of new rust race ended their useful life time. The variety Lerma Rojo 64 had life time of eleven year while others like Yaqui 50, Champingo 52, and Champingo 53 retained their resistance until they were displaced from commercial cultivation by new high-yielding varieties (Borlaug, 1968). The longer life of these genotypes is attributable to their genetic background. They had combination of Hope and Thatcher type and Kenya type resistance.

Despite arguments among researchers on strategies and genetic mechanisms to achieve durable resistance, a reflection of the different host–pathogen systems, they all share the common objective of its utilization for the protection of crops. The association of durable resistance with both major and minor genes depending on different host–pathogen systems and the parasitic behavior of pathogens and their degree of host specialization has been much discussed (Parlevliet, 1993). However, there seems to be a general agreement on the utilization of quantitative resistance controlled by minor genes for achieving durable resistance particularly with heterocyclic fungi that are biotrophic such as rust on cereals.

Johnson & Law (1973) at first proposed the term of durable resistance in the context of the generalized idea that resistance expressed as a low, but positive apparent infection rate "r" was an attribute of horizontal resistance effective against all pathotypes and controlled by polygenes (Van der

Plank, 1968, 1975). However, after the British wheat cultivar Joss Cambier, which showed low infection rate to stripe rust, became highly susceptible in a short time due to a new pathotype of *P. striiformis* (Johnson & Taylor 1972), durable resistance was more specifically redefined as "the resistance that remains effective in a cultivar that is widely grown for a long period of time in an environment favorable to the disease" (Johnson, 1983).

The use of race-specific rust resistance in development of cultivars can routinely be followed with cautious approach in wheat breeding, keeping in mind the continual evolution of new races or biotypes of the pathogen. Race-specific type of resistance is controlled by genes with major to intermediate effects. Race-specific resistance to rusts and powdery mildew is often short lived. There are a number of examples of cultivars with single gene; race-specific resistance became susceptible in a short span of time. At the same time these major genes have provided long lasting resistance and therefore the use of the race-specific resistance should not be discouraged and these genes need not be dumped. Information about the type of resistance is useful in choosing the most appropriate breeding and selection methodology. In this endeavor, both race-specific and non-race-specific resistance could be combined (Tomar et al., 2014). The best known durable resistance genes in wheat are *Sr2*, a stem rust resistance gene, and *Lr34*, a gene that provides resistance to leaf and stripe rust and powdery mildew. *Lr34/Yr18/Sr57/Pm38* gene has been widely used in breeding program at CIMMYT and major wheat breeding organizations. *Sr31/Lr26/Yr9* (original source *Secale cereale* 1BL.1RS translocations in "Veery" lines) and *Sr38/ Lr37/Yr17* (original source *Aegilops ventricosa*, "VPM" lines) have provided durable resistance to many wheat cultivars. However, the resistance of *Yr9* and *Lr26* was broken after two decades of deployment in 1990s by new pathotypes detected in several parts of the world.

Race Ug99 (designated as TTKSK using North American nomenclature) detected in Uganda in 1998 (Pretorius et al., 2000) and its further evolution and spread beyond Eastern Africa pose a new threat to wheat production worldwide. Race Ug99 possesses virulence to the most of the known stem rust resistance genes that are derived from wheat and used in breeding programs worldwide (Singh et al., 2008). In addition Ug99 also possessed virulence to two additional important resistance genes *Sr31* and *Sr38*, stem rust resistant genes used worldwide. The best long-term strategy to mitigate the Ug99 threat is to identify resistant sources among existing materials or develop resistant wheat varieties that can adapt to the prevalent environments in countries under high risk and release them after proper testing while simultaneously multiplying the seed. The "Borlaug Global Rust Initiative" (www.globalrust.org) launched in 2005 is using the following strategies to reduce the possibilities of major epidemics: (a) monitoring the spread of race Ug99 beyond Eastern Africa for early warnings and potential chemical interventions; (b) screening of released varieties and germplasm for resistance; (c) distributing sources of resistance worldwide for either direct use as varieties or breeding; and (d) breeding to incorporate diverse resistance genes and APR into high-yielding adapted varieties and new germplasm. The genes which are still highly effective against Ug99 are *Sr22, Sr26, Sr28, Sr29, Sr32, Sr37, Sr39, Sr40,* and *Sr4*. Genes *Sr33, Sr45,* and *Sr46* derived from *A. tauschii* confer moderate resistance levels that are inadequate under stem rust pressure in screening nurseries in Kenya (Singh et al., 2012). The resistance gene *Sr26* from *Th. elongatum* origin translocated to chromosome 6AL in hexaploid wheat has been used successfully in Australia and remains effective despite its large-scale deployment in the 1970s and 1980s (McIntosh, 1988).Genes *Sr29, Sr32, Sr37, Sr39, Sr40,* and *Sr44* have not been tested widely for their effectiveness to other races and thus are not used in breeding. The size of the alien chromosome segments carrying these genes have to be (and are being) reduced before these genes can be used successfully. The translocation carrying resistance gene *Sr50,* previously known as *SrR,* introduced into wheat from Imperial rye in chromosomes 1BL.1RS and 1DL.1RS is effective against race Ug99 and is being used in an Australian wheat breeding program (Singh et al., 2012).

10.4.2 POWDERY MILDEW RESISTANCE

Powdery mildew resistance has been one of the top four disease resistance priorities in winter and facultative wheat breeding programs worldwide (Braun et al., 1997). The cereal powdery mildew fungi are regarded by the Fungicide Resistance Action Committee (FRAC) as plant pathogens with a high risk of developing resistance to fungicides (FRAC, 2005). Thus breeding and deployment of resistant cultivars is the most economical and environmentally friendly method to avoid fungicide applications and reduction in the yield and avoiding fungicide resistance evolution in the pathogen. The most common breeding strategy has been the use of major genes conferring hypersensitive types of resistance.

Over 45 genes that provide some level of resistance to powdery mildew have been identified. Most of these genes are race specific. Four powdery mildew resistance genes (*Pm1, Pm3, Pm4,* and *Pm5*) have more than one allele conferring resistance. In most cases the effectiveness of race-specific genes is temporal, generally less than five years in commercial cultivars. The presence of a race-specific gene selects for races that are not affected by that gene. In addition some of the highly effective genes are either associated with linkage drag or only available in unadapted backgrounds, thus limiting their usefulness in the breeding programs (Duan et al., 1998; Qiu & Zhang, 2004). Horizontal resistance to powdery mildew also referred to as "slow mildewing," "adult plant resistance," and "partial resistance" has been identified in wheat and provides breeders with a more durable type of resistance (Cowger et al., 2009). However, this type of resistance is controlled by several genes each having additive effects. With multiple genes involved it is difficult to assess and incorporate into new cultivars using traditional breeding and selection techniques.

Wheat cultivars having APR to mildew that remained effective for many years and over broad production areas include Knox and its derivatives Knox 62 (Shaner, 1973) and Massey (Griffey & Das, 1994), Genesee (Ellingboe, 1976), Redcoat (Das & Griffey, 1994), Diplomat (Chae & Fischbeck, 1979), Est Mottin (Zitelli et al., 1982), and Maris Huntsman (Bennett, 1984). The wheat cultivar Knox possesses "general (slow mildewing) resistance" to powdery mildew (Roberts & Caldwell, 1970). Shaner (1973) characterized and described the effect of APR in Knox wheat and its derivatives on powdery mildew development. Subsequently other winter and spring wheat genotypes having partial or APR to powdery mildew have been identified in breeding programs in many countries such as China (Wang et al., 2005), Hungary (Komaromi et al., 2006), Denmark, Finland, Norway, Sweden (Lillemo et al., 2010), Slovakia (Mikulova et al. 2008), and Lithuania (Liatukas & Ruzgas, 2009). Use of cultivar mixtures has been recommended to increase the diversity of resistance gene which can minimize crop loss as well as provide durability to resistance genes deployed. Manthey and Fehrmann (1993) tested the effect of wheat cultivar mixtures on powdery mildew leaf rust and stripe rust development. Infection levels were significantly reduced with the use of cultivar mixtures and the greatest reduction in disease development was observed for powdery mildew. Mixture of near isogenic lines comprising several resistance genes has also been reported to combat fast evolution of new variant of the pathogens. Isogenic lines are identical with respect to all characters except one. Backcross breeding has been used to transfer different disease resistance gene in background of a popular cultivar. Isogenic approach has been applied effectively to combat powdery mildew and rust diseases. Single, major resistance genes against wheat powdery mildew are not very durable because of virulence evolution of the pathogen. Therefore, the sustainable use of the two newly identified

resistance genes will require combination of the two genes in the same genotype or combinations with other mildew resistance genes.

The diploid "A" genome progenitor gene pool of wheat comprising the
three closely related species *T. monococcum* ssp. *monococcum, T. monococcum* ssp. *aegilopoides* (*Triticum boeoticum*), and *T. urartu* harbors useful
variability for many economically important genes including resistance to
diseases (Feldman & Sears, 1981; Singh et al., 2007). Limited backcrossing
should be exercised initially to transfer resistance genes to desirable wheat
background. Once powdery mildew resistance genes have been introgressed
into a cultivated background these can be mobilized to other backgrounds
through marker-assisted selection (MAS). Markers allow breeders to choose
the best resistance gene combinations that will provide stable resistance
against powdery mildew. Identification of DNA markers that are adjacent to
resistance genes and use these markers to select elite wheat lines and cultivars possessing multiple genes conferring durable resistance to powdery
mildew. Currently, molecular markers for more than 35 powdery mildew
resistance genes have been reported (McIntosh et al., 2013).Some of these
markers have been successfully used in map-based cloning (Yahiaoui et al.,
2004), MAS and pyramiding of the resistance genes. Leaf rust and stripe
rust resistance genes from *T. monococcum* and *T. boeoticum* have been transferred to bread wheat following the same strategy.

10.4.3   FUSARIUM HEAD BLIGHT (FHB) RESISTANCE

FHB occurs in all wheat production areas. However, a more powerful occurrence is recorded on warm and humid wheat growing areas. The resistance
is a complex phenomenon; many factors contribute to the final result. This
complexity can be problematic for breeders to consider more than one resistance trait in a program. The following seven components of physiological
head blight resistance have been reported by Mesterhazy (2001):

1. Resistance to invasion
2. Resistance to spreading
3. Resistance to toxin accumulation degradation
4. Resistance to kernel infection
5. Tolerance
6. Resistance to late blighting
7. Resistance to head death above infection site

Host resistance has been considered as the most effective and cost-efficient strategy to combat FHB. However, only limited resources of resistance are available. Resistant sources have been identified in hexaploid wheat including the Chinese cultivar Sumai 3 and its derivatives (Bai et al., 2003), several Japanese accessions (Rudd et al., 2001), the Brazilian cultivar Frontana (Singh et al., 1995), and Eastern European germplasm such as "Praag 8" (Mentewab et al., 2000). Although the durum wheat germplasm collections are generally susceptible to FHB, other tetraploid wheat subspecies within the primary gene pool of durum wheat offer an alternative source of FHB resistance (Oliver et al 2008). In addition to durum wheat tetraploid wheat with AABB genomes has seven other subspecies including Persian wheat (*T.turgidum* ssp. *carthlicum*), wild emmer wheat (*T. turgidum* ssp. *dicoccoides*), cultivated emmer wheat (*T. turgidum* ssp. *dicoccum*), Polish wheat (*T. turgidum* ssp. *polonicum L.*), Oriental wheat (*T. turgidum* ssp. *turanicum*), Georgian emmer wheat (*T. turgidum* ssp. *paleocolchicum*), and Poulard wheat (*T. turgidum* L. ssp. *turgidum*) (van Slageren, 1994). Some *Aegilops* and *Agropyron* spp. have fairly high FHB resistance. Wide-crosses can be made between *T. aestivum* and *Aegilops* spp. or between *T. aestivum* and *Agropyron* spp. to introgress this resistance (Zhuping, 1994).

The resistance to FHB is race non-specific and gives protection against several *Fusarium* species (Mesterhazy, 1999). Breeding for FBH resistance in most cases needs to test the FHB reaction. Breeding is feasible; a number of highly resistant winter wheat lines with resistance to other diseases show this. An indirect way to select plants for FHB resistance cannot be generally exercised via seedling blight as the trait occur in heading phase of the plant. Classical phenotypic selection for resistance is a successful strategy. Multilocation testing of the better advanced lines for FHB resistance is conducted in different regions where epidemics frequently occur. Spikes infected with *Fusarium graminearum* are scored using a 1-to-5 scale based on infection and disease spread as follows.

1. Disease is restricted to the inoculated (initial) spikelet and there are no symptoms on the axis;
2. The axis is infected but the second spikelet adjacent to inoculated spikelet is not;
3. The axis and second spikelet adjacent to inoculated spikelet are infected but the spike holding the inoculated spikelet is not wilting;
4. The axis and additional spikelets are infected; the spike holding the inoculated spikelet is wilting but pathogen moves slowly down the inoculated spikelet;

5. Pathogen moves quickly up and down the axis and entire spike quickly wilts.

The scores are classified as follows:

R = 1.1–2.0   MR = 2.1–3.0   MS = 3.1–4.0   S = 4.1–5.0

In general there are positive relationships between resistance and plant height . Since wheat is less susceptible prior to and after anthesis selecting lines with short flowering periods is recommended (Zhuping, 1994). Oligo-genic as well as polygenic inheritance has been reported in various sources. In case the resistance is quantitatively inherited it is better to delay selection until the F3 or F4; by then homozygotes have formed in an increased delay selection until the F3 or F4; by then homozygotes have formed in an increased proportion of the segregates. However, selection will be successful only if there is uniform and adequate disease pressure in the nursery. Genomics-assisted breeding for improvement of FHB resistance has been implemented in several breeding programs. MAS for relatively large-effect quantitative trait loci (QTL) was an effective approach to introgress known QTL into regionally adapted germplasm (Buerstmayr et al., 2013). Recent advances in genome-wide selection offer great promise to select for even small-effect QTL in breeding populations.

10.4.4 SPOT BLOTCH RESISTANCE

After the leaf rust threat was averted through international breeding efforts spot blotch emerged in the late 1980s as the most damaging foliar wheat disease in the heat-stressed areas of South and South-East Asia. The earliest record on wheat varietal resistance to spot blotch was reported by Nema and Joshi (1971) who found Sonora 64 and NP 884 that are more tolerant to spot blotch compared to other genotypes. The first crosses to incorporate spot blotch resistance were made about three decades ago. These crosses involved moderately resistant cultivars such as BH 1146 from Brazil. However, the level of resistance in progenies was inadequate when tests were carried out at Poza Rica, Mexico. In the mid-1980s, wheat genotypes carrying resis-tance to scab were obtained from the Yangtze River Valley of China also showed varying levels of spot blotch resistance when tested at Poza Rica. These Chinese lines included Suzhoe 1 to 10, Suzhoe 128-OY, Wuhan 1 to 3, Shanghai#1 to 8, and certain Ningmai, Chuanmai18, Ning 8319 Ning 8201,

Longmai, Quangfeng, and Yangmai lines. About the same time the wide-crossing program at CIMMYT produced resistant lines which contained *Thinopyrum curvifolium* in their pedigree (Villareal et al., 1995). Some of these lines and their derivatives are showing good resistance and appear to be promising in Bangladesh, lowland Bolivia, and Nepal. Current strategy followed at CIMMYT is to combine resistances from these diverse sources. Identification of some highly resistant lines from such crosses indicates that resistance is additive in nature. International collaboration contributed to the development of wheat genotypes with improved spot blotch resistance, high grain yield, and acceptable agronomic traits (Sharma & Duveiller, 2007, Khan et al., 2010a), the sources with a high level of resistance seem limited (Duveiller & Sharma, 2009). Since the sources of resistance were identified in the 1990s, genetic control of spot blotch resistance has been investigated extensively during the past 30 years. The published literature suggests that simple to complex genetic control is involved in resistance to spot blotch. A number of studies conducted in seedling stage suggested that both qualitative and quantitative inheritances were involved in conditioning resistance (Khan et al., 2010b).

Studies at seedling stage under controlled conditions do not correlate with field results due to the rapid disease progress and the difficulty in assessing small phenotypic differences among genotypes. Hence breeding and screening for spot blotch resistance must be conducted in the field in regions where the disease occurs every year. Hot spots facilitate regular monitoring for the evaluation of new pathotypes and to identify genetic diversity in potential breeding material. In many locations of South Asia field screening for spot blotch resistance is based on natural infection at hotspot locations such as found in the lowland region or "Terai" of Nepal, the Varanasi area in Uttar Pradesh (India), or in West Bengal (Joshi et al., 2007). A critical step in breeding wheat for spot blotch resistance is to evaluate the resistance traits and their characteristics.

Progress in breeding for resistance to sot blotch was reported to be slow because for various factors including low level of resistance in commercial spring wheat and wide pathogenicity spectrum of pathogen isolates. Therefore there is need to explore wild sources for resistance genes as there is poor genetic diversity for spot blotch resistance in wheat. Mujeeb-Kazi et al. (1996) exploited unique genetic variation which was contributed by different genetic sources for resistance to *B. sorokiniana* and divided the variation under three gene pools (primary, secondary, and tertiary). The most exploited species are D and E1, E2 genome donors belonging to primary and tertiary gene pools. The diploid genome donor belonging to *Triticum tauschii* accessions upon hybridization with durum wheat yielded 570 SHs of which 49 formed a set with high level of resistance to spot blotch (Mujeeb-Kazi et al., 1996). The SHs were produced to be used as bridge cross with susceptible bread wheat to generate material (BW/SH derivatives) for screening for resistance to spot blotch. Among the material categorized as resistant the lines Mayoor, Chiriya 3, and Chiriya 7 were found to be highly resistant against Indian isolates of *B. sorokiniana* (Khan et al., 2011). These lines were developed at CIMMYT, Mexico by introducing the resistance from *T. curvifolium* into synthetic hybrids.

10.5   GENETIC LINKAGES OF DISEASE RESISTANCE GENE

Genetic linkages between genes for rust resistance and gene controlling other characters or resistance to other diseases can be useful in identifying the presence of particular gene for rust resistance in unknown genotypes. In breeding program they can be used as markers for the transfer of a specific resistance gene for rust resistance. Molecular markers linked with rust resistance gene have been widely applied for molecular mapping of the gene and marker-assisted breeding. One of the most popular and perhaps most widely used linked gene segment is that transferred from *S. cereale* spontaneously. The rye segment carries genes for resistance to stem, leaf, and stripe rusts and powdery mildew (*Sr31, Lr26, Yr9, Pm8*). This alien segment is presumed to have genes influencing physiological traits leading to higher yield as evidenced by the number of varieties being cultivated world. The other useful linked gene complexes or pleiotropic are namely *Lr16/Sr23, Lr19/Sr25, Lr20/Sr15/Pm1, Lr21/Sr33/Rg2, Lr23/Sr36, Lr24/Sr24, Lr25/Pm7, Lr34/Yr18/Sr57/Pm38/Ltn, Lr35/Sr39, Lr37/Sr38/Yr17, Yr30/Lr27/Sr2, Lr46/Yr29/Pm39/Ltn2, Yr40/Lr57, Yr46/Lr67/Sr55/Ltn3, Yr47/Lr52, Sr9g/Yr7, Sr34/Yr8, Sr17/Pm5, Sr36/Pm6* (McIntosh et al., 2013). When a large number of markers are segregating simultaneously in a mapping population and these markers are to be placed on a linkage map the first step is to group the markers into linkage groups. Linkage groups are established by considering all estimates of recombination frequencies based on logarithm of odds (LOD) score. If two markers are significantly linked (by LOD value) they belong to the same linkage group. A computerized search through all pairs of markers using a certain threshold value of the LOD score will then produce a grouping of the markers.

10.6   SOURCE OF DISEASE RESISTANCE IN WHEAT

Breeding for disease resistance always requires a steady inflow of novel sources of resistance genes due to the appearance of new virulent pathogen races like stem rust race Ug99. While most resistance genes originate from hexaploid wheat, some have been introduced from related cereal species. A majority of resistance genes catalogued for rusts, powdery mildew, smuts and bunts, and many other facultative parasites appear to be race specific and follow gene-for-gene concept (Flor, 1956). Based on crossability with hexaploid wheat other related species are divided into three major gene pools, the primary gene pool; the secondary gene pool; and the tertiary gene pool (Mujeeb-Kazi & Rajaram, 2002). Many sources of resistance including the alien sources have been used for black rust resistance. Landmark beginning was made by introgression of rye (*S. cereale* L.) gene into bread wheat (1B/1R translocation or substitution) in 1973 (Mettin et al., 1973, Zeller, 1973) which carries *Lr26/Sr31/Yr9* completely linked resistance gene has not only contributed 12–20% yield jump but also imparted resistance to major biotic and abiotic stresses (Cox et al., 1995). Sometimes yield reduction is associated with alien gene introgression in wheat cultivars. For instance*Sr26* showed 9% yield penalty associated with the original 6AS.6AL-6Ae#1L segment originally introgressed into the distal region of the long arm of hexaploid wheat chromosome 6A via an alien segment from *Agropyron elongatum* (syn. *Thinopyrum ponticum*) (Knott, 1961). The *Sr26* is one of the few known major resistance genes effective against the *Sr31*-virulent race Ug99 (TTKSK) and its *Sr24*-virulent derivative (TTKST). Subsequently a number of resistance genes have been introgressed in to wheat from the alien sources (McIntosh et al., 1995 McIntosh et al., 2013).

In wide-crosses (intergeneric and interspecific), classically the self-sterile F1 hybrids, on colchicine treatment, result in fertile amphiploids that may then have practical utility. The fertile amphiploids are sources of backcross I (BCI) derivatives (amphiploid/*Triticum* sources) with eventual production of alien disomic addition lines leading to subtle alien genetic transfers by subsequent cytogenetic manipulation. To minimize yield penalty efforts should be made to transfer only small segment of alien chromosome harboring resistance gene. It is possible by repeated backcross with the recipient parent. In the initial stage crosses, when we are crossing the wild species there is problem of non crossing over which results in the transfer of whole alien chromosome. There are some genes which promote homoeologous pairing between the related genome in wheat. There seems to be no parallel to the chromosome 5B-like manipulative approach that encompasses mono-5B, *phph,* or nulli-tetrasomic stocks as the maternal wheat sources in wide-crosses. These stocks enhance wheat/alien recombinations in the F_1 hybrids and all involve the *ph* system (Sears, 1977; Sharma & Gill, 1983; Mujeeb-Kazi et al., 1984). As an alternative, since a general constraint prevails, it may be appropriate to produce the F_1 hybrid with highly crossable wheat (*PhPh*) and either back-cross or top-cross it with the *phph* stock (Sharma & Gill, 1986). Achieving high recombination is emphasized primarily because of the *T. aestivum* crop species, with its phenomenal cytogenetic flexibility via *Ph* manipulation, offers remarkable opportunities for alien gene transfers and incorporation of homoeologous segments introduced in the best location in the recipient wheat chromosomes. Introgressed segments do not always incur a yield penalty. *Thinopyrum intermedium* translocation, carrying resistance to barley yellow dwarf virus, conferred no significant reduction in grain yield, plant biomass, or grain size (thousand-grain weight) of uninfected plants (Ayala et al., 2001).

10.7   GENE, GENE SYMBOLS, AND DATABASES IN WHEAT

A catalogue of gene symbols and genetic linkages along with marker infor-mation in wheat is published annually and is available on computer desks. Updates to the catalogue with recommended rules are published in the Proceedings of the International Wheat Genetics Symposia, Annual News-letter Cereal Research Communications and Wheat Information Service. A database has been developed to list pedigrees of wheat cultivars registered with the Crop Science Society of America. Similar database of wheat has also been developed by FAO (http://www.fao.org/ag/agp/agpc/doc/field/wheat/data.htm) and CIMMYT sponsored the International Wheat Informa-tion System (IWIS) (http://apps.cimmyt.org/wpgd/Cycles.aspx) for use by wheat breeders, geneticist, and any other interested person. GrainGenes, a database for *Triticeae* and *Avena,* is a comprehensive resource for molecular and phenotypic information for wheat, barley, rye, and other related species including oat (http://wheat.pw.usda.gov/GG3/). Another web resource for wheat database and marker information for resistance gene is MAS Wheat (http://maswheat.ucdavis.edu) and has been developed jointly by USDA and the Durable Rust Resistance in Wheat (DRRW) project of Borlaug Global Rust Initiative (BGRI). Significant progress has been made in sequencing wheat genome by International Wheat Genome Sequencing Consortium (IWGSC). The IWGSC with more than 1100 members in 55 countries is an international collaborative consortium established in 2005 by a group of wheat growers, plant scientists, and public and private breeders. Sequencing of the 21 bread wheat chromosomes with more than 16 gigabases was completed in July 2014 (Eversole et al., 2014). In India wheat database has been developed by ICAR-Indian Institute of Wheat and Barley Research (ICAR-IIWBR), Karnal, India (http://indianwheatdb.com).

10.8   GENE DEPLOYMENT AND WHEAT BREEDING FOR RUST RESISTANCE IN INDIA

Indian National Wheat Rust Survey Program has been conducting rust pathotype surveys for the three rusts at national level since 1930s. The systematic surveys on wheat rusts began with the establishment of "wheat rust laboratory" (presently ICAR-IIWBR Regional Station, Shimla, India) at Shimla by Dr. K.C. Mehta in 1931. It has strengthened the national wheat breeding program regularly by providing diverse sources of resistance for introgression and preventing the release of rust susceptible lines at the national level. Gene deployment strategy has contributed significantly toward management of wheat rusts. Gene deployment has been defined as a centrally planned, properly executed strategic use of vertical resistance genes over a large area to minimize the threat of epidemic or pandemics (Nagarajan, 1984). Race-specific resistance could adequately protect spring wheat in Great Plains of United States and Canada if the same gene was not used in the South (Van der Plank, 1963). Similar schemes were then proposed for India by Reddy and Rao (1979). For deployment of genes for leaf rust resistance they suggested that three regions differing for prevalence of pathotypes of India be planted with different rust resistance genes to stabilize the pathogen. They further suggested that resorting to combination of different genes the possibility of evolving the super race would be minimized.

A breeding program to develop rust resistant varieties was initiated in 1934. After 18 years of work at Indian Agricultural Research Institute, New Delhi variety "NP 809" resistant to all the three rusts was evolved under a planned program. The problem of breeding superior rust resistant varieties by incorporating resistance available only from otherwise agronomically unsuitable exotic wheats was a very complex and long-term project (Pal, 1948). With a view to test the varieties against multiple diseases "plant pathological screening nurseries" are initiated since 1969 and are being conducted at hot spot locations. Since 1967 disease surveillance has been conducted by the All India Co-ordinated Wheat Improvement Program under ICAR-IIWBR, Karnal, India. Vertical resistance governed by single gene expressed in seedling and adult plant stages has been used predominantly in Indian wheat breeding programs due to its effectiveness, qualitative inheritance, ease of introgression and selection.

The work on rust resistance in India began in three stages. Initially the suitable resistant stocks were not available for hybridization, rust losses, were sought to be minimized by introducing rust tolerant varieties. In the second stage the attention was devoted to the production of strains possessing resistance to individual rust and in third stage, attempts have been made to achieve a combination of resistance to all the three rusts. Rust resistant sources such as Khapli, Kenya 184, Kenya Ploughman for black rust, Frontiera, Frontana, Bowie Texas, Mentana, and Tremez Molle for brown rust and Ceres Klein, Klein Cometa, Frondosa, and Spalding's Prolific for yellow rust and exotics like Gabo, Gaza, Yaqui 53, Timstein (*Lr23*) (resistant to black and brown rusts), Rio Negro, La Prevision (for brown and yellow rusts) from other countries were also used in hybridization program to introduce rust resistance in Indian varieties (Tomar et al., 2014). In the first attempt rust tolerant varieties like NP 710, NP 718, and NP 761 were developed and released. Later on the efforts were concentrated to develop genotypes with diverse resistance involving more than two parents. With the available knowledge of different types of resistance Indian wheat breeders have been using these sources and were successful in developing suitable varieties. India has not experienced frequent rust epidemics since last fifty years; it is due to the cultivation of diverse genotypes carrying different resistance genes. The most common method of wheat breeding in India is pedigree method which facilitates selection transgressive segregates for yield components while incorporating of disease resistance. Bulk method, single seed decent, modified pedigree-bulk have also been used by wheat breeders in India (Table 10.2).

In India up to early fifties all the pathotypes of *P. triticina* were avirulent on *Lr3* (Prashar et al., 2006). Thereafter use of Timstein (*Lr10* and *Lr23*) and Sonara 64 (*Lr1+*) in mid fifties paved the way for selection of new pathotypes such as 77 (45R31), 162 (93R7), 162A (93R15), 12 (5R5), and 104 (17R23) which had matching virulence for *Lr3* and/or *Lr1* and partially to *Lr10*. Table 10.3 describes the popular wheat varieties that became susceptible to one or more pathotypes of the rusts a few years after their release. Review of the documented information elucidates that most of these pathotypes have evolved through single step mutation from existing pathotypes. Predominance of virulent pathotypes, susceptible host, and favorable environment led to the epidemic in some part of the India. During 1993 brown rust epidemic occurred in about 4 million hectares in Uttar Pradesh, Punjab, and Haryana. Two best performing cultivars HD 2285 and HD 2329 became

TABLE 10.2 Popular Wheat Varieties Rendered Susceptible due to Changes in Rust Pathogens in India.

Cultivar	Year of deployment / resistance breakdown	Pathotype	Virulence
Stripe rust			
Kalyansona	1967	67S64(31) 70S4(A)	Avirulent on Kalyansona
	1971	66S-64-1(38A)	Virulent on Kalyansona avirulent on Sonalika
Sonalika	1967	66S64(14A) 70S64 (20A)	Resistant to Kalyanasona virulences
		47S102 (K)	Virulent to Sonalika avirulent to CPAN 3004 & HS 240
CPAN 3004 HS 240	1984 1989	46S119	Virulent to CPAN 3004 & HS 240
	1996	-do-	Resistant to all the pts. and avirulent to PBW343
PBW 343	1998 1996	46S119	Resistant to all the pts. and avirulent to PBW343
		78S84 identified in 2002	Virulent to PBW 343, $Yr27$
	2011		
Leaf rust			
HD 2285 HD 2329	1983	109R23 (77A-1)	Avirulent to both
	1984	125R23-1 (77-2)	Virulent to HD 2285 and HD 2329 and avirulent to PBW 299 HUW 206.
	1991	109R63(77-1)	Virulent to PBW 299 and HUW 206 and avirulent to PBW 343 type resistance to brown rust
PBW 343	1996	121R63-1(77-5) and 21R55 (104-2) identified in 1991-92	PBW 343 rendered susceptible
Stem rust			
$Sr24$	1988	62G29-1(40-1)	Resistant to black rust in India
	1989		Virulent to $Sr24$

susceptible to brown rust pts 77-2 (109R31-1) and 77-4 (125R23-1). Similarly, in yellow rust cultivars like UP 2338 and WH 542 (both having *Yr9*) became susceptible to a new pathotype 46S119 in 1996. This pathotype did not pose any challenge in Indian wheat production because PBW 343 varieties found resistant to this pathotype and having wider adaptability, was quickly adopted by farmers of North Western Plain Zone (NWPZ). PBW343 became susceptible to the 78S84 pathotypes first reported in 2000 but large-scale cultivation of this variety continued up to 2009 because of non-favorable weather for yellow rust in NWPZ. Table 10.2 shows the evolution of the new rust pathotypes which lead to withdrawal of popular wheat variety from seed production chain and cultivation. It is evident that the wheat rust pathogens have been fast evolving in India.

Different gene combination was advocated in different parts of India particularly to combat frequent appearance of new pathotypes of leaf rust. It has been suggested that by incorporating different rust resistance gene combination in various ecological area and promoting slow rusting genotypes in the source areas evolution of new pathotypes of wheat rusts can be curtailed and wheat rusts can be managed effectively. However, such an exercise requires tremendous efforts on the part of a breeder who is to cater to the requirement of yield quality, suitability, and acceptability in an area (Prashar et al., 2006). Presently identification of varieties is based partly on its resistance to the major disease in a zone and more to the yield advantage. If a variety adds to the diversity of an area and has more yield it is further promoted for release. Since there are different zones and ecological areas in the existence of pathotypes (Bhardwaj, 2013) deployment of varieties with diverse sources of resistance becomes easy. The popular Indian wheat cultivars Kalyansona, Sonalika, Lok 1, HD 2285, HD 2189, HD 2329, HD 2733, GW 322, PBW 343, PBW 373, PBW 502, RAJ 3765, RAJ 4037, and recently DPW 621-50, HD 2967, DBW 73, and DBW 88 have diverse genes for rust resistance. A list of bread wheat varieties identified by Central Varietal Release Committee (CVRC) for different zones of India during last 15 years along with their resistance is given in Table 10.3.

Some popular wheat cultivars which became susceptible to new pathotypes of resistance are being redeployed by adding additional resistance genes through marker-assisted backcross breeding. The current status of rust resistance breeding involves both conventional and molecular breeding approaches including QTL mapping. Efforts are being made since long time to utilize wild relatives to develop novel germplasm. The diversification of cultivars in a mosaic pattern in each zone intentionally or unintentionally resulted in strategic deployment of resistance genes which contained

TABLE 10.3 Bread Wheat Varieties Released between 2001–2015 in India by CVRC and Their Rust Resistance.

S. No.	Variety	Zone	Year of release	Rust resistance genes postulated		
				Lr	Sr	Yr
1.	AKAW 4627	PZ	2010	24+R	24+	—
2.	CBW 38	NEPZ	2008	10+13+	2+5+	—
3.	DBW 14	NEPZ	2002	23+	2+11+	—
4.	DBW 16	NEPZ	2005	10+13+	2+	—
5.	DBW 17	NWPZ	2006	23+26+	31+	9+
6.	DBW 39	NWPZ	2010	10+13+26+	31+	9+
7.	DBW 71	NWPZ	2012	26+	31+	9+
8.	DBW 88	NWPZ	2013	10+13+	2+11+	4+
9.	DBW 90	NWPZ	2013	3+10+13+	13+	2+
10.	DBW 93	PZ	2013	1+23+26+	2+31+	9+
11.	DBW 107	NEPZ	2014	3+26+	31+	9+
12.	DBW 110	CZ	2014	3+10+1+	-	2+
13.	DPW 621-50	NWPZ	2010	10+13+	2+	—
14.	GW 322	CZ PZ	2002	1+10+13+$	2+11+	—
15.	GW 366	CZ	2006	-	2+	—
16.	HD 2781	PZ	2002	-	2+11+	—
17.	HD 2824	NEPZ	2003	26+23+	31+	9+
18.	HD 2833	PZ	2004	R	R	—
19.	HD 2864	CZ	2004	R	8a+11	2+
20.	HD 2888	NEPZ	2005	24+R	2+24+	2+
21.	HD 2932	CZ PZ	2007	13+	-	2+

TABLE 10.3 (Continued)

S. No.	Variety	Zone	Year of release	Rust resistance genes postulated		
				Lr	*Sr*	*Yr*
22.	HD 2967	NWPZ &NEPZ	2009	*23+*	*2+8a+11+*	*2+*
23.	HD 2985	NEPZ	2009	*10+13+*	*7b+*	*2+*
24.	HD 3043	NWPZ	2011	*23+*		*2+*
25.	HD 3059	NWPZ	2012	*13+*	*2+11+*	
26.	HD 3086	NWPZ	2013	*3+10+13+*	*2+7b+*	*2+*
27.	HD 3090	PZ	2013	*1+26+*	*31+*	*9+*
28.	HD 3118	NEPZ	2014	*13+*	*9b+11+*	*2+*
29.	HI 1500	CZ	2002	*24+R*	*24+*	*2+*
30.	HI 1531	CZ	2005	*24+R*	*2+24+*	–
31.	HI 1544	CZ	2007	*24+R*	*2+24+*	*2+*
32.	HI 1563	CZ	2010	*R*	*2+R*	*2+*
33.	HPW 251	NHZ	2007	*23+26 +*	*2+31+*	*9+*
34.	HPW 349	NHZ	2012	*10z+13+*	*2+*	
35.	HS 375	NHZ	2002	*1+26+34+*	*2+5+31+*	*9+18+*
36.	HS 420	NHZ	2004	*13+10+1+34+#*	*2+*	*18+*
37.	HS 490	NHZ	2008	*23+*	*2+9b*	
38.	HS 507	NHZ	2010	*1+26+*	*31+*	*9+*
39.	HS 542	NHZ	2013	*10+13+*	*2+8a+9b+11+31*	*9+*
40.	HS 562	NHZ	2015	*23+*	*8a+9b+*	*4+*
41.	HUW 510	PZ	2001	*R*	*2+5+*	–

TABLE 10.3 (Continued)

S. No.	Variety	Zone	Year of release	Rust resistance genes postulated		
				Lr	Sr	Yr
42.	HW 2045	NEPZ	2002	–	2+	–
43.	K 0307	NEPZ	2006	1+23+	9b+11+	2+
44.	K 1006	NEPZ	2013	10+23+	8a+9b+11+	2+
45.	MACS 6145	NEPZ	2002	28+	–	–
46.	MACS 6222	PZ	2009	1+26+	2+31+	9+27+
47.	MACS 6478	PZ	2014	1+23+		2+
48.	MP 1203	CZ	2008	1+23+26+	31+	9+
49.	MP 4010	CZ	2002	24+R	2+24+	2+
50.	MP 3288	CZ	2010	24+R	24+	2+
51.	MP 3336	CZ	2012	13+	2+	2+
52.	NIAW 917	PZ	2005	1+26+	2+31+	9+
53.	NIAW 1415	PZ	2010	26+	2+31+	9+
54.	NW 2036	NEPZ	2002	23+26+	31+	9+
55.	NW 5054	NEPZ	2013	23+	7b+	2+
56.	PBW 502	NWPZ	2003	26+8	31+5+2+	9+
57.	PBW 533	PZ	2006	23+26+	31+	9+
58.	PBW 550	NWPZ	2007	26+34+	31+	9+18+
59.	PBW 590	NWPZ	2008	1+23+26	31+	9+
60.	PBW 596	PZ	2008	23+26+	31+	9+
61.	PBW 644	NWPZ	2011	1+13+	2+11+	2+R

TABLE 10.3   *(Continued)*

S. No.	Variety	Zone	Year of release	Rust resistance genes postulated		
				Lr	*Sr*	*Yr*
62.	PBW 660	NWPZ	2013	2+26+	31+	9+R
63.	RAJ 4037	PZ	2003	R	2+R	–
64.	RAJ 4083	PZ	2006	23+	2+11+	2+
65.	RAJ 4229	NEPZ	2012	R	2+5+	2+
66.	RAJ 4238	CZ	2012	24+R	24+	
67.	SKW 196	NHZ	2004	13+34+$	2+R	18+
68.	UAS 347	PZ	2014	10+13+	2+7b+11	2+
69.	VL 804	NHZ	2002	26+34+	2+5+31+	9+18
70.	VL 829	NHZ	2002	1+23+26+34+	2+5+31+	9+18+
71.	VL 832	NHZ	2003	13+34+	2+11+	18+
72.	VL 892	NHZ	2007	10+13+	2+	4+
73.	VL 907	NHZ	2010	10+23+26+34	31+	9+18+27+
74.	WH 1021	NWPZ	2007	1+26+	2+31+	9+
75.	WH 1080	NWPZ	2010	13+	2+9e+	2+
76.	WH 1105	NWPZ	2011	13+	2+11+	2+
77.	WH 1124	NWPZ	2013	10+13+	2+7b	2+
78.	WH 1142	NWPZ	2014	1+23+26+	2+31+	9+

*NHZ = Northern Hill Zone; NWPZ = North Western Plains Zone; NEPZ = North Eastern Plains Zone; PZ = Peninsular Zone; CZ = Central Zone; SHZ = Southern Hills Zone; R = Resistant; $ = adult plant resistance rust infection. The major contributor to Indian wheat breeding program is International Maize and Wheat Improvement Centre (CIMMYT) which is sharing the advanced material.

10.9 MEGA-ENVIRONMENTS (MES) ZONES OF WHEAT

Presently the term ME is used to describe the target global domains. An ME is defined as a broad frequently transcontinental though not necessarily contiguous area occurring in more than one country (Ginkel et al., 1998). It is defined by similar biotic and abiotic stresses, cropping system requirements, consumer preferences, and, for convenience, volume of production. Germplasm generated for a given ME is useful throughout it, accommodating major disease of wheat, though perhaps not all the significant secondary stresses. Over time a set of so-called agro-ecological zones was developed based on a mixture of plant, disease, and edaphic and climatic characteristics. Fifteen different agro-ecological zones had been defined by CIMMYT (Rajaram & Ginkel, 1994):

1. Subcontinent (India, Pakistan, Nepal, and Bangladesh)
2. Eastern Asia (China, Japan, and Korea)
3. Middle East (Turkey to Afghanistan)
4. North Africa and the Iberian Peninsula
5. Nile Valley (Egypt and Sudan)
6. Highlands of Eastern Africa (Ethiopia, Kenya, Tanzania, and Uganda)
7. Southern Africa
8. Northern Mexico and South Western USA
9. Highlands of Central America including Mexico and Guatemala
10. Andean countries (Bolivia, Colombia, Ecuador, and Peru)
11. Southern Cone countries of South America
12. Australia and New Zealand
13. Northern USA and Canada
14. Southern Europe
15. Western Europe

The wheat breeding program of CIMMYT has been structured to serve the germplasm needs of agro-ecological regions of 1–11. However, the breeding program utilized germplasm from all 15 regions in its hybridization efforts including regions 12–15 also. The following briefly describes the general characteristics of the spring wheat MEs with approximate acreages.

1. ME 1: Irrigated temperate; 32 million ha; 99% bread wheat.
2. ME 2: High rainfall temperate environments with an average of more than 500 mm of rainfall during the crop cycle; 10 million ha; 75% bread wheat.
3. ME 3: Acid soils pH below 5.5 with aluminum and possibly manganese toxicity; 1.7 million ha; 100% bread wheat.
4. ME 4: Low rainfall less than 500 mm of rain during crop cycle; 21.6 million ha; 67% bread wheat.
5. ME 5: High temperature; 9 million ha; 100% bread wheat.
6. ME 6: High latitude; 5.4 million ha; 100% bread wheat.

10.10 WHEAT BREEDING METHODOLOGY AT CIMMYT

Breeding efforts in the CIMMYT focus on selecting for minor genes based on APR especially for areas considered to be under high risk and where survival of the pathogen for several years is expected due to the presence of susceptible hosts and favorable environmental conditions. It is thought that this strategy will allow other areas of the world especially facultative and winter wheat growing regions to use race-specific resistance genes more successfully in their breeding programs.

The CIMMYT breeding program develops spring and facultative wheat germplasm targeting major world production zones. The shuttle breeding scheme between Mexican field sites (Ciudad Obregon in North Western Mexico during winter and Toluca or El Batan in the highlands near Mexico City during summer) initiated by Dr. N.E. Borlaug in mid. 1940s is well known for its contribution in developing widely adapted photo-insensitive and rust resistant spring bread wheat varieties that triggered "Green Revolution" in the 1960s. CIMMYT combines "shuttle breeding" and "hot spot" multilocation testing within Mexico and abroad to obtain multiple disease resistance (Ginkel et al., 1998). Shuttle breeding refers to a method where breeding generations from the same crosses are alternately selected in environmentally contrasting locations as a way of combining desirable characters. In Mexico generations are alternated between such locations as Ciudad Obregon, Sonora, an arid environment at 40 masl, and Toluca State of Mexico, a moist highland location at 2650 masl. Hot spots are locations where significant variability for a disease exists, for example highland Ecuador or Ethiopia where there are unique as well as broad virulence gene combinations for stripe rust. Screening generations in these areas and other hot spots expose the genotypes to as broad a range of virulence genes and combinations as possible. This screening together with multilocation testing increases the likelihood of developing durable resistance.

To accommodate selection for APR to Ug99 selection of segregating populations for two generations are generally included in the shuttle breeding of CIMMYT. Two crop seasons per year in both Mexico and Kenya halve the number of years required to generate and test advanced breeding lines. The "single-backcross selected-bulk" breeding approach (Singh & Trethowan, 2007) is applied for transferring multiple minor genes for rust resistance to adapted and high-yielding wheat backgrounds. Parental stocks are derived from many sources worldwide. Generally a set of widely adapted cultivars with high yield performance and stability is crossed with another set of cultivars or genotypes having superior combinations of disease resistance; all are identified through multilocation testing and evaluation in either ME 2 or ME 5. This strategy has resulted in the development of high-yielding wheat lines with high levels of APR.

A breeding procedure known as modified bulk/pedigree is commonly used at CIMMYT. Crosses may be single (A × B), three-way (A × B × C), or limited backcross (A × B × A). Only the best F_1 lines are promoted to F_2. Within the better F_2 populations the best plants are selected by the breeders and experienced research assistants during 3–5 rounds of selection based on good agronomic type, "durable" disease resistance, synchronous tillering, desired spike type, good fertility, and appropriate height and maturity. In F_3 to F_5 lines are identified for phenotypic and disease resistance superiority and then 10 heads are harvested and bulked. In F_6 10 single heads are again selected but planted individually as F_7 lines. Bulk selections are made in F_7 and those with appropriate characteristics are put into yield experiments. The overriding philosophy in the breeding program is a broad-based recurrent selection system where the best genotypes are cycled back into the crossing program based on behavior at many locations. This system retains line resistant or tolerant to important diseases.

Simple and three-way crosses where one or more parents carry APR are being used to breed new high-yielding near-immune wheat materials to all three rusts. High emphasis is given to test a large number of advanced lines for grain yield potential performance in Cd. Obregon, Mexico. Grain yield performance testing during the 2nd year for the selected lines from the 1st year yield trials is done under five diverse environments in Cd. Obregon. While conducting the 1st and 2nd year of yield testing the materials are also characterized for resistance to Ug99 stem rust and yellow rust in Kenya, yellow rust and Septoria leaf blight in Toluca (Mexico), yellow rust in Santa Catalina (Ecuador), leaf rust in El Batan and Cd. Obregon (Mexico), spot blotch (Agua Frias Mexico), FHB (El Batan, Mexico), Karnal bunt in Cd. Obregon (Mexico), etc. End-use quality analyses are also conducted for lines retained after the 1st and 2nd year of yield performance testing. The best entries identified from the above trials are then distributed through various international yield trials such as elite spring wheat trial (ESWYT) and screening nurseries such as International Bread Wheat Screening Nursery (IBWSN). These trials and nurseries carry improved wheat materials that are likely to have adaptation in about 60 million ha of spring wheat growing areas in Africa, Middle East, Asia, and Latin America. The best Ug99 stem rust resistant materials identified through screenings at Njoro (Kenya) are distributed annually through Stem Rust Resistance Screening Nurseries (SRRSN) (Singh et al., 2014).

10.11   BREEDING FOR MULTIPLE DISEASE RESISTANCE

It is assumed that resistant genes effective against prevailing races of a particular region, the chances of yield losses are less than those varieties which are not carrying these resistance genes. Chances of breakdown of resistant varieties are more if only one or two resistance genes are present. Enhanced resistance can be achieved by pyramiding some other resistance genes in addition to the existing one. Incorporation of 3–4 prominent resistant genes conferring resistance against leaf and stem rusts and spot blotch diseases results in less chances of breakdown of resistance and the duration of efficient resistance can be increased and multiple disease resistance can be achieved (Singh *et al.,* 2000). Use of combinations of genes which could be effective at different stages irrespective of whether, they are major or minor, has been suggested as the best method for genetic control of rusts (Roelfs, 1988). It has been suggested that durable resistance to rust may be obtained by combining seedling specific genes with APR. Though this can be achieved by pyramiding effective resistance genes, these are difficult to monitor in the field due to the inability to distinguish the expression of individual resistance genes or due to lack of availability of virulence in the pathogen to differentiate the genes (Fig. 10.1).

Bariana et al. (2007) advocated that three or four lines carrying different minor and major genes for rust resistance can be crossed (three-way and four-way crosses) using conventional and molecular marker technologies. Plants in large segregating populations should be selected under artificially created disease epidemics or at hot spot location. Races of pathogens that have virulence for race-specific resistance genes present in the parents should be used to create the epidemics (Singh *et al.,* 2000). The experience of breeders (Dubin & Ginkel, 1991; Duveiller & Gilchirst, 1994; Dubin & Rajaram, 1996) to achieve partial resistance in breeding populations suggested utilization of polygenic type of resistance for leaf blight complex and tan spot. Breeding for durable resistance based on minor additive genes has been challenging and often slow for following reasons.

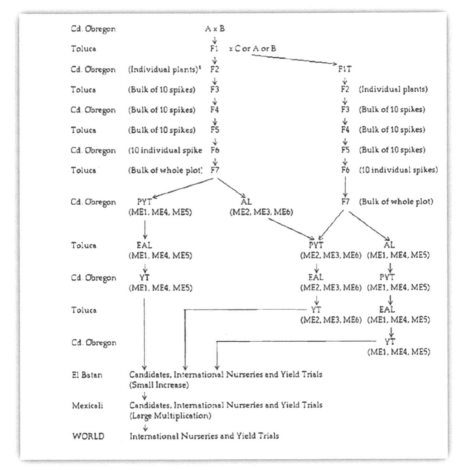

FIGURE 10.1   Scheme of modified pedigree bulk selection practiced at CIMMYT (modified from original source: Guide to the CIMMYT wheat breeding program Ginkel et al., 1998).

1. A sufficient number of minor genes may not be present in a single source genotype.
2. A source genotype may be poorly adapted.

3. There may be confounding effects from the segregation of both major and minor genes in the population.
4. Crossing and selection schemes and population sizes are more suitable for selecting major genes.
5. Reliable molecular markers for several minor genes are unavailable.
6. The cost associated with identifying and utilizing multiple markers is high.

However, germplasm carrying combinations of minor genes would be very useful in transferring these genes to adapted local cultivars. With the advent of molecular marker technology it is now possible to tackle such complex problems. Now selections can be made more efficiently with a process called MAS where molecular markers close to genes of interest are used to assist breeders in selecting the best gene combinations. MAS is seen as complementing the existing screening for yield quality and disease resistance in the breeding program. Molecular marker technologies offer a range of novel approaches to improve the efficiency of a breeding program. It is also important to remember that markers are just one of many screening tools available to plant breeders and that different tools may be appropriate for different traits or stages of the breeding program. DNA-based molecular markers have several advantages over the traditional phenotype based selection especially when disease-escaped-susceptible plants are likely to be confused for resistant plants. Further in the case of APR pyramiding strategy its detection is not possible in the presence of a seedling resistance gene due to masking effect. In India the seedling resistance gene *Lr24/Sr24, Yr5, Yr10, Yr15, Sr26, Sr32, Lr35/Sr39* and the APR gene *Yr17/Lr37/Sr38 Yr18/Lr34/Sr57 Yr29/Lr46 Yr46/Lr67/Sr55, Lr48* and *Sr2* have been reported to be effective against the pathotypes of the sub-continent and molecular markers linked to these genes have been identified and validated.

Availability for tightly linked DNA markers can also be useful in maintaining and diversifying the combinations of additive slow rusting resistance genes in the wheat germplasm and cultivars. DNA markers have been used extensively in wheat to do genome mapping and to construct linkage maps. Linkage maps provide information on chromosomal regions that contain major genes and QTL. Microsatellite or simple sequence repeats (SSR), sequenced-tagged sites (STS), cleaved amplified polymorphic sites (CAPS), sequence characterized amplified region (SCAR), and single nucleotide polymorphisms (SNPs) are now popularly being used because of their robustness. A number of rust resistance genes have been targeted for DNA marker development because of their potential value to wheat breeding.

10.12 CONCLUDING REMARKS

The effective control of the wheat rusts and other important disease has been possible largely due to the exchange of improved germplasm as a vital source of genetic variation. Breeding program for disease resistance still requires strengthening in the perspective of climate change and evolution of further virulent races in the wheat pathogens. Pathogens can mutate to overcome the existing resistance genes especially when common genotypes are released simultaneously in adjoining counties and MEs. Therefore breeding for disease resistance has to be more focused and new sources of resistance in wheat may be used. The maintenance of high levels of resistance involves a continuing search for, and development of, new forms or combinations of resistance to ensure that the varieties in farmers' fields have effective genetic resistance against the current strains of the pathogens. The wheat breeders and pathologists have to work hand in hand to breed new varieties equipped with effective gene combination to encounter any possible threat of newly evolved pathotypes. To ensure that diseases do not cause significant economic losses, the wheat breeders need to have an understanding of degree of cultivar susceptibility, dynamics of epidemiology, emergence of new pathotypes, and the resistance genes available for use. Interdisciplinary approach, where breeders work in tandem with the pathologists, agronomists, physiologist, and social scientists, is the need of the hour. Gene management in field has considerable promise for sustained control of rusts. The impetus on the high-yielding cultivars must continue and the emphasis on disease resistance may be even higher. Appropriate management of the host resistance genes through deployment of resistant cultivars in different geographical regions and gene pyramiding are the suitable strategies to increase the durability of resistance. Use of durable resistance will be an excellent strategy for providing agronomically acceptable option to farmers. Broader application of the genes having minor phenotypic effect would be a way to durable resistance. Further it is most important that a proper balance is created in providing appropriate level of resistance to other important diseases like powdery mildew, leaf blight complex, Karnal bunt, etc. along with rust resistance so that dividends due to containment of rusts are successfully passed to wheat growers and other stakeholders in wheat industry.

KEYWORDS

- disease resistance in wheat
- durable resistance
- breeding
- resistance genes
- rust resistance
- powdery mildew resistance
- spot blotch resistance
- fusarium head blight resistance

REFERENCES

Ayala, L.; van Ginkel, M.; Khairallah, M.; Keller, B.; Henry, M. Expression of *Thinopyrum intermedium*-Derived Barley Yellow Dwarf Virus Resistance in Elite Bread Wheat Backgrounds. *Phytopathology.* **2001,** *91,* 55–62.

Bai, G. H.; Chen, L. F.; Shaner, G. E. Breeding for Resistance to Fusarium Head Blight of Wheat in China. In *Fusarium Head Blight of Wheat and Barley;* Leonard, K. J., Bushnell, W. R., Eds.; APS Press: St. Paul, MN, 2003; pp 296–317.

Bariana, H. S.; McIntosh, R. A. Cytogenetic Studies in Wheat XIV. Location of Rust Resistance Genes in VPM1 and their Genetic Linkage with other Disease Resistance Genes in Chromosome 2A. *Genome.* **1993,** *36,* 476–482.

Bariana, H. S., et al. Breeding Triple Rust Resistant Wheat Cultivars for Australia Using Conventional and Marker-Assisted Selection Technologies. *Aust. J. Agric. Res.* **2007,** *58*(6), 576–587.

Bennett, F. G. A. Resistance to Powdery Mildew in Wheat: A Review of its Use in Agriculture and Breeding Programmes. *Plant Pathol.* **1984,** *33,* 279–300.

Bhardwaj, S. C. *Puccinia -Triticum* Interaction, an Update. *Indian Phytopathol.* **2013,** *66*(1), 14–19.

Biffen, R. H. Mendel's Law of Inheritance and Wheat Breeding. *J. Agri. Sci.* **1905,** *1,* 4–48.

Borlaug, N. E. In *Wheat Breeding and its Impact on World Food Supply,* Proceedings of the 3rd International Wheat Genetics Symposium, Canberra, Aug 5–9,1968; The Australian Academy of Science: Canberra, 1968; pp 1–36.

Braun, H. J., et al. Breeding Priorities of Winter Wheat Programs. In *Wheat, Prospects for Global Improvement;* Braun, F., et al., Eds.; Kluwer Academic Publishers: Dordrecht, The Netherlands, 1997; Vol. 6, pp 553–560.

Buerstmayr, H.; Buerstmayr, M.; Schweiger, W.; Steiner, B. Genomics-Assisted Breeding for Fusarium Head Blight Resistance in Wheat. In *Translational Genomics for Crop Breeding, Biotic Stress;* Varshney, R. K., Tuberosa, R., Eds.; John Wiley & Sons Ltd.: Chichester, 2013; Vol. 1, doi, 10.1002/9781118728475.ch4.

Chae, Y. A.; Fischbeck, G. W. Genetic Analysis of Powdery Mildew Resistance in Wheat Cultivar 'Diplomat'. *Zeitschrift fuer Pflanzenzuechtung.* **1979**, *83,* 272–280.

Cowger, C.; Parks, R.; Marshall, D. Appearance of Powdery Mildew of Wheat Caused by *Blumeria graminis* f. sp. *tritici* on *Pm 1 7*-Bearing Cultivars in North Carolina. *Plant Dis.* **2009**, *93,* 1219.

Cox, T.; Gill, B. S.; Sears, R. G. Notice and Release of KS94WGRC32 Leaf Rust Resistant Hard Red Winter Wheat Germplasm. *Ann. Wheat Newslett.* **1995**, *41,* 241.

Das, M. K.; Griffey, C. A. Heritability and Number of Genes Governing Adult-Plant Resistance to Powdery Mildew in Houser and Redcoat Winter Wheats. *Phytopathology.* **1994,** *84,* 406–409.

Duan, X.;Sheng, B.; Zhou, Y.; Xiang, Q. Monitoring of the Virulence Population of *Erysiphe graminis* f. sp. *tritici. Acta Phytophylac. Sin.* **1998,** *25,* 31–36.

Dubin, H. I.; Ginkel M. V. In *The Strains of Wheat Diseases and Disease Research in Warmer Areas,* Proceedings of the International Conference: Wheat for the Non-Traditional Warm Areas, Mexico, July 29–Aug 3, 1990; Saunder D. A., Ed.; CIMMYT: Mexico DF,1991; pp 125–145.

Dubin, H. J.; Rajaram, S. Breeding Disease Resistant Wheat for Tropical Highlands and Lowlands. *Annu. Rev. Phytopathol.* **1996,** *34,* 503–526.

Duveiller, E.; Gilchrist, L. In *Productions Constraints due to Bipolaris sorokiniana in Wheat, Current Situation and Future Prospects, Proceeding of Wheat in the Warmer Areas Rice/ Wheat Systems,* Nashipur, Dinajpur, Bangladesh, Feb 13–16, 1993; Saunders, D., Hettel, G., Eds.; CIMMYT/UNDP: Mexico DF,1994; pp 343–352.

Duveiller, E.; Sharma, R. C. Genetic Improvement and Crop Management Strategies to Minimize Yield Losses in Warm Non-Traditional Wheat Growing Areas Due to Spot Blotch Pathogen *Cochliobolus sativus. J. Phytopathol.* **2009,** *157,* 521–534.

Ellingboe, A. H. Genetics of Host-Parasite Interactions. In *Physiological Plant Pathology;* Heitefuss R., Williams, P. H., Eds.; Springer-Verlag: Berlin, Heidelberg, 1976; pp 761–778.

Ellis, J. G.; Lagudah, E. S.; Spielmeyer, W.; Dodds, P. N. The Past, Present and Future of Breeding Rust Resistant Wheat. *Frontier Plant Sci.* **2014,** *5*(641), 1–13.

Eversole, K., et al. Slicing Wheat Genome. *Science.* **2014,** *345,* 285–287.

Feldman, M.; Sears, E. R. The Wild Gene Resources of Wheat. *Sci. Am.* **1981,** *2*(44)*,* 98–109.

Flor, H. H. Current Status of the Gene-for-Gene Concept. *Ann. Rev. Phytopath.* **1971,** *9,* 275–296.

Flor, H. H. Host Parasite in Flax Rust- its Genetics and other Implications. *Phytopathology.* **1955,** *45,* 680–685.

Flor, H. H. The Complementary Genic Systems in Flax and Flax Rust. *Adv. Genet.* **1956,** *8,* 2954.

FRAC. *Pathogen Risk List;* Fungicide Resistance Action Committee: Limburgerh of, Germany, 2005; pp 1–5.

Ginkel, M.; Trethowan, R.; Cukadar, B. *A Guide to the CIMMYT wheat Breeding Program;* CIMMYT: Mexico DF, 1998; pp 16–33.

Griffey, C. A.; Das, M. K. Inheritance of Adult-Plant Resistance to Powdery Mildew in Knox 62 and Massey Winter Wheats. *Crop Sci.* **1994,** *34,* 641–646.

Johnson, R.; Law, C. N. Cytogenetic Studies of the Resistance of the Wheat Variety Bersée to *Puccinia striformis. Cereal Rusts Bull.* **1973,** *1,* 38–43.

Johnson, R.; Taylor, A. J. Isolates of *Puccinia striiformis* Collected in England From the Wheat Varieties Maris Beacon and Joss Cambier. *Nature.* **1972,** *23,* 105–106.

Johnson, R. Genetic Background of Durable Resistance. In *Durable Resistance in Crops.* Lamberti, F., et al., Eds.; Plenum Press: New York, 1983; pp 5–26.

Joshi, A. K.; Ortiz-Ferrara, G.; Crossa, J. et al. Associations of Environments in South Asia Based on Spot Blotch Disease of Wheat Caused by *Cochliobolus sativus. Crop Sci.* **2007,** *47,* 1071–1081.

Joshi, L. M.; Srivastava K. D.; Singh D. V. Monitoring of Wheat Rust in Indian Subcontinent. *Proc. Indian Acad. Sci. (Plant Sci.).***1985,** *94*(2), 387–406.

Khan, H.; Chowdhury, S. Identification of Resistance Source in Wheat Germplasm against Spot Blotch Disease Caused by *Bipolaris sorokiniana. Archives Phytopath. Pl. Protec.* **2011,** *44*(9), 840–844.

Khan, H.; Tomar, S. M. S.; Chowdhury, S. Genetic Analysis of Resistance to Spot Blotch (*Bipolaris sorokiniana*) in Wheat. *Indian J. Genet. Pl. Br.* **2010a,** *70*(1), 11–16.

Khan, H.; Tomar, S. M. S.; Chowdhury, S. Inheritance Studies on Spot Blotch of Wheat Caused by *Bipolaris sorokinina. Indian J. Genet. Pl. Br.* **2010b,** *70*(3), 229–233.

Knott, D. R. Genetic Analysis of Resistance. In *The Wheat Rusts-Breeding for Resistance;* Springer-Verlag: Berlin, Heidelberg, 1989; pp 58–82.

Knott, D. R. The Inheritance of Rust Resistance. IV. The Transfer of Stem Rust Resistance from *Agropyron elongatum* to Common Wheat. *Can J. Plant Sci.* **1961,** *41,* 109–123.

Komaromi, J., et al. Identification of Wheat Genotypes with Adult Plant Resistance to Powdery Mildew. *Cereal Res. Comm.* **2006,** *34,*1051–1058.

Liatukas, Z.; Ruzgas, V. *Powdery Mildew Resistance of the Lithuanian Winter Wheat Breeding Material;* Latvian Academy of Sciences: Riga, Latvian, 2009; Section B No. 1/2, (660–661), pp 37–44.

Lillemo, M.; Skinnes, H.; Brown, J. K. M. Race-Specific Resistance to Powdery Mildew in Scandinavian Wheat Cultivars Breeding Lines and Introduced Genotypes with Partial Resistance. *Plant Breed.* **2010,** *129,* 297–303.

Manthey, R.; Fehrmann, H. Effect of Cultivar Mixtures in Wheat on Fungal Diseases, Yield and Profitability. *Crop Prot.* **1993,** *12,* 63–68.

McIntosh, R. A. The Role of Specific Genes in Breeding for Durable Stem Rust Resistance in Wheat and Triticale. In *Breeding Strategies for Resistance to the Rust of Wheat;* Simmonds, N. W., Rajaram, S., Eds.; CIMMYT: Mexico DF, 1988; pp 1–9.

McIntosh R. A., et al. In *Catalogue of Gene Symbols for Wheat, 2013–2014 Supplement,* 12th International Wheat Genetics Symposium, Yokohama, Japan, Sept 8–13,2014; pp 8–9. http,//www.shigen.nig.ac.jp/wheat/komugi/genes/macgene/supplement201–2014.pdf

McIntosh R. A.; Wellings C. R.; Park R. F. *Wheat Rust- an Atlas of Resistance Genes;* CSIRO Publications: Canberra, Australia,1995; pp 9–11.

McIntosh, R. A., et al. In *Catalogue of Gene Symbols for Wheat,*12th International Wheat Genetics Symposium, Yokohama, Japan, Sep 8–13,2013; pp 155–186.

Mentewab, A., et al. Chromosomal Location of Fusarium Head Blight Resistance Genes and Analysis of the Relationship between Resistance to Head Blight and Brown Foot Rot. *Plant Breed.* **2000,** *119,* 15–20.

Mesterhazy, A., et al. Nature of Wheat Resistance to Fusarium Head Blight and the Role of Deoxynivalenol for Breeding. *Plant Breed.* **1999,** *118*(2), 97–110.

Mesterhazy A. Breeding for Fusarium Head Blight Resistance in Wheat. In *Wheat in Global Environment of the Series Developments in Plant Breeding;* Springer: Berlin, Heidelberg, 2001; Vol. 9, pp 353–358.

Mettin, D.; Bluthner, W. D.; Schlegel, R. In *Additional Evidence on Spontaneous 1B/1R Wheat Rye Substitutions and Translocations,* Proceedings of the 4th International Wheat

Genetics Symposium, Missouri, Columbia, Aug 6–11, 1973; Sears, E. R., Sears, L. M. S., Eds.; Agricultural Experimental Station, University of Missouri: Columbia, MO, 1973; pp 179–184.

Mikulova, K.; Bojnanska, K.; Cervend, V. Assessment of Partial Resistance to Powdery Mildew in Hexaploid Wheat Genotypes. *Biologia.* **2008,** *63,* 477–481.

Mujeeb-Kazi, A.; Pena, R. J.; Delgado, R. Registration of Five Wheat Germplasm Lines Resistant to *Helminthosporium* Leaf Blight. *Crop Sci.* **1996,** *36,* 216–217.

Mujeeb-Kazi, A.; Rajaram, S. Transferring Alien Genes from Related Species and Genera for Wheat Improvement. In *Bread Wheat, Improvement and Production. Plant Produced Protection* Series No. 30 Curtis, B. C, Rajaram, S., Gomez, M. H., Eds.; FAO: Rome, 2002.

Mujeeb-Kazi, A.; Roldan, S.;Miranda, J. L. Intergeneric Hybrids of *Triticum aestivum* with *Agropyron* and *Elymus* Species. *Cereal Res. Commun.* **1984,** *12,* 75–79.

Nagarajan, S.; Nayar, S. K.; Bahadur, P. Contemplating the Management of Brown Rust Resistance Genes to Mitigate the Spread of *Puccinia recondita tritici. Indian Phytopathol.* **1984,** *37,* 490–497.

Nema, K. G.; Joshi, L. M. Flag Leaf Susceptibility of Wheat to *Helminthosporium sativum* in Relation to Grain Weight. *Indian Phytopathol.* **1971,** *24,* 526–532.

Oliver, R. P. A Reassessment of the Risk of Rust Fungi Developing Resistance to Fungicides. *Pest Manage. Sci.* **2014,** *70*(11), 1641–1645.

Oliver, R. E., et al. Evaluation of Fusarium Head Blight Resistance in Tetraploid Wheat (*Triticum turgidum* L.). *Crop Sci.* **2008,** *48,* 213–222.

Pal, B. P. The Control of Wheat Rust by Breeding. *J. Indian Bot. Soc.* **1948,** *27,*169–186.

Parlevliet, J. E. What is Durable Resistance a General Outline. In *Durability of Disease Resistance;* Jacobs, T., Parlevliet, J. E., Eds.; Kluwer Academic Publishers: The Netherlands, 1993; pp 23–29.

Person, C. O. Gene-for-Gene Relationships in Parasitic Systems. *Can. J. Bot.* **1959,** *37,* 1101–1130.

Prashar, M.; Bhardwaj, S. C.; Jain, S. K. Gene Deployment in Wheat Rust Management. In *Plant Protection in New Millennium*; Satish Serial Publishing House: New Delhi, India, 2006; pp 165–179.

Pretorius, Z. A.; Singh, R. P.; Wagoire, W. W.; Payne, T. S. Detection of Virulence to Wheat Stem Rust Resistance Gene *Sr31* in *Puccinia graminis* f. sp. *tritici* in Uganda. *Plant Dis.* **2000,** *84,* 203.

Qiu,Y. C.; Zhang, S. S. Researches on Powdery Mildew Resistant Genes and their Molecular Markers in Wheat. *J. Triticeae Crops.* **2004,** *24,*127–132.

Rajaram, S.; Ginkel, M. *A Guide to the Bread Wheat Section*; Wheat Special Report No 5, CIMMYT: Mexico DF, 1998.

Reddy, M. S. S.; Rao, M. V. Resistance Genes and their Deployment for Control of Leaf Rust of Wheat. *Indian J. Genet. Plant Breed.* **1979,** *39,* 359–363.

Roberts, J. J.; Caldwell, R. M. General Resistance (Slow Mildewing) to *Erysiphe graminis* f. sp. *tritici* in 'Knox' Wheat (Abstract). *Phytopathology.* **1970,** *60,* 1310.

Roelfs, A. P. Genetic Control of Phenotypes in Wheat Stem Rust. *Annu. Rev. Phytopathol.* **1988,** *26,* 351–367.

Rudd, J. C., et al. Host Plant Resistance Genes for Fusarium Head Blight, Sources Mechanisms and Utility in Conventional Breeding Systems. *Crop Sci.* **2001,** *41,* 620–627.

Sears, E. R. An Induced Mutant with Homoeologous Pairing in Wheat. *Can. J. Genet. Cytol.* **1977,** *19,* 585–593.

Shaner, G. Evaluation of Slow-Mildewing Resistance of Knox Wheat in the Field. *Phytopathology*. **1973**, *63*, 867–872.

Sharma, H. C.; Gill, B. S. New Hybrids between Agropyron and Wheat. 2. Production, Morphology and Cytogenetic Analysis of F1 Hybrids and Backcross Derivatives. *Theor. Appl. Genet.* **1983**, *66*, 111–121.

Sharma, H. C.; Gill, B. S. The Use of *phI* Gene in Direct Transfer and Search for Ph-like Genes in Polyploid *Aegilops* Species. *Z. Pflanzenzucht.* **1986**, *96*, 1–7.

Sharma, R. C.; Duveiller, E. Advancement toward New Spot Blotch Resistant Wheats in South Asia. *Crop Sci.* **2007**, *47*, 961–968.

Singh, K., et al. Molecular Mapping of Leaf and Stripe Rust Resistance Genes in *Triticum monococcum* and their Transfer to Hexaploid Wheat. In *Wheat Production in Stressed Environments;* Buck, H., Nisi, J. E., Solomon, N., Eds.; Springer: Netherlands, 2007; pp 779–786.

Singh, R. P.; Trethowan, R. Breeding Spring Bread Wheat for Irrigated and Rainfed Production Systems of Developing World. In *Breeding Major Food Staples;* Kang, M., Priyadarshan, P. M., Eds.; Blackwell Publishing: Ames, IA, 2007; pp 109–140.

Singh, R. P., et al. Race Non-Specific Resistance to Rust Diseases in CIMMYT Spring Wheats. *Euphytica.* **2011**, *179*, 175–186.

Singh, R. P., et al. Progress towards Genetics and Breeding for Minor Genes Based Resistance to Ug99 and other Rusts in CIMMYT High Yielding Spring Wheat. *J. Integr. Agric.* **2014**, *13*(2), 255–261.

Singh, R. P.; Huerta-Espino, J.; Rajaram, S. Achieving Near-Immunity to Leaf and Stripe Rusts in Wheat by Combining Slow Rusting Resistance Genes. *Acta Phytopathol. Entomol. Hung.* **2000**, *35*, 133–139.

Singh, R. P.; Ma, H.; Rajaram, S. Genetic Analysis of Resistance to Scab in Spring Wheat Cultivar Frontana. *Plant Dis.* **1995**, *79*, 238–240.

Singh, S.; Singh, R. P.; Huerta-Espino, J. Stem Rust. In *Disease Resistance in Wheat;* CABI Plant Protection Series: Wallingford, CT, 2012; pp 20–32

Singh, R. P., et al. Will Stem Rust Destroy the World's Wheat Crop? *Adv. Agronomy.* **2008**, *98*, 271–309.

Tomar, S. M. S.; Singh, S. K.; Sivasamy, M.; Vinod. Wheat Rusts in India, Resistance Breeding and Gene Deployment-A review. *Indian J. Genet.* **2014**, *74*(2), 129–156.

Van der Plank, J. E. *Principles of Plant Infection;* Academic Press: New York,1975; p 216.

Van der Plank, J. E. *Disease Resistance in Plants;* Academic Press: New York,1968; p 206.

Van der Plank, J. E. *Plant Diseases, Epidemics and Control;* Academic Press: New York,1963; p 349.

van Slageren, M. W. *Wild Wheats, A Monograph of Aegilops L. and Amblyopyrum (Jaub. & Spach) Eig (Poaceae);* Wageningen Agriculture Univ Wageningen: The Netherlands,1994; pp 94–97.

Villareal, R. L.; Mujeeb-Kazi, A.; Gilchrist, L. I.; Del, Toro E. Yield Losses to Spot Blotch in Spring Bread Wheat in Warm Nontraditional Wheat Production Areas. *Plant Dis.* **1995**, *79*, 893–897.

Wang, Z. L., et al. Seedling and Adult Plant Resistance to Powdery Mildew in Chinese Bread Wheat Cultivars and Lines. *Plant Dis.* **2005**, *89*, 457–463.

Yahiaoui, N.; Srichumpa, P.; Dudler, R.; Keller, B. Genome Analysis at Different Ploidy Levels Allows Cloning of the Powdery Mildew Resistance Gene *Pm3b* from Hexaploid Wheat. *Plant J.* **2004**, *137*, 528–538.

Zeller, E. J. In*1B/1R Wheat Rye Chromosome Substitutions and Translocations,* Proceedings of the 4th International Wheat Genetics Symposium,Missouri, Columbia, 1973; Sears, E. R., Sears, L. M. S., Eds.; Agricultural Experiment Station, University of Missouri: Columbia, MO, 1973; pp 209–211.

Zhuping, Y. *Breeding for Resistance to Fusarium Head Blight of Wheat in the Mid- to Lower Yangtze River Valley of China;* Institutional Multimedia Publications Repository, CIMMYT: Mexico DF, 1994; pp 4–7.

Zitelli, G.; Pasquini, M.; Gras, M. A. Resistenza alle Malattie di una Varieta Italiana di Frumento Tenero e Sua Probabile Origine. *Genetica Agraria.* **1982,** *36,* 194–196.

CHAPTER 11

HOST RESISTANCE TO SPOT BLOTCH (*BIPOLARIS SOROKINIANA*) IN WHEAT AND BARLEY

DEVENDRA PAL SINGH*

ICAR-Indian Institute of Wheat and Barley Research, Karnal 132001, Haryana, India

Corresponding author. E-mail: dpkarnal@gmail.com

CONTENTS

ABSTRACT

Spot blotch caused by a hemiobiotorphic fungus, *Bipolaris sorokiniana* is quite important disease of wheat in warmer and humid climate of South Asia and other parts of the world. It is the major cause of leaf blight in India. The disease is capable of causing yield losses up to 50% in susceptible varieties and also in seed viability as well as seed discolouration. The seed infected due to *B. sorokiniana* carries the infection besides crop residue, weeds, and soil. The disease affects the wheat crop from seed germination till seed development and causes symptoms on seed, seedlings, roots, leaves, leaf sheath, and spike. The infected leaves initially develop minute necrotic spot which may or may not be surrounded by yellow halo and presence of halo and its intensity depends on the level of resistance to spot blotch in wheat genotypes. The pathogen also produces "helminthosporol" toxin which is capable to produce necrotic symptoms. Although disease may be managed using fungicidal seed treatment and foliar sprays, but deployment of resistant cultivars remained the most popular method of disease management. During last two decades, research is carried out on pathogenic variability, creation of disease epiphytotics, rating scale, and unveiling the mechanism of host resistance in case of spot blotch. The chapter summarize the information on host resistance against spot blotch caused by *B. sorokiniana*.

11.1   INTRODUCTION

Spot blotch caused by *Bipolaris sorokiniana* (Sacc.) Shoemaker is now treated at par with leaf rust in terms of reducing wheat and barley yields. The disease causes losses in yield up 51% and deteriorate quality of grains in warmer regions of South Asia and other countries of the world in wheat as well as in Barley (Akram et al., 2003; Singh et al., 2002, 2004, 2007, 2011, 2014; Malik et al., 2008b; Singh, 2003, 2004). Besides, spot blotch or leaf blight is also causes foot rot and black point, or seed discoloration (Singh & Kumar, 2008). The pathogen is quite harmful to the seed viability and is the major cause of poor germination in wheat, and barley produced in warmer regions. Although management of spot blotch is worked out using fungicides and botanicals, but it is not much adopted in commercial grain crop (Malik et al., 2008a; Singh et al., 2005, 2007, 2008; Singh, 2014). Therefore, the cultivation of wheat and barley is not possible in warmer and humid climate without proper resistance to spot blotch along with leaf and stem rusts in wheat and barley varieties. The research on looking for sources of resistant (R) and breeding for spot blotch resistance is of recent phenomenon as compared to rusts. It is due to the fact that wheat and barley are traditionally crops of cooler regions and spot blotch occurs in warmer areas in developing countries. In recent years, good advancement in research on resistance to spot blotch took place due to increasing demand of extra wheat from limited land in traditional wheat growing areas and spreading wheat and barley in nontraditional areas (Singh et al., 2007). Besides, India, Nepal, Pakistan, and Bangladesh, good collaborative research on *B. sorokiniana* is taking place in CIMMYT, Belgium, Germany, Brazil, USA, Canada, and Australia. It advanced knowledge generated on the topic is summarized in this chapter.

11.2 *BIPOLARIS SOROKINIANA:* A MAJOR PATHOGEN OF LEAF BLIGHT

B. sorokiniana was found to be associated with majority of leaf blight samples of wheat and barley in India and other South Asian countries during 15 years. The other pathogens isolated from blighted wheat leaves were *Alternaria triticina, A. alternata*, and *Pyrenophora tritici-repentis* (Singh, 2007, 2010, 2011). The pathogen, *B. sorokiniana* is dominated and covered the leaf when inoculated with *A. triticina* (Singh, 2007). It is present in all six agro ecological zones in India. The pathogen dominates at boot leaf stage in relatively cooler regions like northwestern plains zone (NWPZ) thus causing major losses amongst leaf blight pathogens (Fig. 11.1).

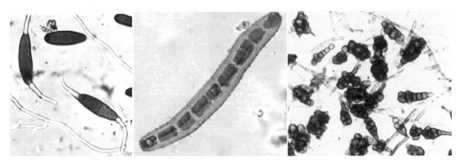

FIGURE 11.1   Spores of *B. sorokiniana* (left), *P. tritici-repentis*, and *A. triticina* (right).

The leaves of susceptible genotypes of wheat and barley are dried prematurely. The use of chemical control is suggested but it may only reduce the disease and more than two sprays are needed since infection may come with seed, soil, plant debris, and air. The infected plant produces thin, shriveled, and discolored grains with low-seed viability. So far no immune wheat and barley is identified against spot blotch.

11.3   MECHANISM AND EVALUATION OF HOST RESISTANCE

The infection process on the leaves usually occurs through natural wounding, stomata, or with the use of an appressorium-like structure through the cell wall (Schäfer et al., 2004), attributed to a biotrophic life style. Plant responses are usually cell wall appositions and HR-like response. If plant responses are insufficient and fail to stop the invasive growth, the fungus starts its necrotrophic way of living (Mattias, 2008). This causes more death and collapse of tissue and further uncontrolled spread of fungus leading to visible necrotic spots (Schäfer et al., 2004). Plant mechanisms involves in the *B. sorokiniana* interaction. The pathogen of spot blotch is also known as hemibiotrophic fungus. The spores are capable to produce germ tube and penetrate in all wheat genotypes irrespective of level of resistance to spot blotch. However, the rate of necrosis of mesophyll tissue represented by light yellow halo around necrotic spot and size of spot after infection varies in different genotypes. It is much slower in resistant genotypes as compared to susceptible genotypes (Singh, 2006). Ibeagha et al. (2005) studied the mechanism of host resistance against *B. sorokiniana* in wheat genotypes, Yangmai 6, M 3 (W7976), Shanghai 4 and Chirya 7, and a susceptible variety Sonalika. After infection, the fungus spreads within the mesophyll tissue (necrotrophic phase) which was restricted in the resistant genotypes. Epidermal cell wall-associated defense, spreading as well as the extent of electrolyte leakage of infected tissue, correlated well with field resistance, and proposed that the cellular host responses, such as formation of cell wall appositions as well as the degree of early mesophyll spreading of fungal hyphae are indicative of the defense potential of the respective host genotype and, therefore, could be used for the characterization of new spot blotch resistance traits in cereals.

Diversity in the isolates of *B. sorokiniana* has been reported from different regions of world (Singh et al., 2004, 2008; Saharan et al., 2008; Chand et al., 2010). It was concluded that the gene-for-gene model is not the principal system operating in barley–*B. sorokiniana* pathosystem, although this plays a role in some interactions (Ghazvini & Tekauz, 2007). The term "pathotype," which is the virulence phenotype of an isolate as elicited on a set of host differentials was introduced in the *Hordeum vulgare–B. sorokiniana* pathosystem when Valjavec-Gratian and Steffenson (1997b) evaluated the virulence diversity of 36 isolates of *B. sorokiniana* collected mostly from North Dakota and detected differential virulence using three-differential lines (DLs). Valjavec-Gratian and Steffenson (1997a) reported that both virulence in *B. sorokiniana* and resistance in host are under monogenic control in barley-*B. sorokiniana* isolate combination.

ND90Pr (pathotype 2) and resistance in the barley line ND 5883 were controlled by a single gene. However, they emphasized that their finding neither confirmed nor refuted a possible gene-for-gene interaction in the *H. vulgare–B. sorokiniana* pathosystem. Nonetheless, the term pathotype was used extensively to describe virulence diversity of *B. sorokiniana* populations in subsequent studies. Differential virulence in *B. sorokiniana* likewise has been reported in several other recent studies (Arabi & Jawhar, 2004; Gamba & Estramill, 2002; Meldrum et al., 2004; Singh, 2006).

11.3.1 EVALUATION OF RESISTANCE BASED ON INFECTION RESPONSES

Singh (2006, 2008) could characterize infection responses (IRs) in wheat against *B. sorokiniana* (Fig. 11.2) both at seedling and adult plant stages. The scoring was proposed as below for IRs.

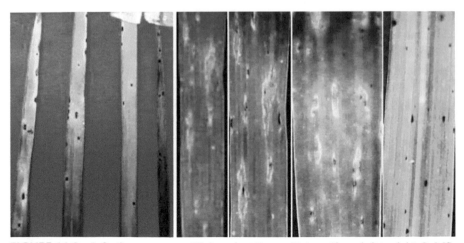

FIGURE 11.2 Infection responses (IRs) against *B. sorokiniana* (from left to right, S, MS, MR, and R in seedlings and adult plant stage).

Resistant (R): No yellow halo around necrotic spot. Small dark brown necrotic spot with not much spread.

Moderately resistant (MR): Necrotic spot comparatively bigger with little yellow halo around.

Moderately susceptible (MS): Necrotic spot surrounded with clear yellow halo. The spots increase in size with time and tend to coalesce with each other.

Susceptible (S): Very distinct and yellow halo around necrotic spot. The spots are elongated and boat shaped, and increases in size quite fast. These coalesce with each other and results larger spots (Singh, 2003).

The IRs against pathogen at seedling were quite similar at adult stage of plants. The seedling resistance test (SRT) based on IRs emerged as valuable technique to evaluate the large numbers of breeding lines against *B. soro-kiniana* under controlled conditions at IIWBR, Karnal. The lines showing no or little halo around necrotic spots during seedling stage were also quite resistant at adult plant stage in field. The genotypes with MS and S behaved almost similar way against spot blotch at flowering stages (Singh, 2006).

11.3.2 EVALUATION OF RESISTANCE BASED ON PER CENT LEAF AREA BLIGHTED

The evaluation of number of genotypes of wheat done (on an average of 1600 genotypes per year) since last more than ten years under Crop Protection, All India Coordinated Wheat & Barley Improvement Project (AICWBIP) identified numbers of confirmed sources of resistance against spot blotch. It was done at 15 hot spot locations situated in northern, central, and peninsular India. These plants were inoculated with spore suspension of *B. sorokiniana* virulent isolates and recording of disease was made at medium milk stage (Singh et al., 2014) using 0–9 double digit scale (Singh & Kumar, 2003) by taking in to account of per cent blighted area on flag (F) and F-1 leaves. The score (0–9) was as follows:

0: No blight, 1: Up to 10% leaf area blighted, 2: 11–20% leaf area blighted, 3: 21–30% leaf area blighted, 4: 31–40% leaf area blighted, 5: 41–50% leaf area blighted, 6: 51–60% leaf area blighted, 7: 61–70% leaf area blighted, 8: 71–80% leaf area blighted, 9: >80% leaf area blighted. The average disease score of multi-location data was calculated by averaging the figures of both digits separately and then rounding up of decimal points 0.5–0.9 to 1 and 0.1–0.4 to 0 (Singh & Kumar, 2005). The further categorization of resistance based on disease score was done as follows:

Average score	Highest score	Category
00–13	35	Highly resistant (HR)
14–35	Up to 57	Resistant (R)
36–57	Up to 68	Moderately resistant (MR)
58–69	Up to 79	Moderately susceptible (MS)
> 69	> 79	Susceptible (S)

The most optimum stage of disease recording was medium milk stage (75 in Zadoks' scale) assessed after comparing the data of multi-locations studies over years and 12 locations (Singh et al., 2014). The rating scale based on per cent leaf area blighted on top two leaves at medium milk stage was proved effective, easy, and relevant in screening of large numbers of breeding lines against leaf blight in wheat and barley (Singh et al., 2006, 2007).

11.3.3   CORRELATION OF INFECTION RESPONSE AND SPOT BLOTCH SCORE

The wheat genotypes were assessed for their resistance against *B. sorokiniana* at both seedling (3–4 leaves) and adult stages (medium milk). Out of 49 genotypes tested, ten numbers of genotypes, HRLSN 14, HRLSN 15, HRLSN 21, HRLSN 23, Chirya 7, Ning 8201, SW 89-5422, 7th EGPSN 101, LBRL 6, and LBRL 13 were showing R type IRs against spot blotch pathogen and adult plant score ranging from 00 to 12 only against 79 score of susceptible checks. The entries showing MR types IR were HRLSN 22, BW/SH 6, BW/SH 16, Chirya 3, Ning 8319, PC-OE-BW-22, YM#6, Shanghai 7, BL 1724, LBRL 11, and LBRL 12 with spot blotch score up to 24. The percentage of dried leaves per tiller was in the range of 8.0–39.1 in case of genotypes showing R and MR types of IRs as compared to 66.6% in case of susceptible checks at mid milk stage. The most promising genotypes showing only 8% dried leaves were SW 89-54422 and LBRL 13. The study indicated a strong relationship between IRs, adult plant score and number of leaves dried at different crop stages, and genotypes showing R and MR types of IRs were showing clear advantage in terms of spot blotch resistance and retaining green leaves till grain filling stage and may be used criteria for selection of promising lines (Singh, 2006). The genotypes and varieties, C 306, DBW 17, HD 2888, HS 277, K 0307, KARAWANI/4NIF/3/SOTY//NAD63/CHRIS, PBW 343, PBW 373, VL 892, NIAW 295, and TL 2942 were found highly resistant (HR) whereas, 15 genotypes were resistant (Singh et al., 2005, 2014). A total of 102 entries, resistance to spot blotch, rusts, and other diseases were shared with wheat breeders contributed in National Genetic Stock Nursery (NGSN) during 1999–2000 till 2009–2010 crop seasons in India. These were utilization from 3 to 64% of centers. The deployment of recently identified spot blotch resistant varieties like HS 490, TL 2942, VL 829, in Northern hills zone (NHZ), DBW 17, and PDW 314 in NWPZ, DBW 39 in North-eastern plains zone (NEPZ), HI 1531, and HI 8498 in central zone (CZ), MACS 6222, UAS 304, and HD 8663 in peninsular zone, HW 5207 in Southern hills zone (SHZ), KRL 213, KRL 210, and MACS 2496 for saline and alkaline soils expected in successful management of spot blotch in field. Resistant genetic stocks like M3 (Harit 1) to spot blotch have been registered in National Gene bank at NBPGR (Singh et al., 2007) in India. Likewise, Verma et al. (2002, 2013) identified barley genotypes showing resistance to spot blotch.

11.3.4　EVALUATION AGAINST HELMINTHOSPOROL TOXIN

The application of purified toxin "Helminthosporol" of *B. sorokiniana* on wheat leaves *in situ* showed low rate of necrosis up to 48 h of application at lower doses of 2.5 and 5.0 mM concentration resistance genotypes tested against *B. sorokiniana* as compared to susceptible genotypes. After 48 h of application, the toxin is causing damage to leaf tissue uniformally in different genotypes. The IR and necrosis up to 36 h of application of toxin of *B. sorokiniana* along with adult plant score of spot blotch on top two leaves at medium milk stage gives good indication of level of host resistance in wheat to spot blotch (Singh, 2006; Singh et al., 2003, 2010, 2015; Åkesson, 1995).

11.4　RESISTANCE TO RUSTS AND SPOT BLOTCH IN WHEAT

The pre-coordinated yield trial entries (about 1300 entries per year) are screened against spot blotch at four hot spot locations in India under artificially inoculated and disease epiphytotic conditions. The per cent spot blotch resistant entries during 2005–2006 till 2007–2008 were highest in NWPZ 49.6%, followed by NEPZ 24.6%, NHZ 18.6%, CZ 2.3%, and SHZ 0.0%. It was possible to get resistance to rusts and spot blotch in same genotype. The entries possessing resistance to spot blotch + three rusts, two rusts, one rust, or no to rusts were, 29, 28, 27, and 14%, respectively. The genotypes with resistance to spot blotch + stem rust, spot blotch + leaf rust in North, spot blotch + leaf rust in South, and spot blotch + stripe rust, were 52, 75, 634, and 61%, respectively (Singh et al., 2009). The spot blotch resistance may be achieved along with rust resistance in wheat and triticales. The possibilities of such combined resistance are higher in case of leaf and stripe rusts as compared to stem rust. The combined resistance to leaf rust and spot blotch will be quite effective in containing the losses in yield and quality in wheat in NEPZ in particular as well as increase the yield in other zones. The supply of data of pre-coordinated entries on status of resistance of entries of wheat and triticale has given choice to the breeders of better selection and proposing entries in coordinated yield trials. A careful analysis of yield trials in coordinated trials of wheat revealed that out of 38 wheat entries possessing *Lr34* leaf rust resistant gene, only four such entries, HS 445, VL 852, SKW 196, and HD 2329 were susceptible, eight were resistant, and rest 26 entries were MR (Singh et al., 2010). The stem rust resistant entries with *Sr2* gene were also found susceptible to spot blotch in majority of cases (Singh, D. P. unpublished). Singh (2014) identified number of barley genotypes with resistance to spot blotch whereas Singh et al. (2010) found multiple resistant genotypes.

11.5 CONCLUDING REMARKS

Spot blotch (*B. sorokiniana*) is the most damaging and largely associated with leaf blight complex of wheat and barley in warmer regions of South Asia and other countries. The association of weaker pathogens like, *A. triticina is* of no or little significance in present day wheat varieties in India. It may be due to changes in genetic constitution and high aggressiveness of *B. sorokiniana.* Therefore, the research work needs to be concentrated to manage spot blotch using effective host resistance. The virulence analysis done indicated presence of groups of isolates differing in their morphology, pathology, and genetic properties. The use of double-digit score based on flag and flag-1 leaf at flowing stage so far proved highly effective, easy to screen large breeding population of wheat and barley, and brought good success in identification of spot blotch resistant varieties. The IR at seedling stage may be used to narrow down population of wheat and barley and lines showing MS and S responses may be discarded early. Likewise use of "Helminthosporol" toxin may be made to evaluate host resistance *in situ* conditions. The spot blotch resistance may be combined with rust resistance.

KEYWORDS

- spot blotch
- wheat
- barley
- *Bipolaris sorokiniana*
- *Alternaria triticina*
- host resistance
- infection response
- helminthosporol toxin
- double-digit score
- resistant genotypes

REFERENCES

Åkesson, H. Infection of Barley by *Bipolaris sorokiniana*: Toxin Production and Ultrastructure. Ph.D. Thesis, Lund University, Sweden, 1995.

Akram, M.; Singh, A.; Singh, D. P.; Singh, S. Spot Blotch of Wheat Caused by *Bipolaris sorokiniana*: An Overview. *Farm Sci. J.* **2003,** *12,* 93–106.

Arabi, M. I. E.; Jawhar, M. Identification of *Cochliobolus sativus* (Spot Blotch) Isolates Expressing Differential Virulence on Barley Genotypes in Syria. *J. Phytopathol.* **2004,** *152,* 461–464.

Chand, R.; Pradhan, P. K.; Prasad, L. C.; Kumar, D.; Verma, R. P. S.; Singh, D. P.; Joshi, A. K. Diversity and Association of Isolates and Symptoms of Spot Blotch Caused by *Bipolaris sorokiniana* in Barley (*Hordeum vulgare* L.). *Indian Phytopathol.* **2010,** *63,* 154–157.

Gamba, F.; Estramill, E. In *Variation in Virulence within Uruguayan Population of Cochliobolus sativus,* Proceedings of Second International Workshop Barley Leaf Blights, Syria, Apr 7–11, 2002; ICARDA: Aleppo, Syria, 2002; pp 59–62.

Ghazvini, H.; Tekauz, A. Host-Pathogen Interactions among Barley Genotypes and *Bipolaris sorokiniana* Isolates. *Plant Dis.* **2007,** *92,* 225–233.

Ibeagha, A. E.; Ralph, H.; Patrick, S.; Singh, D. P.; Kogel, K. H. Model Wheat Genotypes as Tools to Uncover Effective Defense Mechanisms against the Hemibiotrophic Fungus *Bipolaris sorokiniana. Phytopathology.* **2005,** *95,* 528–532.

Malik, V. K.; Singh, D. P; Panwar, M. S. Losses in Yield Due to Varying Severity of Leaf Blight Caused by *Bipolaris sorokiniana* in Wheat. *Indian Phytopathol.* **2008b,** *61,* 526–527.

Malik, V. K.; Singh, D. P.; Panwar, M. S. Management of Spot Blotch of Wheat (*Triticum aestivum*) Caused by *Bipolaris sorokiniana* Using Foliar Sprays of Botanicals and Fungicides. *Indian J. Agric. Sci.* **2008a,** *78,* 646–648.

Mattias, P. Cell Death and Defence Gene Responses in Plant-Fungal Interactions Doctoral Thesis, Swedish University of Agricultural Sciences, Uppsala, 2008; p 41. (Source: http://pub.epsilon.slu.se/1845/1/Mattias_Persson.pdf).

Meldrum, S. I.; Ogle, H. J.; Platz, G. J. Pathotypes of *Cochliobolus sativus* on Barley in Australia. *Aust. Plant Pathol.* **2004,** *33,* 109–114.

Saharan, M. S., et al. Molecular Characterization of Variability in *B. sorokiniana* Isolates Causing Spot Blotch in Wheat in India. *Indian Phytopathol.* **2008,** *61,* 268–272.

Schäfer, P.; Hückelhoven, R.; Kogel, K. H. The White Barley Mutant Albostrians Shows a Super Susceptible but Symptomless Interaction Phenotype with the Hemibiotrophic Fungus *Bipolaris sorokiniana. Mol. Plant-Microbe Interact.* **2004,** *17,* 366–373.

Singh, D. P. Resistance in Wheat to Spot Blotch Pathogen *(Bipolaris sorokiniana)* and its Toxin 'Helminthosporol'. In *Perspective in Plant Health Management;* Abstracts, National Symposium on Dec 16–16, 2010; AAU: Anand, India, 2010.

Singh, D. P.; Gyanendra, S.; Jag, S. Infection Response is an Effective Tool in Selecting Spot Blotch Resistance in Wheat – A Case Study. In *Plant Pathology in the Changing Scenario;* Abstracts of Papers, National Symposium on Feb 27–28, 2008; p 26.

Singh, D. P.; Sharma, A. K.; Singh, S. S. Role of Leaf Rust Resistant *Lr 34* Gene on Severity of Spot Blotch Caused by *Bipolaris sorokiniana* in Wheat. In *Molecular Approaches for Management of Fungal Diseases of Crop Plants;* Abstracts of Papers, National Symposium on Dec 27–30; IIHAR: Bangalore, 2010.

Singh, D. P. Assessment of Losses Due to Leaf Blights Caused by *Bipolaris sorokiniana* (Sacc.) Shoemaker and *Helminthosporium teres* (Sacc.) in Barley. *Pl. Dis. Res.* **2004,** *19,* 173–175.

Singh, D. P., et al. In *Resistance to Spot Blotch Caused by Bipolaris sorokiniana in Pre-coordinated Yield Trial Material of Wheat and Triticales in India, Abstracts of Papers,* National Symposium on "Plant Pathology in the Changing Scenario", New Delhi, Feb 27–28, 2009; pp 26–27.

Singh, D. P. Evaluation of Barley Genotypes against Multiple Diseases. *SAARC J. Agric.* **2008,** *6,* 117–120.

Singh, D. P. First Report of Tan Spot of Wheat in Northern Hills and Northwestern Plains Zones of India. *Plant Dis.* **2007,** *91,* 460.

Singh, D. P. Screening of Wheat Genotypes for Leaf Blight Resistance at Seedling Stage Based on Infection Response against *Bipolaris sorokiniana* – a Rapid and Effective Technique. *Indian Wheat Newslett.* **2003,** *9*(2), 10–11.

Singh, D. P.; Kumar, P. Method of Scoring of Leaf Blight of Wheat Caused by *Bipolaris sorokiniana* (Sacc.) Shoem. on Top Two Leaves at Adult Plant Stage. In *Integrated Plant Disease Management;* Sharma, R. C., Sharma, J. N., Eds.; Scientific Publishers: Jodhpur, India, 2005; pp 289–294.

Singh, D. P. *Bipolaris sorokiniana* Causing Spot Blotch Disease in Wheat and its Management in India. In *Wheat: Productivity Enhancement under Changing Climate;* Singh, S. S., et al., Eds.; Narosa Publishing House: New Delhi, 2011; pp 274–285.

Singh, D. P. Distribution of Foliar Blight Pathogens – The National Scenario and Status of Resistance in Present Day Varieties. Presented in the 49th All India Wheat and Barley Workers' Meet, Aug 27–30, PAU: Ludhiana, 2010.

Singh, D. P., et al. Leaf Blight *(Bipolaris sorokiniana)* Resistant Wheat Genetic Stock (Registered). *Indian Phytopathol.* **2007,** *60,* 118–120.

Singh, D. P., et al. Losses Caused Due to Leaf Blight in Wheat in Different Agroclimatic Zones of India. *Plant Dis. Res.* **2002,** *17,* 313–317.

Singh, D. P., et al. Optimum Growth Stage of Wheat and Triticale for Evaluation of Resistance Against Spot Blotch. *Indian Phytopathol.* **2014,** *67,* 423–425.

Singh, D. P., et al. Assessment of Losses Due to Leaf Blight in Popular Varieties of Wheat under Different Sowing Conditions and Agroclimatic Zones in India. *Indian J. Agric. Sci.* **2004,** *74,* 110–113.

Singh, D. P., et al. In *Comparison of Host Resistance to Bipolaris sorokiniana, the Causal Agent of Leaf Blotches and its Toxin in Wheat,* Proceedings of 4th International Wheat Tan Spot and Spot Blotch Workshop, Bemidji, MN, July 21–24, 2002; Rasmussen, J. B., Friesen, T. L., Ali, S., Eds.; Agricultural Experiment Station, North Dakota State University: Bemidji, MN, 2003; pp 74–78.

Singh, D. P., et al. Leaf Blight Resistant Genotypes in Barley Identified Based on Percent Leaf Area Covered on Flag and a Leaf Below Flag Leaf. In *Plant Pathogens: Exploitation and Management;* Abstracts of Papers, Annual Meet of IPS and National Symposium on Jan 16–18; RDU: Jabalpur, 2007.

Singh, D. P., et al. Management of Spot Blotch of Wheat Caused by *Bipolaris sorokiniana* in Wheat Using Fungicides. *Indian Phytopathol.* **2014,** *67,* 308–310.

Singh, D. P., et al. Multiple Disease and Insect Pests Resistant Genotypes of Wheat and Triticale and Their Utilization in Breeding for Resistance. *Greener J. Agric. Sci.* **2014,** *4,* 150–165.

Singh, D. P., et al. Chemical Control of Leaf Blight of Wheat. *Indian J. Agric. Sci. Res.* **2005,** *39,* 229–231.

Singh, D. P. *Final Report of AP Cess Fund Project on Pathogenic Variability in Bipolaris sorokiniana and Alternaria triticina Causing Leaf Blight of Wheat in Different Agroclimatic Zones of India (3–7/2001-PP);* ICAR: New Delhi, 2006; p 61.

Singh, D. P. In *Leaf Blight of Wheat Caused by Bipolaris sorokiniana in India: Importance, Epidemiology and Management Strategies for Sustaining Wheat Production in Hot and Humid Climate of Eastern India,* Proceedings of 4th International Wheat Tan Spot and Spot Blotch Workshop, Bemidji, MN, July 21–24, 2002; Rasmussen, J. B., Friesen, T. L., Ali, S., Eds.; Agricultural Experiment Station, North Dakota State University: Bemidji, MN, 2003, pp 79–85.

Singh, D. P.; Kumar, P.; Singh, S. K. Resistance in Wheat Genotypes against Leaf Blight Caused by *Bipolaris sorokiniana* at Seedling along with Adult Plant Stage. *Indian Phytopathol.* **2005,** *58,* 344.

Singh, D. P.; Saharan, M. S.; Kumar, P. Mapping of *Bipolaris sorokiniana* in Different Agroclimatic Zones of India and Analysis of Variability. *J. Mycol. Pl. Pathol.* **2004,** *34,* 1008–1009.

Singh, D. P.; Sharma, A. K.; Singh, S. S. *Bipolaris sorokiniana* Causing Spot Blotch Disease in Wheat and its Management. In *Third International Group Meeting on Wheat Productivity Enhancement under Changing Climate;* Abstracts, Feb 9–12, UAS: Dharwad, India, 2011.

Singh, D. P. Status of Tan Spot of Wheat over Years in Different Agro-Ecological Regions in India. In *Mycosphaerella and Stagonospora Diseases of Cereals;* Abstracts of Papers, 8th International Symposium on Sep 11–14; Hotel Sevilla: Mexico City, 2011.

Singh, D. P. Technique For Evaluation of Wheat Genotypes against Helminthosporol Toxin Produced by *Bipolaris sorokiniana* – the Causal Organism of Spot Blotch and Relationship between Host Resistance to Fungus and its Toxin. In *Emerging Plant Diseases, Their Diagnosis and Management*; Abstracts of Papers, National Symposium on Jan 31–Feb 2; UNB: Siliguri, India, 2006.

Singh, D. P., et al. Sources of Resistance to Leaf Blight (*Bipolaris sorokiniana* and *Alternaria triticina*) in Wheat (*Triticum aestivum, T. durum, T. dicoccum*) and Triticale. *Indian Phytopathol.* **2015,** *68,* 221–222.

Singh, D. P.; Kumar, P. In *Validation of Double Digit Method of Scoring of Leaf Blight Caused by Bipolaris sorokiniana of Wheat Based on Per Cent Area Blighted of Flag and a Leaf Below it.* Paper Presented in Annual Meet of Indian Phytopathological Society, YSPUHF: Nauni, Solan, Nov 14–15, 2003.

Singh, D. P.; Kumar, P. Role of Spot Blotch (*Bipolaris sorokiniana*) in Deteriorating Seed Quality in Different Wheat Genotypes and its Management Using Fungicidal Seed Treatment. *Indian Phytopathol.* **2008,** *61,* 49–54.

Singh, D. P.; Singh, S. K.; Pankaj, K. Status of Leaf Blight of Wheat Caused *Alternaria triticina* over Past one Decade in India. In *Climate Change, Plant Protection, and Food Security Interface;* Abstracts of Papers, National Symposium on Dec 17–19, 2009; BCKV: Kalyani, India, 2007; pp 117.

Valjavec-Gratian, M.; Steffenson, B. J. Genetics of Virulence in *Cochliobolus sativus* and Resistance in Barley. *Phytopathology.* **1997a,** *87,* 1140–1143.

Valjavec-Gratian, M.; Steffenson, B. J. Pathotypes of *Cochliobolus sativus* on Barley. *Plant Dis.* **1997b,** *81,* 1275–1278.

Verma, R. P. S.; Singh, D. P.; Sarkar, B. Resistance to Leaf Spot (*Bipolaris sorokiniana* (Sacc.) Shoemaker) and Net Blotch (*Helminthosporium teres* Sacc.) in Barley. *Indian J. Plant Genet. Res.* **2002,** *15,* 17–18.

Verma, R. P. S; Singh, D. P.; Chand, R; Singh, V. K.; Singh, A. K.; Selvakumar, R. Resistance to Spot Blotch (*Bipolaris sorokiniana*) in Barley Germplasm. *Indian J. Plant Genet. Res.* **2013,** *26,* 220–225.

CHAPTER 12

MOLECULAR MARKERS FOR WHEAT IMPROVEMENT: TOOL FOR PRECISION RUST RESISTANCE BREEDING

SUBODH KUMAR, SUBHASH C. BHARDWAJ*, OM P. GANGWAR, PRAMOD PRASAD, HANIF KHAN, and SIDDANNA SAVADI

ICAR-Indian Institute of Wheat and Barley Research, Regional Station, Flowerdale, Shimla 171002, Himachal Pradesh, India

Corresponding author. E-mail: scbfdl@hotmail.com

CONTENTS

ABSTRACT

Wheat rusts are among the major biotic impediments in our efforts to consolidate the genetic gains. The management of wheat rusts through host resistance is the safest and environmental friendly approach. Incorporation of new and diverse sources of rust resistance is a continuous pursuit. So efforts are always going on to incorporate of both major and minor genes to develop durable rust resistant wheat varieties. It is necessary to have tools to determine the resistant genotypes of crop varieties and their confirmation into the target genotype. The molecular marker techniques are sound tools, which enhance efficiency and effectiveness of the genetic analysis of resistance, and the selection of combinations of resistance genes in a crop improvement program. DNA-based molecular markers have acted as versatile tools and found their way into various fields like taxonomy, plant breeding, and genetic engineering. Robust and tightly linked markers can be used to select indirectly for the desirable allele and this represents the simplest form of marker-assisted selection (MAS). MAS not only helps in compressing the population but is also a precise tool to have desirable genotypes through foreground and background selections. Markers can also be used for dissecting polygenic traits into their Mendelian components or quantitative trait loci (QTL). Several types of molecular markers, both Non-polymerase chain reaction (PCR) (restriction fragment length polymorphism) and PCR based (randomly amplified polymorphic DNA marker, amplified fragment length polymorphism, microsatellites etc.) have been utilized in wheat improvement during the recent years and efforts are on to have better and fine type of markers taking the advantage of available genome sequence of wheat.

12.1 INTRODUCTION

Wheat is cultivated in about 215 mha in the world and provides 20% calories and protein requirements for 4.5 billion people in 94 countries. The production of wheat is almost stagnant for the last few years. To improve wheat production, there is a need to overcome the genetic barriers but at the same time, efforts are on to plug the avoidable losses. Wheat rusts are among the major biotic impediments in our efforts to consolidate the genetic gains. The management of wheat rust through host resistance is the safest and environmental friendly approach. The efforts for management through genetic means are always jeopardized by the evolution of new virulent pathotypes, which render resistant wheat varieties susceptible. Therefore, incorporation of new and diverse sources of rust resistance is a continuous pursuit. So efforts are always to incorporate both major and minor genes to develop durable rust resistant wheat varieties. However, resistant genotypes cannot be selected indiscriminately. There are a large number of resistance genes available for many host–pathogen systems with varied response to the biotic constraints both qualitatively and quantitatively. The resistance genes may differ in their specificity, or the pathogen subpopulations that they affect. They may differ in the magnitude of their effects as measured by the extent of disease/damage to the plant or an aspect of pathogen biology. It is necessary to have a sound knowledge of the properties of individual resistance genes and different gene combinations, both within individual genotypes, and in different mixtures of genotypes. In addition, it is necessary to have tools that make it possible to efficiently determine the resistant genotypes of crop varieties and their confirmation into the target genotype.

12.2 MOLECULAR MARKERS AS EFFECTIVE TOOLS IN WHEAT IMPROVEMENT

The molecular marker techniques are sound tools, which enhance efficiency and effectiveness of the genetic analysis of resistance, and the selection of combinations of resistance genes in a crop improvement program. It can also make it possible to analyze the structure of pathogen populations, which helps in the understanding of the interactions between resistance gene and virulence pattern of pathogen populations.

The theoretical advantages of using genetic markers and the potential value of genetic marker linkage maps and direct selection in plant breeding were first reported about 80 years ago (Crouch & Ortiz, 2004). However, it was not until the advent of deoxyribonucleic acid (DNA) marker technology in the 1980s that a large number of environmentally insensitive genetic markers generated to adequately follow the inheritance of important agronomic traits and since then DNA marker technology has dramatically enhanced the efficiency of plant breeding. DNA-based molecular markers have acted as versatile tools and found their own position in various fields like taxonomy, plant breeding, and genetic engineering (Joshi et al., 2011). Molecular markers include biochemical constituents (e.g., secondary metabolites in plants) and macro-molecules, viz., proteins, and DNA. However, analysis of secondary metabolites is restricted to those plants that produce a suitable range of metabolites, which can be easily analyzed and distinguished by varieties (Joshi et al., 2011). These metabolites, which are being used as markers should be ideally neutral to environmental effects or management practices. Hence, amongst the molecular markers used, DNA markers are more suitable and ubiquitous to most of the living organisms.

Diversity based on phenotypic and morphological characters, usually varies with environments and evaluation of traits requires growing the plants to full maturity prior to identification, but now the rapid development of biotechnology allows easy analysis of large number of loci distributed throughout the genome of the plants. Molecular makers have proven to be powerful tools in the assessment of genetic variation and in elucidation of genetic relationships within and among the species (Chakravarthi & Naravaneni, 2006). Molecular markers for classification of genotypes are abundant, but unlike morphological traits, markers are not affected by the environment (Staub et al., 1997). Molecular genetics or the use of molecular techniques for detecting differences in the DNA of individual plants has many applications of value to crop improvement (Jonah et al., 2011). The differences are called molecular markers because they are often associated with specific gene and act as a "sign posts" to those genes and such markers when very tightly linked to genes of interest can be used to select indirectly for the desirable allele and this represents the simplest form of marker-assisted selection (MAS) (Hoisington et al., 2002).

Markers can also be used for dissecting polygenic traits into their Mendelian components or quantitative trait loci (QTL) and this increasing understanding of the inheritance, and gene action for such traits allows the use of markers-selection procedures (Anderson et al., 1993). The discovery of polymerase chain reaction (PCR) was a land mark in this effort and proved to be a unique process that brought about a new class of DNA profiling marker, which has facilitated the development of marker-based gene tags, map-based cloning of agronomically important genes, variability studies, phylogenetic analysis, synteny mapping, market assisted selection of desirable genotypes, etc. DNA markers offer several advantages over traditional phenotypic markers, as they provide data that can be analyzed objectively. Several types of molecular markers are available and each of them has their advantages and disadvantages (Cadalen et al., 1998).

Molecular markers are identifiable DNA sequences, found at specific locations of the genome and associated with the inheritance of a trait, or linked gene (FAO, 2004). Thottappilly et al. (2000) refer to molecular markers as naturally occurring polymorphism, which include proteins and nucleic acids that are clearly different. Markers must be polymorphic (i.e., they must exist in different forms) so that the chromosome carrying the mutant gene can be distinguished from the chromosome with normal gene by form of the marker it carries. Polymorphism can be detected at three levels: Morphological, biochemical, or molecular. The term DNA fingerprinting/profiling is used to describe the combined use of several single locus detection systems and is being used as versatile tools for investigating various aspects of plant genomes. These include characterization of genetic variability, genome fingerprinting, genome mapping, gene localization and analysis of genome, evolution, population genetics, taxonomy, plant breeding, and diagnostics (Joshi et al., 2011).

12.3 CHARACTERS OF EFFECTIVE MOLECULAR MARKERS

According to Joshi et al. (2011), however, an ideal DNA maker should pose the following properties.

i)   Highly polymorphic, which is the simultaneous occurrence of a trait at the same population of two or more discontinues variants or genotypes.
ii)   Co-dominant inheritance—different forms of marker should be detected in a diploid organism to allow discrimination of homozygote and heterozygote.
iii) Frequent occurrence in genome.
iv)   Selective neutral behavior (the DNA sequences of any organism are neutral to environmental conditions or management practices).
v)   Easy access (availability).
vi)   Easy and fast assay.
vii) Reproducible—highly reproducibility, and
viii) Easy exchange of data between laboratories.

12.4 TYPES OF DNA MARKERS

12.4.1 NON-PCR BASED GENETIC MARKERS

12.4.1.1 RESTRICTION FRAGMENT LENGTH POLYMORPHISM

The first and foremost molecular markers system called the restriction fragment length polymorphism (RFLP) was developed in early 1980 (Farooq & Azam, 2002). The RFLPs are simply inherited naturally occurring Mendelian characters. Genetic information is stored in the DNA sequence on a chromosome and variation in this sequence is the basis for the genetic diversity within species. This marker was first reported by Botstein et al. (1980); in the detection of DNA polymorphism (Agarwal et al., 2008). Genomic restriction fragment of different lengths between genotypes can be detected on southern blots and by a suitable probe. In this method, DNA is digested with restriction enzyme like EcoR1, which cut the DNA at specific sequences, electrophoresed, blotted on a membrane, and probed with a labeled clone. RFLP marker provides a way to directly follow chromosome segments during recombination as they follow Mendelian rules and greatly aid in the construction of genetic maps. When an F1 plant undergoes meiosis to produce gametes, its chromosomes will undergo recombination by crossing over and this recombination is the basis of conventional genetic mapping and when used, RFLP markers require hybridization of probe DNA with sampled plant DNA.

12.4.2 POLYMERASE CHAIN REACTION BASED MARKERS

PCR is an *in vitro* method of nucleic acid synthesis by which a particular segment of DNA can be specifically replicated (Mullis & Faloona, 1987). The process involves two oligonucleotide primers that flank the DNA fragment of interest and amplification is achieved by a series of repeated cycles of heat denaturation of the DNA, annealing of the primer to their complementary sequences, and extension of the annealed primers with a thermophilic DNA polymerase. Since the extension products themselves are also complementary to primers, successive cycles of amplification essentially double the amount of the target DNA synthesized in the previous cycle, and the result is an exponential accumulation of the specific target fragment.

12.4.2.1 RANDOMLY AMPLIFIED POLYMORPHIC DNA MARKER

The randomly amplified polymorphic DNA marker (RAPD) detects nucleotide sequence polymorphism in DNA by using a single primer of arbitrary nucleotide sequence (oligonucleotide primer, mostly 10 bases long) (Williams et al., 1991). In this reaction, a single species of primer anneals to the genomic DNA at two different sites on complementary strands of DNA template.

Advantages associated with RAPD analysis include:

i)   Use of small amount of DNA, which makes it possible to work with population that is not accessible with RFLP. It is fast and efficient in analysis having high-density genetic mapping as in many plant species, such as alfalfa (Kiss et al., 1993), faba bean (Torress et al., 1993), and apple (Hemmat et al., 1994).
ii) Non-involvement with radioactive assays (Kiss et al., 1993).
iii) Non-requirement of species-specific probe libraries.
iv) Non-involvement in blotting or hybridization.

Limitations of RAPD markers are:

i)   Its polymorphisms are inherited as dominant or recessive characters causing a loss of information relative to markers which show co-dominance.
ii) Primers are relatively short, a mismatch of even a single nucleotide can often prevent the primer from annealing, hence leads to a loss of band.
iii) Suffers from problems of repeatability in many systems, especially when transferring between populations and laboratories as is frequently necessary with MAS programs (Liu et al., 1994).

12.4.2.2   AMPLIFIED FRAGMENT LENGTH POLYMORPHISM (AFLP)

AFLPS are fragments of DNA that have been amplified using directed primers from restriction of genomic DNA (Metthes et al., 1998). In this approach, the sample DNA is enzymatically cut up into small fragments (as with RFLP analysis), but only a fraction of fragments is studied following selective PCR amplification (Liu et al., 1994). It is a combination of RFLP and RADP methods. AFLP technique shares some characteristic with both RFLP and RAPD analyses (Farooq & Azam, 2002) and combines the specifically of restriction analyses with PCR amplification. AFLP is extremely sensitive technique and the added use of fluorescent primers for automated fragment analysis system and software packages to analyze the biallelic data makes it well suitable for high-throughput analysis.

The major advantages of AFLP techniques (Farooq & Azam, 2002) are:

i)   Generation of a large number of polymorphism.
ii) The PCR technique is fast with high-multiplex ratio which makes the AFLP very attractive choice.

The problems associated with AFLPs are of three types and all are related with practical handling, data generation, and analysis. These problems are not unique to AFLP technology but also associated with other markers systems.

12.4.2.3   SIMPLE SEQUENCE REPEAT OR SHORT TANDEM REPEATS (SSRS) OR MICRO SATELLITES

These are ideal genetic markers for detecting differences between and within species of genes of all eukaryotes (Farooq & Azam, 2002). It consists of tandemly repeated 2–7 base pair units arranged in repeats of mono-, di-, tri-, tetra-, and penta-nucleotides (A, T, AT, GA, AGG, AAAG, etc.) with different lengths of repeat motifs. These repeats are widely distributed throughout the plants and animal genomes that display high level of genetic variation based on differences in the number of tandemly repeating units of a locus. The variation in the number of tandemly repeated units results in highly polymorphic banding pattern (Farooq & Azam, 2002), which is detected by PCR, using locus specific flanking region primers where they are known. The reproducibility of microsatellites is such that they can be used efficiently by different research laboratories to produce consistent and repetitive data (Saghai Maroof et al., 1994).

12.4.3   OTHER MICROSATELLITE BASED PRINCIPLES

12.4.3.1   RANDOMLY AMPLIFIED MICROSATELLITE POLYMORPHISM (RAMP)

This is a micro satellite-based marker, which shows a high degree of allelic polymorphism, but they are labor-intensive (Agarwal et al., 2008). The technique involves a radio-labeled primer consisting of a 5^1 anchor and 3^1 repeats, which is used to amplify genomic DNA in the presence or absence of RAPD primers (Agarwal et al., 2008).

12.4.3.2   THE SEQUENCE CHARACTERIZED AMPLIFIED REGION (SCAR)

The SCARs are PCR-based markers that represent genomic DNA fragments at genetically defined loci that are identified by PCR amplification using sequence specific oligonucleotide primer (McDermoth et al., 1994).

12.4.3.3 SIMPLE PRIMER AMPLIFICATION REACTION (SPAR)

SPAR uses the single SSR oligonucleotide principle.

12.4.3.4 SEQUENCE-RELATED AMPLIFIED POLYMORPHISM (SRAP)

The aim of SRAP technique (Li & Quiros, 2001) is the amplification of open reading frames (ORFs). It is based on two-primer amplification using the AT- or GC-rich cores to amplify intragenic fragment for polymorphism detection (Agarwal et al., 2008).

12.4.3.5 TARGET REGION AMPLIFICATION POLYMORPHISM (TRAP)

The TRAP technique (Hu &Vick, 2003) is a rapid and efficient PCR-based technique, which utilizes bioinformatics tools and expressed sequence tag (EST) database information to generate polymorphic markers, around targeted candidate gene sequences.

12.4.3.6 SINGLE NUCLEOTIDES POLYMORPHISM (SNP)

A single nucleotide differences in the sequence of a gene or segment of the genome. There are typically 10s of 1000s of SNPs and a variety of methods for analyzing them, including highly automated/high-throughput procedures with simultaneous scoring of many markers. Detection of SNPs can be done without gels. SNP marker is just a single base change in a DNA sequence, with a usual alternative of two possible nucleotides at a given position. For such a base position with sequence alternatives in genomic DNA to be considered as an SNP, it is considered that the least frequent allele should have a frequency of 1% or greater SNPs are usually biallelic in practice.

12.4.3.7 SEQUENCE TAGGED SITE (STS)

These markers amplify the unique DNA sequence, which is specific to a locus and does not occur elsewhere in the genome. The markers for rust resistance genes, *Lr9*, *Lr20/Pm1*, and *Lr24* (RAPD-derived); *Lr35* and *Sr22* (RFLP-derived); *Lr19*, *Lr26*, *Lr28*, *Lr37*, *Sr24*, *Sr26*, *Sr31*, *Sr38*, *Sr39*, *Yr5*, and *Yr9* (AFLP-derived); and *Lr34*, *Sr13*, *Sr25*, and *Sr26* (EST-derived) have been reported in the literature.

12.4.3.8   INSERTION SITE-BASED POLYMORPHISM (ISBP)

Transposable elements have unique insertion sites that are highly conserved between different accessions of plants. ISBP markers have been used as PCR based markers by some workers. ISBP markers were also developed from the BAC end sequences of chromosome 3B (Paux et al., 2006).

12.4.3.9   DIVERSITY ARRAYS TECHNOLOGY (DART) MARKERS

DArT marker system provides a cost effective whole-genome finger-printing technique and is efficient for species, which have complex genomes like wheat and others which do not have prior DNA sequence information. Even one DArT assay is capable of typing hundreds to thousands of SNP, and insertion/deletion (indel) polymorphisms distributed throughout the genome. It involves assembly of a group of DNA samples derived from the target germplasm. Details of information about DArT marker system is provided on website (http://www.diversityarrays.com/). DArT marker technique has evolved further and is called as DArTseq. It involves sequencing of the genomic representations on the next generation sequencing (NGS) platforms. Many wheat populations have been mapped using DArT and DArTseq system. DArT markers linked with *Lr34/Yr18/Pm38*, *Lr46/Yr29/Pm39*, *Sr2*, *Sr6*, *Sr25*, and *Yr51* are now available for MAS.

Comparative account of the generally used marker systems in wheat rust resistance improvement is given in Table 12.1.

This information suggests that RFLP, SSR, and AFLP markers are the most effective in detecting polymorphism. However, given the large amount of DNA required for RFLP detection and difficulties in amounting RFLP analysis, AFLP, and SSR are currently the most popular markers in wheat. An increasing amount of sequence information and the determination of the gene function in wheat will lead in near future to the preferred use of new marker type, such as SNPs.

TABLE 12.1   Comparison of the Most Common Marker Systems in Wheat.

Feature	RFLP	RAPD	AFLP	SSR	SNP
Required DNA (μg)	10	0.02	0.5–1.0	0.05	0.05
DNA quality	High	High	Moderate	Moderate	High
PCR based	No	Yes	Yes	Yes	Yes
No. of polymorph loci analyzed	1.0–3.0	1.5–50	20–100	1.0–3.0	1.0
Ease to use	Not easy	Easy	Easy	Easy	Easy
Amenable to automation	Low	Moderate	Moderate	High	High
Reproducibility	High	Unreliable	High	High	High
Development cost	Low	Low	Moderate	High	High
Cost per analysis	High	Low	Moderate	Low	Low

12.5   APPLICATION OF MOLECULAR MARKERS IN RUST RESISTANCE WHEAT BREEDING

Application of these markers for genetic studies of wheat has been so much diverse. Main use includes:

i)   Assessment of genetic variability and characterization of germplasm.
ii) Identification and fingerprinting of genotypes.
iii) Estimation of genetic distance, diversity between population, inbreeds, and breeding materials.
iv) Detection of monogenic and QTL.
v)   MAS.
vi) Identification of sequences of useful candidate genes.

DNA fingerprinting of cereals has a long scientific history. When DNA profiling technology first came in to use RFLP was considered state-of-art. RFLP technology was followed by random amplification of polymorphic DNA(RAPD), followed by AFLP and most recently use of microsatellite markers or SSR.

Advantage of SSR markers are:

i)   The method is relatively simple and can be automated.
ii) Most of these markers are monolocus and show Mendelian inheritance.
iii) SSR markers are highly informative.

iv) A high number of public SSR primers are available.

v)   Effective cost per genotype and primer similar to that for RAPD.

More than 207 genes of resistance to rust fungi: *Puccinia triticina* (74 *Lr* leaf rust resistance genes), *P. striiformis* (75 *Yr* stripe rust resistance genes), and *P. graminis* f. sp. *tritici* (58 *Sr*—stem rust resistance genes) have been identified in wheat (*T. aestivum* L.) and its wild relatives according to recent papers. The number of named and later mapped *Lr*, *Sr*, and *Yr* resistance genes in wheat has increased significantly. However, the wheat genome (17.3 pg per cell) belongs to the largest among crop species and contains nearly 17,000 Mbp per haploid nucleus (Arumuganathan & Earle, 1991). The size of this genome as well as the high percentage (over 90%) of non-coding sequences and three genomes A, B, and D with seven homoeologous chromosomes cause molecular identification and cloning of wheat rust resistance genes to be difficult. Each of 42 hexaploid wheat chromosomes has the average size of about 800 Mbp. Physical distance between crossing-overs (= 1 cm) varies from 0.3 to 3 Mbp (Feuillet et al., 1995). Wild relative species of wheat usually have one genome in common with wheat, which makes them very useful for searching and mapping new resistance genes. Because various wheat-related species carry different genomes (*Triticum* sp., genome B; *Aegilops speltoides*, genome S, similar to B; *T. boeoticum*, genome A; and *Ae. squarrosa*, genome D), they have been and still are used as sources of resistance genes in breeding. A frequent way to transfer the resistance genes is using wheat lines with translocation of a chromosome fragment carrying the genes from a wild species. This was done in the case of genes *Lr19*, *Lr24*, and *Lr29* derived from *Agropyron elongatum* (Procunier et al., 1995; Schachermayr et al., 1995; Prins et al., 2001). The development of new DNA-based assays has led to their application for designing direct and tightly linked markers-restriction fragments length polymorphism (RFLP), random amplified polymorphic DNA (RAPD), cleaved amplified polymorphic sequence (CAPS), sequence characterized amplified regions (SCAR), and sequence tagged sites (STS) to identify individual resistance genes in wheat accessions.

Different categories of molecular markers are available for wheat rust resistance genes (Tables 12.2–12.4). Table 12.2 lists the available markers for *Yr*, Table 12.3 for *Lr*, and Table 12.4 for *Sr* genes in wheat material being used worldwide. It is evident from these tables that most of the rust resistance genes are identifiable through SSR markers.

TABLE 12.2 Markers for *Yr* Genes in Wheat (McIntosh et al., 2004, 2005, 2006, 2007, 2008, 2009, 2010, 2011, 2012, and 2013 (http://maswheat.ucdavis.edu/protocols).

S. No.	Genes	Location	Primers	Type of primers
1	*Yr1*	4AL	Stm673acag	SSR
2	*Yr3*	2B	*Xwmc356*	SSR
3	*Yr4*	3BS	*Xbarc75*	SSR
4	*Yr5*	2BL	*Xbarc349*	SSR
5	*Yr7*	2B	*Xgwm526*	SSR
6	*Yr9*	1BL	*Xgwm582*	SSR
7	*Yr15*	1B	*Xgwm582*	SSR
8	*Yr17*	2AS	VENTRIUP and LN2	RAPD and SCAR
9	*Yr18*	7D	*Xgwm120* and *Xgwm295*	SSR
10	*Yr24*	2B	*Xgwm498*	SSR
11	*Yr26*	1BL	*Xbarc181*	SSR
12	*Yr27*	2B	*Xcdo152*	RFLP
13	*Yr29*	1BL	*Xbac17R*	SSR
14	*Yr32*	2AL	*Xwmc198*	SSR
15	*Yr33*	7DL	*Xgwm111*	SSR
16	*Yr34*	5AL	*Xgwm410*	SSR
17	*Yr35*	6BS	*Xgwm50*	SSR
18	*Yr36*	6B	*Xbarc101*	SSR
19	*Yr40*	5DS	MAS-CAPS16	CAPS marker
20	*Yr41*	2BS	*Xgwm410*	SSR
21	*Yr43*	2BL	*Xwgp110*	
22	*Yr44*	2BL	*Xwgp100*	
23	*Yr45*	3DL	*Xwp118*	
24	*Yr46*	4DL	*Xgwm165* and *Xgwm192*	SSR
25	*Yr47*	5BS	SSR	*Xgwm234*
26	*Yr48*	5AL	*Xwmc727*	SSR
27	*Yr49*	3DS	*Xgwm161*	SSR
28	*Yr50*	4BL	*Xbarc1096*	SSR
29	*Yr51*	4AL	wPt0763	
30	*Yr52*	7BL	*Xbarc182*	SSR
31	*Yr53*	2BL	*Xwmc441*	SSR
32	*Yr54*	2DL	*Xgwm301*	SSR

TABLE 12.2 *(Continued)*

S. No.	Genes	Location	Primers	Type of primers
33	*Yr55*	2DL	*Xmag3385*	
34	*Yr56*	2AS	*Xsun167*	
35	*Yr57*	3BS	*Xgwm389*	
36	*Yr58*	3BL	1121669/3023704	
37	*Yr59*	7BL	*Xbarc32*	SSR
38	*Yr60*	4AL	*Xwmc776*	SSR
39	*Yr61*	7AS	*Xwp5467*	
40	*Yr62*	4BL	*Xgwm192*	SSR
41	*Yr63*	7BS	*IWB33120*	
42	*Yr64*	1BS	*Xgdm33*	SSR
42	*Yr65*	1BS	*Xgwm18*	SSR
43	*Yr66*	3DS	*IWB18087/IWB56281*	SSR
44	*Yr67*	7BL	*–IWB71995*	SSR

*grap= resistance gene-analog polymorphism.

TABLE 12.3 Markers for *Lr* Genes in Wheat (McIntosh et al., 2004, 2005, 2006, 2007, 2008, 2009, 2010, 2011, 2012, and 2013 (Source: http://maswheat.ucdavis.edu/protocols).

S. No.	Genes	Location	Primers	Type of primers
1	*Lr1*	5DL	pTAG621-5 and pTAG621-3	STS and RFLP
2	*Lr3a*	6BL	TaR16	cDNA marker
3	*Lr9*	6BL	J 13/1 and J 13/2	RFLP
4	*Lr10*	1AS	F1-2245 and*Lr10-6/r2*	RFLP and STS
5	*Lr12*	4BS	*Xgwm251*	SSR
6	*Lr13*	2BS	*Xgwm630*	SSR
7	*Lr14a*	7BL	*Xgwm344*	SSR
8	*Lr14b*	7BL	*Xgwm146*	SSR
9	*Lr16*	2BS	*Xwmc764*	SSR
10	*Lr17a*	2AS	*Xbarc212*	SSR
11	*Lr19*	7DL	GbF and GbR	AFLP and STS
12	*Lr20*	7AL	STS638-L and STS638-R	STS
13	*Lr21*	1DL	D14-L and D14-R	RGA
14	*Lr22a*	2DS	*Xgwm455*	SSR

TABLE 12.3 *(Continued)*

S. No.	Genes	Location	Primers	Type of primers
15	*Lr24*	3DL	J9/1 andJ9/2	RFLP
16	*Lr25*	4BS	*Lr25F20 andLr25R19*	RAPD
17	*Lr26*	1BL	J07IF1 andJ07IR1	RAPD
18	*Lr28*	4AL	*Lr28-01*and*Lr28-02*	STS
19	*Lr29*	7DS	UBC219/1 andUBC219/2	RAPD
20	*Lr32*	3D	*Xwmc43*	SSR
21	*Lr34*	7D	*Xgwm120*	SSR
22	*Lr35*	2B	BCD260F1 and 35R2	RFLP and STS
23	*Lr37*	2AS	VENTRIUP and LN2	RAPD and SCAR
24	*Lr39*	2DS	*Xgdm35*	SSR
25	*Lr42*	1D	*Xwmc432*	SSR
26	*Lr46*	1BL	*Xbac17R*	SSR
27	*Lr47*	7AS	PS10R and PS10L2	SCAR
28	*Lr48*	4BL 2B	*Xwmc332*	SSR
29	*Lr49*	4BL2AS	*Xbarc163*	SSR
30	*Lr50*	2BL	*Xgwm382*	SSR
31	*Lr51*	1BL	S30-13L and AGA7-759R	SCAR
32	*Lr52*	5BS	*Xgwm234*	SSR
33	*Lr53*	6BS	*Xgwm50*	SSR
34	*Lr57*	5DS	MAS-CAPS16	CAPS marker
35	*Lr58*	2BL	*Xncw-Lr58-1*	STS
36	*Lr60*	1DS	*Xbarc149*	SSR
37	*Lr61*	6BS	P81/M70269/P87/M75131	
38	*Lr63*	3AS	*Xbarc321*	SSR
39	*Lr64*	6AL	*Xbarc104*	SSR
40	*Lr65*	2AS	*Xbarc212*	SSR
41	*Lr66*	3A	S15-t3	SCAR
42	*Lr67*	4DL	*Xgwm165* and *Xgwm192*	SSR
43	*Lr68*	7BL	*Xgwm146*	SSR
44	*Lr71*	1BL	*Xgwm18*	SSR
45	*Lr72*	7BS	*Xwmc606*	SSR
46	*Lr74*	3BL	*IWB69699/IWB20762*	

TABLE 12.4 Markers for *Sr* Genes in Wheat (McIntosh et al., 2004, 2005, 2006, 2007, 2008, 2009, 2010, 2011, 2012, and 2013 (Source: http://maswheat.ucdavis.edu/protocols).

S. No.	Genes	Location	Primers	Type of primers
1	*Sr*2	3BS	*Xgwm533*	SSR
2	*Sr*6	2DS	*Xwmc453*	SSR
3	*Sr*8b	6A	*Xgwm334*	SSR
4	*Sr*9e	2B	*Xgwm47*	SSR
5	*Sr*9h	2B	*Xgwm47*	SSR
6	*Sr*13	6AL	*Xwmc59*	SSR
7	*Sr*17	7BL	*Xwmc273*	SSR
8	*Sr*22	7AL	CSIH81-BM and CSIH81-AG	STS markers
9	*Sr*24	3DL	*Sr24#12*	STS marker
10	*Sr*25	7DL	Gb	
11	*Sr*26	6AL	BE518379 and *Sr26#43*	PCR based
12	*Sr*30	5DL	*Xcfd12*	SSR
13	*Sr*31	1BL	SCSS30.2 and SsCSS26.1	SCAR markers
14	*Sr*33	1DS	*Xwmc336*	SSR
15	*Sr*35	3AL	*Xbf483299*	
16	*Sr*36	2BS	*Xwmc477*	SSR
17	*Sr*38	2AS	VENTRIUP and LN2	RAPD and SCAR
18	*Sr*39	2B	*Sr39F2 and Sr39R3*	SCAR
19	*Sr*40	2BS	*Xwmc344*	SSR
20	*Sr*42	6D	*Xbarc183*	SSR
21	*Sr*45	1DS	*Xwmc222, Xgwm106*, and *Xbarc229*	SSR
22	*Sr*47	2BL	*Xgwm47*	SSR
23	*Sr*48	2AL	*Xstm673acag*	
24	*Sr*49	5BL	*Xwmc471*	SSR
25	*Sr*50	1DS	AW2-5	
26	*Sr*51	3AL, 3BL, and 3DL	3SS	
27	*Sr*52	6AS	6V3	EST-STS
28	*Sr*53	5D	BE443102/Mbo1	
29	*Sr*54	2D	*Xcfd-283*	SSR
30	*Sr*55	4DL	*Xgwm165* and *Xgwm192*	SSR
31	*Sr*56	5BL	wPt9116	

12.6   MARKER-ASSISTED SELECTION (MAS)

MAS is a tool to confirm the target gene in segregating progenies. It helps in reducing the time and efforts in identifying, and confirming the desired resistance and it has become a very handy and strong technique in rust resistance breeding. It is the identification of DNA sequences located near resistance genes that can be tagged to breed for traits that are difficult or time consuming to confirm through the traditional technologies. The MAS refers to the application of DNA markers that are tight-linked to the target loci as a substitute for or complement the phenotyping data. By determining the allele of a DNA marker, plants that possess particular genes or QTL may be identified on the basis of their genotype whereas supplementary phenotype data makes the selection fool proof. Basic advantages of MAS compared to conventional phenotypic selection which are: (a) Precise, less time consuming, and simple technique as compared to the phenotypic breeding; (b) Selection may be carried out at initial breeding populations and single plants may be tagged with high reliability. In this method, linkages are sought between DNA markers and specific rust resistance genes or QTLs. Therefore, the MAS is being undertaken to develop desirable rust resistant population worldwide.

Another offshoot of this technology is gene mining where we are able to tag novel/new rust resistance genes in landraces, improved germplasm or the one derives in secondary or tertiary gene pools. During the preceding 10 years more than 50 new rust resistance genes or QTLs have been identified and also introgressed in to present day wheat cultivars.

12.7   CONCLUDING REMARKS

Molecular markers have opened new horizons in wheat breeding. We are not only able to compress the population and carry on with the right stuff but are also able to select for the traits which are of complex inheritance or difficult to select. Tightly linked or robust markers are a sure tool to select or reject the breeding material. The marker-assisted technology has become more refined and accurate with the years. With the availability STS, SSR markers, Insertion site based polymorphism, DArT markers, SNPs, we are able to work with more authenticity. When we see even present trend, MAS for resistance to wheat diseases has become more handy, practical, and useful.

KEYWORDS

- **molecular markers**
- **DNA markers**
- **rust resistance genes**
- **marker assisted selection**
- **wheat**
- **stem rust**
- **leaf rust**
- **stripe rust**

REFERENCES

Agarwal, M.; Shrivastava, N.; Padh, H. Advances in Molecular Marker Technique and their Application in Plant Science. *Plant Cell Rep.* **2008,** *27,* 617–631.

Anderson, J. A.; Sorrells, M. E.; Tanksley, S. D. RFLP Analysis of Genomic Regions Associated with Resistance to Preharvest Sprouting in Wheat. *Crop Sci.* **1993,** *33,* 453–459.

Arumuganathan, K.; Earle, E. D. Nuclear DNA Content of Some Important Plant species. *Plant Mol. Biol. Rep.* **1991,** *9,* 208–218.

Botstein, D.; White, R. I.; Skolnick, M.; Davis, R. W. Construction of a Genetic Linkage Map in Man Using Restriction Fragment Length Polymorphisms. *Am. J. Hum. Genet.* **1980,** *32,* 314–331.

Cadalen, T.; Boeuf, C.; Bernard, M. An Inter Varietal Molecular Marker Map in *Triticum aestivum* L. Em. Thell. and Comparison with a Map from a Wide Cross. *Theor. Appl. Genet.* **1998,** *789,* 495–504.

Chakravarthi, B. K.; Naravaneni, R. SSR Marker Based DNA Fingerprinting and Diversity Study in Rice (*Oryza sativa* L.). *Afr. J. Biotechnol.* **2006,** *5*(9), 684–688.

Crouch J. H.; Ortiz, R. Applied Genomics in the Improvement of Crops Grown in Africa. *Afr. J. Biotechnol.* **2004,** *3* (10), 489–496.

FAO. *The State of Food and Agriculture 2003–2004: Agricultural Biotechnology-Meeting the Needs of the Poor?;* FAO: Rome, 2004. http://www.fao.org/docrep/006/Y5160E/Y5160E00.HTM (Accessed on 30 Sep. 2015).

Farooq, S.; Azam, F. Molecular Markers in Plant Breeding-1: Concepts and Characterization. *Pak. J. Biol. Sci.* **2002,** *5* (10), 1135–1140.

Feuillet, C.; Messmer M.; Schachermayr, G.; Keller, B. Genetic and Physical Characterization of the *Lr1* Leaf Rust Resistance Locus in Wheat (*Triticum aestivum* L.) *Mol. Gen. Genet.* **1995,** *248,* 553–562.

Hemmat, M.; Weeden, N. F.; Managanaris, A. G.; Lawson, D. M. Molecular Marker Linkage Map for Apple. *J. Heredity.* **1994,** *85,* 4–11.

Hoisington, D., et al. The Application of Biotechnology. In *Wheat Improvement and Production;* Curtics, B. C., Rajaram, S., Gomez, H., Eds.; FAO: Rome, 2002.

Hu, J.; Vicks, B. A. Target Region Amplification Polymorphism a Novel Marker Technique for Plant Genotyping. *Plant Mol. Boil. Rep.* **2003,** *21,* 289–294.

Jonah, P. M., et al. The Importance of Molecular Markers in Plant Breeding Programmes. *Global J. Sci. Front. Res.* **2011,** *11* (5), 1–9.

Joshi, S. P.; Prabhakar, K.; Ranjekar, P. K.; Gupta, V. S. Molecular Markers in Plant Genome Analysis, 2011; pp 1–19, http/www.ias.ac.in/currsci/jul25/articles 15.html (Accessed on 30 Sep. 2015).

Kiss, G. B., et al. Construction of a Basic Linkage Map of Alfalfa Using RFLP. RAPD, Isozyme and Morphological Markers. *Mol. Gen. Genet.* **1993,** *238,* 129–137.

Li, G.; Quiros, C. F. Sequence-Related Amplified Polymorphism (SRAP), a New Marker System Based on a Simple PCR Reaction: Its Application to Mapping and Gene Tagging in Brassica. *Theor. Appl. Genet.* **2001,** *103,* 455–461.

Liu, C. J., et al. An RFLP-Based Genetic Map of Pearl Millet (*Pennisetum glaucum*). *Theor. Appl. Genet.* **1994,** *8,* 481–487.

McDermoth, J. M., et al. Genetic Variation in Powdery Mildew of Barley: Development of RAPD, SCAR and VNTR Markers. *Phytopathology.* **1994,** *84,* 1316–1321.

McIntosh, R. A., et al. Catalogue of Gene Symbols for Wheat: Supplement. **2004,** http:// www.shigen.nig.ac.jp /wheat/komugi /genes/ macgene/ supplement2004. html.

McIntosh, R. A., et al. Catalogue of Gene Symbols for Wheat: Supplement. **2008,** http:// www.shigen.nig.ac.jp/wheat/komugi/genes/macgene/supplement2008.pdf.

McIntosh, R. A., et al. Catalogue of Gene Symbols for Wheat: Supplement. **2006,** http:// www.shigen.nig.ac.jp/wheat/komugi/genes/macgene/supplement2006.pdf.

McIntosh, R. A., et al. Catalogue of Gene Symbols for Wheat: Supplement. **2005,** http:// www.shigen.nig.ac.jp/wheat/komugi/genes/macgene/supplement2005.pdf.

McIntosh, R. A., et al. Catalogue of Gene Symbols for Wheat: supplement. **2013,** http://www. shigen.nig.ac.jp/wheat/komugi/genes/macgene/ supplement 2013.pdf.

McIntosh, R. A., et al. Catalogue of Gene Symbols for Wheat: supplement. **2011,** http://www. shigen.nig.ac.jp/wheat/komugi/genes/macgene/supplement2011.pdf.

McIntosh, R. A., et al. Catalogue of Gene Symbols for Wheat: supplement. **2010,** http://www. shigen.nig.ac.jp/wheat/komugi/genes/macgene/supplement2010.pdf.

McIntosh, R. A., et al. Catalogue of Gene Symbols for Wheat: supplement. **2009,** http://www. shigen.nig.ac.jp/wheat/komugi/genes/macgene/supplement2009.pdf.

McIntosh, R. A., et al. Catalogue of Gene Symbols for Wheat: supplement. **2012,** http://www. shigen.nig.ac.jp/wheat/komugi/genes/macgene/supplement2012.pdf

McIntosh, R. A., et al. Catalogue of Gene Symbols for Wheat: Supplement. **2007,** http:// www.shigen.nig.ac.jp/wheat/komugi/genes/macgene/supplement2007.pdf.

Metthes, M. C.; Daly, A.; Edwards, K. J. Amplified Fragment Length Polymorphism(AFLP). In *Molecular Tools for Screening Biodiversity;* Karp, A., Isaac, A. P. G., Ingram, D. S., Eds.; Chapman & Hall: London, 1998; pp 183–190.

Mullis, K. B.; Facoona, F. A. Specific Synthesis of DNA *in vitro* via a Polymerase Catalyzed Chain Reaction. *Methods Enzymol.* **1987,** *155,* 335–350.

Paux, E., et al. Characterizing the Composition and Evolution of Homoeologous Genomes in Hexaploid Wheat through BAC-End Sequencing on Chromosome 3B. *Plant J.* **2006,** *48,* 463–474.

Prins, R., Groenewald, J. Z., Marais, G. F., Snape, J. W., Koebner, R. M. D. AFLP and STS Tagging of *Lr19,* a Gene Conferring Resistance to Leaf Rust in Wheat. *Theor. Appl. Genet.* **2001,** *103,* 618–624.

Procunier, J. D., et al. PCR-Based RAPD/DGGE Markers Linked to Leaf Rust Resistance Genes *Lr 29* and *Lr25* in Wheat (*Triticum aestivum* L.). *J. Genet. Breed.* **1995,** *49,* 87–92.

Saghai Maroof, M. A., et al. Extraordinarily Polymorphic Satellite DNA in Barley: Species Diversity, Chromosomal Locations, and Population Dynamics. *Proc. Natl. Acad. Sci. USA.* **1994,** *91,* 5466–5470.

Schachermayr, G., et al. Identification of Molecular Markers Linked to the *Agropyron elongatum*-Derived Leaf Rust Resistance Gene *Lr24* in Wheat. *Theor. Appl. Genet.* **1995,** *90,* 98–990.

Staub, J. C.; Serquen, F. C.; Mccreight, J. A. Genetic Diversity in Cucumber (*Cucumis sativus* L.): An Evaluation of Indian Germplasm. *Genet. Resour. Crop Evol.* **1997,** *44* (4), 315–326.

Thottappilly, G.; Magonouna, H. D.; Omitogun, O. G. The use of DNA Markers for Rapid Improvement of Crops in Africa. *Afr. Crop Sci. J.* **2000,** *8* (1), 99–108.

Torress, A. M.; Weeden, N. F., Martin, A. Linkage among Isozyme, RFLP and RAPD Markers. *Plant Physiol.* **1993,** *101,* 394–452.

Williams, D. A.; Rios, M.; Stephens, C.; Patel, V. P. Fibronectin and VLA-4 in Aematopoietic Stem Cell-Microenvironment Interactions. *Nature.* **1991,** *352,* 438–441.

CHAPTER 13

GENE PYRAMIDING FOR DEVELOPING HIGH-YIELDING DISEASE-RESISTANT WHEAT VARIETIES

M. SIVASAMY[1,*], V. K. VIKAS[1], P. JAYAPRAKASH[1],
JAGDISH KUMAR[2], M. S. SAHARAN[3], and INDU SHARMA[3]

[1]ICAR-Indian Agricultural Research Institute, Regional Station, Wellington 643231, Tamil Nadu, India

[2]ICAR-National Institute of Biotic Stress Management, Raipur 493225, Chhattisgarh, India

[3]ICAR-Indian Institute of Wheat and Barley Research, Karnal 132001, India

*Corresponding author. E-mail:iariwheatsiva@rediffmail.com

CONTENTS

ABSTRACT

In this chapter gene pyramiding for developing high yielding disease resistant wheat varieties, wheat rusts in India, the details of occurrence, epidemiology, and management of rusts particularly the pathotypes of leaf rust pathogen in field and spectrum of dominant pathotypes of wheat rusts in India over the years are given in detail. Further the importance of rust monitoring for the sake of gene deployment by understanding types of rust resistance in wheat to take up the gene deployment strategies to combat rusts are discussed in depth. The usefulness of linked genes used in the gene pyramiding program and the effective introgression of genes through hybridization and marker-assisted selection to develop durable rust resistant varieties particularly the effective genes against stem rust race Ug99 at IARI, Wellington are discussed. The details on origin, chromosomal location, infection type, source stocks, environmental variability, and stocks developed at IARI, RS, Wellington for effective single leaf rust genes viz., *Lr9, Lr19, Lr24, Lr26, Lr28, Lr32, Lr34, Lr35, Lr37, Lr39, Lr44, Lr45, Lr46, Lr47, Lr53, Lr57, Lr67, and Lr68*, stem rust genes, *Sr2, Sr24, Sr25, Sr26, Sr2, Sr30, Sr31, Sr32, Sr33, Sr36/Pm6*, and *Sr38* and stripe rusts *Yr9, Yr10, Yr15*, and *Yr17* are given. For achieving durable resistance in wheat, pyramiding new rust resistance genes is the effective way.

13.1 INTRODUCTION

Wheat is one of the most staple cereal food crop produced in the world followed by maize. Global wheat production reached a record high of 690 MT—4.3% up on 2012 (FAO, 2013). Over the past 50 years in India, the wheat production and productivity gradually increased to reach all time high of 94.88 MT from 30.3 mha area during 2011–2012 (Anonymous, 2012), and it ranks second in the world wheat production next to China, with an average productivity of 3.12 t/ha, which is higher than the global average of 3.1 t/ha (Anonymous, 2013). In India, 95% of wheat produced is *Triticum aestivum*, followed by *T. durum* (4%), and *T. dicoccum* (1%). Over the past, wheat production in India witnessed a moderate growth rate of 2.25%. The growing demand from projected population growth over 110 million tons of wheat is required by 2020 A.D. in India, thus posing a challenging task to the wheat researchers and policy planners. There are several biotic and abiotic constraints toward sustaining and realizing the potential yield in wheat. Among biotic stresses, the rust pathogens remained a threat to wheat production globally and have diverse race profiles. In India also, wheat crop is severely damaged by all the three rusts viz., black or stem rust (*Puccinia graminis* f. sp. *tritici* Erik. & Henn), brown or leaf rust (*P. triticina* Erik.), and yellow or stripe rust (*P. striiformis tritici* West).

13.2 WHEAT RUSTS IN INDIA: OCCURRENCE, EPIDEMIOLOGY, AND MANAGEMENT

Stem rust pathogen *P. graminis* f. sp. *tritici* survives throughout the year only in the Nilgiri hills of South India and the Himalayas are too cold for pathogen to survive during winter. Therefore, consequent upon cessation of wheat cultivation from Nilgiris, the stem rust has drastically disappeared from India owing to debilitated inoculum build up in absence of host. Leaf rust pathogen (*P. triticina*) needs an intermediate temperature and thus survives and spreads from both southern (Nilgiris) and northern (Himalayas) foci and is of importance in all six agro-ecological zones (Nagarajan & Joshi, 1985). It infects both durum and bread wheat. The pathogen is a member of the *P. recondita* complex (*P. triticiduri* V. Bourgin). Subsequently, this rust has ravaged wheat crop several times causing varietal break down of rust resistance in famous post-green revolution wheat cultivars like WL 711, Arjun, WH 147, HD 2329, HD 2285, PBW 343 etc. Stripe rust pathogen (*P. striiformis*) needs low temperature and survives the whole year in the Himalayas. It is an endemic pathogen in cooler parts of the country and its epidemic in early 70s was responsible for withdrawal of a high yielding semi dwarf variety Kalyansona in North India.

Wheat rusts epiphytotics have been recorded from time to time in India. The earliest record of stem rust epiphytotics was in 1827 from central India (Jubbalpore). Later, severe stem rust epiphytotics occurred during 1956–1957 from Pusa (Bihar), in Sanchore (Rajasthan) during 1973–1974, in Narmada valley (Madhya Pradesh) during 1978–1979. High losses in crop yields were reported by rusts from time to time (Asthana, 1948; Barclay, 1890; Joshi et al., 1975; Mehta, 1940; Prasada, 1960). The stem rust problem continued to be at low level after cultivation of Mexican semi dwarf wheat. The stripe rust problem was aggravated with the extensive cultivation of Kalyansona during early 1970s in the north and on *aestivum* wheat in the South hill zone. With the replacement of Kalyansona, stripe rust problem has been appreciably minimized in the north; however, it still continues to be a problem in Nilgiri hills of South India. India in particular has not faced any rust epidemic since last three-and-a-half decades because of proper addition of rust resistance genes in wheat cultivars. However, emergence and spread of Ug99 race of stem rust pathogen, *P. graminis tritici,* has been now recognized as another possible threat to future wheat production in many countries (Singh et al., 2006). Besides Ug99, a new race of yellow rust, *78S84* has also been observed on the most widely grown variety PBW 343. Of the three wheat rusts, leaf rust continues to be important, regularly occurring in all agro-climatic zones. Despite fluctuations in the climate that affect other two rusts, leaf rust continues to occur almost unaffected at many places every year. Continuous efforts aimed at introgression of resistance against this rust have also not really thwarted this pathogen from infecting wheat crop. Remarkably, it has always adapted to new resistance in quick time and this feature has been both worrisome and elusive for wheat scientists. The 77 group of races in leaf rust pathogen have been the most dominant in the country and has produced more than 10 variants since 1950 (Fig. 13.1). Though most of the variants have been picked up in south hills, a few have also been picked up in north hills.

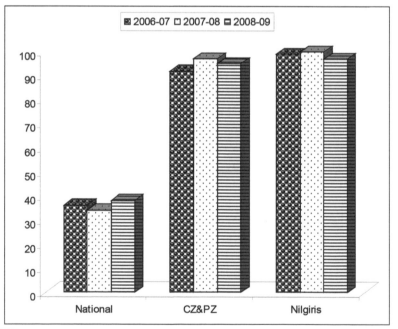

FIGURE 13.1   Percentage of 77 groups of pathotypes of leaf rust pathogen in field samples.

In 1887, Barclay recorded on *Berberisaristata,* the aecidial stages of *Puccinia* and believed that it is that of *P. graminis tritici* described by De

Bary (Barclay, 1890). Then through experimentation, Mehta (1940) demonstrated the link between the aecidial stage on barberries and stem rust uredial stage on a grass *Poa nemoralis*. The *P. graminis tritici* has been demonstrated to exhibit a very large host range and to survive in uredial form on several grasses (Prasada, 1948; Joshi & Manchanda, 1963; Bahadur et al., 1973; Pathak et al., 1979) and self-sown crop or volunteer wheat plants. The urediospores thus surviving the hot summer in hills cause fresh infection on the main November-sown crop in the adjoining plains. The earlier held view that both central Nepal hills and the bunch of hills in South India called the Nilgiris (Mehta, 1952) acting as primary inoculum source was re-examined in the late 1960s when the new high yielding dwarf wheat varieties of International Maize and Wheat Improvement Center (CIMMYT) origin were introduced to India. The urediospores survive through the non-wheat-growing summer months on the off-season crops (plants) available in the hills of Nilgiri and Palni hills. After a detailed study of the weather conditions that influence spore transport and deposition, Nagarajan and Singh (1975) postulated the "Indian Stem Rust Rules" (ISR).

Mehta in 1930 started the annual *P. graminis tritici* race (= virulence) analysis work in India and used the standard differentials (Stakman & Levine, 1922; Mehta, 1940, 1952). Now, *P. graminis tritici* virulences are identified based on the host–pathogen reaction matrix, mapped on a set of near isogenic lines and lines with known *Sr* genes (stem rust resistance genes) (Bahadur et al., 1985). Postulating the *Sr* genes in the about to be released wheat varieties is done by comparing the host–pathogen reaction matrix of a given variety with that of the near isogenic lines. In this matching attempt, it is assumed that when more than one *Sr* gene is present in a given material there all gene expression is additive (Nagarajan et al., 1986; Nagarajan et al., 1987). At the time of wheat variety release, this information is considered to ensure that *Sr* gene combinations are properly deployed over space and time.

13.3   RUST MONITORING FOR THE SAKE OF GENE DEPLOYMENT

A continuous monitoring of wheat rusts pathogens has been regularly done in past, since 1930, to map the prevalence and distribution of black, brown, and yellow rust virulences. Also new virulences emerging on account of cultivating a monogenic genotype or otherwise are also kept under track. Through this exercise, wheat breeders are cautioned and advised to take timely step to breed varieties with more effective resistance genes whenever any shift in pathogen virulences are observed.

The avirulence/virulence genes present in the pathogen are characterized by analyzing more than 2000 rust samples per year and by challenging them on differential possessing different resistance genes. Information over years shows that there is a difference in the brown rust and yellow rust virulence patterns between the wheat-growing zones. On the strength of this information, rests the gene deployment and the varietal diversificationprogram.

During mid-eighties, based on the information on race distribution pattern and the environmental factors, a strategy was developed to minimize the losses due to rusts. The strategy had two major elements, (a) deploying different rust resistance gene combinations in different wheat-growing environments and (b) releasing a large number of wheat varieties of different maturity periods and having shades of difference in the disease severity to rusts. At present, more than 50 varieties representing diverse genetic backgrounds are grown in the country. This varietal diversity and deploying them over space and time brought yield stability and avoided epidemics. Such diversity of host resistance genes in space and time has also buffered the disease spread. For example, earlier, brown rust of wheat used to cause annually about 10% yield loss whereas it has been minimized to less than 2% per year and thereby added to the annual wheat production of 4–5 MT. This has also brought in the necessary stability in wheat production from year to year.

TABLE 13.1   Spectrum of Dominant Pathotypes of Wheat Rusts in India Over the Years.

Period	Stem rust	Leaf rust	Stripe rust
Till 1975	11, 11A, 14, 15, 17, 21, 24, 24A, 34, 34-1, 40, 42	10, 11, 12, 12a, 17, 20, 63, 77, 106, 107, 108, 162	13, 14, 19, 20, 31, 38
1975–1980	21-1, 40A, 21A-2	77A, 104	14A, 20A, 38A, I
1980–1985	117A-1	114A, 104B, 12-2	K
1985–1990	40-1, 117-1	77-1, 77-2, 77-3, 12-1, 12-3, 12-4, 107-1, 108-1	L, N, P
1990–1995	117-2, 117-3, 117-4, 117-5, 117-6	77-4, 77-5, 104-2, 104-3	T, U, CI, CII, CIII
1995–till date	40-2	77-6, 77-7, 77-8	*Yr9* virulence (46S119), 78S84

13.4   TYPES OF RUST RESISTANCE IN WHEAT

During the last two decades, identification of resistance genes and development of near isogenic lines has been accomplished in wheat-*Puccinia* system. In recent compillation, McIntosh (2012) has listed >55 of stem rust, >70 of leaf rust, and >48 of stripe rust that accord resistance against pathogenic populations of *P. recondita tritici, P. graminis tritici,* and *P. striiformis* and some of the linked genes conferring multiple disease resistance which include rusts, powdery mildew, *barley yellow dwarf virus* (BYDV), eye spot, etc. Majority of them are classified as race-specific genes with hypersensitive response to the pathogen.

Although this type of resistance has been very commonly exploited in breeding program, it is observed that this type of resistance is short lived due to evolution of new races (McDonald & Linde, 2002). Based on the host response to the pathogen, Flor (1955) described that race-specific resistance is conditioned by the interaction of specific genes in the host with those of the pathogens. Stakman et al. (1962) described immunity as the absence of visible lesions on the host plant where as Robinson (1976) observed lesions on leaves after the pathogen infection, but no colonization of pustules on host could be recorded. It is generally agreed that plants immune to diseases are immune to infection by the pathogen. Genes, *Lr28* and *Lr24,* showed immunity in India over a very long period, even flecks were not observed on the host leaves. Many of the genes for hypersensitive reaction can produce an immune response to some rust races and visible fleck infection types to other races. Van der Plank (1968) put forwarded the concept of vertical (race-specific) and horizontal resistance (race non-specific). While studying stripe rust of wheat, Johnson (1978, 1981) introduced a term, durable resistance, which refers to resistance that has remained effective in cultivars for a very long period. In other terms, this resistance remains effective during the prolonged and widespread use in an environment favorable to disease (Johnson, 1984). Vertical resistance in wheat is effective from seedling stage onwards, but is short lived, whereas horizontal resistance is often effective at adult plant stage (Robinson, 1976), durable and effective against a wide spectrum of rust races. Parlevliet (1985) described it as partial resistance or slow rusting as it is associated with race non-specificity and is often referred as adult plant resistance (APR).

Slowing effect on disease development may be attributed to resistance controlled by minor genes (Browder, 1973). Some of the genes exhibit moderate or intermediate type of resistance, when the seedling produces 1–3 types of reactions. Such genes minimize the intensity and sporulation at adult plant stage. A *T. timopheevii* derived-gene *Lr18* and *Aegilops ventricosum*-derived genes *Lr37/Sr38/Yr17* are temperature sensitive and produce intermediate type of infection. Usually, genes that give moderate resistance in the seedling stage give the same type of resistance in the adult stage. A combination of several genes may confer durability of resistance on the premise that when components such as increased latent period, smaller pustule size, production of reduced number of uredia per unit area, etc., may contribute toward slow development of rust. Cultivars, viz., HP 1102, HP 1209, HW 741, and HW 971 carrying *Lr13* in combination with *Lr34* or *Lr23* were slow rusters and produced moderate infection type against 77-1, 77-2, and 104 races of leaf rust (Chopra & Kulkarni, 1980). HD2189 showing APR for more than 30 years in peninsular India carries race-specific genes, *Lr1, Lr13, Lr17*, and a race non-specific gene *Lr34/Sr57/Yr18/Bydv/Ltn*. In nature, there are a number of genes that accord resistant reaction to number of brown rust races such as *Lr9, Lr19, Lr24, Lr28, Lr32, Lr35, Lr37, Lr39, Lr44, Lr45, Lr53, Lr57* etc., and exploitation of such genes in commercial agriculture can exert strong selection pressure on *P. recondita tritici* as has been the case with *Lr26* cultivars during the recent past. Such vertical resistance genes result into a situation called "Boom and Bust." The evolution of brown rust virulences 77A (109R31), 12A (5R13), 12-2(1R5), and 104B (29R23) can be attributed to the bust phases. Until 1984, there was no race in India to match gene *Lr26* in pathogenicity. The widespread testing and release of selections possessing *Lr26*(1B/1R) translocation in various combinations involving other *Lr* genes have resulted, all of a sudden, in the arousal of different virulences matching *Lr26*. Similarly, *Yr9* and *Sr31*, the genes for yellow and stem rusts, respectively, tightly linked to *Lr26* fallen prey to a new race of *P. striiformis tritici* evolved in Mediterranean and transferred to India after 1995.

13.5 GENE DEPLOYMENT STRATEGIES TO COMBAT RUSTS

The cultivation of rust resistant wheat varieties is the most economical, efficient, and environmentally advantageous method to reduce yield losses caused by rust diseases. Judicious deployment of rust resistance genes has kept the rust diseases under control. Evolution of new virulent races of rust pathogens is a continuing phenomenon which renders resistance genes ineffective and thus necessitates deployment of newer genes. However, this approach demands a constant effort to identify, characterize, and incorporate resistance genes, mainly due to the great capability of rust populations to change (McCallum & Seto-Goh, 2005).

Lagudah (2011) proposed the effective use of specific rust resistance genes, out of which over 150 resistance genes that confer resistance to leaf rust, stripe rust, or stem rust have been catalogued or introgressed into wheat from related species. Few of these genes are highly influenced by environmental factors as well as genetic background of the host plants (McGregor & Manners, 1985) and few of these genes are the slow-rusting APR type that confers partial resistance in race non-specific manner to one or multiple rust diseases. Slow rusting, race non- specific, and partial resistance were defined by Parlevliet (1975). Slow rusting is a type of resistance where disease progresses at a retarded rate, resulting in intermediate to low disease levels against all pathotypes of a pathogen. The concept of race non-specific, horizontal resistance was widely used by Caldwell (1968) in breeding leaf rust resistant wheat varieties. APR is commonly detected at the post-seedling stage and often as field resistance, although some APR genes are sensitive to varying temperature and light conditions.

It is expected that by increasing genetic diversity in the field, the losses due to rusts are likely to be curtailed. While detailed knowledge of the genetic basis is not essential for an effective resistance breeding (Simmonds, 1985), much can be gained if the genetics of resistance is known. Browning et al. (1969) advocated the use of vertical resistance in different wheat zones along the *Puccinia* path in USA. McIntosh (1985) advocated the use of the combinations of oligogenes which exhibited low coefficient of rust infection. Gene deployment is a promising and effective strategy to curtail the rust epidemics. Like the other countries, the existence of a *Puccinia* path in the Indian subcontinent was established (Nagarajan & Joshi, 1980). It was earlier suggested to deploy specific gene or gene combinations in each zone (Joshi & Nagarajan, 1977). Therefore, strategically devised gene deployment can be followed in central and peninsular India against stem rust of wheat. Similar possibilities exist in the Indo-Gangetic Plain for managing leaf rust (Nagarajan & Joshi, 1980).

The strategy for gene deployment along the "*Puccinia* path" for managing stem and leaf rust of wheat has been proposed by Nagarajan et al. (1984) and Bahadur et al. (1985) following the effectiveness of resistance genes and the distribution of pathotypes. Bahadur et al. (1994) advocated the use of resistance genes, namely, *Sr24, Sr31,* and *Sr36* for stem rust and *Lr9, Lr19, Lr24,* and *Lr34* for the management of leaf rust in different agro-ecological zones in combination of other desirable genes. Stripe rust resistance gene *Yr9* provided resistance for a very long period in India but now has become ineffective due to evolution of new virulences. Deployment of genes such as *Yr5, Yr10,* and *Yr15,* which are individually as well as collectively effective against the prevailing races, would be an effective strategy to check the losses due to stripe rust. Virulence against *Yr5* was reported in Australia (Wellings & McIntosh, 1990) and India (Nagarajan et al., 1986). The gene

Yr10 has also been defeated due to emergence of a new virulence in Canada (Randhawa et al., 2011) but no virulence has been detected yet in India.

India has been deploying rust resistant cultivars from the very beginning. Before 1962, cultivars with tall stature, viz., NP 4, NP 12, NP 789, NP 824, C 591, C 218, K 68, PKD 4, PV 18, and C 306 exhibiting rust resistance at adult plant stage against prevailing rust pathotypes were released. Rust resistant Mexican wheats, Lerma Rojo and Sonora 64, were introduced after 1962 and the cultivars, Sonalika and Kalyansona, selected from segregating materials were grown on large scale shaping the Green Revolution. Through these varieties, rust resistance genes *Lr1, Lr3, Lr10* and *Lr13,Sr2, Sr5, Sr11,* and *Yr2* were deployed (Reddy & Rao, 1977; Nayar et al., 1994). The susceptible cultivars were replaced rapidly by new releases.

A successful, though unintentional, deployment for stem rust resistance is the large-scale cultivation of HD 2189 in Peninsular India which is known to carry *Sr57/Lr34/Yr18* along with *Sr2* and *Sr11*. HD 2189 also carried *Lr13*. Presently, this cultivar is resistant to Indian stem rust pathotypes. Another cultivar DWR 162 carrying *Sr31* was released in 1993 in PZ. Therefore, the inoculum which spreads from Karnataka or Nilgiris is not able to multiply as it lands on the resistant cultivars. Consequently, three popular cultivars of Central Zone, namely, Lok-1, Sujata, and WH 147, though susceptible, are protected due to lack of transported inoculum. In addition, resistant cultivars, HW 2004, HI 1500 (both rain-fed wheats sown early), and DL 788-2 (timely sown) carrying *Lr24/Sr24* are being cultivated on large scale in the Central Zone. To avoid any expected rust epidemic, wheat breeders need to diversify *Lr24*-based leaf rust resistance in Central Zone, as presence of *Lr24* in early sown varieties HW 2004, HI 1500, and HI 1531,timely sown variety HI 1544, and late sown variety DL 788-2 and MP 4010, poses a potential threat due to its continuity in time and space; particularly, in view of occurrence of *Lr24* virulence in neighboring Nepal (Mishra et al., 2014).

13.5.1 INTROGRESSION OF EFFECTIVE GENES THROUGH HYBRIDIZATION AND MARKER-ASSISTED SELECTION

A meticulously planned wheat improvement program employing backcross method to introgress effective rust and powdery mildew resistance genes was initiated during late 1980s and early 1990s. Popular Indian bread wheat and *dicoccum* wheat varieties were taken for theprogram. The reference stocks (RIL's) obtained were initially evaluated for its resistance and the effective genes conferring resistance to existing pathotypes were only taken for theprogram. The effective resistance genes were introgressed initially through conventional backcross hybridization method taking advantage of IARI Wellington where in all three rusts and other foliar diseases occur on a susceptible line all through the year and considered as natural "hot spot." Later, when markers were made available, both conventional and marker-assisted selection (MAS) approaches are undertaken. Initially, a number of backcrosses were 8–9 and now is restricted with BC_3. For molecular confirmation, the mapping population was picked at BC_1F_2 stage.

Following alien rust resistance genes in the backcross program taken up at Wellington:

Stem rust resistance genes: *Sr2* (linked to pseudo black chaff (*Pbc*) on glumes), *Sr22, Sr24, Sr25, Sr26, Sr27* (linked to apical claw on ear), *Sr29, Sr30, Sr31, Sr32, Sr33, Sr35, Sr36, Sr38, Sr39, Sr42, Sr43, Sr44, Sr47,*and *Sr49*.

Leaf rust resistance genes: *Lr9* (now not effective in India), *Lr19* (new virulent reported), *Lr24, Lr26* (not effective in India), *Lr28* (new virulent reported), *Lr32, Lr34* (APR race non-specific and linked to leaf tip necrosis), *Lr35* (APR response), *Lr37, Lr39, Lr40, Lr41, Lr42, Lr45* (linked to pink awn/glumes at milking stage under low temperature), *Lr46* (APR race non-specific), *Lr47, Lr53, Lr57, Lr67* (APR), and *Lr68* (APR).

Stripe rust resistance genes: *Yr9, Yr10, Yr15, Yr16, Yr17, Yr18, Yr24, Yr25, Yr26, Yr29, Yr30, Yr35,* and *Yr40*.

Powdery mildew resistance genes: *Pm6, Pm8, Pm38,* and *Pm39*.

Pleiotropic or closely linked to genes (race non-specific) exploited which are effective to other diseases: *Lr34/Yr18/Pm38/Bdv1/Sr57/Ltn, Lr46/Yr29/Pm39/Ltn, Lr67/Yr46,* and *Sr2/Yr30/(Lr27)/Pbc*.

Other linked genes that are exploited: *Lr19+Sr25, Lr24+Sr24, Lr26+Yr9+Sr31+Pm8, Lr37+Yr17+Sr38, Sr39+Lr35, Lr53/Yr35,* and *Lr57/Yr40*.

Pyramiding of effective stem rust resistance genes currently under progress to overcome threat from Ug99 and its variants of stem rust race virulent on *Sr*31, *Sr*24, and *Sr*36—virulence spectrum of Ug99 (TTKSK).

TABLE 13.2 Effective Rust Resistance Genes/Linked Genes Used in the Gene Pyramiding Program at IARI, Wellington.

Gene	Source	Reference stocks used (RIL's)	Chromosomal location
*Lr9 (not effective at Wellington since 1995)	Aegilops umbellulata	Abe	6BL
*Lr19/Sr25 (pathotype 77-8 reported from PZ, India during 2008)	Thinopyrum ponticum	Cook and now Wheatear	7DL
*Lr24/Sr24 (pathotype40-1 reported from Wellington on Sr24)	Th. ponticum	Tr380-14*7/3Ag#14 Janz, Sunleg, RL6064 Agent	3DL
*Lr26/Sr31/Yr9/Pm8 (pathotype 77-1 reported from Wellington for Lr26)	Secale cereal (Petkus rye)	WH 542(Bucanora)	T1BL1RS
*Lr28(pathotype)	Ae. speltoides	CS 2A/2M 4/2	4AL
*Lr32	Ae. squarrosa/T. tauschii	C86-8/KalyansonaF4/, Thatcher Lr32 later	3DS
*Lr34/Sr57/Yr18/BDV1/Pm 38/Ltn (APR-race non-specific/pleiotropic	T. aestivum, Terenizo	RL 6058, Webster	7DS
*Lr35/Sr39	Ae. speltoides	Thatcher+ Lr 35	2B
*Lr37/Sr38/Yr17	Ae. ventricosa	Thatcher*8/VPM 1, RL 6081	2AS
*Lr39	T. tauschii	KS 92 WGRC 15, EZ 350692	2DS
Lr44	T. spelta	EC 381202 (RL 6147)	1BL
Lr45	S. cereal (Imperial Rye)	EC 381203 (RL 6144)	TAS-2R
*Lr46/Yr29/Pm39/Ltn, pleiotropic(APR)	T. aestivum	PAVON 76, Dimond Bird)	1BL
*Lr47	Ae. speltoides	PAVON 76S, Lr 47, KS 90 H 450	7AS
Lr53/Yr35	T. turgidum–T. dicoccoides	AUS 91388	6BS
*Lr57/Yr40	Ae. geniculata	T291-2(WL 711)	T5DL.5DS-5M^{g}S (0.95)
*Lr67/Yr46pleiotropic(APR)	T. aestivum	RL6077	4DL

TABLE 13.2 (*Continued*)

Gene	Source	Reference stocks used (RIL's)	Chromosomal location
*Lr68	T. aestivum	Parula, Arula1, Arula2	7BL
*Sr2/Lr27/Yr30/(pseudo black chaff)APR	T. aestivum	Hope, Songlen, Arthur and many Indian released cultivars, e.g., Lok-1	3BS
*Sr-22 (APR)	T. boeoticum	Co 1213 HSBVN 163313 with reduced segment	7AL
*Sr24/Lr24	Th. ponticum	Tr380-14*7/3Ag#14	3DL
*Sr25/Lr19	Th. ponticum	Whetear-1	7DL
*Sr26	Th. ponticum	DARF*6/3Ag3/Kite	6AL
Sr27	S. cereale (Imperial)	Kalyanasona'*4/Sr27	3A
Sr29	T. aestivum	Pusa 4/Etoile de choisy	6DL
Sr30	T. aestivum	BtSr30Wst	5DL
*Sr31/Lr26/Yr9/Pm8			
Sr32	Ae. speltoides	CnsSr32 AS	2A, 2B, 2AS
*Sr33	T. tauschii	RL 5405	1DL, 1DS
Sr35	T. monococcum	Mq (2)5*G2919	3AL
Sr36/Pm6	T. timopheevii	Cook*6/C 80-1	2BS
Sr38/Lr37/Yr17	Ae. ventricosa	Thatcher*8/VPM 1, RL 6081	2AS
*Sr39/Lr35	Ae. speltoides (APR)	Thatcher+ Lr 35	2B
Sr42	T. aestivum	EC 381206	6DS
Sr43	Th. ponticum	EC 381210	7DL
*Sr44	Th. intermedium		

TABLE 13.2 (Continued)

Gene	Source	Reference stocks used (RIL's)	Chromosomal location
*Yr9/Lr26/Sr31/Pm8	S. cereal (Petkus rye)	WH 542 (Bucanora)	T1BL.1RS
*Yr10	T. spelta	Moro, Yr10+WH 542	1BS
*Yr15	T. dicoccoides	T. dicoccoidesG-25, V763-2312	1BL
Yr16	Cappelle–Desprez	Cappelle–Desprez	2DS
*Yr17/Lr37/Sr38	Ae. ventricosa	Thatcher*8/VPM 1, RL 6081	2AS
*Yr18/Lr34/Sr57/BDV1/Pm 38//Ltn	T. aestivum, Terenizo	RL 6058	7DS
*Yr30/Sr2/Lr27 (pseudo black chaff)	T. aestivum	Maden	3BS
Yr35/Lr53	T. turgidum-T. dicoccoides		6BS
Yr37/Lr54	Ae. kotschyi		2DL.
Yr38/Lr56	Ae. sharonensis		6A
Yr40/Lr57	Ae. geniculata		T5DL.5DS-5MgS (0.95)
Yr42/Lr62	Ae. neglecta		
Pm6/Sr36	T. timopheevii	Cook*6/C 80-1, Abe	2BS
Pm8/Yr9/Lr26/Sr31	S. cereale	WH 542, Bacanora	T1BL.1RS
Pm 38/Lr34/BDV1/Yr18/Sr resistance/Ltn	T. aestivum, Terenizo	RL 6058	7DS

*Genetic markers available in public domain. Details on genetical markers can be accessed through maswheat.ucdavis.edu/.

Genes that are currently effective against Ug99 (2, 9) are listed below:

Origin of genes	Effective *Sr* genes
Lophopyrum ponticum	*Sr25*
Triticum aestivum	*Sr28*[1], *Sr29*[2], *SrTmp*[1]
T. turgidum	*Sr2, Sr13*[1,2], *Sr14*[1]
T. monococcum	*Sr22, Sr35*
T. timopheevii	*Sr36*[1], *Sr37*
T. speltoides	*Sr32, Sr39*
T. tauschii	*Sr33*[2], *Sr45*
T. araraticum	*Sr40*
Thinopyrum elongatum	*Sr26, Sr43*
Th. intermedium	*Sr44*
Secale cereale	*Sr27*[1], *Sr1A. 1R*[1]

[1]*Virulence for the gene is known to occur in other races.*

[2]*Level of resistance conferred in the field usually not enough.*

13.5.2 INTROGRESSION OF SINGLE (OLIGO GENES) EFFECTIVE GENES INTO COMMERCIAL INDIAN WHEAT CULTIVARS

13.5.2.1 LR9

Triticum umbellulatum-derived gene *Lr9* was transferred to Chinese Spring wheat. Translocation 47 became known as Transfer (Sears, 1956). *Lr9* is located in 6B of the chromosomal arm showing low infection type of 0 and occasionally 1+ and has low environmental variability. However, the virulent for *Lr9* was found in USA in 1971 four years after its use in soft red winter wheat (Shaner et al., 1972). The source stock Abe and Arthur 71 were used for transferring *Lr9* into commercial cultivars of India at IARI, Regional Station, Wellington in India (Tomar & Menon, 1998). Initially, it has the gene *Lr9*-conferred seedling resistance to 13 Indian races of leaf rust (Sawhney et al., 1977) and has consistently displayed immune reaction to leaf in adult plant stage at Wellington for over 25 years (Tomar & Menon, 1998). At IARI, Regional Station, Wellington, authors had developed near isogenic (BC) lines carrying *Lr9* in the background of 12 popular Indian bread wheat cultivars such as HW 2067 (HD 2009), HW 2054 (HD 2285), HW 2055 (HD 2329), HW 2070 (HS 240), HW 2056 (HUW 234), HW 2053 (Kalyansona), HW 2052 (LOK-1), HW 2060 (NI 5439), HW 2057 (PBW 226), HW 2051 (Sonalika), HW 2058 (WH 147), and HW 2059 (WH 542).

However, even before its exploitation for commercial use, the authors for the first time detected virulence for *Lr9* in September 1998 on the Indian bread wheat cultivars Sonalika and HUW 234*5/Abe (HW 2056), both carrying the gene *Lr9*. The virulence was further confirmed on other Indian common wheats with *Lr9* as well as on Transfer, Thatcher*6/Transfer (RL6010), addition line (CS 6U), Abe, Arthur 71, and Oasis (Nayar et al., 2003) and the pathotype has been designated as 77-7. As per the authors' observation, the alien segment with *Lr9* is associated with an enhanced susceptibility to powdery mildew in the backcross derivatives.

13.5.2.2 LR19/SR25

The linked genes *Lr19/Sr25* located in the chromosome 7D transferred from *Th. ponticum* (10x) in the winter wheat Agrus had linkage drag of yellow pigmentation in the flour and hence could not be utilized in wheat breeding because of their undesirable linkages with yellow flour pigmentation which was not commercially acceptable. Agatha, a line resulting in the translocation stock T4, when subjected to irradiation carried *Lr19/Sr25* devoid of yellow pigmentation (Sharma & Knott, 1966) and subsequent independent transfers of *Lr19/Sr25* Chinese spring were used worldwide to exploit this gene Knott (1980, 1984). In one of the lines, Agatha-235, transferred from *A. elongatum* carried only leaf rust resistance gene, presumably *Sr25* was lost along with the gene for yellow pigmentation (Friebe et al., 1994, 1996). However, the authors here have used lines Sunstar*6/C 80-1 and Condor*6/C 80-1 in white-seeded background with highly reduced pigmentation as a donor parent in the backcross program to introgress *Lr19/Sr25* into the susceptible Indian cultivars and developed 20 NIL's carrying *Lr19* [HW 2083 (C 306), HW 2079 (HD 2009), HW 2049 (HD 2285), HW 2046 (HD 2329), HW 2045 (HD 2402), HW 2078 (HD 2687), HW 2048 (HI 1077), HW 2085 (HP 1205), HW 2080 (HS 240), HW 2043 (HUW 234), HW 2084 (J 24), HW 2081 (Kalyansona), HW 2041 (LOK-1), HW 2086 (MACS 2496), HW 2082 (NI 5439), HW 2044 (PBW 226), HW 2050 (Sonalika), HW 2087 (UP 262), HW 2042 (WH 147), and HW 2047 (WH 542)] but later the gene source Sunstar*6/C 80-1 and introgressed lines were confirmed to carry *Lr24/Sr24* and not *Lr19/Sr25* (Prabhu et al., 1998). Out of these, HW 2044 and HW 2045 were released as cultivars in India. Later, the authors used the Australian cultivar "Cook" as donor parent which carry *Lr19/Sr25* and *Sr36/Pm6* and "Wheatear." A wide spectrum of pathotypes of leaf rust prevalent in USA, Canada, Australia, India, and other countries are virulent on *Lr19* (Agatha). Virulence for *Lr19* was detected in *P. recondita* (CBJ/QQ) in Mexico (Huerta-Espino & Singh, 1994) from the cultivar Oasis 86. Since Agatha equals its parent Thatcher in yield, milling, and baking quality, the authors observed enhanced yield in backcross lines [HW 4203 (HD 2285), HW 4204 (HD 2329), HW 4205 (HD 2402), HW 4206 (HD 2687), HW 4207 (HS 240), HW 4208 (J 24), HW 4209 (Kalyansona), HW 4210 (LOK-1), HW 4211 (MACS 2496), HW 4212 (NI 5439), HW 4213 (PBN 51), HW 4218 (WH 147), HW 4219 (WH 542), HW 4220 (RL 6144), HW 3627 (RAJ 3077), HW 3608 (HD 2687), HW 3607 (HD 2402), HW 3601 (C 306), HW 3620 (MACS 2496) etc.] carrying *Lr19* developed using "Cook" as donor which were confirmed to carry either *Lr19/Sr25* alone or *Lr19/Sr25* and *Sr36/Pm6* at molecular level (Sivasamy et al., 2009) although yield reductions associated with alien genetic transfers have been reported in some cases in hexaploid wheat cultivars in Sweden (Sunann) and Mexico (Oasis 86) (McIntosh et al., 1995). However, Singh et al. (2006) reported enhanced yield in the wheat derivatives carry *Lr19/Sr25* gene complex. The gene *Lr19* has been deployed by us for the first time in the Southern hills region of India. The cultivars HW 2044 carrying *Lr19/Sr25* is expected to serve as an effective genetic barrier against the spread of rust from the southern foci of rust inoculum. In addition, they have a greater potential in wheat breeding. The authors also observed that *Lr19* enhances the susceptibility to *Pm*.

13.5.2.3   LR24/SR24

The leaf rust resistance gene *Lr24/Sr24* was originally detected in a bread wheat line possessing spontaneous translocation involving chromosome 3D in agent in the stock Agent derived from *Lophopyrum elongatum* selected from across of this line cv. Triumph is the origin (Gough & Merkle, 1971). The gene present in the genotype Agent is a spontaneous wheat-translocation involving 3Ag and 3DL chromosomes. In Agent, *Lr24* is tightly linked with stem rust resistance gene *Sr24* and with a gene for red seed color. The initial impediment to use of *Lr24/Sr24* from Agent was in association with red grain color but McIntosh and Partridge (un published 1974) were able to recover white-seeded recombinants. One of the white-seeded recombinant lines, *Tr 380-4*7/3 Ag#14* proved to be a valuable genetic stock for wheat improvement, was used to develop BC lines of popular Indian cultivars carrying *Lr24/Sr24* at Wellington which include: HW 2011, HW 2008, HW 2007, HW 2017, HW 2071, HW 2018, HW 2020, HW 2015, HW 2010, HW

2002A, HW 2006, HW 2073, HW 2003, HW 2016, HW 2001A, HW 2012, HW 2072, and HW 2004.

Pathotypes virulent on *Lr24* have been reported from North America (Browder, 1973), Canada (Kolmer, 1991), South America (Singh, 1991), and South Africa (Pretorius et al., 1990) but this gene is highly effective in Australia and India. But, isolates of *P. graminis* virulent to *Sr24* appeared in South Africa (Le Roux & Rijkenburg, 1987) and in India (pathotype 40-1) (Bhardwaj et al., 1990) although the donor *TR 380-14*7/3Ag#14* exhibited a high magnitude of APR to stem rust indicating the presence of some additional factors for resistance. However, *Lr24* still continues to be highly effective in seeding as well as in adult stage to Indian pathotypes of *P. recondita,* and virulence for *Lr24* occurs in low frequencies in most geographical areas (Huerta-Espino, 1992). Bread wheat cultivars carrying *Lr24/Sr24* are widely grown in Australia, North America, and South Africa. The cultivars viz., DL784-3 (Vidisha), HW 2004 (Amar), DL788-2 (Vaishali), HW 2045 (Kausambi), HD 2781 (Aditya), HI 1500 (Amrita), MP 4010, Raj 4037, HD 2851 (Pusa Vishesh), HD 2833 (Tripti), HI 1531, COW(W)-1, HD 2888 (Pusa Wheat), AKAW3722 (Vimal), AKAW 4627, and HW 5207 (Pusa Navagiri) all carrying *Lr24/Sr24* have been released in recent years in India for commercial cultivation. The deployment of this effective gene complex in Indian released cultivars for the last more than a decade, widely across India played pivotal role in checkmating the brown rust (Tomar et al., 2014).

13.5.2.4   LR26/SR31/YR9/PM8

The *S. cereale*-derived linked gene *Lr26/Sr31/Yr9/Pm8* is a spontaneous wheat-rye substitution which was identified by Kattermann (1937) in which the wheat chromosome 1B was replaced by rye chromosome 1R (1BL/1R#1S). Occurrence of 1B (1BL/1RS) are1R (1B) were reported by Mettin et al. (1973) and Zeller (1973). However, some wheat consisting of both substitution and translocation biotypes (Zeller, 1973) were also reported and *Lr26* is completely linked with *Sr31, Yr9*, and *Pm8*. The linkage of *Yr9* for stripe rust resistance and *Pm8* for powdery mildew resistance was reported by (Bartos et al., 1973). Rajaram et al. (1983) reported the significant heterotic effect on grain yield of this translocation. However, genotypes carrying this translocation has the linkage drag of poor grain quality, because dough made from these tend to be sticky with over mixing and earlier reports of tight linkage to red grain. The global importance of 1RS in wheat program has been well documented. Depending on the wheat genotype into which the translocation has been introduced, 1RS may also directly increase yield (Villareal et al., 1997). The impact of this disease resistance gene complex on global wheat production is evident from worldwide exploitation for its yield increase in many of both winter and spring wheat cultivars. However, after the Ug99 pathotype became virulent on *Sr31*, worldwide efforts are on to diversify the genetic basis of resistance; nevertheless, one cannot ignore the yield advantage associated with this gene complex. Although more than 60% of Indian wheat cultivars carrying this gene complex are complementing wheat yields in India, unintentionally the Indian cultivars carry additional gene complexes like *Sr2+, Sr22, Sr14, Lr34+, Lr46+,* and *Lr67+* gene complexes minimize the direct threat from Ug99 stem rust race or *Lr26* virulent race occurring in India.

13.5.2.5   LR28

McIntosh et al. (1982) reported that the *Ae. speltoides*-derived gene located in 4AL mapped at 53 cm from the centromere. This gene was incidentally transferred while attempts were made to transfer the stripe rust resistance gene (*Yr8*) from *Ae. comosa* var. *comosa* to wheat, a high pairing line of *Ae. speltoides* was used by Riley et al. (1968) to induce homoeologous recombination. The gene *Lr28* shows infection type 0; sometimes 1+ to 2+ especially with South American isolates (Huerta-Espino, 1992). The authors used the line CS *2A/2M#4/2* in their BC program which was identified to have a gene for leaf rust resistance, *Lr28* derived from *Ae. speltoides* (McIntosh et al., 1982). The segment carrying *Lr28* is most likely to be derived from the short arm of *Ae. speltoides* chromosome 7S#2 during homoeologous recombination resulting in the translocate in chromosome T4AS.4AL-7S#2 (Friebe et al., 1996).

The gene *Lr28* confers a high degree of resistance in seedling as well as at adult plant stage to all the leaf rust pathotypes prevailing in India. It is also widely effective in Australia, South Asia, and Europe. However, most *P. recondita* isolates in North America are virulent to *Lr28*. No detrimental effects are associated with the presence of *Lr28* but the durability of resistance is likely to be low (McIntosh et al., 1995). Of late, a pathotype virulent on *Lr28* has been reported from Wellington in India (Bhardwaj, 2010), although it was not reported earlier from Indian sub-continent. The cultivar Sunland was registered in New South Wales, Australia as prime with hard texture of seed in 1992. In India, the authors of the compendium have also transferred *Lr28* into 16 Indian cultivars susceptible to leaf rust and developed BC lines viz., HW 2034 (C 306), HW 2064 (HD 2009), HW 2038 (HD 2285), HW 2037 (HD 2329), HW 2065 (HD 2402), HW 2066 (HD 2687), HW 2063 (HS 240), HW 2039 (HUW 234), HW 2036 (J 24), HW 2061 (Kalyansona), HW 2032 (LOK-1), HW 2035 (NI 5439), HW 2040 (PBW 226), HW 2031 (Sonalika), HW 2033 (WH 147), and HW 2062 (WH 542).Out of these, HW 2034 has been released in India as cultivar as MACS 6145. Two backcross lines, HD 2329*7/CS 2A/2M#4/2 and WH147*7/CS 2A/2M#4/2 did not show any yield difference under rust-free conditions as compared to recurrent parents HD 2329 and WH 147, respectively. Some of the backcross lines carrying the gene *Lr28* showed fast rusting to stem rust and reduced susceptibility to powdery mildew as compared to their recurrent parents under natural condition at Wellington indicating the phenomenon of fast rusting to stem rust which appears to be associated with the resistance imparted by *Lr28* (Tomar & Menon, 1999).

13.5.2.6   LR32

Kerber (1987, 1988) confirmed the location of this gene at 3DS which was derived from *Ae. squarrosa* (=*T.tauschii*) which has been found effective in adult plant stage at Wellington as well as in seedling stage with low infection type to the prevalent pathotypes of leaf rust with low environmental variability. Sawhney and Sharma (1990) reported that *Lr32* produces mainly infection type 2 against different pathotypes of leaf rust in seedling stage. Huerta-Espino (1992) and Kerber (1987) also recorded low infection type (0; to 2+) on the lines carrying *Lr32* as well as on donor *Ae. squarrosa*. Pathogenic variability for *Lr32* has not been reported except an isolate from Bulgaria and Turkey. Commercially, this gene has not been exploited much; however, the authors have transferred *Lr32* initially using C86-8/Kalyansona F4 in the background of 14 Indian bread wheat cultivars susceptible to leaf rust namely HW 4001 (C 306), HW 4002 (HD 2285), HW 4003 (HD 2329), HW 4015 (HD 2687), HW 4013 (HS 240), HW 4004 (HUW 234), HW 4005 (Kalyansona), HW 4006 (LOK-1), HW 4007 (NI 5439), HW 4008 (PBW 226), HW 4009 (Sonalika), HW 4012 (UP 262), HW 4010 (WH 147), and HW 4011 (WH 542). Later, Prabhu et al. (1998) confirmed molecularly that all these lines including the donor C86-8/KalyansonaF4 carried only *Lr28* and hence the authors used the donor Thatcher *Lr32* later to transfer this gene into commercial cultivars. The backcross derivatives with *Lr32* also showed fast rusting to stem rust and reduced susceptibility to Pm as compared to their recurrent parents (Tomar & Menon, 1999) similar to that of *Lr28*.

13.5.2.7 LR34/SR57/YR18/BDV1/PM38/LTN

The gene *Lr34* derived from varieties Terenzio showed a moderate level of susceptibility to leaf rust (30–60 MS) at adult plant stage. Sawhney and Sharma (1990) reported that *Lr34* mainly exhibited an infection type 3 to many pathotypes of leaf rust in the seedling. Several researchers have, however, reported that *Lr34* interacted favorably with *Lr13* (Roelfs, 1988) and also with *Lr33* and *LrT3* (Samborski & Dyck, 1982) to confer durable resistance to leaf rust. The transfer of *Lr34*, an APR gene which is associated with *Sr57, Yr18*, BDV1, and leaf tip necrosis (*Ltn*) genes as pleiotropic association if pyramided with major or minor gene(s), expected to give durable rust resistance (Singh & Rajaram, 1992). However when alone, they were observed to be ineffective at Wellington, while wheat varieties, Chris, Frontana, and La Prevision all carrying the genes *Lr13* and *Lr34* exhibited moderate susceptibility with the exception of Era showing resistance to leaf rust. The study indicated that there is little or no interaction between *Lr34* and *Lr13*. The stocks RL 6058, Webster, and Diamondbird carry *Lr34* which are used at Wellington to introgress and pyramid the gene with other APR genes *Lr46, Lr67*, and *Lr68* with the objective of developing durable rust resistant wheat varieties.

13.5.2.8 LR35

The gene *Lr35* located at 2B linked to a gene for seedling resistance to stem rust gene *Sr39* (Kerber & Dyck, 1990). The APR gene *Lr35* was transferred by homoeologous recombination from *Ae. speltoides* chromosome 2S#2 to wheat chromosome 2B (Kerber & Dyck, 1990). The line RL 6082 (TC*6/ RL 5711) carrying *Lr35* which is linked to *Sr39* showed moderate susceptibility with low intensity (30 MR, MS) to leaf rust and high degree resistance to stem rust at Wellington. Seedling tests of line carrying the gene *Lr35* produced an infection type 3 (Sawhney et al., 1994) with selected pathotypes of leaf rust. They explained that *Lr35* imparts races non-specific APR which is likely to be durable. The gene *Lr35* has not been used for cultivar improvement (McIntosh et al., 1995). Although *Lr35* reportedto have yield penalty in the lines introgressed with this gene when compared to recurrent parent, in our experiences careful selection of transgressive segregants facilitates selecting of lines without having any yield penalty. The gene *Lr35* linked to *Sr39* is effective against stem rust races occurring in India and Ug99 which can be well exploited.

13.5.2.9   LR37/SR38/YR17

The gene *Lr37* is derived from *Ae. ventricosa* (Bariana & McIntosh, 1994) and is present in a French winter wheat VPM1 which has been selected for resistance to strawbreaker disease or eye spot disease caused by *Pseudocercosporella herpotrichoides* located on chromosome 7D. The gene *Lr37* was found effective against the virulent Indian pathotypes of *P. recondita* in adult plant stage at Wellington. This gene is located on the short arm of chromosome 2A and is linked with the genes *Sr38* and *Yr17* (Bariana & McIntosh, 1993) both providing moderate degree of resistance to stem rust and stripe rust, respectively, Kloppers and Pretorius (1995) found that the gene *Lr37* conferred slow rusting resistance as they observed fewer uredia smaller uredinium with a low rate of uredinial appearance. In India, the authors have incorporated the gene *Lr37* in 16 commercial Indian bread wheat cultivars susceptible to leaf rust viz., HW 4022 (HD 2285), HW 4023 (HD 2329), HW 4033 (HD 2687) HW 4032 (HS 240), HW 4024 (HUW 234), HW 4025 (Kalyansona), HW 4026 (Lok-1), HW 4034 (MACS 2496), HW 4027 (NI 5439), HW 4035 (PBN 51), HW 4028 (PBW 226), HW 4029 (Sonalika), HW 4030 (WH 147) HW 4031 (WH 542), HW 4036 (HI 1077), and HW 4037(RAJ 3077) effecting five back-crosses. The expressions of resistance in the seedling stage with leaf rust pathotypes 77-5, 77-6, and 77-7 greatly vary (0; 1 to x =) (V.C. Sinha, Personal communication) in different genetic backgrounds. In the adult plant stage, the resistance of *Lr37* was of high magnitude in HD 2285, Kalyansona, NI 5439, and WH 147 backgrounds, while the same level of resistance could not be combined in the backcross derivatives of HD 2329, Lok-1, and Sonalika and it was observed that effectiveness of this gene is highly influenced by temperature and showed susceptible reaction at >25°C day temperature. Moreover, the linked yellow rust resistance gene *Yr17* showed susceptible reaction in certain genetic background, particularly in Kalyansona. The lines carrying *Lr37* seem to be associated with any enhanced susceptibility to Nilgiris flora of *Erysiphe graminis tritici* (Menon & Tomar, 2001) (Score 4) as compared to the recurrent parents (Score 3). Enhanced susceptibility to powdery mildew is also evident on the donor line compared to Thatcher and its derivatives. These linked genes have been used in Australian wheat cultivars, Sunbird, Sun state, and Trident (McIntosh et al., 1995). The gene *Lr37* is usually expressed as an APR gene.

13.5.2.10   LR39

A leaf rust resistance gene designated *Lr39* was transferred from *tauscher* accession TA 1675 to the wheat cultivar Wichita and released as the wheat germplasm KS86WGRC02 (Gill et al., 1988). The gene had been located on the short arm of chromosome 2D by geocentric analysis and the seedlings showed infection type of 0 to; The genetic mapping of leaf rust resistance in KS89WGRC10 on chromosome 2DS, along with similar race specificity observed for *Lr39* and *Lr41* suggests that these two genes are the same. This is also supported by results of our allele studies with line WX93D246R-1 that also has *Lr39*. To date, *Lr39* has been transferred to wheat from at least four accessions of diverse geographic origin as well as an accession of. The closely linked wheat microsatellite marker *Xgdm35* should be useful for MAS for *Lr39* (*Lr41*).

At Wellington, the donor KS92WGRC15 is being used and it is conferring high degree of resistance with infection of 0 to; in the seedling. Introgression of this effective gene is in progress at Wellington and the gene was successfully transferred into commercial Indian cultivar, HD 2285 (HW 3904) and it is being evaluated under the common varietal trial (CVT) of IARI.

13.5.2.11   LR44

Triticum spelta-derived gene *Lr44* located on the 1BL transferred to RL 6147(EC 381202) exhibited moderate resistance to leaf rust at Wellington with the field reaction of 20S, 20S, F and 2 for stem, leaf, stripe, and Pm, respectively. This is being exploited at Wellington to impart yellow rust resistance by pyramiding with other leaf and stem rust genes.

13.5.2.12   LR45

Mukade et al. (1970) using X-rays transferred leaf rust resistance from *S. cereale* cv. Petkus to wheat. The leaf rust resistance gene designated as *Lr45* has been located on the wheat-rye translocation chromosome T2AS-2R#3S. 2R#3L (Friebe et al., 1996). The last nine years of evaluation against leaf rust at Wellington indicated that the gene is effective against prevailing leaf rust pathotypes in the Nilgiris. The gene conferred high degree of leaf rust resistance against all the prevailing leaf rust pathotypes. The authors have successfully transferred this gene into 30 commercial bread wheat culti-vars of India and developed mapping populations and from these materials, Bhojaraja (2012) mapped the *S. cereale*-derived gene *Lr45* on 2A and devel-oped PCR-based co-dominant markers, for the first time. The tightly linked phenotypic marker *pink color on the awns and margin of glumes* which get well expressed during dough stage under cooler weather conditions was well exploited by authors initially to transfer this gene in the absence of genetic marker and developed backcross derivatives, HW 4501 (C306), HW 4502 (GW273), HW 4503 (HD 2189), HW 4504 (HD2285), HW 4505 (HD 2329), HW 4506 (HD 2402), HW 4507 (HD2687), HW 4508 (HD 2733), HW 4509 (HD 2877), HW 4510 (HI 977), HW 4511 (HI 1077), HW 4512 (HP 1205), HW 4513 (HS 240), HW4514 (HUW 234), HW 4515 (J 24), HW 4516 (Kalyansona), HW 4517 (Lalbahadur), HW 4518 (Lok-1), HW 4519 (MACS 2496), HW 4520 (NI 5439), HW4521 (NIAW 34), HW 4522 (PBN 51), HW4523 (PBW 226), HW4524 (PBW 343), HW4525 (PBW 502), HW 4526 (Raj 3077), HW 4527 (UP 2338), HW 4528 (UP 2425), HW 4529 (WH 147), and HW 4530 (WH 542) which carry *Lr45*. Further, these lines were pyramided with *T. timopheevii*-derived stem rust resistance gene *Sr36/Pm6* to develop wheat lines conferring resistance to leaf, stem rusts, and powdery mildew. Another interesting observationwas, in all, the *Lr45* introgressed lines the ear length increased with increased grain weight and lower most spike sterility observed in some recurrent parent restored as fertile ones.

13.5.2.13 LR46

Leaf rust resistance gene *Lr46* is a slow rusting gene. This gene do not provide the host plant with complete immunity against a set of leaf rust (*P. triticina*) races; instead they can delay the infection process or reduce the development of symptoms caused by a wider range of leaf rust races on adult plants. Singh et al. (1998) first described *Lr46* in 1998 in cultivar Pavon 76, and located on chromosome 1B. Martinez et al. (2001) showed that the latency period of infected adult plants was significantly lower in plants carrying *Lr46* compared to the controls without the gene. *Lr46* was also responsible for an increase in the fraction of early aborted fungal colo-nies. The type of resistance conferred by *Lr46* is similar to that of *Lr34*, although with a smaller effect. William (2003) found that *Lr46* was tightly linked or pleiotropic to a stripe rust resistance gene designated *Yr29*. The tight linkage of a slow rusting gene to a stripe rust resistance gene was also found for the pair *Lr34/Yr18*. The authors effectively used this gene complex in gene pyramiding in developing durable rust resistance wheat varieties in selected Indian commercial wheat cultivars HD 2733, UP 2338, PBW 343, and HD 2687 at Wellington.

13.5.2.14 LR47

The leaf rust resistance gene *Lr47* confers resistance to a wide spectrum of leaf rust strains at Wellington. This gene was transferred from chromosome 7S of *T. speltoides* to chromosome 7A of *T. aestivum*. Pavon 76—a donor carrying this gene is used for transferring this useful gene and the marker *Xabc465* is effectively used in the Man and the Biosphere (MAB) program.

13.5.2.15 LR53

The rust resistance genes, *Lr53* and *Yr35*, transferred to common wheat from *T. dicoccoides* were reported previously to be completely linked on chromosome 6B. Australian line AUS 91388 carrying *Lr53/Yr35* is used at Wellington to transfer this gene. The gene showed infection type 0 to; in seedling and immune response in the adult plant stage. However, the authors observed fast rusting for stem rust in the donor and the derived lines.

13.5.2.16 LR57

The wheat, *Ae. geniculata* introgression T5DL.5DS-5MgS (0.95), with stripe rust resistance gene *Yr40* and leaf rust resistance gene *Lr57*, is an effective source of resistance against most isolates of the rust pathogen in Kansas and India. The wheat variety WL 711 carrying *Lr57* developed at PAU, Ludhiana was used for transferring the gene *Lr57* into commercial bread wheat cultivars of India. The gene *Lr57* if effective against prevalent pathotypes of leaf rust in India can be well exploited.

13.5.2.17 LR67

The *Lr67* gene for APR to leaf rust was identified in the common wheat accession PI 250413 (Dyck & Samborski, 1979) and transferred into Thatcher to produce the backcross line RL6077 (Thatcher*6/PI250413). *Lr67* is phenotypically similar to *Lr34* because it could also be associated with resistance to stem rust (Dyck et al., 1994) and stripe rust (Singh, 1992), although *Lr67* confers a lower level of leaf rust resistance than that induced by *Lr34* (Hiebert et al., 2010) and they observed leaf tip necrosis, which is associated with *Lr34* and *Lr46*, was also recorded in segregants carrying *Lr67*. The combinations of *Lr34, Lr46,* and *Lr67* represent an attractive option to breeders for durable multi-pathogen resistance to leaf rust, stripe rust, stem rust, and powdery mildew. Comparison of RL 6077 with Thatcher and RL 6106 (Thatcher + *Lr34*) over four years of field testing showed that both *Lr34* and *Lr67* conditioned improved resistance compared with Thatcher, with *Lr34* conferring a higher level of resistance compared with *Lr67*. Also, field trials showed no significant effect of the APR gene on average yield and other agronomic traits, such as height, maturity, lodging, and kernel weight. Similarly, end-use quality traits (whole wheat, particle size, and sodium dodecyl sulfate (SDS) sedimentation) did not differ significantly for both lines. This gene is being exploited by pyramiding with other APR genes *Lr34* and *Lr46* for developing durable rust wheat varieties in selected commercial bread wheat cultivars at Wellington.

Interestingly, many released Indian wheat cultivars carried the APR genes *Lr34+, Lr46+, and Lr67+* gene complexes. Our observations at Wellington (India) are that the linked APR genes *Lr46/Yr29* present in Pavon76 and Diamond bird produced 10MR-20MS and 10S reactions at adult plant stage to leaf and stripe rusts, respectively. Similarly, RL6077 carrying *Lr67/Yr46* exhibited 30 MRMS reaction to leaf rust and 20S to stripe rust. The genotype, Parula carrying *Lr34, Lr46,* and *Lr68* showed 20 MR to leaf, 20S to stripe, and 40MSS to stem rust against prevailing pathotypes in the Nilgiris. Sivasamy et al. (2013) identified 36 Indian germplasm lines of which five carried *Lr46*, another five possessed *Lr67,* and seven lines carried the combination of *Lr34* and *Lr67*. The genes *Lr46* and *Lr67* have not yet been postulated among Indian cultivars. Many breeding programs are using these genes in developing cultivars with slow rusting resistance for all three rusts.

13.5.2.18   LR68

Lr68 is an APR conferring slow rusting resistance to wheat leaf rust caused by *P. triticina*. This gene, formerly designated *LrP*, was first described in CIMMYT's spring bread wheat Parula (FKN/3/2*Frontana//KENYA 350 AD.9C.2/Gabo 55/4/Bluebird/Chanate). Parula is a line developed at CIMMYT in 1981 that already had *Lr34* and *Lr46* (William et al., 1997,

2007; Herrera-Foessel, 2009). The likely origin of *Lr68* is the Brazilian cultivar Frontana (Herrera-Foessel, 2012). Lillemo (2011) tested the effect of *Lr68*, *Lr34*, and *Lr46* on leaf rust in an F_6 recombinant inbred line population derived from a Avocet-*YrA*×Parula cross in nine field environments in Mexico, Brazil, Argentina, Uruguay, and Chile. The authors were able to confirm additive effects of *Lr68* with the other two slow rusting genes at each site, and also showed that in sites in Argentina and Uruguay, *Lr68* showed a stronger effect than *Lr34*.

Herrera-Fossel (2012) showed in field tests in Mexico that *Lr68*-carrying lines had lower severities to leaf rust infection with *P. triticina* races MCJ/SP and MBJ/SP than the susceptible checks. For most of the tests, the effect of *Lr68* was smaller than those of *Lr34*, *Lr46*, and *Lr67*, and the combined effect of *Lr34*, *Lr46*, and *Lr68* in Parula resulted in near immunity. The line Parula carrying these gene complexes are currently used at Wellington to develop durable rust resistant wheat varieties which are in advance stage of constituting. In the MAB programat Wellington, STS marker, csLV34, and *Xgwm295* for *Lr34*, *Xwmc44* for *Lr46*, "http://wheat.pw.usda.gov/cgi-bin/graingenes/report.cgi?class=locus;name=Xcfd71-4D"\t"GG1" //*Xcfd71-4D*, and"http://wheat.pw.usda.gov/cgi-bin/graingenes/report.cgi?class=locus;name=Xcfd23-4D"\t"GG2"\\ *Xcfd23-4D* for *Lr67* were used to pyramid the genes.

13.5.2.19   SR2 (PSEUDO BLACK CHAFF/PBC)

The APR gene *Sr2* located in 3BS chromosome, shows recessive inheritance and is closely associated with *Lr27* and *Yr30* (Singh & McIntosh, 1984). The *T. turgidum* var. *dicoccum* was initially transferred to Hope and H44-24 (McFadden, 1930). *Sr2* is widely occurring in many wheat varieties in Australia, Canada, Kenya, USA, Mexico, and Indian subcontinent (Luig, 1983; Roelfs, 1988) which include some of the varieties like Songlen (additionally carrying *Sr5*, *Sr6*, *Sr8a*, and *Sr36*), Bluebird series include Nuri70 (additionally carrying *Sr5*, *Sr6*, and *Sr8a*), Lerma Rojo 64 (+ *Sr6*, *Sr7b*, and *Sr9e*). The variety Pavon (+*Sr8a*, *Sr9g*, and *Sr30*) and Sonalika carried *Sr2* (McIntosh, 1988). The *Sr2* is the most important stem rust resistant gene to be deployed in modern plant breeding in wheat (McIntosh, 1988; Rajaram et al., 1988; Roelfs, 1988). This could be attributed to its non-hypersensitive and non-race-specific resistance (APR) and in combination with other genes, it offers durable resistance for the stem rust worldwide (McIntosh, 1979). This gene alone is not effective to the pathotypes of stem rust prevailing in the Nilgiris (Wellington) and in association with other genes through additive gene action, confers high degree of resistance. Several Indian wheat cultivars carry the gene *Sr2* deployed unintentionally and its tight linkage to a phenotypical marker—pseudo black chaff (*Pbc*) offers better scope for the breeders to easily introgress the gene. Through planned breedingprogram, a variety HW 5207 carrying *Sr2* pyramided with *Lr24* and *Yr15* has been developed by the authors which showed lesser intensity of *Pbc*.

13.5.2.20   SR14

The *T. turgidum*-derived gene *Sr14* was first transferred to hexaploid cultivar Steinwedel which resulted in another cultivar Khapstein (Waterhouse, 1933). This gene shows low infection type and low environmental variability but it appears to enhance the distinct necrosis which is very characteristics of this gene (Knott, 1989). The *Khapli* emmer *Sr13* along with *Sr14* is the reference stock along with the source stock Yuma in USA. Although this gene was not widely exploited worldwide, its combined effect with other genes has been lately realized by the breeders especially for durum wheat improvement. The efforts are on at IARI, RS, and Wellington to take up this gene in the gene-pyramidingprogram.

13.5.2.21   SR22

The temperature-sensitive stem rust gene *Sr22* with chromosomal location 7A (Kerber & Dyck, 1973) is more effective at lower temperatures. The *monococcum*-derived gene *Sr22* present in the stock RL5244 is often found in the wild einkorns (The, 1973). The use in agriculture was limited until recent time because of the linkage drag of larger segment of *Sr22* transfer which resulted in yield penalties. However, the authors are currently using this stock *Co 1213 HSBVN 163313* (Bariana and Lagudah, personal communications) with reduced segment in their gene pyramiding program.

13.5.2.22   SR24/LR24

This gene complex has already been discussed elaborately under *Lr24*. Although it has been deployed in number of cultivars worldwide, the virulences for *Sr24* has been reported in South Africa (Le Roux& Rijken-berg, 1987) and in India (Bhardwaj et al., 1990) compelling to use this gene complex with other effective stem rust genes to harness the effectiveness of *Lr24* in India. Number of backcrossed and NIL lines carrying *Sr24/Lr24* has been developed at Wellington (see under *Lr24*).

13.5.2.23   SR25/LR19

Very few cultivars carrying these linked genes have been released for commercial use in the world. *Sr25* in combination with *Sr36* and *Sr6* has exhibited a high degree of resistance in Australian cultivar Cook which indi-cated that *Sr25* could be useful in combination with other genes (Luig, 1983). The line Cook*6/C80-1 exhibited immune reaction to stem rust pathotypes prevailing in the Nilgiris as compared to Sunstar*6/C 80-1 and its deriva-tives. Both these lines carry *Sr25*. Prabhu et al. (1998) confirmed this line (Sunstar*6/C 80-1) carry *Lr24/Sr24* not *Sr25* (see under *Lr19*).

13.5.2.24   SR26

Knott (1961, 1968) used irradiation for transferring stem rust resistance gene *Sr26* from long arm of a group 6 *Agropyron elongatum* chromosome to the long arm of wheat chromosome 6A. The use of *Sr26* has contrib-uted immensely toward cultivar improvement. Martin (1971), for the first time, in Australia, transferred *Sr26* to Eagle variety. Subsequently, the spectacular resistance imparted by this gene has been extensively used in several Australian cultivars grown widely, which competed satisfactorily with contemporary cultivars although it does cause a reduction in yield (The et al., 1988; McIntosh et al., 1995). The gene *Sr26* continues to be very effective in India also and this effective alien stem rust resistance has been introgressed into five well adapted but stem rust susceptible Indian bread wheat cultivars through a judicious backcrossing. The gene *Sr26* is dominant and produces a typical infection type which has served as a good indicator for selection of genotypes in each segregating genera-tion for making subsequent backcrosses. The gene exhibits low infection type and no virulence has been identified anywhere in the world (Huerta-Espino, 1992).

13.5.2.25   SR27

The gene *Sr27* was transferred, by using irradiation treatment, from *S. cereale* (Imperial rye) chromosome 3R to chromosome 3A of Chinese Spring wheat (Acosta, 1962). Rao (1978) confirmed that *Sr27* is derived from short arm of rye chromosome 3R. Sawhney and Goel (1981) reported that *Sr27* is effective in seedling stage against 19 pathotypes of stem rust in India which included the pathotypes commonly occurring in the Nilgiris. The line carrying *Sr27* exhibited very high degree of resistance at the adult stage in Wellington. Virulence for *Sr27* is rare. Harder et al. (1972) isolated an East African culture virulent on a Pembina line with *Sr27*. Initially, *Sr27* was very effective in Australia but later on, isolates of stem rust from triticale variety Coorong were virulent on wheat seedlings with *Sr27*. Cultivar Satu was recommended in Australia as a replacement for Coorong, later mutant of the Coorong pathotype evolved (McIntosh et al., 1983). Although, the gene *Sr27* has not been utilized commercially, we have transferred *Sr27* in the genetic background of Indian wheat varieties like Kalyansona, C306, and Lok-1. The successful transfer and pyramiding of the effective stem rust genes, *Sr26* and *Sr27,* in the adapted Indian bread wheat cultivars, already carrying other linked leaf rust genes *Lr19, Lr24, Lr28,* and *Lr32* which are expected to confer resistance to occurring stem rust pathotypes in India and also Ug99, developed through backcross program at IARI, Regional Station, Wellington are listed in Table 13.3.

13.5.2.26   SR30

Knott and McIntosh (1978) identified *Sr30*, a recessive gene which is located on long arm of 5D chromosome. The Webster gene *Sr30* believed to carry morphogenic resistance to stem rust is the only non-alien gene conferring moderate resistance to stem rust in India. *Sr30* is reported to be effective to 12 cultures of Indian stem rust pathotypes at seedling stage; however, the most prevalent pathotypes viz., 12, 40A, and 117A-1 exhibited virulence on *Sr30* (Sawhney & Goel, 1981). Virulence(s) to *Sr30* have been reported in several countries (Huerta-Espino, 1992). Commercial cultivars with *Sr30* were released in Australia but soon virulent pathotypes increased. Genotype likes Lerma Rojo 64A when introduced in India was initially resistant to stem rust but later, virulent pathotypes developed.

TABLE 13.3 Wheat Genotypes Pyramided with *Sr26, Sr27* Developed at IARI RS, Wellington.

Parent cultivar	Introgressed line	Genes incorporated	Pedigree of improved line	Reaction to stem rust at Wellington
C 306	HW 2023	*Sr24* Sr-26, Lr24*	C 306*7//DARF*6/3AG3/Kite	15R MR
Kalyansona	HW 2021	*Sr-24*Sr-26, Lr24*	Kalyansona*7//DARF*6/3AG3/Kite	20R MR
Lok-1	HW 2094	*Sr24*Sr-26, Lr24*	Lok-1*6//DARF*6/3AG3/Kite	10R MR
NI 5439	HW 2026	*Sr24*Sr-26, Lr24*	NI 5439*7//DARF* 6/3AG3/Kite	20R MR
Sonalika	HW 2027	*Sr24*/Lr24, Sr26*	Sonalika*7//DARF* 6/3AG3/Kite	5R MR
WH 147	HW 2022	*Sr24*/Lr24, Sr26*	WH 147*7//DARF* 6/3AG3/Kite	20R Mr
Kalyansona	HW 2088	*Sr26, Lr28*	Kalyansona*3//CS 2A/2M 4/2/Kite	10R MR
Lok-1	HW 2096	*Sr26, Lr28*	Lok-1*3//CS 2A/2M 4/2/Kite	10R MR
WH 147	HW 2099	*Sr26, Lr28*	WH-147*3//CS 2A/2M 4/2/Kite	20R MR
Kalyansona	HW 2089	*Sr26, Lr32*	Kalyansona *3//C86-8/Kalyansona(F4)/Kite	10R MR
NI 5439	HW 2090	*Sr26, Lr32*	NI 5439*3//C 86-8/Kalyansona(F4)/Kite	20R M
C 306	HW 2091	*Sr27, Sr24*/Lr24*	C 306*3//TR 380-14 *7/3Ag#14/KS *Sr27*	F
Kalyansona	HW 2025	*Sr27, Sr24*/Lr24*	Kalyansona*3//TR380-14*7/3Ag#14/KS *Sr27*	F-TR
Lok-1	HW 2095	*Sr27, Sr24*/Lr24*	Lok-1*3//TR 380-14 *7/3Ag#14/KS *Sr27*	F-TR
C 306	HW 2093	*Sr27, Lr28*	C 306*3//CS 2A/2M 4/2/KS *Sr27*	F
Kalyansona	HW 2024	*Sr27, Lr28*	Kalyansona//CS 2A/2M 4/2/KS *Sr27*	F-TR
HD 2687	HW 2078	*Sr25/Lr19, Sr31*/Lr26/ Yr9/Pm8*	HD 2687*3//Sunstar*6/C 80-1	10R MR

*Individually susceptible to Ug99 but effective in combination with *Sr26, Sr27*, and other genes.

13.5.2.27   SR31/LR26/YR9/PM8

A *S. cereale* cv. Petkus derivative was a spontaneous translocation (see *Lr26*) isolated from Germany in the 1930s (Zeller, 1973). The *Sr31* has been profitably exploited in CIMMYT wheat breeding program and continues to occur at high frequencies and present in cultivars in European Union, China, and USA. Widespread occurrence in Indian subcontinent can be noticed among the recently released cultivars, viz., PAK. 81, Sarhad 82, CPAN 1922, HUM 206, CPAN 3004, UP 2338, WH 542, PBW 343, and HD 2687. The value of *Sr31* as a source of protection against stem rust is difficult to determine (McIntosh et al., 1995). However, use of the gene *Sr31* may reflect the broad agronomic adaptability worldwide rather than the unique contribution of stem rust resistance. The global importance of 1BL/1RS in wheat breeding program has been well documented (Rajaram et al., 1988; Villareal et al., 1997; Kazman et al., 1998). These are potential problems of bread-making characteristics associated with *Sr31* which has restricted its use in Australia. The gene *Sr31* exhibits low infection type in seedling stage and shows moderately resistant to moderately susceptible reaction to stem rust pathotypes in the Nilgiris. The gene *Sr31* in combination with the gene *Sr25* and *Sr24* have shown enhanced resistance to stem rust in the genetic backgrounds of many Indian bread wheats.

13.5.2.28   SR32

Sears (1973) used homoeologous recombination to introgress gene *Sr32* imparting resistance to stem rust from the group 2 *Ae. speltoides* chromosome 2S#1 to wheat chromosomes (McIntosh, 1991) 2A (C 82.1), 2B (C 82.2) and 2D (C 82.3). Although C 82.2 (*Sr32*) is a normal translocation (McIntosh et al., 1995), the reasons for non-utilization of this gene are not known. No virulence on *Sr32* has been found anywhere in the world. The gene *Sr32* exhibited a very high degree of APR to stem rust pathotypes at Wellington. However, seedling reaction to 40A and 40-1 pathotypes was of low infection type (IT; 1+). Patil and Deokar (1996) reported that *Sr32* conferred effective seedling resistance to 18 Indian stem rust pathotypes. Stem rust resistant reactions obtained world over indicate that *Sr32* may be a useful gene for the improvement of wheat cultivars (McIntosh et al., 1995).

13.5.2.29   SR33/LR21

The gene *Sr33* has been transferred from *Ae. squarrosa* (Kerber & Dyck, 1979) and has been located on chromosome 1DS of wheat. It has exhibited moderate resistance to Indian pathotypes of stem rust in the adult plant stage. No virulence on *Sr33* has been reported in the survey made by Huerta-Espino (1992). This gene has not yet been utilized commercially anywhere in the world. The gene *Sr33* is linked to the genes *Lr21, Rg2*, and *Gli-D1* (McIntosh et al., 1995). Genes with moderate intensity of infection like *Sr33* may be quite useful in wheat breeding.

13.5.2.30   SR36/PM6

Allard and Shands (1954), Nyquist (1957) transferred stem rust resistance gene *Sr36* from *T. timopheevii* to common wheat chromosome 2BS. The gene *Sr36* provides a high degree of resistance to stem rust pathotypes in India; it has also exhibited a high degree of APR to pathotypes prevailing in the Nilgiris. Timgalen (*Sr36*), an Australian wheat cultivar was used as one of the parent in the development of cvs., HW 657 and HW 888. The variety, HW 657 was released as a commercial cultivar and exhibited stem rust resistance in peninsular India, while HW 888 showed resistance to stem rust over a period 20 years at multilocations in India. Both these genotypes presumably carry *Sr36* gene The stem rust resistance conferred by *Sr36* has been very valuable in Australia and many cultivars like Mengavi, Mendos, Timgalen, and Cook were released. However, pathotypes of stem rust virulent on *Sr36* were isolated in Australia. Pathogenic variations in most of the major regions have also been reported (Huerta-Espino, 1992).

13.5.2.31   SR38/LR37/YR17

The stem rust resistance gene *Sr38* has been found to be completely linked with leaf rust resistance gene *Lr37* and stripe rust resistance gene *Yr17* (Bariana & McIntosh, 1993). The gene *Sr38* exhibited a moderate degree of resistance at adult stage to stem rust pathotypes prevailing in the Nilgiris. The gene has not been used widely and few varieties in Australia carry linked genes. In India, the authors have also introgressed this useful linkage in several genetic backgrounds (*Lr37*) where *Sr38* exhibited moderate resistance to stem rust. The gene *Sr38* showed 1+ to 2C infection type to Indian stem rust pathotype 40A and 40-1(V.C. Sinha, personal communication). The wheat lines carrying this gene complexes are already listed in *Lr37* (see under *Lr37*).

13.5.2.32 YR9/SR31/LR26/PM8

Many winter and spring wheats either in cultivation or used as genetic stocks possess the gene *Yr9* which is associated with *Lr26* and *Sr31* and also often with *Pm8* (see *Sr31/Lr26*). The 1BL/1RS translocation has not only provided multiple disease resistance that may not always be durable but also provided wide adaptability, particularly in spring wheats. Initially, the *Yr9* was effective worldwide with a very low infection but the fast deployment of wheats with *Yr9* has usually been followed by an increase in pathotypes with virulence for this gene. Virulence occurs in Africa, China, Europe, South America (Stubbs, 1985), New Zealand (Wellings &Burdon, 1992), and in UK (Bayles et al., 1990). The T 1BL/1R#1S translocation is the most successful wheat–alien translocation and is still in use worldwide for the improvement of common wheat (Villareal, 1991). In India, almost every wheat zone has a cultivar carrying *Yr9*. When the area covered by varieties carrying *Yr9* increased in Northwestern zone, a pathotype (46S119) virulent to *Yr9* was reported from Gurdaspur (Punjab) in 1996 (Nayar et al., 1996). Some of the Indian wheat varieties carrying *Yr9* exhibit resistance to the new resistance to *Yr9* virulence. A large number of genotypes including winter wheat cultivars, Kavkaz, Aurora, Clement, and Skorospelka 35 all carrying 1BL/1RS translocation showed resistance to stripe rust at Wellington (more than 20 seasons), indicating that no virulence attacking *Yr9* is present in the Nilgiris.

13.5.2.33 YR10

The gene *Yr10* originating from Turkish bread wheat confers a high degree of stripe rust resistance at Wellington. No virulence for *Yr10* has so far been detected in India. Therefore, this gene appears to be potentially a useful source in wheat improvement. Its use in Indian wheat breeding program has also been advocated (Sharma, 2000). The gene *Yr10* has been utilized in limited stocks in USA (cvs. Moro, Crest, and Jacman) and Australia (cv. Angas); however, virulent pathotypes emerged soon after the deployment of *Yr10* in USA (Line & Qayoum, 1991) whereas, no virulence was reported from UK and Australia. *Yr10* has also been identified in Iranian

T. spelta 415 (Kerma & Lange, 1992). The molecular marker identified for *Yr10* (Hammond-Kosack & Jones, 1997) can also be used in MAS in wheat improvement program. Through the marker-assisted breedingprogram, the gene *Yr10* is being incorporated/pyramided in selected Indian commercial bread wheat cultivars at IARI, New Delhi/Wellington, DWR, Karnal and PAU, Ludhiana under various wheat improvement projects.

13.5.2.34   YR15

The gene *Yr15* has its origin in *T. turgidum* var. *dicoccoides* accession G 25 (Grama & Gerechter, 1974). *Yr15* present in six derivatives in the background of V763-2312 and V763-254 obtained from Israel in 1993 exhibited resistance to the stripe rust pathotypes present in the Nilgiris. The gene *Yr15* also showed APR to the stripe rust races prevailing at Lahaul Spiti (HP) in northern hills. The race specificity of *Yr15* is yet to be detected. Commercially, the gene has not been utilized, although, authors have introduced *Yr15* into some of the Indian wheat cultivars. At Wellington, *Yr15* has been successfully transferred into number of commercial Indian bread wheat cultivars (Vinod et al., 2006). The authors could successfully develop a high yielding wheat variety HW 5207 pyramided with *Yr15*, *Sr24/Lr24*, and *Sr2* through marker-assisted breeding and the marker linked to *Yr15Xgwm* 273 was used (Jagdish Kumar et al., 2012).

13.5.2.35   YR17/SR38/LR37

The gene *Yr17* is closely linked with stem rust resistance gene *Sr38* and leaf rust resistance gene *Lr37* and is located on the short arm of wheat chromosome 2A (Bariana & McIntosh, 1993). The line RL 6081 carrying *Yr17* showed a low infection type with moderate intensity of stripe rust infection on adult plants at Wellington. Bariana and McIntosh (1994) reported that moderately resistant seedlings with *Yr17* become more susceptible at lower temperature and low light intensities. In the backcross lines produced in the background of Indian wheat cultivars, *Yr17* exhibited differential response at adult plant stage to a mixture of stripe rust pathotypes at Delhi during 2000–2001. Preliminary evaluation of the backcross lines carrying *Yr17* indicated that these linked genes are not likely to impose any penalty on yield. The wheat lines carrying this gene complexes are already listed in *Lr37* (see under *Lr37*).

13.6   DURABLE RESISTANCE: THE DESIRED GOAL DURING PYRAMIDING NEW GENES

Durable disease resistance remains effective in a cultivar even though it may be widely grown over a long period of time in an environment that favors disease epidemics. This descriptive term does not provide an explanation to the basis of inheritance of this trait. Durable resistance has the following dimensions; (a) space—coverage of a large area; (b) time grown, for many years; and (c) conditions—high inoculum load and favorable weather. Van der Plank through a series of publications rationalized that non-specific or horizontal resistance will neither lead a variety into boom and bust cycle nor exert any directional selection pressure on the pathogen, and therefore, will be durable. Although Van der Plank considered durable resistance to be polygenically governed trait, he cited a number of examples such as the maize *P. polysora* system in Africa, where a single resistance gene contained the disease for a number of years. Wheat varieties, Thatcher and Lee, have withstood stem rust for 55 and 30 years, respectively. Cappelle–Desprez expresses at adult stage a moderate resistance to yellow rust and this has been maintained for the last 100 years.

Many host resistance genes that are matched by the pathogen survive in the breeding population for a long time, since these genes are not totally overcome by the pathogen and they still carry some amount of residual resistance. In the barley-*Erysiphe graminis hordei* system, they are referred to as defeated genes. In wheat-*P. striiformis* system, segregation for resistance to yellow rust can be obtained through minor gene effects, temperature-sensitive genes, adult plant genes, and various forms of disease resistance. The breeding strategy and selection methodology has to be viewed accordingly.

13.6.1   PYRAMIDING RESISTANCE GENES

When only a single resistance gene is there in a host, soon it tends to become susceptible. Subsequently adding one or a few more resistance gene in the background of that cultivar will make it resistant. But if these genes are brought into one background, as gene action is additive, the host will be resistant to a wide spectrum of the pathogen and the resistance base will last long. Since gaining virulence is at the cost of fitness, a pathotype able to infect all the resistance genes in such a variety is likely to be less fit in nature and hence may not induce epidemic. This approach of pyramiding resistance genes might also prolong the usefulness of the resistance genes.

Pyramiding resistance gene provides greater durability if the pathogen is solely dependent on asexual life cycle and mutation and recombination are less pronounced (Marshall, 1977). Combinations of resistance genes have provided a good field resistance against wheat stem rust in Australia for several years (McIntosh, 1992). In addition, since alternate host of *P. graminis tritici* is non-functional in Australia, pyramiding resistance genes has paid rich dividends. In North America, resistance gene combinations involving *Sr2* has provided durable resistance to stem rust, while *Lr13* and *Lr34* when combined with other leaf rust resistance genes have also provided durable resistance (Kolmer, 1991). Pyramiding resistance genes has provided in some cases durable resistance. For instance, French variety Cappelle–Desprez has durable resistance to eyespot of wheat, caused by *Pseudocercosporella herpotrichoides.* The other source is VPM, derived from a cross involving a wild grass, *Ae. ventricosa.* Molecular markers linked to these genes have been identified (Worland et al., 1988; Koebner & Martin, 1990). Seedlings with both these genes having better eyespot resistance (Doussinaultt & Douaire, 1978) can be selected using molecular markers and this felicitates the plant breeders to select for better resistance to eye spot disease.

There are several varieties in India cultivated as commercial varieties which were pyramided with some genes unintentionally. For example, HD 2189 (Carry *Sr57/Lr34/Yr18* along with *Sr2* and *Sr11Lr13*), Lok-1(*Lr13, Sr2*), C 306 (*Lr34/Sr57/Yr18* and *Lr67/46*), etc., naturally acting as genetic barrier in spread of rust in the *Puccinia* path. Several lines derived from H44 and Hope also exhibit durable stem rust resistance. Varieties such as Thatcher, Lee, Hope, Kenya Page, Africa Mayo, and Selkirk that have been globally used possess the gene *Sr2,* the APR gene. The gene is tightly linked with the *Pbc* gene, and when present in combination with other genes as in Selkirk, produces a durable resistance. In Australia, wheat varieties with five to six different resistance genes are cultivated. Gene *Sr36* derived from *T. timopheevii* (*SrTt-1*) is present in Mengavi, Mendos, Timson, Cook, Timgalen, and Shortim, in various blends, with *Sr5, Sr6, Sr7a, Sr8, Sr9c, Sr11,* and *Sr17.* In race surveys, occurrence of matching virulences was detected for most of the genes either alone or in combinations. However, virulences with combined virulence for *Sr36* were very low in frequency in Australia despite the fact that *Sr36* was released in varietal background as early as 1967. Therefore, like *Sr2,* combining *Sr36* with other resistance genes can render wheat varieties durable.

It is impossible to select for multiple genes under the field in the absence of heavy disease pressure. It is all the more difficult to select for non-race-specific genes. Field selections are often problematic, as they require a certain environment and disease presence that is not consistent from year to year and expertise. The phenotypical markers tightly linked to some genes like *Pbc* to *Sr2*, Apical claw to *Sr27*, pink awns/glume to *Lr45*, fast rusting to stem rust for *Lr28,* etc., offer better chance to pyramid some genes, although some of their expressions are much influenced by temperature and genetical background. Hence, use of genetical markers linked to genes is the only reliable approach.

Identified markers located very close to race-specific and durable rust resistance genes has been identified and used worldwide. Breeders are now using these markers to pyramid genes into the new varieties and with the availability of robust molecular markers linked to various rust resistance genes (Bariana et al., 2007; Prabhu et al., 2009; Singh et al., 2012), two or more genes can be pyramided efficiently in a relatively short time even in the absence of virulence. Several rust resistance genes have been transferred in the genetic background of popular Indian cultivars using conventional backcross breeding (Tomar & Menon, 2001) and through marker-assisted breeding (Samsampour et al., 2009; Sivasamy et al., 2009; Revathi et al., 2010; Kailash et al., 2011; Vinod et al., 2010; Chhuneja et al., 2011) in India. Though, only a few of these backcrossed lines such as HW 2004 (C 306+*Lr24*), HW 2044 (PBW 226+*Lr24*), HW 2045 (HD 2285+*Lr24*), and HW 2034 (C 306+*Lr28*) could be released as a cultivar, nevertheless, these backcross lines are useful genetic resources providing rust resistance genes in the improved and diverse genetic backgrounds. These backcross lines can also be utilized for pyramiding one or more genes.

Combinations of resistance genes have provided a good field resistance against wheat stem rust in Australia for several years (McIntosh, 1992). Moreover, since alternate host of *P. graminis tritici* is non-functional, in Australia, pyramiding resistance genes has paid rich dividends. In North America, resistance gene combinations involving *Sr2* has provided durable resistance to stem rust, while *Lr13* and *Lr34* when combined with other leaf rust resistance genes have also provided durable resistance (Kolmer, 1991). At Wellington, the authors could successfully pyramid some of the leaf and stem rust genes in the recent times as listed in the Table 13.4.

13.6.2   CURRENT STATUS OF GENE PYRAMIDING IN INDIA

If India had not experienced frequent rust epidemics, it is because of the cultivation of diverse genotypes carrying different resistance genes. The

TABLE 13.4 Wheat Varieties Pyramided and Confirmed with Rust Resistance Genes at Molecular Level at Wellington.

S. No.	Gene/pyramided gene complex confirmed to carry	Names of the variety
1	$Sr26+(Sr24+Lr14)$ and $(Sr2+Yr30^a)$	HW 2027
2	$(Sr25+Lr19)+(Lr34+Yr18+BYdv+Ltn^a)+(Sr24+Lr24)$	HW 4220
3	$(Lr34+Yr18+BYdv+Ltn^a)+(Sr24+Lr24)+(Sr2+Yr30^a)$	HW 2093, HW 4057
4	$(Sr25+Lr19)+(Sr2+Yr30^a)$	HW4209,HW4218, HW3627
5	$(Sr24+Lr24)+(Sr2+Yr30^a)$	HW2005, HW2008, HW2015, HW 2016, HW2091, HW2096, HW 5207*, HW2043, HW 2044, HW 2071-1A**, HW 2072**, HW 4029, HW 4066** AND HW 2081
6	$(Sr25+Lr19)+(Lr34+Yr18+BYdv+Ltn^a)$	HW 4205, HW 4219 AND HW 3607
7	$Sr26+(Sr2+Yr30^a)$	HW2099
8	$(Lr34+Yr18+BYdv+Ltn^a)+(Sr2+Yr30^a)$	HW 2017
9	$Sr2+Yr30^a$	HW 3070**, HW 4005, HW 4042, HW 4042, HW 4043, HW 4049, HW 4050
10	$Sr24+Lr14$	HW 2059**, HW 4053, HW 4055, HW 4065A, HW 4055
11	$Sr25+Lr19, Sr36+Pm6$	HW 4204, HW 4206**, HW 4207**, HW 4208, HW 4213**, HW 3608**, HW 3601, HW 3620**, HW 3614**
12	$Sr26$	HW 2021
13	$Lr34+Yr18+BYdv+Ltn^a$	HW 2022, HW 4202, HW 2062**

[a]APR race non-specific minor gene.
*Additionally, confirmed to carry $Yr15$.
**All the recurrent parents carry additional gene complex $Sr31+Lr26+Yr9+Pm8$.

Many of the lines listed above are expected to carry additional rust resistance genes either $Sr27$, $Sr36+Pm6$, $Lr28$, $Lr32$, and $Lr37+$ gene complexes for which we have not evaluated with the markers.

current status of rust resistance breeding involves both conventional and molecular breeding approaches including QTL mapping. Pyramiding of major genes using MAS in agronomically suitable cultivars is being pursued (Samsampour et al., 2009; Revathi et al., 2010; Kailash et al., 2011; Vinod et al., 2010; Chhuneja et al., 2011) to prevent the breakdown of resistance and the pyramided lines are being tested at multi-locations for their yield potential. Resistance genes have also been incorporated through conventional breeding methods, which have later been confirmed molecularly (Sivasamy et al., 2009). Leaf rust resistance genes *Lr19/Sr25* and *Lr28* are pyramided in the background of cv. HD 2687 carrying 1BL/1RS translocation using marker-assisted (back-ground selection) breeding (Kailash et al., 2011). Marker-assisted background selection can result in rapid recovery of recurrent parent genotype in a short span of 2–3 backcross generations (Ribaut et al., 2002).

Cook*6/C80-1, an Australian line, carries *Lr19/Sr25* and a DNA segment carrying *Sr36/Pm6* from *T. timopheevii* was used as donor for *Lr19*. The pyramided line showed more than 90% genomic similarity with recurrent parent, HD 2687. Another popular cultivar HD 2932 has been improved for multiple rust resistance by marker-assisted transfer of genes *Lr19, Sr26,* and *Yr10* (Niharika, 2012).

Similarly, popular cultivars HD 2733 and HD 2967 are being improved by transferring multiple rust resistance genes utilizing MAS. Effective gene for stripe rust resistance (*Yr15*) has been introgressed into a popular but rust susceptible cultivar HD 2329 (Vinod et al., 2006).

Currently, the cultivar HD 2967 is occupying a large area in North Western Plains Zone (NWPZ). It exhibits resistance to all the three rusts at adult plant stage. A high degree of APR against most virulent and prevalent races 77-5 and 104-2 is ascribed to minor gene based durable resistance (Malik et al., 2012), although it has been postulated to carry *Lr13+* which alone is not effective against above-mentioned races of leaf rust. Details on genetical markers can be accessed through *maswheat.ucdavis.edu/*.

13.7 CONCLUDING REMARKS

To effectively combat and checkmate the rust epidemics in the country, the race-specific and non-race-specific resistance genes need to be pyramided with the high yielding varietal background which is expected to confer durable rust resistance by taking advantage of "hot spot" locations and available markers in the public domain, linked to catalogued genes. Identification and mapping of rust resistance genes is crucial for development of effective and durable resistance in wheat. Intense efforts are currently made for worldwide search for APR genes and many of identified ones from Indian bread wheat genotypes offer much scope to exploit the large pool of Indian old varieties and land races expected to carry additional APR genes. Hence, the search for this type of novel or similar type of resistance in the durum or hexaploid wheat should be intensified.

KEYWORDS

- **wheat**
- **rust resistant gene**
- **gene pyramiding**
- **gene deployment strategy**
- **high yielding**
- **disease resistant**
- **wheat varieties**

REFERENCES

Acosta, A. C. The Transfer of Stem Rust Resistance from Rye to Wheat. *Diss. Abstr.* **1962,** *23,* 34–35.

Allard, R. W.; Shands, R. G. Inheritance of Resistance to Stem Rust and Powdery Mildew in Cytologically Stable Spring Wheats Derived from *Triticum timopheevi. Phytopathology.* **1954,** *44,* 266–274.

Anonymous. *Progress Report of All India Coordinated Wheat and Barley Improvement Project 2011–12, Vol. I, Crop Improvement;* Vinod, T., Chatrath, R., Singh, G., Tyagi, B. S., Sareen, S., Raj Kumar, Singh, S. K., Satish Kumar, Charan Singh, Venkatesh, K., Verma, A., Indu Sharma, Eds.; Directorate of Wheat Research: Karnal, India, 2012; p 308.

Anonymous. *Progress Report of All India Coordinated Wheat and Barley Improvement Project 2012–13,*

Vol. I, Crop Improvement; Directorate of Wheat Research: Karnal, India, 2013.

Asthana, R. P. Wheat Rusts and Their Control. *Nagpur Agr. Col. Mag.* **1948,** *22,* 136–143.

Bahadur, P.; Singh, D. V.; Srivastava, K. D. Management of Wheat Rusts- A Revised Strategy for Gene Deployment. *Indian Phytopathol.* **1994,** *47*(1), 41–47.

Bahadur, P.; Nagarajan, S.; Nayar, S. K. The Proposed System for Virulence Analysis II. *Puccinia graminis* f. sp. *tritici* in India. *Proc. Indian Acad. Sci.* **1985,** *95,* 29–33.

Bahadur, P., et al. Impact of Grass Introduction on Cereal Rusts in India. *Indian J. Agric. Sci.* **1973,** *43,* 287–290.

Barclay, A. On Some Rusts and Mildews in India. *J. Bot.* **1890,** *20,* 257–260.

Bariana, H. S., et al. Breeding Triple Rust Resistant Wheat Cultivars for Australia Using Conventional and Marker-Assisted Selection Technologies. *Aust. J. Agric. Res.* **2007,** *58,* 576–587.

Bariana, H. S.; McIntosh, R. A. Cytogenetic Studies in Wheat XIV. Location of Rust Resistance Genes in VPM1 and Their Genetic Linkage with Other Disease Resistance Genes in Chromosome 2A. *Genome.* **1993,** *36,* 476–482.

Bariana, H. S.; McIntosh, R. A. Characterization and Origin of Rust and Powdery Mildew Resistance Genes in VPM1 Wheat. *Euphytica.* **1994,** *76,* 53–61.

Bartos, P.; Valkoun, J.; Kosner, J.; Slovencikova, V. Rust Resistance of Some European Cultivars 'Salzmunder Bartweizen' and Weique. *Euphytica.* **1973,** *20,* 435–440.

Bayles, R. A.; Channell, M. H.; Stigwood, P. L. *Yellow Rust of Wheat;* United Kingdom Cereal Pathogen Virulence Survey, 1989 Annual Report; United Kingdom Cereal Pathogen Virulence Survey Committee: Cambridge, 1990; pp 11–17.

Bhardwaj S. C., et al. A Pathotype of *Puccinia graminis* f. sp. *tritici* on *Sr24* in India. *Cereal Rust. Powder. Mild. Bull.* **1990,** *18,* 35–37.

Bhardwaj, S. C. Virulence of *Puccinia triticina* on *Lr28* and its Evolutionary Relation to Prevalent Pathotypes in India. *Cereal Res. Commun.* **2010,** *38,* 83–89.

Bhojaraja, N. K. Molecular Mapping of Leaf Rust Resistnce Gene *Lr45*in Wheat (*Triticum aestivum L.*). Ph.D. Thesis, Indian Agricultural Research Institute, New Delhi, 2012, p73.

Browder, L. E. Specificity of the *Puccinia recondita* f. sp. *tritici*: *Triticum aestivum* 'Bulgaria 88' Relationship. *Phytopathology.* **1973,** *63,* 524–528.

Browning, J. A., et al. Regional Deployment for Conservation of Oat Crown Rust Resistant Genes. *Spec. Rep. Iowa Agric. Home Econ. Exp. Stn.* **1969,** *64,* 49–56.

Caldwell, R. M. In *Breeding for General and/or Specific Plant Disease Resistance, Proceedings of the Third International Wheat Genetics Symposium,* Canberra, Australia, Aug. 5–9, 1968; Finlay, K. W., Shephard, K. W., Eds.; Australian Academy of Sciences: Canberra, Australia, 1968; pp 263–272.

Chhuneja, P., et al. Marker Assisted Pyramiding of Leaf Rust Resistance Genes *Lr24* and *Lr28* in Wheat (*Triticum aestivum*). *Indian J. Agric. Sci.* **2011,** *81*(3), 214–218.

Chopra, V. L.; Kulkarni, R. N. Slow-Rusting Resistance: Its Components, Nature and Inheritance. *Z. Pflanzenkr. Pflanzenschutz.* **1980,** *87,* 562–573.

Doussinaultt, G.; Douaire, G. Analysis d'vncroisement diallele chez le blatendre pour 1' etude del la Resistance au Pietin-verse (*Cercosporella herpotrichoides*). *Annal. Amelior. Plant.* **1978,** *28,* 479–491.

Dyck, P. L.; Kerber, E. R.; Aung, T. An Inter-Chromosomal Reciprocal Translocation in Wheat Involving Wheat Leaf Rust Resistance Gene *Lr34*. *Genome.* **1994,** *37,* 556–559.

Dyck, P. L.; Samborski, D. J. Adult-Plant Leaf Rust Resistance in PI 250413, an Introduction of Common Wheat. *Can. J. Plant Sci.* **1979,** *59,* 329–332.

FAO. *Wheat Harvest Point to Production Increasing to 690 Million Tonnes-4.3 Per Cent up on 2012;* FAO's Quarterly *Crop Prospects and Food Situation*Report: Rome, Italy, 2013.

Flor, H. H. Host-Parasite Interactions in Flax Rust-Its Genetics and Other Implications. *Phytopathology.* **1955,** *45,* 680–685.

Friebe, B., et al. Characterization of Wheat- Alien Translocation Conferring Resistance to Disease and Pests: Current Status. *Euphytica.* **1996,** *91,* 59–87.

Friebe, B., et al. Compensation Indices of Radiation Induced Wheat-*Agropyron elongatum* translocation Conferring Resistance to Leaf Rust and Stem Rust. *Crop Sci.* **1994,** *34,* 400–404.

Gill, B. S. Registration of KS86WGRC02 Leaf Rust Resistant Hard Red Winter Wheat Germplasm. *Crop Sci.* **1988,** *28*(1), 207.

Gough, F. J.; Merkle, O. G. Inheritance of Stem and Leaf Rust Resistance in Agent and Argus Cultivars of *Triticum aestivum*. *Phytopathology.* **1971,** *61,* 1501–1505.

Grama, A.; Gerechter-Amitai, Z. K. Inheritance of Resistance to Stripe Rust (*Puccinia striiformis*) in Crosses between Wild Emmer (*Triticum dicoccoides*) and Cultivated Tetraploid and Hexaploid Wheats 11, *Triticum aestivum. Euphytica.* **1974**, *23*, 393–398.

Hammond-Kosack, K. E.; Jones, J. D. G. Plant Disease Resistance Genes. *Annu. Rev. Plant Physiol.* **1997**, *48*, 575–607.

Harder, D. E.; Mathenge, Gr.; Mwaura, L. K. Physiologic Specialization and Epidemiology of Wheat Stem Rust in East Africa. *Phytopathology.* **1972**, *62*, 166–171.

Herrera-Foessel, S. A. Characterization and Mapping of a Gene Component for Durable Leaf Rust Resistance in Chromosome Arm 7BL. *Phytopathology.* **2009**, *99*, S53–S53.

Herrera-Foessel, S. A. Lr68: A New Gene Conferring Slow Rusting Resistance to Leaf Rust in Wheat. *Theor. Appl. Genet.* **2012**, *124*, 1475–1486.

Hiebert, C. W., et al. An Introgression on Wheat Chromosome 4DL in RL 6077 (Thatcher*6/ PI 250413) Confers Adult Plant Resistance to Stripe Rust and Leaf Rust (Lr67). *Theor. Appl. Genet.* **2010**, *121*, 1083–1091.

Huerta-Espino J.; Singh, R. P. First Report of Virulence for Wheat Leaf Rust Gene Lr19 in Mexico. *Plant Dis.* **1994**, *78*, 640.

Huerta-Espino J. Analysis of Wheat Leaf Rust and Stem Rust Virulence on a Worldwide Basis. Ph.D. Thesis, University of Minnesota, MI, 1992.

Jagdish Kumar, et al. A High Yielding Bread Wheat Genotype with Potential to Act as Genetic Barrier for Rust Spread in Central India. *Indian Phytopathol.* **2012**, *65*(1), 45–51.

Johnson, R. Practical Breeding for Durable Resistance to Diseases. *Euphytica.* **1978**, *82*, 529–540.

Johnson, R. Durable Resistance: Definition of Genetic Control and Attainment in Plant Breeding. *Phytopathology.* **1981**, *71*, 567–568.

Johnson, R. A Critical Analysis of Durable Resistance. *Annu. Rev. Phytopath.* **1984**, *22*, 309–330.

Joshi, L. M.; Nagarajan, S. In *Regional Deployment of Lr Genes for Brown Rust Management,* Ist National Seminar on Genetics and Wheat Improvement, Ludhiana, India, Feb 22–23, 1977; Gupta, A. K., Ed.; Oxford & IBH Publishing Co.: New Delhi, 1977; pp 87–92.

Joshi, L. M.; Manchanda, W. C. *Bromus japonicus* Thumb, Susceptible to Wheat Rusts under Natural Conditions. *Indian Phytopathol.* **1963**, *16*, 312–16.

Joshi, L. M.; Srivastava, K. D.; Ramanujam, K. An Analysis of Brown Rust Epidemics of 1971–72 and 1972–73. *Indian Phytopathol.* **1975**, *28*, 138.

Kailash B. B., et al. Molecular Marker Assisted Pyramiding of Leaf Rust Resistance Genes Lr19 and Lr28 in Bread Wheat (*Triticum aestivum* L.) Variety HD 2687. *Indian J. Genet. Plant Breed.* **2011**, *71* (4), 304–311.

Kattermann, G. Zur Cytologie halmbehaarter Stämme aus Weizenroggenbastardierung. *Züchter.* **1937**, *9*, 196–199.

Kazman, M. E.; Lein, V.; Robbelen, G. The 1BL-1RS Translocation in Recently Developed European Wheats. In *Current Topics in Plant Cytogenetics Related to Plant Improvement;* Lelly, T., Ed.; WUV-Univ.-Verl.: Vienna, Austria, 1998; pp 33–42.

Kerber, E. R.; Dyck, P. L. Transfer to Hexaploid Wheat of Linked Genes for Adult Plant Leaf Rust and Seedling Stem Rust Resistance from an Amphiploid of *Aegilops speltoides* x *Triticum monococcum. Genome.* **1990**, *33*, 530–537.

Kerber, E. R. Telocentric Mapping in Wheat of the Gene Lr32 for Resistance to Leaf Rust. *Crop Sci.* **1988**, *28*, 178–179.

Kerber, E. R.; Dyck, P. L. Inheritance of Stem Rust Resistance Transferred from Diploid Wheat (*Triticum monococcum*) to Tetraploid and Hexaploid Wheat and Chromosome Location of the Gene Involved. *Can. J. Genet. Cytol.* **1973,** *15,* 397–409.

Kerber, E. R.; Dyck, P. L. In *Resistance to Stem Rust and Leaf Rust of Wheat in Aegilops squarrosa and Transfer of a Gene for Stem Rust Resistance to Hexaploid Wheat,* Proceedings of the Fifth International Wheat Genetics Symposium, New Delhi, India, Feb 23–28, 1978; Ramanujam, S., Ed.; Indian Society of Genetics and Plant Breeding: New Delhi, India, 1979, pp 358–364.

Kerber, E. R. Resistance to Leaf Rust in Hexaploid Wheat, *Lr32* for Resistance to Leaf Rust. *Crop Sci.* **1987,** *28,* 178–179.

Kerma, G. H. J.; Lange, W. Resistance in Spelt Wheat to Yellow Rust II. Monosomic Analysis oftheIranian Accession 415. *Euphytica.* **1992,** *63,* 219–224.

Kloppers, F. J.; Pretorious, Z. A. Histology of the Infection and Development of *Puccinia recondite f. sp. tritici*in a Wheat Line with *Lr37. J. Phytopathol.* **1995,** *143,* 261–267.

Knott D. R. "The Wheat Rusts-Breeding for Resistance." In *Monographs on Theoretical and Applied Genetics;* Springer-Verlag: Berlin, 1989; Vol. 12.

Knott, D. R. The Inheritance of Rust Resistance. VI. The Transfer of Stem Rust Resistance from *Agropyron elongatum* to Common Wheat. *Can. J. Plant Sci.* **1961,** *41,* 109–123.

Knott, D. R. The Inheritance of Resistance to Stem Rust Races 56 and 15F3-1L (Can.) in the Wheat Varieties Hope and H-44. *Can. J. Genet. Cytol.* **1968,** *10,* 311–320.

Knott, D. R.; McIntosh, R. A. The Inheritance of Stem Rust Resistance in the Common Wheat Cultivar Webster. *Crop Sci.* **1978,** *17,* 365–369.

Knott. D. R. Mutation of a Gene for Yellow Pigment Linked to *Lr19* in Wheat. *Can. J. Gen. Cytol.* **1980,** *22,* 651–654.

Knott, D. R. The Inheritance of Resistance to Race 56 of Stem Rust in 'Marquillo' Wheat. *Can. J. Gen. Cytol.* **1984,** *26,* 174–176.

Koebner, R. M. D.; Martin, P. K. Association of Eye Spot Resistance in Wheat Cv. Cappelle Desprez with Endopeptidase Profile. *Plant Breed.* **1990,** *104,* 312–317.

Kolmer, J. A. Physiologic Specialization of *Puccinia recondita* f.sp. *tritici* in Canada in 1990. *Can. J. Plant Pathol.* **1991,** *13,* 371–373.

Lagudah, E. S. Molecular Genetics of Race Non-Specific Rust Resistance in Wheat. *Euphytica.* **2011,** *179,* 81–91.

Le Roux, J.; Rijkenberg, F. H. J. Pathotypes of *Puccinia graminis* f.sp. *tritici* with Increased Virulence for *Sr24. Plant Dis.* **1987,** *71,* 1115–1119.

Lillemo, M. In *Multiple Rust Resistance and Gene Additivity in Wheat: Lessons from Multi-Location Case Studies in the Cultivars Parula and Saar,* Proceedings of Borlaug Global Rust Initiative 2011 Technical Workshop, St. Paul, June 13–16, 2011; pp 111–120.

Line, R. F.; Qayoum, A. *Virulence, aggressiveness, evolution and distribution of races of Puccinia striiformis (the cause of stripe rust of wheat) in North America, 1968–1987;* US Department of Agriculture Technical Bulletin No. 1788; US Department of Agriculture: Washington, 1991; p 44.

Luig, N. H. *A Survey of Virulence Genes in Wheat Stem Rust, Puccinia graminis f. sp. tritici.* Verlag Paul Parey: Berlin, 1983; p 199.

Malik, B. S., et al. Notification of Crop Varieties and Registration of Germplasm: HD 2967, a Rust Resistant Wheat Variety for Timely Sown Irrigated Conditions. *Indian J. Genet.* **2012,** *72,* 101–102.

Marshall, D. R. The Advantages and Hazards of Genetic Homogeneity. *Ann. N. Y. Acad. Sci.* **1977,** 287, 1–20.

Martin, R. H. Eagle: A New Wheat Variety. *Agric. Gaz. NSW.* **1971,** *82,* 206–207.

Martinez, F.; Niks, R. E.; Singh, R. P.; Rubiales, D. Characterization of *Lr46,* a Gene Conferring Partial Resistance to Wheat Leaf Rust. *Hereditas.* **2001,** *135,* 111–114.

McCallum, B. D.; Seto-Goh, P. Physiologic Specialization of Wheat Leaf Rust (*Puccinia triticina*) in Canada in 2002. *Can. J. Plant Pathol.* **2005,** *27,* 90–99.

McDonald, B. A.; Linde, C. The Population Genetics of Plant Pathogens and Breeding Strategies for Durable Resistance. *Euphytica.* **2002,** *124,* 163–180.

McFadden, E. S. A Successful Transfer of Emmer Characters to *Vulgare* Wheat. *J. Am. Soc. Agron.* **1930,** *22,* 1020–1034.

McGregor, A. J.; Manners, J. G. The Effect of Temperature and Light Intensity on Growth and Sporulation of *Puccinia striiformis* on Wheat. *Plant Pathol.* **1985,** *34* (2), 263–271.

McIntosh, R. A. In *Genetical Strategies for Diseases Control,* Proceedings of Fifth Australasian *Plant Pathology Conference, University* Auckland, NZ, 22 May, 1985; pp 39–44.

McIntosh, R. A. Close Genetic Linkage of Genes Conferring Adult Plant Resistance to Leaf Rust and Stripe Rust in Wheat. *Plant Pathol.* **1992,** *41,* 523–527.

McIntosh, R. A.; Wellings, C. R.; Park, F. R. *Wheat Rusts: An Atlas of Resistance Genes.* CSIRO: East Melbourne, Victoria, Australia, 1995; p 200.

McIntosh, R. A.; Dyck, P. L.; Green, G. J. Inheritance of Leaf Rust and Stem Rust Resistances in Wheat Cultivars Agent and Agatha. *Aust. J. Agric. Res.* **1997,** *28,* 37–45.

McIntosh, R. A. In *A Catalogue of Gene Symbols for Wheat,* Proceedings of the Fifth International Wheat Genetics Symposium, New Delhi, India, Feb 23–28, 1978; Indian Society of Genetics and Plant Breeding, Indian Agricultural Research Institute: New Delhi, India, 1979; pp 1299–1309.

McIntosh, R. A., et al. Vulnerability of Triticales to Wheat Stem Rust. *Can. J. Plant Pathol.* **1983,** *5,* 61–69.

McIntosh, R. A. Alien Sources of Disease Resistance in Breed Wheats, In *Nuclear and Organellar Genome of Wheat Species,* Proceedings of Dr. H. Kihara Memorial International Symposium on Cytoplasmic Engineering in Wheat, Yokohama, Japan, July 3–6, 1991; Sakuma, T., Kinoshita, T., Eds.; Kihara Memorial Yokohama Foundation for the Advancement of Life Science: Yokohama, Japan, 1991; pp 320–332.

McIntosh, R. A. The Role of Specific Genes in Breeding for Durable Stem Rust Resistance in Wheat and Triticale. In *Breeding Strategies for Resistance to the Rusts of Wheat;* Simmonds, N. W., Rajaram, S., Eds.; CIMMYT: Mexico, 1988; pp 1–9.

McIntosh, R. A.; Miller, T. E.; Chapman, V. Cytogenetical Studies in Wheat XII. *Lr28* for Reaction to *Puccinia graminis tritici. Z. Pflanzenkr. Pflanzenschutz.* **1982,** *87,* 274–289.

Mehta, K. C. Further Studies on Cereal Rusts in India, Part I. *Sci. Monogr. Coun. Agric. Res. India.* **1940,** *14,* 1–224.

Mehta, K. C. Further Studies on Cereal Rusts in India, Part II. *Sci. Monogr. Coun. Agric. Res. India.* **1952,** *18,* 1–165.

Menon, M. K.; Tomar, S. M. S. *Aegilops*-derived Specific Genes in Common Wheat and Their Introgression into Commercial Indian Bread Wheat Cultivars. *Indian. J. Genet.* **2001,** *61*(2), 92–97.

Mettin, D.; Bluthner, W. D.; Schlegel, G. In *Additional Evidence of Spontaneous 1B/1R Wheat-Rye Substitutions and Translocations,* Proceedings of the Fourth International Wheat Genetics Symposium, University of Missouri, Columbia, Aug 6–11, 1973; Sears, E. R., Sears, L. M. S., Eds.; University of Missouri: CO, 1973; pp 179–184.

Mishra, A. N.; Prakasha, T. L.; Kaushal, K.; Dubey, V. G. Validation of *Lr24* in Some Released Bread Wheat Varieties and its Implications in Leaf Rust Resistance Breeding and Deployment in Central India. *Indian Phytopath.* **2014,** *67*(1), 102–103.

Mukade, K.; Kamio, M.; Hosoda, K. The Transfer of Leaf Rust Resistance from Rye to Wheat by Intergeneric Addition and Translocation. In *'Mutagenesis in Relation to Ploidy Level'*, Gamma field Symposia No. 9, Ibaraki, Japan, July 17–18, 1970; IRB: Japan, 1970; pp 69–87.

Nagarajan, S.; Joshi, L. M. Epidemiology in the Indian Subcontinent. In *The Cereal Rusts Vol. II; Diseases, Distribution, Epidemiology, and Control;* Roelfs, A. P., Bushnell, W. R., Eds.; Academic Press: New York, 1985; pp 371–399.

Nagarajan, S.; Joshi, L. M. Further Investigations on Predicting Wheat Rusts Appearance in Central and Peninsular India. *Phytopathol. Z.* **1980,** *98,* 84–90.

Nagarajan, S.; Bahadur, P.; Nayar, S. K. Occurrence ofa New Virulence 47S102 of *Puccinia striiformis* West in India during Crop Year, 1982. *Cereal Rusts Bull.* **1984,** *12,* 28–31.

Nagarajan, S.; Nayar, S. K.; Bahadur, P.; Bhardwaj, S. C. Evaluation of Some Indian Wheats for *Yr, Lr* and *Sr* Genes by Matching Technique and Genetic Uniformity Observed. *Cereal Rusts Bull.* **1987,** *15,* 53–64.

Nagarajan, S.; Nayar, S. K.; Bahadur, P.; Kumar, J. *Wheat Pathology and Wheat Improvement.* IARI Regional Station: Shimla, India, 1986; p 12.

Nagarajan, S.; Singh, H. The Indian Stem Rust Rules-A Concept on the Spread of Wheat Stem Rust. *Plant Dis. Rep.* **1975,** *59,* 133–136.

Nayar, S. K., et al. *Basis of Rust Resistance in Indian Wheats;* Research Bulletin No. 1; DWR Regional Station: Shimla, India, 1994; p 33.

Nayar, S. K., et al. Distribution Pattern of *Puccinia recondita tritici* Pathotypes in India During 1990–94. *Indian J. Agric. Sci.* **1996,** *66,* 621–630.

Nayar, S. K., et al. Appearance of New Pathotype of *Puccinia recondita tritici* Virulent on *Lr9* in India. *Indian Phytopathol.* **2003,** *56* (2), 196–198.

Niharika, M. Marker Assisted Introgression of Genes for Multiple Resistance in Wheat Variety HD 2932. Ph. D. Thesis, Indian Agricultural Research Institute, New Delhi, 2012, p 70.

Nyquist, N. E. Monosomic Analysis of Stem Rust Resistance of a Common Wheat Strain Derived from *Triticum timopheevi. Agron. J.* **1957,** *49,* 222–223.

Parlevliet, J. E. Partial Resistance of Barley to Leaf Rust, *Puccinia hordei.* I. Effect of Cultivar and Development Stage on Latent Period. *Euphytica.* **1975,** *24,* 21–27.

Parlevliet, J. E. Resistance of the Non-Specific Type. In *The Cereal Rusts Vol. 2; Diseases, Distribution, Epidemiology and Control;* Roelfs, A. P., Bushnell, W. R., Eds.; Academic Press: Orlando, FL, 1985; pp 501–525.

Pathak, K. D.; Joshi, L. M.; Chinnamani, S. Natural Occurrence of *Puccinia graminis tritici* on *Brachypodium sylvaticum* during off Season in Nilgiri Hills. *Indian Phytopathol.* **1979,** *32,* 308–309.

Patil, J. V.; Deokar, A. B. Host Parasite Interactions between Lines and Varieties of Wheat with Known *Sr* Genes and Races of Stem Rust. *Cereal Rust. Powder. Mild. Bull.* **1996,** *24,* 91–97.

Prabhu, K. V., et al. Molecular Markers Linked to White Rust Resistance in Mustard *Brassica juncea. Theor. Appl. Genet.* **1998,** *97,* 865–870.

Prabhu, K. V. Marker Assisted Selection for Biotic Stress Resistance in Wheat and Rice. *Indian J. Genet. Plant Breed.* **2009,** *69,* 305–314.

Prasada, R. Studies on Rusts of Some ofthe Wild Grasses Occurring in the Neighbourhood of Simla. *Indian J. Agric. Sci.* **1948,** *18,* 165–170.

Prasada, R. Fight the Wheat Rust. *Indian Phytopathol.* **1960,** *13,* 1–5.

Pretorius, Z. A.; LeRoux, J.; Drijenpondt, S. C. Occurrence and Pathogenicity of *Puccinia recondita* f.sp. *tritici*on Wheat in South Africa During 1988. *Phytophylactica.* **1990,** *22,* 225–228.

Rajaram, S.; Singh, R. P.; Toress, E. Current CIMMYT Approaches in Breeding Wheat for Rust Resistance. In *Breeding Strategies for Resistance to the Rusts of Wheat;* Simmonds, N. W., Rajaram, S., Eds.; CIMMYT: Mexico, 1988; pp 101–118.

Rajaram S.; Mann, C. E.; Ortis Ferrara, G.; Muzeeb-Kazi, A. In *Adaptation, Stability and High Yield Potential of Certain 1B/1R CIMMYT Wheats,* Proceedings of the Sixth International Wheat Genetics Symposium, Kyoto, Japan, Nov 28–Dec 3, 1983; Sakamoto, S., Ed.; Kyoto University: Kyoto, Japan, 1983; pp 613–621.

Randhawa, H. S.; Gaudet, D. A.; Puchalski, B. J.; Graf, R. In *New Virulence Races of Stripe Rust in Western Canada,* Ist Canadian Wheat Symopsium, Winnipeg, MB, Nov 30–Dec 2, 2011.

Rao, M. V. Varietal Improvement. In *Wheat Research in India 1966–1976;* Jaiswal, P. L., Tata S. N., Gupta, R. S., Eds.; Delhi Press: Jhandewalan Estate, New Delhi, 1978; p 244.

Reddy, M. S. S.; Rao, M. V. In *Genetic Classification and Control of Leaf Rust Pathogen of Wheat in India,* Proceedings of Ist National Seminar on Genetics and Wheat Improvement, Ludhiana, Feb 22–23, 1977; Gupta, A. K., Ed.; Oxford & IBH Publishing Co.: New Delhi, India, 1977; pp 78–85.

Revathi P.; Tomar S. M. S.; Vinod; Singh N. K. Marker Assisted Gene Pyramiding of Leaf Rust Resistance Genes *Lr24 and Lr28* along with Stripe Rust Resistance Gene *Yr15* in Wheat (*Triticum aestivum* L.). *Indian J. Genet. Plant Breed.* **2010,** *70*(4), 349–354.

Ribaut, J. M.; Jiang, C.; Hoisington, D. Efficiency of a Gene Introgression Experiment by Backcrossing. *Crop Sci.* **2002,** *42,* 557–565.

Riley, R.; Chapman, V.; Johnson, R. Introduction of Yellow Rust Resistance of *Aegilops comosa* into Wheat by Genetically-Induced Homoeologous Recombination. *Nature.* **1968,** *217,* 383–384.

Robinson, R. A. *Plant Pathosystems.* Springer-Verlag: Berlin, 1976; p 184.

Roelfs, A. P. Resistance to Leaf and Stem Rusts in Wheat. In *Breeding Strategies for Resistance to the Rusts of Wheat;* Simmonds, N. W., Rajaram, S., Eds.; CIMMYT: Mexico, 1988; pp 10–12.

Samborski, D. J.; Dyck, P. L. Enhancement of Resistance to *Puccinia recondita* by Interactions of Resistance Genes in Wheat. *Can. J. Plant Path.* **1982,** *4,* 152–156.

Samsampour, D., et al. Marker Assisted Selection to Pyramid Seedling Resistance Gene *Lr24* and Adult Plant Resistance Gene *Lr48* for Leaf Rust Resistance in Wheat. *Indian J. Genet.* **2009,** *69*(1), 1–9.

Sawhney, R. N.; Goel, L. B. Race Specific Interaction between Wheat Genotypes and Indian Culture of Stem Rust. *Theor. Appl. Genet.* **1981,** *60,* 161–166.

Sawhney, R. N.; Sharma D. N. Identification of Sources for Rust Resistance and for Durability to Leaf Rust in Common Wheat. *SABRAO J.* **1990,** *22,* 51–55.

Sawhney, R. N.; Nayar, S. K.; Singh, S. D.; Chopra, V. L. Virulence Pattern of the Indian Leaf Rust Races on Lines and Varieties of Wheat with *Lr* genes. *SABRAO J.* **1977,** *9,* 13–20.

Sawhney, R. N.; Sharma, J. B.; Sharma, D. N. Nonspecific Adult Plant Resistance to Leaf Rust with Potential for Durability. *Cereal Rust. Powder. Mild. Bull.* **1994,** *22,* 9–13.

Sears, E. R. The Transfer of Leaf Rust Resistance from *Aegilops umbellulata* to Wheat. *Brookhaven Symp. Biol.* **1956,** *9,* 1–22.

Sears, E. R. In *Agropyron-Wheat Transfers Induced By Homoeologous Pairing,* Proceedings of the 4th International Wheat Genetics Symposium, MI, Columbia, Aug 6–11, 1973; University of Missouri: MI, 1973; pp 191–199.

Shaner, G.; Roberts J. J.; Finney, R. E. A Culture of *Puccinia recondita* Virulent to the Wheat Cultivar 'Transfer'. *Plant Dis. Report.* **1972,** *56,* 827–830.

Sharma, D.; Knott, D. R. The Transfer of Leaf Rust Resistance from *Agropyron* to *Triticum* by Irradiation. *Can. J. Genet. Cytol.* **1966,** *8,* 137–143.

Sharma, A. K. In *Resistance Breeding for New Virulent Pathotypes of Rust,* Proceedings of the 39th All India Wheat Workers' Workshop, BHU, Varanasi, Aug 27–30, 2000.

Simmonds, N. W. A Plant Breeder's Perspective of Durable Resistance. *FAO Plant Protect. Bull.* **1985,** *33,* 13–17.

Singh, A.; Pallavi, J. K.; Gupta Promila; Prabhu, K. V. Identification of Microsatellite Markers Linked to Leaf Rust Resistance Gene *Lr25* in Wheat. *J. Appl. Genet.* **2012,** *53*(1), 19–25.

Singh, R. P.; Huerta-Espino J.; Sharma R.; Joshi, A. K. In *High Yielding Spring Bread Wheat Germplasm for Irrigated Agro-Ecosystem,* Proceedings of the International Symposium on Wheat Yield Potential: Challenges to International Wheat Breeding, Ciudad Obregon, Sonora, Mexico, March 20–24, 2006; CIMMYT: Mexico, 2006; p 16 (Abstract).

Singh R. P; McIntosh R. A. Complementary Genes for Resistance to *Puccinia recondita tritici* in *Triticum aestivum.* Genetic and Linkage Studies. *Can. J. Genet. Cytol.* **1984,** *26,* 723–735.

Singh R. P. Pathogenicity Variations of *Puccinia recondita* f. sp. *tritici* in Wheat Growing Areas of Mexico during 1988 and 1989. *Plant Dis.* **1991,** *75,* 790–794.

Singh, R. P. Genetic Association of Leaf Rust Resistance Gene *Lr34* with Adult-Plant Resistance to Stripe Rust in Bread Wheat. *Phytopathology.* **1992,** *82,* 835–838.

Singh, R. P.; Rajaram, S. Genetics of Adult-Plant Resistance to Leaf Rust in 'Frontana' and Three CIMMYT Wheats. *Genome.* **1992,** *35,* 24–31.

Singh, R.; Mujeeb-Kazi, A.; Huerta-Espino, J. Lr46: A Gene Conferring Slow-Rusting Resistance to Leaf Rust in Wheat. *Phytopathology.* **1998,** *88,* 890–894. 10.1094/PHYTO. 1998.88.9.890

Sivasamy, M. Phenotypic and Molecular Confirmation of Durable Adult Plant Leaf Rust Resistance (APR) Genes *Lr34, Lr46* and *Lr67* Linked to Leaf Tip Necrosis (*Ltn*) in Select Registered Indian Wheat (*Triticum aestivum* L.) Genetic Stocks. *Cereal Res. Commun.* **2013,** DOI 10.1556/crc.2013.0054

Sivasamy, M., et al. Introgression of Useful Linked Genes for Resistance to Stem Rust, Leaf Rust and Powdery Mildew and Their Molecular Validation in Wheat. *Indian J. Genet.* **2009,** *69,* 17–27.

Stakman, E. C.; Stewart D. M.; Loegering W. Q. *Identification of Physiological Races of Puccinia graminis var. tritici;* USDA Agricultural Research Service E617: Washington DC, 1962, p 53.

Stakman, E. C.; Levine, M. N. *The Determination of Biologic Forms of Puccinia graminis on Triticum sp;* Technical Bulletin 10: *University of Minnesota,* Agricultural Experiment Station, MI, 1922; pp 1–10.

Stubbs, R. W. Stripe Rust. In *The Cereal Rusts. Vol. 2; Diseases, Distribution, Epidemiology and Control;* Roelfs, A. P., Bushnell, W. R., Eds.; Academic Press: Orlando, FL, 1985; pp 61–101.

The, T. T., et al. In *Grain Yields of Near Isogenic Lines with Added Genes for Stem Resistance,* Proceedings of the 7th International Wheat Genetics Symposium, Cambridge, July 13–19, 1988; Miller, T. E., Koebner, R. M. D., Eds.; IPSR: Cambridge, England, 1988; pp 901–906.

The, T. T. Transference of Resistance to Stem Rust from *Triticum monococcum* L. to Hexaploid Wheat. Ph.D. Thesis, The University of Sydney, 1973.

Tomar, S. M. S.; Singh, Sanjay K.; Sivasamy, M.; Vinod. Wheat Rusts in India: Resistance Breeding and Gene Deployment-A Review. *Indian J. Genet. Plant Breed.* **2014,** *74*(2), 129–156.

Tomar, S. M. S.; Menon, M. K. Adult Plant Response of Near-Isogenic Lines and Stocks of Wheat Carrying Specific *Lr* Genes against Leaf Rust. *Indian Phytopathol.* **1998,** *51,* 61–67.

Tomar, S. M. S.; Menon, M. K. Fast Rusting to Stem Rust in Indian Bread Wheat Cultivars Carrying the Genes *Lr28* and *Lr32*. *Wheat Inf. Serv.* **1999,** *88,* 32–36.

Tomar, S. M. S.; Menon, M. K. Genes for Disease Resistance in Wheat. Indian Agricultural Research Institute, New Delhi, 2001, pp. 152.

Van der Plank, J. E. *Disease Resistance in Plant Infection.* Academic Press: New York, 1968; p 206.

Villareal, R. L.; Banuelos, O.; Muzeeb-Kazi, A. Agronomic Performance of Released Durum Wheat (*Triticum turgidum* L.) Stocks Possessing the Chromosome Substitution T1BL. 1RS. *Crop Sci.* **1997,** *37,* 1735–1740.

Villareal, R. L. The Effect of Chromosome 1B/1R Translocation on the Yield Potential of Certain Spring Wheats (*Triticum aestivum L.*). *Plant Breed.* **1991,** *106,* 77–81.

Vinod, et al. In *Molecular Marker Assisted Pyramiding of Rust Resistance Genes to Counter the Threat Posed by Evolution of New Virulences in Common Wheat,* Proceedings of National Seminar on "Checkmating Evolution of Race Group 77 of Wheat Leaf Rust Pathogen", IARI Regional Station, Wellington, The Nilgiris, March 14, 2010; IARI: New Delhi, India, 2010.

Vinod, et al. Evaluation of Stripe Rust Resistance Genes and Transfer of *Yr15* into Indian Wheats. *Indian J. Agric Sci.* **2006,** *76*(6), 362–366.

Waterhouse, W. L. On The Production of Fertile Hybrids from Crosses between Vulgare and Khapli Emmer Wheats. *Proc. Linn. Soc. NSW.* **1933,** *58,* 99–104.

Wellings, C. R.; McIntosh, R. A. *Puccinia striiformis* f. sp. *tritici* in *Australia. Vort. Pflanz.* **1990,** *24,* 114.

Wellings, C. R.; Burdon, J. J. Variability in *Puccinia striiformis* f. sp. *tritici* in Australia. *Vort. Pflanz.* **1992,** *24,* 114.

William, H. M. Molecular Marker Mapping of Leaf Rust Resistance Gene *Lr46* and Its Association with Stripe Rust Resistance Genes *Yr29* in Wheat. *Phytopathology.* **2003,** *93,* 153–159.

William, H. M., et al. Detection of Quantitative Trait Loci Associated with Leaf Rust Resistance in Bread Wheat. *Genome.* **1997,** *40,* 253–260.

William, H. M., et al. Characterization of Genes for Durable Resistance to Leaf Rust and Yellow Rust in CIMMYT Spring Wheats. *Wheat Production in Stressed Environments*; Developments in Plant Breeding Vol. 12; Buck, H. T., Nisi, J. E., Salomon, N., Eds.; Springer: Dordrecht, The Netherlands, 2007; pp 65–70.

Worland, A. J., et al. Location of a Gene for Resistance to Eyespot (*Pseudocercosporella herpotrichoides*) on Chromosome 7D of Wheat. *Plant Breed.* **1988,** *101,* 43–51.

Zeller, F. J. In *1B/1R Wheat-Rye Chromosome Substitutions and Translocations,* Proceedings of the 4th International Wheat Genetics Symposium, MI, Columbia, Aug 6–11, 1973; Sears, E. R., Sears, L. M. S., Eds.; Agricultural Experiment Station, University of Missouri: Columbia, MO, 1973; pp 209–221.

CHAPTER 14

BREEDING STRATEGIES AND PROSPECTS OF WHEAT IMPROVEMENT IN NORTHWESTERN HIMALAYAS

DHARAM PAL* and MADHU PATIAL

ICAR-Indian Agricultural Research Institute, Regional Station (Cereals & Horticulture Crops), Tutikandi Facility, Shimla 171004, Himachal Pradesh, India

Corresponding author. E-mail: dpwalia@rediffmail.com

CONTENTS

ABSTRACT

Wheat is an important winter cereal of North Western Himalayan region of India. The average productivity of this region is low as compared to Indo-Gangetic plains. Low productivity is primarily due to rain fed cultivation and losses due to biotic stresses. Among biotic stresses, stripe rust has emerged as an important disease not only in Himalayan region but also in North-Western plains, because it receives primary inoculum of rusts from off-season crop and volunteer plants grown in different altitudes of Himalaya. Breeding strategy is focused to utilize diverse rust resistance gene sources with reliable yield and moisture stress tolerance for developing high yielding wheat varieties fitting to Himalayan ecology. With prime aim to provide semi-dwarf high yielding wheat varieties to hill farmers and keeping national interest to contain wheat rusts in the hills, a number of wheat varieties were released and deployed in Himalayan region.

14.1 INTRODUCTION

Wheat is a staple diet and calorie base for a major part of the world's population. It is an important winter cereal of humid Western Himalayan Region (Jammu & Kashmir, Himachal Pradesh, Uttarakhand, hilly regions of West Bengal, and northeastern states) occupying 1.39 mha area (Gupta & Kant, 2012). This area has a wider climatic variation and has wheat cultivation ranging from tropical to temperate weather. The average productivity of Northern hill states is about 1.87 tons/ha as compared to the national average of 3.06 tons/ha (Anonymous, 2014). Low productivity of this region is primarily due to rain fed cultivation and losses due to biotic stresses.

14.1.1 RAIN FED CULTIVATION

More than 83% of area of wheat in northern hills zone is under rain fed cultivation. No rains during the sowing period of the crop generally results in poor germination of seed. The moisture stress experienced during the flowering and gain-filling period also affects wheat yields. The success of wheat crop is dependent on the availability of residual moisture after harvesting of *Kharif* crop and distribution of rainfall during the crop season.

14.1.2 BIOTIC STRESSES

In the Indian sub-continent all the three types of rusts viz., leaf rust/brown rust (*Puccinia triticina*), stripe rust/yellow rust (*Puccinia striiformis* sp. *tritici*), stem rust/black rust (*Puccinia graminis* sp. *tritici*) occurs and causes huge loss to the wheat crop. Leaf rust, remained the widely distributed rust disease followed by stripe rust. However, among the three rusts, stripe rust has become threat to the wheat crop in North-Western Himalayan region and the adjoin North-Western Plains Zone. The summer heat in the plains destroys the local sources of infections of the rusts on wheat but survives during summer at the higher altitudes of the Himalaya and remains prevalent throughout the year (Mehta, 1940). It occurs under cool and moist weather conditions or at high altitudes (Lupton & Macer, 1962). Joshi (1976) reported that stripe rust appears by the end of December and early January along the foothills of Punjab, Haryana, and Western region of Uttar Pradesh. From this region it spreads and establishes in the Northern plains of the country by the end of February. Based on this usual dissemination pattern of spread of inoculum from foci of infection to plains was termed as *Puccinia* path (Nagarajan & Joshi, 1980).

In Uttarakhand hills, stripe rust severity reached up to 80S with high prevalence on most of the widely cultivated wheat varieties during *Rabi* season 2010–2011. In different parts of Himachal Pradesh, the severity of yellow rust was observed 10–80S (Kant & Jain, 2011). Wheat cultivars derived from "Veery" germplasm carrying *Yr9/Lr26/Sr31/Pm8* gene complex located on 1B/1R translocation, provided protection against losses due to leaf and stripe rusts till 1995. However, the evolution of pathotype (pt) 46S119 virulent on *Yr9* and *Yr2* has changed the whole scenario (Nayar et al., 1996). After the breakdown of *Yr9* based resistance, some protection against stripe rust was offered by "Attila" derived wheat varieties that had *Yr27* effective against pt. 46S119. The breakdown of *Yr27* with new variant 78S84 (Prashar et al., 2007) created a fear of stripe rust epidemic due to wide cultivation of PBW343 or varieties with similar resistance. The prevalence of *Yr9, Yr2*, and *Yr27* virulences and recently identified five new pathotypes viz., 46S117, 110S119, 238S119, 110S247, and 110S84 which rendered *Yr11, Yr12*, and *Yr24* susceptible (Gangwar et al., 2015) is of grave concern and therefore, it is imperative to develop and deploy high yielding rust resistant wheat varieties in high altitude areas of North-Western Himalayas where wheat is grown in summer/off season and in foot hills the crop is grown as regular/*Rabi* crop.

14.2   HISTORICAL PERSPECTIVE OF WHEAT IMPROVEMENT

Breeding work for the improvement of wheat for North-Western Himalaya was started in India at Shimla in 1934. Breeding program at that time was aimed at the hybridization involving certain exotic wheat lines and some of the leading Indian wheats. A cross involving this type of combination (Thatcher (E124)/NP 165) yielded a total of 30 hybrid strains with fixed high degree of resistance and desirable agronomic characters of which NP 788, NP 789, and NP 790 were developed as elite genotypes (Pal et al., 1956). Wheat varieties developed and released from 1952 to 2014 are given in Table 14.1.

TABLE 14.1   Wheat Varieties Possessing Rust Resistant Genes Recommended for Cultivation during Post Green Revolution Era.

Variety	Rust resistance gene combination
HS 86	*Lr13+Yr2+Sr11*
HS 207	*Lr26+34+1+ Yr9+18+ Sr2+5+31*
HS 240	*Lr26+34+1+ Yr 9+18+ Sr2+31+*
HS 277	*Lr26+34+ Yr9+18+ Sr5+31*
HS 295	*Lr23+34+Yr2ks+18+ Sr2+8b+11*
HS 365	*Lr26+1+ Yr9+Sr5+31+*
HS 375	*Lr26+34+1+ Yr 9+18+ Sr2+5+31*
HS 420	*Lr13+10+34+Yr18*
HS 490	*Lr23+ Sr2+9b+*
HS 507	*Lr26+1+ Yr9+Sr31+*
HS 542	*Lr13+10+ Yr2+ Sr5+8a+9b+11+*

14.2.1   PRE-GREEN REVOLUTION ERA

With prime aim to provide better wheat varieties to hill farmers and keeping national interest to contain wheat rusts in the hills, five tall wheat varieties (NP770, NP 809, Ridley, NP 829, and NP 818) with high degree of resistance against rusts were released during 1951–1965. These varieties helped significantly in checking spread of rust inoculum from hills to the plains.

14.2.2 GREEN REVOLUTION ERA

With the advent of semi-dwarf spring wheat in India and consequently release of "Sonalika" and "Safed Lerma," resulted significant enhancement in productivity of hilly region. With available rust resistant plant genetic resources two single dwarf wheat varieties, "Girija" for higher hills and "Shailaja" for lower hills were released in the year 1975.

14.2.3 POST-GREEN REVOLUTION ERA

Green revolution brought new challenges to sustain its growth under new challenges of genetic vulnerability to diseases and disease management with environmental considerations. Major emphasis of the breeding programs in the hilly areas is on stripe and leaf rusts. Detailed studies were carried out on genetics of rusts resistance with targeted Indian pathotypes. These studies provided framework to formulate new breeding strategies to achieve mandate of wheat improvement with focus on rust resistance and deploying different rust resistant genes in the hills. Considering these objectives along with changing need of hill farmers, wheat varieties viz., HS 86, HS 207, HS 240, HS 277, HS 295, HS 342 (Mansarovar), HS 365, HS 375 (Himgiri), HS 420 (Shivalik), HS 490 (Pusa Baker), HS 507, and HS 542 were released for cultivation in hills.

14.3 DEPLOYMENT OF RUST RESISTANT VARIETIES TO CONTAIN WHEAT RUSTS IN THE HIMALAYAS

The most common, very effective and environment safe method of rust management is the deployment of rust resistant wheat varieties in the hills. Therefore, breeding strategy is focused at utilizing diverse rust resistance, besides reliable yield and moisture tolerance for developing wheat varieties in Northern hills. Resistance based on single dominant gene that produce a hypersensitive response is generally considered to be vulnerable to genetic changes in pathogen virulence (Wilcoxon, 1981). Although the life of effective race-specific resistance genes can be prolonged by using gene combinations, an alternative approach being implemented at international maize and wheat improvement center (CIMMYT) is to deploy varieties that possess adult plant resistance (APR) based on combinations of minor, slow rusting genes (Singh et al., 2014). APR genes individually provide low levels of resistance and combinations of three or more genes are essential to express commercially adequate levels of resistance (Bariana & McIntosh, 1993). Hence, deploying wheat varieties possessing diverse gene combinations could be a feasible strategy to delay the evolution of pathotypes with matching virulence genes. Durable resistance in wheat is rarely found, and the basis for the most durable resistance to wheat leaf rust has been combination of different genes (Samborski, 1985). The cultivation of wheat varieties with gene combinations of *Lr29+Lr15, Lr9+Lr10, Lr10+Lr13* and *Lr15+Lr20* as a management strategy for controlling leaf rust in India was recommended (Nagarajan et al., 1986). *Lr10* in combination with other genes has been found to be an effective source of resistance against leaf rust (Nayar & Bhardwaj, 1998). Among adult plant resistance genes, *Lr34* is the only gene from *Triticum aestivum* that was observed to be partially effective against leaf rust in India (Sawhney & Sharma, 1990). Resistance gene *Lr34* involved in gene combinations conferring durable resistance to leaf rust and expresses enhanced resistance in combinations with other resistance genes (German & Kolmer, 1992). Under heavy leaf rust inoculum pressure, the loss in the *Lr34* deficient genotype ranged between 60 and 84% whereas it ranged between 14 and 18% when *Lr34* was present (Singh et al., 1994). Slow rusting impact of *Lr34* and its interaction with other genes has made it very useful gene for rust resistance in India (Nayar et al., 1999). The short life span of resistant wheat varieties warrants intensification of research toward general resistance due to slow rusting. Slow rusting resistance is expressed in the field with slower disease progress and judged in green house by its components viz., longer latent, incubation period, fewer and small uredinia, and lower spore production per unit area (Bjarko & Line, 1988; Singh & Rajaram, 1992; Prabhu et al., 1993). The gene *Lr34* is reported to be either linked with *Yr*18 or under pleiotropic genetic control with the presence of leaf tip necrosis and characteristic broken stripes of yellow rust on flag leaf (Dyck, 1987). The wheat varieties viz., HS 365 and HS 295 developed for North-Western Himalaya have shown slower progress of leaf rust due to longer latent period and lower uredopustules/cm^2 at seedling stage against the predominant and virulent pt 121R63-1 of brown rust. The variety HS 365 has also shown lowest number of uredo pustules/cm^2 on leaf surface at adult plant stage (Khurana et al., 2003).

Therefore, the presence of *Lr34* in wheat varieties is extremely beneficial in rust resistance breeding program. Besides, *Lr34*, two new adult plant resistance genes *Lr48* and *Lr49* identified in cultivars CSP44 and VL404, respectively, are now available with the plant breeders for combating leaf rust pathogen (Saini et al., 2002). The non-hypersensitive adult plant resistance conferred by the gene $Yr18$ is believed to be durable (McIntosh, 1992) but level of resistance conferred by $Yr18$ alone is not adequate for commercial exploitation (Singh et al., 2001). The line CSP 44 carrying $Yr18$ has shown terminal disease severity to stripe rust ranging from TR-20MR, whereas in RL 6058 carrying $Yr18$ the severity varied between 30 and 40S. The line CSP44 carry new stripe rust resistance gene in addition to $Yr18$ which offer long lasting resistance and thus increase diversity for genes conferring durable resistance (Khanna et al., 2005).

The cultivation of wheat varieties as illustrated in Table 14.1 may provide suitable genetic barriers for minimizing leaf and stripe rusts inoculum at the source. It can be deduced that the mosaic of rust resistance genes viz., $Lr1+10+13+23+26+34$, $Yr2+2ks+9+18$ and $Sr2+5+8a+9b+11+31$ as deployed in the Himalayas has proved effective for protecting the wheat crop of adjoining plains of North-Western India from epidemics of wheat rusts . Now it has become an accepted national strategy to deploy Lr gene combination in North India to contain leaf rust epidemic (Nagarajan et al., 2006). The presence of rust resistance genes in various combinations along with unidentified genes in the gene pool of released varieties is expected to provide considerable protection against leaf and stripe rusts pathogens to keep the yield losses below the damaging level. Marker aided selection being an effective breeding strategy has been employed to introgress effective leaf rust resistance gene $Lr19$ and $Lr24$ in wheat varieties HS 240 and HS 295. Breeding lines validated for leaf rust resistance genes using sequence characterized amplified regions (SCAR) and simple sequence repeat (SSR) markers are under testing in preliminary yield trials. The strategic use of resistance would not only reduce the initial inoculum but would also discourage the selection of new pathotypes.

14.4 BREEDING METHODOLOGY

14.4.1 GERMPLASM COLLECTION, SCREENING, AND UTILIZATION

The northwestern hills are the hidden treasures of germplasm resources and found to be one of the biggest repositories of landraces of winter wheat and rye (Partap et al., 2001). However, these genetic resources have considerably eroded from the traditional wheat growing areas but still some of the land races/farmers' varieties are in cultivation in high hills and tribal areas of Jammu & Kashmir and Himachal Pradesh (Table 14.2). Although, such

varieties and landraces have low yielding capacity with disease susceptibility but they do possess yield stability, which is important for subsistence
farming in the tribal areas and some novel rust resistance gene pool. Moreover, they are good sources of genes for traits such as drought tolerance (as
most of them are grown in rainfed conditions) and for nutrition and dough
quality (selected by the farmers with very long experience and local preferences) by virtue of which they are still existing and or competing with their
superior counterparts (Pal et al., 2007). The impact of high yielding varieties
is clearly visible in this region also, but it is more in irrigated valley areas
than in rain fed remote and tribal areas. About 55% loss of wheat genetic
resources has been recorded in this region over the last 30 years (Rana et
al., 2000). In view of the above, a systematic germplasm collection program
was initiated in the genetic diversity rich areas especially the *changar* and
kandi areas (highly rainfed) and tribal belt of North-Western Himalaya for
strengthening our wheat germplasm. Genetic diversity including farmers'
varieties/landraces was observed more in remote (Pal et al., 2007). Some of
these land races have been used in wheat improvement programs.

TABLE 14.2 Characteristic Features of Wheat Landraces Present in Northwestern Himalaya.

Land race	Characteristic feature(s)
Rundal	Awnless, tall, thin stem, makes good bread, good yield
Jhuldi	Awned, long ears, bold grain size
Sirohun	Awned, gives good yield in rotation with paddy but less with maize
Kathuan	Pink pigmentation in the ear, poor yielder, very good quality
Kishal	Awned, good bread making quality, average yielder
Kankoo	Good plant vigor, awnless, tall, average yield, non-shattering, bread tasty and does not dry quickly, flour white
Dharmouri	Red grained, high tillering ability, drought tolerant
Desi mundal	Awnless, bread tasty, disease resistant, non-shattering type
Brad kanak	Awned, more straw, drought resistant, flour brown but bread tasty, very high tillering, shattering type, lodging resistant
Bharadoo	Late maturing, more tillers, good dough quality
Ralieun	Easy threshable, flour white, high yield, bread tasty
Mundu, Misri	Cold tolerant, amber color, soft bread
Ridley	Very tall, red grained, late maturing
Shruin	Awnless, tall, small grain, good dough quality

14.4.2 PRE-BREEDING RUST RESISTANT GENETIC STOCKS

Identifying rust resistance sources effective against virulent pathotypes of rust provides opportunities to the plant breeders for incorporating viable genes into germplasm pools and permit the system to release the cultivars carrying diverse resistance genes. Hence, pre-breeding genetic stocks are developed and being used for wheat improvement in Northern hills zone (Table 14.3).

TABLE 14.3 Wheat Germplasm Possessing Rust Resistance against Prevalent Pathotypes of Rusts in Northern Hills Zone.

Germplasm	Pedigree	Trait
WBM 1587	MILAN/SHA	Resistance source against stripe rust pt 46S119
WBM 1591	PYN/BAU/MILAN	Resistance source against stripe rust pt 46S119
HS 424	CPAN 3004//HPW(DL) 30/HS286	Resistant against all the prevalent pathotypes of leaf and stem rusts
HS 431	V 81623//BUC/PVN	Resistant against all the prevalent pathotypes of leaf and stem rusts
HS 492	HPW42/CPAN2032// UNATH KS	Resistant against leaf rust

14.4.3 EVALUATION OF WHEAT SEGREGATING MATERIALS FOR DIRECT INTRODUCTIONS

Wheat breeding materials from various segregating international nurseries are selected either for direct introduction or used as donors in wheat improvement program. The breeding materials, after at least three years of evaluation as advanced bulks in preliminary trials, qualify for testing under All India Coordinated trials of Northern hills zone.

14.4.4 CROSSING BLOCKS

14.4.4.1 SPRING CROSSING BLOCK

Promising wheat materials from various observation nurseries, national genetic stock nursery, and elite wheat entries reached at the level of advanced varietal trials are the source materials to constitute spring crossing block. In order to synchronize the flowering for attempting crosses, same set of genotypes are sown in different dates. Depending upon the need, 10–15% of the source materials of the crossing block are replaced every year. Simple, three-way crosses and very limited backcrosses are attempted depending upon the requirement of the areas.

14.4.4.2   WINTER CROSSING BLOCK

Facultative wheat genotypes having blend of spring and winter gene pools with stay green habit, profuse root biomass, efficient grain filling rate and short grain filling duration fit effectively in the Himalayan ecology. This block consists of genotypes having good agronomic features along with semi-spreading and spreading type of growth habit.

14.4.4.3 F_1 AND THREE-WAY F_1 GENERATION

F_1s and three-way F_1s are grown under good fertility and irrigated conditions to get sufficient F_2 seed. Seed of each F_1 is bulked. In case of three-way F_1 individual plants are harvested separately. The segregating generations are handled using modified pedigree-bulk method of breeding (Van Ginkel et al., 2000).

14.4.4.4   HANDLING OF F_2 GENERATION

Seed of F_2 generation is space planted under good fertility and irrigated conditions at rust hot spot. Infector rows are planted on both sides of F_2 and artificial conditions are created by inoculating the plants with a mixture of inoculum of prevalent pathotypes of leaf and stripe rusts. Poor performing F_2s are rejected. Single plants are selected based on disease reaction, number of effective tillers, plant height, lodging resistance, maturity, spike length, spike type, spike fertility, etc. The grains are examined for color, size, texture, and luster.

14.4.4.5   SELECTION IN F_3 AND SUBSEQUENT GENERATIONS

The seed from individually selected plants is planted in two rows of 2 m length under good agronomic conditions at commercial seeding rate.

Artificial epiphytotic conditions for creating rusts are maintained. Ten spikes from each of the best lines judged on the basis of desirable agronomic features and disease resistance are harvested and bulked. The methodology of selection of lines repeated in F_4 and F_5. In F_6 generation the selected 10 spikes are harvested individually to raise spike to row progenies in F_7. The best F_7 lines showing synchrony in plant height (80–100 cm), spike type, maturity, and adult plant resistance against rusts are selected, harvested in bulk to advance it to F_8 as advanced bulks.

14.4.4.6 EVALUATION OF ADVANCED BULKS UNDER YIELD TRIALS

The best performing advanced bulks are evaluated along with checks representing different production conditions following optimum agronomic practices. The advanced bulks showing significant superiority against respective checks are selected to constitute preliminary yield trials for different production conditions. For one more year the best performing genotypes which have shown consistent better performance over checks, disease resistance in preliminary disease screening nurseries across different rust hot spots are evaluated under common varietal trials before their entry to All India Coordinated Trials to find out their suitability for different climatic zones. The entries entered in All India Coordinated Trials are tested in multilocations for three years. The consistently performing genotypes showing significant grain yield superiority over best check, disease resistance and quality are proposed for identification followed by its release to the farmers in the appropriate area of adaptation.

14.4.4.7 DISTRIBUTION OF BREEDER SEED

Breeder seed of newly released wheat varieties is made available to the State Departments of Agriculture. The State Departments of Agriculture multiply the seed at large scale in their state farms and distribute to the farmers.

14.5 CONCLUDING REMARKS

The most common, very effective and environment safe method of rust management is to exploit genetic resistance. There is urgent need to identify and explore novel rust resistance genes for developing rust resistance wheat varieties. The breeding strategy needs to be focused at utilizing diverse and new sources of rust resistance, besides reliable yield and moisture tolerance from spring and winter gene pools, possessing semi-spreading and stay green habit, profuse root biomass, efficient grain filling rate and short grain filling duration for developing wheat varieties fitting well in the Himalayan ecology. Therefore, developing and deploying wheat varieties possessing diverse rust resistance gene combinations with reliable grain yield levels and drought tolerance could be a feasible strategy to delay the evolution of new pathotypes and enhancing wheat productivity of North-West Himalayan region.

KEYWORDS

- **wheat varieties**
- **breeding strategies**
- **rust resistance**
- **gene deployment**
- **germplasm**
- **North-West Himalayas**

REFERENCES

Anonymous. Progress Report of the All India Coordinated Wheat and Barley Improvement Project. Project Director's Report. Indu Sharma, Ed.; DWR: Karnal, 2014; pp 1–120.

Bariana, H. S.; McIntosh, R. A. Cytogenetic Studies in Wheat XIV. Location of Rust Resistance Genes in VPM1 and Their Genetic Linkage with Other Disease Resistance Genes in Chromosome 2A. *Genome.* **1993,** *36,* 476–482.

Bjarko, M. E.; Line, R. F. Heritability and Number of Genes Controlling Leaf Rust Resistance in Four Cultivars of Wheat. *Am. Phytopathol. Soc.* **1988,** *78,* 457–461.

Dyck, P. L. The Association of a Gene for Leaf Rust Resistance with the Chromosome 7D Suppressor of Stem Rust Resistance in Common Wheat. *Genome.* **1987,** *29,* 467–469.

Gangwar, O. P.; Prasad, P.; Khan, H.; Bhardwaj, S. C. Pathotype Distribution of *Puccinia* Species on Wheat and Barley during 2014–15. *Mehtaensis.* **2015,** *35,* 1–29.

German, S. E.; Kolmer, J. A. Effect of Gene *Lr*34 in the Enhancement of Resistance to Leaf Rust of Wheat. *Ther. Appl. Genet.* **1992,** *84,* 97–105.

Gupta, H. S.; Kant, L. Wheat Improvement in Northern Hills of India. *Agric. Res.* **2012,** *1,* 100–116 (DOI 10.1007/s400003-012-012-z).

Joshi, L. M. Recent Contribution towards Epidemiology of Wheat Rusts in India. *Indian Phytopath.* **1976**, *29*, 1–16.

Kant, L.; Jain, S. K. Wheat Yellow Rust Resurgence in the Northern Hills and Tarai Regions. *ICAR News.* **2011**, *17*, 14.

Khanna, R.; Bansal, U. K.; Saini, R. G. Genetics of Adult Plant Stripe Rust Resistance in CSP44, a Selection from Australian Wheat. *J. Genet.* **2005**, *84*, 337–340.

Khurana, R.; Nayar, S. K.; Lakhanpal, T. N. Slow-Rusting Behaviour of Some Wheat Cultivars against Brown Rust. *Indian Phytopath.* **2003**, *56*, 470–472.

Lupton, F. G. H.; Macer, R. C. F. Inheritance of Resistance to Yellow Rust (*Puccinia glumarum* F. Rikss. & Henn.). *Trans. Br. Mycol. Soc.* **1962**, *45*, 21–45.

McIntosh, R. A. Close Genetic Linkage of Genes Conferring Adult Plant Resistance to Leaf Rust and Stripe Rust in Wheat. *Plant Pathol.* **1992**, *41*, 323–327.

Mehta, K. C. *Further Studies on Cereal Rusts in India.* Imperial Council of Agricultural Research: India, 1940; Vol.14, pp 1–244.

Nayar, S. K.; Prashar, M.; Bhardwaj, S. C. Occurrence of *Yr9* Virulence of Yellow Rust and its Management -An Indian Context. 2nd International Crop Science Congress: New Delhi, India, Nov. 17–24, 1996.

Nagarajan, S.; Joshi, L. M. Further Investigations on Predicting Wheat Rusts Appearance in Central and Peninsular India. *Phytopathol. Z.* **1980**, *98*, 84–90.

Nagarajan, S.; Nayar, S. K.; Kumar, J. Checking the *Puccinia* Species that Reduce the Productivity of Wheat (Triticum spp.) in India-A Review. *Proc. Indian Natl. Sci. Acad.* **2006**, *72*, 239–247.

Nagarajan, S.; Nayar, S. K.; Bahadur, P.; Kumar, J. Wheat Pathology and Wheat Improvement, IARI Regional Station: Shimla, India, 1986.

Nayar, S. K.; Bhardwaj, S. C. Management of Wheat Rusts in India. In *IPM System in Agriculture;* Upadhyay, R. K., Mukherji, K. G., Rajak, R. L., Eds.; Aditya Books Pvt. Ltd: New Delhi, India, 1998; Vol. 3, pp 305–324.

Nayar, S. K.; Bhardwaj, S. C.; Prashar, M. Characterization of *Lr34* and *Sr2* in Indian Wheat (*Triticum aestivum*) Germplasm. *Indian J. Agri. Sci.* **1999**, *69*, 718–721.

Pal, B. P.; Vasudeva, R. S.; Kohli, S. P. *Breeding Rust Resistant Hill Wheats in India.* ICAR Research Series: India, 1956; Vol. 5, pp 1–56.

Pal, D.; Kumar, S.; Rana, J. C. Collection and Characterization of Wheat Germplasm from North – West Himalaya. *Indian J. PlantGenet. Resour.* **2007**, *20*(2), 170–173.

Partap, A.; Singh, S.; Choudhary, H. K.; Sethi, G. S. Bioresource Potential of North Western Himalayas. *Indian J. Plant Genet. Resour.* **2001**, *14*, 134–136.

Prabhu, K. V.; Luthra, J. K.; Nayar, S. K. Slow Rusting Resistance in Wheat (*Triticum aestivum*) to Leaf Rust (*Puccinia recondita*) in Northern Hills of India. *J. Agri. Sci.* **1993**, *63*, 354–357.

Prashar, M.; Bhardwaj, S. C.; Jain, S. K.; Datta, D. Pathotypic Evolution in *Puccinia striiformis* in India during 1995–2004. *Australian J. Agric. Res.* **2007**, *58*, 602–604.

Rana, J. C.; Sharma, B. D.; Gautam, P. L. Agri-diversity Erosion in the North-West Indian Himalayas- Some Case Studies. *Indian J. Plant Genet. Resour.* **2000**, *13*, 252–258.

Saini, R. G., et al. *Lr48* and *Lr49* Novel Hypersensitive Adult Plant Leaf Rust Resistance Genes in Wheat (*Triticum aestivum*). *Euphytica.* **2002**, *124*, 365–370.

Samborski, D. J. Wheat Leaf Rust. In *The Cereal Rusts, Volume II. Diseases, distribution, epidemiology and control*; Roelfs, A. P., Bushnell, W. R., Eds.; Academic Press: Orlando,1985; pp 39–59.

Sawhney, R. N.; Sharma, D. V. Identification of Sources for Rust Resistance and for Durability to Leaf Rust in Common Wheat. *SABRAO J.* **1990,** *22,* 51–55.

Singh. R. P.; Espino, J. H.; William, M. In *Slow Rusting Genes based Resistance to Leaf and Yellow Rust in Wheat: Genetics and Breeding at CIMMYT*, Proceeding 10th Assembly of Wheat Breeding Society of Australia, Mildura, Australia, Sep 16–21, 2001; Wheat Breeding Society of Australia Inc.: Australia, 2001; pp 103–108.

Singh, R. P., et al. Progress towards Genetics and Breeding for Minor Genes based Resistance to Ug99 and other Rusts in CIMMYT High Yielding Spring Wheat. *J. Integr. Agric.* **2014,** *13*(2), 255–261.

Singh, R. P.; Hong, M.; Huerta, J. Rust Diseases of Wheat. In *Guide to the CIMMYT Wheat Crop Protection Sub-Program*; Saari, E. E., Hettel, G. P., Eds.; CIMMYT Wheat Special Report No.24: Mexico,1994; pp 19–33.

Singh, R. P.; Raja Ram, S. Genetics of Adult Plant Resistance in Frontana and Three CIMMYT Wheats. *Genome.* **1992,** *35,* 24–31.

Van Ginkel, M.; Trethowan, R.; Cukardar, B. A Guide to the CIMMYT Bread Wheat Program (rev), Wheat Special Report No.5. CIMMYT: Mexico, 2000.

Wilcoxon, R. D. Genetics of Slow Rusting in Cereals. *Phytopathology.* **1981,** *71,* 989–993.

PART II
Diseases in Diverse Agroecological Conditions

CHAPTER 15

DISEASES OF WHEAT IN BRAZIL AND THEIR MANAGEMENT

S. P. VAL-MORAES*

*Technology Department, UNESP-Universidade Estadual Paulista/
College of Agricultural and Veterinarian Sciences, Jaboticabal,
Sro Paulo 14884900, Brazil*

Corresponding author. E-mail: valmoraes.silvana@gmail.com

CONTENTS

ABSTRACT

Wheat is one of the very important aspects of the agribusiness trade balance in Brazil. This chapter gives an overview of this sector, providing the general information about wheat imports. Two of Brazil's coldest states, Paraná and Rio Grande do Sul, account for over 90% of wheat production and imports around US$700 million in wheat every year. However, wheat is very susceptible to mycotoxin contamination caused by fungal infection. The southern region is characterized by the cultivation of soft wheat cultivars; its greatest obstacle is the high rainfall during harvest months, which favors pre-harvest sprouting, leaf rust, and *Fusarium* head blight. In the south-central and central regions, water deficiency, excessive heat, and wheat blast (*Magnaporthe oryzae* pathotype *Triticum*) are the main limiting factors. The south-central and central regions are marked to produce bread wheat. The quality profile of Brazilian wheat varies according to the region where it is produced. The southern Brazil has a semi-temperate climate, higher rainfall, soil that is more fertile, advanced technology and high input use, appropriate infrastructure, and experienced farmers. This region produces most of Brazil's cereal grains and oilseeds. The aim of this chapter is to present the main diseases of wheat in Brazil. Emphasis has been put on the characteristics of the diseases, their epidemiological aspects, management using common control measures.

15.1   INTRODUCTION

Wheat is one of the very important commodities of the agribusiness trade in Brazil. The national wheat production is not enough for domestic consumption, and there is a deficit of 5.8 million tons, that is, there is a strong dependence for wheat on foreign markets (Leivas et al., 2014). The import of wheat in Brazil is around US$ 700 million every year. In Brazil, wheat is grown in the south especially the Rio Grande do Sul state, southeast, and mid-west regions in Figure 15.1 (MAPA, 2015). Varieties grown in the Rio Grande do Sul belong to bread wheat (*Triticum aestivum* L.), which is the most widely cultivated worldwide (Scheeren, 1986). Two of Brazil's coldest states, Paraná and Rio Grande do Sul, account for over 90% of wheat production. Wheat crop produced in Brazil is susceptible to mycotoxins contamination caused by fungal infection. The south-central and central regions are marked to produce bread wheat. The quality profile of Brazilian wheat varies according to the region where it is produced. The Southern Brazil

FIGURE 15.1   Wheat production in south, southeast, and mid-west regions of Brazil.

has a semi-temperate climate, higher rainfall, fertile soils, advanced technology, high input use, appropriate infrastructure, and experienced farmers. This region produces most of Brazil's cereal grains and oilseeds. The southern region is characterized by the cultivation of soft wheat cultivars and its greatest obstacle is the high rainfall during harvest months, which favors pre-harvest sprouting, leaf rust, and *Fusarium* head blight (FHB). In the south-central and central regions, water deficiency, excessive heat, and wheat blast (*Magnaporthe oryzae* pathotype *Triticum*) are the main limiting factors. The most important diseases of wheat in Brazil in general are, leaf rust (*Puccinia recondita* f. sp. *tritici*), head blight (*Gibberella zeae* and *Fusarium graminearum*), powdery mildew (*Blumeria graminis* f. sp. *tritici*), soil- borne wheat mosaic vírus—SBWMV, Septoria blight (*Mycosphaerella graminicola*—teleomorph *Septoria tritici*), spot blotch (*Bipolaris sorokiniana*), and blast disease caused by *Magnaporthe oryzae* pathotype *Triticum* (Bacaltchuk et al., 2006). Spraying with fungicides is usually practiced to manage these diseases. Wheat is also attacked by bacterial diseases in Brazil. The main bacterial diseases reported are, leaf streak (*Xanthomonas campestris* pv. *undulosa*), and leaf burning, bleaching or blight (*Pseudomonas syringae* pv. *syringae*). These are mainly managed by following crop rotation. The information on the recent progress toward unraveling the principal diseases of wheat crop in Brazil is compiled here. Emphasis in this chapter has been put on the characteristics of the diseases, their epidemiological aspects, and management using common control measures, which are followed in Brazil.

However, in Brazil, yields are low and unstable due to strong phosphorous fixation and acid soils with high levels of soluble aluminum; severe disease pressures, variable rainfall, often excessive in South Brazil and short in Central Brazil; and unseasonable frosts. In Northern Brazil, early, mid, and late-season heat and mid and late season drought adversely affect the crop. Frosts at flowering and excess rain at harvest are common. In Southern Brazil, frosts at flowering and rains at harvest can severely reduce production (Kohli & McMahon, 1988). The subtropical climate of Southern Brazil associated with the non-adoption of direct planting systems can favor the onset of grave epidemics, such as scab, a principal fungal disease that attacks wheat grain, leading to serious problems including low quality of grain and production losses (Astolfi et al., 2011).

15.2   WHEAT IN BRAZIL

Wheat was one of the first domesticated food crops and for 8000 years, has been the basic staple food of the major civilizations of Europe, West Asia, and North Africa (Curtis, 2015). Martim Afonso de Sousa first brought the wheat to Brazil in 1534 and grain was planted in the south, southeast regions initially. Hot weather hampered the expansion of wheat. It was only in the second half of the eighteenth century that the wheat crop began to grow in Rio Grande do Sul. Nevertheless, in the early nineteenth century, rust wiped out the wheat fields.

The planting was only resumed in the 20s of last century. From the 40s, wheat cultivation began to expand in Rio Grande do Sul and Parana, which has become the main producing state in Brazil. Research on seeds has increased the planted area and wheat crop yield.

Today, Brazil produces about 6 million tons of wheat, and rest over 4 million tones is imported to meet demand of consumer. Brazil did show 7.9% growth rate in wheat yield 1976–1985, which in later years fallen slightly negative. Soil degradation and erosion are problems throughout the region. Conservation tillage practices, especially zero-tillage and incorporation of green manure crops in the rotation, are spreading rapidly in the region (Kohli & McMahon, 1988).

In the period, 1993–1995, between seven and 10 mha of wheat were grown in the large southern cone region, largely in Argentina and Brazil. During that period, however, wheat area has reduced since 1985 by 12% (Curtis, 2015).

15.3 MAJOR DISEASES

Fungi, bacteria, and viruses cause several diseases in wheat. The main diseases affecting wheat root system are the common rot caused by *B. sorokiniana* and take-all disease caused by *Gaeumannomyces graminis* var. *tritici.* The main fungal foliar diseases are leaf rust, FHB, blast, powdery mildew, tan spot, spot and glume blotch, caused by *Puccinia triticina*, *G. zeae*, *Magnaporthe oryzae* pathotype *Triticum*, *Blumeria graminis* f. sp. *tritici*, *Pyrenophora tritici-repentis*, *B. sorokiniana,* and *Stagonospora nodorum*, respectively (Table 15.1).

The two main bacterial diseases are bacterial blight caused by *Xanthomonas campestris* (Pam.) Dow pv. *undulosa* (Mohan & Mehta, 1985) and bacterial blight or halo blight caused by *Pseudomonas syringae* pv. *coronafasciens* (Elliot) Young, Dye, and Wilkie (Wiese, 1998). The main diseases caused by viruses are common mosaic of wheat; and barley yellow stunting (Casa et al., 2000; Wiese, 1977).

15.4 LIFE CYCLE AND CONTROL OF WHEAT DISEASES

The diseases are among the main factors that limit the productivity and the expansion of wheat in Brazil. Diseases caused by fungi, bacteria, and viruses cause significant crop damage. Infection by these agents may occur at different stages of plant development.

The rusts and powdery mildew are responsible for significant damage to grain yield, which implies increasing efforts to develop resistant cultivars (Barcelloos et al., 1997). In wheat, the damage caused by powdery mildew can reach as high as 62% (Fernandes et al., 1988; Reis et al., 1996) and leaf rust can reach as high of the up to 63% (Reis et al., 2000). These diseases are managed using resistant cultivars, although the resistance is not durable due to frequent genetic variability in these pathogens.

TABLE 15.1   Wheat Diseases in Brazil.

Pathogens (Fungi)	Diseases	References
Gaeumannomyces graminis var. *tritici* (Sacc.)	Take all of wheat, *Ophiobolus* or root rot	Reis et al. (1983)
B. sorokiniana (Sacc.) Shoemaker (*Cochliobolus sativus* (Ito et Kurib.) Drechsler ex Dastur.	Spot blotch	Reis (1998)
F. graminearum Schwabe (teleomorph *G. zeae* (Schwein.) Petch.)	Leaf spot	Reis (1998)
Blumeria graminis f. sp. *tritici*	Powdery mildew	Reis et al. (1997)
P. recondita Rob. Ex. Desm. f. sp. *tritici* Eriks. & Henn.	Leaf rust	Barcellos et al. (1997)
Puccinia graminis f. sp. *tritici* Eriks & Henn.	Stem rust	Hu and Rijkenberg (1998)
Pyrenophora tritici repentis (Died) Drechs. [*Drechslera tritici repentis* (Died) Schoem.]	Tan spot	Reis and Casa (1996)
Leptosphaeria nodorum Müller [*Septoria nodorum* (Berk.) Berk.]	Glume blotch	Reis (1998)
M. graminicola (Fuckell) Schroeter [*S. tritici* Rob. in Desm.]	Septoria blotch or blight	Duncan and Howard (2000)
Fusarium nivale (Fr.) Ces. (*Calonectria nivale* Schaffn.)	Snow mold	Reis (1991)
Ustilago tritici (Pers.)	Loose smut	Reis (1991)
G. zeae Petch (*F. graminearum* Schwab.	Scab	Reis et al. (1996)
Magnaporthe oryzae pathotype *Triticum* B.C. Couch and L.M. Kohn	Blast	Igarashi et al. (1986)

15.4.1 TAKE ALL OF WHEAT

Wheat take all is caused by is of *Gaeumannomyces graminis tritici* (Ggt). It is distinguished from *G. graminis* var. *avenae*, in virulence on oats, and *G. graminis* var. *graminis*, which occurs on grasses, including rice. The take all caused by *G. graminis* var. *graminis* has been reported in rice in Brazil (Prabhu & Filippi, 2002). Take all is common in wheat grown in monoculture, tillage, soils with pH > 5.5, in regions with higher rainfall 450 mm (Reis & Bacaltchuk, 1979). Ammonium fertilizer can decrease disease levels and infection cycles of take-all in wheat (Colbach et al., 1996).

Thus, there is a real need to determine the effect of soil nutrient supply on disease development and biocontrol activities of biocontrol agents. The dramatic effect of take-all is higher in low fertility soils or those that have nutrients in unbalanced amounts. Smiley and Cook (1973) demonstrated that intensity of take all disease was declined by a high availability of NH_4^+ N: NO_3^- N, especially where nitrification was reduced. The NH_4^+ effect can be attributed to a combination of many factors, including increased root growth, increased host resistance, altered pathogenicity of the causative agent and due to the change in soil microflora. Moreover, NH_4^+ N favors the development of *Pseudomonas putida*, a bacterium antagonistic of Ggt. The most effective measures of control of take all are crop rotation with not susceptible species such as oats (*Avena sativa*), oilseed radish (*Raphanus sativus*), and canola (*Brassica napus* L.).

15.4.2 ROOT ROT

The root rot of wheat is caused by *B. sorokiniana*. It is responsible for the marked reduction in the ability of root to absorb water and nutrients from soil. This leads to the plants weaker and prone lodging and attack of other diseases. Environmental factors play an important role in the severity of disease caused by *B. sorokiniana* and other root rot fungi. Warm soil temperatures favor growth of *B. sorokiniana*. It occurs between 16 and 40°C, with optimal soil temperatures of 28–32°C. Moist soil conditions during planting also favors infection and colonization by pathogen (Murray et al., 1998).

Any movement of soil by wind, water, and implements can help in dissemination of inoculum. Infested seed can also serve as a means of dissemination of the pathogen over long distances (Mathre, 1987). Dissemination of secondary inoculum is not important for continued disease development below ground, but provides inoculum for subsequent crops (Murray et al., 1998). The practice of crop rotation is the best way to break the life cycle of pathogen.

15.4.3 LEAF RUST

Leaf rust is frequent in wheat crop, being present in all places where the crop is grown. It may cause yield losses to the extent of 63% (Reis & Casa, 2007). Leaf rust is caused by *P. triticina* (Menezes & Oliveira, 1993). The causal agent of wheat leaf rust is biotrophic parasite, and therefore, only survives in the parasitic phase in volunteer plants present in crops, the long paths, roads and highways (Reis & Casa, 2007). Cycle of host-pathogen relationships: *P. triticina* Eriks. × *T. aestivum* is shown in Figure 15.2.

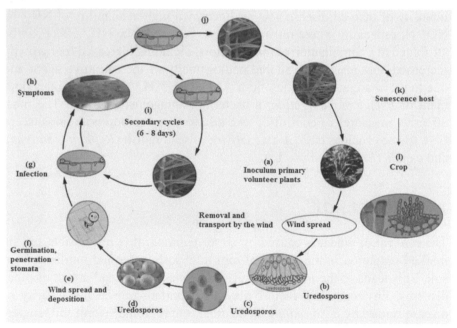

FIGURE 15.2   Wheat leaf rust cycle (Source: Reis et al., 2015a, with permission).

The uninterrupted monitoring of the wheat rust pathogenic population in Brazil in past helped in understanding of the evolution and virulence of races. The use of international nomenclature adopted by some programs has allowed the comparison of the variability of leaf rust pathogen in Brazil with that in other countries, especially where frontiers are not barriers for spore transportation. It has been confirmed by the occurrence of the similar races all over in one region (Bianchin et al., 2012).

Control measures include the use of cultivars with adult plant resistance. The application of triazole fungicides and mixtures of strobilurin at recommended doses. The indicator of the first application is the threshold economic damage (LDE). The fungicide may be sprayed at a gap of 15–20 days between applications. The volunteer plants (grass) may be eliminated (Reis et al., 2015a).

15.4.4   HEAD BLIGHT

FHB is a fungal disease of great importance for small-grain crops worldwide (McMullen et al., 1997). It is mostly caused by members of the *F. graminearum* species complex (Fg complex), teleomorph: *G. zeae* (Goswami & Kistler, 2004).

The main sources of inoculum are infected seeds and crop residues. Optionally, inoculum remains viable between the crops by parasitizing volunteer wheat plants along paths, roads, and highways and as free conidia dormant in the soil (Fig. 15.3). The inoculum of FHB survives in crop debris and infects wheat crops from flowering to grain filling stages when weather conditions are favorable (McMullen et al., 1997). Although yield losses are associated with reduced kernel plumpness and weight, the fungus produces mycotoxins that may accumulate to high levels, making harvested grain and their by-products inadequate for human and animal consumption (Creppy, 2002).

Integrated management practices include crop rotations, resistant wheat varieties, and fungicide applications that reduce fungal infection and subsequently mycotoxin production (Edwards, 2004). Deoxynivalenol (DON) is one of several mycotoxins produced by certain *Fusarium* species that frequently infect corn, wheat, oats, barley, rice, and other grains in the field or during storage (Sobrova et al., 2010). A survey of wheat grain in Brazil during 2006–2008 was conducted for FHB. It was concluded that the concentrations and spatial distribution of two trichothecenes were the major concern and was associated with FHB damage in commercial grain samples from a major wheat-growing area in Brazil (Del Ponte et al., 2012). Further studies are needed to identify agronomic and biological factors related to the variability in DON and nivalenol (NIV) levels due to FHB in wheat grains (Del Ponte et al., 2012).

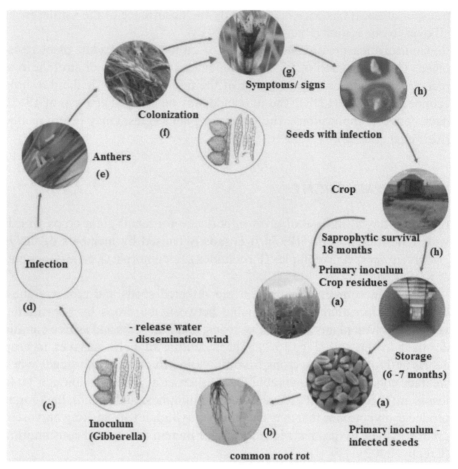

FIGURE 15.3   Life cycle of *Fusarium* spp. in wheat (Danelli et al., 2015 with permission).

15.4.5   WHEAT BLAST

The fungus *Magnaporthe oryzae* pathotype *Triticum* has a wide host range with emphasis on wheat, rice, and barley. Numerous cultivated grasses, native and invasive, are mentioned as hosts of this pathogen. In Brazil, *Pyricularia* occurs in Poaceae commonly found in rice and wheat crops, such as *Brachiaria plantaginea, Cenchrus echinatus, Digitaria horizontalis, D. sanguinalis, Eleusine indica, Hyparrhenia rufa, Pennisetum setosum, Rhynchelytrum roseum, Setaria geniculata* (grasses), *Echinochloa colonum, E. crus-galli,* and *Lolium multiflorum* (Urashima & Kato, 1994; Urashima & Bruno, 2001; Nunes et al., 2002; Prabhu & Filippi, 2006).

According to Agrios (2005), the chain of events that involve in the development of disease, including the phases and sub-phases of development of causal agent and the effect of disease on the host, constitutes also the cycle of corresponding disease and cycle of host-pathogen relationships (Fig. 15.4). The blast is characterized as polycyclic disease of wheat leaves, showing several cycles of the pathogen therefore considered difficult to control (Reis et al., 2015a).

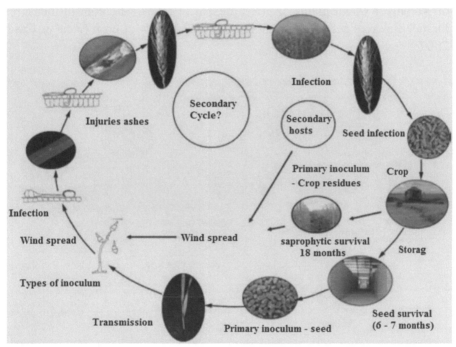

FIGURE 15.4 Wheat blast disease caused by *Magnaporthe oryzae* pathotype *Triticum* (Source: Reis et al., 2015b with permission).

15.4.6 POWDERY MILDEW

Wheat powdery mildew is a common disease occurring on wheat crops. It can cause damage of up to 62% in grain yield (Reis & Casa, 2007). Powdery mildew is caused by fungus *Blumeria graminis* Speer f. sp. *tritici* Em. Marchal (Menezes & Oliveira, 1993). In parasitic phase, powdery mildew is an ectoparasite. Its large mass of mycelia, conidia, and conidiophores develops on the plant surface. The spores are called conidia. Another type of fruiting body is cleistothecia that appears in senescence tissues has asci and ascospores (Wiese, 1977).

In Brazil, the presence of ascospores inside this fruiting body are not yet reported (Reis et al., 2015b). This fungus produces virulent races and distribution of this is not clearly understood in Brazil (Costamilan & Linhares, 2002).

The cycle of host-pathogen relationships can be seen in Figure 15.5. The pathogen is biotrophic in nature and only survives in the parasitic phase on volunteer wheat plants or grass present in crops. The control measures include the application of fungicides on aerial parts of crop (Reis & Casa, 2007).

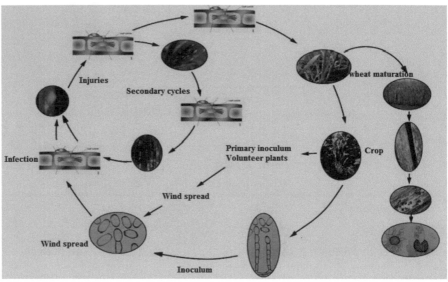

FIGURE 15.5   Wheat powdery mildew cycle (Source: Reis et al., 2015d with permission).

15.4.7   TAN SPOT

Tan spot of wheat is of frequent occurrence and is present in where untreated seeds are sown in monoculture and tillage. This disease may cause 48% losses in grain yield (Reis & Casa, 2007). A fungus, *Pyrenophora tritici repentis*, causes the disease (Menezes & Oliveira, 1993).

The disease cycle (Fig. 15.6) consists of the survival of the pathogen, sporulation, release, removal, transportation, deposition, germination, and penetration, colonization (parasitism) of pathogen, symptom expression, and sporulation on dead tissue. Thus, at the end of the primary cycle, process is repeated several times resulting in the growth of disease by secondary cycles. Control strategies always aim to interrupt one or more stages the disease cycle. The most effective control measures are crop rotation with non-susceptible plant species, with oats, radish, and canola.

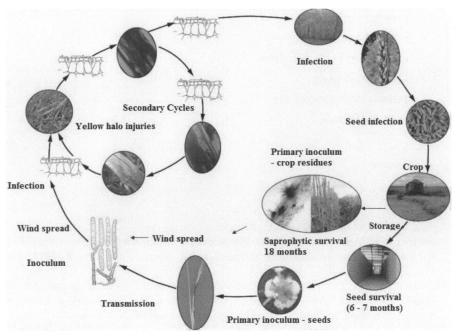

FIGURE 15.6 Wheat tan spot or yellow stain-cycle caused by *Pyrenophora tritici-repentis* (Source: Reis et al., 2015c with permission).

15.5 CONCLUDING REMARKS

Diseases are the most important yield-limiting factors in wheat production in Brazil due to favorable climate and growing of soft bread wheat. The fungicides are generally used to control these diseases. The important diseases are leaf rust, spot blotch and root rot, bacterial diseases, blast, powdery mildew, tan spot, head scab, and take-all in Brazil. The studies on the physiology of the wheat plant and host-pathogen relationships of different pathogens that attack it are needed for proper disease management in Brazil. The breeding for disease resistance is the need of the hour to reduce yield losses due to diseases.

KEYWORDS

- wheat diseases
- Brazil
- disease management
- powdery mildew
- wheat blast
- mycotoxin contamination
- bacterial diseases
- semi-temperate climate
- crop rotational

REFERENCES

Agrios, G. N. *Plant Pathology;* 5th ed.; Elsevier Academic Press: Burlington, 2005.

Astolfi, P., et al. Molecular Survey of Trichothecene Genotypes of *Fusarium graminearum* Species Complex from Barley in Southern Brazil. *Intern. Food Microb.* **2011,** *148,* 197–201.

Bacaltchuk, B., et al. *Características e Cuidados Com Algumas Doenças do Trigo;* Embrapa Trigo: Passo Fundo, Rio Grande do Sul, 2006; p 11. html Documentos Online, 64, http://www.cnpt.embrapa.br/biblio/do/p_do64.html (accessed Sep 03, 2015).

Barcellos, A. L., et al. Ferrugem da Folha do Trigo (*Puccinia recondita*): Durabilidade da Resistência. *Summa Phytopathol.* **1997,** *23,* 10–17.

Bianchin, V.; Barcellos, A. L.; Reis, E. M.; Turra, C. Genetic Variability of *Puccinia triticina* Eriks. in Brazil. *Summa Phytopathol.* (online). **2012,** *38*(2), 113118. ISSN 0100-5405. http://dx.doi.org/10.1590/S0100-54052012000200001 (accessed Sep 03, 2015).

Colbach, N.; Maurin, N.; Huet, P. Influence of Cropping System on Foot Rot of Winter Wheat in France. *Crop Prot.* **1996,** *15,* 295–305.

Casa, R. T.; Reis, E. M.; Schons, J. *Vírus do Nanismo Amarelo da Cevada–VNAC;* UPF: São Paulo, 2000; p 22.

Costamilan, L. M.; Linhares, W. I. Efetividade de Genes de Resistência de Trigo a Oídio. *Fitopatol. Bras.* **2002,** *27,* 621–625.

Creppy, E. E. Update of Survey, Regulation and Toxic Effects of Mycotoxins in Europe. *Toxicol. Lett.* **2002,** *127*(3), 19–28.

Curtis, B. C. *Wheat in the World;* FAO: Rome, 2015. http://www.fao.org/docrep/006/y4011e/y4011e04.html (accessed Sep 03, 2015).

Danelli, A. L. D.; Zoldan, S.; Reis, E. M. Giberela-Ciclo da Doença., http://www.orsementes.com.br/sistema/anexos/artigos/20/Ciclo%20giberela.pdf (accessed Sep 30, 2015).

Del Ponte, E. M.; Garda-Buffon, J.; Badiale-Furlon E. Deoxynivalenol and Nivalenol in Commercial Wheat Grain Related to *Fusarium* Head Blight Epidemics in Southern Brazil. *Food Chem.* **2012,** *132*(2), 1087–1091.

Duncan, K. E.; Howard, R. J. Cytological Analysis of Wheat Infection by the Leaf Blotch Pathogen *Mycosphaerella graminicola*. *Mycol. Res.* **2000,** *104*(09), 1074–1082.

Edwards, S. G. Influence of Agricultural Practices on *Fusarium* Infection of Cereals and Subsequent Contamination of Grain by Trichothecene Mycotoxins. *Toxicol. Lett.* **2004,** *153,* 29–35.

Fernandes, J. C. M.; Rosa, O. S.; Picinini, E. C. Perdas no Potencial de Rendimento de Linhas Quase-Isogênicas de Trigo Devidas ao Oídio. *Fitopatol. Bras.* **1988,** *13,* 131.

Goswami, R. S.; Kistler, H. C. Heading for Disaster: *Fusarium graminearum* on Cereal Crops. *Mol. Pl. Pathol.* **2004,** *5*(6), 515–525.

Hu, G.; Rijkenberg F. H. Subcellular Localization of Beta-1, 3-Glucanase in *Puccinia recondita* f. sp. *tritici*-Infected Wheat Leaves. *Planta.* **1998,** *204*(3), 324–34.

Igarashi, S., et al. *Pyricularia* sp. em Trigo. I. Occurencia de *Pyricularia* sp. no Estado do Paraná. *Fitopatol. Bras.* **1986,** *11,* 351–352.

Kohli, M. M.; Mc Mahon, M. A. A Perspective of Research Needs for Non-Irrigated Tropical Countries. In *Wheat Production Constraints in Tropical Environments;* Klatt, A. R., Ed.; CIMMYT: Mexico, 1988.

Leivas, J. F., et al. Biophysical Parameters in a Wheat Production Region in Southern Brazil. 2014. http://www.alice.cnptia.embrapa.br/alice/bitstream/doc/1001291/1/4203.pdf (accessed Sep 03, 2015).

MAPA (Ministério Da Agricultura, Pecuária E Abastecimento). *Trigo;* 2015. http://www.agricultura.gov.br/vegetal/culturas/trigo (accessed Sep 03, 2015).

Mathre, D. E. *Compendium of Barley Diseases;* American Phytopathological Society: St. Paul, MN, 1987.

McMullen, M.; Jones, R.; Gallemberg, D. Scab of Wheat and Barley: A Re-Emerging Disease of Devastating Impact. *Plant Dis.* **1997,** *81,* 1340–1348.

Menezes, M.; Oliveira, S. M. A. *Fungos Fitopatogênicos;* UFRPE: Recife, 1993; p 227.

Mohan, S. K.; Mehta, Y. R. Estudos Sobre *Xanthomonas campestris* pv. Ondulosa em Trigo e Triticale no Estado do Paraná. *Fitopatol. Bras.* **1985,** *10,* 447–453.

Murray, T. D.; Parry, D. W.; Cattlin, N. D. *A Color Handbook of Diseases of Small Grain Cereal Crops;* Iowa State University Press: Ames, IA, 1998.

Nunes, C. D. M.; Brancão, N.; Rodrigues, R. S.; Reis, J. C. Ocorrência de Brusone no Azevém em Diferentes Locais do RS Brasil. *Fitopatol. Bras.* **2002,** *27,* 231.

Prabhu, A. S.; Filippi, M. C. Ocorrência do Mal-do-pé Causado por *Gaeumannomyces graminis* var. Graminis, Uma Nova Enfermidade em Arroz no Brasil. *Fitopatol. Bras.* **2002,** *27*(4), 417–419.

Reis, E. M., et al. Ferrugem da Folha do Trigo-Ciclo da Doença. 2015a. http://www.orsementes.com.br/sistema/anexos/artigos/16/Ciclo%20da%20ferrugem%20da%20folha%20do%20trigo.pdf (accessed Sep 30, 2015).

Reis, E. M. *Doenças do Trigo IV: Septorioses;* Ciba Geigy: São Paulo, 1998; p 29.

Reis, E. M.; Bacaltchuk, B. O mal-do-pé do Trigo. *Trigo soja.* **1979,** *45,* 12–15.

Reis, E. M.; Casa, R. T. *Doenças do Trigo VI: Mancha Amarela da Folha;* Bayer: Passo Fundo, São Paulo, 1996; p 16.

Reis, E. M.; Blum, M. M. C.; Casa, R. T.; Medeiros, C. A. Grain Losses Caused by the Infection of Wheat Heads by *Gibberella zeae* in Southern Brazil, from 1984 to 1994. *Summa Phytopathol.* **1996,** *22,* 134–137.

Reis, E. M.; Casa, R. T. *Doenças dos Cereais de Inverno-Diagnose,* Epidemiologia e Controle: Graphel Lages: Lages, Santa Catarina, 2007; p 176.

Reis, E. M.; Casa, R. T.; Hoffmann, L. L.; Mendes, C. Effect of Leaf Rust on Wheat Grain Yield. *Fitopatol. Bras.* **2000**, *25*, 67–71.

Reis, E. M.; Danielli, A. L. D.; Zoldan, S. Brusone do Trigo-Ciclo da Doença. 2015b, http://www.orsementes.com.br/sistema/anexos/artigos/15/Ciclo%20brusone.pdf (accessed Sep 30, 2015).

Reis, E. M.; Zodan, S.; Danielli, A. L. D.; Tonin R. B. A Mancha-Amarela-da-Folha do Trigo-Ciclo da Doença. 2015c. http://www.orsementes.com.br/sistema/anexos/artigos/22/Ciclo%20mancha-amarela.pdf (accessed Sep 30, 2015).

Reis, E. M.; Zoldan S.; Danielli, A. L. D.; Avozani, A. Oídio do Trigo-Ciclo da Doença. 2015d. http://www.orsementes.com.br/sistema/anexos/artigos/23/Ciclo%20o%C3%ADdio%20trigo.pdf (accessed Sep 30, 2015).

Reis, E. M. *Doenças do Trigo V: Ferrugens;* Bayer: Passo Fundo, São Paulo, 1991; p 20.

Reis, E. M.; Casa, R. T.; Hoffmann, L. L. E Efeito de Oídio, Causado Por *Erysiphe graminis* f. sp. Tritici, Sobre o Rendimento de Grãos de Trigo. *Fitopat. Bras.* **1997**, *22*, 492–495.

Reis, E. M.; Lhamby, J. C. B.; Santos, H. P.; Kouchhann, R. A. Efeito de Calcário e de Sistemas de Semeadura na Incidência de *Gaeumannomyces graminis* var. Tritici e de *Helminthosporium sativum* em Raizes de Trigo (*Triticum aestivum*). *Summa Phytopathol.* **1983**, *8*(3–4), 56–64.

Scheeren, P. L. *Informações Sobre o Trigo (Triticum spp.);* Embrapa – CNPT: Passo Fundo, 1986; Documentos 2, p 34.

Smiley, R. W; Cook, R. J. Relationship between Take-All of Wheat and Rhizosphere pH in Soils Fertilized with Ammonium vs. Nitrate-Nitrogen. *Phytopathol.* **1973**, *63*, 882–890.

Sobrova, P., et al. Deoxynivalenol and its Toxicity. *Interdisc. Toxicol.* **2010**, *3*(3), 94–99. doi:10.2478/v10102-010-0019-x (accessed Sep 30, 2015).

Urashima, A. S.; Bruno, A. C. Interrelação Sexual de *Magnaporthe grisea* do Trigo com a Brusone de Outros Hospedeiros. *Fitopatol. Bras.* **2001**, *26*, 21–26.

Urashima, A. S.; Kato, H. Varietal Resistance and Chemical Control of Wheat Blast Fungus. *Summa Phytopathol.* **1994**, *20*, 107–112.

Wiese, M. V. *Compendium of Wheat Diseases;* APS Press: St. Paul, MN, 1977; p 106.

Wiese, M. V. *Compendium of Wheat Diseases;* 2nd ed.; APS Press: St. Paul, MN, 1998; pp 8–9.

CHAPTER 16

STATUS OF WHEAT DISEASES AND THEIR MANAGEMENT IN GUJARAT STATE OF INDIA

S. I. PATEL[1], V. A. SOLANKI[2], B. M. PATEL[1], and R. S. YADAV[3*]

[1]Wheat Research Station, S. D. Agricultural University, Vijapur 382870, Gujarat, India

[2]Department of Plant Pathology, N. M. College of Agriculture, Navsari Agricultural University, Navsari 396450, Gujarat, India

[3]ICAR-Directorate of Groundnut Research, Junagadh 362001, Gujarat, India

*Corresponding author. Email: yadavrs2002@gmail.com

CONTENTS

ABSTRACT

Biotic stresses are the conspicuous inhibitions for realizing the attainable productivity in wheat. The diseases like leaf and stem rusts are potentially most dreadful when the variety is susceptible and conditions are favorable for their development. Foliar blight and black point occur in punctuated pockets. The wheat produced in Gujarat state is totally free from Karnal bunt which is an important disease from sanitary and phyto-sanitary (SPS) point of view in international wheat grain trade. This chapter deals with the status of important diseases of wheat in Gujarat state of central zone of India, their symptomatology, epidemiology, and management.

16.1 INTRODUCTION

India harvested record wheat production of 95.85 million ton during 2013–2014 from an area of 30.47 million ha (Anonymous, 2015). During 2014–2015, a quantum fall of 5.07 million ton in comparison to the previous year is mainly attributed due to unexpected hailstorms, erratic, and unseasonal precipitation in the major wheat growing regions of the country. Besides this, fluctuations in ambient temperature during cropping season had been the major contributing factor for variability in productivity of the wheat in India and especially in Gujarat state and may influence further as the consequences of global warming and climate change in future. Gujarat is important wheat producing state in central zone of India and wheat produced is considered of high quality and fetches higher prices in market.

16.2 WHEAT SCENARIO IN GUJARAT

Traditionally, wheat is cultivated in an area ranging from 5-7 lakh ha in Gujarat. However, during the last decade, a conspicuous three-folds increase in production and acreage is discernible. This trends in production of 4.7 million ton from 1.5 million ha with productivity of 3143 kg/ha during 2013–2014 was attributable to high yielding varieties, improved technology, back up with irrigation facilities, congenial administrative policies and favorable weather conditions especially prolonged winter. The wheat productivity during different triennium in Gujarat is presented in Figure 16.1.

Unlike other parts of India, Gujarat is known for diversified animal husbandry based agriculture with entrepreneurial penchant for cash crops dominating non-cereal crop husbandry. The high end technologies particularly the new varieties and micro-irrigation systems (MIS) have picked up in the state mainly because of good backward linkages by the state Government. The enviable achievement in wheat production (~3000 kg/ha) during last 3–4years was ushered despite narrow temperature window of around 14 weeks as compared to over 25 weeks elsewhere in wheat bowl of North India.

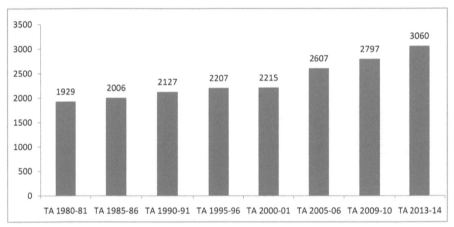

FIGURE 16.1   Average triennium productivity (kg ha⁻¹) of wheat in Gujarat.
TA; triennium ending year.

It is impressive to note that per day productivity of wheat in Gujarat is comparable to the best global per day productivity of wheat. There are ample chances of enhancement of area under indicated that the area, production, and productivity of wheat in Gujarat have been increased by 106, 181, and 27%, respectively, irrigated durum wheat for which the successful feasibility endeavors were taken along the Narmada linkage network. Interestingly, the productivity of new durum varieties is higher than aestivum varieties in most of the areas.

Bread wheat (*Triticum aestivum* L.) is the overarching (85%) species followed by durum or macaroni wheat (*Triticum durum* Desf.). The acreage is almost equal in three major wheat-growing regions *viz*; Saurashtra (37%), North Gujarat (31%), and Middle Gujarat (30%). South Gujarat has hemmed acreage of less than 2% (Fig. 16.2).

TABLE 16.1 The Decadal Status of Area (Mha), Production (million ton), and Productivity (kg/ha) of Wheat in Gujarat (Source: DOA, GOG, Gandhinagar).

Year	Total wheat			Irrigated wheat			Rain fed wheat		
	Area (million ha)	Production (million ton)	Productivity (kg/ha)	Area	Production	Productivity	Area	Production	Productivity
2004–05	0.73	1.81	2482	0.65	1.75	2697	0.08	0.06	674
2005–06	0.86	2.32	2701	0.79	2.27	2882	0.07	0.05	652
2006–07	1.07	2.79	2603	0.96	2.72	2823	0.11	0.07	646
2007–08	1.27	3.84	3013	1.19	3.78	3182	0.09	0.06	690
2008–09	1.09	2.59	2375	1.02	2.54	2500	0.08	0.05	691
2009–10	0.88	2.35	2678	0.85	2.34	2740	0.02	0.01	527
2010–11	1.59	5.01	3156	1.50	4.94	3297	0.09	0.07	804
2011–12	1.35	4.07	3015	1.27	4.00	3158	0.08	0.07	847
2012–13	1.02	2.94	2876	0.98	2.91	2975	0.04	0.03	720
2013–14	1.50	4.71	3143	1.43	4.66	3253	0.07	0.05	755
2014–15	1.15	3.22	2810	–	–	–	–	–	–

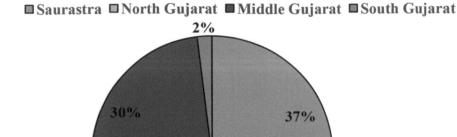

FIGURE 16.2 Distribution of bread wheat (*T. aestivum* L.) acreage in Gujarat state of India

Durum (macaroni)wheat is predominantly grown in two pockets, Bhal and Godhra. Dicoccum wheat (*Triticum dicoccum*) is also grown in insignificant punctuated pockets in Saurashtra along the basins of Ojat River. All bread wheat is irrigated while macaroni wheat is grown both under rain fed as well as irrigated conditions. Ironically, the rain fed macaroni wheat can be ascribed as the most stressed wheat exposed to triple stress conditions of water, heat and dreaded soil hazards. The area under dry wheat has diminished after the major water conservation and development endeavors in Gujarat have culminated in extending irrigation facilities. The state has wide range of cropping systems *viz.* Cotton-Wheat-Pearl millet, Mung-Wheat-Pearl millet, Cotton-Wheat-Mung bean, Cotton-Wheat, Groundnut-Wheat, Paddy-Wheat, Pearl millet-wheat, and so forth. Amber and hard grain wheat is preferred locally and is consumed as chapatti for which bread wheat is more suitable. Similarly, durum wheat with hard vitreous grain is more appropriate for semolina, pasta, and local products like laddu, bhakhari, and sheera. Lesser yellow berry and high β-carotene contents are the most desirable traits of durum wheat. Dicoccum is hard textured too but is more appropriate for paste or gruel preparations like upma. It is needless to mention that Gujarat wheat has no sanitary and phyto-sanitary (SPS) issues from quarantine point of view and therefore, holds great potential for exports.

16.3   CLIMATE AND WEATHER CONDITIONS

The sowing of wheat in Gujarat is normally done on a higher temperatures notch (46–48 meteorological week for normal sowing and 48–50 meteorological week for late sowing) during which average maximum temperature (average of 25 years) would remain above 30°C. The maximum temperature would remain around 26–28°C from second fortnight of December till first fortnight of February. During this period, the anthesis and early dough stage of normal sown wheat crop is almost completed. The temperature rises to above 30°C thereafter which forces grain maturity. The difference between maximum and minimum temperature also play pivotal role in the development of wheat crop. On long-term basis (average of 25 years), difference was ranging from 11.1 to 18.5°C which had significant impact on wheat productivity. On the other hand, slightly declined trend on short-term basis (average of five years) as well as during 2014–2015 especially during anthesis to grain maturity stages is very evident that it could have manifested in higher wheat productivity.

16.4   IMPORTANT DISEASES OF WHEAT IN GUJARAT

Biotic stresses are the conspicuous inhibitions for realizing the attainable productivity in wheat. Among these, like diseases leaf and stem rusts are potentially most important. Foliar blight and black point occur in punctuated pockets. Wheat produced in Gujarat is free of from Karnal bunt which is important from SPS point of view. In pursuant to extraordinary grain quality and absence of SPS issues the state has enormous opportunities for wheat export particularly that of durum wheat.

16.4.1   LEAF RUST

Leaf rust caused by *Puccinia triticina* Eriks occurs primarily on the leaf blades although leaf sheaths can also be infected under favorable conditions at temperatures (10–30°C, high inoculum densities and susceptible cultivars. The initial symptoms of leaf rust appear as flecks, or tiny water soaked light-colored spots, 7–10 days after inoculation (McCallum et al., 2012). These progress into small sporulating pustules with chlorosis or necrosis surrounding the pustule in the incompatible or resistant interactions, or to larger sporulating pustules without chlorosis or necrosis within 10–14 days after inoculation (Fig. 16.3) (Roelfs et al., 1992; McIntosh et al., 1995). It generally lacks the abundant teliospore production at the end of the season, resulting in a brown leaf lesion. However, when teliospores are produced, they usually remain covered by the leaf epidermal cells. Losses in grain yield are primarily attributed to reduced floret set and grain shriveling.

(A)   (B)

(C)   (D)

FIGURE 16.3 Leaf rust of wheat at different stages of its development. A and B; flecks and/or tiny light-colored spots, C; small to large sporulating pustules, D; brown leaf lesion.

16.4.2 STEM RUST

Stem rust is caused by *Puccinia graminis* Pers. f. sp. *tritici* Eriks & E. Henn. It is considered most dreadful due to its rapid spread particularly where the growing season is warmer (>15°C). Severe infection of stem rust interrupts nutrient flow to the developing panicle resulting in shriveled grains and stem weakened by rust infection are prone to lodging (Roelfs et al., 1992). Stem rust has not caused severe yield losses in Gujarat since last four decades mainly due to adequate diversified resistance in the varieties commonly grown in the state.

(A)   (B)   (C)   (D)

FIGURE 16.4   Stem rust of wheat at different plant parts. A, B and C; elongate blister-like pustules/uredinia in leaf, leaf sheath, glumes and awns, D; dark brown to black telia in infected mature plants.

The stem rust disease appears as elongate blister-like pustules, or uredinia, most frequently on the leaf sheaths but also on stem tissues, leaves, glumes, and awns. Stem rust pustules on leaves develop mostly on the dorsal side, but may penetrate and make limited sporulation on the ventral side. The initial macroscopic symptom is typically a small chlorotic fleck, which appears a few days after infection. On the leaf sheath and glumes, pustules rupture the epidermis and give a shabby appearance. Masses of uredinio-spores produced on the pustules are brownish red in color and easily shaken off the plants. As infected plants mature, uredinia change into telia, altering color from red to dark brown to black; thus, the disease is also called black rust (Fig. 16.4). Teliospores are attached tightly to plant tissue. The pathogen on the alternate host barberry produces raised yellow-orange lesions on leaves, petioles, blossoms, and fruits. Symptoms include clusters of orange spore mass. (Singh et al., 2012).

16.4.3 DISEASE CYCLE OF LEAF AND STEM RUSTS

Rust pathogens are biotrophs in nature (Mendgen & Hahn, 2002), which produces large number of uredospores during crop season and these are transmitted through wind on the same or new host plant. Typically, most spores will be deposited close to the source (Roelfs & Martell, 1984). However, long distance dispersal is also well documented. In India, both leaf and stem rusts appear in wheat growing belt during *Rabi* crop season. The uredospores/teliospores survive and the hills during off-season on self-sown wheat plant crop or volunteer hosts, and act as a source of inoculum. There is no evidence so far that the alternate host for both stem and leaf rust pathogens are functional in India. The fungus is inhibited by high temperatures (>40°C) during summer followed by rainy season makes the survival of teliospores of wheat rusts difficult in Gujarat. The complete cycle from infection to the production of new spores can take as little as seven days during ideal conditions. The uredospores acts as repeating spores during wheat crop season. The disease cycle may therefore be repeated many times in one season. Mehta (1952) opined that dissemination of rusts results from wheat crop on hills and plains. He also visualized that wheat crop is cultivated round the year in the Nilgiri and Pulney hills in south India and Himalayan ranges in North India carried infection of rusts.

16.4.4 DYNAMISM OF RUST PATHOTYPES IN GUJARAT

Rust pathogens are dynamic in nature and evince innumerable oddities. These have many pathotypes and emergence of new pathotype results a resistant variety susceptible to rusts. Hence, dynamism of pathotypes has tangible repercussions.

The pattern of dynamics of brown rust pathotypes in Gujarat during the last decade is depicted in Figure 16.5. Two pathotype *viz*; 21R55 (104-2) and 121R63-1 (77-5) were overarching in their presence indicating these pathotypes establish, multiply and spread very fast under prevailing environmental conditions. The other pathotypes *viz*., 77-11, 77, 104-B, 104-3, 77-6, 77-1, 77-2, and 77-8 were reported in the range of 1–18% of the samples. These findings were based on analyses of rusts samples collected from Gujarat through rigorous surveillance and analyzed at IIWBR, Regional station, Flowerdale, Shimla for different pathotypes.

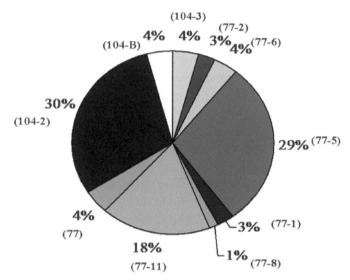

FIGURE 16.5   Pathotypes of leaf rust detected in Gujarat and their percentange.

16.4.5   *MANAGEMENT OF RUSTS IN GUJARAT*

16.4.5.1   *HOST RESISTANCE*

The principle mechanism of control of the cereal rusts has been through the use of resistant cultivars (Johnson, 1981) because other remedial measures usually are not pragmatic. Breeding for rust resistance and cultivation of such varieties is quite viable proposition to manage wheat rusts. However, frequently occurrence of new variants and shift in virulence pattern render resistant varieties susceptible (Bhardwaj et al., 2005). As a follow up action of pathotype situation, deployment of varieties with diverse resistance is put into practice for management of wheat rusts. Since rusts are obligate parasites, any resistance genes in host cultivars that curtail or eliminate rust reproduction will place tremendous selection pressure on variants that are virulent to the resistance genes. The widespread use of wheat cultivars with different rust resistance genes may results in different rust races in the major wheat production regions. In view of this large numbers of breeding lines were screened every year at Wheat Research Station, Vijapur, Gujarat by creating artificial rusts or epiphytotic conditions. A total number of 11,683 lines were screened against stem and leaf rusts during last five years (2010– 2014) as summarized in Table 16.2. The genetically diversified materials having resistant genes for rusts were involved using breeding program. Some of the wheat cultivars evolved in Gujarat exhibited wide genetic diversity for rusts resistance in wheat (Table 16.3).

TABLE 16.2   Genotypes Screened for Rust Resistance under Epiphytotic Conditions during 2010–14 crop Seasons.

Year	No. of entries						
	Coordinated yield trials	NBPGR	CIMMYT	Vijapur	Juna-gadh	Arnej	Dhan-dhuka
2010–11	489	1097	–	338	286	–	206
2011–12	621	1179	123	302	298	42	199
2012–13	654	979	150	327	279	35	171
2013–14	511	863	200	324	300	35	172
2014–15	526	–	–	370	281	49	177
Total	2801	4118	473	1661	1544	161	925

TABLE 16.3   Genetic Diversity for Rusts Resistance in Released Wheat Cultivars.

Wheat cultivars	Postulated gene(s)		
	Stem rust *(Sr)*	Leaf rust *(Lr)*	Yellow rust *(Yr)*
Lok 1	*11+2+*	*13+*	*2(KS)+*
GW 173	*2+9b+11+*	*23+34+*	*2(KS)+18+*
GW 190	*2+31+*	*26+23+1+*	*9+*
GW 273	*2+31+*	*13+10+*	*9+*
GW 322	*11+2+*	*13+*	–
GW 1139	*2+*	*13+*	–
GW 366	*2+*	–	–
GDW 1255 (d)	*9e+*	–	–
GW 451	*2+7b*	–	*2+*

(d) represents to *Triticum durum*.

Most of the released varieties possess an $Lr13+$ gene which provides the resistance against brown rust because of introduction of pathotypes 121R63-1 and 21R55 which are virulent on $Lr23+$ and $Lr26+$. Postulation of genes for black rust in released varieties clearly noticed the $Sr2+$ and $Sr11+$ genes. Gene $Sr2$ provides the durable source of resistance. Whereas $Sr11+$, $Sr9b+$, $Sr31+$ are also conferring varying degree of resistance. The varieties released from the center evinced inordinate resistance to both black and brown rusts and hence no incidence of rusts was reported from the state for the last few years.

The status of resistance in breeding materials developed by the center during 2009–2010 to 2013–2014 was assessed using multiplication screening against leaf and stem rusts. It clearly indicated that nearly 90% of the breeding lines developed were either resistant or moderately resistant against leaf rust (Fig. 16.6) whereas, nearly 75% are resistant to moderately resistant against stem rust (Fig. 16.7).

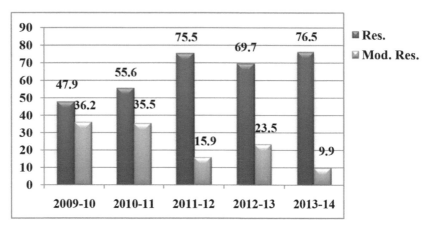

FIGURE 16.6   Status of resistance in wheat genotypes of Gujarat against leaf rust during 2009-2014.

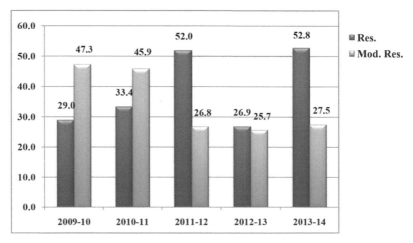

FIGURE 16.7   Status of resistance (% Entries, R, and MR) in wheat genotypes against stem rust during 2009-2014 in Gujarat.

16.4.5.2 CHEMICAL CONTROL

The rust pathogens are air-borne and are capable of causing damage over a large area, the chemical control may not be feasible as a routine practice. However foliar sprays of fungicides can be undertaken at the disease initiation to contain its spread. The effective fungicides for the control of wheat rusts are propiconazole @ 0.1%, mancozeb @ 0.2%, and tebuconazole @ 0.1%.

16.4.5.3 CONFRONTING WHEAT STEM RUST PATHOTYPE UG99

Preemptive strategy to screen existing varieties and advanced breeding lines at hot spots in Kenya with the help of ICAR-IIWBR Karnal has been adopted. During 2007–2008 screening program, two released varieties *viz.* GW 273 and GW 322 of Gujarat were found resistant to Ug99. Screening of advance breeding lines from Gujarat was carried out in Ethiopia and Kenya during 2013 against stem rust pathotype Ug99. Among these, GW 432 in Ethiopia and GW 1276, GW 1277, and GW 1280 in Kenya were reported resistant with (Act up to 10.0 only.

16.4.6 FOLIAR BLIGHT

Foliar blight is a complex disease and is caused by *Helminthosporium sativum* Pammel. king and Bakke [Syn. *Bipolaris sorokiniana* (Sacc.) Shoem] appears to be major pathogen along with *Alternaria triticina* Prasada & Prabhu and *A. alternata* (Fr.) Keissler (Misra, 1973; Joshi et al., 1974, 1978). The disease causes significant yield losses in warmer wheat growing areas. It is the major biotic constraint in wheat in the Gangetic Plains of South Asia, especially in the rice-wheat cropping system (Duveiller et al., 1998; Minh et al., 1998). Grain yield losses due to foliar blight vary greatly, depending on wheat crop husbandry. In affected areas, yield losses depend on genotype, sowing time, location, and stress conditions (Villareal et al., 1995; Sharma & Duveiller, 2004; Rosyara et al., 2005). Under resource constraint farming conditions, leaf blight can cause substantial grain yield reductions (Sharma & Duveiller, 2006). Heat stress conditions coupled with the low use of inputs lead to increased foliar blight severity. Change in climatic conditions resulted reduced window of cool winter and increased heat stress, which gradually causes higher levels of foliar blight disease (Rosyara et al., 2010).

16.4.6.1   SYMPTOMS OF FOLIAR BLIGHT

B. sorokiniana causes spot blotch as well as head and seedling blight (Zillinsky, 1983). Typically, the first foliar symptom appears as small, light brown blotches that develop into oval-shaped, light brown, necrotic lesions bordered with a yellow halo (Schilder & Bergstrom, 1993). In susceptible genotypes, these lesions extend very quickly in oval to elongated blotches having light brown to dark brown color. They may reach several centimeters before coalescing and inducing the death of the leaf. Fruiting structures develop readily under humid conditions and are generally easily observed on old lesions.

The infection of *A. triticina* is initially evident as small, oval, discolored lesions, irregularly scattered on the leaves and become dark brown to grey and irregular in shape. Some are surrounded by a bright-yellow marginal zone. The lesions vary in size, reaching a diameter of 1 cm or more. As the disease progresses, several lesions coalesce to cover large areas, resulting in the death of the entire leaf. The infected leaf begins to dry from the tip, prematurely, when lesions appear. The lowermost leaves are the first to show the signs of infection; the fungus gradually spreads to the upper leaves (Fig. 16.8). Heavily infected fields present a burnt appearance and can be identified from a distance (Prasada & Prabhu, 1962; Prabhu & Prasada, 1966; Singh, 1990).

FIGURE 16.8   Foliar blight of wheat in Gujarat.

16.4.6.2 EPIDEMIOLOGY OF FOLIAR BLIGHT

The leaf blight disease is favored by moderate to warm temperatures (18–32°C) and particularly by humid weather. The role of infected seed as a primary source of inoculum appears to be important (Shaner, 1981). The efficiency pathogen transmission from infected seeds to plumules and coleoptile tips may be as high as 87% for *B. sorokiniana* (Reis & Forcelini, 1993). The fungus invades the moistened seed after sawing and just after emergence, as early as the first leaf stage, sporulation is induced in the presence of direct sunlight (Spurr & Kiesling, 1961). The disease severity is directly related to the duration of leaf moisture and high temperature in the field (Reis, 1991). The post inoculation exposure of plants to higher temperature (35°C) and relative humidity 95% resulted early development of symptoms than those exposed 24 h before inoculation. However, plant kept at 25°C and relative humidity of 90% developed symptoms earlier than exposed plants and had higher score of leaf blight (Singh et al., 2011). The effects of environmental conditions may allow a better measurement of small, quantitative differences of resistance among wheat genotypes (Duveiller et al., 1998).

A forecast model on significance of biological and meteorological parameters to predict foliar blight disease under natural conditions was designed based on experiments conducted (1998–1999 to 2006–2007) at Wheat Research Station, Vijapur, Gujarat. The significant independent variables *viz.* crop age in terms of days after sowing (DAS) and temperature (minimum, maximum, mean, morning, and afternoon) were found crucial for disease forecast. The results indicated that temperature played significant role in disease development and hence thermal time (degree-days) in consonance with crop age was more important in designing prediction equation rather than maximum, minimum, morning, and afternoon temperatures. Prediction equation based on pooled analysis with crop age (i.e., DAS) and cumulative degree-days (CDD) revealed that an increase of one unit in independent variable like DAS and CDD increased foliar blight intensity and area under disease progress curve (AUDPC) to the tune of 0.72 and 0.08% and 4.46 and 0.64%, respectively. Following conclusions were drawn:

1. The late sown crop is more vulnerable to foliar blight disease development than timely sown conditions in Gujarat.
2. Crop age (DAS) and temperatures (minimum, maximum, mean, morning, and afternoon) exhibited positive correlation; whereas relative humidity (morning, afternoon, and mean) showed negative correlation with percent disease intensity and AUDPC.

3.   Foliar blight prediction equation was significantly correlated up to 1% level of significance. (Patel, Personal Communication) (Table 16.4).

TABLE 16.4   Foliar Blight Prediction Equations for Percent Disease Index (PDI) and Area Under Disease Progressive Curve (AUDPC) at 1 & 5% Level of Significance.

Parameter	Predicted equation	Regression co-efficient (R^2)
PDI**	Y = −25.7544 + 0.7298 DAS + 0.0878 CDD	0.7709
AUDPC**	Y = −196.2409 + 4.4636 DAS + 0.6421 CDD	0.7779

**$p = 0.05$.

16.4.6.3   MANAGEMENT OF FOLIAR BLIGHT

Host Resistance

Host resistance is the most important and effective method to manage the leaf blight in the field conditions. During every year, a number of entries are evaluated at hot spot locations under artificially inoculated conditions. The percent leaf blight resistant entries in different zones of India were in the range of 0.0–49.6 (Singh, 2011). Similarly, under Gujarat conditions, the disease appearance was reported negligible in released varieties, GW 322, GW 496, GW 173, and GDW 1255.

Chemical Control

Seed treatment with carboxin, carboxin + thiram (1:1) @ 0.25% and thiram @ 0.3% was effective in controlling seed borne infection and increased seed germination percentage (Singh, 2011). Based on chemical control experimental results under Gujarat conditions, three foliar sprays of mancozeb @ 0.25% at 15 days' interval were recommended as a prophylactic measures after eight weeks of germination. Likewise three curative sprays of thiophanate methyl @ 0.035% at 15 days' interval on the appearance of disease were also recommended (Patel, Personal Communication).

16.4.7   BLACK POINT

The disease produces dark brown to black discoloration at the embryonic end of the seed and surrounding areas. The entire seed coat may be discolored under severe infection. The species of *Alternaria, Cladosporium, C. sativus, Fusarium, Helminthosporium,* and *Pyrenophora* are found associated with black point. Some of these fungi are pathogenic. Black point symptoms may also be caused due to the enzyme, peroxidase also (Jacobs & Rabie, 1987; Ellis et al., 1996; Williamson, 1997). Susceptible genotypes exhibited a higher level of ferulic acid. Weather plays a crucial role in the development of black point. The disease is favored by high relative humidity, rainfall. High humidity from milk to dough stage, late season irrigation, and lodging often trigger infection. Seed is susceptible to infection during filling or maturation, particularly at the milk and soft dough stages. Diseased kernels are discolored, weathered, and black at the embryo end extending to the ventral surface (Conner & Davidson, 1988). The embryos may be shriveled, brown to black in color and will have less germination ability. Black point can affect grain quality and food products made from it can have a displeasing odor and color (Sharma, 2012). It may not reduce yield-but presence of black point in harvested grain can reduce grade and quality of produce.

To ascertain the extent of infection of black point, survey is conducted every year and nearly 500 samples were collected from different marketing yards and inspected for intensity and severity of infection. The survey conducted from 2001–2002 to 2013–2014 revealed that about 15–50% infection for black point were observed having varied degree of

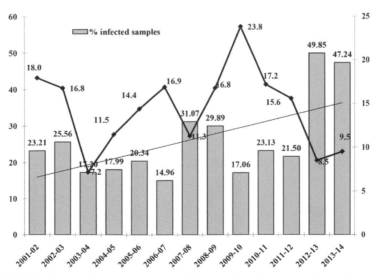

FIGURE 16.9   Temporal status and extent of black point infection in wheat of Gujarat. (Left side) Y axis: per cent infected samples and (right side) Y axis: percent range of infection.

infection in collected samples from Gujarat (Fig. 16.9). The data further indicated that among the samples, the range of infection varied from 7.2 (2003–2004) to 23.8% (2009–2010). The high black point infection observed during last couple of years was due to off-seasonal rain received during wheat season from flowering to early dough stage. (Patel, Personal communication).

In general, the percent grain infected with black point per spike were more in the middle and bottom portion of the spike in comparison to top portion. Moreover, 1000-grain weight of black point affected grains was higher than healthy grains. Significantly positive correlation of 1000-grain weight with disease incidence exhibited that bolder grains in spike were most vulnerable to infection. Disease did not affect to the grain yield adversely.

16.4.7.1   EFFECT OF BLACK POINT OF WHEAT ON QUALITY AND MARKET PRICE

The surveys conducted in different agricultural produce marketing corporation (APMC) in North Gujarat revealed that about 3.7 to 12.5% market prices were reduced due to black point infected grain lot (5–50%) compared to healthy looking grain lots (Fig. 16.10). Black point affects the color and luster of the wheat grains. The flour color also becomes dull whitish which in turns affects the chapatti quality. However, there were no effects on protein contents as well as quality of the wheat.

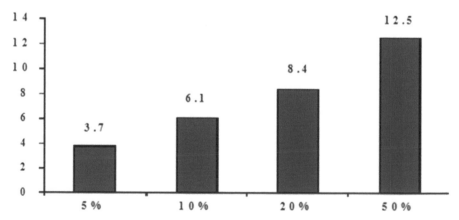

FIGURE 16.10   Effect of different levels of black point (%) on market prices (% reduction).

16.4.7.2   DISEASE MANAGEMENT

The agro-ecological situation of Gujarat mainly falls under arid and semi-arid climate where irrigation water is in short supply which leads to adoption of MIS like sprinkler or drip irrigation systems in wheat cultivation also. The continuous maintenance of high humidity in the immediate vicinity of plants favors the development of black point. Since, these MIS including both sprinkler as well as drip are not usual practices for irrigation of wheat crop in Gujarat, hence more research is needed to quantify as to when and how the system is to be adopted for wheat crop under prevailing conditions. One spray of mancozeb @ 0.25% or chlorothalonil @ 0.2% at hard dough stage is recommended for effective and economical management. (Patel, Personal Communication).

16.4.8   KARNAL BUNT

The climatic conditions of Gujarat are not favourable for infection and spread of Karnal bunt disease caused by *T. indica* (Nagarajan et al. 1997). The wheat produced in the state is therefore declared Karnal bunt free. Free movement of seed materials in the adjoining states limits the effectiveness of quarantine regulations. The post-harvest seed health surveillance is conducted every year to keep vigil on the introduction of this disease in Gujarat state. More than 500 samples per year are collected from the various marketing yards as well as from the farmers' fields and inspected for Karnal bunt infection. Till date, none of the samples showed any infection of Karnal bunt in Gujarat. The locally produced wheat possesses high quality and therefore has great potential for export. (Patel, Personal Communication).

16.5   INTEGRATED MANAGEMENT OF WHEAT DISEASES IN GUJARAT STATE

Stem and leaf rusts are major diseases of wheat in Gujarat state. Foliar blight occurs in punctuated areas. The integrated package consists of the following components:

- Old and rust susceptible varieties like Lok 1 should be discouraged.
- Rust resistant varieties, GW 322, GW 366, GW 451, GDW 1255, GW 11, and so forth, are to be grown for disrupting *Puccinia* path.

- All the recommended varieties possess very high degree of resistance against stem and leaf rusts.
- If the disease is observed under field conditions, two sprays of propiconazole (Tilt 25 EC) @ 0.1% significantly reduced the rusts diseases.
- The Gujarat state is free from Karnal bunt infection. Hence, no need to control it.
- Black point is observed especially in bold seeded varieties if congenial environment prevails during ear emergence. Hence, sprinkler irrigation should be avoided. One spray of mancozeb @ 0.25% or chlorothalonil @ 0.2% at hard dough stage is recommended for effective and economical management.
- The sprays of 0.035% thiophanate methyl or 0.25% mancozeb or 0.2% Ziram at the appearance of leaf blight disease followed by next spays at the intervals of 15 days are recommended for effective management.

16.6   CONCLUDING REMARKS

Keeping in view the change in climatic conditions, monitoring and surveillance need to detect any shift in spectrum and incidence of diseases in Gujarat state of India. So far wheat in Gujarat is free from Karnal bunt. The stem and leaf rusts are continuous challenge to both plant breeders and plant pathologists in developing durable rust resistant wheat cultivars in the state. Due to evolution and selection of new rust races, new and effective resistance genes must be added into future wheat varieties on regular basis to manage rusts effectively. Leaf blight is becoming important in the state due to sprinkler irrigated conditions. The scenario may further aggravate in the light of global.

KEYWORDS

- **wheat diseases**
- **Gujarat**
- **disease management**
- **leaf rust**
- **stem rust**
- **foliar blight**

- **black point**
- **epidemiology**
- **host resistance**

REFERENCES

Anonymous. *Progress Report of All India Coordinated Wheat and Barley Improvement Project 2014–15, Director's Report.* Indu Sharma, Ed.; Indian Institute of Wheat and Barley Research: Karnal, India, 2015; p 120.

Bhardwaj, S. C.; Prashar M.; Kumar Subodh; Jain, S. K.; Datta, D. *Lr19* Resistance Becomes Susceptible to *Puccinia triticina* in India. *Plant Dis.* **2005,** *89,* 1360.

Conner, R. L.; Davidson, J. G. N. Resistance in Wheat to Black Point Caused by *Alternaria alternata* and *Cochliobolous sativus. Can. J. Plant Sci.* **1988,** *68,* 351–358.

Duveiller, E., et al. In *Evaluation of Resistance to Spot Blotch of Wheat in Mexico: Improvement of Disease Assessment in the Field and under Controlled Conditions,* Proceedings of the International Workshop Helminthosporium Diseases of Wheat: Spot Blotch and Tan Spot, CIMMYT, El Batan, Mexico, Feb 9–14, 1997; Duveiller, E., Dubin, H. J., Reeves, J., McNab, A., Eds.; CIMMYT: El Batan, Mexico, 1998; pp 171–181.

Ellis, S. A.; Gooding, M. J; Thompson, A. J. Factors Influencing the Relative Susceptibility of Wheat Cultivars to Black Point. *Crop Prot.* **1996,** *15,* 69–76.

Jacobs, B.; Rabie, C. J. The Correlation between Mycelium Presence and Black Point in Barley. *Phytophylactica.* **1987,** *19,* 77–81.

Johnson, R. *Durable Disease Resistance.* In *Strategies for Control of Cereal Diseases;* Jenkyn, J. F., Plumb, R. T., Eds.; Blackwell: Oxford, 1981; pp 55–63.

Joshi L. M., et al. Some Foliar Diseases of Wheat during 1969–70 Crop Season. *Indian Phytopathol.* **1974,** *27,* 178–181.

Joshi L. M., et al. *Annotated Compendium of Wheat Diseases in India;* ICAR: New Delhi, 1978; p 332.

Joshi, L. M.; Singh, D. V.; Srivastava, K. D.; Wilcoxon, R. D. Karnal Bunt-a New Minor Disease that is Now a New Threat to Wheat. *Bot. Rev.* **1983,** *43,* 309–338.

McCallum, B.; Hiebert, C.; Huerta-Espino, J.; Cloutier, S. Wheat Leaf Rust. In *Disease Resistance in Wheat;* Indu Sharma, Ed.; CABI Plant Protection Series, CAB International: Oxfordshire, 2012; pp 33–62.

McIntosh, R. A.; Wellings, C. R.; Park, R. F. *Wheat Rusts: An Atlas of Resistance Genes;* CSIRO Publications: East Melbourne, Australia, 1995.

Mehta K. C. *Further Studies on Cereal Rusts in India;* Imperial Council of Agricultural Research: New Delhi, India, 1952; Vol. 18, p 165.

Mendgen, K.; Hahn, M. Plant Infection and the Establishment of Fungal Biotrophy. *Trends Plant Sci.* **2002,** *7,* 352–356.

Minh, T. D. Long, T. D.; Ngu, T. T. In *Screening Wheat for Bipolaris sorokiniana Resistance in Vietnam,* Proceedings of the Helminthosporium Blights of Wheat: Spot Blotch and Tan

Spot Workshop, CIMMYT, El Batan, Mexico, Feb 9–14, 1997; Duveiller, E., Dubin, H. J., Reeves, J., McNab, A., Eds.; CIMMYT: El Batan, Mexico DF, 1998; pp 213–217.

Misra A. P. *Helminthosporium Species Occurring on Cereal and other Graminae.* Final Report of PL 480 Project, Tirhut College of Agriculture: Dholi, Bihar, 1973; p 289.

Nagrajan, S.; Aujla, S. S.; Nanda, G. S.; Sharma, I.; Goel, L. B.; Kumar, J.; Singh, D. V. Karnal Bunt (*Tilletia indica*) of Wheat – a Review. *Rev. Plant Pathol.* **1997,** *76,* 1207–1214.

Prabhu, A. S.; Prasada, R. Pathological and Epidemiological Studies on Leaf Blight of Wheat Caused by *Alternaria triticina. Indian Phytopathol.* **1966,** *19,* 95–111.

Prasada, R.; Prabhu, A. S. Leaf Blight of Wheat Caused by a New Species of *Alternaria. Indian Phytopathol.* **1962,** *15,* 292–293.

Reis, E. M. In *Integrated Disease Management: The Changing Concepts of Controlling Head Blight and Spot Blotch,* Proceedings of the International Conference: Wheat for Non-traditional Warm Areas, *Foz do Iguacu,* Brazil, Jul 29–Aug 3, 1990; Saunders, D. A., Hettel, G., Eds.; UNDP/CIMMYT: Mexico, DF, 1991; pp 165–177.

Reis, E. M.; Forcelini, C. A. Transmissao de *Bipolaris sorokiniana* de Sementes Para Orgaos Radicularese Aereos do Trigo. *Fitopatol. Bras.* **1993,** *18,* 76–81.

Roelfs, A. P.; Singh, R. P.; Saari, E. E. *Rust Diseases of Wheat: Concepts and Methods of Disease Management;* CIMMYT: Mexico DF, 1992.

Roelfs, A. P.; Martell, L. B. Uredospore Dispersal from a Point Source within a Wheat Canopy. *Phytopathology.* **1984,** *74,* 1262–1267.

Rosyara, U. R., et al. Yield and Yield Components Response to Defoliation of Spring Wheat Genotypes with Different Level of Resistance to Helminthosporium Leaf Blight. *J. Inst. Agric. Anim. Sci.* **2005,** *26,* 43–50.

Rosyara, U. R.; Subedi, S.; Duveiller, E.; Sharma, R. C. The Effect of Spot Blotch and Heat Stress in Variation of Canopy Temperature Depression, Chlorophyll Fluorescence and Chlorophyll Content of Hexaploid Wheat Genotypes. *Euphytica.* **2010,** *174,* 377–390.

Schilder; Bergstrom, G. *Tan spot.* In *Seedborne Diseases and Seed Health Testing of Wheat*; Mathur, S. B., Cunfer, B. M., Eds.; Jordburgsforlaget: Copenhagen, Denmark, 1993; pp 113–122.

Shaner, G. Effect of Environment on Fungal Leaf Blights of Small Grains. *Ann. Rev. Phytopathol.* **1981,** 19, 273–296.

Sharma, I. *Diseases in Wheat Crops-An Introduction.* In *Disease Resistance in Wheat;* Indu Sharma, Ed.; CABI Plant Protection Series, CAB International: Oxfordshire, 2012; pp 1–17.

Sharma, R. C.; Duveiller, E. Effect of *Helminthosporium* Leaf Blight on Performance of Timely and Late-Seeded Wheat under Optimal and Stressed Levels of Soil Fertility and Moisture. *Field Crops Res.* **2004,** *89,* 205–218.

Sharma, R. C.; Duveiller, E. Spot Blotch Continues to Cause Substantial Grain Yield Reductions under Resource-Limited Farming Conditions. *J. Phytopathol.* **2006,** *154,* 482–488.

Singh, D. P. *Bipolaris sorokiniana* Causing Spot Blotch Disease in Wheat and its Management in India. In *Wheat: Productivity Enhancement under Changing Climate;* Singh, S. S., et al., Eds.; Narosa Publishing House: New Delhi, 2011; pp 274–285.

Singh R. S. *Plant Diseases.* Oxford IBH Publishing Co. Pvt. Ltd: New Delhi, India, 1990; p 512.

Singh, S.; Singh R. P.; Huerta-Espino, J. Stem Rust. In *Disease Resistance in Wheat;* Indu Sharma, Ed.; CABI Plant Protection Series, CAB International: Oxfordshire, 2012; pp 18–32.

Singh, D. V. Bunts of Wheat in India. In *Problems and Progress of Wheat Pathology in South Asia;* Joshi, L. M., Singh, D. V., Srivastava, K. D. Eds.; Malhotra Publishing House: New Delhi, 1986; pp 31–40.

Spurr, H. W. Jr.; Kiesling, R. L. Field Studies of Parasitism by *Helminthosporium sorokiniana. Pl. Dis. Rep.* **1961,** *45,* 941–943.

Villareal, R. L.; Mujeeb-Kazi, A.; Gilchrist, L. I.; Del Toro, E. Yield Loss to Spot Blotch in Spring Bread Wheat Production Areas. *Plant Dis.* **1995,** *79,* 1–5.

Williamson, P. M. Black Point of Wheat: *In Vitro* Production of Symptoms, Enzymes Involved and Association with *Alternaria alternata. Austral. J. Agric. Res.* **1997,** *48,* 13–19.

Zillinsky, F. *Common Diseases of Small Grain Cereals, a Guide to Identification.* CIMMYT: Mexico DF, 1983, p 141.

CHAPTER 17

DISEASE SPECTRUM ON BARLEY IN RAJASTHAN AND INTEGRATED MANAGEMENT STRATEGIES

P. S. SHEKHAWAT*, S. P. BISHNOI, and R. P. GHASOLIA

Rajasthan Agricultural Research Institute SKN Agriculture University, Durgapura, Jaipur, Rajasthan 302018, India

Corresponding author. E-mail: pdsingh87@yahoo.in

CONTENTS

ABSTRACT

Barley (*Hordeum vulgare*) is a primitive sacred cereal grain, which contributes nearly 12% of the global coarse cereal production. Globally, it occupied fourth rank amongst cereals. It is considered a crop of marginal farmers due to its low input requirement and better adaptability on marginal lands, tolerance to drought, salinity, and alkalinity. It is an important winter cereal crop in Rajasthan and rank on top in area and production of barley amongst the different barley growing states of India. The agro-climatic situations of Rajasthan are quite suitable for barley cultivation. During the last decade demand of malt barley in domestic market for industrial utilization has increased. Water availability may be crucial and a limiting factor for growing those crops which need more water. As a result of it, area under barley cultivation is expected to increase in Rajasthan due to climate change. Diseases are important constraints besides water availability to barley production and quality. Due to the change in climate conditions, cropping pattern and adoption of new technologies particularly use of micro irrigation system in Rajasthan, the diseases like stripe rust, leaf rust, leaf stripe, loose smut, foliar blight, covered smut, and cereal cyst nematode are causing significantly yield losses in Rajasthan. In this chapter, important barley diseases prevailing in Rajasthan during past one decade have been described in detail and integrated management practices are outlined.

17.1   INTRODUCTION

Barley (*Hordeum vulgare* L.) is an ancient cereal grain crop, which upon domestication has evolved from largely a food grain to a feed and malting grain (Baik & Ullrich, 2008). It is fourth largest cereal crop in world next to rice, wheat and maize with a share of 7% of global cereal production (Pal et al., 2012). Barley can grow in a wide range of environments including extremes of latitude, longitude, and high altitude (Vangool & Vernon, 2006). It is frequently being described as the most cosmopolitan of the crops and also considered, as poor man's crop because of its low input requirements and better adaptability to harsh environment like drought, salinity, alkalinity, and grow well on marginal lands (FAO, 2002). In India, barley is an important coarse cereal crop, being grown in rabi season in Northern and North Eastern Plains Zones. In India, barley occupied an area of 0.67 million ha with total production 1.6 million tons with a productivity of 24.2 q/ha (Table 17.1) (Anonymous, 2015). The major barley growing states in India are

Rajasthan, Uttar Pradesh, Haryana, Punjab, Madhya Pradesh, Uttarakhand and Himachal Pradesh (Kumar et al., 2014).

TABLE 17.1   Recent Trends in the Area and Production of Barley in India.

Year	Area (000ha)	Production (000 tons)	Productivity (q/ha)
2010–2011	699.0	1564.0	22.40
2011–2012	643.4	1618.0	25.1
2012–2013	695.0	1743.2	25.1
2013–2014	674.0	1830.0	27.1
2014–2015	672.0	1626.3	24.2

It is an important cereal crop next to wheat in acreage and production in the state of Rajasthan, in North-Western India. The state ranked first in barley area and production with 3.4 lac ha and 9.62 lac tons, respectively (Table 17.2). Barley grain is used as feed for animals, malt for industrial uses and for human food. Barley straw is used as animal fodder in many developing countries including India. It is also used for green forage. Due to the liberalization trade policies recently, the use of barley for malt and beer preparation has increased in India. The industrial requirement of barley is about 3.5–4.0 lac metric tons and it is growing at annually the rate of 10%. About 20–30% of the total barley production is utilized for malt preparation. Many malt companies used contract farming for production of the two-row malt barley in the states of Rajasthan, Punjab, Haryana, Utter Pradesh, and Uttarakhand (Kumar et al., 2014). In India, its utilization as food crop (mainly hull type) is restricted to the tribal areas of hills and plains. The barley products like "sattu" (in summers because of its cooling

TABLE 17.2   Trends in Area, Production and Productivity of Barley in Rajasthan.

Year	Area (Lac ha)	Production (Lac tons)	Productivity (q/ha)
2010–2011	3.28	8.58	26.2
2011–2012	2.78	7.89	28.4
2012–2013	3.07	8.53	27.7
2013–2014	3.09	9.42	30.4
2014–2015	3.43	9.62	28.03

Source: www.RajasthanKrishi.gov.in

effect on human body) and "Missi roti (for its better nutritional quality) and "Daliya" mixed with butter have been traditionally used in India including Rajasthan. The data on percent utilization of barley in different industries and barley malt for brewing and other purposes are presented in Figures 17.1 and 17.2 (Kumar et al., 2014).

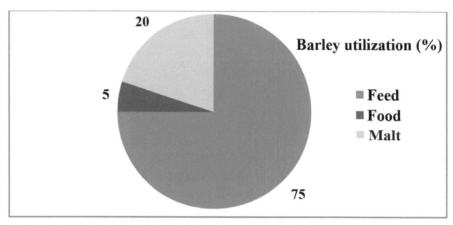

FIGURE 17.1   Barley utilization for feed, food, and malt in India.

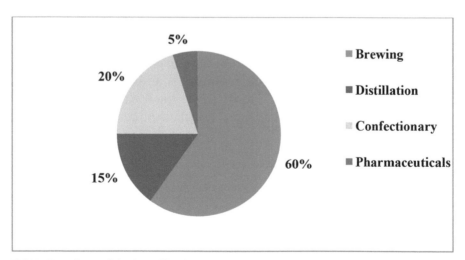

FIGURE 17.2   Malt barley utilization in India.

17.2 CLIMATE AND WEATHER CONDITIONS

The mean maximum temperature of past ten crop years (2004–2005 to 2013–2014) was in the range of 21.1–40.4°C whereas minimum temperature range was 7.3–26.3°C. During 2014–2015 crop season the maximum and minimum temperature ranges were 17.7–40.8 and 5.3–16.1°C, respectively, which were slightly lower than average values of earlier 10 years. The relative humidity I and II values were also at lower side.

17.3 BARLEY SCENARIO IN RAJASTHAN

The major area of Rajasthan state falls in North Western Plains Zone (NWPZ), except Kota and Udaipur regions which are in Central Zone (CZ). At state level, it is further divided into ten agro-climatic zones based on soil types and rainfall distribution. About 75% area of six zones falls in NWPZ and remaining 25% area of four zones is in CZ. The productivity in different barley growing districts in Rajasthan varies depending on the soil fertility status. The barley growing districts are broadly classified into three categories (high, medium, and low) are based on productivity pattern. These are:

High productivity districts: Hanumangarh, Pratapgarh, Sikar, Alwar, Jaipur, Jhunjhunu, Dausa, and Sri Ganganagar.

Medium productivity districts: Dholpur, Karoli, Sawai Madhopur, Jaisalmer, Jodhpur, Barmer, Jalore, Sirohi, Kota, Baran, Bundi, Jhalawar, Dungarpur and Bharatpur.

Low productivity districts: Ajmer, Pali, Tonk, Nagaur, Bikaner, Churu, Banswara, Udaipur, Bhilwara, Chittorgarh, and Rajsamand.

The Rajasthan Agricultural Research Institute (RARI), Durgapura, Jaipur, took up research on development of varieties for rain-fed and irrigated cultivation under low fertility conditions of marginal soils. A number of barley varieties suitable to different cultural and agro-climatic conditions have been developed for feed and malt purposes (Table 17.3). New varieties for specific conditions like saline-sodic soils in Northern plains and cereal cyst nematode (*Heterodera avenae*) resistance for cereal cyst nematode (CCN) infested soils of Rajasthan and Haryana were also developed. Systematic research efforts have resulted in gradual improvement in

TABLE 17.3 Barley Varieties Released in Rajasthan.

Variety	Cross/parentage	Year of release	released for	Salient features
RD 2052	Api-CM-67/SO-727// PL101	1987	Rajasthan for nematode infested soils, irrigated conditions	Six-row hulled barley with high yield, resistant to CCN (Molya disease)
RD 2035	RD137/PL101	1994	North Western Plains Zone (NWPZ) for irrigated timely sown conditions	Six-row hulled barley with high yield, resistant to CCN (Molya disease)
RD 2503	RD103/BH153// RD2046	1997	NWPZ for irrigated timely sown conditions	Six-row barley, good grain for malting
RD2508	RD2035/P409	1997	NWPZ for rain-fed irrigated late sown conditions	Six-row hulled barley with high yield
RD 2552	RD2035/DL472	1999	NWPZ and North Eastern Plains Zone (NEPZ) for irrigated timely sowing and also for saline soils	Six-row hulled barley with high yield under normal and saline soils
RD 2624	Bilara2/RD2508	2003	NWPZ for rain-fed timely sowing	Six-row hulled barley with high grain yield and suitable for low fertility soils
RD 2592	RD2503/UBL9	2003	Rajasthan for irrigated conditions	Six-row hulled barley with high grain yield and suitable for high management conditions
RD 2660	RD2052/RD 2566	2006	NWPZ and NEPZ under rain-fed	Six-row hulled barley for rain-fed
RD 2668	RD2035/BCU73	2007	NWPZ for irrigated timely sowing	Two-row barley in irrigated timely sown for malt
RD 2715	RD387/BH602// RD2035	2008	Central Zone (CZ) for irrigated timely condition	Six-row hulled dual purpose barley for green fodder and grain
RD 2786	RD2634/NDB1020// K425	2013	CZ for irrigated timely condition	Six-row hulled barley with high yield
RD 2794	RD2035/RD2683	2013	NWPZ and NEPZ for irrigated timely sowing and also for saline soils	Six-row hulled barley with high yield alkaline soils.
RD 2849	DWRUB52/PL705	2014	NWPZ irrigated timely sowing	Two row barley in irrigated timely sown for malt

productivity of the varieties under different production conditions/areas (Table 17.3). The gain in yield potential under different production conditions is quite evident through new varieties, where a gradual increase in grain yield of the new cultivars has been achieved with the continuous breeding efforts.

The state has wide range of cropping system viz. Pearl-millet-Barley, Groundnut-Barley, Green-gram-Barley, Cowpea-Barley, Maize-Barley, Sorghum-Barley, Cotton-Barley, and Cluster-bean-Barley. The crop season 2014–2015 was not favorable for barley due to adverse climatic conditions and strong wind at the time of crop maturity, led to decline in the production and productivity over last year. Rajasthan being the hub of barley cultivation, MNC's malting and brewing industries are the major importers of barley. The varieties developed by RARI are being adopted and cultivated at the farmers' field. The breeder seed is in demand and RARI varieties are figure in 72% of the total indent of the country (Anonymous, 2015). Most of the barley varieties developed recently are highly resistant to yellow rust.

17.4   IMPORTANT DISEASES OF BARLEY IN RAJASTHAN

The intensity of diseases is increasing due to intensive cultivation of barley in Rajasthan under higher input condition. Among these, stripe and leaf rusts, stripe disease, loose smut, covered smut, and cereal cyst nematode are causing significantly higher losses in yield.

17.4.1   STRIPE RUST

It is caused by *Puccinia striiformis* f. sp. *hordei* and is an important disease of barley in India and several parts of world (Safavi et al., 2012). Stripe rust is traditionally a disease of a cool and moist climate. However, in recent years, stripe rust has emerged as serious threat in warmer areas where the disease was previously considered of little importance (Hovmoller et al., 2008). The high incidence of stripe rust may result high losses in susceptible varieties. Stripe rust in barley may cause losses up to 60% (Park et al., 2007). In general, stripe rust is more destructive than stem and leaf rusts. In the epidemic years it may implicit considerable losses due to premature death of foliage and sometimes by sterility of spikelet and shriveling of grains.

17.4.1.1 SYMPTOMS

The most clear symptoms appears as uredo-pustules in the form of narrow, orange-yellow, linear stripes on upper side of leaf, and on sheath, neck, and glumes. The pustules coalesce to form long stripes on leafater. The orange pustules contain spores which re-infect green plants. Glumes are often infected. Stripe rust symptoms usually appear earlier in the season than stem and leaf rusts. The uredo pustules turn in to dark brown to black teliosori with increase in temperature toward maturity of crop (Fig. 17.3 A–C).

A   B

C

FIGURE 17.3   Stripe rust infected (A) crop, (B) spikes, and (C) telial stage on stem.

Puccinia striiformis is a heteroecious, with uredial and telial stages occurring on barley. The uredospores initiate germination at 9–15°C with free water. The germtube growth takes place at 10–15°C. An optimal temperature 8–13°C favors formation of appressorium and sub-stomata vesicle under humid conditions.

17.4.1.2   DISEASE CYCLE

The pathogen over-summers under cool climate in the inner valley of Himalayas on barley and on volunteer plants in uredial stage. When favorable temperature returns, the sori bust and release abundant uredospores these move toward foothills of Northern India during December and January, which the crop is about a month old. The inoculum then spreads by katabatic winds to sub-mountainous parts of NWPZ. Primary infection foci occur in this region along the mouth of rivers, viz. Tavi, Ravi, Beas, Satluj, and Yamuna. Later on the katabatic winds carry the uredospores to adjoining plains and cause infection. The pathogen becomes partially systemic in leaf and later produces several uredo cycles and infect other plants. At maturity of the crop, the temperature begins to rise and it is not congenial for uredospore production. The spread of the rust is checked above 25°C and telial stage develops mostly on the lower side on leaf as black dots arranged in rows. The teliospores apparently serve no function in the absence of alternate host and recurrence of the disease takes place through air borne uredospores from primary foci of infection. Physiological races of *P. striiformis* causing yellow rust of barley in India are, 0S0-1 (24), 0S0 (57), 1S0 (M), 4S0 (G), 4S0-3(G-1), 4S0-3(G-1), and 5S0 (Q) (Table 17.4).

The rusted plants are stunted and they produce small spike in extreme disease situation (Figs. 17.3). There is impairment of fertility of florets to reduce the number of seed set per spike (Srivastava, 2010). Severe infection results in pre-mature drying of the plants. Thus, the major cause of loss in barley yield is on account of producing shriveled seed with reduced size and grain weight (Fig. 17.4). The uredispores loose viability rapidly at temperature above 15°C. Disease development is more rapid when temperature is between 10 and 15°C when intermittent rain or dew is present.

TABLE 17.4 Pathotypes Spectrum of Stripe Rust of Barley in India.

Crop year	Stripe rust pathotypes detected in samples (numbers)							Total samples
	0S0-1 (24)	0S0 (57)	1S0 (M)	4S0 (G)	4S0-3 (G-1)	4S0-3 (G-1)	5S0 (Q)	
2003–2004	–	–	24	–	–	–	–	24
2004–2005	14	7	50	5	7	–	–	83
2005–2006	–	–	–	–	–	–	·	–
2006–2007	1	–	17	5	–	–	–	23
2007–2008	–	3	29	2	–	–	–	34
2008–2009	6	2	19 (Raj.5)*	3	–	–	–	30
2009–2010	26	4 (Raj.3)	8 (Raj.8)	–	–	–	–	38
2010–2011	3	2	28	3	–	–	–	36
2011–2012	3	1	13 (Raj.1)	–	–	–	1	18
2012–2013	–	1	–	–	–	–	–	1
2013–2014	–	3 (Raj.1)	7	–	–	–	–	10
Total	53	23	195	18	7	0	1	297
Percent	17.8	7.7	65.7	6.06	2.4	0.0	0.3	

*Number of samples from Rajasthan.

FIGURE 17.4   (A) Healthy grain and (B) shriveled grain due to yellow rust.

17.4.2   LEAF RUST

The leaf rust in barley is caused by *Puccinia hordei.* It is widely distributed disease in barley in the Indian sub-continents. It is the most common rust prevalent in North-Western and North-Eastern, Central, and Peninsular regions of India.

17.4.2.1   SYMPTOMS

Disease appears as small, orange-brown, circular pustules, scattered mainly on upper surface of leaf blade and sometimes on leaf sheath, peduncles, internodes, and ear heads also. Heavily infected leaves die prematurely.

17.4.2.2   DISEASE CYCLE

Leaf rust of barley is macro-cyclic rust. In absence of alternate hosts, the pathogen perpetuates in the form of uredial stage on self-sown barley plants and collateral hosts in both Himalayas and South Indian hills. The rust inoculums in the form of uredospores spread from hilly region into plains where they cause infection on regular barley crop.

   Leaf rust develops rapidly between 15 and 22°C when moisture is not limiting. Following infection, new pustules and spores may be produced within eight days. Uredospores are wind-borne and can spread long distances. Teliospores develop later in the season either within uredial sori or within separate telial sori. In India, where rust cannot overwinter and alternate host is not present, the fungus is re-introduced annually through repeating spores called uredospores. The pathotypes of leaf rusts are H_1, H_2, H_3, H_4, and H_5.

17.4.3   STEM RUST

The stem rust is resulted due to infection of *Puccinia graminis* f. sp. *tritici*. It mainly occurs in Central and Peninsular Zones of India. It is basically a disease of warmer climate. The uredospores of stem rust from CZ to Northern region never arrive in time usually does not cause much damage. Isolated but severe epidemics of stem and leaf rusts was recorded in Jalore district of Rajasthan in 1973. The yield losses implicated by stem rust have been enormous in stem rust prone areas of the country.

17.4.3.1   SYMPTOMS

The large pustules are oval to elongated and often surrounded by characteristic to margin. Pustules are full of reddish brown spores which fall away easily. They can occur on stems, leaf surfaces, sheaths, and heads. Remnants of the epidermis are visible on the margins of the pustule, giving it a ragged appearance. As a plant mature the pustules produce black teliosori. They are conspicuous, linear, oblong, dark to black and often merging with one another to cause linear patches of black lesions. *Puccinia graminis* is a macrocyclic, and heteroecious fungus. Both uredial and telial stages are produced on barley and other two stages, pycnial and aecial are developed on alternate hosts like Berberies and Mahonia. Since alternate hosts are not functional in India, the uredospores, play major role in the survival and perpetuation of stem rust.

17.4.3.2   DISEASE CYCLE

Survival of the stem rust in the Indian sub-continent is only through uredospores at high hills. In hilly areas, stem rust survives in the summer mainly on volunteer wheat, barley, triticale, and some grasses including common wheat grass and barley grass in the form of uredospores, wet summers cause growth of self-sown wheat, barley and other host of stem rust. These plants can become heavily infected with stem rust in the autumn, and act as source of rust inocula for the new barley crop in the season. The disease is favored by warm, and moist weather. The disease develops optimally between 15 and 24°C and seriously hampered below 15°C and above 40°C temperatures. Physiological races of *P. graminis* recorded in India are, 79G31(11), 62G29 (40A), 37G19 (117-6), 75G5 (21A-2), and 7G43(295).

17.4.4 EPIDEMIOLOGY OF RUSTS IN INDIA

17.4.4.1 LEAF RUST

Mehta (1940, 1952) suggested that leaf rust of wheat spread both from South and North Indian hills. This view is supported by recent studies. It has been demonstrated that the first buildup of leaf rust like stem rust takes place in plains of Karnataka in South India, generally in the last week of December. At the same time the infection also establishes in Bihar in North-Eastern region of the country. The rust population from Southern region moves Northwards toward Maharashtra and Madhya Pradesh and population moves from Northern region toward South. Finally both the populations, moving in opposite direction merge into each other (Joshi et al., 1974). Nagrajan et al. (2006) explored that the uredospores of leaf rust from northern region are also carried away by western disturbance. This observation suggests that if more western disturbances accompanied by frequent rains occur in North India, there is apparently good chance for the spread and built up of leaf rust in North-Western region to lead to an epidemic.

17.4.4.2 STRIPE RUST

Stripe rust pathogen prefers low temperature for infection and symptoms expression so it survives in uredial form at several locations in Himalayan ranges in the absence of alternate host (Mehta, 1940). By December/January when the crop is about a month old, the rust appears along the hills in NHZ/ NWPZ along the mouth of rivers such as Tawi, Ravi, Beas, Sutlej, and Yamuna where temperatures are low and dew is abundant. The disease can survive in Nilgiri and Pulney hills but it cannot spread even to the foothills of Nilgiri due to unfavorable warm weather (Joshi et al., 1984). It is a major problem of North and North-Western region in India.

17.4.4.3 STEM RUST

Joshi et al. (1974) reported that stem rust uredospores spread primarily from Nilgiri and Pulney hills and Himalayas play a minor role in recurrence of epidemics of the disease in northern India. The sporulation of stem rust over naturally infected fields in southern hills is faster under warm weather. Hence, many places in peninsular India get early infection of stem rust in December and January. In contrast, low temperature prevailing in Himalayan region during winter months hampers the sporulation and multiplication of *P. graminis*. The South Indian hills are considered the main foci of stem rust infection in India (Joshi, 1976). Spread of stem rust from central India to other parts of northern India is favored by the repeated passage of western disturbance, linking both areas. The associated winter precipitation helps countrywide spread of stem rust (Bhardwaj et al., 2010).

17.4.5 MANAGEMENT OF BARLEY RUSTS

17.4.5.1 HOST RESISTANCE

Growing of resistant varieties is the most practical approach to manage the rusts of barley. The stripe rust resistant varieties developed at RARI, Durgapura so far are RD 2052, RD 2552, RD 2624, RD 2592, RD 2660, RD 2668, RD 2715, RD 2786, RD 2794, and RD 2849. The donors used for different traits are given in Table 17.5.

TABLE 17.5 Donor Lines Used in Breeding Programs at RARI, Durgapura, Jaipur.

Traits	Potent donor lines
All the three rusts	RD 2552, RD 2550, RD2849, RD 2900, RD, 2904, RD 2909, RD 2913 and RD 2915
Stripe rust	RD 2560, RD 2620, RD 2655, RD2662, RD 2665, RD 2669, RD 2670, RD 2674, RD 2676, RD 2696, RD 2712, RD 2899, RD 2910
Stripe and stem rusts	RD 2891, RD 2903, RD 2908, RD 2914
Stripe and leaf rusts	RD 2786, RD 2901, RD 2905, RD 2907

TABLE 17.5 *(Continued)*

Traits	Potent donor lines
Stem and leaf rusts	RD 2919
Stripe rust+ cereal Cyst nematode	RD 2552. RD 2585, RD 2624, RD 2657
Leaf blight	RD 2670, RD 2677, RD 2683, RD 2696
Cereal cyst nematode	Rajkiran, RD 2035, RD2052, RD 2508, RD 2606, RD 2624
Foliar aphids	RD 2630
Straw quality (dual purpose)	RD 2035 RD 2552, RD2618, RD 2670, RD 2696, RD 2715, RD 2717
Malt quality	RD 2503, RD 2511, RD 2560, RD 2606, RD 2651, RD 2666, RD 2668, RD 2719, RD 2722
Rainfed	RD31, RD2521, RD 2615, RD 2624, RD 2660, RD 2675, RD 2685
Salinity/sodicity	BL-2, RD 2552, RD 2556, RD 2588, RD 2692, RD 2693

17.4.5.2   FUNGICIDES

Brown et al. (2002) reported that the foliar application of propiconazole, tebuconazole, Folicur, or Bayleton applied at the first sign of stripe rust infection was effective in controlling the disease.

Singh et al. (2010) observed that foliar spray of propiconazole (Tilt 25 EC) @ 0.1%, reduced the incidence of barley rust. Selvakumar et al. (2014) reported that the foliar application of triademefon 25%WP @ 0.1% gave the best rust control to 89.33% followed by tebuconazole 25.9% EC @ 0.1% (87.23%).

17.4.6   STRIPE DISEASE

The disease is incited by *Drechslera graminea* (Rabenh.) Shoemaker (formerly *Helminthosporium gramineum*). It is worldwide in occurrence with considerable damage to the crop from seedling to maturity. The disease reduces the yield and quality of barley and causes economically important yield losses in many countries. Yield losses from 3.3 to 15% have been reported in infected spikes.

17.4.6.1   SYMPTOMS

Small pale colored spots appear on lower leaves and sheaths on about two months old plants. These spots fuse together and form long, parallel, brownish stripes. These stripes are continuous from the base to the tip of the leaf. The infection is systemic and symptoms can be seen on all tillers of a plant and on all leaves. The yellow stripes soon become brown as tissue necrosis progresses and finally the tissues dry out and the leaf blade shredded (leaf shredding). A grey to olive grey mass of conidiophores and conidia develop on lesions/spots. The spikes are blighted, twisted, compressed, and brown in color. Infected plants are stunted and produce sterile spikes (Fig. 17.5).

FIGURE 17.5   Stripe disease infected crop.

The mycelium is endophytic and intercellular, branched, sub-hyaline to light yellow and septate. The conidiophores are thick, erect, grey to olivaceous in color, emerges through stomata in groups (usually 3–5). Basal segment is enlarged and distal portion is slender. The conidia are yellowish-brown, straight, sub-cylindrical to slightly tapering having rounded ends, 1–7 septate. Germination takes place from all the cells (mostly from both the end cell).

17.4.6.2   DISEASE CYCLE

The pathogen survives through mycelium in seed and in crop debris lying in soil. The infection is systemic. Primary infection is caused by hyphae which penetrate through the coleoptiles, coleorhizae, or root and grow up ward through the seedling. Abundant conidia are produced on infected leaves during periods of high humidity. The conidia are wind-blown to nearby plants where they lodge on flowers and infect seeds at kernel development stage. Secondary infection occurs by means of conidia formed on conidiophores after primary infection of the crop. Most favorable condition is heavy dew or rainfall during flowering. Low soil temperature (around 12°C) and moderate soil moisture are favorable for systemic invasion. Infection greatly reduced at soil temperature >15°C. Cool and moist soil conditions enhance the disease in seedlings. Disease is favored by deep sowing, (4–5 cm) as compared to 2–3 cm depth. Infection can occur at temperatures ranging from 10 to 33°C.

17.4.6.3   MANAGEMENT

- Use only thoroughly cleaned, certified barley seeds from disease-free fields.
- Deep ploughing and burying of crop residues soon after harvest.
- Where practical, keep down susceptible grasses and volunteer (small grains) by cultural or chemical means.
- Rotate small grains and grasses with non-host crops, preferably with legumes.
- Produce seed in dry areas.
- Adopt shallow sowing (2–3 cm depth).
- Treat the seeds with carboxin +thiram @ 0.2% or tebuconazole 2% DS @ 0.1%.
- Soaking of seeds in solutions of 0.3% $CuSO_4$ or 0.5% $FeSO_4$ or $ZnSO_4$ (0.5%).
- Grow resistant or tolerant varieties.

17.4.7 LOOSE SMUT

Under the changing climatic scenario incidence of loose smut has increased in NWPZ in India during the last one decade. None of the barley varieties under cultivation in India are resistant to this disease.

Loose smut of barley has a long history of its existence. The disease occurs throughout the world wherever barley is cultivated. It is prevalent in all barley-growing states of India, however, the incidence of loose smut is relatively more in cool and moist areas of northern plains and hills than dry peninsular region of the country (Goel et al., 1977). In recent years, the incidence of loose smut in barley is in increasing trend due to climatic change. Loose smut is very destructive as almost every panicle of the affected plant is converted into black powdery mass of smut spores and there is no grain formation. The yield loss is, therefore, proportion to percentage of infected panicles.

17.4.7.1 SYMPTOMS

The fungus infects open flowers and establishes in the embryo of developing seed. Infected panicles emerge as a mass of dark brown spores. Masses of the olive-brown smut spores replace the entire head of plants with little development of floral bracts and awns. Smutted heads often emerge earlier than healthy heads. Spores are dislodged and scattered by wind after delicate membranes surrounding break (Fig. 17.6). These panicles normally stand taller in the crop and spores are blown onto adjacent healthy heads by wind. The bare stalk is left and this may be the only sign of the disease at later stage in the season. Infected seeds appear healthy and but carry a minute dormant infection inside the embryo. Hence the disease is internally seed-borne. Infected plants appear normal until the emergence of the heads. The disease is most common in cool, high rainfall areas and may be more common in the year following a wet spring, which promotes seed infection. In Rajasthan use of sprinkler irrigation is quite common in major barley growing areas which favor the disease. The pathogen survives from one season to the next only a dormant mycelium within the embryo of infected barley seeds.

The systemic infection proceeds soon after germination of infected seed and symptoms are visible only at the time of heading. The spikes emerged contain dark black powdery mass of smut spores. The spore mass is initially covered by a smooth, delicate, greyish membrane which soon ruptures and releases the spores. Subsequently, the spores are dislodged by wind leaving the naked rachis. In diseased plants, a few or all the spikes are affected but in most of the cases entire spike gets converted into smut sori, though partially infected spikes can often be seen.

FIGURE 17.6   Loose smut of barley.

17.4.7.2   DISEASE CYCLE

The infection of loose smut is internally seed-borne. The fungus perpetuates inside the embryo of infected barley seeds as the dormant mycelium which resumes its activity, and invades all part of the plant, without causing injury to it. The active hypha just behind the growing point keeps pace inter-cellularly through the young seedlings until it reaches apex of the shoot.

17.4.7.3   MANAGEMENT OF LOOSE SMUT

Seed Certification: Use certified seeds from authenticated agencies. The permissible maximum limits of loose smut infection are 0.1% and 0.5% for foundation and certified seeds respectively (Tunwar & Singh, 1988).

Heat Therapy: For eradication of internally seed-borne infection of loose smut, soaking of seed is done in normal water for 4–6 h and then dipping in hot water at 49°C for 2 min followed by drying before sowing (Srivastava, 2010). The solar heat treatment is effective during bright summer day in the months of May–June in India. Duhan and Beniwal (2006) proposed soaking of seed in water (1:1 w/v) and drying in sun.

Chemical Control: Seed treatment with systemic fungicides is one of the most promising practices for control of seed borne diseases. Seed treatment with tebuconazole 2% DS @ 1 g/kg seed or carboxin 75%WP @ 2 g/kg seed was found best for controlling loose smut of barley.

Biological Control: Seed treatment with *Trichoderma viride* @ 4 g/kg seed with half dose of carboxin 75% WP @ 1.25 g seed is effective for the control of loose smut of wheat (Singh et al., 2000).

Cultural control: The regular inspection of crop at the early heading stage and careful rouging of smutted panicles helps to prevent spread of the pathogen. The smutted panicles should be covered with a paper or plastic bag and plucked with scissor.

17.4.8 COVERED SMUT

It is caused by *Ustilago hordei* (Pers.) Lagerh. The disease is found in all barley growing counties and it is more extensively distributed than loose smut of barley (Martens et al., 1984; Mathre, 1997). In India, it is common in northern parts where it takes heavy toll of yield in areas where susceptible varieties are grown. Other than barley, it is also appeared on oat (*Avena sativa*), rye (*Secale cereale*), brome grass (*Bromus* sp.), wheat grass (*Agropyron* sp.) and wild rye (*Elymus* sp.). In the Kurgan region of Russia, 7.7–14.1% yield losses have been estimated. In Rajasthan (India), the incidence of covered smut of barley was reported up to 46% (Mathur & Bhatnagar, 1986). Smut spores/teliospores are mostly developed in the floral organs (most attacked part is ovary).

17.4.8.1 SYMPTOMS

It becomes visible when the panicle emerge. The smutted panicles emerge at the same time as those of healthy plants but remain shorter and are usually retained within the sheath for a longer time or may sometimes fail to emerge at all. Every panicle in a diseased stool and all grains in a diseased panicle are affected. The sori are formed in the ovaries and covered with a silvery semi-persistent membrane until the plants are fully matured. The grains are replaced by black powdery mass of teliospores. The spore masses are held together due to deposition of a fatty substance. This feature renders treatment of seeds difficult, unless the fat is removed (Fig. 17.7). During threshing, the spores get attached to the seed surface. Thus, pathogen becomes externally seed-borne in nature. Besides these, infected plants show reduction in height, tillers/plant, tiller length, ears/plant, and ear length (Jain et al., 1997).

FIGURE 17.7   Covered smut of barley.

Mycelium is branched and septate. Teliospores or smut spores are round to elliptical, brown in color and 6–9 μm or sometimes up to 11 μm. If the smut spores are in mass then they appear as black. Smut spores are lighter in color on one side. The spore wall is smooth. The optimum temperature for spore germination on medium is 25°C. On germination, the spores produce a typical four-celled promycelium on which uni-nucleate, ovate to oblong sporidia are produced. Due to inter-sporidial anastomoses and/or hyphal anastomoses, a dikaryotic hypha (n+n) is arises that causes infection in the host.

17.4.8.2   DISEASE CYCLE

The pathogen survives through the externally seed-borne teliospores (smut spores) which have attached on the surface of healthy seeds during threshing or in infested soil. The primary infection is caused by "dikaryotic mycelium" (n + n) only. When infested seeds are sown, the smut spores germinate on seed coat simultaneously with seeds to produce four meiotic, haploid sporidia/basidiospores. Fusion of two sporidia of opposite mating type results in obligate parasitic and dikaryotic mycelia (Bakkeren & Kronstad, 1994). This dikaryon penetrates the host plant through young coleoptiles of the seedling causing primary infection. This leads to systematically infected plants that sporulate exclusively in male or female inflorescence (Laurie et al., 2012). During plant colonization, the dikaryotic mycelium grows in or just below the shoot meristem. The infection proceeds mostly symptomless until differentiation of the colonized meristem into floral tissue, which cues the fungus to proliferate and sporulate in the spikelets of the inflorescence. Mycelium of the pathogen branches rapidly in the crown buds and floral structures. Smut spores are produced by the cells of the hyphae. The hyphae collect in the ovaries and transform into teliospores, which replace the grains. No infection occurs when the primary shoot of the host has grown out above the soil surface. There is no secondary infection and the disease is monocyclic. The optimum temperature for spore germination is 20°C and the maximum is 35°C. Deep sowing enhances the period of susceptibility to the infection.

17.4.8.3   DISEASE MANAGEMENT

- Use healthy seeds, in healthy soil.
- Uproot and burn affected plants as early as possible.
- Adopt shallow sowing (2.5 cm), as it helps shoots to come up early thus reducing the chances of infection.
- Adopt early sowing.
- Follow crop inspection and certification of seed crops.
- Seed treatment with carboxin or thiram or zineb @ 2.5 g/kg seeds (Bedi & Singh, 1974; Singh, 2009). Complete disease control can be achieved by treating the seeds with Sprint 75 WP (mancozeb 50% + carbendazim 25%) @ 0.3% or Raxil 2DS (tebuconazole 2%) @ 0.15% or Raxil 60 FS (tebuconazole 60%) @ 0.1% or Vitavax power 75 WP (carboxin 37.5% + thiram 37.5%) @ 0.15% (Kaur et al., 2014).
- Cultivar C-163 had been found immune to covered smut while BJ 13, BJ 14, BJ 24, BJ 26, BJ 28, and BJ 29 are resistant (i.e., showing less than 1% infection) (Singh, 2009).

17.4.9 MOLYA DISEASE

Cereal cyst nematode, *Heterodera avenae* was first recorded as a parasite of cereals in Germany by Kuhn (1874), as merely a biological form of *H. schachtii* Schm., the beet eelworm. Wollenweber (1924) clearly demonstrated morphological and physiological differences by which cereal cyst nematode could be separated from beet eelworm. Schmidt (1930) observed that larvae from oats were larger than those from sugar beets and designated as *H. schachtii* sub sp. *major*. Researchers in different countries continued to use *Heterodera major* instead of *Heterodera avenae* (Franklin, 1940). Franklin et al. (1959) reviewed the nomenclature and proposed the adoption of *H. avenae*.

H. avenae is the principal species on temperate cereals (Rivoal & Cook, 1993), while it has been detected in many countries including Australia, Canada, Israel, South Africa, Japan, and most European countries (Kort, 1972), as well as India (Vasudeva, 1958) and countries within North Africa and West Asia, including Morocco, Tunisia, Pakistan and Libya (Sikora, 1988), Algeria (Mokabli et al., 2001), and Saudi Arabia (Ibrahim et al., 1999). Although its distribution is global, much of the research has been confined to Europe, Canada, Australia, and India.

Meagher (1972) suggested that *H. avenae* originated in Europe and disseminated to other parts by activities of men and wind. Since its detection from Sikar, in 1958 by Vasudeva, it has also been reported from different parts of India. The disease caused by the nematode is known as "Molya" (Molya in local language of Sikar and Jaipur denotes "deformed" and disease gets its name due to characteristically deformed root system), It is present in Ajmer, Alwar, Bhilwara, Bikaner, Bharatpur, Churu, Chittorgarh, Dausa, Dholpur, Hanumangarh, Jaipur, Jhunjhunu, Pali, Sirohi, Tonk, and Udaipur districts of Rajasthan (Koshy & Swarup, 1971; Mathur et al., 1975; Bishnoi & Bajaj, 2003); Ambala, Bhiwani, Faridabad, Gurgaon, Mahendergarh, Rohtak, and Sirsa districts of Haryana (Bhatti et al., 1980; Bajaj & Walia, 1985); Faridkot, Hoshiarpur, Jalandhar, Kapurthala, Ludhiana, Patiala, and Sangrur districts of Punjab (Koshy & Swarup, 1971; Chhabra, 1973); Jammu and Kashmir (Singh et al., 1976); Himachal Pradesh (Koshy & Swarup, 1971); Delhi and Aligarh, Badaun, Bulandshahar, and Ghaziabad districts of Uttar Pradesh (Swarup et al., 1982; Siddiqui et al., 1986). It is speculated that this nematode is spreading slowly and gradually.

17.4.9.1 PATHOTYPES

Eleven pathotypes have been detected from various countries on the basis of international test assortment. Five pathotypes of *H. avenae* have been reported in Rajasthan and were different from European pathotypes (Mathur et al., 1974). Bhatti et al. (1977) found that the behavior of Haryana population was different from Rajasthan population and speculated the possibility of *H. avenae* being a species complex. Swarup et al. (1979) reported that population of Rajasthan (Jaipur and Udaipur) belonged to pathotype-1 and distinct from population of Punjab (Ludhiana and Hoshiarpur) which formed pathotype-2. Siddiqui and Hussain (1989) showed that Ghaziabad (Utter Pradesh) population was different from Aligarh (Utter Pradesh) population on the basis of host differential test. After scanning the literature Andersen and Andersen (1982) assigned Indian populations to pathotype Ha 21, Ha 31, and Ha 41. Bekal et al. (1998) attributed Najafgarh (Delhi) population to Ha 71 pathotype. Bishnoi and Bajaj (2004) reported that Jaipur, Udaipur, Narnaul, Sirsa, and Delhi populations belong to pathotype Ha 21 while Ludhiana (Punjab) and Ambala (Haryana) populations belong to pathotype (Ha 41) and Una (Himachal Pradesh) population belong to Ha 31.

17.4.9.2 SYMPTOMS

The above-ground symptoms of damage can be seen within a month after sowing, becoming quite marked by the end of January. During this period, second stage juveniles are abundant in the soil. By mid-February, white females can be seen attached to roots. The disease is characterized by patchy growth of stunted and yellow plants. The symptoms associated with *H. avenae* damage are characterized by uneven patches of poorly growing plants, randomly distributed throughout a field. It may vary in size from 1 to 100 m² or more. Size and number of patches are directly related to nematode population levels as well as nematode distribution in the field. Under monoculture, the patches coalesce and damage may uniformly cover the entire field within 3–4 years. Infested plants exhibit stunt, thin and narrow leaves, reduced number of tiller, delayed emergence of ears, reduced number of spikelets and grains (Fig. 17.8). In France, successful detection of *H. avenae* in wheat fields was achieved with the use of radio thermometry (Nicolas et al., 1991; Lili et al., 1991). It is possible that this technique could be extended to thermography which could improve the detection of cereal cyst nematode attacks in large areas. Severely infected plants remain stunted, 30–60 cm high. Panicles, if formed, have very few grains.

FIGURE 17.8   Cereal cyst nematode infested wheat crop at Durgapura (Jaipur).

Presence of cysts on the roots is the only confirmatory indication of nematode infestation (Kort, 1972). Symptoms produced on roots are different, depending on the host. Root system becomes bushy, bunchy, and shallow due to their proliferation and has slight swelling near the root-tip. Such plants can be easily pulled out of the ground. Wheat attacked by *H. avenae* shows increased root production such that the roots have a "bushy-knotted" appearance with several females visible at each knot (Fig. 17.9), (Rivoal & Cook, 1993). Oat roots are shortened and thickened, while barley roots appear same as in wheat affected. Other species of *Heterodera* also appear to produce host-specific symptoms on the roots of cereals. For example, *Heterodera latipons* did not produce knotted roots like *H. avenae* in Israel (Mor et al., 1992). Under European conditions root division takes place at the points of juvenile invasion, giving an appearance of a knotted root system. In Australia, a much-branched root system is characteristic of infested wheat and barley and to a lesser extent oat. Tufting of roots may not be noticeable during field examination due to adhering soil.

FIGURE 17.9   Cereal cyst nematode infested roots of barley.

17.4.9.3   DISEASE CYCLE

The life cycle of *H. avenae* involves only one generation during a cropping season, irrespective of geographic region and the host range of this
nematode is restricted to graminaceous plants. There is sexual dimorphism,
with males remaining vermiform, whereas females become lemon shaped
and spend their -life inside or attached to a root. Eggs are retained within
the female's body and, after the female has died, the body wall hardens to
protect the eggs and juveniles. The eggs within a cyst remain viable for
several years. In India, juveniles emerge in November while in Europe emergence occurs during mid-March to mid-July, and in Australia in June–July.
Second-stage juvenile penetrates the host root just behind the growing tip,
moving through the cortical cells and stellar region where the larvae remain
parallel to the stellar tissues and start feeding on the specialized cells called
syncytium surrounding the head (Johnson & Fushtey, 1966). The development of nematode inside root involves three moults. The larvae that become
females swell to lemon-shape with the enlargement of body and protrude

out of the root tissues. Those larvae which are destined to become males remain vermiform and move out of the root tissues into the soil and fertilize the females. Females start forming eggs which fill up the whole body. At this stage, females die and body cuticle changes color from white to brown. In Australia, cysts may be seen during August–October (Meagher, 1972), in England in May–July, and in India white females may be seen in February. On an average, a cyst may contain 200–250 eggs and larvae. A bold cyst may contain as many as 600 eggs.

Second-stage larvae penetrate the roots usually in the meristematic region behind the root cap and other parts of the root also. The larvae move through the cortex both intercellularly and intracellularly and establish for feeding in the stellar region and lie parallel to the stele. The cells surrounding the head begin to enlarge and their protoplasm becomes dense. Cell wall dissolution and merging of protoplasm results in the formation of elongate multinucleate syncytia with thickened outer boundaries (Williams, 1969). The cells involved in syncytial formation include those of endodermis, pericycle, phloem, and protoxylem. Johnson and Fushtey (1966) observed the syncytia within six days after inoculation in oat. Cotten (1970) found the initiation of syncytium in resistant barley cv. KVL 91, but subsequently these cells become vacuolated with less dense granular cytoplasm than in susceptible roots. Initiation of syncytia in resistant spring barley cv. Sabrlis were found but differed in size and appearance from those in susceptible cultivar.

Few and isolated reports are available regarding the interaction of *H. avenae* with other pathogens. Gill and Swarup (1977) found *H. gramineum* and *Fusarium moniliforme* on barley plants infested with *H. avenae*. Both the fungi singly suppressed the nematode multiplication but in the presence of both, nematode multiplication was significantly greater than their individual effects. Meagher et al. (1978) reported that the combination of *H. avenae* and *Rhizoctonia solani* caused more reduction in wheat growth than infestation with either pathogen alone. In plants infested with *H. avenae* ear emergence was delayed for 10 days where *H. avenae* and *R. solani* complex delayed it for 13 days.

Andersen (1961) and Mathur et al. (1975) observed complete suppression of larval emergence from newly produced dark brown cyst until several months have elapsed and indicated that cysts require a period of maturation or dormancy. Larval emergence is possible from white, yellow, and light brown cysts only if exposed to proper temperature conditions. The dormancy appears to be facultative since it is induced by high temperature.

The dormancy in India extends from April to October, though availability of proper temperature (4–50°C) may shorten this period by two months (Swarup & Gill, 1969). Johnson and Fushtey (1966) stated that a minimum period of eight weeks of low temperature was necessary before substantial hatching could occur. The induction of dormancy appears to be correlated with the change in cyst color as well as with increases in temperature. Juvenile emergence from eggs in brown cysts requires a period of dormancy of two or more months and is strongly regulated by an increase in temperature. Emergence of juveniles may take place spontaneously when suitable temperature prevails. Often the periods of mass emergence from cysts coincide with a cropping season. Temperatures, availability of moisture and root diffusate are important determinants of juvenile emergence. Emergence of *H. avenae* can take place at temperatures 10–25°C, with the optimum between 20 and 22°C (Swarup & Gill, 1969). The optimum temperature for the Australian *H. avenae* population is 10°C (Brown, 1981). Fluctuating temperatures or alternate exposure of cysts to low and high temperatures stimulates *H. avenae* emergence; release of juveniles at lower temperatures (10–15°C) can be obtained with wheat and barley root diffusate. Root diffusive from one-week-old barley seedlings stimulates emergence of juveniles from the cysts (Williams & Siddiqui, 1972). Ecotypes of *H. avenae* show variation in hatching cycles from induction or suppression of dormancy (diapause) at different temperatures (Rivoal & Cook, 1993).

17.4.9.4   LOSSES DUE TO MOLYA

H. avenae can cause economic damage irrespective of soil type when the intensity of cereal cropping exceeds a certain limit (Kort, 1972). However, greater damage occurs in light soils. The nematode cause severe disease (Molya) on temperate cereals, such as barley and wheat in India, while tropical cereals, sorghum and maize, are non-hosts (Gill & Swarup, 1971; Sharma & Swarup, 1984). In the North-Western part of India, between four- and sixteen-fold increase in yield of wheat and barley, respectively, have been obtained after Nematicidal treatments (Swarup et al., 1976). Staggering annual yield losses of three million pounds sterling in Europe, 72 million Australian dollars in Australia and nine million U.S. dollars in India have been calculated as being caused by *H. avenae* (Wallace, 1965; Brown, 1981; Van Berkum & Seshadri, 1970). The losses in Australia are now greatly reduced due to control of the disease with resistant and tolerant cultivars. The extent of losses depends upon nematode inoculum present in the soil. Handa

(1983) estimated yield losses upto 87% in heavily infested nematode field in Rajasthan, but Mathur et al. (1986) reported avoidable loss in wheat ranging from 32 to 67% inocula varying from 4.6 to 10.6 eggs/ml soil. Mukhopadhyay (1972) reported that 10 cysts/kg soil caused 10% loss which may go up to 64% with 1250 cysts/pot. Under most severe infection the losses may be upto 100% (Seshadri, 1970). In term of rupees Van Berkum and Seshadri (1970) estimated loss of 2.55 crores rupees for barley in three districts of Rajasthan. Yield losses due to this nematode to barley are 17–77% in Saudi Arabia (Ibrahim et al., 1999) and 20% in Australia (Meagher, 1972).

Little is known about the economic importance of the species *H. latipons*, even though it was first described in 1969 (Sikora, 1988). Field studies in Cyprus indicated a 50% yield loss to barley (Philis, 1988). Because the cysts are similar in size and shape, it is possible that previous findings of this recently described nematode species were erroneously attributed to the *H. avenae* (Kort, 1972). In West Asia and North Africa, *H. latipons* has been found on wheat and barley in four countries (Sikora, 1988). It has also recently been confirmed in Turkey (Nicolas et al., 1991). It has also been reported from several Mediterranean countries associated with the poor growth of wheat (Kort, 1972). Other cyst nematodes, such as *P. punctata* and *H. hordecalis*, have been described from roots of cereals in several countries, but their distribution and economic importance is unknown.

17.4.9.5   MANAGEMENT

Attempts have been successful to manage the population through *H. avenae* growing non-host crops because CCN possesses narrow host range being specific to cereals. Handa (1983) found that nematode population decreased by 70% with continued rotation of mustard, carrot, fenugreek, and gram or by fallowing. This led to 87% increase in the yield. The barley yield increased by 56% with two years rotation of non-host crops (Handa et al., 1975).

It has been known for the past several years that organic manuring influences the development of *Heterodera* species. On the basis of pot and field trials, Mathur (1969) reported that oil cakes, farm yard manure, compost and saw dust applications resulted in improved plant growth and subdued multiplication of the CCN. The application of nitrogen resulted in better plant growth and more nematode multiplication, phosphorus and potash treatments did not cause any significant effect. Handa (1983) reported that increased dose of nitrogen significantly improved the plant height, ear length, and fodder and grain weight of barley whereas increase in nematode population was non-significant but certainly there was significant increase over initial population. Plant growth promotion and nematode multiplication, however, depends on the initial population. Better plant growth can be obtained only in lower populations up to 5 L/ml soil but above it, results may be affected.

Mixed cropping of wheat and barley (Gojra) is common practice in Rajasthan. Handa et al. (1980) reported the beneficial effect of resistant variety of barley (Rajkiran) with susceptible variety of wheat Kalyansona for increase in grain yield and decrease in nematode population as compared to susceptible crop of wheat/barley. They further indicated the possibility of use of combination of different crops with varying susceptibility to nematode to decrease the population and obtaining optimum yield. While studying the survival of *H. avenae* under controlled temperatures, a marked reduction in survival of cyst contents at 28°C was recorded (Mathur, 1969). Handa et al. (1975) observed decrease in population of cereal cyst nematode with summer ploughings and subsequent increase in the yield of cereals. Soil temperature during May and June generally remain more than 30°C and go up to 48°C in India. Deep summer ploughings may expose deeper soil layer (up to 20–30 cm) to sun. The moisture of the soil evaporates and cysts are subjected to dry heat. The cyst contents start drying and mortality to eggs and juveniles occurs. The success of the ploughing depends on high temperatures and no rainfall during May and June. Summer ploughings (3–5) shall bring down the initial population and subsequently better yield of barley could be obtained.

Use of chemicals is costly and non-eco-friendly but its application becomes indispensable where nematode populations have reached very high levels. Nematicides are applied through different ways. Research work on control of CCN in cereals with soil incorporation of chemicals has been attempted by a large number of workers. Swarup et al. (1976) and Handa et al. (1980) found that application of DD, dibromochloropropane (DBCP) or granular formulations effectively decreased the soil population of *H. avenae* and increasing the yield of barley. Granular formulations of carbofuran or aldicarb greatly increased the yield of barely and reduced the nematode population. This treatment provided additional and complete control of aphids in barley. Mathur et al. (1986) and Handa et al. (1980) found drilling of granular formulations better than broadcast.

Field experiments conducted with DD, DBCP (EC), aldicarb and carbofuran have revealed high degree of effectiveness against *H. avenae* on barley (Yadav & Singh, 1975; Handa et al., 1980; Mathur et al., 1986; Handa et al., 1975; Singh & Yadav, 1978; Handa et al., 1980, 1985) They observed that carbofuran and aldicarb when applied through drill @1.0 kg a.i./ha were most effective. Use of these chemicals is recommended when the initial nematode population is very high (>5 Juveniles/g soil) and other alternatives are not available.

Use of seed dressing chemicals is quite common and practiced for control of many fungal diseases. The main advantage of seed dressers is (a) use of less quantity of chemical, (b) reduction in the cost of application, and (c) easy to apply. Much work for management of CCN through seed treatment was not done as the availability of seed dressing nematicides is limited. Among some of the seed dressing nematicides tested at different doses, carbofuran and aldicarb sulfone @ 4 kg a.i. /ha were found effective in increasing grain yield of barley with no change in final population. Integration of summer ploughings + irrigation, summer ploughings + nitrogenous fertilizers + seed treatment or soil application of nematicides was found effective in increasing the barley yield and suppressing nematode population (Handa, 1965). Handa et al. (1985) and Mathur et al. (1986) has tried combination of summer ploughings + nitrogenous fertilizers and found that nitrogen at the highest dose (90 kg N/ha or) increased grain yield in tall barley variety in unploughed plots when ploughings were combined with nitrogen, smaller doses of nitrogen gave equal or better results.

The production of resistant cultivars is the most promising method for managing *H. avenae* on other pathogens. Breeding for disease resistance is complicated because of the presence of different pathotypes of *H. avenae* with different levels of virulence. Nevertheless, several cultivars of barley have been developed in our country by incorporating resistance in the local popular cultivars. Looking to the seriousness of the nematode problem in the country which is increasing day by day especially in sandy soils; the best way left out to get rid of nematode is to evolve resistant varieties. Morocco, Maroccaine, PI 253826, PL 101, and C-164 were most effective resistant varieties covering wide range of pathotypes of *H. avenae* in India but it is not effective against *H. filipjevi* of Punjab and Ambala population of Haryana. Any one of these can be utilized as resistant donor parents (Bishnoi et al., 2004) since resistance is monogenic and dominant in nature.

17.5 CONCLUDING REMARKS

Barley in India is preferred due to low input and better adaptability to fragile ecology like drought, salinity-alkalinity, and marginal lands. The agro-climatic conditions for barley cultivation are more suitable in Rajasthan. Rajasthan has maximum area and production of barley in India. Once considered as coarse cereal, barley now has become quality cereal due to its better nutritive quality and industrial utilization for malting and brewing. During the last decade, demand of malt barley in domestic market for industrial utilization has increased. In the era of climate change, where water availability will be becoming a limiting factor for growing crops, area under barley cultivation will be increased in the states like Rajasthan. A number of rusts resistant barley varieties suiting to different cultural and agro-climatic conditions of Rajasthan have been developed. Due to change in climate and cropping pattern and adoption of new technologies particularly, use of micro irrigation system as well as application of more fertilizers under irrigated condition resulted more foliage growth thus favoring buildup of pathogens. Rust pathotypes are evolving in more virulent forms. A resistant cultivar becomes susceptible due to new pathotype. Incorporation of varietal resistance for prevailing diseases has been the major thrust in high yielding varieties. Resistance to rusts will continue to be primary requirement for new varieties. Recently with introduction of irrigated barley cultivation in NWPZ, leaf blight incidence has increased. Though the available barley varieties are showing resistance to rusts but they lack good resistance to leaf blight. Thus there is a need for incorporation of blight resistance in barley varieties in Rajasthan.

KEYWORDS

- barley
- diseases
- management
- rust
- smuts
- Rajasthan
- stripe disease
- molya disease
- Rajasthan

REFERENCES

Andersen, S. Resistance Mod Havreal, *Heterodera Avenae*, Med. 68 Kge.Vet. Land Haisk, Kobenhaven, 1961; p 179.

Andersen, S.; Andersen, K. Suggestion for Determination and Terminology of Pathotypes and Genes for Resistance in Cyst Forming Nematode Especially *Heterodera avenae*. *EPPO Bulletin*. **1982**, *12*, 379–386.

Anonymous. Annual Progress Report of Barley Network 2014–15; ICAR-Indian Institute of Wheat and Barley Research, Karnal, India, 2015; p 275.

Baik, B. K.; Ullrich, S. E. Barley for Food: Characteristics, Improvement and Renewed Interest. *J. Cereal Sci.*, **2008**, *48*, 233–242.

Bajaj, H. K.; Walia, R. K. A New Report on the Occurrence of Cereal Cyst Nematode *Heterodera avenae* Woll. 1924 in Ambala District (Haryana). *Geobios New Reports*, **1985**, *4*, 169–170.

Bakkeren, G.; Kronstad, J. W. Linkage of Mating Type Loci Distinguishes Bipolar from Tetrapolar Mating in Basidiomycetous Smut Fungi. *Proc. Natl. Acad. Sci.* **1994**, *91*, 7085–7089.

Bedi, P. S.; Singh, G. Control of Covered Smut of Barley by Seed Treatment. *Indian Phytopathol.* **1974**, *27*, 379–381.

Bekal, S.; Jahier, J.; Rivoal, R. Host Response of Different *Triticeae* to Species of the Cereal Cyst Nematode Complex in Relation to Breeding Resistant Durum Wheat. *Fund. Appl. Nemat.* **1998**, *21*, 359–370.

Bhardwaj, S. C.; Prashar, M.; Singh, S. B.; Kumar, S.; Sharma, Y. P. Strategies to Combat Wheat Rust in India. *Indian Farming*. **2010**, *60*, 5.

Bhatti, D. S.; Dalal, M. R.; Dahia, R. S. Some Factorial Involvement in Non Parasitisation of Maize by *Heterodera avenae*. *Indian J. Nematol*. **1977**, *7*, 112–116.

Bhatti, D. S.; Dahia, R. S.; Gupta, D. C.; Madan, I. Plant-Parasitic Nematodes Associated with Various Corps in Haryana. *Haryana Agric. Uni. J. Res.* **1980**, *10*, 413–414.

Bishnoi, S. P.; Singh, S.; Metha, S.; Bajaj, H. K. Isozyne Pattern of *Heterodera avenae* and *H. filipjevi* Population of India. *Indian J. Nematol.* **2004**, *34*(1), 33–36.

Bishnoi, S. P.; Bajaj, H. K. On the Species and Pathotypes of *Heterodera avenae* Complex of India. *Indian J. Nematol.*, **2004**, *34*(2), 147–152.

Bishnoi, S. P.; Bajaj, H. K. Development of *Heteredera avenae* on Resistant and Susceptible Cultivars of Barley and Oat. *Indian J. Nematol.* **2003**, *33*(1), 9–12.

Brown, R. A. Nematode Diseases. In *Economic Importance and Biology of Cereal Root Diseases in Australia. Report to Plant Pathology;* Subcommittee of Standing Committee on Agriculture: Australia, 1981.

Brown, W. M.; Velasco, V. R.; Johanson, R. In *Chemical Suppressive of Barley Yellow Rust. Meeting the Challenges of Yellow Rust in Cereals Crop*, Proceedings of the First Regional Conference on Yellow Rust in the Central and West Asia and North Africa Region, Karaj, Iran, May 8–14, 2001; ICARDA: Aleppo, Syria, 2002, pp, 261–264.

Chhabra, H. K. Distribution *Heterodera avenae* Woll., Cereal Cyst Nematode in Punjab. *Curr. Sci.* **1973**, *42*, 441.

Cotton, J. Some Aspects of Breeding for Cereal Cyst Eelworm (*H. avenae* Woll.) Resistance in Barley. Zesz, Probl. Postep. Neuk. Roln. No. 92. Proceedings of the 9th International Nematology Symposium, Warsa, 1970, pp 231–237.

Duhan, J. C.; Beniwal M. S. Improved Solar Energy Treatment for the Control of Loose Smut, Flag Smut and Karnal Bunt Diseases of Wheat. *Seed Res.* **2006**, *32*, 184–188.

FAO. Food Barley Improvement, **2002**. (http://www.fao.org/ag/AGP/AGPC/doc/Field/other/ act.html).

Franklin M. T. On the Specific Status of the so Called Biological Strains of *Heterodera schachtii* Schmidt. *J. Helminth.* **1940**, *18*, 193–208.

Franklin, M. T.; Thorne, G.; Oostenbrink, M. Proposal to Stabilize the Scientific Name of the Cereal Cyst Eelworm (Class Nematoda) Z. M. (S) 375. *Bull. Zool. Nomencl.* **1959**, *17*, 76–85.

Gill, J. S.; Swarup, G. Effect of Interaction between *Heterodera avenae* Woll. 1924, *Fusarium moniliforme* and *Helminthosporium gramineum* on Barley Plants and Nematode Reproduction. *Indian J. Nematol.* **1977**, *7*(1), 42–45.

Gill, J. S.; Swarup, G. On the Host Range of Cereal Cyst Nematode, *Heterodera avenae*, the Causal Organism of 'Molya' Disease of Wheat and Barley in Rajasthan, India. Indian. *J. Nematol.* **1971**, *1*, 63–67.

Goel, L. B.; Singh, D. V.; Srivastava, K. D.; Joshi, L. M.; Nagrarajan, S. *Smuts and Bunts of Wheat in India;* Indian Agricultural Research Institute: New Delhi, India, 1977; p 38.

Handa, D. K.; Mathur, R. L.; Mathur, B. N.; Yadav, B. D. Estimation of Losses in Barley due to Cereal Cyst Nematode in Sandy and Sandy Loam Soils. *Indian J. Nematol.* **1985**, *15*(2), 163–166.

Handa D. K. Studies on 'Molya' Disease (*Heterodera avenae*) of Wheat and Barley with Special Emphasis on its Control. Ph.D. Thesis, University of Rajasthan, Jaipur, 1983; p 340.

Handa, D. K.; Mathur, B. N.; Sharma, G. L.; Yadav, B. D. Evaluation of Barley Varieties for Resistance to Molya Disease Incited by *Heterodera avenae*. *Indian. J. Mycol. Pl. Pathol.* **1980**, *10*(2), 9.

Handa, D. K.; Mathur, R. L.; Mathur, B. N. Studies on the Effect of Deep Summer Ploughings on the Cereal Cyst Nematode (*Heterodera avenae)*, and Yield of Wheat and Barley. (Abstr.). *Indian J. Mycol. Pl. Pathol.* **1975**, *5*(1), 18.

Hovmoller, M. A.; Yahyaoni, E.; Milus, E. A.; Justesen, A. Rapid Global Spread of Two Aggressive Strains of Wheat Rust Fungus. *Mol. Ecol.* **2008**, *17*, 3818–3826.

Ibrahim, A. A. M., et al. Damage Potential and Reproduction of *Heterodera avenae* on Wheat and Barley under Saudi Field Conditions. *Nematology.* **1999**, *1*(6) 625–630.

Jain, A. K., et al. Morphological Characters as Affected by Covered Smut of Barley. *Adv. Plant Sci.* **1997**, *10*(1), 237–239.

Johnson, P. W; Fushtey, S. G. The Biology of Oat Cyst Nematode, *Heterodera avenae* in Canada. VI. Nematode Development and Related Anatomical Changes in Roots of Oat and Corn. *Nematologica.* **1966**, *12*, 630–636.

Joshi, L. M. Recent Contributions towards Epidemiology of Wheat Rusts in India. Presidential Address of Indian Phytopathological Society. *Indian Phytopathol.* **1976**, *29*, 1–5.

Joshi, L. M., et al. Survey and Epidemiology of Wheat Rust in India. In *Current Trends in Plant Pathology;* Raychaudhuri, S. P., Verma, J. P., Eds.; 1974; pp 150–159.

Joshi, L. M., et al. *Wheat Diseases News Letter;* 1984; *17*, 1–20, IARI: New Delhi.

Kaur, A., et al. Management of Covered Smut of Barley through Cultural, Chemical and Biological Methods. *J. Mycol. Pl. Pathol.* **2014**, *44*(4), 432–437.

Kort, J. Nematode Diseases of Cereals of Temperate Climates. In *Economic Nematology*; Webster, J. M., Ed.; Academic: New York, 1972; pp 97–126.

Koshy, P. K.; Swarup, G. Distribution of *Heterodera avenae*, *H. zeae*, *H. cajani* and *Anguina tritici* in India. *Indian J. Nematol.* **1971**, *1*, 106–111.

Kuhn, J. Uber das Kommen von Rubbennematoden an den Wurzehdra Halmfruchle, Landav. *JBR.* **1874,** *3,* 47–50.

Kumar, V., et al. Barley Research in India. Retrospect and Prospects. *J. Wheat Res.* **2014,** *6*(1) 1–20.

Laurie, J. D., et al. Genome Comparison of Barley and Maize Smut Fungi Reveals Targeted Loss of RNA Silencing Components and Species-Specific Presence of Transposable Elements. *Plant Cell.* **2012,** *24*(1), 1733–1745.

Lili, Z., et al. Detection Infrarouge Thermique des Maladies du blé d'hiver. *EPPO Bull.* **1991,** *21,* 659–672.

Martens, J. W., et al. *Diseases of Field Crops in Canada;* Canadian Phytopathological Society: Baltimore, MD, 1984; p 160.

Mathre, D. E. *Compendium of Barley Diseases;* American Phytopathological Society: Baltimore, MD, 1997; p 120.

Mathur, A. K.; Bhatnagar, G. C. Studies on the Compatibility of Vitavax with Aldrin, E.C. *Pesticides.* **1986,** *20*(9), 53.

Mathur, A. K.; Bhatnagar, G. C. Sources of Resistance in Barley against Stripe Caused by *Helminthosporium gramineum. Indian Phytopathol.* **1992,** *45*(1), 115–116.

Mathur, B. N., et al. The Occurrence of Biotypes of the Cereal Cyst Nematode (*Heterodera avenae)* in the Light Soils of Rajasthan and Haryana, India. *Indian J. Nematol.* **1974,** *16,* 152.

Mathur, B. N., et al. On the Symptoms and Distribution of Molya Disease of Wheat and Barley Caused by *Heterodera avenae* in the Light Soils of Rajasthan and Haryana, India. *Nematologica.* **1975,** *20,* 19–26.

Mathur, B. N. Studies on Cereals Root Eelworm (*Heterodera avenae*) with Special Reference to 'Molya' Disease of Wheat and Barley in Rajasthan. Ph.D. Thesis, University of Rajasthan, Jaipur, 1969, p 233.

Mathur, B. N., et al. On the Loss Estimation and Chemical Control of 'Molya' Disease of Wheat caused by *Heterodera Avenae* in India. *Indian J. Nematol.* **1986,** *16,* 152.

Meagher, J. W. Cereal Cyst Nematode *(Heterodera avenae* Woll). Studies *on Ecology and Content in Victoria;* Technical Bulletin No. 24; Department of Agriculture: Victoria, Australia, 1972; p 50.

Meagher, J. W.; Brown, R. H.; Rovira, A. D. The Effects of Cereal Cyst Nematode (*Heterodera avenae*) and *Rhizoctonia solani* on the Growth and Yield of Wheat. *Aust. J. Agric. Res.* **1978,** *29,* 1127–1137.

Mehta, K. C. Further Studies on Cereal Rusts in India; Scientific Monograph. No. 14; Imperial Council Agricultural Research: New Delhi, India, 1940, p 224.

Mehta, K. C. Further studies on Cereal Rusts in India; Scientific Monograph. No. 18; Imperial Council Agricultural Research: New Delhi, India, 1952, p 165.

Mokabli, A., et al. Influence of Temperature on the Hatch of *Heterodera avenae* Woll. Populations from Algeria. *Nematologica.* **2001,** *3*(2), 171–178.

Mor, M.; Cohn, E.; Spiegel, Y. Phenology, Pathogenicity and Pathotypes of Cereal Cyst Nematodes, *Heterodera avenae* Woll. and *H. latipons* (Nematoda: Heteroderidae) in Israel. *Nematologica.* **1992,** *38,* 444–501.

Mukhopadhyay, M. C., et al. Studies on the 'Molya' Disease of Wheat and Barley. *Indian J. Nematol.* **1972,** *2,* 11–20.

Nagrajan, S.; Nayar, S. K.; Kumar, J. Checking the *Puccinia* Species that Reduce the Productivity of Wheat (*Triticum* spp.) in India-A Review. *Proc. Indian Natl. Sci. Acad.* **2006,** *72,* 239–247.

Nicolas, H.; Rivoal, R.; Duchesne, J.; Lili, Z. Detection of *Heterodera avenae* Infestations on Winter Wheat by Radiother-Mometry. *Rev. Nematol.* **1991**, *14*, 285–290.

Pal, D.; Kumar, S.; Verma, R. P. S. Pusa Losar (BH 380) the First Dual-Purpose Barley Variety for Northern Hills of India. *Indian J. Agric. Sci.* **2012**, *82*, 164–165.

Park, R. F.; Bariana, H. S.; Wellings, C. S. Preface. *Aus. J. Agric. Res.* **2007**, *58*, 469.

Philis, I. Occurrence of *Heterodera latipons* on Barley in Cyprus. *Nematol. Med.* **1988**, *16*, 223.

Rivoal, R.; Cook, R. Nematode Pests of Cereals. In *Plant Parasitic Nematodes in Temperate Agriculture;* CAB International: Wallingford, CT, 1993; pp 259–303.

Safavi, S. A.; Atahusaini, S. M.; Ebrahimnejad, S. Effective and Ineffective Resistance Gene and Resistance Reaction of Promosing Barley Lines to *Puccinia Striiformis* f. sp. *hordei* in Iran. *Asian J. Plant Sci.* **2012**, *11*, 52–57.

Schmidt, O. Sind Ruben-und Hafer-Nematoden Identische Wiss Arch. *Landwirtsch (Abt A).***1930**, *3*, 420–461.

Selvakumar, R., et al. Management of Stripe Rust of Barley Using Fungicides, *Indian Phytopathol.* **2014**, *67* (2), 138–142.

Seshadri, A. R. New Vistas in Crops Yields. In *Agricultural Book;* ICAR: New Delhi, India, 1970; pp 370–411.

Sharma, S. B.; Swarup, G. *Cyst Forming Nematodes of India;* Cosmo Publications: New Delhi, India, 1984; p 150.

Siddiqui, M. R.; Hussain, S. I. Studies on Occurrence of Biotype of *Heterodera avenae* in Uttar Pradesh, India. *Intl. Nematol. Network Newsl.* **1989**, *6*, 8–11.

Siddiqui, Z. A.; Siddiqui, M. R.; Hussain, S. I. First Report on the Occurrence of *Heterodera avenae* Wollenweber, 1924 in Uttar Pradesh. India. *Intl. Nematol. Network Newsl.* **1986**, *3*, 11.

Sikora, R. A. Plant Parasitic Nematodes of Wheat and Barley in Temperate and Temperate Semi-Arid Regions -a Comparative Analysis. In *Nematodes Parasitic to Cereals and Legumes in Temperate Semi-Arid Regions;* Saxena, M. C., Sikora, R. A., Srivastava, J. P. Eds.; ICARDA: Aleppo, Syria, 1988, pp 46–48.

Singh, D. P., et al. Efficacy of *Trichoderma viride* in Controlling the Loose Smut of Wheat caused by *Ustilago segetum* var. *tritici* at multilocations. *J. Biol. Control.* **2000**, *14*, 35–38.

Singh, D. P., et al. Integrated Pest Management in Barley (*Hordeum vulgare*). *Indian J. Agric. Sci.* **2010**, *80*, 437–442.

Singh, H.; Yadav, B. S. Control of Molya Disease of Wheat caused by *Heterodera avenae* (Abstr.). *Indian J. Mycol. Pl. Pathol.* **1978**, *10*(2), 15.

Singh, I., et al. Distribution of Cereal Cyst Nematode *H. avenae* Woll. on Wheat in Ludhiana (Punjab). *J. Res. PAU.* **1976**, *14*, 314–317.

Singh, R. S. *Plant Diseases;* 9th ed.; Oxford & IBH Publishing Co. Pvt. Ltd.: New Delhi, India, **2009**, pp 346–349.

Srivastava, K. D. *Biology and Management of Wheat Pathogens;* Stadium Press Pvt. Ltd.: New Delhi, India, 2010, p 270.

Swarup, G., et al. Response of Wheat and Barley to Soil Fumigation by DD and DBCP against "Molya" Disease Caused by *Heterodera avenae. Indian J. Nematol.* **1976**, *6*, 150–155.

Swarup, G., et al. Distribution of *Heterodera avenae* the causal Organism of 'Molya' Disease of Wheat and Barley in India. *Curr. Sci.* **1982**, *51*, 896–897.

Swarup, G.; Sethi, C. L.; Seshadri, A. R.; Kaushal, K. K. On the Biotype of *Heterodera avenae*, the Causal Organism of 'Molya' Disease of Wheat and Barley in India. *Indian J. Nematol.* **1979**, *9*, 164–168.

Swarup, G.; Gill, J. S. Varietal Response of Wheat and Barley to Infestation by *Heterodera avenae*, First All India Nematology Symposium, New Delhi, 1969; pp 64–65.

Tunwar, N. S.; Singh, S. V. *Indian Minimum Seed Certification Standards;* The Central Seed Certification Board, Department of Agriculture and Cooperation, Ministry of Agriculture: Govt. of India, New Delhi, India, 1988, p 388.

Van Berkum, J. A.; Shesadri, A. R. In Some Important Nematode Problems in India (abstract), 10th International Nematology Symposium, Pescara, Italy, 1970, pp 136–137.

Vangool, D.; Vernon, L. Potential Impact of Climate Change on Agricultural Land Use Suitability: Barley; Report, No. 302, Department of Agriculture: Western Australia, 2006.

Vasudeva, R. S. Annual Report of Indian Agricultural Research Station: New Delhi, India, June 30, 1958.

Wallace, H. R. The Ecology and Control of the Cereal Root Nematode. *J. Austr. Inst. Agric. Sci.* **1965,** *31*, 178–186.

Williams, T. D.; Siddiqi, M. R. *Heterodera avenae.* In *C.I.H. Descriptions of Plant Parasitic Nematodes;* Wilmott, S., Gooch, P. S., Siddiqi, M. R., Franklin, M., Eds.; Commonwealth Agricultural Bureau: Farnham, Slough, UK; Set 1, No. 2, 1972.

Yadav, B. S.; Singh, H. Effect of Certain Chemicals for the Control of Molya Disease of Wheat Caused by *Heterodera avenae* (Abstr.) *Indian J. Mycol. Plant Pathol.* **1975,** *5*(1), 17.

CHAPTER 18

RESOURCE CONSERVATION AGRICULTURE PRACTICES, RHIZOSPHERE, AND DISEASES OF WHEAT UNDER WHEAT–RICE CROPPING SYSTEM**

ANJU RANI[1] and DEVENDRA PAL SINGH[2*]

[1]*Department of Bioscience & Biotechnology, Banasthali University, Vanasthali 304022, Rajasthan, India*

[2]*ICAR-Indian Institute of Wheat and Barley Research, Karnal 132001, Haryana, India*

Corresponding author. E-mail: dpkarnal@gmail.com

**Part of M.Sc thesis of senior author*

CONTENTS

ABSTRACT

Rice–wheat is the most important cropping system in Indo Gangetic region of India and South Asia. It is in practice since last four decades and there is no immediate alternate to it in this region. Being two different soil systems (aerobic and unaerobic), it is of much importance to keep track of effect of rice-wheat cropping system on soil and crop health for its sustainability and profitability. There had been studies indicating decline of organic carbon under rice-wheat cropping system in states like Punjab and Haryana in India especially under situation of non-incorporation of crop residues in soil. The results of studies conducted at ICAR-IIWBR Karnal, India on retention and incorporation of crop residues of either rice or wheat or both had positive effect on soil microbial population, root health and had lower diseased root volume as compared to burning of crop residues.

18.1   INTRODUCTION

Rice (*Oryza sativa* L.)–wheat is the most important two-crop combination per year in the intensive rice-based cropping system of Asia. It accounts one third of the irrigated rice and half of the irrigated wheat in South Asia. Available estimates show that 12 million hectares (mha) of this cropping system exist in four countries of South Asia: 9.4 mha in India, 1.5 mha in Pakistan, 0.6 mha in Bangladesh, and 0.5 mha in Nepal. There are about 9 mha of this cropping system in China. In this cropping system, rice is grown during rainy season (June to November) and wheat during the winter season (November to April). This is an intensive cropping system and its productivity varies from 5 to 10 metric tons of grains per hectare per year (Singh et al., 2014). However, concerns are expressed about the sustainability of the cropping system, keeping in view of different types of ecological systems of rice and wheat and its impact on soil properties and plant health (André & Lagerlof, 1983). It has been estimated that system covers more than 23.5 mha in different countries (Bimb et al., 1994). In Punjab and Haryana states of India, almost 95–98% of the rice–wheat cropping area is practiced under irrigated conditions, with over 90% of water requirement sourced from groundwater (Ambast et al., 2006). This has contributed to a fall in water table levels in central Punjab (the main area for rice–wheat cropping) by more than 33 cm per year from 1979 to 1994 in 46% of the area. The conventional method of puddling rice is also damaging the soil. In puddling, the soil for rice, a thick hardpan can develop, which restricts root growth in crops grown in rotation with rice. The constant flooding of the soil also leads to greater losses of soil microbes and nitrogen fertilizer. The practice of burning the rice stubble has a substantial negative effect on the environment. More than 90% of the 17 Mt of rice stubble in Punjab are burnt each year, resulting in thick smoke blanketing the region, since the burn-off occurs over a short period. The air pollution caused by this burn-off has serious adverse health effects on humans and animals; it has been blamed for causing road accidents and the closing of airports due to poor visibility. Furthermore, the burning results in the loss of nutrients and organic matter from the soil. Recently it has been seriously viewed and legal action is initiated to stop this practice in India in the northern states.

Much of this work has focused on developing resource-conserving technologies (RCTs) helping to produce cereals at a lower cost while attempting to improve soil health through reduced tillage and stubble retention. These RCTs are being adopted according farmers' needs and situations. In zero tillage, wheat seeds are drilled into unplowed fields which retain the residues from the rice crop. In reduced tillage, the seeds are surface sown onto rota-tilled soil. Farmers save on tillage costs. The RCTs are also more sustainable as they use less energy and are more environmentally friendly. In spite of above advantages, information on the effect of RCTs on soil health, flora, and diseases of wheat is scanty. Conservation tillage is a set of practices that leaves crop residues on the surface which increases water infiltration and reduces erosion. It is a practice used in conventional agriculture to reduce the effects of tillage on soil erosion. The increasing popularity of the techniques has led to the need for research into their effects on soil health. The conventional mode of agriculture through intensive agricultural practices was successful in achieving goals of production, but simultaneously led to degradation of natural resources (Sturz et al., 1997). Conservation agriculture (CA) is an approach for designing and management of sustainable and resource-conserving agricultural systems (Cartwright et al., 1996; Pankhurst, 1997). It is currently practiced on more than 80 mha worldwide in more than 50 countries and the area is expanding rapidly (Rothrock, 1992). In the 2004–2005 wheat season, zero tillage is estimated to have been used on nearly 2 mha of sown area in India. Taken together, area under zero tillage/ CA was estimated for 2000–2001 to be of the order of 60 mha, most of it in the USA, Australia, Canada, Brazil, and Argentina (Roper & Gupta, 1995; Sumner et al., 1981). Soil health studies were carried out at ICAR-Indian Institute of Wheat and Barley Research (IIWBR), Karnal and CCS HAU Rice Research Station, Kaul in Haryana, North India.

18.2 RESOURCE CONSERVATION AGRICULTURE AND RHIZOSPHERE OF WHEAT

The studies were conducted at ICAR-IIWBR Karnal, India to know the effect of CA on soil health, and soil micro biota profiles under rice–wheat cropping system in wheat crop (Anju Rani, 2012). The size and composition of microbial populations were used to assess the changes in the soil biota happened in response to CA. Soil microflora is important for sustainable agriculture as its activity contributes to increasing agricultural production. Disease susceptibility of crops may be reduced considerably by better understanding the interactions between pathogens and crop residue and then modifying local environmental conditions, crop rotations, tillage practices, and antagonistic microflora accordingly. These are often good indicators of the biological status of soil. High levels of soil borne pathogens often indicate poor soil health because of the increased threat of root disease. Different types of cropping systems and tillage practices can also influence soil health for crop growth as they will influence the number of detrimental and beneficial organisms in the rhizosphere. CA maintains a permanent or semi-permanent organic soil cover resulted due to growing crop or incorporation of dead mulch. Its function is to protect the soil physically from sun, rain, and wind and it is a food to soil biota. The soil micro-organisms and soil fauna take over the tillage function and soil nutrient balancing. Mechanical tillage disturbs this process. Therefore, zero or minimum tillage and direct seeding are important elements of CA. A varied crop rotation is also important to avoid disease and pest problems.

The studies were conducted on micro biota profiles in the soil, root architecture, and dehydrogenate activity. The main treatments were, removal of rice and wheat crop residues, incorporation of rice and wheat crop residues, incorporation of rice residue and removal wheat residue, burning of rice and wheat residues, burning of rice residue and removal wheat residue, retention of rice and wheat residues, and retention of only rice residue and removal wheat residue. The sub treatments were nitrogen applications @ 100, 150, and 200 kg/ha during 2011–2012 crop season. The sampling of wheat plants was done using standard procedure from field and the rhizosphere soil was assessed for microbiological population and activity. Soil samples were processed within 1 h. Soil sample (1 g) was diluted up to 10^{-6} adding 9 ml sterilized water in the test tubes. Out of it, 0.1 ml of each dilution (10^{-1}, 10^{-3}, 10^{-5}, and 10^{-6}) was spread on nutrient agar, potato dextrose agar, *Pseudomonas* agar base + 1% (v/v) glycerol (Table 18.1) (Bridge et al., 1999).

TABLE 18.1   Media Used for Assay of Soil Microbes.

Ingredients	g/l	Ingredients	g/l
Agar medium for *Actinomyces*		**Nutrient agar**	
Sodium caseinate	2.0	Peptic digest of animal tissue	5.0
L Asparagine	0.1	Beef extract	1.5
Sodium propionate	4.0	Yeast extract	1.5
Dipotassium phosphate	0.5	Sodium chloride	5.0
Magnesium sulphate	0.1	Agar	15.0
Ferrous sulphate	0.001	pH	7.4
Agar	15.0	***Pseudomonas* isolation agar base**	
pH	8.1	Peptone	20.0
Potato dextrose agar (PDA)		Agar	13.6
Infusion of potatoes	200.0	Potassium sulphate	10.0
		Magnesium chloride	1.4
Dextrose	20.0	Irgasan (triclosan)	25.0 mg.
Agar	15.0	Glycerol	20.0

The agar plates were incubated at 28°C up to seven days. Morphologically different microbial colonies were selected after 48 and 96 h of incubation, assessed broadly and cultured on respective media for further studies. The number of cultivable bacteria in the original gram of soil by averaging the results from each countable plate was counted according to procedure of Florida International University (1996). The bacterial isolates were identified by Institute of Microbial Technology (IMTECH) (Chandigarh).

Biological oxidation of soil organic compounds is generally a dehydrogenation process carried out by specific dehydrogenases involved in the oxidative energy transfer of microbial cells. This activity is a measure of microbial metabolism and thus of the oxidative microbial activity in soil. The technique involves the incubation of soil with 2, 3, 5-triphenyltetrazolium chloride (TTC) either in the presence or in the absence of added electron donating substrate. Microbial dehydrogenase activity during this incubation results in reduction of water soluble colorless TTC to water insoluble red 2, 3, 5-triphenyltetrazolium formazan which was extracted from soil and read calorimetrically for quantification.

The root samples were taken from different plots and were thoroughly washed with distilled water until the adhering soil is completely removed. The root samples were then surface dried and analyzed by scanning and using Win RHIZO 2012A software for different root parameters like total length, surface area, root volume, etc.

18.2.1 MICROBIAL POPULATION

The predominant fungal species found in the wheat rhizosphere under rice–wheat cropping system and CA were *Aspergillus terreus, Aspergillus heteromorphus, Fusarium* spp., *Penicillium* spp., *Alternaria triticina,* and *Bipolaris sorokiniana.* Bacterial counts were more than fungal and actino-mycetes counts. The predominant actinomycetes spp. were *Streptomycetes* spp. which have biocontrol properties (Table 18.2). Amongst different treat-ments, total colony forming units (CFU) were highest in plots where practice of retention of residues of both rice and wheat crops was followed. Differ-ences were also found in terms of kind of microbial populations amongst different treatments of RCTs. Significantly higher counts of CFU were in treatments provided with N150 kg/ha (Table 18.2).

18.2.2 EFFECT ON ROOT ARCHITECT

The root parameters (analyzed region width, height, area, and diseased root area) differed in different treatments of RCA. The effect of N doses was not significant. Better root architect was recorded in plots having wheat and rice crop residue. The burning of crop residues had negative effect on root health (Table 18.3). Root architecture shows different surface area, root volume, healthy condition of root. These results show higher surface area is respon-sible for higher number of micro biota in soil, more lengths of root help plant in water logging conditions (Table 18.3).

18.2.3 DEHYDROGENASE ACTIVITY

The dehydrogenase activity of soil biomass was low in different RCT treat-ments as compared to plots where rice residue was burnt and wheat residue was incorporated (Table 18.4).

The results of these studies indicated beneficial effect of CA on soil microbes, their activities, and root architect which may in turn help the rice–wheat cropping system to sustain effectively over long period without dete-riorating soil health.

TABLE 18.2 Effect of Different Resource Conservation Agricultural Practices on Microbial Profiles in Wheat.

Main treatments	CFU/Plate (total)	Total colony counts/Petri plate (9 cm diameter)					
		Streptomycetes spp.		Bacterial colony color		Aspergillus heteromorphus
		Chalky white colony	Dark yellow	Orange	Cream	Black colony
Removal of both rice and wheat crop residue	52.1	1.7	14.8	8.0	25.3	2.0
Incorporation of both rice and wheat crop residue	34.0	2.2	14.0	1.3	16.4	0.0
Incorporation of rice residue and removal of wheat residue	32.1	1.4	10.4	1.7	18.4	0.0
Burning of both rice and wheat residues	44.2	8.5	6.5	3.6	25.4	0.0
Burning of rice residue and removal of wheat residue	38.3	1.4	11.0	2.8	23.0	0.0
Retention of both rice and wheat residues	54.1	1.3	13.4	1.5	37.4	0.0
Retention of rice residue and removal of wheat residue	25.9	1.3	2.0	1.5	20.3	0.0
MEAN	40.1	2.5	10.3	2.9	23.8	0.3
CD (5%)	3.3	1.2	2.5	1.1	2.1	0.3
Sub treatments						
N 100 kg/ha	39.0	4.2	5.9	2.2	26.0	0.5
N 150 kg/ha	50.3	1.7	16.6	3.2	28.4	0.3
N 200 kg/ha	31.0	2.0	8.4	3.5	16.9	0.0
Mean	40.1	2.6	10.3	3.0	23.8	0.3
CD (5%)	2.2	0.8	1.7	0.7	1.4	0.3

TABLE 18.3 Effect of Different Resource Conservation Agricultural Practices on Root Architect in Wheat.

Main treatments	Root parameters			
	Analyzed region width (cm)	Analyzed region height (cm)	Analyzed region area (cm^2)	Root volume diseased (cm^3)
Removal of both rice and wheat crop residues	8.6	12.7	109.5	1.4
Incorporation of rice and wheat crop residues	9.3	12.1	112.8	2.7
Incorporation of rice residue and removal wheat residue	9.5	12.0	113.7	2.5
Burning of both rice residue and wheat residue	9.9	12.6	123.6	3.0
Burning of rice residue and removal wheat residue	9.3	11.7	109.1	1.8
Retention of both rice and wheat residues	11.0	11.5	126.6	2.1
Retention of rice residue and removal wheat residue	8.9	10.3	91.9	2.0
Mean	9.5	11.8	112.5	2.2
CD (5%)	0.8	0.8	11.0	0.6
Sub treatments				
N 100 kg/ha	9.8	11.4	111.9	2.3
N 150 kg/ha	9.1	11.9	109.3	2.1
N 200 kg/ha	9.6	12.2	116.2	2.3
Mean	9.5	11.8	112.5	2.2
CD (5%)	0.5	0.5	7.2	0.4

TABLE 18.4 Effect of Different Resource Conservation Agricultural Practices on Dehydrogenase Activity in Wheat.

Main treatments	Dehydrogenase activity		
	Concentration ppm/ (4 g)	Reading/g	Reading/dry weight
Removal of both rice and wheat crop residues	178.9	44.7	50.9
Incorporation of both rice and wheat crop residues	222.5	55.6	62.8
Incorporation of rice residue and removal wheat residue	118.3	29.6	33.7
Burning of both rice and wheat residue	252.6	63.2	71.4
Burning of rice residue and removal wheat residue	305.5	76.4	87.0
Retention of both rice and wheat residues	209.4	52.4	58.8
Retention of rice residue and removal wheat residue	124.7	31.1	34.9
Mean	201.7	50.4	57.1
CD (5%)	12.6	3.1	3.5
Sub treatments			
N 100 kg/ha	225.5	56.3	63.5
N 150 kg/ha	191.2	47.8	54.3
N 200 kg/ha	188.3	47.0	53.4
Mean	201.7	50.4	57.1
CD (5%)	8.2	2.0	2.3

18.3   RESOURCE CONSERVATION AGRICULTURE AND DISEASES OF WHEAT

Due to tillage practices, physical and chemical properties of the soil, root growth, nutrient uptake, and microbial population are altered and these indirectly affect the viability and activity of the plant pathogens as well as host response to these. The changes also occur in temperature, moisture, aeration, compaction, porosity, plant nutrients, pH, and organic matter of soils due to tillage practices. Several diseases are more damaging in high than low-residue seedbeds, and in crops planted during early autumn to reduce soil erosion during winter, especially unirrigated winter wheat in rotation with summer fallow in low rainfall zones of 250–400 mm rain fall (Smiley, 1996).

CA may reduce pests and diseases by integrating crop rotation, which breaks the cycles that perpetuate crop diseases such as wheat rust and pest infestations (ICARDA, 2016). Fujisaka et al. (1994) reported higher losses by grassy weeds due to rice–wheat cultivation. Subsequent plant residue decomposition may result in phytotoxin release and the stimulation of toxin producing micro-organisms. It predisposed the plants to pathogens (Sturz et al., 1997). Relatively high soil microbial activity can lead to competition effect that may ameliorate pathogen activity and survival. Microbial antagonism in root zone can lead to the formation of disease suppressive soil. The losses in yield due to foliar and root pathogens (Septoria blotch, glume blotch, Rhizoctonia root rot, seed and root rot, powdery mildew, and crown rot) have been reported higher under CA (Belmar et al. 1987). The incidence of spot blotch and Fusarium common root rot was partially or completely controlled by reduced tillage. The others, however, reported contrasting findings (Ram Singh et al., 2005). In the experiments conducted at ICAR-IIWBR Karnal, India, the incorporation and retention of crop residues of rice and wheat had significantly low diseased root volume (Table 18.3).

18.4   CONCLUDING REMARKS

The effect of RCT on rhizospheric microbes of wheat under rice–wheat cropping system was observed in different treatments involving crop residue incorporation in soil after harvesting of rice and wheat, removal and burning residue in field, and their combinations along with sub treatments of nitrogen doses (100, 150, and 200 kg/ha). Differences in populations of bacteria, actinomycetes, and fungi were found in different treatments. These may have biocontrol properties against soil borne diseases. Likewise, root architect changed due to resource conservation tillage thus affecting vigor of plants. The dehydrogenase activity also differed in treatments of RCT. Amongst different RCT treatments, retention of crop residue in soil after harvest of both wheat and rice crops was best for soil and plant health. The work conducted on effect of RCT on wheat diseases was not conclusive, keeping in view results of different reports.

ACKNOWLEDGEMENT

Thanks to Dr. R.K. Sharma PI (RM) and his team for their cooperation in conducting the studies at ICAR-IIWBR Karnal.

KEYWORDS

- **cropping system**
- **conservation agriculture**
- **crop residue**
- **dehydrogenases**
- **soil health**
- **rice-wheat system**
- **diseases**
- **wheat**

REFERENCES

Ambast, S. K.; Tyagi, N. K.; Raul S. K. Management of Declining Groundwater in the Trans Indo-Gangetic Plain (India): Some Options. *Agr. Water. Manage.* **2006,** *82,* 279–296.

André, O.; Lagerlof, J. Soil Fauna (Microarthropods, Enchytraeids, Nematodes) in Swedish Agricultural Cropping Systems. *Acta Agric. Scand.* **1983,** *33,* 33–52.

Anju Rani. *Effect of Resource Conservation Agriculture Practices on Rhizosphere of Wheat under Wheat–Rice Cropping System;* Project Report Submitted for Partial Fulfillment of M.Sc. Degree: Department of Bioscience & Biotechnology, Banasthali University, Rajasthan, 2012, p 20.

Belmar, S. B.; Jones, R. K.; Starr, J. L. Influence of Crop Rotation on Inoculum Density of *Rhizoctonia solani* and Sheath Blight Incidence in Rice. *Phytopathology.* **1987,** *77,* 1138–1143.

Bimb, H. P.; Dubin, J. H. *Studies of Soil Borne Diseases and Foliar Blight of Wheat at the National Wheat Research Experiment Station, Bhairahawa, Nepal;* Wheat Special Report, 36: CIMMYT, Wheat Programme, 1994.

Bridge, J., et al. *Technical Protocols for Crop Disease Assessment, Plant and Soil Sampling, Isolation and Extraction of Fungi and Nematodes;* DFID Rice-Wheat Soil Health Project: CABI Bioscience, UK Centre, Egham, Surrey, TW20 9TY, UK, 1999; p 20.

Cartwright, R. D., et al. Conservation Tillage and Sheath Blight of Rice in Arkansas. *Ark. Expt. Sta. Res. Ser.* **1996,** *456,* 83–88.

Florida International University. *Isolation of Soil Bacteria: Viable Isolation and Pure Culture Cornell University: Enrichment and Isolation* "Lab Manual for Soil Microbiology"; David Zuberer, Ph.D. Thesis, 1996.

Fujisaka, S.; Harrington, L.; Hobbs, P. Rice-Wheat in South Asia: Systems and Long-Term Priorities Established Through Diagnostic Research. *Agric. Sys.* **1994,** *46* (2), 169–187.

ICARDA. *Conservation Agriculture: Opportunities for Intensified Farming and Environmental Conservation in Dry Areas.* International Center for Agricultural Research in Dry Areas (ICARDA): Syria, 2016. http://reliefweb.int/report/world/conservation-agriculture-opportunities-intensified-farming-and-environmental.

Pankhurst, C. E. Biodiversity of Soil Organisms as Indicators of Soil Health. In *Biological Indicators of Soil Health;* Pankhurst, C. E. et al., Eds.; CAB International: Wallingford, 1997; pp 297–324.

Ram Singh, et al. *Long-Term Response of Plant Pathogens and Nematodes to Zero-Tillage Technology in Rice-Wheat Cropping System;* Technical Bulletin No. 7; Regional Research Station, Kaul, Department of Nematology and Directorate of Extension Education, CCS Haryana Agricultural University, Hisar, 2005. http://www. hau.ernet.in/hisar_admin/newspdf/1421384428plantprokaul.pdf.

Roper, M. M.; Gupta, V. V. S. R. Management-Practices and Soil Biota. *Aust. J. Soil Res.* **1995,** *33,* 321–339.

Rothrock, C. S. Tillage Systems and Plant Disease. *Soil Sci.* **1992,** *154,* 308–315.

Singh, D. K., et al. Evaluation of Agronomic Management Practices on Farmers' Fields under Rice-Wheat Cropping System in Northern India. *Int. J. Agron.* **2014,** http://dx.doi.org/10.1155/2014/740656.

Smiley, R. W. Diseases of Wheat and Barley in Conservation Cropping Systems of the Semi-arid Pacific Northwest. *Am. J. Alternative Agr.* **1996,** *11* (2–3), 95–103.

Sturz, A. V., et al. A Review of Plant Disease, Pathogen Interactions and Microbial Antagonism under Conservation Tillage in Temperate Humid Agriculture. *Soil. Till. Res.* **1997,** *41,* 169–189.

Sumner, D. R., et al. Effects of Reduced Tillage and Multiple Cropping on Plant Diseases. *Annu. Rev. Phytopathol.* **1981,** *19,* 167–187.

PART III
Pathogenic Variability and Its Management

CHAPTER 19

SURVEY AND SURVEILLANCE OF WHEAT BIOTIC STRESSES: INDIAN SCENARIO

M. S. SAHARAN*

Division of Plant Pathology, ICAR-Indian Agricultural Research Institute, New Delhi 110012, India.

Corresponding author. E-mail: mssaharan7@yahoo.co.in

CONTENTS

ABSTRACT

Biotic stresses are the main constraints in wheat production worldwide. Realizing the importance of crop health monitoring, regular wheat disease surveys were initiated during 1967 under All India Co-ordinated Wheat Improvement Programme (AICWIP). In India, the concerted efforts of wheat surveillance have contributed significantly in keeping vigil on new pathogens and devising strategy for developing new resistant varieties ahead the pathogen can cause loss to the crop. A systematic plant protection program emphasizes the need for identifying the potentially important pests. Crop health monitoring and surveillance activities help in identifying these crop health problems early enough, so that timely management measures can be undertaken to keep the pests at the lowest possible level, or to have information that the pests, feared by the farmer, may not warrant intensive crop protection measures. These surveys were carried out through mobile units and trap nurseries which generated considerable information on the appearance of rusts in different parts of the country. With the advancement in technology, the machine driven intelligent system has come to existence to facilitate remote agriculture monitoring in developed nations. Earlier, emphasis the working of earlier robot was based upon radio frequency (RF), which faced the problem of range. This problem is now overcome by intelligent robot which employs global system for mobile communication (GSM) technique. The robot captures the visual image of plant leaves, compares its color and height with reference value set in the system, and correspondingly indicates the plant health. In future, remote based technique either through remote sensing or through intelligent robot can play important role in surveillance of wheat diseases in India. This will help in getting information in real time and necessary management options will be followed well in time to avoid any major loss to food grain production in the country if any new pest/pathogen appears in severe form due to changing climatic scenario.

19.1   INTRODUCTION

Pest surveillance is the systematic monitoring of biotic and abiotic factors of the crop ecosystem in order to predict the pest outbreak. By the surveillance programs, the pest population dynamics under field conditions can be known which in turn helps in devising the appropriate management strategies. Disease surveying is basic to all effective disease control and research programs. These surveys are equally important throughout the research work as a means of assessing the effectiveness of control measures. Surveys are required to maintain claims of "pest-free" status of an area, to detect new populations of quarantine pests, and to delimit populations of quarantine pests. Pest surveys are also an integral part of control and eradication programs. Surveys are helpful to know how a pest/disease is multiplying in an area and when it is expected. Pest/disease control measures can be initiated in time due to advance forecasting.

The main objectives for surveying pests are listed below.

1. To develop a list of pests or hosts present in an area.
2. To demonstrate a pest-free area (the absence of a particular pest in an area) or places of low pest prevalence for trade purposes.
3. To develop a baseline list of pests before ongoing monitoring for changes in pest status for pest management and control.
4. For early detection of exotic pathotypes.
5. For early detection of established organisms becoming pests.
6. To delimit the full extent of a pest following an incursion.
7. To monitor progress in a pest eradication campaign.
    For achieving these objectives, the following courses of action in envisaged:

1. Mobile or roving surveys
2. Trap plot nurseries

Crop protection program at ICAR – Indian Institute of Wheat and Barley Research (IIWBR), Karnal usually has two major objectives, viz., (a) maximum or optimum income per unit land or input and (b) safe supply of food and livelihood for a family from harvest to harvest. The latter is the dominant objective under subsistence farming in developing countries. There is dire need to keep vigil on the occurrence or appearance of pests so that the losses could be avoided. A systematic plant protection program emphasizes the need for identifying the potentially important pests. Crop health monitoring and surveillance activities help in identifying these crop health problems early enough, so that timely management measures can be undertaken to keep the pests at the lowest possible level, or to have information that the pests, feared by the farmer, may not warrant intensive crop protection measures.

Realizing the importance of crop health monitoring, regular wheat disease surveys were initiated during 1967 under "All India Co-ordinated Wheat Improvement Programme (AICWIP)." These surveys were carried out through mobile units and trap nurseries which generated considerable information on the appearance of rusts in different parts of the country. This also helped in epidemiological studies. These surveys further confirmed the findings of Dr. K. C. Mehta on the movement and occurrence of rusts in the country. The information generated was published in the form of bulletins that were issued by the Division of Mycology and Plant Pathology, Indian Agricultural Research Institute (IARI), New Delhi from time to time. The survey and surveillance program was strengthened during 1995 through an AP-cess fund project on survey and surveillance for pests and diseases with IIWBR as the nodal center and four other zonal centers viz., Ludhiana (North Western plain zone (NWPZ)), Kanpur (North Eastern plains zone (NEPZ)), Powarkheda (central zone (CZ)), and Pune (peninsular zone (PZ)). Extensive surveys were conducted and pest profile was prepared. During 1995, a "Wheat Crop Health Newsletter" was published from Directorate of Wheat Research (DWR), Karnal on monthly basis during the crop season and is continuing. The latest information on crop health status is disseminated quickly through this newsletter. All issues of it are available on DWR website since the year 2000 (Sharma et al. 2000-12).

Crop health surveillance, apart from providing a decision-making tool for taking up the plant protection measures, also provides information on the occurrence and spread of the diseases and other pests, already occurring in an area or the ones newly introduced to the area due to the change in cropping sequences and cultural practices, introduction of newer and resistant genotypes, and similar factors. Monitoring of diseases of wheat crop was systematically conducted under the AICWIP during late sixties to early eighties in the post-green revolution era, when new high yielding varieties were introduced in the country. This helped in enlisting and prioritizing the health problems of wheat for developing relevant crop protection strategy. Now, under the changing cropping sequences, like introduction of rice–wheat rotation, more area coming under irrigation, and the increase in cropping intensity, etc., there is every likelihood of a changed disease scenario too. In other words, the pests of less importance during yester years may become the pests of major concern tomorrow.

Crop health monitoring and surveillance is being done with following objectives.

1. To monitor the occurrence and spread of the diseases/pests.
2. To monitor the inoculum in the environment for purpose of prediction or forewarning.

3. To monitor or keep vigil on the entry of new pests or their races/ pathotypes, etc.

   For achieving these goals or objectives, the following course of action in envisaged:

   1. Trap plot nurseries
   2. Mobile or roving surveys

Rapid progress has been made toward the goal of establishing a global cereal rust monitoring system (Hodson, 2009). The system has reached the point where it can now be regarded as a fully operational global disease monitoring system. New technologies are playing an increasingly important role in rust tracking. Through Indian Council of Agricultural Research (ICAR)–Borlaug Global Rust Initiative (BGRI) collaboration, a robust and functional data management system—the wheat rust toolbox—is now in place in India. This includes extensive rust surveillance and race databases. New web resources are providing access to a wealth of information regarding rust surveillance and monitoring in ways not previously possible.

19.2 WHEAT AND BIOTIC STRESSES

Overall India is the second largest producer of wheat after China. Wheat is the next most important food crop in the country following rice, both in area, production, and consumption. Wheat growing regions in India represent the diverse agro-climatic conditions. The country has witnessed an increase in wheat area from 10 mha in early sixties to around 30.37 mha in 2014–2015. The major wheat growing states of India are Punjab, Haryana, Uttar Pradesh, Madhya Pradesh, Rajasthan, and Bihar. These states account for about 90% of India's total wheat produce. The success story of enhanced wheat production during last 40–45 years by keeping the rusts at its lowest ebb demonstrates the strength of not only the consolidated efforts of the technology developed by the scientists across the country but the government's policies for making them adaptable by the farmers through different countrywide programs. The era of cultivation of semi-dwarf wheat which began in 1960s ushered a phenomenal success as these genotypes along with spurt in inputs revolutionized the wheat production in the following years, leading to "Green Revolution." The wheat production in India, ever since, has increased many folds from 6.4 mt in 1950 to 95.91 mt during 2013–2014 with productivity to the tune of 3.07 t/ha. Sustaining this level of productivity is a big challenge and efforts are on to break the yield barriers.

A host of biotic stresses affects wheat crop leading to huge losses in yield. Since wheat is grown in different agro-climatic conditions in our country, the constraints to its production vary from one zone to other. Biotic stresses are the main constraints in wheat production worldwide. The black or stem rust (*Puccinia graminis* Pers. f. sp. *tritici* Erikss. & Henn) is important in warmer areas whereas the brown or leaf rust (*Puccinia triticina* Eriks.) in the entire country and the yellow or stripe rust (*Puccinia striiformis* Westend.) in cooler areas.

19.3 TRACKING RUSTS

As a national strategy under the overall umbrella of ICAR, IIWBR is coordinating All India Wheat and Barley Improvement Project through which rust pathotypes are being monitored, strategically resistance-genes incorporated with major emphasis on adult plant resistance (APR) and slow rusting. The importance of crop health monitoring in Indian wheat program was realized as early as 1920s and random rust surveys were conducted by Dr. K.C. Mehta and colleagues. From time to time there was integration of mobile surveys, trap nurseries, use of satellite for disease survey and evaluation of spore dissemination, infrared image (s), analysis of rain samples for uredospores to monitor the disease situation. In addition, mobile surveys are undertaken at various crop growth stages, along the specific routes. These roving or mobile surveys also help in keeping a vigil on the entry of a new pest or pathotype. A very good example is the detection of a new virulence (*Yr*9 virulence) of stripe rust in 1996 from the bordering areas of Punjab and identification of stripe rust resistant germplasm within two months of its confirmation. Subsequent surveys carried out through mobile units and trap nurseries generated considerable information on the appearance and spread of rusts in different parts of the country became an integral component of wheat improvement. During 1967–1968 under the AICWIP monitoring of diseases of wheat crop was systematically introduced in all wheat growing states to know the prevalence, spread of wheat diseases, and performance of cultivated wheat varieties to rust. Co-operating scientists in different states monitor the wheat crop at 15 days' interval starting from either December end or after the crop is more than 45 days of maturity.

Each state has different agro-climatic regions and the extensive and intensive monitoring is done in disease prone areas, however, other areas also monitored on alternate trips. A team of 2–3 scientists at fortnightly intervals moves by car in different directions in a state and halt at a distance of 30 to 50 km to monitor 2–4 wheat fields on both sides of the road. Within wheat field randomly move around to look for diseases, pests, weeds, and general crop health. For rust prevalence, intensity and spread within and surrounding fields is monitored. In case, there are 2–3 patches of the disease in a range of 100–200 m having traces to 40S it is considered of limited occurrence but if there are more than five such patches having infection varying from traces to 80S it is considered spreading and then the adjoining fields are monitored intensely more so if the field is grown with a susceptible variety. Sometimes the spread could be visualized just standing outside the fields, such fields are considered to be severely infested where farmers are immediately apprised of the fact that if the crop is not sprayed losses could occur. The losses may vary from < 1 to 100% depending upon the varietal susceptibility, stage at which infection occurs, and the severity of infection on flag leaf up to grain formation stages. Many a times under severe situations of the rust infection the sporulation occurs on glumes and pericarp of the grain. Under any situation of rust prevalence from mild to severe form scientists of agricultural universities, Krishi Vigyan Kendra (KVK) staff, and officers of department of agriculture are contacted for awareness campaigns to apprise the farmers of rust detection and spraying the crop with fungicide (s) and making it available timely through government agencies (Sharma et al., 2013).

Besides mobile surveys another strategy had been planting of Wheat Disease Trap Plot Nursery (TPN). To specify the exact purpose, the nursery was later on designated as Wheat Disease Monitoring Nursery (WDMN). The nursery has become an integral part of the crop protection program and has assumed so much significance that it is regularly planted at various strategic locations and constant watch is kept on the appearance and spread of wheat diseases both in regular as well as off season crop. Such nurseries usually contain varieties or entries with known genetic constitution so that the occurrence or appearance of a new pathotype or pest could be identified. Under the AICWIP, the WDMN/TPN is planted at multi-locations including those all along the Western border. At present, it is planted at more than 60 locations. Looking into the spurt of stripe rust in the last five years more locations have been added during 2013–2014, where stripe rust is noticed early and appears regularly, every year. It contains a common set comprising of 15 lines having popular varieties, resistant varieties, a susceptible variety, and some resistant lines. In addition, five predominantly cultivated zone specific varieties are also part of this 20 line nursery. It is serving as an important tool to know the wheat disease situation, progress, and appearance of new variants or races on resistant materials. It also helps to keep a vigil on the occurrence of wheat diseases along the Western border, especially for the stripe rust. Every year, the crop health is monitored and the field samples of the rusts are analyzed for their virulence (pathotype analysis). Distribution pattern of the rust virulences (pathotype) provides much-needed orientation for the wheat breeding programs and executing the resistance-gene/varietal deployment.

19.3.1   COMMON SET OF VARIETIES OF WDMN

WL 711, HD 2329, Agra Local, HD 2160, Lal Bahadur, WL 1562, HW 2021(*Sr26/Sr24*), HD 2204, C 306, WH 147, HW 2008 (*Sr24/Lr24*), Kharchia mutant, HP 1633, DL 784-3, and RNB 1001.

19.3.2   ZONE-SPECIFIC VARIETIES

NWPZ: WH 1105, WH 542, PBW 343, DPW 621-50, and WH 896

NEPZ: K 8804, HD 2402, HP 1102, HUW 468, and NW 1014

CZ: HI 8381, DL 803-3, Lok -1, GW 273, and GW 322

PZ and SHZ: MACS 2496, Bijaga Yellow, HW 971, HD 2501, and HW 2022 (*Sr24/Lr24*)

North hills zone (NHZ) and high altitude zone: HPW 349, VL892, HS 420, Sonalika, and HS 507

On the same pattern as that of WDMN/TPN, a regional WDMN, called as "SAARC Nursery" is planted and conducted in Afghanistan, Bangladesh, India, Nepal, and Pakistan in collaboration with CIMMYT, Kathmandu. This nursery was started in 1989–1990 and first planted in 1990. It is composed of 20 varieties viz. Annapurna 1, WL 1562, HD 2204, PBW 343, HD 2687, HD 2189, HP 1633, RAJ 3765, PBW 373, Pak 81, Punjab 65, Chakwal 86, Faisalabad 85, Inquilab 91, Faisalabad 83, Rawal 87, Kohsar, Bakhtawar 94, Gourab, and susceptible check drawn from the SAARC member countries. This nursery is also constituted at ICAR-IIWBR Regional Station, Shimla. The activity is also coordinated since 2012–2013 crop season at CIMMYT, Nepal. The disease appearance in the nursery provides an evidence of comparative wheat disease situation and virulences prevalent in the SAARC countries. The nursery is monitored by the competent co-operators at the respective locations. These observations help a breeder to re-cast the breeding strategy toward breeding for resistance to the new or likely to emerge virulences before these (pathotypes or virulences) become a real threat to wheat production in SAARC region.

19.4 OFF-SEASON SURVEYS

Frequent surveys are being taken during the off season in the high hills of Himachal Pradesh (Lahaul, Spiti, and Kinnaur) and Jammu and Kashmir (Ladakh). Major focus is on the occurrence of yellow rust and surveillance for the stem rust pathotype, Ug99. Survey is also being carried out in hills of Nepal having off season wheat, selfsown wheat, and alternate/collateral hosts to understand the epidemiology of wheat rusts. Extensive surveys are conducted in the country to monitor the occurrence of stem rust pathotype, Ug99. Till today, there has not been any report from anywhere in the country (Saharan et al. 2013-15; Singh et. al. 2016).

19.5 WHEAT CROP HEALTH NEWSLETTER

Wheat Crop Health Newsletter was started during 1995–1996 for faster dissemination of crop health information. This newsletter is being issued every month in every crop season regularly. Newsletter is also made available on DWR web site (www. dwr.res.in).

19.6 WHEAT RUST EPIDEMICS AND ECONOMIC LOSSES IN INDIA

In India, epidemics due to stripe rust have occurred in one or the other region of North hills and North Western plains about once in every 10 years. Complete losses in wheat yields may occur in fields of highly susceptible cultivars and the pathogen has a narrow spectrum of virulence in these zones on wheat (Saari & Prescott, 1985). It assumes great economic significance, especially in the parts of areas with cool and wet environmental conditions spread over 10 mha. It has been stated that the rust caused huge losses during 1904–1905, in parts of Punjab due to damp weather and lack of sunshine. Between 1967 and 1974, brown and yellow rusts occurred every year, but only twice caused accountable yield losses due to the isolated epidemics which occurred in some pockets of Western Uttar Pradesh during 1971–1972.

A pandemic of the same two rusts occurred during 1972–1973 in Western Uttar Pradesh, Haryana, and Punjab (Nagarajan & Joshi, 1975).

Cultivation of resistant varieties is the most effective, eco-friendly, and economically viable method of managing wheat stripe rust. However, rust resistant cultivars of wheat turn as a result of evolution of new variations in stripe rust pathogen. A rust resistant cultivar generally becomes susceptible in about five years (Sawhney, 1995). Chemicals are generally avoided and deployed under emergency situations mainly to curtail the initial inoculums of rusts. Varietal development with use of diverse resistance to diseases and strategic deployment of resistant varieties is practiced in India. But rust virulences also keep pace with this change and the varieties which are grown over large areas gradually succumb to rusts with the appearance of new virulence. In spite of climate change as reflected by fluctuations in weather conditions and occurrence of stripe rust in high proportions in parts of India, an increasing trend in wheat production was recorded for six years in a row from 2006 to 2015 with some decline during 2012–2013 and 2014–2015 due to higher range in minimum temperature and heavy rainfall in a very short spell of time resulting in water logging.

19.7   STRIPE RUST STATUS DURING 2000–2015

Crop health of wheat is monitored thoroughly during the main crop season as well during the off season. Major focus was on the occurrence of stripe rust as it was noticed in high proportions in some of the areas of Northern hills and plains zones since 2006. The extensive surveys were conducted by the scientists of wheat crop protection programme at different cooperating centers including IIWBR, Karnal. Special teams of scientists were constituted every year during the annual meet of "All India Wheat Workers." Information on wheat crop health was disseminated through the "Wheat Crop Health Newsletter," which was issued on monthly basis during the crop season. The newsletter was also put on IIWBR website (http://www. dwr.res.in). Advisories for deployment of right varieties and use of protection technologies were issued to the farmers for stripe rust management, as and when required.

In 2001, a new pathotype 78S84 of *P. striiformis* with virulence on widely grown wheat variety, PBW 343 (*Yr* 3, *Yr* 9, *Yr* 27 virulences) was detected. Its inoculum got built up steadily due to the cultivation of PBW 343 over a large area (Prashar et al., 2007). Due to high yields, PBW 343 became popular amongst farmers in the hilly regions and Eastern India as well besides main wheat growing zone of North Western plains. It once occupied as high as 9-mha area in India. The stripe rust pathogen survives in Northern India and due to continuous growing of susceptible varieties in the major wheat belt of NWPZ, the inoculum built up also increased in the hilly areas. Since 2006–2007, the stripe rust is occurring in high intensity in one or the other parts of NHZ and NWPZ. Emphasis was laid for growing rust resistant wheat varieties such as PBW 550, DPW 621-50, WH 1021, HD 3043, and WH 542 of bread wheat in disease prone areas of NWPZ whereas HS 375, HS 490, HS 507, VL 892, and VL 907 in NHZ. Due to congenial weather for stripe rust, two pathotypes, 78S84 (Yr27 virulence on PBW 343) and 46S119 (Yr9 virulence), were most prevalent during 2010–2011 crop season and most of the varieties grown in North Western plain zone became susceptible to it. During 2010–2011, due to congenial weather, stripe rust appeared in severe form in the plains of Jammu and Kashmir, foot hills of Punjab and Himachal Pradesh, parts of Haryana, and *tarai* region of Uttarakhand states wherever susceptible varieties were grown. In Punjab, especially in the districts of Ropar, Nawan Shahar, and Hoshiarpur, the disease was well spread over a large area on most of the varieties being grown by the farmers. In Haryana, the disease was severe in Yamuna belt. During 2011–2012, stripe rust incidence was less as compared to 2010–2011. But timely, coordinated efforts by IIWBR (ICAR), State Agricultural Universities (SAUs), Department of Agricultural and Cooperation (DAC), and state agriculture departments helped in averting the epidemics and India had record harvests of wheat.

19.8 STRATEGIC PLANNING

There had been recurrence of stripe rust in some of the years and the situation demanded planning not only at the scientific level but also at the involved policy issues which could only be carried out through the department of agriculture and cooperation. During 2010–2011, strategic plan was envisaged to limit its occurrence and spread of stripe rust in India. Major emphasis was on bringing in farmers' awareness in replacing susceptible varieties with resistant ones, early detection of the rust/initial foci of infection by regular monitoring after 40 days of planting and immediately spraying wheat with recommended fungicides to limit its spread. Special cards were devised for raising the awareness in farmers and circulated in large numbers in all the affected areas. In disease prone areas, trap nursery was planted for early detection of stripe rust. The advanced variety trials were also planted in affected areas to identify resistant varieties. Varietal development and deployment of

new varieties is a regular feature in the country which was further strengthened and led to create genetic diversity at farmers' field through release and
popularization of several stripe rust resistant varieties viz., HD 2967, WH
1105, HD 3086, DBW 88, HD 3059, WH 1021, WH 1080, HD 3043, DBW
71, DBW 90, HS 507, HPW 349, and HS 542.

19.9   POST-HARVEST SURVEYS

The post-harvest grain sampling and analysis for various seed-borne diseases
have been regular activities. Analysis of grain samples was done for monitoring the status of Karnal bunt (KB) disease and to identify the disease
free or low disease regions. Other seed-borne maladies like black point and
discoloration (which may be confused by traders with KB), and ear cockle
nematode were also monitored. Based on these, the "low risk zone," "high
risk zone," and "no risk zone" have been identified for KB. The states of
Maharashtra, Karnataka, Gujarat, and parts of Madhya Pradesh have been
declared as KB free. Parts of Northern Madhya Pradesh and Rajasthan have
been in the low risk zone, while the NWPZ comprising of Punjab, Haryana,
Western Uttar Pradesh, foot hills of Uttarakhand, and Himachal Pradesh has
been identified as endemic or "high risk zone" for KB. There is scope to
delimit the low KB areas even in the endemic zone itself. Some areas in
Punjab and Haryana have been identified as low KB areas. These studies
have great importance in view of the global trade.

19.10   CONCLUDING REMARKS

In India, the concerted efforts of wheat surveillance have contributed significantly in keeping vigil on new pathogens and devising strategy for developing new resistant varieties ahead the pathogen can cause loss to the crop.
But now with the advancement in technology, agriculture system is not
confined to traditional human centric approach. The machine driven intelligent system has come to existence to facilitate remote agriculture monitoring. Earlier, emphasis the working of earlier robot was based upon radio
frequency (RF), which faced the problem of range. This problem is now
overcome by intelligent robot which employs global system for mobile
communication (GSM) technique. The robot captures the visual image of
plant leaves, compares its color and height with reference value set in the
system, and correspondingly indicates the plant health. With the use of

GSM, the robot can be operated even from a remote area. In future, remote based technique either through remote sensing or through intelligent robot can play important role in surveillance of wheat diseases. This will help in getting information in real time and necessary management options will be followed well in time.

KEYWORDS

- **survey and surveillance**
- **biotic stresses**
- **wheat**
- **disease trap nursery**
- **rusts**
- **India**
- **SAARC**

REFERENCES

Hodson, D. P.; Cressman, K.; Nazari, K.; Park, R. F.; Yahyaoui, A. In *The Global Cereal Rust Monitoring System,* Proceedings of Oral Papers and Posters, Technical Workshop, BGRI, Cd. Obregón, Mexico, March 17–20, 2009; McIntosh, R., Ed.; pp 35–46. http://www.globalrust.org/db/attachments/resources/737/10/BGRI_2009_proceedings_cimmyt_isbn.pdf (Last accessed on 21 December 2015).

Nagarajan, S.; Joshi, L. M. An Historical Account of Wheat Rust Epidemic in India and their Significance. *Cereal Rust Bull.* **1975,** *3,* 29–33.

Prashar, M.; Bhardwaj, S. C.; Jain, S. K.; Datta, D. Pathotypic Evolution in *Puccinia striiformis* in India during 1995–2004. *Aust. J. Agric. Res.* **2007,** *58,* 602–604.

Saari, E. E.; Prescott, J. M. World Distribution in Relation to Economic Losses. In *The Cereal Rusts;* Roelfs, A. P., Bushnell, W. R., Eds.; Academic Press: New York, 1985; Vol. 2, pp 257–292.

Saharan, M. S., et al. *AICW&BIP Crop Protection Reports*; IIWBR/DWR, 2013–2015.

Singh, D.P., et al. AICW&BIP crop protection Report, IIWBR, 2016, p 221.

Sawhney, R. N. Genetics of Wheat–Rust Interaction. In *Plant Breeding Reviews;* 1995; Vol. 13, pp 293–343.

Sharma, A. K., et al. *AICW&BIP Crop Protection Reports*; IIWBR/DWR, 2000–2012.

Sharma, I.; Saharan, M. S.; Bhardwaj, S. C. *Stripe Rust Status and Management in India;* Technical Bulletin No. 14; 2013, p 20.

CHAPTER 20

EVOLUTION OF WHEAT RUST PATHOGENS IN THE INDIAN SUBCONTINENT

SUBHASH C. BHARDWAJ[1]*, SUBODH KUMAR[1], T. R. SHARMA[2], OM. P. GANGWAR[1], PRAMOD PRASAD[1], HANIF KHAN[1], and SIDDANNA SAVADI[1]

[1]ICAR-Indian Institute of Wheat and Barley Research, Regional Station, Flowerdale, Shimla 171002, Himachal Pradesh, India

[2]ICAR-NRCPB, Lal Bahadur Shashtri Building, Pusa, New Delhi 110012, India

*Corresponding author. E-mail: scbfdl@hotmail.com

CONTENTS

ABSTRACT

Wheat rusts are dynamic pathogens. There is a tug of war between the rusts and wheat scientists. Rust resistant wheat varieties become susceptible due to shift in virulence which is influenced by the wheat cultivars in the fields. New virulences emerge which render rust resistant wheat varieties susceptible to a within 4–5 years of their cultivation. Over the years more virulent forms go on appearing, however, only fittest one rules the flora for a long period. New virulences arise due to sexual recombinants on alternate hosts. However, in countries where alternate hosts are not functional, new pathotypes arise mostly through mutation and somatic hybridization in few cases. These mutations occur as a natural phenomenon; however, resistant varieties in the field facilitate their selection. Few rust resistance genes are more prone to the mutations than the others. Mutations are generally forward (gain in virulence), however, reverse (loss of virulence) mutations have been observed in few cases. Loss of virulence is specific to few rust resistance genes only. Our experience shows that wheat rusts have even overcome the transgenes and other alien introgressions undertaken for rust resistance in wheat. It means we cannot eradicate wheat rusts but better we learn to live with these pathogens. Intelligent management is to be practiced. A vigil has to be kept for the occurrence of new virulences. More resistance sources, preferably diverse and varieties with more than two effective rust resistance genes will have to be given preference. Taking into consideration the racial patterns, skilful deployment of varieties can help in keeping wheat rusts below the threshold levels.

20.1   INTRODUCTION

Rusts are devastating pathogens of wheat worldwide. All three rusts occur on the wheat crop grown in India. While brown (leaf) rust is the most widespread, stem rust has assumed significance in view of emergence of virulence on *Sr31*(Ug99) which has rendered 40% of wheat material susceptible throughout the world (Pretorius et al., 2000). Stripe rust is a major constraint in Northern India. Yield losses due to brown rust under favorable conditions could be up to 40% whereas those by black and yellow rusts could be as high as 100% (Anonymous, 1992). In India, alternate hosts do not appear to play any role in the recurrence of wheat rusts (Mehta, 1940, 1952). High summer temperatures during teleutospore formation followed by the rainy season, availability of *Berberis* species only in the hills, non-synchronization of young barberry leaves with the basidiospores coupled with inoculation studies led to the conclusion that *Berberis* species are non-functional as alternate hosts in India (Mehta, 1940, 1952). Consequent upon the confirmation of *Berberis* species as alternate host for stripe rust (Jin et al., 2010), further exhaustive studies on the role of *Berberis* under Indian conditions were conducted using the recently available information. However, aecial infections from *Berberis* did not cause any infection on wheat. Therefore, it can be said fairly that *Berberis* does not act as alternate host in this part of the world. Under natural conditions there is a continuity of wheat plants between the summer crops, supplemented with self-sown wheat to the regular crop which causes the recurrence of wheat rusts year after year (Mehta, 1940, 1952; Nagarajan & Joshi, 1985). However, keeping into account the different racial patterns in source areas and the various target areas, it can be said fairly that there is more than we know about the recurrence of wheat rusts in India. In this context, role of collateral hosts is also under the scanner and is being worked out. It has been envisaged that in absence of functional alternate hosts, new pathotypes originate mainly through mutation (Bhardwaj, 2012).

Variability in *Puccinia striiformis* (stripe rust of wheat) is minimal in India. This might be due to its occurrence in a limited area ranging from hills to foothills of Himalayan region. But recent studies in Pakistan indicate high diversity in hills in South Asia. Though *Puccinia graminis* Pers. f. sp. *tritici* (black rust of wheat) is quite variable, however, *Puccinia triticina* Eriks. (brown rust of wheat) is the most variable and 23 new pathotypes have been identified during the last 21 years (Bhardwaj, 2012). Many effective *Lr* genes *viz. Lr9, Lr19,* and *Lr28* have become susceptible during the last 14 years (Bhardwaj, 2011). Presently three pathotype groups (12, 77, and 104) of *P. triticina* (Bhardwaj et al., 2010a), two (40, 117) of *P. graminis tritici* (Bhardwaj et al., 2006b), and two lineages of *P. striiformis* are predominant in India.

In Australia, new pathotypes of *P. graminis tritici* were derived primarily from existing pathotypes by single (or rarely double) step mutation (Luig, 1977). Likewise seven variants of pathotype 104-2, 3, (6), (7), 11 that differed by single virulences were detected during 1984–1992. Pathotype 104-2, 3, (6), (7), 11 and its derivative with virulence to *Lr20* increased rapidly in frequency and later spread to New Zealand as exotic incursion (Park et al., 1995). The studies reported here aim at finding out the possible evolutionary route of leaf, stem rust, and stripe pathogens of wheat over the years in India.

20.2 SURVEILLANCE FOR WHEAT RUSTS

Leaf and stem samples infected with leaf, stripe, and stem rusts were collected from wheat growing areas of India and Nepal. Systemic and random surveys were conducted in regular as well as summer crops in Tamil Nadu and Himachal Pradesh.

20.3 VIRULENCE STRUCTURE/PATHOTYPE DESIGNATION

The rust infected samples were inoculated with the help of a lancet needle on one-week old susceptible wheat variety A-9-30-1 (for stripe rust) or Agra Local (for leaf and stem rusts). The uredospores produced in sufficient quantity in 15 days on these inoculated plants were used to inoculate the revised Indian 0, A, and B sets of differential hosts (Table 20.1) and near-isogenic lines (NILs). The inoculated plants were sprayed with a fine mist of water and placed overnight in dew chambers at $16 \pm 2°C$ (stripe rust), $22 \pm 2°C$ (leaf and stem rusts), 100% relative humidity, and 12 h daylight. The plants were then transferred on to the greenhouse benches and kept at $16 \pm 2°C$ (stripe rust) and $22 \pm 2°C$ (leaf and stem rusts) in relative humidity of 40–60%, and illuminated at about 15,000 lux for 12 h (Nayar et al., 1997). Infection types (resistant or susceptible) on the test lines were recorded 14 days after inoculation following modified method of Stakman et al. (1962). Infection types were characterized as 0 = no visible infection, 0; = small hypersensitive flecks, 1 = uredia minute, surrounded by necrotic areas, 2 = small to medium uredia surrounded by chlorotic area, 3 = uredia small to medium in size and chlorotic areas may be present, and 3+ = uredia large with or without chlorosis, sporulating profusely. Infection type 3, 3+ is classified when both 3 and 3+ pustules occur together. To find out the similarities and differences among the pathotypes of *Puccinia* species on wheat in India, all the pathotypes of a group were inoculated simultaneously under optimum conditions on the sets of differentials (Table 20.1), and known *Lr, Sr,* and *Yr* genes under same set of conditions. Based on avirulence/virulence structure of the known pathotypes and chronology, possible routes of evolution (tree) were constituted. Pathotypes were designated on the basis of the binomial nomenclature as given by Nagarajan et al. (1983, 1984) and Bahadur et al. (1985) and designated as standard pathotypes. To facilitate international communication, leaf rust pathotypes were also designated as per Kolmer et al. (2007) and stem rust pathotypes were named as by Jin et al. (2008).

TABLE 20.1 The Revised Composition of Sets of Differentials for the Identification of Pathotypes of *Puccinia* Species on Wheat in India.

Set-0	Set-A	Set-B
Leaf rust (*Puccinia triticina*)		
IWP 94 (*Lr23+*)	*Lr14a*	Loros (*Lr2c*)
Kharchia Mutant	*Lr24*	Webster (*Lr2a*)
Raj 3765	*Lr18*	Democrat (*Lr3*)
PBW 343	*Lr13*	Thew (*Lr20*)
UP 2338	*Lr17*	Malakoff (*Lr1*)
K 8804	*Lr15*	Benno (*Lr26*)
Raj 1555	*Lr10*	HP1633 (*Lr9+*)
HD 2189	*Lr19*	
Agra Local	*Lr28*	
Stem rust (*Puccinia graminis tritici*)		
Sr24	*Sr13*	Marquis (*Sr7b+*)
NI 5439	*Sr9b*	Einkorn (*Sr21+*)
Sr25	*Sr11*	Kota (*Sr28+*)
DWR 195	*Sr28*	Reliance (*Sr5+*)
HD 2189	*Sr8b*	Charter (*Sr11+*)
Lok 1	*Sr9e*	Khapli (*Sr7a+*)
HI 1077	*Sr30*	Tc*6/*Lr26* (*Yr9*)
Barley Local	*Sr37*	
Agra Local		
Stripe rust (*Puccinia striiformis*)		
WH 147	Chinese 166 (*Yr1*)	Hybrid *46* (*Yr4*
Barley Local	Lee (*Yr7*)	Heines VII (*Yr2+*
WH 416	Heines Kolben(*Yr6*)	Compair (*Yr8*)
PDW 215	Vilmorin 23(*Yr3*)	*T. spelta album* (*Yr5*)
HD 2329	Moro (*Yr10*)	Tc*6/*Lr26* (*Yr9*)
HD 2667	Strubes Dickkopf	Sonalika (*Yr2+)*
PBW 343	Suwon92XOmar	Kalyansona *Yr2*(KS)
HS 240	Riebesel47/51(*Yr9+*)	
Anza		
A–9-30-1		

20.4 PHYLOGENETIC STUDIES

There had been frequent changes in races of *P. triticina, P. graminis tritici* in the Indian subcontinent. Three groups of pathotypes namely 12, 77, and 104 were common in *P. triticina* whereas two groups, that is, 40 and 117 of *P. graminis tritici* were predominant. In *P. striiformis* two lineages have been observed. International designations of these pathotype groups are given in Table 20.2. The evolution of variants had been frequent in these groups. The sequential evolution of these groups of pathotypes is presented in the following sections.

20.5 PUCCINIA TRITICINA

About 9600 samples of *P. triticina* were analyzed during 2002–2015. Overall, 99% of the population was observed belonging to 12, 162, 77, and 104 groups. Among these four pathotype groups, 77 group was predominant and constituted more than 54% of the samples analyzed. Thirty seven percent of the samples were in the 104 group whereas around 7% of the population was either 12 or 162 groups.

In India, three predominant pathotypes of *P. triticina* contributed for more than 75% of the samples analyzed during 2002–2011. Among the three predominant pathotypes of *P. triticina* in India, pathotype 77-5 was present in 39% of the samples hence considered as most frequent followed by pathotypes 104-2 (26%) and 104-3(10.3%).

20.5.1 RACE GROUP 77

Race 77 was identified in 1954 (Vasudeva et al., 1955) and it has been the fastest evolving race and has evolved 13 biotypes during the last 61 years. The specific features of race 77 are that it is virulent on eight resistance genes *viz., Lr1, Lr2a, Lr2c, Lr3, Lr14a, Lr15, Lr18,* and *Lr20.* Mostly the new pathotypes have emerged with gain in virulence, appearing to a mutation for one resistance gene. In general there has been gain in virulence; however, there are instances of reverse or backward mutation resulting in loss of virulence. In 1974 it gained virulence on resistance gene *Lr10* which resulted in pathotype 77A (Nayar et al., 1975) and loss of virulence on *Lr20* led to 77A-1 (Nayar et al., 1980). Subsequently acquisition of virulence in pathotypes 77A on *Lr26, Lr23,* and *Lr19* resulted in

TABLE 20.2 North American Equivalents of Predominant Pathotype Groups of *P. triticina* and *P. graminis* f. sp. *tritici* of Indian Subcontinent.

S. no.	Designations of *P. triticina* pathotypes		Designations of *P. graminis* f. sp. *tritici* pathotypes	
	Indian binomial (vernacular)	N. American[1]	Indian binomial (vernacular)	N. American[2]
1	5R5 (12)	FGTTL	104G13(40)	PHPSC
2	5R37 (12-1)	FHPNM	62G29(40A)	PTHSC
3	1R5 (12-2)	FGTTL	62G29-1(40-1)	PTHSM
4	49R37 (12-3)	FHTRL	58G13-1(40-2)	PKRSC
5	69R13 (12-4)	FGTRM	127R29(40-3)	PTTSF
6	29R45 (12-5)	FHTPM	37G3(117)	KRCSC
7	5R45 (12-6)	FHRPM	36G2(117A)	JRCSC
8	93R45 (12-7)	FHTTL	38G18(117A-1)	JRHSC
9	49R45 (12-8)	FHRSL	166G2(117-1)	JRHSC
10	93R37 (12-9)	FHTTM	33G2(117-2)	KHCSC
11	5R13 (12A)	FGTTL	167G3(117-3)	KRHSC
12	45R31 (77)	TGTPB	166G3(117-4)	KRHSC
13	109R63 (77-1)	THTTB	166G2-2(117-5)	JRHSC
14	109R31-1 (77-2)	TGTTL	37G19(117-6)	KRHSC
15	125R55 (77-3)	THTTB		
16	125R23-1 (77-4)	TGTTL		
17	121R63-1 (77-5)	THTTM		
18	121R55-1 (77-6)	THTTL		
19	121R127 (77-7)	TRTTL		
20	253R31 (77-8)	TGTTL		
21	121R60-1 (77-9)	MHTKL		
22	377R60-1 (77-10)	MHTKQ		
23	125R28 (77-11)	MGTTL		
24	109R31 (77A)	TGTTB		
25	109R23 (77A-1)	TGTTL		
26	17R23 (104)	PGTNL		
27	21R31-1 (104-1)	PGTSL		
28	21R55 (104-2)	PHTTL		
29	21R63 (104-3)	PHTTL		
30	93R57 (104-4)	NHKTL		
31	21R31 (104A)	MGTDC		
32	29R23 (104B)	MGTDL		

[1]Leaf rust races designated as described by Kolmer et al. (2007).

[2]Stem rust races designated as described by Jin et al. (2008).

pathotypes 77-1, 77-2 (Nayar et al., 1987), and 77-8 (Bhardwaj et al., 2005), respectively. Loss of virulence to *Lr20* in 77-1 and 77-2 resulted in the evolution of pathotype 77-3 (Nayar et al., 1991) and 77-4 (Prashar et al., 1991), respectively. Virulence of pathotype 77-2 on resistance gene *Lr26* led to the evolution of pathotype, 77-5 (Nayar et al., 1994), which further evolved with loss of virulence for *Lr20, Lr2* to 77-6 (Jain et al., 1999) and 77-9, respectively (Bhardwaj et al., 2010a, 2010b). Pathotype 77-5 is a predominant pathotype of *P. triticina* for the last 15 years and its mutant on *Lr9* is named as 77-7 (Nayar et al., 2003). Pathotype 77-9 subsequently gained virulence on *Lr28* and resulted in pathotype 77-10 (Bhardwaj et al., 2010b), whereas pathotype 77-11 (Bhardwaj et al., 2011b) is a result of loss of virulence on Thew (*Lr20*) in pathotype 77-9. Gain in virulence has been on *Lr9, Lr10, Lr23, Lr26,* and *Lr28* whereas loss in virulence was on *Lr2a,* and *Lr20.* Both *Lr23* and *Lr26* loci appear to be more vulnerable for the mutation by the pathogen.

20.5.2 RACE GROUP 12

Race 12 of *P. triticina* was identified in 1966 and 10 pathotypes have been identified during these 49 years (Ahmad et al., 1969). This group is characterized based on its avirulence on leaf rust resistance genes *Lr1, Lr2a.* Race 12 acquired virulence on *Lr26* and *Lr23* the mutant 12-1 and 12-2 (Nayar et al., 1991), respectively, got selected. Gain of virulence on *Lr20* in 12-2 resulted in 12A (Goel et al., 1979). Further gain of virulence on *Lr15* and *Lr20* in pathotype 12-1 led to the emergence of pathotypes 12-3 (Nayar et al., 1991) and 12-6 (Jain et al., 2008), respectively. Likewise a mutant of 12 got selected on *Lr10* and *Lr20* was designated as 12-4 (Prashar et al., 1991). Our studies have indicated that the pathotype 12-5 is a putative hybrid between 12-1 and 12A (Bhardwaj et al., 2006a). Pathotypes 12-5 and 12-3 also acquired virulence on *Lr10* and *Lr20* and named as 12-7 (Bhardwaj et al., 2006a) and 12-8 (Bhardwaj et al., 2009), respectively. Similarly, the pathotype 12-7 lost virulence on *Lr20* and resulted in a new pathotype 12-9 (Bhardwaj et al., 2011a, 2011b). In general, it can be said that in most of the cases, there was a gain in virulence; however, loss of virulence was again observed on *Lr20* when pathotype 12-9 evolved. Again, *Lr26* and *Lr23* appear to be more prone for the mutations by the 12 group of the pathotypes.

20.5.3 RACE GROUP 104

Race 104 was identified in 1972 (Nayar et al., 1974) and has evolved into seven pathotypes during the past 43 years. It is the third most diverse group of pathotypes of *P. triticina* in India. Mutants of 104 got selected on *Lr20* and *Lr23* and evolved into new pathotypes 104A (Nayar et al., 1977) and 104B (Nayar et al., 1983), respectively. The pathotype 104A further gained virulence on *Lr23* and was designated as 104-1 (Nayar et al., 1988). Pathotype 104B gained virulence on *Lr26* to result in 104-2 (Nayar et al., 1994), which further gained virulence on *Lr20* and pathotype 104-3 (Nayar et al., 1998) got selected. Loss of virulence in 104-3 on *Lr3* resulted in a new pathotype 104-4 (Bhardwaj et al., 2011a). In this case also, most of the mutations were for *Lr23* and *Lr26*.

20.6 P. GRAMINIS TRITICI

The stem (black) rust pathogen *P. graminis tritici* is more prevalent in Peninsular and Central India where biotypes belonging to the pathotypes 40 and 117 are predominant. The evolution of virulences in these groups is as given below.

20.6.1 PATHOTYPE GROUP 117

This is the most variable group in *P. graminis tritici*. It was identified in 1945 from Madhya Pradesh (Prasada & Lele, 1952) and now evolved into seven biotypes during the past 70 years. Pathotype 117 lost virulence for *Sr7b* and *Sr13* and pathotype 117A was identified in 1960 (Patil et al., 1963). Pathotype 117A gained virulence to *Sr9b* and *Sr* Charter which resulted in 117A-1 (Sharma et al., 1979). Another offshoot 117-1 was the result of gain in virulence to *Sr37* in pathotype 117A (Bhardwaj et al., 1989). Pathotype 117-1 further gained virulence to *Sr7b*, NI5439 to become 117-4, 117-5, respectively (Bhardwaj et al., 1994). Pathotype 117-3 evolved as a result of gain in virulence on *Sr7b*, *Sr13* in pathotype 117-1 (Bhardwaj et al., 1994). Pathotype 117-2 evolved because 117 lost virulence to *Sr11* (Bhardwaj et al., 1990); however, when 117 gained virulence on *Sr* Charter, it was named as pathotype 117-6 (Bhardwaj et al., 1994). In most of the cases, there was gain in virulence except for the evolution of 117A, which was the result of loss of virulence on *Sr7b* and *Sr13*.

20.6.2   RACE GROUP 40

P. graminis tritici pathotype 40 was identified as early as 1932 from Poona (Mehta, 1940) whereas another independent pathotype 40A was identified in the Nilgiri hills in 1974 (Sharma et al., 1975). One-step gain in virulence in 40A on stem rust resistance gene *Sr24* resulted in 40-1 (Bhardwaj et al., 1990) whereas loss of virulence to *Sr11* and simultaneous gain in virulence to *Sr25* resulted in pathotype 40-2 (Jain et al., 2009). Pathotype 40-3 emerged as a result of 40-2 gaining virulence on the genes *Sr13* and *Sr30* (Jain et al., 2013).

20.7   PUCCINIA STRIIFORMIS

Evolution of variants in stipe (yellow) rust of wheat in India has followed two lineages.

20.7.1   YR2 (KALYANSONA) GROUP

In 1993, pathotype analysis of stripe rust samples from Leh Ladakh in Jammu and Kashmir started and three pathotypes CI, CII, and CIII were identified based on their differential reaction. This area has predominance of wheat land races under cultivation and all these pathotypes are very simple and virulent to wheat variety Kalyansona only. Pathotype CI is virulent on *Yr2* (Kalyansona), *Yr3, Yr6,* and *Yr7.* Pathotype CII emerged from CI by gaining virulence on *Yr1.* Further gain of virulence on Suwon92XOmar in pathotype CI resulted in the emergence of CIII (Kumar et al., 1996). In 2001 another pathotype 78S84 evolved from CIII by gaining virulence on *Yr8* and *Yr9,* because of this change the popular wheat variety PBW343 became susceptible to yellow rust (Prashar et al., 2007).

20.7.2   YR (SONALIKA) GROUP

First race designated as I was identified in 1973 when virulence on rust resistance genes *Yr2, Yr6, Yr7, Yr8*, Strubes Dickkopf, and *YrA* but avirulent to *Yr1, Yr3,* and *Yr4* was obtained (Sharma et al., 1973). This pathotype has remained confined to the Nilgiri hills for the last 42 years as a stable population and has not evolved further. In Northern India which is separated from the Nilgiri hills by about 2500 km, mostly Kalyansona type virulence occurred prior to 1982. In 1982, an independent pathotype which was avirulent to Suwon92XOmar and *Yr4* but virulent to *Yr1, Yr2, Yr3, Yr6, Yr7, Yr8*, Strubes Dickkopf, and *YrA* was named as pathotype K (Nagarajan et al., 1983). Subsequently 1988 loss of virulence on *Yr1* resulted in the evolution of race N (Kumar et al., 1989). The Acquisition of virulence on *Yr4* resulted in the emergence of pathotype P in 1990 (Kumar et al., 1991) and when it gained virulence on *Yr1* it became T (Kumar et al., 1994). First pathotype with virulence to *Yr9* was identified as 46S119 in 1996 which is a single step gain of virulence on *Yr9* in pathotype P (Prashar et al., 2007).

In most of the cases there has been gain in virulence; however, loss of virulence was observed on *Yr1* locus. There have been frequent claims that variability in yellow rust pathogen in India is through the introductions from Western Asia or may be South Africa. However, Western Asian population of yellow rust is virulent to Clement whereas that of South Africa on *Yr11, Yr14*, but Indian population is avirulent on all these wheat accessions. Based on these observations and the unique virulence structure of Indian pathotypes of *P. striiformis*, it can be concluded that mutation, selection, and local adaptation to resistant genotypes are the prime reasons for evolution of new virulences in India and exotic incursions are ruled out.

Mutation is probably the most important source of variation and evolution in the rust pathogens (Kolmer, 1989; Knott, 1989). Some of the loci are more prone to mutational changes in *P. triticina*. Mutations have been frequent to resistance genes *Lr2a, Lr2b, Lr2c, Lr10, Lr20*, and *Lr26* (Statler 1985; 1987). Gain in virulence (forward mutation) on the resistance genes *p1, p3, p16, p17*, and *p27* and loss of virulence (reverse mutation) on *p2a, p2c, p23*, and *p28* have also been reported (Statler, 1985, 1987). Induced mutations with N-Methyl-N-nitrosoguanidine lead to induction variation in *P. triticina* with some genes mutation observed to be more common than others (Statler, 1987; Watson, 1981). The virulences on *Lr26* have been frequently observed throughout the world (Long et al., 1989; Nayar et al., 1991). In Australia, virulences on *Lr26* and *Lr20* have been very common (Park et al., 1995). It has been reported that virulences on resistance genes like *Lr9, Lr19*, and *Lr28* have been the result of single step mutation in India (Nayar et al., 2003; Bhardwaj et al., 2005, 2010b). From the literature gain in virulence is very common in *P. graminis tritici*; however, loss of virulence has also been recorded in some new races in Australia (Luig, 1983). In *P. graminis* f. sp. *tritici*, resistance genes *Sr11* and *Sr37* appear more prone to mutation hence these genes have been reported to be more susceptible to 17 and 5 pathotypes, respectively. In India, pathotype 11A was reported as virulent on *Sr37* long before the genes deployment in the wheat varieties (Sinha et al., 1976). Subsequently, virulence of the 117 group on *Sr37* has also been reported (Bhardwaj et al., 1994). Similarly, pathotype 62G29-1 with virulence for *Sr24* was detected in Nilgiri hills in 1989, much before the release of variety HW 2004 which contains stem rust resistance gene *Sr24* in 1995 and is considered as one step mutation in pathotype 40A (Bhardwaj et al., 1990). These examples illustrate the significance of systematic annual survey for the presence of wheat rust pathogens in different parts of the world. This type of studies help us in identification of most effective resistance genes to be deployed in specific epidemiological regions. In stripe rust, two lines of evolution have been observed in India. One has virulence for resistance genes *Yr2* and *YrA* which has led to the evolution of virulent pathotype 46S119. It appears to be a mutant pathotype as virulence for *Yr9* which did not exist earlier. In second type, the simple races which were virulent on Kalyansona (*Yr2*) evolved to 78S84 which is also virulent on *Yr9* in India. Because of emergence of this pathogen the most popular wheat variety PBW34 became susceptible to *P. striiformis* in India.

The rate of mutation in fungi has been estimated to be 10^{-6} (Parlevliet & Zadoks, 1977). Since wheat rusts produce an enormous number of spores, mutation could be the common cause of evolution of new pathotypes. Gassner and Straib (1932) were the first to describe the role of mutation in the formation of new races in *P. striiformis* and estimated a mutation frequency of 1–6 per 100,000–200,000 uredospores. In some cases, however, somatic hybridization has also been found to be the cause of new pts (Park et al., 1999; Bhardwaj et al., 2006a). Generation of new races by the fusion of germ tubes of uredospores was first shown by Little and Manners (1969).

Mutation is independent of the host; however, presence of the rust resistance genes in field favors selection of new pathotypes. Some loci like *Lr26* on *P. triticina* and now *Sr31* on *P. graminis tritici* in Uganda and adjoining countries are more prone to mutation. The resistance gene pyramids with durable resistance discourage the selection of virulent mutants in a pathogen population as mutation to virulence at different loci is equal to the product of rates of mutation to virulence for each avirulence/virulence locus (Mundt, 1990). Though mutations for single gene virulence are selected when single genes are in use, however, virulence for multiple genes is also possible (Mundt, 1991). These have been attributed to radical mutation (Green, 1975). This phenomenon can explain the gain for multiple virulence in pathotypes 40A, 40-2, and 40-3 of *P. graminis tritici* and pathotypes 12-5, 104-1, and 104-2 of *P. triticina*. Mutants on resistance genes *Yr1, Yr2,* and *Yr3* have been found to occur frequently (Stubbs, 1968). Mutation rate for *Yr7* and *Yr9* is high whereas for *Yr5* is rare (Stubbs, 1985). Johnson et al. (1978) indicated high mutability for *Yr8*. Somatic recombinants have been shown by Taylor (1976), who found one recombinant with decrease in virulence on Chinese 166. In South Africa single step mutation events have resulted in many new pathotypes in *P. striiformis* (Boshoff et al., 2002).

Point mutations have also been found to alter the avirulence/virulence structure in *Melampsora lini*. A dominant gene in *M. lini* that inhibits expression of five avirulence genes has been reported (Jones, 1988). A single mutation from non-inhibition to inhibition could result in the simultaneous inactivation of five different avirulence alleles (Jones, 1988). This phenomenon can explain the gain of virulence in the pathotypes 40A, 40-2, and 40-3 of *P. graminis tritici*; 12-5,104-1, and 104-2 of *P. triticina*. Similarly, Griffiths and Carr (1961) irradiated the population of *Puccinia coronata* f. sp. *avenae* with UV light and found that one mutant was virulent on six differential cultivars to which the original culture was avirulent and avirulent on a seventh to which the original was virulent. These differentials, however, possessed different factors of resistance. Flor (1960) studied the mutations caused by X-ray treatment for a culture of *M. lini*. Sixteen of these progeny were virulent to cultivars Abyssinian and Leona, whereas original isolate was avirulent. A single deletion mutation that affected more than one avirulence/virulence locus could result in simultaneous, independent mutations from avirulence to virulence (Mundt, 1990) and pathotypes K, N, P, P-1, CII, CIII, and CIII-1 in *P. striiformis* might be following a similar phenomenon. Most of the new pathotypes in the species of *Puccinia* in India have evolved a single step gain in virulence for one gene. Mutation on multiple rust resistance genes in wheat can be explained in the light of earlier work (Flor, 1960; Griffiths & Carr, 1961; Mundt, 1990, 1991). Molecular basis of variation in Indian population of rust fungi is under investigation and it may give further leads. With the advent of next generation sequencing technology this type of genome wide sequence analysis is possible. The genome sequence analysis of Indian population of leaf rust fungi, *P. triticina* has been done under a multi-institutional project funded by the Department of Biotechnology, Govt. of India.

20.8   CONCLUDING REMARKS

In India, no evidence exists that alternate hosts play any functional role in the recurrence of wheat rusts. Wheat rusts get propagated clonally in the form of uredospores. Wheat grown in summer (off-season) and the regular crop provides opportunity for continuity of the rust pathogens throughout the year. Role of collateral or other auxiliary hosts is also under investigations. Rusts produce enormous number of uredospores which facilitate the evolution of new virulences possibly through single step mutation for virulence. Though there are 50 pathotypes of *P. triticina* reported from India, within the last decade the 77 group of pathotypes is predominated followed by the 104, 12, and 162 groups. Studies on avirulence/virulence structure of the predominant pathotypes showed closeness to earlier known pathotypes and their chronology of appearance have led to the conclusion that in India, wheat rust pathogens evolve mainly through mutation. Putative hybrid has been observed only in pathotype 12-5. Mutation for gain in virulence (forward mutation) appears to be more commonly observed for *Lr10, Lr23,* and *Lr26* whereas mutations for avirulence (reverse mutation) were observed for *Lr2a, Lr3,* and *Lr20,* in *P. triticina.* Single step gain in virulence was observed on *Lr9, Lr19, Lr23, Lr26,* and *Lr28.* Some of the loci like *Lr26* and *Lr23* appear to more prone for the mutation than others. In the case of *P. graminis tritici,* gain in virulence has been mainly on *Sr11, Sr37* and in one case each on *Sr24,* and *Sr25.* Single instances of loss of virulence in one case each was observed on *Sr7b, Sr13,* and *Sr11.* In *P. striiformis* gain in virulence has been observed to *Yr1, Yr2, YrA,* and *Yr9.* Loss of virulence was observed to *Yr1* only. There have been frequent claims that variability in yellow rust pathogen in India is through the introductions from Western Asia or may be South Africa. However, Western Asian population of yellow rust is virulent to Clement whereas that of South Africa on *Yr11* and *Yr14,* but Indian population is avirulent on all these wheat accessions. Based on these observations and the unique virulence structure of Indian pathotypes of *P. striiformis,* it can be concluded that mutation, selection, and local adaptation to resistant genotypes are the prime reasons for evolution of new virulences in India and exotic incursions have not been observed. Available evidences suggests that evolution of new virulences of *Puccinia* species in India has been mainly through mutation. However, dynamic evolution of *Puccinia* genome will be better understood by the decoding of their genomes.

KEYWORDS

- leaf rust
- stem rust
- stripe rust
- *Puccinia graminis tritici*
- *P. triticina*
- *P. striiformis*
- races
- pathotypes
- evolution
- Indian subcontinent

REFERENCES

Ahmad, S. T.; Singh, S. Addition to the Wheat Rust Races in India. *Indian Phytopathol.* **1969**, *22*, 524–525.

Anonymous. The Wheat Rust Patrol Striking a Fast-moving Target. In *Partners in Research for Development;* ACIAR: Canberra, Australia, 1992; Vol. 5, pp 20–21.

Bahadur, P.; Nagarajan, S.; Nayar, S. K. A Proposed System of Virulence Designation in India II. *Puccinia graminis* f. sp. *tritici. Proc. Indian Acad. Sci. (Plant Sci.).* **1985**, *95*, 29.

Bhardwaj, S. C., et al. Occurrence of a New Virulent form 117-1 in Race 117 of *Puccinia graminis* f. sp. *tritici. Cereal Rusts Powdery Mildews Bull.* **1989**, *17*, 1–5.

Bhardwaj, S. C., et al. A Pathotype of *Puccinia graminis* f. sp. *tritici* on *Sr24* in India. *Cereal Rusts Powdery Mildews Bull.* **1990**, *18*, 35–37.

Bhardwaj, S. C. Wheat Rust Pathotypes in Indian Subcontinent Then and Now. In *Wheat-Productivity Enhancement under Changing Climate;* Singh, S. S., et al., Eds.; Narosa Publishing House Pvt. Ltd.: New Delhi, India, 2012; pp 227–238.

Bhardwaj, S. C.; Kumar, S.; Gupta, N.; Sharma, S. A New Pathotype 93R57 of *Puccinia triticina* on Wheat from India. *J. Wheat Res.* **2011a**, *3* (1), 24–25.

Bhardwaj, S. C., et al. Further Variation in 117 Group of Pathotypes in *Puccinia graminis tritici* in India. *Pl. Dis. Res.* **1994**, *9* (1), 91–96.

Bhardwaj, S. C., et al. Physiologic Specialization of *Puccinia triticina* on Wheat (*Triticum* Species) in India. *Indian J. Agric. Sci.* **2010a**, *80* (9), 805–811.

Bhardwaj, S. C., et al. Virulence on *Lr28* in Wheat and Its Relation to Prevalent Pathotypes in India. *Cereal Res. Commun.* **2010b**, *38* (1), 83–89.

Bhardwaj, S. C., et al. Two New Pathotypes 29R45 and 93R39 of *Puccinia triticina* from India. *Indian Phytopathol.* **2006a**, *59* (4), 486–488.

Bhardwaj, S. C., et al. Occurrence of Two New Pathotypes 93R45 and 49R45 of *Puccinia triticina* Causing Brown Rust of Wheat in India. *Pl. Dis. Res.* **2009**, *24* (1), 6–8.

Bhardwaj, S. C., et al. Two New Pathotypes 125R 28 and 93R45 of *Puccinia triticina* on Wheat from India and Sources of Resistance. *Indian Phytopathol.* **2011b,** *64* (3), 240–243.

Bhardwaj, S. C., et al. Physiologic Specialization of *Puccinia graminis tritici* on Wheat in India during 2002–04. *Indian J. Agric. Sci.* **2006b,** *76,* 386–388.

Bhardwaj, S. C.; Prashar, M.; Jain, S. K.; Datta, D. *Lr19* Resistance in Wheat Becomes Susceptible to *Puccinia triticina* in India. *Plant Dis.* **2005,** *89,* 1360.

Bhardwaj, S. C. Strategic Centres. In *100 Years of Wheat Research in India;* Singh, S. S., et al., Eds.; Directorate of Wheat Research (ICAR): Karnal, India, 2011; pp 243–262.

Boshoff, W. H. P.; Pretorius, Z. A.; Van Niekerk, B. D. Establishment, Distribution and Pathogenicity of *Puccinia striiformis* f. sp. *tritici* in South Africa. *Plant Dis.* **2002,** *68,* 485–492.

Flor, H. H. The Inheritance of X-Ray Induced Mutations to Virulence in a Urediospore Culture of Race 1 of *Melampsora lini. Phytopathology.* **1960,** *50,* 603–605.

Gassner, G.; Straib, W. Uber Mutationen in Einer Biologischen Rasse Von *Puccinia glumarum tritici* (Schmidt) Erikss. u. Henn. *Z. Indukt. Abstammu. Vererbungsl.* **1932,** *63,* 154–160.

Goel, L. B., et al. A New Virulent Biotype of Race 12 of Leaf Rust of Wheat and Sources of Resistance to it. *Indian Phytopathol.* **1979,** *32* (1), 152–154.

Green G. J. Virulence Changes in *Puccinia graminis* f. sp. *tritici* in Canada. *Can. J. Bot.* **1975,** *53,* 1377–1386.

Griffiths, D. J.; Carr, J. H. Induced Mutation for Pathogenicity in *Puccinia coronata avenae. Trans. Br. Mycol. Soc.* **1961,** *44,* 601–607.

Jain, S. K., et al. Physiologic Specialization and New Virulences of *Puccinia graminis* f. sp. *tritici* Causing Black Rust of Wheat in India during 2005–2009. *Indian J. Agric. Sci.* **2013,** *83* (10), 1058–1063.

Jain, S. K., et al. Two New Pathotypes 5R45 and 29R7 of *Puccinia triticina* from India. *J. Mycol. Plant Pathol.* **2008,** *38* (1), 123–125.

Jain, S. K., et al. A New Pathotype 77-6(121R55-1) of *Puccinia recondita tritici* with Virulence for *Lr23* and *Lr26* in India. *Pl. Dis. Res.* **1999,** *14* (1), 7–10.

Jain, S. K., et al. Emergence of Virulence to *Sr25* of *Puccinia graminis tritici* on Wheat in India. *Plant Dis.* **2009,** *93* (8), 840.

Jin, Y., et al. Detection of Virulence to Resistance Gene *Sr24* within Race TTKS of *Puccinia graminis* f. sp. *tritici. Plant Dis.* **2008,** *92,* 923–926.

Jin, Y., et al. Century-Old Mystery of *Puccinia striiformis* Life History Solved with the Identification of *Berberis* as an Alternate Host. *Phytopathology.* **2010,** *100,* 432–435.

Jones, D. A. Genetic Properties of Inhibitor Genes in Flax Rust that Alter Avirulence to Virulence on Flax. *Phytopathology.* **1988,** *78,* 342–344.

Johnson, R.; Priestdey, R. H.; Taylor, E. C. Occurrence of Virulence in *Puccinia striiformis* for Compair Wheat in England. *Cereal Rusts Bull.* **1978,** *6,* 11–13.

Knott, D. R. Genetic Analysis of Resistance. In *The Wheat Rusts- Breeding for Resistance.* Springer-Verlag: Berlin, 1989; pp 58–82.

Kolmer, J. A. Virulence and Race Dynamics of *Puccinia recondita* f. sp. *tritici* in Canada during 1956–1987. *Phytopathology,* **1989,** *79,* 349–356.

Kolmer, J. A.; Long, D. L.; Hughes, M. E. Physiologic Specialization of *Puccinia triticina* on Wheat in the United States in 2007. *Plant Dis.* **2007,** *93,* 538–544.

Kumar, J.; Nayar, S. K.; Prashar, M.; Nagarajan, S. Mapping the *Puccinia striiformis* Race Flora in Ladakh and Other Valleys of the Interior Himalaya. *Cereal Rusts Powdery Mildews Bull.* **1996,** *24* (1 & 2), 98–103.

Kumar, J.; Nayar, S. K.; Prashar, M.; Bhardwaj, S. C.; Bhatnagar, R.; Singh, S. B. New Forms M (1S0) and N (46S102) of *Puccinia striiformis* Westend in India. *Indian J. Mycol. Pl. Pathol.* **1989**, *19* (3), 322–324.

Kumar, J.; Nayar, S. K.; Prashar, M.; Bhardwaj, S. C.; Bhatnagar, R. New Variants 47S103 and 102S100 of *Puccinia striiformis* West. *Indian Phytopathol.* **1994**, *47* (3), 258–259.

Kumar, J.; Sinha, V. C.; Nayar, S. K.; Prashar, M.; Bhardwaj, S. C.; Bhatnagar, R. Monitoring New Virulence Patterns of *Puccinia striiformis* West. *Indian J. Mycol. Pl. Pathol.* **1991**, *21* (2), 161–163.

Little, R.; Manners, J. G. Somatic Recombination in Yellow Rust of Wheat (*Puccinia striiformis*):II. Germ Tube Fusion, Nuclear Number and Nuclear Size. *Trans. Br. Mycol. Soc.* **1969**, *53*, 259–267.

Long, D. L.; Shafer, J. F.; Roelfs, A. P.; Roberts, J. J. Virulences of *Puccinia recondita* f. sp. *tritici* in the United States in 1987. *Plant Dis.* **1989**, *73*, 294–297.

Luig, N. H. The Establishment and Success of Exotic Strains of *Puccinia graminis tritici* in Australia. *Proc. Ecol. Soc. Aust.* **1977**, *10*, 89–96.

Luig, N. H. *A Survey of Virulence Genes in Wheat Stem Rust, Puccinia graminis f. sp. tritici;* Parey: Berlin, 1983; p 199.

Mehta, K. C. *Further Studies on Cereal Rusts in India*; Imperial Council of Agricultural Research, Scientific Monograph No. 14: New Delhi, 1940; pp 1–201.

Mehta, K. C. *Further Studies on Cereal Rusts in India*; Imperial Council of Agricultural Research, Scientific Monograph No. 18: New Delhi, 1952; pp 1–165.

Mundt, G. C. Probability of Mutation to Multiple Virulence and Durability of Resistance Gene Pyramids. *Phytopathology.* **1990**, *80* (3), 221–223.

Mundt, G. C. Probability of Mutation to Multiple Virulence and Durability of Resistance Gene Pyramids: Further Comments. *Phytopathology.* **1991**, *81* (3), 241–243.

Nagarajan, S.; Joshi, L. M. *Epidemiology in the Indian Subcontinent*. In *The Cereal Rusts, Diseases, Distribution, Epidemiology and Control;* Roelfs, A. P., Bushnell, W. R., Eds.; Academic Press: Orlando, FL, 1985; Vol. 2, pp 371–402.

Nagarajan, S.; Nayar, S. K.; Bahadur, P. The Proposed Brown Rust of Wheat (*Puccinia recondita* f. sp. *tritici*) Virulence Monitoring System. *Curr. Sci.* **1983**, *52* (9), 413–416.

Nagarajan, S.; Bahadur, P.; Nayar, S. K. Occurrence of a New Virulence, 47S102 of *Puccinia striiformis* West., in India during Crop Year, 1982. *Cereal Rusts Bull.* **1984**, *12* (1), 28–31.

Nayar, S. K.; Nagarajan, S.; Bahadur, P.; Bhatnagar, R. A New Virulence 104-1(21R31-1) of *Puccinia recondita tritici* in India. *Curr. Sci.* **1988**, *57*, 1005–1006.

Nayar, S. K., et al. Race 104, a New Virulence of Leaf Rust Attacking Dwarf Wheats, and its Sources of Resistance. *Indian J. Agric. Sci.* **1974**, *44* (8), 547–549.

Nayar, S. K., et al. Appearance of New Pathotype of *Puccinia recondita tritici* Virulent on *Lr*9 in India. *Indian Phytopathol.* **2003**, *56* (2), 196–198.

Nayar, S. K., et al. Occurrence of New Virulent Pathogenic Forms in Race 77 of *Puccinia recondita tritici* in India. *Curr. Sci.* **1987**, *56* (16), 844–845.

Nayar, S. K.; Prashar, M.; Bhardwaj, S. C. *Manual of Current Techniques in Wheat Rusts*; Research Bulletin No. 2; Regional Station, DWR: Flowerdale, Shimla. 1997, p 32.

Nayar, S. K.; Prashar, M.; Basandarai, A. K. Two New Pathotypes of *Puccinia recondita tritici* with Combined Virulence for *Lr23* and *Lr26*. *Pl. Dis. Res.* **1994**, *9* (2), 122–126.

Nayar, S. K., et al. Pathotype 104-3(21R63) of *Puccinia recondita tritici* with Combined Virulence for *Lr23* and *Lr26*. *Indian Phytopathol.* **1998**, *51* (3), 290–291.

Nayar, S. K., et al. Pathotypes of *Puccinia recondita* f. sp. *tritici* Virulent on *Lr26* (1BL.1RS translocation) in India. *Cereal Res. Commun.* **1991,** *19* (3), 327–331.

Nayar, S. K.; Singh, S.; Sharma, S. K. A Virulent Biotype of Race 77 of Leaf Rust of Wheat and Its Sources of Resistance. *Curr. Sci.* **1975,** *44* (20), 742.

Nayar, S. K., et al. A New Biotype of Race 77 of Leaf Rust of Wheat and Sources of Resistance. *Indian Phytopathol.* **1980,** *33,* 623–624.

Nayar, S. K.; Singh, S.; Sharma, S. K.; Chatterjee, S. C.; Goel, L. B. A New Virulence of Brown Rust Culture 104 of Leaf Rust of Wheat and Its Sources of Resistance. *Curr. Sci.* **1977,** *46,* 61.

Nayar, S. K.; Srivastava, M.; Nagarajan, S.; Bahadur, P. Occurrence of a New Biotype 104B (29R23) of *Puccina recondita tritici* and Sources of Resistance. *Cereal Rusts Powdery Mildew Bull.* **1983,** *11* (2), 48–52.

Park, R.; Burdon, J. J.; Jahoor, A. Evidence for Somatic Hybridization in Nature in *Puccinia recondita* f. sp. *tritici,* the Leaf Rust Pathogen of Wheat. *Mycol. Res.* **1999,** *103* (6), 715–723.

Park, R. F.; Burdon, J. J.; McIntosh, R. A. Studies on the Origin, Spread and Evolution of an Important Group of *Puccinia recondita* f. sp. *tritici* Pathotype in Australasia. *Eur. J. Plant Pathol.* **1995,** *101,* 613–622.

Parlevliet, J. E.; Zadoks, J. C. The Integrated Concept of Disease Resistance: A New View Including Horizontal and Vertical Resistance in Plants. *Euphytica,* **1977,** *26,* 5–21.

Patil, B. P.; Mutkekar, M. L.; Sulaiman, M. On the Occurrence of Biotype 117A of *Puccinia graminis* (Pers.) Eriks. and Henn. in India. *Indian Phytopathol.* **1963,** *16,* 244–245.

Prasada, R.; Lele, V. C. New Physiologic Races in India. *Indian Phytopathol.* **1952,** *5,* 128–129.

Prashar, M.; Nayar, S. K.; Kumar, J.; Bhardwaj, S. C. Two New Pathotypes 12-4 and 77-4 in *Puccinia recondita* f. sp. *tritici* in India. *Indian J. Pl. Pathol.* **1991,** *9,* 14–16.

Prashar, M.; Bhardwaj, S. C.; Jain, S. K.; Datta, D. Pathotypic Evolution in *Puccinia striiformis* in India during 1995–2004. *Austr. J. Agric. Res.* **2007,** *58,* 602–604.

Pretorius, Z. A.; Singh, R. P.; Wagorie, W. W.; Payne, T. S. Detection of Virulence to Stem Rust Resistance Gene *Sr31* in *Puccinia graminis* f. sp. *tritici. Plant Dis.* **2000,** *84,* 203.

Sharma, S. K.; Singh, S. D.; Nayar, S. K. A New Virulence of Stem Rust of Wheat Attacking Chhoti Lerma. *Curr. Sci.* **1975,** *44,* 486.

Sharma, S. K.; Ruikar, S. K.; Goel, L. B.; Kulkarni, S. A New Virulence 117A-1 of Stem Rust of Wheat and Its Sources of Resistance. *Indian Phytopathol.* **1979,** *32,* 498–499.

Sharma, S. K.; Singh, S.; Goel, L. B. Note on a New Record of Sonalika Infecting Race of Yellow Rust in India and Sources of Its Resistance. *Indian J. Agric. Sci.* **1973,** *43,* 964–965.

Sinha, V. C.; Upadhyaya, Y. M.; Bahadur, P. A New Virulence in Race 11 of Stem Rust of Wheat and Some Sources of Resistance. *Indian J. Genet. Pl. Br.* **1976,** *36* (3), 345–346.

Stakman, E. C.; Stewart, D. M.; Loegering, W. Q. *Identification of Physiological Races of Puccinia graminis* var. *tritici*; United States Department of Agriculture, Agricultural Research Service: Washington, DC, 1962; p 53.

Statler, G. D. Mutations Affecting Virulence in *Puccinia recondita. Phytopathology.* **1985,** *75,* 565–567.

Statler, G. D. Mutation Studies with Race 1- *Puccinia recondita. Can. J. Plant Pathol.* **1987,** *9,* 200–204.

Stubbs, R. W. Artificial Mutation in the Study of the Relationship between Races of Yellow Rust of Wheat. *Proc. Eur. Mediterr. Cereal Rusts Conf.* **1968,**60–62.

Stubbs, R. W. Stripe Rust. *Cereal Rusts*; Academic Press: New York, 1985; Vol. 1, pp 546.

Taylor, B. C. In *The Production and Behavior of Somatic Recombinants in Puccinia striiformis*, *Proceedings* of the Fourth *European* and *Mediterranean* Cereal Rusts Conference, Interlaken, Switzerland, Sep 5–10, 1976; EMCRF: Wageningen, The Netherlands, 1976; pp 436–438.

Vasudeva, R. S.; Lele, V. C.; Misra, D. P. A New Physiologic Race of *Puccinia triticina* Eriks. in India. *Indian Phytopathol.* **1955,** *8,* 3.

Watson, I. A. Wheat and Its Rust Parasites in Australia. In *Wheat Science-Today and Tomorrow*; Evans, L. T., Peacock, W. J., Eds.; Cambridge University Press: New York, 1981; pp 129–147.

CHAPTER 21

STEM RUST PATHOTYPE UG99: AN INDIAN CONTEXT

PRAMOD PRASAD, SUBHASH C. BHARDWAJ*, HANIF KHAN, OM P. GANGWAR, SIDDANNA SAVADI, and SUBODH KUMAR

ICAR-Indian Institute of Wheat and Barley Research, Regional Station, Flowerdale, Shimla 171002, Himachal Pradesh, India

Corresponding author. E-mail: scbfdl@hotmail.com

CONTENTS

ABSTRACT

Stem rust caused by *Puccinia graminis tritici* is the most feared diseases of wheat and known to cause severe epidemics throughout the world. Stem rust resistance gene *Sr31* in combination with other stem rust resistance genes kept the stem rust fungus under control until the emergence of *Sr31* virulence; designated as TTKSK and popularly known as Ug99, in Uganda during 1998. The emergence of Ug99 is considered as a highly significant event, having far-reaching consequences not only for Africa but also for the global wheat production due to susceptibility of majority of the wheat cultivars to Ug99. Ug99 has got the ability to defeat the resistance imparted by most of the stem rust resistance genes. Since 1998 eleven variants in Ug99 lineage have been detected from different African countries. In India the stem rust prone area is less than 25% of the total area under wheat and commendable diversity for stem rust resistance has been reported in Indian wheat material. Nevertheless the consequences of entry of Ug99 race into the country or independent mutation for *Sr31* or other *Sr* genes cannot be overlooked. Indian advanced wheat material is being continuously screened at Ug99 hot spot locations *viz.*, Kenya, Ethiopia etc. Besides this, strong vigil is being kept on the evolution and spread of new *Puccinia graminis tritici* pathotypes in India and neighboring countries.

21.1   INTRODUCTION

Bread wheat (*Triticum aestivum* L.), the third largest cereal produced in the world, provides 21% of the food calories and protein to at least 4.5 billion people living in developing countries (Braun et al., 2010). Wheat is the host to many microbes causing a number of diseases such as rusts, bunts, smuts, powdery mildew, leaf blight, etc. which causes huge losses to the quality and quantity of the produce. Among the biotic stresses, rusts are broadly prevalent and notorious pathogens, posing serious threat to wheat production worldwide. Wheat rusts are caused by *Puccinia* spp., which are different from most other plant pathogens in the sense that they are biotrophic and require a living plant host for their growth and reproduction. However, there are few reports of axenic culture of rust pathogens (Williams et al., 1967). Of the wheat rusts, stem rust (Sr)/black rust caused by the *Puccinia graminis* f. sp. *tritici* (*Pgt*) is one of the most dangerous plant diseases worldwide and is known to cause famines and economic crisis.

Recent examples of wheat diseases/pathotypes include the worldwide emergence of new aggressive races, that is, *Yr9* virulence of *Puccinia striiformis* (stripe rust), *Sr31* virulences (Ug99 group) of *Pgt* (Sr) (Pretorius, 2000), and wheat blast caused by *Magnaporthe oryzae* in Brazil. The Ug99 emergence in East Africa in 1998 was once considered as a challenging threat for global wheat production due to susceptibility of more than 20% of the global wheat cultivars to this pathotype. It has been estimated that the area under the risk of Ug99 amounts to around 50 mha of wheat grown globally, that is, about 25% of the world's wheat area (Singh et al., 2008). The Ug99 race, as other pathotypes of the wheat rust pathogens is evolving rapidly, till date 11 variants have been documented as the member of the Ug99 pathotype lineage (Prasad et al., 2016; Patpour et al., 2015). In Indian context, though the Sr prone area is <25% of the total area, but as the history repeats, the possible implications of independent mutation for *Sr31* cannot be ignored (Bhardwaj et al., 2014). Stem rust resistance screening conducted in Kenya and Ethiopia against Ug99 group of pathotypes has revealed that there is sufficient resistance available in Indian wheat material to avoid epidemic caused by *Sr31* type of virulences.

21.2 WHEAT RUSTS

Among the wheat diseases, rusts are the most important from historical perspective. Wheat rusts have appeared in epiphytotic form from time to time in many countries. These diseases have forced the farmers to change their cropping pattern in many parts of the world. In the southern parts of the United States, people consume corn rather than wheat which is consumed in the northern United States. This is due to the poor growth of wheat in southern part because of rust disease. In ancient Roman religion, the "Robigalia" festival was celebrated to protect wheat fields from rust disease. In India, rusts had been considered to cause a loss of more than $0.6 million every year before the introduction of Mexican dwarf wheat varieties, even in these varieties farmers may lose about 10% of yield due to rusts. Wheat rusts, caused by *Pgt* (black/Sr), *P. triticina* (brown/leaf rust), and *P. striiformis* (yellow/stripe rust), are really dangerous because they multiply geometrically on the wheat plants and complete one disease cycle within mere 7–14 days' time depending on prevailing temperature (Roelfs, 1985b). They produce enormous number of uredospores (over two trillion per hectare) at reasonable infection levels (Rowell & Roelfs, 1971), and frequently travel thousands of kilometers by wind (Roelfs, 1985a). These pathogens are highly

specific for host selection and are termed as *formae speciales* or f. sp. of that
pathogen infecting a particular crop. This was first described by Eriksson
in 1894 (Anikster, 1984). For example, wheat rust cannot infect corn and
corn rust cannot infect wheat, but some species such as *P. graminis* have
many f. sp. that can infect many host genera (Cummins, 1971). Cereal rust
pathogens are further specialized as they can infect a particular genotype of
a species but not the other and vice-versa. This type of specialization was
originally described by Stakman and Piemeisel (1917) as strains of a particular pathogen but later the term strain was replaced by physiological race,
which have the ability to infect specific resistance genes or cultivars but not
others (Stakman et al., 1962). This variation among pathogen strains or races
was summarized by Flor (1971) in his gene-for-gene hypothesis which says
"For each gene that conditions reaction in the host, there is a corresponding
gene in the parasite that conditions pathogenicity." A highly virulent race
can break down resistance governed by many resistance genes and produce
compatible reaction with its hosts; at the same time a less virulent or avirulent race can infect the genotypes with lesser number of resistance genes.
Rust pathogens undergo change in their virulence pattern through sexual
and asexual mechanisms. Sexual recombination of virulence in the cereal
rusts involves alternate host species (*Berberis* species for black and yellow
rusts; *Mahonia* and *Mahoberberis* species for black rust; *Thalictrum* and
Anchusa species for brown rust) to complete their life cycle (Jackson &
Mains, 1921; Craigie, 1927; Jin et al., 2010). Role of barberry is highly
significant in generating new Sr pathotypes (Roelfs, 1982) and causing
major epidemics. Sr epidemics in the United States and Canada in early
1900s were the outcome of evolution of new pathotypes through the sexual
recombination of the pathogen in Barberry. This initiated a major *Berberis*
eradication campaign in the United States and Canada (Campbell & Long,
2001). Moreover, mutation is the primary source of new variation in virulence (Groth, 1984) in the areas including India, where role of alternate hosts
in the pathogens' life cycle is not proved till today.

Wheat rusts are widely distributed throughout India. Yellow rust is mostly
restricted to northern states and southern hills while black rust is known
to effect wheat crop in central and southern part of the country. However,
recently (2013–2014) black (stem) rust of wheat was observed on indige-
nous wheat material planted at ICAR-NBPGR, Bhowali and ICAR-VPKAS,
Almora and Pantnagar (Uttarakhand). Although brown rust is evenly distributed in almost all the wheat-growing regions of the country, yet it is not
known to cause perceptible damage. The historical account of major wheat
rust epidemics in India (Nagarajan & Joshi, 1975) suggests that a number of

epidemics took place in Jabalpur during 1786, 1805, 1827, 1828–1829, and 1831–1832. Later on rust epidemics was reported in Central India during 1879 and 1894–1895. There was severe Sr outbreak on Einkorn wheat during 1907. Again the Sr epidemic occurred during 1946–1947 in Central India and during 1948–1949 in South India. Rust epidemics were also reported from major wheat growing areas of India and they include the epidemics at Delhi (1843), Allahabad, Banaras and Jhansi (1884 and 1895), Punjab and sub mountainous regions of Gorakhpur (1905), and Indo Gangetic plains (1910–1911). Both brown and yellow rusts caused huge losses in Western Uttar Pradesh during 1971–1973. Another epidemic of brown rust occurred in 1993 in about 4 mha of Northwestern India. Since the green revolution there are few reports of any major epidemics of wheat rusts in India.

Butler and Hayman (1906) estimated ~INR 40 million annual losses due to wheat rusts in India. Mehta (1952) reported a total loss of about INR 60 million annually in wheat and barley due to rust diseases. The magnitude of losses due to rusts can be assumed from the fact that due to rust epidemic in 1956–1957 in Bihar, only half maund of grain per acre could be harvested as against the lowest average yield of nine maunds per acre. Avoidance of yellow rust epidemic alone in Northern India during 1994–1996, which devastated wheat crop in Pakistan, saved INR 41,750 millions in that crop season.

21.3   WHEAT STEM RUST (*PUCCINIA GRAMINIS* F. SP. *TRITICI*)

Sr of wheat (*Pgt*) also known as black/summer rust is historically the most damaging disease of wheat worldwide. The Sr damage to wheat is an ongoing menace dating back to biblical times where it was reported to cause "mildewing and blasting of wheat" (Arthur, 1929). Devastating Sr epidemic of wheat from Tuscany, Italy (1766) is one of the first reports on Sr epidemic published by Fontana (1767). Since then several studies on pathogens' life cycle, taxonomy, cytology, physiology, epidemiology, virulence, and host–plant interaction have improved understanding of wheat black rust. Black rust can completely damage the healthy looking susceptible wheat crop within few weeks of initial infection under congenial environmental conditions for disease development (Fetch et al., 2011). Sr epidemics have been reported from almost all the continents where wheat is grown and major epidemics have been documented from Eastern Europe (Spehar, 1975; Spehar et al., 1976), Australia and North America (Stakman & Harrar, 1957; Saari & Prescott, 1985), Canada and the United States (Roelfs, 1978), and India (Nagarajan & Joshi, 1975).

Pgt (Eukaryota, Fungi, Basidiomycota, Pucciniomycotina, Puccini-omycetes, Pucciniales, and Pucciniaceae) is a basidiomycetous rust fungus having the ability to infect more than 360 species of plants (Anikster, 1984). *P. graminis* has been divided into subspecies and *formae speciales* based on its morphology, host range, and fertility crosses. Subspecies *graminicola* is known to infect non-cereal grasses whereas subspecies *graminis* causes black rusts on cereal crops. Subspecies *graminis* has been further divided into variety *stakmanii* (effecting barley, oat, and rye) and *graminis*, causing disease mostly on wheat. The subspecies concept is not in agreement with new *formae speciales* designations, which is based on fertility crosses. Crosses between *formae speciales tritici* and *secalis* (rye rust) and between *avenae* (oat rust) and *poae* (*poa* rust) were found to produce viable offspring (Johnson, 1949). *Pgt* infects wheat through uredospores (dikaryotic: n + n) during warmer periods. Uredospores are also known as repeating spores and they move from one plant to the other or from one region to the other and infect the susceptible cereal host. This process can result in rapid spread of the disease. Near late summer when temperatures decreases pathogen start producing bicelled teliospores. Before over wintering karyogamy takes place in bicelled teliospores (Boehm et al., 1992), which germinate during early spring by producing one promycelium from each cell of the teliospore. Each promycelium produces two pairs of haploid basidiospores. Basidiospores are then transported to nearby *Berberis* species through insects or winds. On the upper leaf surface of *Berberis* species, the basidiospores penetrate the tissue and form pycnia. Thereafter pycnia start producing different mating types of pycniospores and receptive hyphae (Johnson & Newton, 1946). The pycnio-spores can disperse through wind, water, or by insects (Leonard & Szabo, 2005). Pycniospores of one mating type are then transferred to compatible mating type receptive hyphae of a pycnium, where spermatization or sexual recombination takes place between compatible pycniospore and recep-tive hyphae, which is followed by formation of aecia below the pycnium (Johnson & Newton, 1946). In aecia diploid aeciospores are produced, which after coming in contact with wheat plant starts infection and produce uredia (Roelfs, 1985b). Numerous uredospores are produced in a uredium over a period of time. A fully developed uredium produces ~23 μg (Katsuya & Green, 1967) of uredospores daily. One microgram contains about 4.5×10^2 uredospores (Rowell & Olien, 1957). Thus in a single uredium about 10,000 uredospores are produced per day (Rowell & Roelfs, 1971). On the surface of wheat plant uredospores produce a germ tube, which grows along the epidermis up to stoma, where it develops appressorium. A penetration peg is formed at the base of the appressorium, which moves through substomatal cavity and forms a vesicle. Subsequently this vesicle elongates and form hyphae, that penetrate into and grows inside the plant tissue. On contact with the host cell wall, hyphae develop haustorial mother cell to release host cell wall degrading enzymes. This results in the formation of haustorium for nutrient uptake of from host, required by the pathogen for its growth and successful infection.

The genome of *Pgt* is made up of 18 chromosomes (Boehm et al., 1992) with 88.6 Mb genome size (Duplessis et al., 2011). There is no evidence for whole genome duplication but larger than expected genome seems to be a result of a large number of transposable elements, which form 45% of the genome (Duplessis et al., 2011). About 17,773 protein-coding genes are predicted in *Pgt*, of which only 35% showed significant homology to known proteins (Duplessis et al., 2011).

21.3.1   WHEAT STEM RUST EPIDEMIOLOGY IN INDIA

Role of alternate hosts in the life cycle of *Pgt* has not been proved in India till date. Even after several efforts by following the proven methodology, we have not been able to infect wheat plants by aeciospores taken from *Berberis* species present in northern or southern hills of the country. This suggests that these *Berberis* species are harboring the *Puccinia* species which are not at all infectious to wheat. Wheat rusts in India are supposed to survive and initiate infection through uredospore, which are being carried forward either by self-sown wheat or some unknown grass(es) or other hosts. It is a widely known fact that wheat rusts survive in off-season on self-sown wheat plants in Himalayas and Nilgiri and Palani hills of Southern India. However, recent studies have indicated that some of the grasses growing in the plains might be acting as collateral hosts for wheat rust pathogens. With the precise and accurate procedures for the identification of pathotypes of *Puccinia* species on wheat in India (Nagarajan et al., 1983, 1985, Bahadur et al., 1985) and its modifications (Bhardwaj, 2012), it has become possible to track the dissemination routes of these pathogens. Both Himalayas and South Indian hills act as the primary foci of infection of *Pgt* (Mehta, 1952, Joshi et al., 1974). However, extensive monitoring and surveys carried out during the past few decades advocate that black rust uredospores spread primarily from Nilgiri and Pulney hills or some other foci in Peninsular India rather than the Himalayas, which play minor role in disease occurrence in plains. One of the reasons for this could be the differences between incubation periods of the pathogen at both the locations. The pathogen under cool climate of

Himalaya takes 40–50 days to sporulate in winter months. Contrarily, it takes 8–10 days for sporulation in Nilgiris; thus the inoculum multiplication takes place at faster rate (Joshi, 1976).

Uredospores of *Pgt* move from south to north and get deposited with rain drops. It is observed that the cyclonic disturbances that occur during October/November in the Bay of Bengal cross the Coromandal coast and then re-curve toward Central and Peninsular region of the country (Nagarajan & Singh, 1973). Examination of the rain drops along with satellite television cloud photographs (STCPs) and trajectories proved that due to cyclonic wind circulations the uredospores of *Pgt* from higher elevation of Nilgiris are lifted to conviction currents and transported to hundreds of kilometers toward Central and North India, where persistent damp weather favors early infection of black rust. Based on this study, Nagarajan and Singh (1975) formulated three synoptic rules called the Indian stem rust rules (ISR) to describe the likelihood of occurrence of Sr in India. The rules are:

ISR1: Storm of depression should be formed either in the Bay of Bengal or in the Arabian Sea demarcated by longitude 65–81° East and 8–15° North.

ISR2: A persistent high-pressure cell over the south part of Central India (not far from the Nilgiris) must be present.

ISR3: Appearance of the deep trough extending up to southern India and caused by onward movement of western disturbance and associated rainfall over Central India.

Data on pathotype analysis during the last 35 years indicate that the Nilgiri hills act as a source of inoculum for Maharashtra, Madhya Pradesh, Gujarat, and to some extent to Karnataka. Sr pathotypes in 117 group do not occur in Nilgiri hills, however, are more prevalent in Karnataka, Maharashtra, and in some years in Madhya Pradesh also. It is indicative of the fact that in addition to Nilgiri hills, Karnataka and Maharashtra are also contributing as a source of Sr inoculum (Bhardwaj et al., 1989) to Central and Peninsular India. In Northern India Sr is observed in Ladakh area of Jammu and Kashmir, Himachal Pradesh, and Uttarakhand. Ladakh plays no role in the epidemiology of Sr in rest of India as the Sr pathotype 34-1, which is prevalent in Ladakh region is not reported elsewhere in India. In Himachal Pradesh and Uttarakhand 21, 21-1, and 21A-2 pathotypes are prevalent but generally these pathotypes do not occur in the plains of India. However, few years back, 21A-2 pathotype was detected in Madhya Pradesh and Gujarat.

21.3.2 STEM RUST RESISTANCE IN INDIAN WHEAT

Genetic resistance is one of the most cost effective, ecofriendly, and successful means of wheat rust management. Availability of varieties with adequate level of resistance to particular rust does not require any additional prophylactic measures for rust management. However, developing a cultivar with resistance to all/most of the pathotypes present in an area and maintaining adequate resistance for long period is a difficult task. A few historic cultivars, such as Thatcher and Hope (Hare & McIntosh, 1979) remain resistant to black rust for many years. In general, a particular cultivar remains resistant for about five years or more, which is also the average agronomic life span of a cultivar. Information on genetics of wheat rust resistance is widely available (Tomar et al., 2014; Walia & Kumar, 2008; Bhardwaj, 2011). Based on the available information a number of *Sr* resistance genes *viz.*, *Sr2*, *Sr5*, *Sr6*, *Sr7a*, *Sr7b*, *Sr8a*, *Sr8b*, *Sr9b*, *Sr9e*, *Sr11*, *Sr12*, *Sr13*, *Sr17*, *Sr21*, *Sr24*, *Sr25*, *Sr30*, and *Sr31* have been characterized in Indian wheat material. Among these *Sr2*, *Sr11*, and *Sr31* were very common in bread wheat whereas *Sr7b*, *Sr9e*, and *Sr11* conferred Sr resistance in many durum lines. *Sr26*, *Sr27*, *Sr31*, *Sr32*, *Sr33*, *Sr35*, *Sr39*, *Sr40*, *Sr43*, and *SrTt3* (Jain, 2013) confer resistance against Indian population of *Pgt*. Maintaining the diversity of resistance genes or cultivars with different resistance genes is one of the key aspects of effective wheat rust management. But tendency of the farmers to grow a single variety often leads to monoculturing over larger area. Cultivation of two mega cultivars such as PBW 343 on seven and Inquilab 91 on six mha in India and Pakistan, respectively (Joshi, 2008), is the best example of monoculture. Cultivation of fewer cultivars that carry less diverse/major/race-specific resistance genes often leads to higher genetic uniformity and subsequently to greater disease vulnerability or boom and bust cycle.

Resistance gene *Sr2*, transferred into hexaploid wheat from *Triticum turgidum* L. f. sp. *dicoccum* Schrank ex. Schübler (cv. Yaroslav), along with other minor genes transferred from cultivars Hope and H-44 have provided durable Sr resistance in modern day cultivars. The *Sr2* gene confers inadequate slow rusting when present alone under heavy disease pressure, but in association of other slow rusting or minor genes its effect is many folds. However, sufficient information is unavailable on other minor genes of *Sr2* complex and their interactions. Knott (1989) is of the opinion that adequate levels of multigenic resistance by minor genes to Sr can be accomplished by accumulating at least five minor genes. Moreover, pyramiding of multiple major genes promises prolonged resistance durability and little chances for the evolution of the pathogen. Thus pyramiding multiple major genes in a genotype for which no virulence is available in the pathogen population could prolong the effectiveness of that particular genotype.

21.4 EVOLUTION IN WHEAT RUST PATHOGENS

E.C. Stakman was the first man to introduce the terms pathogens physiological forms or races for *Pgt* (Stakman & Piemeisel, 1917). The physiological forms or races differ in their virulence pattern for particular resistance genes or their combinations in the host. These physiological forms or races can be differentiated by comparing their avirulence/virulence pattern on standard set of wheat differentials. The nomenclature system for designating wheat rust races have evolved since first differential set developed by Stakman and Levine (1922). Since then a number of *Pgt* nomenclature systems like the North American nomenclature system (Roelfs & Martens, 1988) have been developed in different parts of the world. These systems are based on different infection types produced by different races on wheat differentials. Broadly the infection types are classified in to five classes, that is, 0: no infection or flecking; 1: small uredia often surrounded by necrosis; 2: small to medium size uredia surrounded by necrosis or chlorosis; 3: medium uredia associated with chlorosis; 4: large uredia without chlorosis or necrosis (Roelfs & Martens, 1988). The resistant reaction is assigned to infection types 0, 1, or 2 while the infection types 3 or 4 are assigned for susceptible reaction. In India, binomial system (Bahadur et al., 1985) having 24 differentials, is followed for designating *Pgt* pathotypes. This system has been updated from time to time (Bhardwaj, 2012).

Plant pathogens and their hosts have got unique property to co-evolve naturally to sustain diverse environments. A collection of complex ecological interactions existing under natural ecosystem forces such co-evolution to take place. Plants resist pathogen attack through a range of preformed and induced pathogen barriers. Simultaneously pathogens try to defeat plant defenses for successful disease development through diverse virulence mechanism. Pathogen associated molecular patterns (PAMPs) triggered immunity is assumed to be the first level of induced defenses in plants, which involves recognition of PAMPs. As a result the pathogen evolves by producing pathogenicity effectors to conquer the host immunity. Such survival pressure guides both host and its pathogen for an antagonistic host–pathogen co-evolution (Stahl & Bishop, 2000). The gene-for-gene hypothesis (Flor, 1955) determines the magnitude of disease development in the host plant to understand the plant–pathogen co-evolution by way of different assumptions and prophecies. The gene-for-gene hypothesis has been proved in a number of host pathogen systems, especially where obligate parasites are involved.

Evolution or adaptation of a pathogen to changing ecosystem is strongly linked its mode of reproduction. Both sexual and asexual reproduction play pivotal role in deciding the adaptation and evolutionary potential of the pathogen (Fisher, 2007). Sexual reproduction produces the genetic variations required for evolution of newer forms of the pathogen to defeat hosts' defense barriers, while asexual reproduction allows rapid multiplication of newly evolved or fittest individuals. The other means by which the plant pathogens undergo evolution include mutation, selection, parasexuality, somatic hybridization, introduction, etc. Leonard (1969) was one of the earliest workers who studied quantitative adaptation in genetically heterogeneous population of *P. graminis* f. sp. *avenae* on two different host genotypes. He recorded 10–15% increase in the mean infection efficiency of the pathogen population on the host on which they were maintained, but not on other hosts.

Introgression of 1B.1R (*Lr26/Sr31/Yr9*) resistance to wheat from rye (Mettin et al., 1973; Zeller, 1973) revolutionized the wheat production system. Besides the resistance to three rusts of wheat, it conferred tolerance to many abiotic stresses. However, within few years of its commercialization the leaf rust resistance gene *Lr26* became ineffective to new pathotypes of *P. triticina* in Europe (Bartos et al., 1984), India (Nayar et al., 1987), North America (Kolmer, 1991), and elsewhere. Probably this was due to the independent mutations in *P. triticina* for gain in virulence for *Lr26*. Since first identification of *Lr26* virulence in 1987, there are 19 pathotypes in India with *Lr26* virulence (Bhardwaj, 2012). Likewise, virulence to *Sr* resistance gene *Sr24* was first identified in South Africa (Roux & Rijkenberg, 1987) whereas in India there was an independent single step mutation for gain in virulence in pathotype 40A with virulence to *Sr24* and this resulted in the emergence of a new *Pgt* pathotype 40-1 (Bhardwaj et al., 1990). Likewise, there are numerous examples where independent mutation in wheat rust pathogens having virulence against one gene of the host are reported, example include *Yr9* (McIntosh et al., 1995; Prashar et al., 2007). One of the alarming examples of pathogen evolution is the emergence of Ug99 pathotype (*Sr31* virulence) of *Pgt*. It was first identified in Uganda in 1998. Before that *Sr31* (linked to *Lr26* and *Yr9*) and other resistance genes kept the Sr disease under control for about 60 years and people believed that genetic resistance had overcome the wheat Sr (Singh et al., 2011).

21.4.1 UG99: EMERGENCE, EVOLUTION, AND SPREAD

Developing rust resistant varieties has been a continuous exercise over the years. One of the achievements of the Green Revolution of the 1960s was to reduce yield losses due to wheat rusts, as many resistance genes introduced in wheat during that period conferred resistance to most of the rust pathotypes of that time. A number of rust resistant sources including the alien ones have been used to combat wheat Sr. Transfer of *Secale cereale* L. (rye) gene (1B/1R translocation or substitution), carrying wheat rust resistance genes *Lr26/Sr31/Yr9* into bread wheat (Mettin et al., 1973; Zeller, 1973), has not only increased yield potential by 12–20% but also protected plant from key biotic and abiotic stresses (Cox et al., 1995). *Sr31* in combination with other *Sr* resistance genes kept the *Sr* fungus under control for more than two decades. Sr of wheat was thought to be a disease of past until 1998, when a new pathotype of *Pgt* was observed in Uganda. This new pathotype, designated as TTKSK in 1999, popularly known as Ug99, has the ability to overcome the resistance conferred by majority of the *Sr* resistance genes including *Sr31*. Since then 11 variants in Ug99 lineage (also confirmed by molecular markers) have been detected in different African countries (Prasad et al., 2016, Patpour et al., 2015). The arrival of Ug99 and magnitude of its disease causing potential was a major event in history of plant pathogen evolution and was treated as a big fear for global wheat cultivation as most of the wheat cultivars of that time were susceptible to Ug99. The estimation advocated that 50 mha of wheat grown globally is under the risk of Ug99, which is ~25% of the world's wheat area (Bhardwaj et al., 2014).

The variants in Ug99 lineage are designated based on North American nomenclature system (Jin et al., 2008). This system of nomenclature assigns a five-letter code to each pathotype of *Pgt*. TTKSK, the earliest pathotype in Ug99 lineage, was detected in 1999 with virulence to *Sr31*. It is now reported to spread to other countries like Kenya (2001), Ethiopia (2003), Sudan (2006), Yemen (2006), Iran (2007), Tanzania (2009), and Egypt (2014). Other variants, that is, TTKSF (virulent to *Sr21* and avirulent to *Sr31*) from South Africa (2000) and Zimbabwe (2009), TTKST (virulent to *Sr31, 21* and *24*) from Kenya (2006), Tanzania (2009), and Eritrea (2010), TTTSK (virulent to *Sr31* and *36*) from Kenya (2007) and Tanzania (2009), TTKSP (virulent to *Sr21* and *24*) from South Africa (2007), PTKSK (virulent to *Sr31* but avirulent to *Sr21, 24,* and *36*) from Ethiopia (2007) and Kenya (2007), PTKST (virulent to *Sr31* and *24* but avirulent to *Sr21*) from Ethiopia (2007), Kenya (2008), South Africa (2009), Eritrea (2010), Mozambique (2010), and Zimbabwe (2010), and TTKSF (virulent to *Sr21* but avirulent to

Sr31) have been reported from South Africa (2000) and Zimbabwe (2009) (Prasad et al., 2016, Singh et al., 2011). All the *Pgt* pathotypes belonging to Ug99 group may differ based on their avirulence/virulence profiles but they are closely interrelated genetically, having nearly related DNA fingerprints.

During 2014 three new variants in Ug99 lineage, that is, TTKTT, TTKTK, and TTHSK were detected, making total members to 11 in Ug99 lineage (Table 21.1). TTKTT (virulent on *Sr24, 31, and Tmp*) from Kenya, TTKTK (virulent on *Sr31 and Tmp*) from Kenya, Egypt, Eritrea, Rwanda, and Uganda, and TTHSK (avirulent on *Sr30* and virulent on *Sr31*) from Kenya have been confirmed (http://rusttracker.cimmyt.org/?page_id=22). All variants are supposed to be the result of single step mutations. Acquired virulence to *Sr* genes *Sr24* and *Sr36* is known from Kenya (*Sr24, Sr36*), Egypt (*Sr24*), Eritrea (*Sr24*), Ethiopia (*Sr24, Sr36*), Mozambique (*Sr24*), Rwanda (*Sr36, Sr24*), South Africa (*Sr24*),Tanzania (*Sr24, Sr36*), Uganda (*Sr24, Sr36*), and Zimbabwe (*Sr24*). The Ug99 variants with combined virulence to *Sr31* and *Sr24* have been detected throughout Africa and their further spread in near future to other countries is very much expected (Singh et al., 2015). Resistance contributed by *SrTmp* gene has now been broken down by the latest variants, that is, TTKTT and TTKTK detected in Kenya. The confirmed detection of TTKTK from Kenya, Egypt, Eritrea, Rwanda, and Uganda during 2014 suggests its rapid spread in Africa (Patpour et al., 2015).

Ethiopia experienced severe Sr epidemics in the southern wheat production region during 2013 and 2014 (Singh et al., 2015). This epidemic was caused by a non-Ug99 group pathotype of *Pgt* and designated as TKTTF. Pathotype TKTTF is virulent to Digalu variety in Ethiopia carrying resistance gene *SrTmp* that was effective to the known variants of Ug99 group in Ethiopia. TKTTF was first detected from Ethiopia in August 2012 but remained at a low frequency until October 2013 when it caused grain loss up to 100% in more than 10,000 ha area (Singh et al., 2015).

21.4.2 UG99: GLOBAL CHALLENGES AND SUCCESS

It has been estimated that Ug99 prone area accounts for about 50 mha globally which is about 1/4th of the world's wheat area (Singh et al., 2008). Rapid evolution and potential of long distance dispersal of *Pgt* pathotypes especially pathotypes belonging to Ug99 lineage make them capable of causing explosive epidemics. They produce millions of uredospores and these spores could move from one plant or area to the other. The primary mode of *Pgt* dispersal is through wind by step-wise range expansion. The

TABLE 21.1 Distribution of Variants of *Puccinia graminis tritici* in Ug99 Group Reported from Various Countries till 2014, with Avirulence or Virulence Status (Source: http://rusttracker.cimmyt.org/?page_id=22, modified from Singh et al., 2015).

Pathotype*	Common alias	Key virulence (+) or avirulence (−)	Year of identification	Confirmed countries (year detected)
TTKSK	Ug99	+Sr31	1999	Uganda (1998/9), Kenya (2001), Ethiopia (2003), Sudan (2006), Yemen (2006), Iran (2007), Tanzania (2009), Eritrea (2012), Rwanda (2014), Egypt (2014)
TTKSF		−Sr31	2000	South Africa (2000), Zimbabwe (2009), Uganda (2012)
TTKST	Ug99	+Sr31, +Sr24	2006	Kenya (2006), Tanzania (2009), Eritrea (2010), Uganda (2012), Egypt (2014), Rwanda (2014)
TTTSK	Ug99	+Sr31, +Sr36	2007	Kenya (2007), Tanzania (2009), Ethiopia (2010), Uganda (2012), Rwanda (2014)
TTKSP		−Sr31, +Sr24	2007	South Africa (2007)
PTKSK		+Sr31, −Sr21	2007	Uganda (1998/9), Kenya (2009), Ethiopia (2007), Yemen (2009)
PTKST		+Sr31, +Sr24, −Sr21	2008	Ethiopia (2007), Kenya (2008), South Africa (2009), Eritrea (2010), Mozambique (2010), Zimbabwe (2010)
TTKSF+		−Sr31, +Sr9h	2012	South Africa (2010), Zimbabwe (2010)
TTKTT		+Sr31, +Sr24, +SrTmp	2015	Kenya (2014)
TTKTK		+Sr31, +SrTmp	2015	Kenya (2014), Egypt (2014), Eritrea (2014), Rwanda (2014), Uganda (2014)
TTHSK		+Sr31, −Sr30	2015	Kenya (2014)

*Pathotype designation follows the North American nomenclature system described by Jin et al. (2008).

stepwise movement of *Yr9* virulent pathotype of *P. striiformis* is one of the best example of such type step-wise range expansion. *Yr9* virulent pathotype of *P. striiformis* evolved in Eastern Africa and migrated to Middle East, West Asia, and finally to South Asia in a step-wise manner in about 10 years, and caused severe outbreak throughout its path (Singh et al., 2004). Ug99 variants are projected to migrate in the similar fashion. TTKSK has been detected from Uganda, Kenya, Ethiopia, Sudan, Yemen, Iran, Tanzania, Eritrea, Rwanda, and Egypt (Fig. 21.1). The migration of Ug99 up to Egypt (2014) is particularly important to mention here, as its migration route presents strong indication of its future migration in to the major wheat growing areas of the Middle East and South Asia. Moreover; East Africa with West and South Asia shares single epidemiologic zone for migration of rust pathotypes, which supports the probable migration of rust pathotypes from East Africa to Southwest Asia (Singh et al., 2004).

Interestingly, *geographic information system* (GIS) data predicts two potential migration routes for Ug99 into Southwest Asia (Hodson et al., 2005). The first and the most likely route is as described by Singh et al.

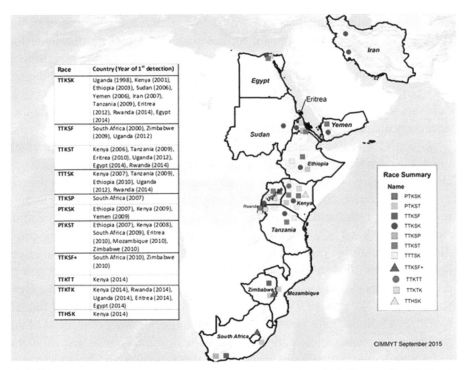

FIGURE 21.1   Ug99 Pathotypes Group Summary, September 2015. (Source: Ravi P. Singh, CIMMYT, Mexico, modified).

(2004) for the migration *Yr9*-virulence for yellow rust. The second route, which is highly speculative and less probable, connects East Africa directly with Southern Pakistan/Western India (Joshi, 2008). Rumors on detection of Ug99 came from Pakistan during 2009, but pathotyping based on North American Nomenclature System and DNA marker analysis ruled out the same (Mirza et al., 2010). Till date Ug99 is following the above-mentioned first route and have reached up to Iran (2007) and Egypt (2014). If somehow TTKSK, the only variant of Ug99 reported outside Africa, or any other variant from Ug99 lineage reaches these regions, they are definitely a threat to the wheat crop in stem rust prone areas where the crop is protected by *Sr* resistance gene *Sr31* or any other ineffective gene for migrating pathotypes. The detailed studies on prediction of further migration of Ug99 in future is being undertaken using the model HYbrid Single-Particle Lagrangian Integrated Trajectory (HYSPLIT), an airborne particle trajectory model developed by National Oceanic and Atmospheric Administration (NOAA), USA (Draxler & Rolph, 2003).

To fight the global wheat Sr threat Nobel laureate N. E. Borlaug started Borlaug Global Rust Initiative (BGRI) earlier Global Rust Initiative (GRI) in 2005 to monitor the spread of wheat Sr pathotype Ug99, to screen the released varieties and germplasm for resistance to Ug99, to distribute the sources of resistance worldwide, breeding to incorporate diverse resistance genes and adult plant resistance (APR) gene into high-yielding adapted varieties. The success of BGRI lies in a timely replacement of susceptible cultivars with resistant ones having equal or better yield potential and other necessary characteristics.

Global Cereal Rust Monitoring System (GCRMS) has been implemented under the collaboration of BGRI, Consultative Group on International Agricultural Research (CGIAR) centers, advanced research labs, national agricultural programs, and UN-FAO to integrate and disseminates up-to-date information on Sr incidence as well as emergence, evolution and spread of *Pgt* pathotypes. GCRMS has been successful in developing a strong coordinated international rust surveillance network. A collaborative global effort begun in April 2008 to mitigate the threat posed by Ug99 and other potential pathotypes causing wheat rust with the implementation of Durable Rust Resistance in Wheat (DRRW) Project, led by Cornell University and funded by Bill & Melinda Gates Foundation (BMGF). The primary objectives of the project are to replace susceptible varieties with durably resistant varieties, to harness recent advances in genomics for introducing non-host resistance (immunity) into wheat and to develop strong international collaboration in wheat research.

There was increased emphasis toward phenotyping and breeding for stem-rust-resistance in wheat varieties. BGRI with the support of DRRW Project contributed a lot in this. More than 350,000 wheat varieties, germplasm collections including advanced breeding materials from wheat producing countries of Africa and Asia were screened for resistance to Sr pathotype Ug99 and its derivatives during past one decade at Kenya Agricultural and Livestock Research Organization (KALRO), Njoro, Kenya. Similarly at Kulumsa and Debre Zeit (Ethiopia) over 87,000 wheat accessions were screened. More than 85% of the materials from different countries screened at these two locations were susceptible to pathotypes of Ug99 lineage. After the vigorous phenotyping and breeding for Sr resistance, the proportion of resistant material over the years increased significantly till 2014. Nearly 20% of entries have shown good level while other 20% entries from all the countries were showing intermediate levels of resistance to Ug99 pathotype group. Screening of advanced lines and wheat varieties at Njoro, Kenya from India, Pakistan, and Ethiopia indicated that 38.4, 56.7, and 69.4% of entries, respectively, was highly to moderately resistant to Ug99 group of pathotypes (Singh et al., 2015). This proportion of resistance is supposed to be adequate enough for Sr prone areas of India and Pakistan.

More than 59 *Sr* resistance genes have been designated in common wheat and its wild relatives. Many of the *Sr* genes are single-locus major genes (McIntosh et al., 2011) sometimes, conferring varying age-dependent resistance. In wheat rusts two types of resistance, that is, seedling or all stage resistance and adult plant stage resistance are reported. Seedling resistance is governed by race specific major genes. While APR, conferred by race non-specific genes, is expressed in adult plant stage only. Most of the APR genes are minor genes and act as quantitative trait loci (QTL). Resistance conferred by multiple minor genes contributes individual small effects but in group their contribution is quite useful.

Minor genes complex generate high levels of resistance in adult plants (Singh et al., 2011). *Sr2,* the most utilizes APR gene linked to *Pbc* gene, that conferred pseudo-black chaff (PBC) was transferred to hexaploid wheat "Hope" and "H44-24a" from tetraploid emmer wheat "Yaroslav" (McFadden, 1939). This is the most effective resistance gene when used in combination with other minor genes. The "*Sr2*-Complex" comprised of *Sr2* and up to five additional minor genes have contributed a lot to international Sr breeding efforts (Singh et al., 2011). Pyramiding multiple slow-rusting APR genes in one agronomically superior variety to get high levels of durable Sr resistance is particularly important to restrict or trim down the emergence and evolution of new virulent pathotypes of Sr in different wheat growing areas. Recently three pleiotropic APR genes *Sr55* (linked to *Lr67/Yr46/Pm46*), *Sr57* (linked to *Lr34/Yr18/Pm38/Sb1/Bdv1*), and *Sr58* (linked to *Lr46/Yr29/ Pm39*) were identified (Yu et al., 2014). At present race-specific *Sr* resistance genes *Sr22, Sr25, Sr26, Sr33, Sr35, Sr45,* and *Sr50* are reported to be effective to all the identified pathotypes of Ug99 group and possibly their use in combinations would be quite useful. *Sr25* when present in combinations with some minor genes confers high level of stem rust resistance in Misr 1 and Muquawin 09 in Egypt and Afghanistan, respectively, having gene combination *Sr2* and *Sr25*, are the example of such varieties. In India, *Sr25* virulent pathotype (PKTSC) was identified in 2009 (Jain et al., 2009). This diminished the usefulness of *Sr25* for future breeding programs in India.

21.4.3 IS UG99 A REAL THREAT TO WHEAT PRODUCTION IN INDIA?

At the time of Ug99 emergence it was estimated that 50 mha of wheat grown globally was under Ug99 risk, that is, about 25% of the world's wheat area (Singh et al., 2008). Varieties of Indian subcontinent carrying *Sr31* gene like PBW343, PBW373, and others are susceptible to Ug99 group of pathotypes. These pathotypes are expected to travel further out of East Africa to other major wheat growing areas in years to follow. Nevertheless the sources of Sr resistance specifically to Ug99 group are available in many countries including India to defend possible outbreak of Sr epidemics (Singh et al., 2011).

In India North West Plains Zone (NWPZ), the food basket of the country with 10.0 mha under wheat is the major wheat growing area. If somehow any production threat like Ug99 detected in NWPZ, it will have serious implications on food self-sufficiency of the country. In this zone the maximum temperature remains below 2°C for weeks during December and January. The mean daily temperature reaches 18°C only after mid-March, which later increases further and make conditions favorable for development of *Pgt*. Due to sudden increase in mean temperature beyond 30°C during late April wheat crop matures and harvesting starts. As a result *Pgt* do not get enough time to develop and cause damage beyond threshold level even during most favorable seasons for Sr development (Nagarajan & Joshi, 1985). The summer temperature during May–July goes beyond 40°C which is too high for the survival of *Pgt*. After the detection of Ug99 in Uganda, it was widely hypothesized that Ug99 could be a serious threat to Indian wheat production. However, the detailed epidemiological studies on the relevance of Ug99 demonstrate that it is not a threat to the NWPZ of India (Nagarajan, 2012; Bhardwaj et al., 2014). The central and peninsular zones of the country are the only areas where any pathotype of Sr can cause infection but the available data suggests that varieties like Lok1, HI 9498, WH 147, GW322, HI 1531, HI 8627, HD 4672, MACS 2846, NI 5439, DL 788-2, MPO 1215, NIDW 295, HI 8663, UAS 321, and UAS 431, which are in cultivation in Central and Peninsular India, possess considerable resistance to most of the predominant pathotypes of that region or Ug99 variants (Bhardwaj et al., 2014). Thus if somehow Ug99 enters these zones we would have sufficient resistant cultivars available with us to defend it.

21.4.4 INDIAN INITIATIVE TO MEET UG99 THREAT

Keeping the serious threat of Ug99 to wheat production in India under consideration, Indian wheat program initiated Ug99 mitigation activities in collaboration with CIMMYT, BGRI, DRRW, and other national and international organizations. These activities include rapid identification and development of Ug99 resistant wheat cultivars for their quick deployment in Sr prone areas of the country. Central and peninsular zones are the most Sr prone areas in India, which accounts for about 25% of total area and practically, it may not be a threat in the main wheat belt (NWPZ), yet the vigil is being kept on the possible implications of the entry of Ug99 type of pathotypes into the country or independent mutation for *Sr31* or any other more virulent pathotypes. Among more than 1000 Indian germplasm screened against Ug99 type of pathotypes in Kenya and Ethiopia, wheat variety HW1085 and others (Table 21.2) for central and peninsular zone and three genetic stocks, that is, FLW 2 (PBW 343 + *Sr24*), FLW6 (HP 1633 + *Sr24*), and FLW 8 (HI 1077 + *Sr25*), developed for rust resistance breeding were found to be resistant to Ug99 group of pathotypes under natural conditions in Kenya and Ethiopia. After the detection of TTKST (2007), TTKSP (2007), PTKST (2008), and TTKTT (2015), the new variants in Ug99 group from East Africa, the usefulness of the gene *Sr24* was reduced. In one study commendable diversity for Sr resistance was observed in Indian wheat genotypes (Bhardwaj et al., 2003). Few Indian wheat cultivars carrying *Sr25/Lr19* gene, which is still effective to Ug99 group of pathotypes have been developed. These newly developed lines also carry *Sr36* derived from *Triticum timopheevii* and powdery mildew resistance gene *Pm6*. *Sr26* another effective *Sr* gene has been incorporated into WH 147 and NI 5439, one of the widely adopted cultivars of Sr prone central and peninsular zones, respectively.

TABLE 21.2   Stem Rust Resistance in Indian Genotypes to Ug99 Group of Pathotypes at Njoro, Kenya and Kulumsa & Debre Zeit, Ethiopia.

Average coefficient of infection	Genotypes
0–10	A 9-30-1, AKDW 4021, DDK 1037, DDK 1038, DDW 14, DL 153-2, GW 1250, HD 2781, HD 3014, HD 4720, HDR 77, HI 8381, HI 8498, HUW 234, K 9107, MACS 1967, MACS 2846, MACS 2988, MACS 2998, MACS 3742, MACS 5009, MPO 1215, WH 147, NDW 1020, NDW 940, NI 5439, NIDW 295, PBW 315, PBW 612, PDW 274, PDW 316, PDW 317, RSP 561, *Sr22*, *Sr32*, *Sr35*, *Sr39*, *Sr42*, TL 2942, TL 2963, and TL 2966
10.1–15.0	HUW 234, K 508, AKAW 4627, AKDW 2997-16, COW (W) 1, DDK 1001, DDK 1009, DDK 1025, DWR 1006, GW 1139, GW 1251, GW 406, HD 4671, HUW 464, K 0710, PL172, PL 419, RD 2503, *Sr21*, UPD 85, and WHD 943
15.1–20.0	HS 512, J 858, Jagriti, K 409, and MACS 2971

Studies suggest that durum wheat is a good source of rust resistance as these pro different durum varieties appears to be driven by genes different than the known ones. Thus the deployment of such varieties will enhance much needed gene diversity and reduce the vulnerability of cultivated wheat to new virulences (Joshi, 2008). Therefore, the cultivation of durum wheat varieties like MACS 2846, HD 4672, HI 8498, etc. showing moderate to good resistance to Ug99 group should be encouraged in bordering and Sr prone areas. Recently, Sharma et al. (2015) have reported that wheat lines A-9-30-1, AKDW 4021, DDK 1037, DDK 1038, DDW 14, DL153-2, GW 1250, HD 2781, HD 3014, HD 4720, HDR77, HI 8381, HI 8498, HUW 234, HW 5211, K 9107, MACS 1967, MACS 2846, MAC S2988, MACS 2998, MACS 3742, MACS 5009, MPO 1215, NI 5439, NIDW 295, PBW 315, PBW 612, PDW 274, PDW 316, PDW 317, WH 147, *Sr22*, *Sr32*, *Sr35*, *Sr39*, and *Sr42* and triticale varieties TL 2942, TL 2963, and TL 2966 were found resistant to Ug99 and its variants on evaluation in Kenya and Ethiopia during 2006–2011 (Table 21.2).

Occurrence of wheat rust pathotypes is being critically monitored in India and neighboring countries for detection of any new variants of *Pgt* in initial stages. During the green revolution, it was speculated that Mexican dwarf wheat varieties carrying Norin-10 gene might be rendered susceptible by rust diseases. Therefore, a close wheat diseases monitoring network, in the form of wheat disease monitoring nurseries (WDMN (erstwhile trap plot nursery)) was started way back in 1967. Wheat disease monitoring nursery works as an efficient indicator for monitoring the evolution of rust pathotypes, occurrence, and spread of wheat diseases especially rusts across the different wheat growing zones of India. During 2014–2015 crop year, the WDMN was planted at more than 40 locations across India (Gangwar et al., 2015). Mobile surveys are also conducted at every 20 km distance in Northern and Central India by survey teams of different zones. Survey teams follow fixed route to track the distribution of wheat rust pathotypes. Thus the availability of Ug99 resistance sources along with strong attentive vigilance on the evolution and spread of new Sr pathotypes in India and neighboring countries makes Indian wheat program fully prepared to mitigate any possible Sr epidemic due to Ug99 group of pathotypes or others in the country.

21.5   CONCLUDING REMARKS

Since the detection of Ug99 (TTKSK) in Uganda during 1998, 11 variants in Ug99 lineage have been reported till date from East Africa. These variants carry virulence to several *Sr* resistant genes commonly present in leading global spring and winter wheat germplasm. The pathotypes of Ug99 group have defeated the resistance conferred by several key genes including *Sr9h, Sr24, Sr31, Sr36, SrTmp,* etc. During 2014 three new pathotypes, that is, TTKTT, TTKTK, and TTHSK belonging to Ug99 group were detected in East Africa. TTKSK, the first pathotype in Ug99 lineage, is migrating rapidly and has been detected from several countries like Kenya, Ethiopia, Sudan, Yemen, Iran, Tanzania, and Egypt. It is the only pathotype of Ug99 group reported beyond African periphery and may spread to other areas. So far *Sr31* remains resistant to black rust population in India and neighboring countries, however, there is complete preparedness to combat any threat. More than 95% of the wheat black rust prone area in Central and Peninsular India is under the cultivation of Ug99 resistant wheat material. Ethiopia experienced severe Sr epidemics during 2013 and 2014 due to the *SrTmp* virulence (TKTTF), a non-Ug99 variant of *Pgt*. It caused 100% grain loss in more than 10,000 ha area of Ethiopia and is being considered as another challenging threat to global wheat production. Significant progress has been made to tackle the Sr menace at global level. Developments made in breeding high-yielding wheat varieties with high level of race-specific resistance or APRs, field screening of global wheat germplasm at hot spot locations, pathogen surveillance, and cloning of rust resistance genes have been instrumental. Besides having sufficient resistant genotypes for Sr prone areas, India is keeping strong vigil on the evolution and spread of new

Sr pathotypes in India and neighboring countries. This is indicative of the fact that India is fully prepared to counter the Ug99 threat.

Future strategies to tackle the risk associated with the variants of Ug99 group or any other virulent variant of *Pgt* include replacement of currently popular susceptible varieties with high yielding resistant cultivars in the areas where these variants are already established or bordering areas, diversification of rust resistance to Ug99 variants among wheat cultivars, international collaboration for screening and breeding of rust resistant varieties, and rapid multiplication and adoption of resistant varieties.

KEYWORDS

- *Triticum aestivum*
- stem rust
- rust epidemiology
- resistant genotypes to Ug99
- *Puccinia graminis*
- stem rust race Ug99
- India

REFERENCES

Anikster, Y. The Formae Speciales. In *Cereal Rusts;* Bushnell, W. R., Roelfs, A. P., Eds.; Academic Press: Orlando, FL, 1984; Vol. 1, pp 115–130.

Arthur, J. C. *The Plant Rusts (Uredinales);* Wiley: New York, 1929.

Bahadur, P.; Nagarajan, S.; Nayar S. K. A Proposed System for Virulence Designation in India to *Puccinia graminis* f. sp. *tritici. Proc. Indian Acad. Sci. (Plant Sci.).* **1985,** *95,* 29–33.

Bartos, P.; Stuchlikova, E.; Kubova, R. Wheat Leaf Rust Epidemics in Czechoslovakia in 1993. *Cereal Rusts Bull.* **1984,** *12,* 40–41.

Bhardwaj, S. C., et al. Occurrence of a New Virulent Form 117–1 in Race 117 of *Puccinia graminis* f. sp. *tritici* in India. *Cereal Rusts Powdery Mildews Bull.* **1989,** *17*(1), 1–5.

Bhardwaj, S. C., et al. Diversity of Resistance for *Puccinia graminis tritici* in Wheat (*Triticum aestivum*) and Triticale Mutant. *Indian J. Agric. Sci.* **2003,** *73*(12), 676–679.

Bhardwaj, S. C., et al. A Pathotype of *Puccinia graminis* f. sp. *tritici* on *Sr24* in India. *Cereal Rusts Powdery Mildew Bull.* **1990,** *18,* 35–38.

Bhardwaj, S. C.; Prashar, M.; Prasad, P. Ug99-Future Challenges. In *Future Challenges in Crop Protection against Fungal Pathogens, Fungal Biology;* Goyal, A., Manoharachary, C., Eds.; Springer Science and Business Media: New York, 2014; pp. 231–248.

Bhardwaj, S. C. *Resistance Genes and Adult Plant Rust Resistance of Released Wheat Varieties of India;* Research Bulletin No. 5; Regional Station, IIWBR: Flowedale, Shimla, 2011, pp. 24–28.

Bhardwaj, S. C. Wheat Rust Pathotypes in Indian Subcontinent Then and Now. In *Wheat-Productivity Enhancement under Changing Climate;* Singh, S. S., et al., Eds.; Narosa Publishing House: New Delhi, India, 2012; pp. 227–238.

Boehm, E. W. A., et al. An Ultrastructural Pachytene Karyotype for *Puccinia graminis* f. sp. *tritici. Can. J. Bot.* **1992,** *70,* 401–413.

Braun, H. J.; Atlin, G.; Payne, T. Multi-Location Testing as a Tool to Identify Plant Response to Global Climate Change. In *Climate Change and Crop Production;* Reynolds, M. P., Ed.; CABI: London, 2010; pp. 115–138.

Butler, E. J.; Hayman, J. M. Indian Wheat Rusts with a Note on the Relation of Weather to Rust on Cereals. *Mem. Dep. Agric. India Bot. Serv.* **1906,** *1,* 1–57.

Campbell, C. L.; Long, D. L. The Campaign to Eradicate the Common Barberry in the United States. In *Stem Rust of Wheat: From Ancient Enemy to Modern Foe*; Peterson, P. D., Ed.; APS Press: St. Paul, MN, 2001; pp. 16–50.

Cox, T. S.; Gill, B. S.; Sears, R. G. Notice and Release of KS94WGRC32 Leaf Rust Resistant Hard Red Winter Wheat Germplasm. *Ann. Wheat Newslett.* **1995,** *41,* 241.

Craigie, J. H. Discovery of Function of Pycnia of the Rust Fungi. *Nature.* **1927,** *120,* 765–767.

Cummins, G. B. *The Rust Fungi of Cereals, Grasses, and Bamboos;* Springer: Berlin. 1971.

Draxler, R. R.; Rolph, G. D. HYSPLIT (Hybrid Single-Particle Lagrangian Integrated Trajectory). http://www.arl.noaa.gov/ready/hysplit4.html. Accessed 29 November 2007. NOAA Air Resources Laboratory, Silver Spring, MD, 2003.

Duplessis, S., et al. Obligate Biotrophy Features Unraveled by the Genomic Analysis of Rust Fungi. *Proc. Natl. Acad. Sci. USA.* **2011,** *108,* 9166–9171. 10.1073/PNAS.1019315108.

Fetch, T., et al. Rust Diseases in Canada. *Prairies Soils and Crops.* **2011,** *4,* 86–96. http://www.prairiesoilsandcrops.ca/

Fisher, M. C. The Evolutionary Implications of an Asexual Lifestyle Manifested by Penicillium marneffei. In *Sex in Fungi: Molecular Determination and Evolutionary Implications;* Heitman J. K., et al., Eds.; ASM Press: Washington, DC, 2007; p 542.

Flor, H. H. Current Status of the Gene-for-Gene Concept. *Annu. Rev. Phytopathol.* **1971,** *9,* 275–296.

Flor, H. H. Host–Parasite Interactions in Flax-Its Genetic and Other Implications. *Phytopathology.* **1955,** *45,* 680–685.

Fontana, F. Observations on the Rust of Grain. In *Phytopathological Classics;* American Phytopathological Society: Washington, DC, 1767; pp. 1–38.

Gangwar, O. P., et al. *Mehtaensis: Six- Monthly Newsletter;* Regional Station, ICAR-IIWBR: Shimla, 2015, *35*(2), pp 1–29.

Groth, J. V. Virulence Frequency Dynamics of Cereal Rust Fungi. In *The Cereal Rusts;* Bushnell, W. R., Roelfs, A. P., Eds.; Academic: Orlando, FL, 1984; pp. 231–252.

Hare, R. A.; McIntosh, R. A. Genetic and Cytogenetic Studies of the Durable Adult Plant Resistance in 'Hope' and Related Cultivars to Wheat Rusts. *Zeitschrift fur Pflanzenzuchtung.* **1979,** *83,* 350–367.

Hodson, D. P.; Singh, R. P.; Dixon, J. M. In *An Initial Assessment of the Potential Impact of Stem Rust (Race Ug99) on Wheat Producing Regions of Africa and Asia Using GIS,* Absts of the 7th International Wheat Conference, Mar del Plata, 2005, p. 142.

Jackson, H. S.; Mains, E. B. Aecial Stage of the Orange Leaf Rust of Wheat, *Puccinia triticina* Erikss. *J. Agric. Res.* **1921,** *22,* 151–172.

Jain, S. K. Physiologic Specialization and New Virulences of *Puccinia graminis* f. sp. *tritici* Causing Black Rust of Wheat (*Triticum aestivum*) in India during 2005–2009. *Indian J. Agric. Sci.* **2013,** *83*(10), 1058–1063.

Jain, S. K., et al. Emergence of Virulence to *Sr25* of *Puccinia graminis* f. sp. *tritici* on Wheat in India. *Plant Dis.* **2009,** *93,* 840.

Jin, Y.; Szabo, L. J.; Carson, M. Century-Old Mystery of *Puccinia striiformis* Life History Solved with the Identification of *Berberis* an Alternate Host. *Phytopathology,* **2010,** *100,* 432–435.

Jin, Y., et al. Detection of Virulence to Resistance Gene *Sr24* within Race TTKS of *Puccinia graminis* f. sp. *tritici*. *Plant Dis.* **2008,** *92,* 923–926.

Johnson, T.; Newton, M. The Occurrence of New Strains of *Puccinia triticina* in Canada and Their Bearing on Varietal Reaction. *Sci. Agric. (Washington, D.C.).* **1946,** *26,* 468–478.

Johnson, T. Intervarietal Crosses in *Puccinia graminis*. *Can. J. Res.* **1949,** *27,* 45–65.

Joshi, A. K. Ug99 Race of Stem Rust Pathogen: Challenges and Current Status of Research to Sustain Wheat Production in India. *Indian J. Genet. Plant Breed.* **2008,** *68*(3), 231–241.

Joshi, L. M. Recent Contributions towards Epidemiology of Wheat Rusts in India. *Indian Phytopathol.* **1976,** *29,* 1–16.

Joshi, L. M.; Saari, E. E.; Gera, S. D.; Nagarajan, S. Survey and Epidemiology of Wheat Rust in India. In: *Current Trends in Plant Pathology;* Raychaudhuri, S. P., Verma, J. P., Eds.; Professor S. N. Das Gupta Birthday Celebration Committee: Lucknow, India, 1974; pp. 150–159.

Katsuya, K.; Green, G. J. Reproductive Potentials of Races 15B and 56 of Wheat Stem Rust. *Can. J. Bot.* **1967,** *45,* 1077–1091.

Knott, D. R. Genetic Analysis of Resistance. In *The Wheat Rusts- Breeding for Resistance*; Springer-Verlag: Heidelberg, Berlin, 1989; pp. 58–82.

Kolmer, J. A. Physiologic Specialization of *Puccinia recondite* f. sp. *triticina* in Canada in 1990. *Can. J. Plant Pathol.* **1991,** *13,* 371–373.

Leonard, K.; Szabo, L. J. Stem Rust of Small Grains and Grasses Caused by *Puccinia graminis*. *Mol. Plant Pathol.* **2005,** *6,* 99–111.

Leonard, K. J. Selection in Heterogeneous Populations of *Puccinia graminis* f. sp. *avenae*. *Phytopathology.* **1969,** *59,* 1851–1857.

McFadden, E. S. Brown Necrosis, a Discoloration Associated with Rust Infection in Certain Rust Resistant Wheats. *J. Agric. Res.* **1939,** *58,* 805–819.

McIntosh, R.; Wellings, R. C.; Park, R. *Wheat Rusts, an Atlas of Resistance Genes*; CSIRO: East Melbourne. 1995.

McIntosh, R. A., et al. Catalogue of Gene Symbols for Wheat: 2011 Supplement. *Annual Wheat Newsletter.* **2011,** *57,* 303–321.

Mehta, K. C. *Further Studies on Cereal Rusts in India*; Imperial Council of Agricultural Research, Scientific Monograph No. 18: New Delhi, 1952; p. 165.

Mettin, D.; Bluthner, W. D.; Schlegel, R. In *Additional Evidence on Spontaneous 1B/1R, Wheat Rye Substitutions and Translocations*, Proceedings of the 4th International Wheat Genetics Symposium, Columbia, MO, Aug 6–11, 1973; Sears, E. R., Sears, L. M. S., Eds.; Agricultural Experimental Station, University of Missouri: Columbia, MO, 1973; pp. 179–184.

Mirza, J. I., et al. In *Race Analysis of Stem Rust Isolates Collected from Pakistan in 2008–09,* Proceedings of BGRI 2010 Technical Workshop, St. Petersburg, May 30–31, 2010; McIntosh, R., Pretorius, Z., Eds.; BGRI: Ithaca, NY, 2010; p 5.

Nagarajan, S.; Joshi, L. M. Historic Account of Wheat Rust Epidemics in India and Their Significance. *Cereal Rusts Bull.* **1975,** *3,* 25–33.

Nagarajan, S.; Joshi, L. M. Epidemiology in the Indian Subcontinent. In *The Cereal Rusts, Diseases, Distribution, Epidemiology and Control;* Roelfs, A. P., Bushnell, W. R., Eds.; Academic Press: Orlando, FL, 1985; Vil. 2, pp 371–402.

Nagarajan, S.; Singh, H. Satellite Television Photography as a Possible Tool to Forecast Plant Disease Spread. *Curr. Sci.* **1973,** *42,* 273–274.

Nagarajan, S.; Singh, H. The Indian Stem Rust Rules-an Epidemiological Concept on the Spread of Wheat Stem Rust. *Plant Dis. Rep.* **1975,** *59,* 133–136.

Nagarajan, S. Is *Puccinia graminis* f. sp. *tritici*-Virulence Ug99 a Threat to Wheat Production in the North West Plain Zone of India? *Indian Phytopath.* **2012,** *65*(3), 219–226.

Nagarajan, S.; Nayar, S. K.; Bahadur, P. The Proposed Brown Rust of Wheat (*Puccinia recondita* f.sp. *tritici*) Virulence Monitoring System. *Curr. Sci.* **1983,** *52*(9), 413–416.

Nagarajan, S.; Nayar, S. K.; Bahadur, P. The Proposed Systems of Virulence Analysis III. *Puccinia striiformis* West. *Kavaka.* **1985,** *13*(1), 33–36.

Nayar, S. K., et al. Occurrence of New Virulent Pathogenic Forms in Race 77 of *Puccinia recondita* f. sp. *tritici* in India. *Curr. Sci.* **1987,** *56,* 844–845.

Patpour, M., et al. Emergence of Virulence to *SrTmp* in the Ug99 Race Group of Wheat Stem Rust, *Puccinia graminis* f. sp. *tritici,* in Africa. **2015,** *100,* 522. http://apsjournals.apsnet.org/doi/abs/10.1094/PDIS-06-15-0668-PDN.

Prasad, P., et al. Ug99: Saga, Reality and Status. *Curr. Sci.* **2016,** *110 (9),* 1614-1616.

Prashar, M., et al. Pathotypic Evolution in *Puccinia striiformis* in India during 1995–2004. *Aust. J. Agric. Res.* **2007,** *58,* 602–604.

Pretorius, Z. A. Detection of Virulence to Wheat Stem Rust Resistance Gene *Sr31* in *Puccinia graminis* f. sp. *tritici* in Uganda. *Plant Dis.* **2000,** *84*(2), 203.

Roelfs, A. P.; Martens, J. W. An International System of Nomenclature for *Puccinia graminis* f. sp. *tritici. Phytopathology,* **1988,** *78,* 526–533.

Roelfs, A. P. Epidemiology in North America. In *The Cereal Rusts;* Roelfs, A. P., Bushnell, W. R., Eds.; Academic Press, Orlando, **1985a,** pp 403–434.

Roelfs, A. P. *Estimated Losses Caused by Rust in Small Grain Cereals in the United States-1918–76;* U.S. Department of Agriculture, Agricultural Research Service: Washington, DC, 1978; Vol. 1363, pp. 1–85.

Roelfs, A. P.; Long, D. L.; Casper, D. H. Races of *Puccinia graminis* f. sp. *tritici* in the United States and Mexico in 1980. *Plant Dis.* **1982,** *66,* 205–207.

Roelfs, A. P. Wheat and Rye Stem Rust. In *The Cereal Rusts;* Roelfs, A. P., Bushnell, W. R., Eds.; Academic Press: Orlando, FL, 1985b; pp. 3–37.

Roux, L.; Rijkenberg, F. H. J. Pathotypes of *Puccinia graminis* f. sp. *tritici* with Increased Virulence for *Sr24. Plant Dis.* **1987,** *77,* 1115–1119.

Rowell, J. B.; Olien, C. R. Controlled Inoculation of Wheat Seedlings with Urediospores of *Puccinia graminis* var. *tritici. Phytopathology.* **1957,** *47,* 650–655.

Rowell, J. B.; Roelfs, A. P. Evidence for an Unrecognized Source of Overwintering Wheat Stem Rust in the United States. *Plant Dis. Rep.* **1971,** *55,* 990–992.

Saari, E. E.; Prescott, J. M. World Distribution in Relation to Economic Losses. In: *The Cereal Rusts;* Roelfs, A. P., Bushnell, W. R., Eds.; Academic Press: Orlando, FL, 1985; Vol. 2, pp. 259–298.

Sharma, A. K., et al. Evaluation of Wheat (*Triticum aestivum*) Germplasm and Varieties against Stem Rust (*Puccinia graminis* f.sp. *tritici*) Pathotype Ug99 and Its Variants. *Indian Phytopathol.* **2015,** *68*(2), 134–138.

Singh, R. P., et al. The Emergence of Ug99 Races of the Stem Rust Fungus is a Threat to World Wheat Production. *Annu. Rev. Phytopathol.* **2011**, *49*, 465–481.

Singh, R. P., et al. Will Stem Rust Destroy the World's Wheat Crop? *Advances in Agron.* **2008**, *98*, 271–309.

Singh, R. P., et al. Emergence and Spread of New Races of Wheat Stem Rust Fungus: Continued Threat to Food Security and Prospects of Genetic Control. *Phytopathology.* **2015**, *105*(7), 872–884.

Singh, R. P., et al. In *Wheat Rust in Asia: Meeting the Challenges with Old and New Technologies, New Directions for a Diverse Planet,* Proceedings of the 4th International Crop Science Congress, Brisbane, Australia, 26 Sept–1 Oct, 2004; p 209.

Spehar, V. *Epidemiology of Wheat Rust in Europe,* Proceedings of the 2nd International Winter Wheat Conference, Zagreb, Yugoslavia, June 9–19, 1975; Agricultural Institute: Zagreb, Yugoslavia, 1975; pp 435–440.

Spehar, V.; Vlahovic, V.; Koric, B. *The Role of Berberis sp. on Appearance of Virulent Physiologic Races of Puccinia graminis f. sp. tritici,* Proceedings of the 4th European and Mediterranean Cereal Rusts Conference, Interlaken, Switzerland, 1976; EMCRC: Wageningen, pp 63–67.

Stahl, E. A.; Bishop, J. G. Plant–Pathogen Arms Races at the Molecular Level. *Curr. Opin. Plant Biol.* **2000**, *3*, 299–304.

Stakman, E. C.; Stewart, D. M.; Loegering, W. Q. *Identification of Physiological Races of Puccinia graminis var. tritici;* United States Department of Agriculture, Agricultural Research Service: Washington, DC, 1962; p. 53.

Stakman, E. C.; Harrar, J. G. *Principles of Plant Pathology;* Ronald Press: New York, 1957.

Stakman, E. C.; Levine, M. N. *The Determination of Biologic Forms of Puccinia graminis on Triticum sp.* Technical Bulletin No. 10; University of Minnesota Agriculture Experiment Station: St. Paul, MN, 1922, pp. 1–10.

Stakman, E. C.; Piemeisel, F. J. Biological Forms of *Puccinia graminis* on Cereals and Grasses. *The J. Agric. Res.* **1917**, *10*, 429–495.

Tomar, S. M. S., et al. Wheat Rusts in India: Resistance Breeding and Gene Deployment-A Review. *Indian J. Genet Plant Breed.* **2014**, *74*(2), 129–156.

Walia, D. P.; Kumar, S. Genetics of Rust Resistance in Wheat-An Update. In *Crop Improvement Strategies and Applications;* Setia, R. C., Nayyar, H., Setia, N. I. K., Eds.; International Publishing House Pvt. Ltd.: New Delhi, India, 2008; pp. 119–131.

Williams, P. G., et al. Sporulation and Pathogenicity of *Puccinia graminis* f. sp. *tritici* Grown on an Artificial Medium. *Phytopathology,* **1967**, *57*, 326–327.

Yu, L. X., et al. A Consensus Map for Ug99 Stem Rust Resistance Loci in Wheat. *Theor. Appl. Genet.* **2014**, *127*, 1561–1581.

Zeller, E. J. In *1B/1R Wheat Rye Chromosome Substitutions and Translocations,* Proceedings of the 4th International Wheat Genetics Symposium, Columbia, MO, Aug 6–11, 1973; Sears, E. R., Sears, L. M. S., Eds.; Agricultural Experiment Station, University of Missouri: Columbia, MO, 1973; pp. 209–211.

CHAPTER 22

BARLEY STEM RUST RESISTANCE MECHANISMS: DIVERSITY, GENE STRUCTURE, AND FUNCTION SUGGEST A RECENTLY EVOLVED HOST-PATHOGEN RELATIONSHIP

ROBERT SAXON BRUEGGEMAN* and SHYAM SOLANKI

Department of Plant Pathology, North Dakota State University, Fargo 58108, ND, USA

Corresponding author. E-mail: robert.brueggeman@ndsu.edu

CONTENTS

ABSTRACT

Stem rust caused by *Puccinia graminis* f. sp. *tritici* was historically one of the most devastating diseases of barley and wheat. The barley resistance gene *Rpg1* provided remarkably durable resistance against diverse stem rust races in the United States until *P. graminis* f. sp. *tritici* race QCCJF emerged in the US, which was virulent on barley lines containing *Rpg1*. Of higher concern for cereal production worldwide were the highly virulent race TTKSK (A.K.A. Ug99) and its variants, which are virulent on most of world's commercial wheat varieties and barley cultivars, including those containing *Rpg1*. Cloning and characterization of the *Rpg1* gene demonstrated that the resistance responses are activated within minutes of avirulent spores landing on the leaf surface, indicative of an early cell surface receptor mediated resistance response, possibly representing early non race specific PAMP triggered immunity. The emergence of *P. graminis* f. sp. *tritici* races QCCJF and TTKSK prompted the identification of new wheat stem rust resistance genes in barley effective against these new threats. As a result the barley *rpg4/Rpg5* resistance locus was identified on chromosome 5H and cloning efforts showed that the locus harbors three tightly linked genes (*Rpg5, HvRga1* and *HvAdf3*) that function together to provide resistance against a broad spectrum of stem rust races including QCCJF and TTKSK. *Rpg5* and *HvRga1* are tightly linked nucleotide binding site-leucine rich repeat (NLR) *R*-genes with head-to-head genome architecture typical of dual NLR *R*-genes that function together for race specific resistances. *Rpg5* and *HvRga1* possibly follow the "integrated decoy" model where one NLR with an integrated domain functions as the pathogen "sensor and activator" and the second NLR functions as a "repressor" of resistance signaling. *Rpg5* has the typical NLR domains and a C-terminal serine threonine protein kinase domain that is indispensable for resistance function and possibly acts as the pathogen sensory domain. Interestingly, *Rpg5* alleles evolved that contain a phosphatase domain instead of kinase domain that compromise stem rust resistance and act as dominant susceptibility factors explaining the recessive nature of the *rpg4*-mediated wheat stem rust resistance. The requirement of the actin depolymerization factor, *HvAdf3*, for the *rpg4/Rpg5* mediated resistance response is also not surprising as actin cytoskeleton remodeling during pathogen attack is crucial, not only to provide a physical barrier to block pathogen ingress, but also working as an early pathogen sensing mechanism to prepare for later robust *Rpg5/HvRga1* mediated hypersensitive responses. Parallels drawn from the model pathosystem *Arabidopsis-Pseudomonas syringe* are aiding in our understanding of the complex and intertwined molecular orchestra involving cytoskeleton dynamics and cellular signaling operating in barley-stem rust pathosystem. Mounting evidence shows that the early immunity responses effective against broad races of wheat stem rust in barley, that are elicited before haustorial development, probably represent non-host resistance mechanisms suggesting that barley may represent a recent host of the wheat stem rust pathogen.

22.1   INTRODUCTION

Rust fungi infect and damage most crop plants but those infecting the small grain crops such as wheat, barley, rye, and oats historically caused the largest economic losses and pose the biggest threats to world food security. One of the most important has been wheat stem rust caused by *Puccinia graminis* Pers.: Pers. f. sp. *tritici* Eriks. & E. Henn., which infects both wheat and barley (Steffenson, 1992). Wheat stem rust is a major concern as losses in wheat (*Triticum* spp.) can have huge economic implications and also pose food security issues because it is a staple crop in populated regions of the world (Pretorius et al., 2000). Wheat stem rust also infects barley (*Hordeum vulgare* L.), but the losses due to epidemics, although similar in magnitude relative to the yield losses in wheat, are not of such high economic and societal consequences. Barley is a relatively lower acreage crop and is not a staple food except in some small geographic regions, including the highlands of Ethiopia, Eritrea, Yemen, Tibet, Nepal, Ecuador, and Peru (Steffenson et al., 2013). However, barley is very important for the brewing and distilling industries, which has huge value added implications to the economies of countries in every corner of the globe. Thus, its economic importance due to brewing and distilling gives this small acreage commodity tremendous influence to ensure that its end users and stakeholders have a secure supply that meets the quality parameters of the industry.

Historically, wheat stem rust also known as "black rust" was one of the all-time most devastating diseases of the small grains. The earliest record of plant diseases is found in the Bible where it was stated that while still in the desert of Sinai, some thirty-five centuries ago, Moses had warned the people of Israel that if they failed to observe the commandments of Jehovah, "they would be punished by their field crops being destroyed by smut and mildew" (Deut, 28:22). Now it is assumed that, in part, it was due to stem rust (Chester, 1946) because *Puccinia graminis* spores were identified in archeological sites in Israel containing grain bins that date as far back as 1300 B.C. (Kislev, 1982), approximately the same period as this first historic record of disease on cereals. Wheat stem rust was also a serious disease in ancient Greece and during the Roman Empire. It was observed and recorded by Aristotle (384–322 B.C.) that "the grain rust was brought on by the warm vapors," indicating that ancient scientists had made the correlation between a warm moist environment and severe disease onset (Chester, 1946). Historical weather records, indicates a period of unusually high rainfall years that suggest rust epidemics may have caused a food crisis contributing to the fall of the Roman Empire.

In modern times stem rust caused major epidemics to wheat and barley production in the Midwestern United States and Prairie Provinces of Canada until the mid-twentieth century (Steffenson, 1992). Efforts to develop rust resistance in wheat varieties, initiated in Kansas in 1911 then followed by other states, managed the disease, but the use of single resistance genes (R-genes) in wheat varieties grown across vast acreages in the Midwest began the "Boom and Bust" cycles of wheat breeding and production (Peterson, 2001). Single dominant R-genes deployed in wheat varieties protected the crop, "the boom," until selection pressures by these R-gene specificities enriched pathogen populations with advantageous virulence or lack of avirulence genes (Avr-genes). The selection pressure toward individuals carrying these advantageous genes or combinations of genes gave rise to the predominance of R-gene virulent pathogen races, once again causing large-scale stem rust epidemics, "the bust." Stem rust epidemics prior to the 1950s during a single growing season resulted in wheat yield losses as high as 200 million bushels in the United States alone (Steffenson, 1991). The losses on barley during these periods of massive epidemics are not well documented but are believed to be similar to those recorded for wheat (Steffenson, 1991). Then in the 1950s, wheat breeders began pyramiding several R-genes into their varieties. This practice, in combination with the eradication of the P. graminis f. sp. tritici secondary host barberry, the sexual stage of the pathogen, through the barberry eradication program provided more durable resistance ending the occurrence of major stem rust epidemics in wheat (Peterson, 2001).

In the 1950s, Henry H. Flor's groundbreaking genetic analysis of compatible and incompatible interactions in the model flax-flax rust pathosystem using comprehensive genetic analysis of both the pathogen and its host determined that for every dominant R-gene in the resistant host genotype there is a dominant cognate Avr-gene in the pathogen (Flor, 1956). It is this gene-for-gene interaction that is the bases for pathogen recognition by the resistant genotypes, which elicit resistance responses. Flor's "gene-for-gene" hypothesis becomes the main tenet in the study of host–parasite genetics and was later translated into the receptor–ligand functional model. This direct interaction model predicted that R-proteins function as receptors that directly interact with the corresponding Avr protein ligand and this direct interaction acts as the trigger initiating the defense response signaling cascades.

In the last few decades major advancements in molecular techniques and tools facilitated the cloning of nearly a hundred plant *R*-genes from diverse genera of plants. However, only a relatively limited number of the cognate *Avr*-genes have been identified from the pathogens. The *R*-genes have been grouped into a limited number of different classes according to their protein domain structure (Martin et al., 2003) with the vast majority containing an N-terminal nucleotide-binding site (NBS) and a C-terminal leucine rich region (LRR) designated the NOD-like receptors (NLR) family of *R*-genes. Many of the NLR R-proteins also contain an N-terminal domain, of which most fall into one of two classes, the Toll and interleukin-1 receptors (TIR) or the putative leucine-zipper (LZ) motifs (Hammond-Kosack & Jones, 1997). On the pathogen side the *Avr*-genes and/or effectors are very diverse and typically do not fall into major functional classes. However, recent research showed that although effectors may not share primary amino acid similarity there is conserved tertiary structure shared among effector classes, which probably provides functional conservation (De Guillen et al., 2015).

The cloning of *R*-genes and their cognate *Avr*-genes determined that in many instances there is not the expected direct interaction as proposed by the gene-for-gene direct receptor–ligand binding model. The lack of data showing that R-proteins directly interact with their corresponding Avr ligands coupled with the identification of plant proteins that directly interact with both Avr and R-proteins as interaction intermediates led to an alternate hypothesis: the "guard hypothesis" (Van der Biezen & Jones, 1998). In this model, the R-proteins ("the guards") surveil other host proteins ("the guardees") that are targets of modifications by pathogen effector proteins. These effectors are hypothesized to have initially evolved as virulence factors and their host susceptibility targets are modified to manipulate host cell physiology to provide the pathogen with better access to nutrients or to evade host immunity responses (Qi et al., 2011). The host counter-evolved to detect these modifications of the "guardee" proteins by these effectors through surveillance by the R-protein "guards." Resistance protein-mediated detection of these modifications elicits a strong effector triggered immunity (ETI) defense response. This model suggests that R-proteins are present in complexes containing the guardee and typically do not have a direct protein–protein interaction with the Avr effector proteins.

Based on extensive research of plant innate immune systems, much of the work performed in the model *Pseudomonas syringae-Arabidopsis* and *P. syringae*-tomato pathosystems, the plant innate immunity system was delineated into two distinct layers (Dodds & Rathjen, 2010). The first layer of defense responses is typically considered early non-host resistance mediated by the detection of pathogen associated molecular patterns (PAMPs) by plasma membrane bound receptor complexes that contain extracellular pattern recognition receptors (PRRs). These host–parasite interactions occur very early, almost immediately after the pathogen is encountered, and represent the first layers of induced defense responses. The PRRs fall into two receptor classes, the transmembrane receptor-like kinases (RLKs) and the receptor-like proteins (RLPs). The RLKs contain an extracellular receptor domain, a transmembrane domain and a cytosolic kinase signaling domain. The RLPs are similar in structure to the RLKs yet lack the cytosolic signaling domains. Characterization of the PRR receptor complexes has determined that they typically contain hetero duplexes of RLKs and RLPs (Monaghan & Zipfel, 2012). The PAMPs that elicit the PRR signaling complexes are highly conserved molecular patterns found on or secreted by a wide range of microbes. The best-characterized PRR and subsequent signaling mechanism is *FLS2* in *Arabidopsis*, which recognizes the flg22 peptide, a highly conserved PAMP found as a subunit of motile bacteria flagellin (Zipfel et al., 2004). The recognition of flg22 by FLS2 triggers PAMP-triggered immunity (PTI) responses, which among typical PTI responses including *Mitogen-activated protein kinase* (MAPK) pathway induction, and also results in stomata closure to stop the pathogen from entering the host. These resistance mechanisms as well as other bacterial PAMP responses have been extensively studied and in depth reviews have been published (Nicaise et al., 2009; Ali & Reddy, 2008). In fungi the best characterized PAMP is the chitin subunits that make up part of the fungal cell walls, which have also been extensively studied and reviewed (Malinovsky et al., 2014; Wan et al., 2008).

ETI responses are typically elicited by race specific cytoplasmic localized NLR R-proteins, previously discussed. The ETI response is characterized by a strong-programmed cell death (PCD) activation designated as the hypersensitive response (HR). These higher amplitude ETI responses kill the host cells at and adjacent to the infection site effectively stopping the fungus from acquiring nutrients via its haustoria or feeding structure successfully sequestering and halting further colonization. The ETI resistance mechanisms, including the role NLRs play in pathogen perception and understanding of the up and down stream signaling events, have been extensively reviewed (Cui et al., 2015; Gassmann & Bhattacharjee, 2012; Rafiqi et al., 2009).

Until recently PTI and ETI were considered as two separate yet complimentary layers of the plant immune system, however, recent research has indicated that there is an early intimate relationship between the two (Thomma
et al., 2011). It was demonstrated that the FLS2 receptor physically interacts
with three distinct *Pseudomonas syringae* race specific *R*-genes *Rpm1*, *Rps2*,
and *Rps5*, and it was speculated that these *R*-genes guard components of
the FLS2 mediated PTI mechanisms, which are *P. syringae* effector targets.
These effectors have evolved to target the PTI components to suppress the
early non-host resistance mechanisms (Qi et al., 2011). These findings in the
model *P. syringae-Arabidopsis* pathosystem have given rise to the zigzag
model of co-evolution between hosts and parasites, where both are continually counter evolving to gain the upper hand (Jones & Dangle, 2006). The
zigzag model originally developed based on this model pathosystem has been
criticized as being over simplistic (Pritchard & Birch, 2014), yet similar to
Flor's gene-for-gene model, although simplistic and not accounting for the
diversity, many deviations and complexities in host–parasite genetic interactions, it does provide the best model to describe the co-evolutionary arms
races that are occurring in many pathosystems.

Proteins that contain at least one serine/threonine protein kinase (STPK)
domain have also been identified as another class of receptor-like *R*-proteins
(Martin et al., 2003). The involvement of these STPK domain proteins,
with phosphorylation function, suggests that these proteins function in ETI
signaling pathways (Hanks et al., 1988). However, recent functional characterization of the cytoplasmic localized STPK *R*-genes suggest that most are
not components of ETI signaling mechanisms but rather are mimics of the
early PTI pathway signaling components which are targeted by pathogen
effector proteins. These data supported a new twist on the "Guard hypothesis" which is now termed as the "decoy model" (van der Hoorn & Kamoun,
2008). For example, the Tomato *Pto R*-gene conferring resistance to the bacterial pathogen *P. syringae* (Swiderski & Innes 2001; Martin et al., 1993) has
been hypothesized to be a decoy of STPK proteins underlying the FLS2 PRR
complex (van der Hoorn & Kamoun, 2008). In this mechanism the alteration
to the STPK decoy proteins are hypothesized to be under surveillance by
the NLR, PRF. The action of the pathogen to suppress the PTI responses by
targeting the STPK domains underlying these immunity-signaling pathways
with the effectors Avr PtoB is detected through the presence of biologically
irrelevant STPK decoy proteins. These decoys are also targeted by the effectors whose actions elicit ETI through the NLR guard proteins. The involvement of the functional kinases in the ETI signaling mechanisms, as has been
hypothesized for decades, is being challenged by the long-standing inability

to identify these pathways. Also NLR functional data coupled with studies showing nuclear localization and function post pathogen interaction (Mang et al., 2012; Cheng et al., 2009; Shen et al., 2007; Wirthmueller et al., 2007; Bruch-Smith et al., 2007; Deslandes et al., 2003) implies a direct, or very short signaling pathway from NLR pathogen recognition to HR activation. These short direct ETI signaling pathways may be more efficient immunity mechanisms, which do not provide vulnerable targets or hubs for pathogen effectors to suppress. Thus, ETI responses, which have been considered to be less durable than the PTI "nonspecific" resistance responses, functionally may have a more direct and efficient signaling pathways that are not as vulnerable to pathogen effector manipulation as the PTI phosphorylation pathways involving MAPK signal transduction.

It has been shown in many ETI mechanisms that dual NLRs are required for resistance and it appears that tightly linked dual NLR proteins belonging to diverse gene families are required together for pathogen recognition and defense signaling (Eitas & Dangl, 2010; Sinapidou et al., 2004; Gassmann et al., 1999; Deslandes et al., 2003; Birker et al., 2009; Narusaka et al., 2009; Loutre et al., 2009; Wang et al., 2013; Brotman et al., 2012; Ashikawa et al., 2008; Lee et al., 2009; Okuyama et al., 2011; Yuan et al., 2011; Zhai et al., 2014). The NLR pairs have been shown in monocotyledonous and dicotyledonous plants to be required for the recognition of Avirulence proteins from bacterial, viral, oomycete, and fungal pathogens. Thus this mechanism of dual NLRs is a common and widespread mechanism in plant immunity. For eight of these dual NLR resistance mechanisms described in diverse host-pathogen systems, one of the NLRs contains a variable domain that has been hypothesized to be an "integrated decoy" where the variable decoy domain is integrated as part of the NLR. This new twist on defense mechanisms has given rise to the "integrated decoy model" (Cesari et al., 2014). Interestingly the barley *rpg4/Rpg5* stem rust resistance mechanism is an example of an integrated decoy mechanism and will be extensively described and discussed here.

22.2   *RPG1*-MEDIATED RESISTANCE

Similar to the management of stem rust in wheat, stem rust was also managed using genetic resistance in barley, but contrary to wheat, only a single source of resistance was deployed and has protected Midwestern and Canadian Prairie Province barley cultivars (cvs) since 1942. Major stem rust epidemics in barley were controlled since the 1940s by fixing six- and two-rowed barley varieties with the single durable *R*-gene *Rpg1* (Steffenson, 1992). This single *R*-gene has remained remarkably durable in the field considering the "boom and bust" that occurred with single race specific *R*-genes that were deployed in wheat. The *Rpg1* gene confers broad-spectrum resistance against a diversity of races, but these races are designated by their reactions on wheat differentials containing 20 specific wheat stem rust *R*-genes (http://www.fao.org/agriculture/crops/rust/stem/stem-pathotypetracker/stem-differentialsets) following the North American nomenclature system developed by Roelfs and Martens (1988). If known effective barley *R*-genes (*Rpg1*, *rpg4/Rpg5*, and *rpg6*) were used to develop differentials for wheat stem rust virulence on barley, there would only be three differentials compared to the numerous possibilities in wheat (>70 *R*-genes). This limited number of race specific *R*-genes and the broad resistance provided by the limited genes present in the primary barley germplasm pool suggests that these resistances may represent holdovers of non-host resistance mechanisms. We posit that there may have been a relatively short period of host–parasite co-evolution in the barley–wheat stem rust pathosystem, thus the limited history of co-evolution has resulted in a limited number of resistance specificities.

In an effort to understand how these limited stem rust *R*-genes function in barley, especially the exceptional durability provided by the *Rpg1* gene, efforts were undertaken to clone and characterize the *Rpg1*, *rpg4*, and *Rpg5* genes. From the period of 1990 to 2010 this research in the lab of Dr. Andris Kleinhofs at Washington State University culminated in the cloning of the *Rpg1* wheat stem rust *R*-gene (Brueggeman et al., 2002) and the *Rpg5* rye stem rust *R*-gene (Brueggeman et al., 2008) via positional cloning efforts. The identification of both of these barley stem rust *R*-genes uncovered atypical *R*-gene structures. The *Rpg1* *R*-gene encodes a unique protein kinase domain structure with tandem STPK domains, one predicted to be an active kinase and the second a pseudo non-functional kinase (Brueggeman et al., 2008; Nirmala et al., 2010). To validate the dual kinase gene as *Rpg1* the gene from the moderately resistance cv Morex was transformed into the susceptible cv Golden Promise under the control of its endogenous promoter. It was shown that the transformation of a single copy of the *Rpg1* gene converted Golden Promise from a susceptible genotype to near immunity against *Pgt* pathotype MCCF (Horvath et al., 2003). This unexpected result suggested that there are other interacting loci in barley genotypes that enhance these broad-spectrum wheat stem rust resistance mechanisms in barley. Interestingly, transcript analysis shows that it is not expression differences in the *R*-genes contributing to the differential resistance responses, thus other genetic factors are contributing to the enhancement of resistance. This was further substantiated by Expression quantitative trait loci (eQTL) analyses that identified loci/or a locus on chromosome 2H, which enhances resistance mediated by *Rpg1* (Druka et al., 2008) and the same or possibly a closely linked yet distinct locus also enhances *rpg4/Rpg5*-mediated resistance against TTKSK (Moscou et al., 2011). Druka et al. (2008) identified a Histidine Kinase as the putative candidate enhancer of resistance but the genes' function was not validated.

Functional characterization of the RPG1 protein determined that both the functional kinase and pseudokinase domains are essential to confer resistance (Nirmala et al., 2006). Interestingly, it was shown that inoculation with the avirulent isolates specifically elicited a very rapid RPG1 protein phosphorylation event, within 5 min of inoculation, followed by degradation of the RPG1 protein beyond Western blot detection at 24 h post inoculation through the ubiquitin pathway (Nirmala et al., 2010; Nirmala et al., 2007). This very rapid response to the avirulent pathogen suggested that the avirulence proteins may be present in the spore and is immediately released or detected upon leaf contact. It is likely that this early pathogen detection occurs via a plasma membrane bound receptor, possibly a PRR, which transmits the signal across the plasma membrane quickly eliciting phosphorylation of the RPG1 proteins. The RPG1 specific effector proteins were later identified and shown to be two unexpectedly large proteins, a RGD-binding protein and a VPS9 protein, which were likely present in the spore (Nirmala et al., 2011). When these effector proteins were expressed in yeast and co-infiltrated on 10 days old barley seedlings, they elicited a PCD response similar to HR on genotypes containing *Rpg1* but showed no response on barley genotypes lacking *Rpg1* (Nirmala et al., 2011). However, these data are somewhat contradictory to histology data generated by Zurn et al. (2015) that showed inoculation of *P. graminis* f. sp. *tritici* on *Rpg1* containing barley apparently does not elicit early prehaustorial resistance response because the infection process proceeds to haustoria formation in both *Rpg1* resistant lines and susceptible lines lacking *Rpg1* (Zurn et al., 2015). It could be hypothesized that the early recognition event resulting in the phosphorylation (<5 min post pathogen inoculation) and degradation of RPG1 protein (~24 h post inoculation) primes the plant for the later detection of specific avirulence/effectors which may be haustorial secreted. This post haustorial recognition of secreted effectors would results in the strong ETI defense response. Since the infiltration of the two *P. graminis* f. sp. *tritici* effectors, RGD-binding and VPS9, elicits both early phosphorylation and later protein degradation responses, it suggests that the later degradation of the RPG1 proteins must be independent of the later pathogen infection process. This suggests that the protein degradation defense response is set into motion by the early RGD/VPS9 duel effector detection. Since the strong HR response apparently does not occur until after haustoria formation and is associated with the areas of haustoria formation (Zurn et al., 2015) we speculate that the early response may represent a non-host resistance mechanism but that the early recognition events by a putative PRR or plasma membrane localized receptor complex is coupled to later responses by priming the host for this later ETI response, that is elicited by a haustorial-produced effector. The best model for such a mechanism is the FLS2/RPS5 resistance mechanism in *Arabidopsis* against *P. syringae* carrying the Avr PphB effector. Where recognition of FLG22 flagellar subunits by the FLS2 PRR receptor primes the Rps5 resistance mechanism for the later recognition of the secreted Avr PphB effector, which targets the PBS1 kinase guarded by the NLR Rps5 (Porter et al., 2012). However, this *Rpg1*-mediated resistance model does not explain the rapid HR-like response that is elicited when *Rpg1* containing barley lines are infiltrated with the RGD and VPS9-like effector proteins. Yet, some PRR receptors do elicit programmed cell death responses similar to HR (Nicaise et al., 2009). Possibly, the infiltration of the RGD-binding and VPS9 proteins into the apoplast results in a biologically irrelevant strong HR like response because the host receptor complex did not evolve to recognize these effectors in the apoplast. It has been shown that the *Rpg1* gene has much higher expression in the epidermal cells suggesting that these initial recognition complexes involving *Rpg1* may have evolved to have higher levels of expression at the leaf surface as spore recognition is very important to elicit and set the priming of the resistance mechanism into motion. Thus, it is very important to have much higher levels of sensitivity at the leaf surface to detect the pathogen as early as possible and infiltrating a high concentration of these proteins in to the apoplast results in an amplified HR response. We have generated data showing that there is a weak HR response that occurs around the point of spore contact yet the germ tube still grows to the stomata and forms haustoria (Solanki & Brueggeman, unpublished data). Interestingly, it was shown that several transgenic multi copy *Rpg1* lines (Horvath et al., 2003) had higher transcript levels and RPG1 protein levels than wild-type *Rpg1* lines and single copy transgenic lines, yet were susceptible to *P. graminis* f. sp. *tritici* race MCCF. Post pathogen infection the multi copy transgenic lines showed an inability to degrade all the RPG1 protein suggesting the requirement that the RPG1 protein must be degraded during the resistance response and failure to do so leads to loss of resistance activation. Thus, it could be speculated that RPG1 functions to suppress the resistance signaling responses and the degradation releases this suppression. These pieces of information in the resistance responses mediated by *Rpg1* are part of a complex puzzle, which will require extensive further research to unravel.

22.3 *RPG4/RPG5*-MEDIATED RESISTANCE

Although *Rpg1* has been remarkably durable in the Midwestern United States and Canadian Prairie Provinces there have been wheat stem rust races detected and minor epidemics caused by races that show high virulence on barley lines containing the *Rpg1* gene. These races include, but are not limited to, *P. graminis* f. sp. *tritici* race QCCJF that was first identified in North Dakota in 1988 (Roelfs et al., 1991). Then, in 1999 and 2001 a new race of *P. graminis* f. sp. *tritici* race TTKSK (A.K.A. Ug99) was detected in Uganda and Kenya, respectively, that showed virulence on important wheat resistances (Pretorius et al., 2000). *P. graminis* f. sp. *tritici* race TTKSK is also virulent on barley lines containing *Rpg1* (Steffenson & Jin, 2006). Our recent analysis of stem rust races collected from the 1970s to 1990s in North Dakota identified several other isolates that are virulent on *Rpg1* and based on PCR-GBS marker analyses are quite divers from one another suggesting that virulence on *Rpg1* is not as uncommon as previously believed (Sharma Poudel & Brueggeman, unpublished). However, in the absence of suscep-tible wheat varieties, as was the case with race QCCJF, these rust races do not become predominant races in the North American stem rust populations as it is virulence on deployed wheat resistances that is more important in the epidemiology of stem rust epidemics in North America. It is also possible that this lack of virulence on *Rpg1* may have an epistatic effect. The *Avr Rpg1* effector/s may have virulence function that when missing imposes a fitness penalty on the pathogen in both wheat and barley. However, the TTKSK race and its lineage that emerged in Africa although also highly virulent on *Rpg1* containing barley as well as 97% of the world's commercial barley varieties tested (Steffenson et al., 2013) appears to be highly virulent and robust without apparent fitness penalties. As these new races of wheat stem rust threatened epidemics in barley, comprehensive searches of the primary barley germplasm pool, including *H. vulgare* and *Hordeum spontaneum*, were conducted to identify additional sources of resistance to counter these new rust races (Jin et al., 1994; Steffenson et al., 2013). The only effective source of race QCCJF and TTKSK resistance identified to date was origi-nally detected in the unimproved barley line Q21861 and it was designated as the *rpg4* gene. The *rpg4* gene was shown to be an atypical rust resistance because it is recessive in nature and temperature sensitive (Jin et al., 1994). The *rpg4* gene was shown to be effective against the wheat stem rust races QCCJF and TTKSK, and these resistances mapped and co-segregated to the sub telomeric region of barley chromosome 5H (Steffenson et al., 2009), very tightly linked to the region harboring the barley *Rpg5* gene which confers dominant resistance against rye stem rust isolates (Brueggeman et al., 2008). Another, QCCJF resistance source was also identified from the interspecific barley breeding line (212Y1) with introgressed chromatin from *Hordeum bulbosum* (Fetch et al., 2009). This resistance source, designated as the *rpg6* gene, confers resistance to *P. graminis* f. sp. *tritici* races MCCF and QCCJF and was also determined to be recessive. Preliminary seedling data suggested that *rpg6* is not effective against *P. graminis* f. sp. *tritici* race TTKSK suggesting that it has different resistance specificities than both *Rpg1* and *rpg4/Rpg5*. The *rpg6* gene mapped to the telomeric end of barley chromosomes 6HS suggesting that it was a novel *R*-gene. Four other stem rust *R*-genes have been described in barley including *Rpg2*, identified from Hietpas-5 (CIho 7124) (Patterson et al., 1957), *Rpg3*, from PI382313 (Jedel, 1990), *rpgBH* from Black Hulless (CIho 666) (Steffenson et al., 1984), and *RpgU* from the barley cv Peatland (Fox & Harder, 1995). However, these four genes confer low levels of stem rust resistance making them difficult to phenotype and characterize genetically, therefore, to date these resistance loci have not been mapped (Sun & Steffenson, 2005).

The cloning of the *Rpg5* gene uncovered a new R-protein domain structure encoding the typical NLR R-protein and STPK protein domains, discussed earlier, however, the domains are encoded by the single *Rpg5* transcript. This structure was previously an unknown structure among plant *R*-genes (Brueggeman et al., 2008). The *rpg4* wheat stem rust *R*-gene and *Rpg5* rye stem rust *R*-gene were originally determined to be distinct yet closely linked genes and in the original cloning efforts the best candidate *rpg4* gene, *HvADf2*, was determined to encode an actin depolymerizing factor (ADF) (Brueggeman et al., 2008; Brueggeman et al., 2009; Kleinhofs et al., 2009). However, the candidacy was based on restriction fragment length polymorphism (RFLP) marker data and the subsequent high-resolution genetic analysis utilizing new single nucleotide polymorphism (SNP) and cleaved amplified polymorphic sequences (CAPs) markers designed based on comparative genome sequence analysis in the *rpg4/Rpg5* region deter-mined that *HvAdf2* was not *rpg4* (Wang et al., 2013). Utilizing post tran-scriptional gene silencing via barley stripe mosaic virus–virus induced gene silencing (BSMV-VIGS) it was shown that *rpg4*-mediated wheat stem rust resistance is the concerted action of three genes at the *Rpg5* locus including the two NLRs, *Rpg5* and *HvRga1*, and the actin depolymerization factor *HvAdf3* (Wang et al., 2013). A fourth gene tightly linked to the *Rpg5* locus designated *Rme1* (*rpg4* modifier element 1), which only segregated in the Steptoe × Q21861 population was also shown to be required for resistance. It was the recombinants from this bi-parental population that led us to hypothesize that *rpg4* and *Rpg5* were distinct. However, in two other populations (Harrington × Q21861 and MD2 × Q21861) *P. graminis* f. sp. *tritici* races QCCJF and TTKSK resistance mapped to the *Rpg5* locus suggesting that the functional polymorphism at *Rme1* between Steptoe and Q21861 is not present between Harrington, MD2, and Q21861 (Wang et al., 2013).

In depth allele analysis of the three genes at the *Rpg5* locus required for *rpg4*-mediated stem rust resistance in barley determined that the NLR *HvRga1* and actin depolymerization factor *HvAdf3* are conserved and no polymorphisms in primary amino acid sequences correlated with resistance and/or susceptibility. However, all QCCJF and TTKSK susceptible lines can be attributed to four different classes of *Rpg5* susceptible alleles with predicted truncated proteins, missing STPK domains and a new susceptible allele with an E1287A amino acid substitution (Arora et al., 2013). These data suggest that *Rpg5* is the polymorphic R-protein at the *rpg4/Rpg5* locus that contributes to *rpg4*-mediated stem rust resistance, thus *rpg4*-mediated wheat stem rust resistance and *Rpg5*-mediated rye stem rust resistance cannot be considered distinct resistance mechanisms, despite the differences in recessive vs dominant resistance, respectively. This intriguing discontinuity between the two resistance specificities relying on the same disease *R*-gene is intriguing and will be further discussed.

Recent genetic characterization of 10 NLR mediated resistance mechanisms from diverse plant species effective against diverse taxa of pathogens has shown that several resistances against pathogens containing specific *Avr*-genes require two unrelated NLR genes with little or no sequence homology (Eitas & Dangl, 2010). The functional characterization of some of the R-proteins in these dual NLR mechanisms have determined that the protein domains are modular and protein fragments are sufficient for homo and hetero dimerization and to initiate defense signaling (Cesari et al., 2015). During defense signaling the proteins act together in defense complexes by interaction of different combinations of intermolecular protein domains possibly guarding a variable domain found in one of the NLR pairs, recently proposed in the "integrated decoy model" (Cesari et al., 2014). The model proposes that after the decoy domain is targeted by the pathogen effector, the heterodimeric NLR molecular switch is tripped by this pathogen perception and one or both of the NLR proteins act as signaling components by possibly relocalizing to or accumulating in the nucleus to initiate signaling processes and defense gene expression (Cesari et al., 2014).

The STPK *R*-genes have been cloned from diverse taxa of plant species and confer resistance to bacterial and fungal pathogens (Brueggeman et al., 2008; Brueggeman et al., 2002; Fu et al., 2009; Sun et al., 2004; Swiderski & Innes, 2001; Song et al., 1995; Martin et al., 1993). Many early models suggested that the STPK domain functions in plant defense signal transduction pathways involving phosphorylation cascades (Hanks et al., 1988), however, it was shown that some STPK R-proteins directly interact with Avr proteins, demonstrating their ability to function as receptors that can recognize pathogen challenge through protein–protein interactions (Tang et al., 1996). For some bacterial pathosystems it has been shown that NLR and protein kinase genes are required together for resistance and the STPK component actually serves as a "guardee" or "decoy," which is under surveillance by the NLR (van der Bizen & Jones, 1998). Resistance against *P. syringae* strains that carry *Avr PphB* requires two *Arabidopsis* genes, the *RPS5* NLR *R*-gene (guard) and the *PBS1* gene (guardee), a STPK sensory domain gene (Mucyn et al., 2006). As previously mentioned, the dual NLR proteins have shown to be a common theme to resistance mechanisms, with one of the NLRs containing a variable domain also referred to as the "integrated decoy" or "sensory domain." The *rpg4/Rpg5* mechanism is an excellent system for elucidating and validating the STPK as a sensory domain (SD) acting as an integrated decoy. The *Rpg5* NB-LRR-STPK and the requirement of the second NLR, *HvRga1,* fit the model and we have begun work to validate the integrated decoy model in this important crop pathosystem.

As previously described the *Arabidopsis-P. syringae* pathosystem, requiring the *Rps5, Pbs1,* and *AtAdf4* genes, is an excellent model system showing that a phosphorylation pathway links PTI recognition with the priming of an *R*-gene that guards components of PTI signaling pathway/s (Porter et al., 2012). The protein subunits of the *P. syringae* flagella, flg22, is recognized by the PAMP recognition receptor *Fls2* which elicits the PTI responses and triggers a phosphorylation cascade, resulting in the phosphorylation of the AtADF4 protein, which is required as a signaling intermediate that results in the putative priming of the NLR *Rps5 R*-gene. The RPS5 protein guards the PK PBS1, which is targeted for degradation by the later secretion of the Avr PphB effector, which targets PBS1 as a putative decoy of the actual effector susceptibility target BIK1 (Porter et al., 2012). The HvAdf3 protein, which has homology to the AtADF4 protein, was shown to be required for the *rpg4/Rpg5* mediated resistance response and may represent a component of the intermediate signaling pathway linking an early *Pgt* recognition event, detected by the early regulation of *HvAdf3* (Solanki & Brueggeman, unpublished data), and the later ETI response that we hypothesize occurs via a haustoria produced effector that is secreted into the cytoplasm and acts on the RPG5 STPK domain or a recently discovered protein which may act as the scaffold that links the *Rpg5-HvRga1* resistance complex together in the inactive state (Solanki & Brueggeman, unpublished data).

22.4   CYTOSKELETON DYNAMICS AND STEM RUST RESISTANCE MECHANISMS

Resistance in *Arabidopsis* against *P. syringae* strains that carry *Avr PphB* requires at least three host genes, *RPS5*, a NBS-LRR gene, *PBS1*, a serine threonine protein kinase gene, and the ADF *AtAdf4* (Tian et al., 2009). It has been shown in this model *P. syringae- Arabidopsis* pathosystem that RPS5 the NLR race specific *R*-gene responsible for triggering the ETI response directly interact with a PRR and that the NLR R-protein probably guards the PRR STPK domain by detecting pathogen effector/avirulence protein mediated modifications or interactions (Qi et al., 2011). The barley *Rpg5* gene combines the NBS-LRR domain containing 23% amino acid identity and 40% similarity to *Rps5* with a protein kinase containing 60% amino acid identity and 71% similarity to PBS1 in the same gene. *Rpg5* also requires the *Hv*Adf3 gene, which contains 52% identity and 76% similarity to *AtAdf4* for resistance against several *Pgt* pathotypes (Wang et al., 2013). These similarities to the *Rps5* mechanism of resistance and the commonality of early responses to the pathogen have prompted us to use the *Rps5* system as a model for the development of our barley stem rust resistance model. These three protein components may be common among resistance complexes where the *R*-genes are guarding decoys of PTI components, which may be the case with wheat stem rust resistance in barley. However, in the *Rpg5* mechanism it is a putative integrated decoy. The *rpg4/Rpg5* mechanism may provide an excellent system for elucidating the function of PTI/ETI coupled resistance against rust pathogens in a cereal crop to help elucidate the link between PTI plasma membrane localized receptor complexes and the later cytoplasmically localized NLR receptors that elicit the strong ETI responses.

Plants have unique signaling mechanisms that react to environmental cues resulting in responses that involve changes in cytoplasmic organization and cell shape. This system of innate cellular responses to the environment is required of an organism that cannot respond to environmental stresses by evading the perpetrator. The rapid reorganization of the cell's actin cytoskeleton is triggered by environmental cues and is modulated by different biochemical action through a pool of actin binding proteins (Hussey et al., 2006; Staiger & Blanchoin, 2006). ADFs constitute a family of proteins important in these processes (Maciver & Hussey, 2002).

Plant ADFs have been shown to play roles in biotic stress responses (Tian et al., 2009; Miklis et al., 2007); however, information regarding the molecular mechanisms underlying these responses is limited. Plant genomes have small numbers of *Adf* genes as demonstrated by the presence of 12 *Adf* genes in the sequenced genomes of *Arabidopsis* and rice (Feng et al., 2006). This low redundancy of proteins with functional diversity suggests that individual *Adfs*, although having concise function, are differentially modulated by diverse biotic, abiotic, and developmental stimuli.

Plant cells respond to surface colonization of both bacteria and fungi by rapid cell polarization of the actin cytoskeleton and formation of large actin bundles beneath the pathogen entry sites (Schmelzer, 2002; Takemoto et al., 2003). Cytoskeleton polarization in response to pathogen challenge is well documented (Schmelzer, 2002; Lipka & Panstruga, 2005) and is ADF dependent. Over expression of exogenous ADF and drug mediated blockage of actin filament assembly in plants allowed increased fungal pathogen entry by non-host pathogens (Miklis et al., 2007). Cytochalasin E, an actin cytoskeleton inhibitor, and genetic interference by over expression of barley *Adf* (*Hv*Adf3) were utilized in a study of the *mlo* non-race specific resistance system conferring resistance to *Blumeria graminis* f. sp. *hordei* in barley. Actin cytoskeleton inhibition demonstrated that cytoskeleton function was required for basal defense against an appropriate mildew pathogen and for *mlo*-mediated non-race specific resistance at the cell wall. It was also demonstrated that actin filament disruption led to increased fungal penetration by non-host pathogens and was not required for several race specific immune responses (Miklis et al., 2007).

Substantial data suggest that polarization of the cytoskeleton is important in cytoplasmic streaming and vesicle transport to direct the deposition of cell wall fortifying and antimicrobial compounds at the cell periphery (Kwon et al., 2008). Chemical and genetic disruption of actin polymerization in tobacco indicated that cytoskeleton perturbation also primes the cell for HR (Kobayashi & Kobayashi, 2007). It was proposed that the host detects disruption of the actin cytoskeleton similar to the guard hypothesis, thus triggering the HR response. Evidence is mounting that the actin cytoskeleton is the target of plant pathogen effectors to suppress non-race specific resistance or PTI responses, which may signal to induce a second line of defense mediated by the pathogen specific *R*-genes that elicit a stronger ETI response (Jones & Dangl, 2006). Evidence suggests that non-host pathogens that evade the first line of defense and enter the cell are met by an HR mediated resistance that could be due to several simultaneously acting *R*-genes (Thordal-Christensen, 2003).

Previous research suggested that cytoskeleton reorganization responses are characteristic of basal resistance or non-host resistance mechanisms (Kwon et al., 2008) and only limited circumstantial evidence supported its role in specific or *R*-gene mediated resistance. However, recent research indicates that ADF proteins do play a role in some specific *R*-gene mediated resistance pathways against fungal and bacterial pathogens (Wang et al., 2013; Tian et al., 2009). Tian et al. identified an *Arabidopsis thaliana* gene, *AtAdf4* that mediates *RPS5 R*-gene specific resistance against *P. syringae* harboring the effector AvrPphB. They demonstrated that *AtAdf4* was not involved in resistance against pathogen entry at the cell periphery indicative of non-host resistance but that *AtAdf4* mediates *RPS5* initiated race specific defense signaling by modifying the actin cytoskeleton.

Our research characterizing the *rpg4/Rpg5* barley stem rust *R*-gene locus has revealed an *R*-gene (*Rpg5*) mediated resistance mechanism that requires *HvAdf3* indicating reliance on actin reorganization. Based on current data we hypothesize a mechanism whereas the early stem rust resistance responses in barley are intimately linked to an extracellular receptor, which recognizes the pathogen at the surface of the leaf and transduces the signal across the plasma membrane. This is possibly through an early PRR type receptor, which recognizes a conserved PAMP as early differential regulation of *HvAdf3* occurs yet it occurs in both the resistant (rpg4/Rpg5+) and susceptible (Rpg4/RPG5-) barley lines. However, at 6 h post inoculation HvAdf3 is down regulated in the resistant lines and upregulated in the susceptible lines (Solanki & Brueggeman, unpublished data). We have developed a model where the *Rpg5* and *Hv*Rga1 resistance proteins are possibly primed for a more sensitive detection of a later specific haustorial expressed Avr protein through HvAdf3 signaling yet further research is required to understand these complex interactions.

22.5   RECESSIVE NATURE OF *RPG4*-MEDIATED RESISTANCE

An important regulatory process of phosphorylation signaling cascades triggered by biotic or abiotic stimuli in plants is the ability to reverse or turn the signaling process off. One mechanism of *R*-gene initiated signal transduction in plant cells is achieved through protein phosphorylation cascades mediated by protein kinase activity. A hypothesized regulatory mechanism of phosphorylation cascades is through opposing protein kinase and protein phosphatase activities (Rodriguez, 1998). Genome sequence information has determined that PP2Cs are a major class of serine/threonine phosphatases in plants and the recurrent theme is that the super family of plant PP2Cs negatively-regulate STPK kinase signaling pathways through dephosphorylation activity. Presently, there is little information on specific PP2C substrate binding in plant signaling but the large number of PP2Cs in plant genomes suggests that individual phosphatases may have tight specificity in substrate binding (Kerk et al., 2002). We have identified an *Rpg5* allele which contains a PP2C domain in place of the STPK domain, *rpg5*-PP2C, that may negatively regulate signaling processes activated by the *Rpg5/Hv*Rga1 mediated detection of the wheat stem rust races QCCJF and TTKSK. Genetic analysis indicated that the recessive nature of the *rpg4*-mediated stem rust resistance is determined by the presence of the rpg5-PP2C allele in several susceptible genotypes and absence from the resistant genotypes (Solanki et al., unpublished data). Protein phosphatases are known to be antagonistic of phosphorylation pathways and there are two examples of PP2C proteins that negatively regulate resistance pathways in plant systems. These PP2C proteins negatively regulate responses elicited by PAMP receptors. In Arabidopsis the kinase-associated protein phosphatase (KAPP) physically interacts with the FLS2 receptor and over expression of KAPP results in insensitivity to the PAMP flg22 and reduced FLS2/flg22 binding (Gomez-Gomez et al., 2001). In rice the PAMP receptor kinase Xa21 physically interacts with the PP2C protein XB15 that was shown to negatively regulate the innate immune response elicited by the receptor kinase Xa21 against *Xanthomonas oryzae* (Park et al., 2008). The *rpg5*-PP2C allele present at the *rpg4/Rpg5* locus in many susceptible barley varieties appears to act as a dominant susceptibility factor determining the recessive nature of the *rpg4*-mediated wheat stem rust resistance. The *rpg5*-HvPP2C appears to negatively regulate stem rust resistance signaling activated by the *Rpg5/HvRga1 R*-genes against the wheat stem rust races QCCJF and TTKSK but not the rye stem rust isolates. This intriguing twist on the integrated decoy model, the replacement of the STPK domain with a putatively functional PP2C with antagonistic function, is truly remarkable. It is hard to imagine the evolutionary unlikely hood that these integrated decoy mechanisms link with the NLRs unless the plants innate immunity systems have some directed evolutionary process that we have yet to discover or appreciate.

22.6 CONCLUDING REMARKS

After over half a century of managing stem rust through genetic resistance it was believed by many that stem rust was a thing of the past. Norman Borlaug's statement sums this up quite well: "Isn't it quite possible that the greatest ally of the pathogen is our short memory of the disastrous stem rust epidemics from 1951-1954 across much of the wheat producing regions of North America" (Peterson, 2001). The misconception that stem rust has been defeated has fostered complacency and currently restricts research and breading for resistance against this shifty enemy. Stem rust research has received less and less funding, leading to the dependency on too few resistant sources in the United States and the major wheat and barley producing regions worldwide. The recent emergence of a new highly virulent race of stem rust in Eastern Africa and its migration through Middle Eastern Asia has sent plant scientist scrambling to identify and deploy new sources of resistance in both wheat and barley.

The barley–wheat stem rust pathosystem is remarkable because there appears to be very limited resistance sources in barley and the two well-characterized mechanisms confer broad resistance and appear to have common downstream signaling mechanisms. This was elucidated by the identification of fast neutron generated mutants that compromise both *Rpg1* and *rpg4/ Rpg5* mediated resistance (Solanki & Brueggeman, unpublished data). It can be hypothesized that the lack of resistance specificity in the barley— *P. graminis* f. sp. *tritici* pathosystem is due to a limited time of co evolution between barley and wheat stem rust. Thus, this limited co-evolutionary timescale has not given rise to numerous specific *R*-genes, as is seen in the wheat–wheat stem rust pathosystem. Also, the broad resistances provided by the two characterized *R*-genes, *Rpg1* and the *rpg4/Rpg5* complex, suggest that these resistance complexes may rely on an early recognition event that occurs via a plasma membrane signaling complex, possibly a PAMP recognition receptor (PRR) that elicits an early PTI-like non-host resistance response that is coupled to the NLR mediated ETI responses that occur post haustoria formation. The hypothesis that *P. graminis* f sp. *tritici* resistance may be more of an early cell surface receptor complex recognition system has been further substantiated by the fact that the *Rpg1*-mediated resistance mechanism has a very early response elicited < 5 min post inoculation with *Rpg1* avirulent isolates (Nirmala et al., 2010). These responses are measured as RPG1 protein phosphorylation within 5 min of host–parasite interaction, and subsequent RPG1 protein degradation at 24 h post pathogen inoculation (Nirmala et al., 2008). The pathogen proteins that elicit the early response are large molecules that are apparently present in the spore coat, required together to elicit the defense responses, possibly representing PAMPs. Thus, a major question in these plant resistance mechanisms is whether these R proteins are coupled to priming by an early non-host PRR-like receptor and how do they switch from a functionally "inactive" state to "active" state. Although the answers are still largely unknown our functional characterization of these resistance mechanisms is getting us ever so much closer to the answers. The new insights into the wheat stem rust resistance mechanisms in barley are beginning to resolve the complex interactions and suggest that these mechanisms of resistance may be remnants of non-host resistance mechanisms suggesting that *P. graminis* and barley may have had short host–parasite co-evolutionary time period suggesting that barley is a relatively recent host of this devastating pathogen.

KEYWORDS

- **Ug99**
- **NLR**
- **sensory domain**
- **Integrated decoy**
- ***R*-genes**
- **ETI**
- **pseudokinase**
- ***Avr*-genes.**

REFERENCES

Ali, G. S.; Reddy, A. S. N. PAMP-Triggered Immunity. *Plant Signal. Behav.* **2008**, *3*(6), 23–426.

Arora, D.; Gross, T.; Brueggeman, R. Allele Characterization of Genes Required for *rpg4*-Mediated Wheat Stem Rust Resistance Identifies *Rpg5* as the R-Gene. *Phytopathology.* **2013**, *103*, 1153–1161.

Ashikawa, I., et al. Two Adjacent Nucleotide-Binding Site-Leucine-Rich Repeat Class Genes are Required to Confer Pikm-Specific Rice Blast Resistance. *Genetics.* **2008**, *180*, 2267–2276. doi: 10.1534/genetics.108.095034.

Birker, D., et al. A Locus Conferring Resistance to *Colletotrichum higginsianum* is Shared by Four Geographically Distinct Arabidopsis Accessions. *Plant J.* **2009,** *60,* 602–613. doi: 10.1111/j.1365-313X.2009. 03984.x.

Brotman, Y., et al. Dual Resistance of Melon to *Fusarium oxysporum* Races 0 and 2 and to Papaya Ring-Spot Virus is Controlled by a Pair of Head-to-Head- Oriented NB-LRR Genes of Unusual Architecture. *Mol. Plant.* **2012,** *6,* 235–238. doi: 10.1093/mp/sss121.

Brueggeman, R.; Steffenson, B. J.; Kleinhofs, A. The *rpg4/Rpg5* Stem Rust Resistance Locus in Barley: Resistance Genes and Cytoskeleton Dynamics. *Cell Cycle.* **2009,** *8*(7), 977–981.

Brueggeman, R., et al. The Stem Rust Resistance Gene *Rpg5* Encodes a Novel Protein with Nucleotide Binding Site, Leucine-Rich and Protein Kinase Domains. *Proc. Natl. Acad. Sci. USA.* **2008,** *105,* 14970–14975.

Brueggeman, R., et al. The Barley Stem Rust-Resistance Gene *Rpg1* is a Novel Disease-Resistance Gene with Homology to Receptor Kinases. *P. Natl. Acad. Sci. USA.* **2002,** *99,* 9328–9333.

Burch-Smith, T. M., et al. A Novel Role for the TIR Domain in Association with Pathogen-Derived Elicitors. *PLoS Biol.* **2007,** *5,* e68.

Cesari, S., et al. A Novel Conserved Mechanism for Plant NLR Protein Pairs: The "Integrated Decoy" Hypothesis. *Front. Plant Sci.* **2014,** *5,* 606. doi:10.3389/fpls.2014.00606.

Cheng, Y. T., et al. Nuclear Pore Complex Component MOS7/Nup88 is Required for Innate Immunity and Nuclear Accumulation of Defense Regulators in *Arabidopsis. Plant Cell.* **2009,** *21,* 2503–2516.

Chester, K. S. *The Nature and Prevention of the Cereal Rusts as Exemplified in the Leaf Rust of Wheat;* Chronica Botanica: Waltham, MA, 1946.

Cui, H.; Tsuda, K.; Parker, J. E. Effector-Triggered Immunity: From Pathogen Perception to Robust Defense. *Annu. Rev. Plant Biol.* **2015,** *66,* 487–511.

De Guillen, K. Structure Analysis Uncovers a Highly Divers but Structurally Conserved Effector Family in Phytopathogenic Fungi. *PLoS Pathog.* **2015,** *11*(10), e1005228. DOI:10.1371/journal.ppat.1005228.

Deslandes, L., et al. Physical Interaction between RRS1-R, a Protein Conferring Resistance to Bacterial Wilt, and PopP2, a Type III Effector Targeted to the Plant Nucleus. *Proc. Natl. Acad. Sci. USA.* **2003,** *100,* 8024–8029.

Dodds, P. N.; Rathjen, J. P. Plant Immunity: Towards an Integrated View of Plant-Pathogen Interactions. *Nature Rev. Genet.* **2010,** *11,* 539–548.

Druka, A. Exploiting Regulatory Variation to Identify Genes Underlying Quantitative Resistance to the Wheat Stem Rust Pathogen *Puccinia graminis* f. sp. *tritici* in Barley. *Theor. Appl. Genet.* **2008,** *117*(2), 261–272.

Eitas, T. K.; Dangl, J. L. NB-LRR Proteins: Pairs, Pieces, Perception, Partners, and Pathways. *Curr. Opin. Plant Biol.* **2010,** *13,* 472–477.

Flor, H. H. The Complementary System in Flax and Flax Rust. *Adv. Genet.* **1956,** *8,* 29–54.

Feng, Y.; Liu, Q.; Xue, Q. Comparative Study of Rice and Arabidopsis Actin-Depolymerizing Factors Gene Families. *J. Plant Physiol.* **2006,** *163,* 69–79.

Fetch, T.; Johnston, P. A.; Pickering, R. Chromosomal Location and Inheritance of Stem Rust Resistance Transferred from *Hordeum Bulbosum* into Cultivated Barley (*H. vulgare*). *Phytopathology.* **2009,** *99*(4), 339–343.

Fox, S. L.; Harder, D. E. Resistance to Stem Rust in Barley and Inheritance of Resistance to Race QCC. *Can. J. Plant. Sci.* **1995,** *75,* 781–788.

Fu, D. A Kinase-START Gene Confers Temperature-Dependent Resistance to Wheat Stripe Rust. *Science.* **2009,** *323,* 1301–1302.

Gassmann, W.; Bhattacharjee, S. Effector-Triggered Immunity Signaling: From Gene-for-Gene Pathways to Protein-Protein Interaction Networks. *Mol. Plant Microbe Interact.* **2012**, *25*(7), 862–868.

Gassmann, W.; Hinsch, M. E.; Staskawicz, B. J. The Arabidopsis RPS4 Bacterial-Resistance Gene is a Member of the TIR-NBS-LRR Family of Disease- Resistance Genes. *Plant J.* **1999**, *20*, 265–277.

Gomez-Gomez, L.; Bauer Z.; Boller, T. Both the Extracellular Leucine-Rich Repeat Domain and the Kinase Activity of FLS2 are Required for Flagellin Binding and Signaling in Arabidopsis. *Plant Cell.* **2001**, *13*, 1155–1163.

Hammond-*Kosack*, K. E.; *Jones*, J. D. G. Plant Disease Resistance Genes. *Annu. Rev. Plant Physiol. Plant Mol. Biol.* **1997**, *48*, 573–605.

Hanks, S.; Quinn, A.; Hunter, T. The Protein Kinase Family: Conserved Features and Deduced Phylogeny of the Catalytic Domains. *Sci.* **1988**, *241*, 42–52.

Horvath, H., et al. Genetically Engineered Stem Rust Resistance in Barley Using the *Rpg1* Gene. *Proc. Natl. Acad. Sci. USA.* **2003**, *100*, 364–369.

Hussey, P. J., Ketelaar, T.; Deeks, M. J. Control of the Actin Cytoskeleton in Plant Cell Growth. *Annu. Rev. Plant Biol.* **2006**, *57*, 109–125.

Jedel, P. E. A Gene for Resistance to *Puccinia graminis* f. sp. *tritici* in PI 382313. *Barley Genet. Newsl.* **1990**, *20*, 43–44.

Jin, Y.; Steffenson B. J.; Miller J. D. Inheritance of Resistance to Pathotypes QCC and MCC of *Puccinia graminis* f. sp. *tritici* in Barley Line Q21861 and Temperature Effects on the Expression of Resistance. *Phytopathology.* **1994**, *84*, 452–455.

Jones, J. D. G.; Dangl, J. L. The Plant Immune System. *Nature.* **2006**, *44*, 323–329.

Kislev, M. E. Stem Rust of Wheat 3300 Years Old Found in Isreal. *Science.* **1982**, *216*, 993–994.

Kleinhofs, A., et al. Barley Stem Rust Resistance Genes: Structure and Function. *Plant Genome.* **2009**, *2*, 109–120.

Kobayashi, Y.; Kobayashi, I. Depolymerization of the Actin Cytoskeleton Induces Defense Responses in Tobacco Plants. *J. Gen. Plant Pathol.* **2007**, *73*, 360–364.

Kwon, C.; Bednarek, P.; Schulze-Lefert, P. Secretory Pathways in Plant Immune Responses. *Plant Physiol.* **2008**, *147*, 1575–1583.

Lee, S. K., et al. Rice Pi5-Mediated Resistance to *Magnaporthe oryzae* Requires the Presence of Two Coiled-Coil-Nucleotide-Binding-Leucine-Rich Repeat Genes. *Genetics.* **2009**, *181*, 1627–1638.

Lipka, V.; Panstruga, R. Dynamic Cellular Responses in Plant-Microbe Interactions. *Curr. Opin. Plant Biol.* **2005**, *8*, 625–631.

Loutre, C., et al. Two Different CC-NBS-LRR Genes are Required for Lr10-Mediated Leaf Rust Resistance in Tetraploid and Hexaploid Wheat. *Plant J.* **2009**, *60*, 1043–1054.

Maciver, S. K.; Hussey, P. J. The ADF/Cofilin Family: Actin-Remodeling Proteins. *Genome Biol.* **2002**, *3*, Reviews3007.

Malinovsky, F. G.; Fangei, J. U.; Willats, W. G. T. The Role of the Cell Wall in Plant Immunity. *Front. Plant Sci.* **2014**, *6*(5), 178. doi:10.3389/fpls.2014.00178.

Mang, H. G., et al. Abscisic Acid Deficiency Antagonizes High-Temperature Inhibition of Disease Resistance through Enhancing Nuclear Accumulation of Resistance Proteins *SNC1* and *RPS4* in Arabidopsis. *Plant Cell.* **2012**, *24*(3), 1271–1284.

Martin, G. B.; Bogdanove, A. J.; Sessa, G. Understanding the Function of Plant Disease Resistance Proteins. *Annu. Rev. Plant Biol.* **2003**, *54*, 23–61.

Martin, G. M., et al. Map-Based Cloning of a Protein Kinase Gene Conferring Disease Resistance in Tomato. *Science.* **1993,** *262,* 1432–1436.

Miklis, M. Barley *MLO* Modulates Actin-Dependent and Actin-Independent Antifungal Defense Pathways at the Cell Periphery. *Plant Physiol.* **2007,** *144,* 1132–1143.

Monaghan, J.; Zipfel, C. Plant Pattern Recognition Receptor Complexes at the Plasma Membrane. *Curr. Opin Plant Biol.* **2012,** *15,* 349–357.

Moscou, M. J.; Lauter, N.; Steffenson, B.; Wise, R. P. Quantitative and Qualitative Stem Rust Resistance Factors in Barley are Associated with Transcriptional Suppression of Defense Regulons. *PLoS Genetics.* **2011,** *7*(7), e1002208.

Mucyn, T. S.; The Tomato NBARC-LRR Protein Prf Interacts with Pto Kinase in Vivo to Regulate Specific Plant Immunity. *Plant Cell.* **2006,** *18*(10), 2792–2806.

Narusaka, M., et al. RRS1 and RPS4 Provide a Dual Resistance-Gene System against Fungal and Bacterial Pathogens. *Plant J.* **2009,** *60,* 218–226.

Kerk, D., et al. The Complement of Protein Phosphatase Catalytic Subunits Encoded in the Genome of *Arabidopsis. Plant Physiol.* **2002,** *129,* 908–925.

Nicaise, V.; Roux, M.; Zipfel, C.; Recent Advances in PAMP-Triggered Immunity against Bacteria: Pattern Recognition Receptors Watch over and Raise the Alarm. *Plant Physiol.* **2009,** *150,* 1638–1647.

Nirmala, J. Concerted Action of Two Avirulent Spore Effectors Activates Reaction to *Puccinia graminis* 1 (Rpg1)-Mediated Cereal Stem Rust Resistance. *Proc. Natl. Acad. Sci. USA.* **2011,** *108*(35), 14676–14681.

Nirmala, J.; Drader, T.; Chen, X.; Steffenson, B.; Kleinhofs, A. Stem Rust Spores Elicit Rapid RPG1 Phosphorylation. *MPMI.* **2010,** *23*(12), 1635–1642.

Nirmala, J., et al. Proteolysis of the Barley Receptor-Like Protein Kinaserpg1 by a Proteasome Pathway is Correlated with *rpg1*-Mediated Stem Rust Resistance. *Proc. Natl. Acad. Sci. USA.* **2007,** *104*(24), 7518–7523.

Nirmala, J., et al. Subcellular Localization and Functions of the Barley Stem Rust Resistance Receptor-like Serine/Threonine-Specific Protein Kinase Rpg1. *Proc. Natl. Acad. Sci. USA.* **2006,** *103,* 7518–523.

Okuyama, Y., et al. A Multifaceted Genomics Approach Allows the Isolation of the Rice Pia-Blast Resistance Gene Consisting of Two Adjacent NBS-LRR Protein Genes. *Plant J.* **2011,** *66,* 467–479.

Park, C. J., et al. Rice XB15, a Protein Phosphatase 2C, Negatively Regulates Cell Death and XA21-Mediated Innate Immunity. *PLoS Biology.* **2008,** *6*(9), e231. doi:10.1371/journal.pbio.0060231.

Patterson, F. L.; Shands, R. G.; Dickson, J. G. Temperature and Seasonal Effects on Seedling Reactions of Barley Varieties to Three Races of *Puccinia graminis* f. sp. *tritici. Phytopathology.* **1957,** *47,* 395–402.

Peterson, P. D. *Stem Rust of Wheat from Enemy to Modern Foe;* APS press: St Paul, MN, 2001.

Porter, K.; Shimono, M.; Tian, M.; Day, B. Arabidopsis Actin-Depolymerizing Factor-4 Links Pathogen Perception, Defense Activation and Transcription to Cytoskeletal Dynamics. *PLoS Pathog.* **2012,** *8*(11), e1003006.

Pretorius, Z. A.; Singh, R. P.; Wagoire, W. W.; Payne, T. S. Detection of Virulence to Wheat Stem Rust Resistance Gene *Sr31* in *Puccinia graminis.* f. sp. *tritici* in Uganda. *Plant Dis.* **2000,** *84,* 203.

Pritchard, L.; Birch, P. R. J. The Zigzag Model of Plant-Microbe Interactions: Is it Time to Move on? *Mol. Plant Pathol.* **2014,** *15*(9), 865–870.

Qi, Y.; Tsuda, K.; Glazebrook, J.; Katagiri, F. Physical Association of Pattern-Triggered Immunity (PTI) and Effector-Triggered Immunity (ETI) Immune Receptors in Arabidopsis. *Mol. Plant Pathol.* **2011,** *12*(7), 702–708.

Rafiqi, M.; Bernoux, M.; Ellis, J. G.; Dodds, P. N. In the Trenches of Plant Pathogen Recognition: Role of NB-LRR Proteins. *Seminars Cell Dev. Biol.* **2009,** *20,* 1017–1024.

Rodriguez, P. L. Protein Phosphatase 2C (PP2C) Function in Higher Plants. *Plant Mol. Biol.* **1998,** *38,* 919–927.

Roelfs, A. P.; Casper, D. H.; Long, D. L.; Roberts, J. J. Races of *Puccinia graminis* in the United States in 1989. *Plant Dis.* **1991,** *5,* 1127–1130.

Roelfs, A. P.; Martens, J. W. An International System of Nomenclature for *Puccinia graminis* f. sp. *tritici. Phytopathology.* **1988,** *78,* 526–553.

Schmelzer, E. Cell Polarization, a Crucial Process in Fungal Defense. *Trends Plant Sci.* **2002,** *7,* 411–415.

Sinapidou, E., et al. Two TIR:NB:LRR Genes are Required to Specify Resistance to *Peronospora parasitica* Isolate Cala2 in Arabidopsis. *Plant J.* **2004,** *38,* 898–909. doi: 10.1111/j.1365- 313X.2004.02099.x.

Shen, Q. H., et al. Nuclear Activity of MLA Immune Receptors Links Isolate-Specific and Basal Disease-Resistance Responses. *Science.* **2007,** *315,* 1098–1103.

Song, W. Y. A Receptor Kinase-Like Protein Encoded by the Rice Disease Resistance Gene, *Xa21. Science.* **1995,** *270,* 1804–1806.

Staiger, C. J.; Blanchoin, L. Actin Dynamics: Old Friend Friends with New Stories. *Curr. Opin. Plant Biol.* **2006,** *9,* 554–562.

Steffenson, B. J.; Zhou, H.; Chai Y.; Grando, S. Vulnerability of Cultivated and Wild Barley to African Stem Rust Race TTKSK. In *Advance in Barley Science;* Springer: Netherlands, 2013; pp 243–255.

Steffenson, B. J.; Jin, Y.; Brueggeman, R. S.; Kleinhofs, A; Sun, Y. Resistance to Stem Rust Race TTKSK Maps to the *rpg4/Rpg5* Complex of Chromosome 5H of Barley. *Phytopathology.* **2009,** *99,* 1135–1141.

Steffenson, B. J.; Jin, Y. Resistance to Race TTKS of *Puccinia graminis* f. sp. *tritici* in Barley. *Phytopathology.* **2006,** *96,* S110.

Steffenson, B. J. Analysis of Durable Resistance to Stem Rust in Barley. *Euphytica.* **1992,** *63,* 153–167.

Steffenson, B. J. In *Stem Rust of Barley: History, Calamity, and Control,* Proceeding of the 28th Barley Improvement Conference, Minneapolis, MN, 1991, pp 3–8.

Steffenson, B. J.; Wilcoxson, R. D.; Roelfs, A. P. Inheritance of Resistance to *Puccinia graminis* f. sp. *secalis* in Barley. *Plant Dis.* **1984,** *68,* 762–763.

Sun, Y.; Steffenson, B. J. Reaction of Barley Seedlings with Different Stem Rust Resistance Genes to *Puccinia graminis* f. sp. *tritici* and *P. g.* f. sp. *secalis. Can. J. Plant Pathol.* **2005,** *27,* 80–89.

Sun, X.; Cao, Y.; Yang, Z.; Xu, C.; Li, X.; Wang, S; Zhang, Q. *Xa26,* a Gene Conferring Resistance to *Xanthomonas oryzae* pv. *oryzae* in Rice, Encodes an LRR Receptor Kinase-Like Protein. *Plant J.* **2004,** *37,* 517–527.

Swiderski, M. R.; Innes, R. W. The *ArabidopsisPBS1* Resistance Gene Encodes a Member of a Novel Protein Kinase Subfamily. *Plant J.* **2001,** *26,* 101–112.

Takemoto, D.; Jones, D. A.; Hardham, A. R. GFP-Tagging of Cell Components Reveals the Dynamics of Subcellular Re-Organization in Response to Infection of Arabidopsis by Oomycete Pathogens. *Plant J.* **2003,** *33,* 775–792.

Tang, X.; Frederick, R. D.; Zhou, J.; Halterman, D. A.; Yia, Y. Initiation of Plant Disease Resistance by Physical Interaction of Avrpto and Pto Kinase. *Science.* **1996,** *274,* 2060–2063.

Tian, M., et al. Arabidopsis Actin-Depolymerizing Factor AtADF4 Mediates Defense Signal Transduction Triggered by the *Pseudomonas syringae* Effector AvrPphB. *Plant Physiol.* **2009,** *150,* 815–824.

Thomma, B. P. H. J.; Nurnberger, T.; Joosten, M. H. A. J. Of PAMPs and Effectors: The Blurred PTI-ETI Dichotomy. *Plant Cell.* **2011,** *23,* 4–15.

Thordal-Christensen, H. Fresh Insights into Processes of Nonhost Resistance. *Curr. Opin. Plant Biol.* **2003,** *6,* 351–357.

van der Bizen, E. A.; Jones J. D. G. Plant Disease-Resistance Proteins and the Gene-for-Gene Concept. *Trends Biochem. Sci.* **1998,** *23,* 454–456.

van der Hoorn, R. A. L.; Kamoun, S. From Guard to Decoy: A New Model for Perception of Plant Pathogen Effectors. *Plant Cell.* **2008,** *20,* 2009–2017.

Wan, J.; Zhang, X. C.; Stacey, G. Chitin Signaling and Plant Disease Resistance. *Plant Signal. Behav.* **2008,** *3*(10), 831–833.

Wang, X., et al. The rpg4-Mediated Resistance to Wheat Stem Rust (*Puccinia graminis*) in Barley (*Hordeum vulgare*) Requires *Rpg5*, a Second NBS-LRR Gene, and an Actin Depolymerization Factor. *Mol. Plant Microbe Interact.* **2013,** *26,* 407–418. doi: 10.1094/MPMI-06-12-0146-R.

Wirthmueller, L.; Zhang, Y.; Jones, J. D. G.; Parker, J. E. Nuclear Accumulation of the *Arabidopsis* Immune Receptor RPS4 is Necessary for Triggering EDS1-Dependent Defense. *Curr. Biol.* **2007,** *17,* 2023–2029.

Yuan, B., et al. The Pik-p Resistance to *Magnaporthe oryzae* in Rice is Mediated by a Pair of Closely Linked CC-NBS-LRR Genes. *Theor. Appl. Genet.* **2011,** *122,* 1017–1028. doi: 10.1007/s00122- 010-1506-3.

Zhai, C., et al. Function and Interaction of the Coupled Genes Responsible for Pik-h Encoded Rice Blast Resistance. *PLoS ONE.* **2014,** 9, e98067. doi: 10.1371/journal.pone.0098067.

Zipfel, C. Bacterial Disease Resistance in Arabidopsis Through Flagellin Perception. *Nature.* **2004,** *428,* 764–767.

Zurn, J. D.; Dugyala, S.; Borowicz, P.; Brueggeman, R.; Acevedo, M. Unraveling the Wheat Stem Rust Infection Process on Barley Genotypes through Relative qPCR and Fluorescence Microscopy. *Phytopathology.* **2015,** *105*(5), 707–712.

CHAPTER 23

INVERSE GENE-FOR-GENE: NECROTROPHIC SPECIALIST'S *MODUS OPERANDI* IN BARLEY AND WHEAT

JONATHAN RICHARDS, GAZALA AMEEN, and ROBERT BRUEGGEMAN*

Department of Plant Pathology, North Dakota State University, Fargo 58108-6050, ND, USA

Corresponding author. E-mail: robert.brueggeman@ndsu.edu

CONTENTS

ABSTRACT

Throughout their evolutionary history, plants have developed an intricate immune system, including a basal defense response that provides resistance toward a broad range of pathogens, as well as a species or isolate specific recognition response. Biotrophic pathogens, needing living plant tissue to subsist, are successfully sequestered via the initiation of a localized cell death response following the recognition of pathogen avirulence proteins by a dominant resistance gene in the host. This mechanism is classically known as a "gene-for-gene" interaction and has been observed in many biotrophic pathosystems. Intriguingly, necrotrophic pathogens, requiring dead plant tissue for survival, have evolved means to intentionally activate host cell death responses. Inappropriate elicitation of host defense responses by targeted action of pathogen produced necrotrophic effectors on a host dominant susceptibility factor results in programmed cell death, providing a source of nutrients for the necrotroph resulting in a compatible reaction. At the molecular level, this interaction essentially mirrors a "gene-for-gene" model, but the outcome shifts from resistance to susceptibility due to the lifestyle of the phytopathogen. Additionally, as "gene-for-gene" interactions are generally qualitative in nature, these "inverse gene-for-gene" interactions in necrotrophic pathosystems often have an additive or quantitative effect. Characterization of host susceptibility factors show the presence of homologous domains to those of identified resistance genes, further implicating them in typical host defense responses. Necrotrophic effectors, generally observed to be small, secreted proteins with a high cysteine content, evolved to manipulate pieces of the plant immune system for benefit of the pathogen. Four economically important necrotrophic plant pathogens of wheat and barley (*Pyrenophora tritici-repentis*, *Parastagonospora nodorum*, *Pyrenophora teres*, and *Cochliobolus sativus*) have been observed to cause disease in an "inverse gene-for-gene" manner. Discussed in this chapter, thorough investigation of these interactions at both the genetic and molecular level has shed light on these intricate pathosystems, illustrating the complex evolutionary relationships occurring between plants and necrotrophic pathogens. As resistance to necrotrophic pathogens can be facilitated through the lack of susceptibility genes rather than the presence of resistance genes, the current research of "inverse gene-for-gene" interactions will allow wheat and barley breeders to improve their elite breeding germplasm by selecting against host susceptibility factors.

23.1 INTRODUCTION

Plant pathogens are often classified into two general categories, biotrophs and necrotrophs, based upon their lifestyle and the mechanisms in which they obtain nutrients from the hosts they colonize. Biotrophic pathogens require living tissue to subsist and complete their life cycle and often form feeding structures such as haustoria to accomplish this feat in a hostile environment, which requires staying undetected by the plant's immune system (Mendgen & Hahn, 2002). Conversely, the necrotrophic pathogens utilize dead host tissue to complete their lifecycle and tailor this environment by eliciting the host's immune responses by letting their presence be known (van Kan, 2006). A third class, the hemibiotrophs, represents a melding of biotrophic and necrotrophic lifestyles. These organisms begin with a biotrophic phase of variable length, the latent period, followed by a necrotrophic phase which they induce through effector mediated immune system elicitation and manipulation of host cellular physiology (Mendgen & Hahn, 2002; Perfect & Green, 2001). Additionally, necrotrophs can be further classified into necrotrophic generalists or specialists. Necrotrophic generalists have wide host ranges and the ability to invade the host utilizing enzymes or toxins that can effectively target conserved barriers that evolved early in plant speciation to resist pathogen colonization, thus, are present in a broad range of plant species. An example of general virulence factors is the production of a repertoire of cell wall degrading enzymes which can be seen as a brute force infection mechanism utilized by many necrotrophic generalists (van Kan, 2006). Necrotrophic specialists on the other hand have a much narrower host range, often restricted to a single or several closely related host species. The specialists are known to utilize a much more intricate means of infection which involves the production of specific necrotrophic effectors (NEs) that target host immunity mechanisms, purposefully eliciting programmed cell death (PCD) pathways with spatial and temporal specificity. Thus, they specifically modify the environment within the host making it conducive to colonization and completion of their specialized lifecycle (Oliver & Solomon, 2010; Hammond-Kossack & Rudd, 2008). Additionally, resistance or susceptibility to necrotrophs is often quantitative in nature, with multiple dominant susceptibility loci throughout the host genome targeted by unique NEs, also spread throughout the pathogen genome, which quantitatively contribute to disease severity or infection types (Friesen et al., 2008b). Therefore, the hallmark characteristic of this subset of pathogens is the production of proteinaceous or secondary metabolite NEs, which have the ability to subvert the host's immune response leading to different levels of PCD and disease (Stergiopoulos et al., 2013;

Wolpert et al., 2002). Classic characteristics of the proteinaceous NEs are small molecular weight, predicted secretion signals, and a relatively high cysteine content contributing to protein stability (Tan et al., 2010).

Plants evolved different immune responses, ranging from general defense responses that react to a wide range of potentially harmful microbes (non-host resistance) to highly specific recognition of species-specific pathogens (host specific resistance). Pathogen associated molecular patterns (PAMPs), or more thoughtfully considered microbe associated molecular patterns (MAMPs), are conserved motifs in potentially harmful microbes, necessary for survival and/or maintenance of a high level of fitness in their environment (Zipfel, 2008). Examples of well-investigated PAMPs include chitin, a major component of fungal cell walls, and flagella subunits, the structures utilized for motility in many plant pathogenic bacteria (Shibuya & Minami, 2001; Gomez-Gomez & Boller, 2002). During the coevolution with these pathogens, the host evolved PAMP recognition receptors (PRRs) as a major component of their innate immune system to successfully detect pathogen challenge. The main class of PRRs are the receptor like kinases (RLKs) which contain extracellular leucine-rich repeats (LRRs), the receptor domain that recognizes PAMPs, a transmembrane domain, and an intracellular kinase domain that functions in immunity signaling cascades (Zipfel et al., 2006; Chisholm et al., 2006; Nurnberger & Kemmerling, 2006; Coll et al., 2011). Upon PRR recognition of PAMPs, PAMP triggered immunity (PTI) is initiated, resulting in a quick defense response to arrest the pathogen's attempt at colonization. The receptor activation elicits responses including reactive oxygen and nitric oxide production, ion fluxes, altered levels of plant hormones and the activation of the mitogen activated protein kinase (MAPK) signaling cascades. This activation leads to transcriptional activation of defense-associated gene expression (Numberger et al., 2004). A classic example of PAMP recognition by a PRR is the detection of flg22, an N-terminal 22 amino acid sequence in the flagellin protein of many bacteria, by FLS2, an RLK. Upon direct recognition of flg22, BAK1, an additional RLK, is recruited to form a complex with FLS2 to induce the PTI response and effectively reduce the growth and spread of the invading organism (Gomez-Gomez & Boller, 2000; Chinchilla et al., 2006; Chinchilla et al., 2007). Due to the fact that flagella are necessary structures for motility and fitness in most environments including the lifestyle of a phytopathogenic bacteria, the amino acid sequence remains conserved, which allows a pre-formed plant defense response that recognizes the conserved molecular pattern to protect itself from a wide range of bacterial species. Through selection and mutation over time, the pathogen evolved secreted effectors that block components of the host PTI responses, often by altering the phosphorylation signaling cascades underlying PTI and thereby inhibiting the early defense responses, effectively eliciting effector triggered susceptibility (ETS). Returning to the aforementioned FLS2 mediated PTI response, it has been observed that the plant pathogen *Pseudomonas syringae* evolved to produce a protein secreted through the type III secretion system (T3S), AvrPto, which blocks the kinase activity of FLS2, thereby inhibiting the PTI response (Xiang et al., 2008). Following subversion of the broad innate immune response by the pathogen, the host counter or co-evolved to respond to this ETS through the evolution of a second level of immunity that relies on cytoplasmic resistance proteins (R-proteins) that recognize, directly or indirectly, the presence of the ETS virulence effectors (Jones & Dangl, 2006). The cloning of nearly one hundred of these resistance genes (R-genes) from diverse plant species conferring resistance to a wide taxa of plant pathogens have shown that the majority encode R-proteins with the common nucleotide binding sites (NBS) and LRR domains and are referred to as the nucleotide-binding leucine rich repeat receptor (NLRs) (Coll et al., 2011; Belkhadir et al., 2004). The NLRs are the major class of plant immunity receptors, yet a second minor class of immunity receptors are the proteins that contain at least one serine/threonine protein kinase (STPKs) domain, such as the barley stem rust R-gene *Rpg5* (Brueggeman et al., 2008). Via the R-protein receptors, the recognition of the ETS virulence effectors results in a higher amplitude of resistance compared to PTI which is characterized by the hypersensitive response (HR), which is a strong yet tightly regulated PCD response. This second layer of the plant innate immune system is known as effector triggered immunity (ETI). This co-evolutionary process is illustrated by the "zigzag" model, where the evolutionary arms race between the host and the pathogen is illustrated by constant counter measures to gain the upper hand in the host-parasite interaction (Jones & Dangl, 2006).

Within a biotrophic pathosystem, host resistance to plant pathogens is mediated by the recognition of the pathogen through PRR and NLR receptors and the induction of subsequent immunity responses. This recognition often follows Flor's gene-for-gene model, where the host R-protein recognizes a pathogen virulence effector eliciting the defense responses (Flor, 1956). These virulence effectors that evolved in the pathogen to perpetuate colonization, once recognized by the host, elicit defense responses, thus becoming avirulence genes (*Avr*). Typical plant defense responses elicited by Avr protein recognition include the production of reactive oxygen species (ROS), upregulation of pathogenesis-related (PR) genes, and induction of MAPKs, acting in a concerted effort resulting in a regulated and localized PCD (Coll et al., 2011). These physiological responses effectively function to intentionally kill the cells surrounding the point of infection and successfully sequester a biotrophic pathogen by restricting its access to nutrients, arresting any further colonization. However, these effective mechanisms of immunity against a biotroph provide unwanted opportunity to an invading necrotrophic pathogen. Since necrotrophs thrive on dead tissue, the same cellular responses that comprise the HR and successfully defend the host from a biotroph, actually favor the necrotrophic life cycle. Thus, necrotrophic specialists have evolved a repertoire of effectors that mimic biotrophic Avr proteins to purposefully initiate ETI to aid host colonization and cause disease (Coll et al., 2011). This realization prompted the coining of the "inverse gene-for-gene theory" (Friesen et al., 2007), which will be the primary focus of this chapter.

Prominent pathosystems in wheat and barley have been studied that follow the "inverse gene-for-gene" model including the *Pyrenophora tritici-repentis* (*Ptr*)-wheat, *Parastagonospora nodorum*-wheat, *Pyrenophora teres*-barley, and *Cochliobolus sativus*-barley pathosystems. Within these pathosystems, it has been observed that pathogen isolates produce unique NEs (Ciuffetti et al., 2010; Friesen et al., 2008a; Liu et al., 2015), which may be recognized directly or indirectly by host susceptibility genes that have been shown to be similar to different classes of biotroph immunity receptors (Faris et al., 2010, 2014; Richards et al., 2016; Ameen et al., 2016). The presence of both the dominant NE and the dominant host susceptibility gene results in a compatible reaction, or susceptibility, which is inverse to the gene-for-gene model. The genetic characterization, cloning of NEs, and host susceptibility genes as well as their functional characterization have shown diversity in the genes involved in the host-necrotrophic parasite genetic interactions yet a high level of continuity to their pertinence to the inverse gene-for-gene model. Thorough investigation of these pathosystems, including host and pathogen genetic components that determine compatible or incompatible interactions will provide insight into developing effective and possibly durable resistance to these necrotrophic pathogens in elite wheat and barley cultivars (cv).

23.2   DOMINANT SUSCEPTIBILITY TO NECROTROPHIC PATHOGENS

23.2.1   *PYRENOPHORA TRITICI-REPENTIS*

Ptr is the causal agent of tan spot of wheat and is an economically important disease in wheat producing regions of the world. Tomas and Bockus

(1987) demonstrated that *Ptr* culture filtrates contained toxic compounds that induce necrosis, and at times chlorosis, when infiltrated into sensitive wheat lines. Additionally, it was observed that toxic culture filtrate sensitive wheat lines were also susceptible to the pathogen, whereas toxic culture filtrate insensitive lines were resistant (Tomas & Bockus, 1987). The results of this work explicitly illustrate the role of host selective toxins (HSTs), now more commonly referred to as NEs, in the role of host genotype specific disease susceptibility. This work paved the way for the further investigation of the weaponry that necrotrophs possess to incite disease and epidemics in the field. Presently, three NEs produced by *Ptr* have been identified: ToxA, ToxB, and ToxC. Isolates of *Ptr* are classified into eight races based on specific combinations of these three NEs (Table 23.1) (Lamari et al., 2003; Strelkov & Lamari, 2003). The two NEs, ToxA, and ToxB, have been extensively characterized and the ToxA corresponding host dominant sensitivity gene *Tsn1* has been cloned and partially characterized (Ciuffetti et al., 2010; Faris et al., 2010). The ToxB host sensitivity gene, *Tsc2*, has yet to be identified, but the genetic nature of host sensitivity has been analyzed and falls into the inverse-gene-for-gene model (Friesen & Faris, 2004; Abeysekara et al., 2010). The investigation of these host susceptibility genes and corresponding NEs has facilitated the elucidation of the unique manner in which necrotrophs subvert the host's immunity system for their own benefit and the implications that these complex yet intricate genetic interactions have when breeding for resistance, or more accurately, lack of susceptibility.

TABLE 23.1 *Pyrenophora tritici-repentis* Race Structure and Phenotypic Reactions of Differential Lines.

Line	Race							
	1	2	3	4	5	6	7	8
Glenlea	N	N	R	R	R	R	N	N
6B662	R	R	R	R	C	C	C	C
6B365	C	R	C	R	R	C	R	C
Salamouni	R	R	R	R	R	R	R	R

R = resistant, N = necrotic reaction, and C = chlorotic reaction (adapted from Lamari et al., 2003; Strelkov & Lamari, 2003).

23.2.1.1   TOXA

The discovery of toxin production by *Ptr* prompted further research to identify and characterize the pathogen molecule causing this toxicity. Three independent groups identified and purified a proteinaceous toxin from *Ptr* and designated them as Ptr toxin, Ptr necrosis toxin, and ToxA (Tomas et al., 1990; Ballance et al., 1989; Tuori et al., 1995). The toxin purified from these groups showed nearly identical properties, such as similar molecular weights and amino acid content. Further research demonstrated that these were in fact the same protein and are further referred to as ToxA. ToxA is a small, secreted protein, ~13.2 kDa in size, and has been demonstrated to confer a cv specific compatible reaction in wheat. The expression of ToxA in a nonpathogenic *Ptr* isolate results in the restoration of pathogenicity and the induction of necrotic symptoms on a sensitive host, confirming that the production of ToxA is primarily responsible for host cell death (Ciuffetti et al., 1997). Additionally, ToxA is present in the *Ptr* genome as a single copy, which is sufficient to cause significant necrotic symptoms. It was also observed that internalization of ToxA into the host cell is required to initiate cell death, even when biolistically bombarded into ToxA insensitive wheat lines, and is mediated through a necessary arginine-glycine-aspartic acid (RGD) domain (Manning & Ciuffetti, 2005; Manning et al., 2008). Upon internalization, ToxA is trafficked to the chloroplast and binds to ToxABP1, the interaction of which is presumed to disrupt the function of photosystem II resulting in the production of ROS in the chloroplast, ultimately instigating cell death (Manning et al., 2007; Manning et al., 2009). ToxA sensitive and insensitive lines tested are predicted to produce identical ToxABP1 proteins which are regulated at similar levels, indicating that ToxA specificity is not localized to this chloroplastic interaction (Manning et al., 2007). In addition to the apparent function of ToxA acting as an inhibitor of photosystem function, it was observed that cell death induced by infiltration of ToxA required the presence of light and that necrosis was nearly entirely eliminated when infiltrated plants were grown under dark conditions (Manning et al., 2005). The extensive investigation of the biochemical properties and cellular action of Ptr ToxA indicates that a conserved domain of the effector protein mediates entry into the host cell, which upon entry alters the function of photosynthetic processes of the plant to induce the production of ROS contributing to cell death, aiming to manufacture a beneficial environment for the necrotroph.

23.2.1.2 TSN1

Following the identification of ToxA from *Ptr*, the corresponding host gene was thoroughly characterized. Utilizing culture filtrates of *Ptr* isolate 86–124, which was known to produce the ToxA necrosis inducing toxin, sensitivity to ToxA was mapped to wheat chromosome 5BL. Segregation analysis indicated a single recessive gene was conferring insensitivity to ToxA and was designated *Tsn1* (Faris et al., 1996). The recessive nature of this insensitivity gene may now be more accurately described as a dominant susceptibility gene, as seen by the cloning of *Tsn1*. The isolation of *Tsn1* revealed it to be a functional gene harboring domains normally associated with resistance to a biotrophic pathogen, including STPK, NBS, and LRR (Faris et al., 2010). A similar structure to the *Rpg5* gene, which confers resistance to wheat and rye stem rust races and isolates (Wang et al., 2013; Arora et al., 2013). Although *Tsn1* has been shown to be the ToxA corresponding host sensitivity gene, yeast two-hybrid experiments demonstrated that the ToxA and Tsn1 proteins do not directly interact (Faris et al., 2010), indicating either an indirect perception of ToxA or that an alternative function exists which is necessary for the induction of cell death. Interestingly, it was also observed that transcriptional regulation of *Tsn1* follows a circadian rhythm, being at the highest expression level at the beginning of a 12 h photoperiod, whereas expression in plants placed continually in the dark decreased and remained at lesser levels. Additionally, following infiltration with ToxA, expression of *Tsn1* steadily decreased and remained at a significantly low level of expression relative to control plants (Faris et al., 2010). As previously mentioned, the ToxA induced necrosis requires the presence of light, making the observation of correlated *Tsn1* expression and the photoperiod exceptionally intriguing. Potentially, *Tsn1* may have been evolutionarily conditioned to follow a circadian rhythm for the purpose of mediating transport of a beneficial molecule into the cell during the light cycle. Antagonistically, *Ptr* may have hijacked this system to disrupt photosynthetic processes at a perfectly opportune time.

23.2.1.3 TOXB

Certain races of *Ptr* were observed to induce chlorosis on sensitive wheat lines, accompanied either by necrosis or by itself. *Ptr* race five induced chlorotic symptoms on wheat line Katepwa without necrosis and culture filtrates were hypothesized to contain a host specific toxin, designated *Ptr* chlorosis toxin (*Ptr* ToxB) that caused the observed symptoms on susceptible host genotypes. Segregation analysis of F_2 progeny derived from a cross of Katepwa (susceptible/sensitive) × ST15 (resistant/insensitive) determined that susceptibility to the fungus and sensitivity to culture filtrates were both governed by a single dominant gene (Orolaza et al., 1995). The purification of *Ptr* ToxB and the cloning of the corresponding pathogen gene determined that ToxB is a 6.61 kDa, heat stable protein and is 64 amino acids in length (Strelkov et al., 1999; Martinez et al., 2001). Copy number of ToxB in the *Ptr* genome also appears to affect the extent of host symptoms, with isolates containing a greater copy number of ToxB and increased toxin production being able to induce a substantially greater magnitude in symptoms (Strelkov et al., 2002; Strelkov & Lamari, 2003; Martinez et al., 2004). Additionally, whereas the ToxA-wheat interaction results in a rapid formation of necrosis, the ToxB-wheat interaction appears to be a slower response (Ciuffetti et al., 2010).

23.2.1.4   TSC2

As previously mentioned, sensitivity to culture filtrates of *Ptr* race five isolates containing ToxB appeared to be conferred by a single dominant gene in the host (Orolaza et al., 1995). Later studies localized the ToxB sensitivity locus (*Tsc2*) to the short arm of wheat chromosome 2B and through saturation mapping refined its position to an ~3.3 cM interval (Friesen & Faris, 2004; Abeysekara et al., 2010). The *Tsc2* gene is another example of the inverse gene-for-gene model within the *Ptr* pathosystem; however, it has yet to be cloned. Thus, the putative function of this host sensitivity gene is unknown. Interestingly, ToxA and ToxB have no significant homology with each other and induce distinctly different symptoms on sensitive hosts (Ciuffetti et al., 2010). In addition, ToxB appears to act in a dose-dependent manner, whereas a single copy of ToxA is sufficient to induce necrotic symptoms (Ballance et al., 1996; Ciuffetti et al., 1997; Martinez et al., 2004). Both *Tsn1*-ToxA and *Tsc2*-ToxB interactions induce the upregulation of molecular mechanisms associated with defense responses, indicating that *Ptr* potentially has the ability to exploit host cellular pathways in different manners with the same intended outcome of PCD and successful infection (Pandelova et al., 2009; Adhikari et al., 2009; Pandelova et al., 2012).

23.2.2   PARASTAGONOSPORA NODORUM

P. nodorum (formerly *Stagonospora nodorum*) is another example of a necrotrophic specialist fungal pathogen of wheat that causes the disease *S. nodorum* blotch. The labs of Dr. Timothy Friesen and Dr. Justin Faris have conducted thorough investigation on the identification of *P. nodorum* NEs as well as the corresponding wheat genes with which they interact. Within the wheat-*P. nodorum* pathosystem, a total of nine individual interactions between a host sensitivity gene and a specific NE have been genetically characterized (Liu et al., 2004; Faris et al., 2010; Friesen et al., 2006a, 2006b, 2007, 2012; Liu et al., 2009; Zhang et al., 2011; Abeysekara et al., 2009; Gao et al., 2015). Interestingly, one of these effectors, SnToxA, interacts with the aforementioned *Tsn1* gene in wheat. Following the sequencing of the *P. nodorum* genome, a nearly identical ortholog of ToxA from *Ptr* was identified and its presence in *Ptr* is hypothesized to have arisen from a horizontal gene transfer event with *P. nodorum* being the donor (Friesen et al., 2006a, 2006b). It was shown that the gene appears to only exist in these two fungi within a common DNA fragment that is ~11 kb in length, containing the ToxA gene as well as flanking repetitive sequences. A high level of sequence variability was found in SnToxA alleles from 600 *P. nodorum* isolates, whereas allele sequencing of *Ptr ToxA* revealed very little polymorphism. These data suggest that *ToxA* originated from *P. nodorum* and was transferred relatively recently to *Ptr* (Friesen et al., 2006a, 2006b). This horizontal gene transfer across species is an extraordinary example of how diverse pathogens may acquire DNA fragments containing foreign NEs that allow the pathogen to have increased virulence or broader host range. Additionally, due to its seemingly recent transfer, it also illustrates the remarkable effect that the transfer of a NE to another fungal species may have in the rapid increase of fitness of new pathogen specificities and implications for managing crop disease. In addition to the ToxA-*Tsn1* interaction, which was previously discussed, eight other novel interactions have been reported (Table 23.2). For the purpose of this section, focus will be placed on the well-characterized interactions of SnTox1-*Snn1*, SnTox3-*Snn3-B1/D1*, and SnTox6-*Snn6*.

TABLE 23.2   *P. nodorum*-Wheat Interactions Reported to Date and Known Characteristics.

Host sensitivity gene	Gene class	Pathogen effector	Effector type
Tsn1 Faris et al. (2010)	NBS-LRR/ STPK	SnToxA Friesen et al. (2006a, 2006b)	Protein
Snn1 Faris et al. (2014)	WAK	SnTox1 Liu et al. (2012)	Protein
Snn2 Friesen et al. (2007)	N/A	SnTox2	Protein
Snn3-B1 Friesen et al. (2008)	N/A	SnTox3 Liu et al. (2009)	Protein
Snn3-D1 Zhang et al. (2011)	N/A	SnTox3 Liu et al. (2009)	Protein
Snn4 Abeysekara et al. (2009)	N/A	SnTox4	Protein
Snn5 Friesen et al. (2012)	N/A	SnTox5	Protein
Snn6 Gao et al. (2015)	N/A	SnTox6	Protein
Snn7 Shi et al. (2015)	N/A	SnTox7	Protein

23.2.2.1   SNTOX1-SNN1

The discovery of the SnTox1-*Snn1* interaction was the first reported in the wheat-*P. nodorum* pathosystem and was identified via the mapping of sensitivity to a partially purified NE isolated from *P. nodorum*. Utilizing culture filtrates of *P. nodorum* isolate Sn2000, sensitivity was mapped to wheat chromosome 1BS in the International Triticeae Mapping Initiative (ITMI) population. Segregation analysis determined that sensitivity was likely conferred by a single dominant host gene designated *Snn1* (Liu et al., 2004). A positional mapping approach to identify Snn1 resulted in the identification of a wall associated kinase (WAK) gene as the best candidate *Snn1* gene (Faris et al., 2014). This class of susceptibility genes will be further discussed in the *rcs5 C. sativus* R-gene section.

A genomics approach was taken to identify the corresponding NE, which elicited a *Snn1* specific sensitivity response by examination of annotated *P. nodorum* genes using established criteria to prioritize genes as SnTox1 candidates. The main criteria were the absence from isolate Sn79-1087 and presence in mass spectrometry analysis of culture filtrates. The criteria used to further prioritize candidates was based on the characteristics of previously identified NEs including predicted small secreted proteins (<30 kDa), containing RGD or Arginine-X-Leucine-Arginine (RXLR) domains, genomic position near repetitive elements, and detected expression *in planta*. The top candidate genes were expressed in yeast and culture filtrates were infiltrated onto a set of differential lines. It was determined that infiltration of one of these candidate proteins elicited necrosis on *Snn1* containing lines, indicating the successful identification of *SnTox1* (Liu et al., 2012). SnTox1 is predicted to be an ~10.33 kDa protein that is secreted into the apoplast, contains a large proportion of cysteine residues, and harbors a chitin-binding domain. Further evidence indicating that this gene is SnTox1 included the absence in avirulent isolates, gain of pathogenicity when transformed into an avirulent isolate, and loss of pathogenicity when deleted from Sn2000. Additionally, when sensitivity to Sn2000 *SnTox1* knockout mutants were mapped in the ITMI population, the previously identified quantitative trait locus (QTL) on wheat chromosome 1BS completely disappeared (Liu et al., 2012). It has been previously seen that interactions between a NE and a host susceptibility factor trigger various host defense responses. Similar results were seen with SnTox1 infiltrations of sensitive wheat lines, including the upregulation of PR genes (chitinase, PR-1-A1, and thaumatin), production of ROS, and DNA laddering (Liu et al., 2012). Further affirming a connection with photosynthetic pathways, formation of necrosis induced by SnTox1 requires light, as seen in other interactions such as ToxA-*Tsn1* (Liu et al., 2012; Manning et al., 2005).

23.2.2.2   *SNTOX3-SNN3-B1/SNTOX3-SNN3-D1*

Continuing the search for additional NEs, culture filtrates of *P. nodorum* isolate Sn1501 were used to map sensitivity the BR34 × Grandin RIL population. A sensitivity locus on wheat chromosome 2DS was identified and appeared to be the previously identified *Snn2* locus, indicating that SnTox2 is produced by isolates Sn1501. However, a novel locus was also identified on chromosome 5BS, designated *Snn3* (Friesen et al., 2008a). Further analysis utilizing isolate Sn4 and a population of 117 insensitive F_2 plants derived from a cross of TA2377 (sensitive) × AL8/78 (insensitive) mapped sensitivity to SnTox3 to chromosome 5DS at a similar position to the previously identified *Snn3* on chromosome 5BS (Zhang et al., 2011). Map comparisons revealed that these are most likely homoeologous loci and are now designated *Snn3-B1* and *Snn3-D1*. Additionally, it was determined that individuals with only the *Snn3-D1* locus were significantly more sensitive to SnTox3 than individuals with only the *Snn3-B1* locus (Zhang et al., 2011).

Following the identification of the *Snn3*-SnTox3 interaction, efforts were placed in the identification of the gene encoding SnTox3 as well as its physiological properties. Briefly, culture filtrates of *P. nodorum* isolate Sn4 were subjected to size exclusion chromatography and subsequent mass spectrometry of selected protein bands to identify the corresponding genomic sequences of two candidate genes, *SNOG_16063* and *SNOG_08981*. Due to the absence of *SNOG_08981* in the avirulent isolate Sn79-1087, it was selected as a strong effector candidate. Transformation of avirulent isolate Sn79-1087 with a functional copy of *SNOG_08981* caused this isolate to become virulent on *Snn3* harboring lines and subsequent mapping analysis showed that susceptibility to the transformant mapped to the *Snn3* locus, confirming this interaction. Additionally, mutagenesis of endogenous *SNOG_08981* significantly reduced virulence and QTL analysis using the disruption mutants displayed the absence of any significant association at the *Snn3* locus confirming *SNOG_08981* as *SnTox3*. SnTox3 consists of a single 693 base pair exon and the predicted 230 amino acid protein does not have homology to proteins from other organisms. It is present in the genome as a single copy embedded in a region containing repetitive sequences and contains six cysteine residues, similar to previously identified effectors, such as SnToxA (Liu et al., 2009; Friesen et al., 2006a, 2006b).

23.2.2.3   SNTOX6-SNN6

The SnTox6 NE is the most recent to be characterized from *P. nodorum*. Sensitivity to Sn6 culture filtrates in the ITMI population was mapped to wheat chromosomes 5BS and 6A (Gao et al., 2015). The locus on 5BS had been previously identified as conferring sensitivity to SnTox3 (Friesen et al., 2008a). However, the presence of a novel locus on 6A indicated that a previously unidentified interaction was occurring. Utilizing infiltrations from differential wheat lines known to harbor sensitivity to the various *P. nodorum* NEs and previous research, it was determined that isolate Sn6 produced SnToxA, SnTox1, SnTox2, and SnTox3 (Gao et al., 2015). Additionally, in the ITMI population, only sensitivity to SnTox3 segregated, which prompted the creation of a Sn6 SnTox3 deletion mutant to Mendelize the SnTox6 effects. Also, it was seen that transcription levels of *SnTox1* were non-existent *in planta,* which explained why the *Snn1* locus was not detected in QTL analysis. Infiltrations of the ITMI population with the Sn6 deletion mutant culture filtrates indicated that sensitivity segregated as a single host gene, designated *Snn6*. Similar to other identified NEs, further analysis revealed SnTox6 to be an ~12 kDa secreted protein and only induces necrosis under light conditions (Gao et al., 2015). These results illustrate yet another example of a small, secreted protein implicated in necrotrophic effector triggered susceptibility (NETS), further exemplifying the vast repertoire that this necrotrophic specialists has evolved to successfully perturb host immunity machinery to persist.

23.2.3   PYRENOPHORA TERES

Pyrenophora teres is a destructive foliar pathogen of barley and the causal agent of net blotch of barley. Two forms of net blotch exist in nature, net form net blotch (NFNB) and spot form net blotch (SFNB) caused by *P. teres* f. *teres* and *P. teres* f. *maculata*, respectively. These two forms of the disease are morphologically identical and can only be differentiated based on the symptoms they induce on the host and molecular analysis (Smedegard-Petersen, 1971; Rau et al., 2007). NFNB is characterized by the presence of transverse and longitudinal striations on foliar tissue, often accompanied by chlorosis. The SFNB differs from NFNB by the formation of elliptical necrotic lesions, surrounded by a chlorotic halo (Liu et al., 2011). To date the cloning of host genes conferring susceptibility to net blotch or the corresponding effectors have not been reported in the literature, but the genetics of resistance/susceptibility and virulence/Avr of both the host and the pathogen have been well characterized. This work has culminated in the identification of candidate NFNB susceptibility genes (Richards et al., 2016) as well as candidate NEs (Wyatt et al., 2016).

Resistance/susceptibility to net blotch has been mapped to every barley chromosome (reviewed in Liu et al., 2011) and the host-pathogen genetic interactions appear to be exceedingly complex. Dominant resistances, incomplete resistances, and dominant susceptibility (recessive resistances) have been identified in the host (Steffenson et al., 1996; Raman et al., 2003; Cakir et al., 2003; Ma et al., 2004; Manninen et al., 2006; Friesen et al., 2006a, 2006b; Abu Qamar et al., 2008; St. Pierre et al., 2010; Gupta et al., 2011). As described earlier for the mechanisms underlying the wheat-*P. nodorum* and wheat-*Ptr* pathosystems, it is hypothesized that the dominant NFNB and SFNB susceptibility loci in barley harbor susceptibility genes that inappropriately perceive the pathogens triggering NETS. Recently, a proteinaceous effector was identified in *Ptt* isolate 0–1 that appears to interact with a single gene on barley chromosome 6H (Liu et al., 2015). Intracellular wash fluids (IWF) were isolated from infected foliar tissue and used to infiltrate a population derived from a cross between the susceptible parent Hector and the resistant parent NDB 112. The susceptibility mapped to the centromeric region of barley chromosome 6H, a common locus in which resistance/susceptibility has been mapped many times in diverse sets of barley bi-parental mapping populations challenged with unique *Ptt* isolates collected from distinct pathogen populations from around the globe (Liu et al., 2015; Steffenson et al., 1996; Raman et al., 2003; Cakir et al., 2003; Ma et al., 2004; Manninen et al., 2006; Friesen et al., 2006a, 2006b; Abu Qamar et al., 2008; St. Pierre et al., 2010; Gupta et al., 2011). It was shown that this putative NE, designated as PttNE1, present in these IWF was proteinaceous and was estimated to be 6.5–12.5 kDa. As previously mentioned, when the host recognizes a pathogen, often times a chain of defense responses is initiated, such as the release of ROS known as an oxidative burst as well as electrolyte leakage. It was determined that following the interaction between PttNE1 and the corresponding host susceptibility factor, a significant accumulation of hydrogen peroxide (H_2O_2) was detected in the compatible reaction with susceptible parent Hector, but not in the resistant parent NDB 112 (Liu et al., 2015). Additionally, substantial electrolyte leakage was detected in the compatible reaction. These results indicated that the same molecular mechanisms that successfully control a biotrophic pathogen are also being hijacked by this necrotrophic pathogen to proliferate and infect the host via the interaction with a single dominant host susceptibility factor.

P. teres f. *teres* also appears to have a strong repertoire of weapons. Unique loci have been identified in the *P. teres* f. *teres* genome corresponding to virulence on different barley cv. *P. teres* f. *teres* isolates 6A and 15A are virulent on barley cv Rika and Kombar, respectively. The creation of a mapping population derived from a cross between *P. teres* f. *teres* isolates 6A and 15A facilitated the discovery of two unique loci in each pathogen genome correlated with virulence on Rika and Kombar, respectively (Shjerve et al., 2014). Two virulence loci were identified for each isolate, indicating the potential of four unique NEs being produced by these two isolates. Progeny isolates containing a single virulence locus were used to inoculate a Rika × Kombar population and susceptibility to these progeny isolates mapped to the previously identified centromeric 6H locus (Shjerve et al., 2014). Additionally, susceptibility to isolate 15A has also been mapped to barley chromosomes 1H and 2H, and susceptibility to isolate 6A has been identified on chromosomes 3H and 5H (Liu et al., 2015). This indicates that the host possess several other susceptibility loci and that these two isolates harbor other unique NEs, warranting further investigation into this complex pathosystem.

Considerable efforts have been put forth toward the characterization of the centromeric dominant susceptibility locus on barley chromosome 6H in respect to *P. teres* f. *teres* isolates 6A and 15A. As previously stated, barley cv Rika and Kombar exhibit differential susceptibility to isolates 6A and 15A, respectively, and exists in repulsion. Utilizing a map-based cloning approach and marker saturation via the exploitation of the regional synteny with *Brachypodium distachyon* chromosome 3, susceptibility was mapped at high-resolution to ~0.29 cM. Analysis of the syntenic region in *Brachypodium distachyon* revealed 59 annotated genes within the newly delimited region, of which, orthologous barley genes were mined for further marker development and candidate gene identification (Richards et al., 2016a). A predicted extracellular LRR receptor like gene designated *Spt1*, was identified as a strong candidate gene, and extensive allele analysis of diverse barley lines displayed a strong correlation with three diverse alleles, Rika-, Kombar-, and Morex-like alleles, and susceptibility to *P. teres* f. *teres* isolates 6A, 15A, and TA5/TD10, respectively (Richards et al., 2016b). Currently, work toward validation of this candidate gene and its corresponding alleles is underway via attempts through post-transcriptional gene silencing using barley stripe mosaic virus-virus induced gene silencing (BSMV-VIGS) and the development of deletion mutants utilizing CRISPR/CAS9. The results of this research will reveal yet another piece of this complex puzzle and advance our knowledge of the manner in which necrotrophic specialists invade the host at the molecular level in an inverse gene-for-gene system.

23.2.4 COCHLIOBOLUS SATIVUS

C. sativus is an ascomycete fungus, which causes economical losses of barley and wheat in North and South America, Europe, and several countries of Asia (Mathre, 1997; Grey & Mathre, 1984; Weise, 1987). It attacks various plant parts such as roots as a part of common root rot complex; spikes, causing kernel blight or black point; and leaves as spot blotch (Duczek & Jones-Flory, 1994; Grey & Mathre, 1984). Among various symptoms caused by this pathogen, foliar spot blotch of barley and wheat is the most devastating and can cause significant yield loss in barley and wheat (Duveiller & Gilchrist, 1994; Almgren et al., 1999).

As with most foliar diseases of wheat and barley, the most practical disease management strategy is by the use of genetic resistance. Resistance to spot blotch disease in barley can be traced back to CI 7117-77, which is an unimproved line selected from North Dakota State University (Wilcoxson et al., 1990). A cross between CI-7117-77 × Kindred resulted in the line NDB112, which was selected and deployed for Rcs5-mediated seedling and adult plant resistance and has remained effective for over 50 years in the Upper Midwest of the United States (Steffenson et al., 1996). Various studies for mapping spot blotch resistance in barley reported six resistance loci, namely *Rcs1-6* (Bilgic et al., 2006). However, *Rcs5*, have been shown to be responsible for the majority of seedling and some adult resistance in current cv (Søgaard & von Wettstein-Knowles, 1987; Steffenson, et al., 1996; Bilgic et al., 2005; Bovill et al., 2010; Zhou, 2012).

Spot blotch resistance in the barley cv Morex, which has been utilized for many of the genetic and genomic tools in barley, as well as most of the six-rowed barley varieties currently grown in the Midwestern United States, harbor the *Rcs5* locus. Steffenson et al. (1996) mapped *Rcs5*-mediated resistance in a Steptoe × Morex double haploid population using isolate ND85f, to the short arm of Chr. 1(7H). Utilizing a high-resolution mapping population and syntenic genomic regions between grass species *Rcs5* was delimited to an ~0.32 cM region (Drader et al., 2009; Drader & Kleinhofs, 2010; Drader, 2010). Barley genome sequence of cv Morex, revealed four WAK genes in the *Rcs5* region designated *HvWAK2-5*. In efforts to functionally validate these WAK genes, analysis of Steptoe × Morex F_2 individuals showed a 1 (resistant):3 (susceptible) segregation ratio suggesting that *rcs5* is actually recessive in nature (Ameen et al., 2016). Further allele analysis and BSMV-VIGS constructs specific to each WAK gene revealed two of these WAK genes are required for susceptibility and gene silencing of any of these two genes shifted the phenotype from susceptibility to resistance in susceptible cv Steptoe and Harrington (Ameen et al., 2016) suggesting that they represent dominant susceptibility genes that also function in an inverse gene-for-gene manner.

The WAKs make up a large family of transmembrane proteins with a cytoplasmic serine/threonine kinase domain, a transmembrane domain with extracellular galacturonan-binding (GUB) and calcium-binding epidermal growth factor (EGF_CA) domains (He et al., 1996, 1999). Members of the WAK gene families have been implicated in fundamental processes that depend on cell wall and plasma membrane interactions, such as cell elongation and development (He et al., 1998, 1999; Kohorn et al., 2000; Anderson et al., 2001; Lally et al., 2001). Genes encoding WAKs have been found to be upregulated during tolerance to abiotic stress such as cold, heavy metal (aluminum) resistance, salt stress, and more recently to wounding and pathogen challenge. These WAK genes playing a role in disease resistance respond to external stimuli eliciting the MAPK signaling pathways to induce stress response thus should be constitutively expressed receptors. However, there must be a feedback loop resulting in the upregulation of these putative extracellular receptors in response to stress stimuli possibly priming the plant for further pathogen challenge (Sivaguru et al., 2003; He et al., 1999; Lally et al., 2001; Wagner, & Kohorn, 2001; Zhang et al., 2005; Liu et al., 2006; Kohorn & Kohorn, 2012).

In maize, a biotrophic soil-borne fungus *Sporisorium reilianum*, causes head smut that can result in complete loss of seed on infected plants (Wang et al., 2002). Head smut resistance is quantitative; however, *ZmWAK* is a major gene for this quantitative resistance (Zuo et al., 2015; Chen et al., 2008). The resistance provided by *ZmWAK* works early in the mesocotyl of maize seedlings, rather than in the ear or tassel where typical symptoms occur at later growth stage, suggesting *ZmWAK*, monitor the pathogen at early stage and arrest its biotrophic growth before reaching aerial tissues, suggesting a spatiotemporal resistance strategy (Zuo et al., 2015). Recently, a very strong candidate for the dominant wheat susceptibility gene *Snn1*, conferring sensitivity to the necrotrophic fungal pathogen, *P. nodorum* was identified and characterized (Shi et al., 2016). The *Snn1* belongs to the WAK gene family and shown to be the susceptibility target of *P. nodorum* carrying the NE, SnTox1 (Shi et al., 2016). As discussed earlier this NE-WAK interaction also leads to pathogen recognition eliciting NETS.

Considering the evidence of WAK roles in NETS, we hypothesize that the barley *rcs5* recessive R-gene would more appropriately be called a dominant susceptibility gene. Thus the appropriate nomenclature would be *Scs5* (susceptibility to *C. sativus* 5) as the dominant susceptibility protein is targeted by *C. sativus* to induce PCD in an inverse-gene-for-gene manner facilitating disease in the susceptible lines.

23.3 INVERSE GENE-FOR-GENE THEORY

A seminal theory in the field of the plant pathology was the ideation of the gene-for-gene theory by H. H. Flor. Within the flax rust pathosystem, Flor dissected the genetics of both the host (*Linum usitatissimum*) and the pathogen (*Melampsora lini*). By doing so, he discovered that resistance to the biotrophic flax rust not only depends on the presence of a dominant R-gene in the host, but upon the production of a dominant effector, *Avr* gene in the pathogen (Flor et al., 1956). Recognition of this Avr gene product by the host triggers a wide array of host cellular processes comprising PCD known as the HR. During HR, plant cells release ROS known as an oxidative burst, as well as accumulate the plant hormone salicylic acid. Additionally, regulatory changes occur within the cell, such as the synthesis of antimicrobial compounds and the transcriptional induction of MAPK phosphorylation signaling cascades (Coll et al., 2011). These responses, among others, ultimately result in cell death and were initiated from the recognition of the pathogen. In a biotrophic pathosystem, HR effectively sequesters the pathogen and severs its nutrient source of living tissue, thereby restricting any further growth. Additionally, whereas PTI grants a general form of immunity to all isolates of a particular pathogen, ETI may be isolate specific, due to the production of specific effectors that are recognized by specific resistance proteins. This phenomenon now poses a selection pressure on the biotrophic pathogen to evade host detection and avoid the triggering of HR. However, this plant immune response also occurs when being invaded by necrotrophic pathogens. The induction of host defense responses upon inappropriate recognition of a NE produced by a necrotrophic specialist pathogen elicit PCD responses that are detrimental because they facilitate disease caused by this class of pathogen. As necrotrophs thrive on dead tissue, the very same mechanisms that are advantageous as a defense to biotrophs become a detriment to the host when encountering a necrotrophic pathogen. In a necrotrophic system, the roles of selection pressure have now been reversed. The host now attempts to avoid the recognition of the pathogen and subsequently avoid the activation of its intricate PCD pathway, because the pathogen now thrives when it is detected.

A key point to understand about both the gene-for-gene and inverse gene-for-gene models is that both are essentially the same. The plant host recognizes the invading pathogen triggering its elaborate immunity pathways that result in PCD responses. The differences in these models do not exist in the host response, but in the outcome of the interaction, which is ultimately determined by the lifestyle of the pathogen. Immunity responses against a biotrophic pathogen are successful when the interaction results in pathogen recognition resulting on PCD responses that sequester the pathogen resulting in resistance/avirulence. However, the outcome of detection of a necrotrophic pathogen resulting in immunity responses that elicit PCD responses may result in susceptibility/virulence.

The molecular interactions involved in the inverse gene-for-gene model may occur in several unique ways. Since LRRs, one of the primary gene classes associated with plant disease resistance or susceptibility, function as cytoplasmic receptors, it is often speculated that pathogens produced effectors that may interact directly with the LRR recognition domain and subsequently induct a cascade of signaling events that result in HR. Although this direct interaction has yet to be shown in a wheat or barley necrotrophic system, it has been observed to occur in the biotrophic flax rust pathosystem. Flax genes L5, L6, and L7 at the *L* locus were shown to directly interact with seven diverged *AvrL567* variants in the pathogen through yeast two-hybrid assays and interactions detected directly correlated with the formation of necrosis and subsequent Avr in the host. An additional five *Avr567* variants were identified but did not interact with L5, L6, or L7 and also correlated with virulence on flax (Dodds et al., 2006). The results of this study indicate the direct interaction of a pathogen effector with a host receptor triggers PCD and in this case, the successful sequestration of a biotrophic pathogen. Additionally, this work illustrates the selection pressure placed on the pathogen in a classical gene-for-gene model. With a total of 12 allelic variants identified, obvious diversifying selection had occurred in the pathogen population with the aim to alter effector sequence enough to evade perception from the host but maintain any alternative function the effector protein may have. In the hypothetical case of a necrotrophic pathogen directly interacting with a host receptor, the diversifying selection may then be placed on the host receptor, attempting to eradicate recognition and purifying selection placed on the pathogen effector to maintain the interaction. Indirect interactions likely occur in the inverse gene-for-gene model as well, as evidenced by the *Tsn1-ToxA* interaction. It was seen that *Tsn1* and *ToxA* do not directly interact, but *Tsn1* is explicitly needed for the induction of host defense responses and the translocation of ToxA into the cell (Faris et al., 2010). *Tsn1* may be operating as a "guard" to the target of ToxA. Upon modification of the effector target of ToxA, Tsn1 may recognize the alteration of the target protein and begin cellular signaling pathways. Evolutionarily, this interaction may have arisen due to a previous biological function of *Tsn1*, which *Ptr* then exploited. Potentially, Tsn1 may have functioned to monitor another protein(s) essential for host fitness. By the production of toxic compounds with varying functions, *Ptr* may have thrown a barrage of molecules at the host. Eventually, either a pathogen molecule or pathogen induced modification of a host protein was detected by the host and an advantageous triggering of cell death occurred for *Ptr*. A similar event may have occurred with the SnTox1-*Snn1* interaction. As SnTox1 contains a chitin-binding domain, *Snn1* may detect the chitin-SnTox1 complex or SnTox1 alone, directly or indirectly, and begin various defense response mechanisms. The detection of a chitin-bound complex or other classes of molecules may have already pre-existed and was under surveillance by *Snn1*, but *P. nodorum* was able to produce a molecule to mimic this complex by binding free chitin or by the production of a protein similar to the primary *Snn1* ligand. As with a potential direct interaction with a receptor, indirect interactions may then impose a positive selection on the pathogen effector while causing a diversifying selection on the host R-gene. The initial interaction between a pathogen effector and a host R-gene may arise from the multitude of potential interactions between pathogen and host molecules with many of the secreted effectors from the pathogen not resulting in any effect or recognition, but once a beneficial reaction takes place, that molecule/interaction may then be selected for. Direct or indirect interactions may not be the only mechanism being selected for or against within a gene-for-gene or inverse gene-for-gene system, but the shuffling of genomic sequences may also be involved. As with PtrToxB, an increase in copy number of the *PtrToxB* gene strongly correlates with an increase in virulence. Potentially, an increasing number of *PtrToxB* copies were being selected for, increasing its virulence on the host with each addition (Strelkov et al., 2002; Strelkov & Lamari, 2003; Martinez et al., 2004). Also, the evidence of horizontal gene transfer of ToxA from *P. nodorum* to *Ptr* represents an astonishing mechanism in which a pathogen can diversify its weaponry by borrowing arms from an organism living within the same ecosystem (Friesen et al., 2006a, 2006b). By simply adding genomic components from pre-existing genes through an increase in copy number of a beneficial factor or by acquiring one from a fellow plant pathogen, necrotrophs appear to have effective means to diversify themselves to become more successful pathogens.

A multi-allelic susceptibility gene may also exist in the barley-NFNB pathosystem. As previously mentioned, susceptibility to *Ptt* isolates 6A and 15A map to the same locus on barley chromosome 6H and have been shown to harbor unique virulence loci with the hypothesis that they produce distinct NEs underlying these virulence QTL (Shjerve et al., 2014). Preliminary allele analysis data of a candidate *P. teres* f. *teres* susceptibility gene suggests that a single gene with at least three identified diverged alleles may condition susceptibility to different isolates of *P. teres* f. *teres* (Richards et al., 2016b). The presence of a corresponding allele in the host is strongly correlated with susceptibility to particular *P. teres* f. *teres* isolates. Additionally, susceptibility to progeny isolates of 15A × 6A map to the same locus, indicating that a single gene with multiple alleles could be directly or indirectly recognizing unique NEs produced by these two isolates (Shjerve et al., 2014). Alternatively, a single gene could be recognizing multiple effectors or a tightly linked cluster of susceptibility factors may be forming a "susceptibility island" and condition susceptibility to a broad range of pathogen isolates. In the previously mentioned flax rust pathosystem, it has been observed that diverged alleles of a pathogen effector may be recognized by alleles at a single host locus (L) (Ellis et al., 2007) and the same phenomenon may be occurring with the barley-NFNB system as well, with the major differences being the lifestyle of the pathogen and the reversal of selection pressures. More work is needed to validate these hypotheses, but nevertheless, it appears that within the inverse gene-for-gene model lies an enormous amount of complexity, encompassing a wide array of molecules produced by the host and pathogen, and the interactions between them.

23.4 IMPLICATIONS FOR WHEAT AND BARLEY BREEDING

Necrotrophic specialists impose a unique perspective on the selection for genetic resistance in breeding programs. Traditionally, when breeding for resistance to a particular pathogen, the strategy is the selection for dominant R-gene/genes. This approach is appropriate when targeting resistance to a biotrophic pathogen operating in a gene-for-gene system because incorporation of functional R-gene/genes confers effective dominant resistance. However, the paradigm shifts when breeding for resistance toward a necrotrophic specialist pathogen. Although dominant resistances toward necrotrophic pathogens have been identified, in most cases as previously discussed, resistance is not necessarily conferred by a dominant functional gene, but as a lack of several functional susceptibility factor/s, requiring the selection against functional genes or toward null alleles. Necrotrophic specialists commonly harbor multiple effectors that target unique loci through the host genome with great diversity in the pathogen populations, thereby necessitating the elimination of several host loci to successfully achieve "resistance."

23.5 CONCLUDING REMARKS

Successful breeding for resistance to necrotrophic specialists including pathogens such as *Ptr*, *P. nodorum*, *P. teres* f. *teres* and *maculata, and C. sativus* requires the knowledge of the diversity of pathogen isolates in the region as well as the host susceptibility targets. It is imperative to understand and characterize the diversity of NEs in the pathogen to be able to effectively breed against NETS. Since particular pathogen isolates harbor different combinations of NEs, that is, *Ptr* with eight known races each with a different array of effectors, it is important to know which race types are prevalent in the region of interest. Equally as important is the knowledge of the known host susceptibility targets. Understanding the potential molecular interactions that may occur between pathogen isolates and adapted germplasm will allow the successful cleansing of potential susceptibility loci from breeding materials and effective introduction of resistance.

KEYWORDS

- **inverse gene for gene theory**
- **necrotrophic pathogen**
- *Modus operandi*
- **pathosystems**
- *Pyrenophora tritici repentis*
- *Pyrenophora teres*
- *Parastagonospora nodorum*
- *Cochliobolus sativus*
- **wheat**
- **barley**

REFERENCES

Abeysekara, N. S.; Friesen, T. L.; Keller, B.; Faris, J. D. Identification and Characterization of a Novel Host-Toxin Interaction in the Wheat-*Stagonospora nodorum* Pathosystem. *Theor. Appl. Genet.* **2009,** *130,* 117–126.

Abeysekara, N. S., et al. Marker Development and Saturation Mapping of the Tan Spot Ptrtoxb Sensitivity Locus *Tsc2* in Hexaploid Wheat. *Plant Genome.* **2010,** *3,* 179–189.

Abu Qamar, M., et al. A Region of Barley Chromosome 6H Harbors Multiple Major Genes Associated with the Net Type Net Blotch Resistance. *Theor. Appl. Genet.* **2008,** *117,* 1261–1270.

Adhikari, T. B., et al. *Tsn1*-Mediated Host Responses to Toxa from *Pyrenophora tritici-repentis. Mol. Plant Microbe Interact.* **2009,** *22*(9), 1056–1068.

Almgren, I.; Gustafsson, M.; Falt, A. S.; Lindgren, H.; Liljeroth, E. Interaction between Root and Leaf Disease Development in Barley Cultivars after Inoculation with Different Isolates of *Bipolaris sorokiniana. Phytopathology.* **1999,** *147,* 331–337.

Ameen, G.; Drader, T.; Sager, L.; Steffenson, B.; Kleinhofs, A.; Brueggeman, R. Death be not Proud *rcs5* is a Wall Associated Kinase Gene that Functions as a Dominant Susceptibility Factor in the Barley *Cochliobolus sativus* Interaction to Produce Necroptosis. IS-MPMI XVII Congress, Portland, Oregon. 2016, Abstract Number: P17-486.

Anderson, C. M., et al. WAKs: Cell Wall-Associated Kinases Linking the Cytoplasm to the Extracellular Matrix. *Plant Mol. Biol.* **2001,** *47,* 197–206.

Arora, D.; Gross, T.; Brueggeman, R. Allele Characterization of Genes Required for *rpg4*-Mediated Wheat Stem Rust Resistance Identifies *Rpg5* as the R Gene. *Phytopathology.* **2013,** *103,* 1153–1161.

Ballance, G. M.; Lamari, L.; Bernier, C. C. Purification and Characterization of a Host-Selective Necrosis Toxin from *Pyrenophora tritici-repentis. Physiol. Mol. Plant Path.* **1989,** *35,* 203–213.

Ballance, G. M.; Lamari, L.; Kowatsch, R.; Bernier, C. C. Cloning, Expression and Occurrence of the Gene Encoding the Ptr Necrosis Toxin from *Pyrenophora tritici-repentis. Mol. Plant Pathol.* 1996. Online. http://bspp.org.uk/mppol/1996/1209ballance/ (accessed December 15, 2015).

Belkhadir, Y.; Subramaniam, R.; Dangl, J. L. Plant Disease Resistance Protein Signaling: NBS-LRR Proteins and Their Partners. *Curr. Opin. Plant Biol.* **2004,** *7,* 391–399.

Bilgic, H.; Steffenson, B. J.; Hayes, P. M. Comprehensive Genetic Analyses Reveal Differential Expression of Spot Blotch Resistance in Four Populations of Barley. *Theor. Appl. Genet.* **2005,** *111,* 1238–1250.

Bilgic, H.; Steffenson, B. J.; Hayes, P. M.; Molecular Mapping oOf Loci Conferring Resistance to Different Pathotypes of the Spot Blotch Pathogen in Barley. *Phytopathology.* **2006,** *96*(7), 699–708.

Bovill, J., et al. Mapping Spot Blotch Resistance Genes in Four Barley Populations. *Mol. Breeding.* **2010,** *26,* 653–666.

Brueggeman, R., et al. The Stem Rust Resistance Gene *Rpg5* Encodes a Protein with Nucleotide-Binding-Site, Leucine-Rich, and Protein Kinase Domains. *Proc. Natl. Acad. Sci. USA.* **2008,** *105*(39), 14970–14975.

Cakir, M., et al. Mapping and Validation of the Genes for Resistance to *Pyrenophora teres* f. *teres* in Barley (*Hordeum vulgare* L.). *Aust. J. Agric. Res.* **2003,** *54,* 1369–1377.

Chen, Y., et al. Identification and Fine-Mapping of a Major QTL Conferring Resistance against Head Smut in Maize. *Theor. Appl. Genet.* **2008,** *117,* 1241–1252.

Ciuffetti, L. M.; Tuori, R. P.; Gaventa, J. M. A Single Gene Encodes a Selective Toxin Causal to the Development of Tan Spot of Wheat. *Plant Cell.* **1997,** *9,* 135–144.

Ciuffetti, L. M.; Manning, V. A.; Pandelova, I.; Figueroa Betts, M.; Martinez, J. P. Host-Selective Toxins, PtrToxA and PtrToxB, as Necrotrophic Effectors in the *Pyrenophora tritici-repentis*-Wheat Interaction. *New Phytol.* **2010,** *187,* 911–919.

Chinchilla, D.; Bauer, Z.; Regenass, M.; Boller, T.; Felix, G. The *Arabidopsis* Receptor Kinase fls2 Binds flg22 and Determines the Specificity of Flagellin Perception. *Plant Cell.* **2006,** *18*(2), 465–476.

Chinchilla, D., et al. A Flagellin-Induced Complex of the Receptor FLS2 and BAK1 Initiates Plant Defence. *Nature.* **2007,** *448*(7152), 497–500.

Chisholm, S. T.; Coaker, G.; Day, B.; Staskawicz, B. J. Host-Microbe Interactions: Shaping the Evolution of the Plant Immune Response. *Cell.* **2006,** *124,* 803–814.

Coll, N. S.; Epple, P.; Dangl, J. L. Programmed Cell Death in the Plant Immune System. *Cell Death Differ.* **2011,** *18,* 1247–1256.

Dodds, P. N., et al. Direct Protein Interaction Underlies Gene-for-Gene Specificity and Co Evolution of the Flax Resistance Genes and Flax Rust Avirulence Genes. *Proc. Natl. Acad. Sci. USA.* **2006,** *103*(23), 8888–8893.

Drader, T. Cloning of the Seedling Spot Blotch Resistance Gene *Rcs5*. Ph.D. Dissertation, Washington State University, Pullman, 2010.

Drader, T.; Kleinhofs, A. A Synteny Map and Disease Resistance Gene Comparison between Barley and the Model Monocot *Brachypodium distachyon. Genome.* **2010,** *53*(5), 406–417.

Drader, T.; Johnson, K.; Brueggeman, R.; Kudrna, D.; Kleinhofs, A. Genetic and Physical Mapping of a High Recombination Region on Chromosome 7H (1) in Barley. *Theor. Appl. Genet.* **2009,** *118*(4), 811–820.

Duczek, L. J.; Jones-Flory, L. L. Relationship between Common Root Rot, Tillering and Yield Loss in Spring Wheat and Barley. *Can. J. Plant Pathol.* **1994,** *15,* 153–158.

Duveiller, E.; Gilchrist, L. Production Constraints due to *Bipolaris sorokiniana in* Wheat: Current Situation and Future Prospects. In *Wheat in Heat-Stressed Environments: Irrigated Dry Areas and Ice-Wheat Farming Systems;* Saunder, D. A., Hettel, G. P., Eds.; International Maize and Wheat Improvement Center: Mexico, DF, 1994; pp 343–352.

Ellis, J. G.; Dodds, P. N.; Lawrence, G. J. Flax Rust Resistance Gene Specificity is Based on Direct Resistance-Avirulence Protein Interactions. *Annu. Rev. Phytopathol.* **2007,** *45,* 289–306.

Faris, J. D.; Anderson, J. A.; Francl, L. J.; Jordahl, J. G.; Chromosomal Location of a Gene Conditioning Insensitivity in Wheat to a Necrosis-Inducing Culture Filtrate from *Pyrenophora tritici-repentis. Phytopathology.* **1996,** *86,* 459–463.

Faris, J. D., et al. A Unique Wheat Disease Resistance-Like Gene Governs Effector-Triggered Susceptibility to Necrotrophic Pathogens. *Proc. Natl. Acad. Sci. USA.* **2010,** *107*(30), 13544–13549.

Faris, J. D., et al. In *Map-Based Cloning of the Wheat Snn1 Gene Reveals a Wall-Associated Kinase Hijacked by the Necrotrophic Pathogen Stagonospora nodorum to Cause Disease,* Plant and Animal Genome XXII Conference, San Diego, CA, Jan 11–15, 2014.

Flor, H. H. The Complementary Genetic Systems in Flax Aa Flax Rust. *Adv. Genet.* **1956,** *8,* 29–54.

Friesen, T. L.; Faris, J. D. Molecular Mapping of Resistance to *Pyrenophora tritici-repentis* Race 5 and Sensitivity to PtrToxB in Wheat. *Theor. Appl. Genet.* **2004,** *109,* 464–471.

Friesen, T. L., et al. Emergence of a New Disease as a Result of Interspecific Virulence Gene Transfer. *Nature Genet.* **2006a,** *38*(8), 953–956.

Friesen, T. L.; Faris, J. D.; Lai, Z.; Steffenson, B. J. Identification and Chromosomal Location of Major Genes for Resistance to *Pyrenophora Teres* in a Barley Doubled Haploid Population. *Genome*. **2006b,** *409*, 855–859.

Friesen, T. L.; Meinhardt, S. W.; Faris, J. D. The *Stagonospora nodorum*-Wheat Pathosystem Involves Multiple Proteinaceous Host-Selective Toxins and Corresponding Host Sensitivity Genes that Interact in an Inverse Gene-for-Gene Manner. *Plant J.* **2007,** *51*(4), 681–692.

Friesen, T. L., et al. Characterization of the Interaction of a Novel *Stagonospora nodorum* Host-Selective Toxin with a Wheat Susceptibility Gene. *Plant Physiol.* **2008a,** *146*(2), 682–693.

Friesen, T. L.; Faris, J. D.; Solomon, P. S.; Oliver, R. P. Host-Specific Toxins: Effectors of Necrotrophic Pathogenicity. *Cell Microbiol.* **2008b,** *10*(7), 1421–1428.

Friesen, T. L.; Chu, C.; Xu, S. S.; Faris, J. D. Sntox5-*snn5*: A Novel *Stagonosporano-dorum* Effector-Wheat Gene Interaction and its Relationship with the SnToxA-*Tsn1* and SnTox3-*Snn3-B1* Interactions. *Mol. Plant Pathol. Interact.* **2012,** *13*(9), 1101–1109.

Gao, Y., et al. Identification and Characterization of the SnTox6-*Snn6* Interaction in the *Parastagonospora nodorum*-Wheat Pathosystem. *Mol. Plant Microbe. Interact.* **2015,** *28*(5), 615–325.

Gomez-Gomez, L.; Boller, T. FLS2: An lrr Receptor-Like Kinase Involved in the Perception of the Bacterial Elicitor Flagellin in *Arabidopsis*. *Mol. Cell.* **2000,** *5*, 1003–1011.

Gomez-Gomez, L.; Boller, T. Flagellin Perception: A Paradigm for Innate Immunity. *Trends Plant Sci.* **2002,** *7*(6), 251–256.

Grey, W. E.; Mathre, D. E. Reaction of Spring Barleys to Common Root Rot and its Effects on Yield Components. *Can. J. Plant Sci.* **1984,** *64*, 245–253.

Gupta, S.; Li, C.; Loughman, R.; Cakir, M.; Westcott, S.; Lance, R. Identifying Genetic Complexity of 6H Locus in Barley Conferring Resistance to *Pyrenophora teres* f. *teres*. *Plant Breed.* **2011,** *130*(4), 423–429.

Hammond-Kossack, K. E.; Rudd, J. J. Plant Resistance Signaling Hijacked by a Necrotrophic Fungal Pathogen. *Plant Signal Behav.* **2008,** *3*(11), 993–995.

He, Z. H.; Cheeseman, I.; He, D.; Kohorn, B. D. A Cluster of Five Cell Wall-Associated Receptor Kinase Genes, Wak1–5, are Expressed in Specific Organs of Arabidopsis. *Plant Mol. Biol.* **1999,** *39*, 1189–1196.

He, Z.-H.; Fujiki, M.; Kohorn, B. D. A Cell Wall Associated Receptor-Like Protein Kinase. *J. Biol. Chem.* **1996,** *271*, 19789–19793.

He, Z. H.; He, D.; Kohorn, B. D. Requirement for the Induced Expression of a Cell Wall Associated Receptor Kinase for Survival during the Pathogen Response. *Plant J.* **1998,** *14*, 55–64.

Jones, J. D. G.; Dangl, J. L. The Plant Immune System. *Nature*. **2006,** *444*(16), 323–329.

Kohorn, B. D. Plasma Membrane-Cell Wall Contacts. *Plant Physiol.* **2000,** *124*, 31–38

Kohorn, B. D.; Kohorn, S. L. The Cell Wall-Associated Kinases, WAKs, as Pectin Receptors. *Front. Plant Sci.* **2012,** *3*, 88.

Lally, D.; Ingmire, P.; Tong, H. Y.; He, Z. H. Antisense Expression of a Cell Wall–Associated Protein Kinase, *WAK4*, Inhibits Cell Elongation and Alters Morphology. *Plant Cell.* **2001,** *13*, 1317–1331.

Lamari, L., et al., The Identification of Two New Races of *Pyrenophora tritici-repentis* from the Host Center of Diversity Confirms a One-to-One Relationship in Tan Spot of Wheat. *Phytopathology*. **2003,** *93*(4), 391–396.

Liu, Y., et al. Isolation and Characterization of Six Putative Wheat Cell Wall-Associated Kinases. *Funct. Plant Biol.* **2006,** *33*, 811–821.

Liu, Z. H., et al. Genetic and Physical Mapping of a Gene Conditioning Sensitivity in Wheat to a Partially Purified Host-Selective Toxin Produced by *Stagonospora nodorum*. *Phytopathology*. **2004**, *94*(10), 1056–1060.

Liu, Z.; Faris, J. D.; Oliver, R. P.; Tan, K.; Solomon, P. S.; McDonald, M. C.; McDonald, B. A.; Nunez, A.; Lu, S.; Rasmussen, J. B.; Friesen, T. L. SnTox3 Acts in Effector Triggered Susceptibility to Induce Disease on Wheat Carrying the *Snn3* Gene. *PLoS Pathog*. **2009**, *5*(9), e1000581.

Liu, Z.; Ellwood, S. R.; Oliver, R. P.; Friesen, T. L. *Pyrenophora teres:* Profile of an Increasingly Damaging Barley Pathogen. *Mol. Plant Pathol*. **2011**, *12*(1), 1–19.

Liu, Z.; Zhang, Z.; Faris, J. D.; Oliver, R. P.; Syme, R.; McDonald, M. C.; McDonald, B. A.; Solomon, P. S.; Lu, S.; Shelver, W. L.; Xu, S.; Friesen, T. L. The Cysteine Rich Necrotrophic Effector Sntox1 Produced by *Stogonospora nodorum* Triggers Susceptibility of Wheat Lines Harboring *Snn1*. *PLoS Pathog*. **2012**, *8*(1), e1002467.

Liu, Z.; Holmes, D. J.; Faris, J. D.; Chao, S.; Brueggeman, R. S.; Edwards, M. C.; Friesen, T. L. Necrotrophic Effector-Triggered Susceptibility (NETS) Underlies the Barley-*Pyrenophora teres* f. *teres* Interaction Specific to Chromosome 6H. *Mol Plant Pathol*. **2015**, *16*(2), 188–200.

Ma, Z. Q.; Lalpitan, N. L. V.; Steffenson, B. QTL Mapping of Net Blotch Resistance Genes in a Doubled-Haploid Population of Six-Rowed Barley. *Euphytica*. **2004**, *137*, 291–296.

Manninen, O.; Kalendar, R.; Robinson, J.; Schulman, A. H. Application of BARE-1 Retrotransposon Markers to the Mapping of a Major Resistance Gene for Net Blotch in Barley. *Mol. Genet. Genomics*. **2006**, *264*, 325–334.

Manning, V. A.; Ciufetti, L. M. Localization of Ptr ToxA Produced by *Pyrenophora tritici-repentis* Reveals Protein Import into Wheat Mesophyll Cells. *Plant Cell*. **2005**, *17*, 3203–3212.

Manning, V. A.; Hardison, L. K.; Ciuffetti, L. M. Ptr ToxA Interactions with a Chloroplast-Localized Protein. *Mol. Plant Microbe Interact*. **2007**, *20*(2), 168–177.

Manning, V. A., et al. The Arg-Gly-Asp-Containing, Solvent-Exposed Loop of Ptr ToxA is Required for Internalization. *Mol. Plant Microbe Interact*. **2008**, *21*(3), 315–325.

Manning, V. A., et al. A Host-Selective Toxin of *Pyrenophora tritici-repentis*, PtrToxA, Induces Photosystem Changes and Reactive Oxygen Species Accumulation in Sensitive Wheat. *Mol. Plant Microbe Interact*. **2009**, *22*(6), 665–676.

Martinez, J. P.; Ottum, S. A.; Ali, S.; Francl, L. J.; Ciuffetti, L. M. Characterization of the ToxB Gene from *Pyrenophora tritici-repentis*. *Mol Plant Microbe Interact*. **2001**, *14*(5), 675–677.

Martinez, J. P.; Oesch, N. W.; Ciuffetti, L. M. Characterization of the Multiple-Copy Host-Selective Toxin Gene, *ToxB*, in Pathogenic and Nonpathogenic Isolates of *Pyrenophora tritici-repentis*. *Mol. Plant Microbe Interact*. **2004**, *17*(5), 467–474.

Mathre, D. E. *Compendium of Barley Diseases;* 2nd ed.; APS Press: St Paul, MN, 1997.

Mendgen, K.; Hahn, M. Plant Infection and the Establishment of Fungal Biotrophy. *Trends Plant Sci*. **2002**, *7*(8), 352–356.

Numnberger, T.; Brunner, F.; Kemmerling, B.; Piater, L. Innate Immunity in Plants and Animals: Striking Similarities and Obvious Differences. Immunol. Rev. **2004**, *198, 249–266.*

Nurnberger, T.; Kemmerling, B. Receptor Protein Kinases- Pattern Recognition Receptors in Plant Immunity. *Trends Plant Sci*. **2006**, *11*(11), 519–522.

Oliver, R. P.; Solomon, P. S. New Developments in Pathogenicity and Virulence of Necrotrophs. *Curr. Opin. Plant Biol*. **2010**, *13*(4), 415–419.

Orolaza, N. P.; Lamari, L.; Balance, G. M. Evidence of Host-Specific Chlorosis Toxin from *Pyrenophora tritici-repentis*, the Causal Agent of Tan Spot of Wheat. *Phytopathology.* **1995,** *85*(10), 1282–1287.

Pandelova, I., et al. Analysis of Transcriptome Changes Induced by Ptrtoxa in Wheat Provides Insights into the Mechanisms of Plant Susceptibility. *Mol. Plant.* **2009,** *2*(3), 1067–1083.

Pandelova, I., et al. Host-Selective Toxins of *Pyrenophora tritici-repentis* Induce Common Responses Associated with Host Susceptibility. *PLoS One.* **2012,** 10.1371/journal. pone.0040240.

Perfect, S. E.; Green, J. R. Infection Structures of Biotrophic and Hemibiotrophic Fungal Plant Pathogens. *Mol. Plant Pathol.* **2001,** *2*(2), 101–108.

Raman, H., et al. Mapping of Genetic Regions Associated with Net Form of Net Blotch Resistance in Barley. *Aust. J. Agric Res.* **2003,** *54,* 1359–1367.

Rau, D., et al. Phylogeny and Evolution of Mating-Type Genes from *Pyrenophora teres*, the Causal Agent of Barley "Net Blotch" Disease. *Curr. Genet.* **2007,** *51,* 377–392.

Richards, J.; Chao, S.; Friesen, T.; Brueggeman, R. Fine Mapping of the Barley Chromosome 6H Net Form Net Blotch Susceptibility Locus. *G3.* **2016a,** *6,* 1809–1818.

Richards, J.; Friesen, T.; Brueggeman, R. High-Resolution Mapping and Candidate Gene identification of Unique *Pyrenophora teres* f. *teres* Necrotrophic Effector Host Sensitivities on Barley Chromosome 6H. IS-MPMI XVII Congress, Portland, Oregon. **2016b,** Abstract Number: P17-597.

Shi, G., et al. The Wheat *Snn7* Gene Confers Susceptibility on Recognition of the *Parastagonospora nodorum* Necrotrophic Effector SnTox7. *Plant Genome.* **2015,** *8,* 1–10.

Shi, G., et al. The Hijacking of a Receptor Kinase-Driven Pathway by a Wheat Fungal Pathogen Leads to Disease. *Sci. Adv.* **2016,** *2,* e1600822.

Shibuya, N.; Minami, E. Oligosaccharide Signaling for Defence Responses in Plant. *Physiol. Mol. Plant Path.* **2001,** *59,* 223–233.

Shjerve, R. A.; Faris, J. D.; Brueggeman, R. S.; Yan, C.; Koladia, V.; Friesen, T. L. Evaluation of a *Pyrenophora teres* f. *teres* Mapping Population Reveals Multiple Independent Interactions with a Region of Barley Chromosome 6H. *Fungal Genet. Biol.* **2014,** *70,* 104–112.

Sivaguru, M., et al. Aluminum-Induced Gene Expression and Protein Localization of a Cell Wall–Associated Receptor Kinase in Arabidopsis. *Plant Physiol.* **2003,** *132,* 2256–2266.

Smedegard-Peterson, V. *Pyrenophora teres* f. *maculata* f. nov. and *Pyrenophora teres* f. *teres* on Barley in Denmark. *Kgl. Vet. Landbohojsk. Arsskr.* **1971,** 124–144.

Søgaard, B.; von Wettstein-Knowles, P. Barley: Genes and Chromosomes. *Carlsberg Res. Commun.* **1987,** *52,* 123–196.

St. Pierre, S.; Gustus, C.; Steffenson, B.; Dill-Macky, R.; Smith, K. P. Mapping Net Form Net Blotch and Septoria Speckled Leaf Blotch Resistance Loci in Barley. *Phytopathology.* **2010,** *100*(1), 80–84.

Steffenson, B. J.; Hayes, H. M.; Kleinhofs, A. Genetics of Seedling and Adult Plant Resistance to Net Blotch (*Pyrenophora teres* f. *teres*) and Spot Blotch (*Cochliobolus sativus*) in Barley. *Theor. Appl. Genet.* **1996,** *92,* 552–558.

Stergiopoulos, I.; Collemare, J.; Mehrabi, R.; de Wit, P. J. G. M. Phytotoxic Secondary Metabolites and Peptides Produced by Plant Pathogenic *Dothideomycete* Fungi. *FEMS Microb. Rev.* **2013,** *37,* 67–93.

Strelkov, S. E.; Lamari, L.; Balance, G. M. Characterization of a Host-Specific Protein Toxin (PtrToxB) from *Pyrenophora tritici-repentis*. *Mol. Plant Microbe Interact.* **1999,** *12*(8), 726–732.

Strelkov, S. E.; Lamari, L.; Sayoud, R.; Smith, R. B. Comparative Virulence of Chlorosis-Inducing Races of *Pyrenophora tritici-repentis*. *Can. J. Plant Pathol.* **2002**, *24*, 29–35.

Strelkov, S. E.; Lamari, L. Host-Parasite Interactions in Tan Spot (*Pyrenophora tritici-repentis*) of Wheat. *Can. J. Plant Pathol.* **2003**, *25*, 339–349.

Tan, K.; Oliver, R. P.; Solomon, P. S.; Moffat, C. S. Proteinaceous Necrotrophic Effectors in Fungal Virulence. *Funct. Plant Biol.* **2010**, *37*, 907–912.

Tomas, A.; Bockus, W. W. Cultivar-Specific Toxicity of Culture Filtrates of *Pyrenophora tritici-repentis*. *Phytopathology.* **1987**, *77*(9), 1337–1340.

Tomas, A., et al. Purification of a Cultivar-Specific Toxin from *Pyrenophora tritici-repentis*, Causal Agent of Tan Spot of Wheat. *Mol. Plant Microbe Interact.* **1990**, *3*(4), 221–224.

Tuori, R. P.; Wolpert, T. J.; Ciuffetti, L. M. Purification and Immunological Characterization of Toxic Components from Cultures of *Pyrenophora tritici-repentis*. *Mol. Plant Microbe Interact.* **1995**, *8*, 41–48.

Van Kan, J. A. Licensed to Kill: The Lifestyle of A Necrotrophic Plant Pathogen. *Trends Plant Sci.* **2006**, *11*(5), 247–253.

Wagner, T. A.; Kohorn, B. D. Wall-Associated Kinases are Expressed throughout Plant Development and Are Required for Cell Expansion. *Plant Cell.* **2001**, *13*, 303–318.

Wang, Z., et al. Research Advance on Head Smut Disease in Maize. *J. Maize Sci.* **2002**, *10*, 61–64.

Wang, X., et al. The *rpg4*-Mediated Resistance to Wheat Stem Rust (*Puccinia graminis*) in Barley (*Hordeum vulgare*) Requires *Rpg5*, a Second NBS-LRR Gene, and an Actin Depolymerization Factor. *Mol. Plant Microbe Int.* **2013**, *26*, 407–418.

Weise, M. V. *Compendium of Wheat Diseases;* 2nd ed.; APS Press: St Paul, MN, 1987.

Wilcoxson, R. D.; Rasmusson, D. C.; Miles, M. R. Development of Barley Resistant to Spot Blotch and Genetics of Resistance. *Plant Dis.* **1990**, *74*, 207–210.

Wolpert, T. J.; Dunkle, L. D.; Ciufffetti, L. M. Host-Selective Toxins and Avirulence: What's in a Name? *Annu. Rev. Phytopathol.* **2002**, *40*, 251–285.

Wyatt, N.; Brueggeman, R.; Faris, J.; Friesen, T. Identification of *Pyrenophora teres* f. *teres* Candidate Effector Genes in the *VR1* and *VR2* Genomic Regions. IS-MPMI XVII Congress, Portland, Oregon. **2016**, Abstract Number: P9-323.

Xiang, T., et al. *Pseudomonas syringae* Effector AvrPto Innate Immunity by Targeting Receptor Kinases. *Curr. Bio.* **2008**, *18*, 74–80.

Zhang, S., et al. Evolutionary Expansion, Gene Structure, and Expression of the Rice Wall-Associated Kinase Gene Family. *Plant Physiol.* **2005**, *139*, 1107–1124.

Zhang, Z., et al. Two Putatively Homoeologous Wheat Genes Mediate Recognition of SnTox3 to Confer Effector-Triggered Susceptibility to *Stagonospora nodorum*. *Plant J.* **2011**, *65*, 27–38.

Zhou, H.; Muehlbauer, G.; Steffenson, B. Population Structure and Linkage Disequilibrium in Elite Barley Breeding Germplasm from the United States. *J. Zhejiang. Univ. Sci. B.* **2012**, *13*, 438–451.

Zipfel, C., et al. Perception of the Bacterial PAMP EF-Tu by the Receptor EFR Restricts *Agrobacterium*-Mediated Transformation. *Cell.* **2006**, *125*, 749–760.

Zipfel, C. Pattern-Recognition Receptors in Plant Innate Immunity. *Curr. Opin. Immunol.* **2008**, *20*(1), 10–16.

Zuo, W., et al. A Maize Wall-Associated Kinase Confers Quantitative Resistance to Head Smut. *Nat. Gen.* **2015**, *47*(2), 151–157. doi: 10.1038/ng.3170. Epub 2014 Dec 22, PMID:25531751.

CHAPTER 24

WHEAT BLAST CAUSED BY *MAGNAPORTHE ORYZAE* PATHOTYPE *TRITICUM*: PRESENT STATUS, VARIABILITY, AND STRATEGIES FOR MANAGEMENT

DEVENDRA PAL SINGH*

ICAR-Indian Institute of Wheat and Barley Research, Karnal 132001, Haryana, India

*Corresponding author. E-mail: dpkarnal@gmail.com

CONTENTS

ABSTRACT

Wheat blast caused by *Magnaporthe oryzae* pathotype *Triticum* (MoT) is a recent disease of wheat occurring in South American countries like Brazil, Bolivia, Uruguay, Paraguay and Argentina and only during early 2016 has been reported to occur in Bangladesh (South Asia). Disease causes seed discoloration and leaf spots as well blight of spike. The spike blight is most damaging stage of disease and results in to sterility or poor seed setting. The disease likes humid and warm climate and increases if heading stage coincides with intermittent rains. Not many studies are done on epidemiology, pathogenic variability, seed health testing, and control including host resistance in wheat blast. The pathogen is more equipped for mutation and develops resistance to fungicides faster than other wheat pathogens. Use of combined dose of two fungicides is therefore recommended for proper management of wheat blast. Keeping in view of climate change, the pathogen may attain most damaging status in wheat in countries where wheat is cultivated in warm and humid climate. The combined impact of wheat blast with spot blotch, head scab and brown rust will be quite devastating in these regions.

24.1   INTRODUCTION

Magnaporthe oryzae is a species complex that causes blast disease on more than 50 species of poaceous plants. *M. oryzae* has a worldwide distribution as a rice (*Oryza* spp.) pathogen and in the last century emerged as an important wheat (*Triticum*) pathogen in southern Brazil and during 2016 in Bangladesh (S. Asia). Presently, *M. oryzae* pathotype *Oryza* is considered the rice blast pathogen, whereas *M. oryzae* pathotype *Triticum* (MoT) is the wheat blast pathogen. The *Oryza* and *Triticum* pathotypes of *M. oryzae* are distinct at the species level (Castroagudin, 2017; Couch & Kohn, 2002). The wheat blast is considered quite important disease due to the facts that its pathogen strikes directly to wither and deform wheat grains, leaving little time to growers to manage it, it spreads easily via commercial grain shipments or farmer-to-farmer seed exchanges. It is seed-borne and the conidia are airborne in nature. The outbreaks of wheat blast are occasional and difficult to predict, have other cereal crops hosts like barley, maize, oat, and foxtail millet, and produces new races rapidly, making it difficult to develop a wheat cultivar with durable resistance and fungicides only provide partial control of wheat blast. The losses in yield may vary from 10 to 100% depending on weather conditions and degree of susceptibility of a wheat cultivar (Igarashi et al.,

1991; Urashima et al., 1993). Goulart et al. (2007) reported average damage due to wheat blast, in 20 tested materials, was 387 kg/ha or 10.5% of the yield, in the Embrapa Agropecuária Oeste experiment in Brazil. In Indápolis, the damages were greater, reaching, in average 609 kg/ha or 13% of yield. The head weight loss was greater (63.4%) with early infection than with late infection (46.0%). The grains below the infection point in the rachis were larger than the normal ones, thus compensating to some extent the presence of the empty spikelets.

Initially wheat blast was found in Brazil from the north of the Mato Grosso do Sul, São Paulo and Rio Grande do Sul, federal district of Brasília, and the state of Goiás, as well as in the Cerrados in centralwestern Brazil, under irrigated cultivated conditions of wheat crop (Igarashi et al., 1986; Prabhu et al., 1992). The disease is also reported from Paraguay (CRIA), Uruguay, Argentina (Perelló et al., 2015), and Bolivia.

Wheat blast has not been reported in India in spite of its presence in Bangladesh during 2016 and 2017 (Singh et al., 2016, 2017; Malaker et al., 2016). Effective quarantine, survey and surveillance, and integrated management practices are being implemented in India to prevent entry and establishment of wheat blast in India. Collaborative research is being carried out with CIMMYT and other countries in evaluation of genotypes and exchange of information and skills.

24.2 SYMPTOMS

Wheat blast pathogen can infect all aerial plant parts. Foliar symptoms appear as grey-green and water-soaked leaf lesions with dark green borders; these become light tan eye shaped spots with necrotic borders, once they have completely expanded. The most notable symptoms of wheat blast are however, appear on spike which shows partly or complete bleaching and blackened rachis at the point of infection. These are often confused with head scab symptoms. Grains from blast-infected heads are usually small, wrinkled, deformed, and have low-test weight. The severe yield losses occur when spikes are infected at flowering or early grain formation (Fig. 24.1). The leaves of such tillers may not dry and remain green inspite of bleaching and sterility of spike.

FIGURE 24.1 (a) Wheat blast infected field (Source: Dr. P. Malaker/ CIMMYT), (b) Leaf spots (Source: Dr. T. Islam), (c) Spike infection

24.3 FAVORABLE CLIMATE

Under most conducive environmental conditions (18–30°C and >80% relative humidity (RH)) during ear emergence or grain filling, the disease can become an epidemic and devastate wheat crop within a week (CABI, 2017). Most severe wheat blast outbreaks have coincided with wet years; warm temperatures and high humidity. Epidemic years are characterized by several days of continuous rains and average temperatures between 18–25°C during flowering, followed by sunny, hot, and humid days. More precise information on conducive conditions for development and spread of wheat blast is crucial and presently not available. The disease is a major limiting factor in wheat cultivation in areas such as the Cerrados (tropical savanna ecoregion of Brazil) and Bolivian lowlands. Since the disease is seed-borne and able to survive on alternative hosts, with uncertainties related to climate and the absence of satisfactory genetic resistance, more information is needed to forecast epidemics and protect other regions and cropping systems from infection (Ha et al., 2017).

24.4 DISEASE CYCLE

Wheat blast is seed and airborne besides it's pathogen survives on collateral hosts and on crop residue. Seed transmission is most important for spread of disease across continents besides human errors. It also plays a limited role in epidemiology, where spikes are infected mainly by airborne conidia from secondary host grasses. The pathogen is believed to survive between wheat crops on wild plants at field borders and in open grasslands, but the plant species that harbor MoT have yet to be conclusively determined. Several grasses and weeds occur commonly in wheat fields and are secondary hosts. However, information on their role in the epidemiology of wheat blast is scanty. The potential role of lower and older wheat leaves in inoculum build-up before ear emergence needs to be clarified (Cruz et al., 2015). Likewise, the survival of MoT as mycelium in crop residues has to be investigated under conservation agriculture in South Asia (CIMMYT, 2017). The lower leaves act as source of conidia for infection of developing spikes. The symptoms on the leaf are however, difficult to distinguish due to combined infection of other pathogens like *Bipolaris sorokiniana* and *Fusarium* spp.

24.5 VARIABILITY IN PATHOGEN

Wheat blast pathogen is now considered as a separate species of *M. oryzae* (Couch & Kohn, 2002). *M. oryzae* (anamorph *Pyricularia oryzae*) is divided into host-specific subgroups or pathotypes specialized for infecting rice (*Oryza* pathotype), wheat (*Triticum* pathotype), ryegrass (*Lolium* pathotype), foxtail millet (*Setaria* pathotype), and many other plant species. The isolates from wheat are distinct from other host-specific subgroups and it has been proved based on host range, sexual fertility, and fingerprinting with repetitive DNA elements. The origin of the wheat pathogen is still debated. The MoT population has high evolutionary potential and evolving. Wheat blast populations exhibit a mixed reproductive system in which sexual reproduction is followed by the local dispersion of clones. A thorough dissection of the fungal biology, including characterization of more isolates and mating types from Bangladesh, is critical to develop durable disease management approaches (CIMMYT, 2017; Maciel et al., 2014; Tosa, et al., 2006; Urashima et al., 2005). Pieck et al. (2016) used a genomics-based approach to identify molecular markers unique to the *Triticum* pathotype of *M. oryzae*. MoT3 was used for a polymerase chain reaction (PCR)-based diagnostic assay. Conventional PCR primers were designed to amplify a 394 bp product and the protocol consistently amplified from as little as 0.1 ng of purified DNA. The specificity of the MoT3-based assay was also evaluated using *Fusarium* spp. DNA, from which no amplicons were detected.

24.6   MANAGEMENT OF WHEAT BLAST

An integrated approach for disease management is advocated since the pathogen cycle and epidemiology of the wheat blast is not fully understood. The most effective strategy will be to create genetic resistance, coupled with fungicidal seed treatment and foliar sprays on crop with strobilurin (quinone outside inhibitor (QoI)) fungicides to reduce severity of disease. Fungicides in combination of triazoles with strobilurins have proven effective at heading, especially in moderately resistant wheat varieties under low-to-moderate disease pressure (Rios et al., 2016; Paula et al., 2014). Wheat blast fungus can be transmitted efficiently through wheat seed and remain viable and infectious for up to 22 months. Therefore, seed treatment with fungicides like thiram plus carboxin or benomyl is recommended. Oliveira et al. (2015) reported resistance in wheat blast pathogen to QoI fungicides azoxystrobin and pyraclostrobin in Brazil. Eliminating alternative hosts from the field will minimize the presence of wheat blast. A strong surveillance, by monitoring disease appearance, movement, and evolution; in coordination with the state governments, non-governmental organizations (NGOs), research institutes, and private organizations will minimize losses. Strengthening plant quarantine measures governing seed movement, increasing knowledge and awareness about wheat blast amongst scientists, extension workers, decision makers, and farmers are important issues to tackle the disease effectively. Screening of wheat germplasm, identification of resistant wheat lines with durable resistance, development of varieties with better blast resistance, and their effective deployment will be quite fruitful. The novel resistance genes and use of DNA markers associated with resistance genes for marker-assisted selection will be helpful.

So far no confirmed sources of durable blast resistance have been found in wheat. Several Brazilian, Bolivian, and Paraguayan cultivars showed field resistance, but this has not been confirmed under artificial inoculation or multi-location (countries) studies. Cultivars such as BR18, IPR85, and CD113 have shown some resistance at many locations. Likewise, cultivars and advanced lines derived from the CIMMYT line "Milan" have shown a high level of resistance in South America (Urashima et al., 2004). Both qualitative and quantitative genetic resistance to wheat blast have been identified in wheat. Qualitative resistance has been validated only at the seedling stage. The 2NS/2AS translocation from *Aegilops ventricosa* gave some genetic backgrounds in USA.

However, the 2NS/2AS-based resistance appears to be breaking down under field infections in case of virulent pathotypes (Cruz et al., 2016). So far, eight blast resistance genes (*Rmg1 to Rmg8*) have been identified in wheat. Only *Rmg2, Rmg3, Rmg7,* and *Rmg8* provide host-plant resistance against MoT (Anh et al., 2015). Resistance genes must be expressed both at seedling and heading stages for proper control of disease. The identification of corresponding avirulence genes is also a prerequisite. *M. oryzae* fungal/cereal host interactions bear similarities to those of the wheat rusts. Long-term studies on gene-for-gene reactions in rice blast systems and host resistance provide opportunities to combat MoT more effectively (CIMMYT, 2017). .

24.7 CONCLUDING REMARKS

Wheat blast (*M. oryzae* pathotype *Triticum*) is a serious disease which has a potential to wipe wheat cultivation in warmer and humid agroecological zones in world. The nature of it's survival, pace of spread, mutation, attack on most productive stage (spike emergence and grain formation) of wheat, lack of host resistance and effective chemical control poses direct threat to resource poor farmers of South Asia and South American countries. Since pathogen does not recognize political boundaries therefore, requires a joint and collaborative action in terms of prevention of its spread, establishment and losses in wheat. Wheat blast if occurs with others diseases like brown rust, spot blotch and head scab may poss greater threats to wheat cultivation in warmer climate.

KEYWORDS

- **wheat blast**
- ***Magnaporthe oryzae* pathotype *Triticum***
- **wheat**
- **status**
- **variability**
- **disease management**

REFERENCES

Anh, V. L., et al. *Rmg8*, a New Gene for Resistance to *Triticum* Isolates of *Pyricularia oryzae* in Hexaploid Wheat. *Phytopathology.* **2015,** *105* (12), 1568–1572.

CABI. Invasive Species Compendium *Magnaporthe oryzae Triticum* Pathotype (Wheat Blast). 2017. (Source: http://www.cabi.org/isc/datasheet/121970)

Castroagudin, V. L., et al. Wheat Blast Disease Caused by *Pyricularia graminis-tritici* sp. nov. BioRxiv beta. 2017. (Source: http://biorxiv.org/content/early/2016/04/30/051151)

CIMMYT. Understanding and Managing the Threat of Wheat Blast in South Asia, South America, and Beyond. 2017. (Source: http://libcatalog.cimmyt.org/Download/cim/57941. pdf)

Couch, B. C.; Kohn, L. M. A Multilocus Gene Genealogy Concordant with Host Preference Indicates Segregation of a New Species, *Magnaporthe oryzae*, from *M. grisea. Mycologia.* **2002,** *94* (4), 683–693.

Cruz, C. D.; Peterson, G. L.; Bockus, W. W.; Kankanala, P.; Dubcovsky, J.; Jordan, K. W.; Akhunov, E.; Chumley, F.; Baldelomar, F. D.; Valent, B. The 2NS Translocation from *Aegilops ventricosa* Confers Resistance to the *Triticum* Pathotype of *Magnaporthe oryzae. Crop Sci.* **2016,** *56* (3), 990–1000.

Cruz, C. D.; Kiyuna, J.; Bockus, W. W.; Todd, T. C.; Stack, J. P.; Valent, B. *Magnaporthe oryzae* Conidia on Basal Wheat Leaves as a Potential Source of Wheat Blast Inoculum. *Plant Pathol.* **2015,** *64,* 1491–1498.

Goulart, A. C. P.; Sousa, P. G.; Urashima, A. S. Damages in Wheat Caused by Infection of *Pyricularia grisea. Summa Phytopathol.* **2007,** *33* (4), 358–363.

Ha, X.; Wei, T.; von Tiedemann, A.; Duveiller, E. *Epidemiological and Phytopathological Studies on Wheat Blast (Magnaporthe Grisea)-Characterisation of Pathotypes and Resistance in Wheat*; Georg August Universitat Gottingen Fachgebiet für Pflanzenpathologie und Pflanzenschutz: Gottingen, Germany, 2017. (Source: http://www.phytopathology.uni-goettingen.de/?id=445).

Igarashi, S. In *Update on Wheat Blast (Pyricularia oryzae) in Brazil.* Proceedings of the International Conference-Wheat for Nontraditional Warm Areas; Saunders, D. A., Ed.; CIMMYT: Mexico, 1991, pp. 480–483.

Igarashi, S., et al. *Pyricularia* in wheat.1. Occurrence of *Pyricularia* sp. in Paraná State. *Fitopatol. Bras.* **1986,** *11,* 351–352.

Maciel, J. L. N., et al. Population Structure and Pathotype Diversity of the Wheat Blast Pathogen *M. oryzae* 25 Years after Its Emergence in Brazil. *Phytopathology.* **2014,** *104,* 95–107.

Malaker, P. K., et al. First Report of Wheat Blast Caused by *Magnaporthe oryzae* Pathotype *Triticum* in Bangladesh. *Plant Dis.* **2016,** *100,* 2330. (Online: http://dx.doi.org/10.1094/PDIS-05-16-0666-PDN).

Oliveira, S. C. D., et al. Cross-Resistance to QoI Fungicides Azoxystrobin and Pyraclostrobin in the Wheat Blast Pathogen *P. oryzae* in Brazil. *Summa Phytopathol.* **2015,** *41,* 298–304.

Paula, A., et al. Management of Wheat Blast with Synthetic Fungicides, Partial Resistance and Silicate and Phosphite Minerals. *Phytoparasitica.* **2014,** *42* (5), 609–617.

Perelló, A.; Martinez, I.; Molina, M. First Report of Virulence and Effects of *Magnaporthe oryzae* Isolates Causing Wheat Blast in Argentina. *Plant Dis.* **2015,** *99* (8), 1177.

Pieck, M. L., et al. Genomics-Based Marker Discovery and Diagnostic Assay Development for Wheat Blast. *Plant Dis.* **2016,** *101,* 103–109.

Prabhu, A. S.; Fillipi, M. C.; Castro, N. Pathogenic Variation among Isolates of *Pyricularia oryzae* Affecting Rice, Wheat, and Grasses in Brazil. *Trop. Pest Manag.* **1992,** *38,* 367–371.

Rios, J. A., et al. Fungicide and Cultivar Effects on the Development and Temporal Progress of Wheat Blast under Field Conditions. *Crop Prot.* 2016, *89,* 152–160.

Singh, D. P., et al. *Wheat Crop Health Newsletter;* **2016,** *22* (2), 1–10. (Online: www.dwr. res.in)

Singh, D. P., et al. *Wheat Crop Health Newsletter.* **2017,** *22* (4), 1–26. (Online: www.dwr. res.in)

Tosa, Y., et al. Genetic Analysis of Host Species Specificity of *Magnaporthe oryzae* Isolates from Rice and Wheat. *Phytopathology.* **2006,** *96,* 480–484.

Urashima, A. S., et al. DNA Finger Printing and Sexual Characterization Revealed Two Distinct Populations of *Magnaporthe grisea* in Wheat Blast from Brazil. *Czech J. Genet. Plant Breed.* **2005,** *41,* 238–245.

Urashima, A. S.; Igarashi, S.; Kato, H. Host Range, Mating Type and Fertility of *Pyricularia grisea* from Wheat in Brazil. *Plant Dis.* **1993,** *77,* 1211–1216.

Urashima, A. S.; Lavorent, N. A.; Goulart, A. C. P.; Mehta, R. Resistance Spectra of Wheat Cultivars and Virulence Diversity of *Magnaporthe grisea* Isolates in Brazil. *Fitopatol. Bras.* **2004,** *29,* 511–518.

INDEX

T - #0789 - 101024 - C682 - 229/152/30 - PB - 9781774636619 - Gloss Lamination